Topics in

ORGANIC POLAROGRAPHY

Topics In
ORGANIC POLAROGRAPHY

Edited by

P ZUMAN

Department of Chemistry
University of Birmingham
Birmingham, England

PLENUM PRESS
LONDON · NEW YORK · 1970

Plenum Publishing Company Ltd.,
Donington House,
30 Norfolk Street,
London, W.C.2.

U.S. Edition published by
Plenum Publishing Corporation,
114 Fifth Avenue,
New York, New York 10011

Library of Congress Catalog Card Number: 77-104331

ISBN-13: 978-1-4684-1814-9   e-ISBN-13: 978-1-4684-1812-5
DOI: 10.1007/978-1-4684-1812-5

Preface

Even though the number of requests for reprints and the number of quotations in the Science Citation Index has indicated an ever-increasing interest in topics of organic polarography, I have often felt that the reason that some work is less known may well be because the papers were published in less accessible journals. Therefore, I was pleased when I was asked to prepare a selection of my papers on organic polarography for reprinting. This collection of papers may indicate some of the possibilities offered by polarography in the study of properties of organic compounds. The fact that the papers are published in one volume, not only makes the information more easily accessible for the reader, but also enables a direct comparison of related topics.

The mode of selection is discussed in the Introduction. The papers reprinted in this volume are mostly based on work carried out in the J. Heyrovský Institute of Polarography of the Czechoslovak Academy of Sciences in Prague, in cooperation with my co-workers. I would like to take this opportunity of thanking all of them for the pleasure I got from this cooperation on the solution of varying problems of organic electrochemistry.

I would also like to thank the following for permission to reprint papers initially published in various Journals: Academia – the Publishing House of the Czechoslovak Academy of Sciences – and the Johnson Reprint Corporation, for papers in the Collection of Czechoslovak Chemical Communications; the Elsevier Publishing Company, for papers in the Journal of Electroanalytical Chemistry; Macmillan (Journals) Ltd., for papers in Nature; Professor Dr. K. Schwabe for a paper in Zeitschrift für Physikalische Chemie (Leipzig) and to the Editor and Birkhäuser Verlag for a note in Experientia. It is my pleasure to acknowledge the pleasant cooperation offered by Plenum Press, particulary their London office headed by R. H. Leech.

Birmingham   P. Z.
May, 1969

Contents

viii

Part I

Introduction

The electrolysis of organic compounds has been carried out for almost a century, but for a long time there was a rather limited understanding of the processes involved. Often, suggested reaction schemes were deduced only from a knowledge of the initial and final compositions of the electrolysed solution. Polarography has contributed more to the understanding of processes involved in the electrolysis of organic substances, than any other method.

Polarography — the study of current-voltage curves obtained with mercury electrodes with periodically-renewed surfaces — was first developed by Professor J. Heyrovský as a physicochemical method. The study of polarographic electrolysis from this point of view contributed considerably to our understanding of transport phenomena, of the role of potential control in electrolysis, and of effects of chemical reactions antecedent, parallel or consecutive to the electrode process proper.

After the theoretical relationships between the limiting current and concentration had been derived by D. Ilkovič, polarography found wide application as an analytical method. The sensitivity, the small volume of sample and, in particular, the selectivity of polarographic methods were the main factors leading to the present situation, in which the great majority of papers published on polarography (over 20,000) deal with analytical applications.

Only over the past decade has it been recognized that polarography can be useful in solving some fundamental problems of both inorganic and organic chemistry. In organic chemistry, polarography can be used in the determination of equilibrium and rate constants, in studies of reaction mechanisms, in the search for optimal conditions for some preparative reactions, in studies and comparisons of reactivities of organic compounds, and in correlations of structure with polarographic data. For all these purposes, a partial understanding of the course of the electrode process involved is necessary. The type of electrode process involved and the nature of accompanying chemical reactions must be known, for effective use of polarographic data in studying kinetics, equilibria, and mechanisms. Comparisons of reactivities or structural effects are restricted to systems that follow the same type of electrode process. Therefore, before using polarographic data it is necessary to show to which type the process studied belongs, and that it is the same type for all the systems compared.

Once it is understood that the course of the electrode process must be elucidated, it is possible to discuss the methods available for the purpose. At an early stage in the development of organic polarography, it was assumed that processes taking place at the dropping mercury or similar electrodes could be elucidated from current — voltage curves alone. However, during the past decade this opinion has had to be considerably modified. It is now realized that, even though polarographic curves offer a wealth of information to those familiar with their interpretation, it is only in rare and very simple cases that all the information about the electro-active system can be obtained solely from such curves.

1

Even though studies of the electrochemistry of organic compounds in strictly non-aqueous systems have some theoretical interest, and have contributed to the solution of some fundamental thermodynamic problems, their importance in solving the above-mentioned problems is restricted. Practically, it is of greater value for organic chemists to learn about the behaviour of organic compounds under conditions used in practical organic chemistry, where it is rarely possible to exclude traces of water, not to mention the high cost of organic solvents when used in large volumes. Therefore, it is very important to study organic compounds in the presence of varying amounts of water in the solvent, as information can be obtained about the actual reactions (including complications due to reactions with solvent) encountered in practical organic chemistry. In addition to this aspect, organic electrode processes are often accompanied by acid-base equilibria. However, whereas in aqueous solutions there is a good deal of information available on such equilibria and their position can be controlled by buffering, so as to prevent changes in conditions at the electrode surface during electrolysis, current knowledge about acid-base equilibria and acidity control in non-aqueous media is still limited. Before electrochemistry in nonaqueous media can be fully exploited, a deeper understanding of the chemistry in such solvents is essential. Because of this, the following discussion of the techniques for elucidating organic electrode processes is restricted to aqueous and water-containing solutions.

Polarographic curves, identification of electrolysis products and intermediates, structural effects, and combination with other methods may be used for the elucidation of organic electrode processes at the dropping mercury electrode.

Polarographic curves and their dependence on mercury pressure, concentration of the electroactive species and temperature allow us to distinguish the type of process governing the limiting current. It is possible to distinguish diffusion-controlled current from currents involving chemical reactions, catalytic processes or adsorption phenomena. The number and relative heights of waves offer information about the number of electrons transferred and about some intermediates formed. Because many organic electrode processes are accompanied by proton transfers, the effect of pH in well buffered solutions on both wave-heights and half-wave potentials is of particular importance. Generally, the pH-dependence of either half-wave potentials or limiting currents has been applied in the study of electrode processes. Only recently has it been recognized that a combination of both is the most powerful tool in interpretating the acid-base equilibria accompanying the electrode reaction proper and in identifying the structure of the electroactive species. The use of the wave-shape by means of logarithmic analysis has more restricted applications in the identification of the individual steps of the electrode process, even when information can be obtained about the electrode process proper. It should be stressed that polarography offers the valuable possibility of using limiting current as an important tool in the elucidation of electrode processes, and that it has been proved that limiting current offers more and clearer information about the individual steps and participating species than logarithmic analysis or Tafel plot. Furthermore, differences between the logarithmic analyses for different types of electrode processes are usually smaller than those between limiting currents, and experimental verification of logarithmic analyses is subject to greater experimental errors, effects of adsorption, etc. Logarithmic analysis is thus used more as a confirmatory than a diagnostic tool. In some cases, the effects of drop-time and concentration of the electroactive form on half-wave potentials can be used for identifying certain reaction schemes. Similarly, the effects of buffer composition and concentration, as well as the effects of the nature and concentration of the supporting electrolyte can be useful in certain cases. Addition of acids or bases to unbuffered solutions can help to determine the number of protons transferred before the potential-determining step.

In the identification of electrolysis products and intermediates, it has proved very useful — apart from the information obtained from polarographic curves mentioned above — to carry out controlled potential electrolysis, preferably with the dropping mercury electrode. When a mercury pool electrode is used, the products obtained may occasionally differ from those obtained with the dropping mercury electrode. This comparison is discussed in detail in Part II. To achieve considerable conversion within a reasonable time, small volumes of the electrolysed solution must be used. Identification and determination of the yield must then be carried out by sensitive methods e.g., chromatographic, in particular gas-liquid, ultraviolet spectral or submicro techniques.

Comparisons of polarographic behaviour of structurally related organic substances make it possible, analogously to mechanistic studies of homogeneous reactions, to draw conclusions about the type of the electrode reaction involved, and to decide whether the suggested reaction scheme is in accordance with the observed structural effects or not. Structural effects also indicate the composition of the transition state and thus contribute to the study of the electrode process proper.

It has proved valuable to combine polarographic investigations with those carried out by other electrochemical methods. In particular, rectangular voltage polarization, single and multi-sweep methods, and current-voltage curves with a hanging mercury drop electrode offer information on the electrolysis products formed. It is nevertheless important to stress that techniques which are not potentiostatic, and where the voltage sweep is applied to an unrenewed surface, can give products different from those formed under polarographic conditions, e.g., because intermediates formed at one potential can interact with species generated at another potential.

Useful information about chemical reactions, in particular consecutive reactions, accompanying the electrode process can also be obtained from chronopotentiometric curves. Adsorption phenomena can be detected by a.c. polarography, from the change of the instantaneous current at the dropping mercury electrode with time ($i - t$ curves), from electrocapillary curves, and from differential capacity measurements. Finally, many of these methods offer information on reversibility of the electrode process.

Recently, several new and modified electrochemical techniques have been developed. Several of these techniques look promising, but most of them have one serious limitation: so far they have been applied only to a few (if any) organic systems, so that it is difficult to assess their scope. One essential advantage of polarography is the wide range of systems which have been investigated over the past forty years. Chronopotentiometry and a.c. polarography seem to come second with respect to the number of useful applications. The real utility of other techniques will become known only after more extensive use.

This selection of reprints includes examples of studies of organic systems by means of the techniques just discussed. The degree of sophistication naturally changed with time – the number of techniques used and the amount of information gained in the more recent papers is greater than in the older ones. Reprints of papers dealing solely with substituent effects have been excluded from this collection, because such information has been summarized elsewhere.* Papers dealing with anodic waves corresponding to the formation of mercury compounds have also been omitted, because the chemical processes involved are still not sufficiently understood. In order to keep the size of this book within practical limits, papers dealing with catalytic processes have been left out. Various papers have dealt with problems of kinetics of reactions in the solution, where polarography has been applied in the study of reaction mechanisms as an analytical method; of such papers, only those that deal with compounds discussed from some other point of view in this volume are included. References can be found in the attached bibliography.

Most of the experimental work reported is connected one way or another with the polarographic reduction of carbonyl compounds. The author first became associated with carbonyl compounds in 1948, at which time he studied bismuth complexes under Professors J. Heyrovský and R. Brdička at the Department of Physical Chemistry at Charles University, Prague. During that work, two unexpected waves were observed in the presence of lactic acid. The more negative was proved to correspond to the reduction of pyruvic acid which was present in lactic acid; the more positive was proved to belong to a ketimine formed by the reaction of pyruvic acid with ammonia, which had diffused into the polarographic cell from a washbottle in series, where it had accumulated during preceding experiments. This led to an investigation of the reaction of carbonyl compounds with amines, and subsequent studies.

The papers in this collection are not arranged chronologically, but rather logically, according to the electroactive group involved.

Reductions of some simpler carbonyl compounds, especially aryl alkyl ketones and aldehydes, are discussed in Part III. Their behaviour was investigated because they were intermediates in some more complex electrode processes discussed later, and because polar and steric effects on their reduction were compared.

Discussion of aromatic systems bearing two carbonyl groups in the para-position, which are reduced in a reversible two-electron step, is included. p-Diacetylbenzene was first investigated to show whether the two carbonyl groups in this acetophenone derivative are reduced simultaneously or in consecutive steps. Experimental evidence has shown that neither of these paths is followed and that the system shows entirely different properties from those of the corresponding monocarbonyl compounds.

* P. Zuman: Substituent Effects in Organic Polarography, Plenum Press, New York, 1967.

In β-substituted carbonyl compounds, which are discussed in Part IV, the carbonyl group is still reduced first. The substituent affects the reduction potential and the acid-base equilibria involved. Polarographic waves of β-substituted carbonyl compounds have made it possible to study reactions in which these compounds are formed or undergo elimination.

3-Thianaphthenone, which possesses a grouping $COCH_2X$ in which the carbonyl group is reduced and the CH_2 group has acidic properties, was included in this group. For such compounds, polarography offers information about equilibria and rates of dissociation leading to an ambident anion. Several further systems of that type, such as ethyl benzoyl acetate, ω-cyanoacetophenone, benzoylacetone etc. have been studied recently.

For $a, β$-unsaturated carbonyl compounds, discussed in Part V it has been confirmed that the C=C double bond is reduced before the reduction of the carbonyl group. It has been shown that this sequence is followed not only in the reduction of $a, β$-unsaturated ketones but also in the reduction of $a, β$-unsaturated aldehydes. The author's suggestion (1954) that in the reduction of $a, β$-unsaturated ketones the electron transfer is preceded by protonation of the ketone, now appears to be generally accepted. The effects of interposed protonation of the ambident anion formed in the first two-electron step or interposed hydration – dehydration equilibria have been investigated.

Comparison of a- and $β$-aminoketones led to the observation that in a-aminoketones the $C - N$ bond is reduced before the carbonyl group. These reductions of the single $C - N$ bond together with reductions of activated single $C - S$ and $C - O$ bonds are discussed in Part VI. These studies have recently been extended to some cyclic compounds such as saccharine and its derivatives.

Reduction of carbonyl compounds derivatives containing a C=N bond are described in Part VII. The waves have also been used in the study of formation of various Schiff bases.

An investigation of substituted benzonitriles was prompted by the anomalous behaviour of the p-cyano derivative observed in a study of substituted benzophenones, together with deviations from linear Hammett plots for p-cyano-substituted benzene derivatives and the reducibility of p-cyanobenzoic acid. The reduction of substituted benzonitriles resulted either in the nucleophilic substitution of the cyanide ion or the reduction of the cyano group to CH_2NH_2. These modes of reduction are discussed in Part VIII.

Finally, Part IX discusses the reduction of some non-benzenoid aromatic substances, in particular sydnones, azulenes, and tropylium ions. Apart from the reduction course, attention has also been paid to accompanying acid-base equilibria and adsorption phenomena.

The papers reprinted in this collection are indicated in the Bibliography by the numbers in square brackets at the end of the relevant entry.

Bibliography*

1. P. Zuman: Formation of reducing substances during the heating of sugar solution (in Czech): *Listy cukrovar.* **62**, 81–7 (1945/46).
2. P. Zuman: The influence of the purification methods on the content of reducing substances in sugar solution (in Czech): *Listy cukrovar.* **62**, 89–93 (1945/46).
3. P. Zuman: Artificially prepared reductones (in Czech): *Listy cukrovar.* **62**, 97–9 (1945/46).
4. P. Zuman: Polarographic behaviour of acetone: *Nature* **175**, 585–6 (1950). **[30]**
5. P. Zuman: The reaction of carbonyl compounds with primary amines (in German): *Collection* **15**, 839–74 (1950). **[31]**
6. R. Brdička, and P. Zuman: Polarographic behaviour of reductones and coumarindiol. *Collection* **15**, 766–78 (1950); (in Russian) *Chimija* **1**, 289–301 (1951).
7. P. Zuman: A list of half-wave potentials. *Collection* **15**, 1107–1208 (1950).
8. P. Zuman: Polarographic maxima of second kind (in Czech). *Chem. Listy* **44**, 162–4 (1950).
9. P. Zuman: The reaction of carbonyl compounds with primary amines. Proceedings Ist Internat. Polarograph. Congr., Prague 1951, Part I (1951) in Russian: 704–11; in English 711–18; Part III (1952) in Czech: 520–29.
10. P. Zuman: The reaction of pyruvic acid with the imidazole derivatives (in Czech). *Chem. listy* **45**, 40–2 (1951).
11. P. Zuman: The reaction of carbonyl compounds with amines I. The reaction of pyruvic acid with ammonia (in Czech). *Chem. listy* **45**, 65–72 (1951).
12. P. Zuman: The determination of sulphydryl substances in some fruits (in Czech). *Chem. listy* **46**, 73–8 (1952); (in German) *Collection* **16**, 510–25 (1951); Proc. Ist Internat. Polarograph. Congr. Part III (1952) (in Czech) 582–601; (in Russian) 601–3; (in English) 603–6.
13. P. Zuman: A contribution to the polarography of aluminium and beryllium (in Czech). *Chem. listy* **46**, 326–28 (1952).
14. P. Zuman: Polarographic behaviour of anthocyanines I (in Czech). *Chem. listy* **46**, 328–31 (1952); (in German) *Collection* **18**, 36–42 (1953).
15. P. Zuman: Polarographic analysis of organic substances. Proc. Ist Internat. Polarograph. Congr. Part III (1952). (in Czech) 146–60; (in Russian) 160–76; (in German) 177–93.
16. P. Zuman: The determination of pyruvic acid in lactic acid (in Czech). *Chem. zvesti* **6**, 191–99 (1952).
17. J. Koryta, and P. Zuman: The polarography of barbituric acid derivatives I. Barbituric acid (in Czech). *Chem. listy* **46**, 389–93 (1952); (in German) *Collection* **18**, 197–205 (1953).
18. P. Zuman, and F. Šantavý: The polarography of heart glycosides bearing an aldehydic group (in Czech). *Chem. listy* **46**, 293–96 (1952); (in Russian) *Collection* **18**, 28–35 (1953).
19. P. Valenta, and P. Zuman: A contribution to the polarography of tetravalent germanium (in Czech). *Chem. listy* **46**, 478–79 (1952).
20. P. Zuman: The reaction of carbonyl compounds with amines II. The reaction of pyruvic acid with glycine, alanine and colamine (in Czech). *Chem. listy* **46**, 516–20 (1952).
21. P. Zuman: The reaction of carbonyl compounds with amines III. The reaction of the oxidation product of ascorbic acid with ammonia (in Czech). *Chem. listy* **46**, 521–25 (1952).
22. P. Zuman, and M. Březina: The reaction of carbonyl compounds with amines IV. The reaction of acetone with ammonia and glycine (in Czech). *Chem. listy* **46**, 599–603 (1952).
23. P. Zuman: Aldimines, ketimines, Schiff's bases (a review in Czech). *Chem. listy* **46**, 688–708 (1952).

* Abbreviation "Collection" is used for the Collection of Czechoslovak Chemical Communications.

24. P. Zuman, and Ž. Procházka: The bound form of ascorbic acid IV. The polarographic determination of ascorbic acid in ascorbigen concentrates (in Czech). *Chem. listy* 47, 357–62 (1953); (in Russian) *Collection* 18, 442–49 (1953).

25. P. Zuman, R. Zumanová, and B. Souček: Polarographic determination of carbon disulfide using the anodic wave (in Czech). *Chem. listy* 47, 178–88 (1953); (in German) *Collection* 18, 632–47 (1953).

26. P. Zuman, and F. Šantavý: Polarographic study of cyanhydrine reaction in alkaline medium (in Czech). *Chem. listy* 47, 267–69 (1953); (in German) *Collection* 19, 174–76 (1954).

27. P. Zuman, J. Koryta, and R. Kalvoda: The polarography of barbituric acid derivatives II. Veronal (in Czech). *Chem. listy* 47, 345–56 (1953); (in German) *Collection* 18, 350–65 (1953).

28. M. Březina, and P. Zuman: The reaction of carbonyl compounds with amines V. The polarographic study of the reaction of cyclanones with primary amines. Equilibrium states (in Czech). *Chem. listy* 47, 975–91 (1953).

29. P. Zuman, J. Tenygl, and M. Březina: The polarography of steroids I. Directly reducible ketosteroids (in Czech). *Chem. listy* 47, 1152–61 (1953); (in German) *Collection* 19, 46–57 (1954). [20]

30. P. Zuman: The significance of Hammett's equation in polarography (in Czech). *Chem. listy* 47, 1234–36 (1953); (in German) *Collection* 19, 599–602 (1954).

31. P. Zuman, R. Zumanová, and B. Souček: The polarographic determination of hydrogen sulfide in the presence of carbon disulfide. (in Czech) *Chem. listy* 47, 1409–10 (1953).

32. P. Zuman, R. Zumanová, and B. Souček: The polarographic irreversibility of the system diethyldithiocarbamate-bis(diethylthiocarbamyl)disulfide. (in Czech) *Chem. listy* 47, 1522. (1953).

33. P. Zuman: The polarographic behaviour of ferrocyanide, ferricyanide and the Prussian blue. (in Czech) *Chem. listy* 47, 1523–25 (1953); (in German) *Collection* 19, 602–5 (1954).

34. J. Blatná, J. Fragner, V. Šanda, P. Zuman, and D. Žuffová: The determination of l-ascorbic acid in foods. (in Czech) *Prumysl potravin* 4, 402 (1955); Czechoslovak Ministry for Food Production, Recommended analytical methods No. 4, Prague 1953.

35. P. Zuman: Polarography in pharmacy (a review in Czech). *Čsl. Farmacie* 2, 413–18 (1953); (in German) *Pharmazie* 8, 903–09 (1953).

36. P. Zuman: The advantages of polarographic analysis. (in Czech) Proc. 1st Conf. Anal. Chem. Prague 1953, 296–99.

37. P. Valenta, and P. Zuman: A note to the article: "Polarographic Reduction of Germanium" by A. K. Das Gupta and C. K. N. Nair, *Anal. Chim. Acta* 10, 591–93 (1954).

38. P. Zuman: The influence of structure on polarographic behaviour of organic compounds. (a review in Czech) *Chem. listy* 48, 94–150 (1954).

39. P. Zuman, and M. Kabát: Polarographic study of nitroprusside. (in Czech) *Chem. listy* 48, 368–77 (1954); (in German) *Collection* 19, 873–84 (1954).

40. P. Zuman: Polarographic behaviour of juglone and its determination in the presence of ascorbic acid in fruits and green parts of the walnut (Juglans regia). (in Czech) *Chem. listy* 48, 525–32 (1954); (in German) *Collection* 19, 1140–49 (1954).

41. R. Zumanová, and P. Zuman: The determination of 2,3-dimercaptopropanol (BAL) in pharmaceutical preparations using polarography and potenciometry. (in German) *Pharmazie* 9, 554–56 (1954).

42. P. Zuman: The polarography of barbituric acid derivatives III. Thiobarbituric acid derivatives. (in Czech) *Chem. listy* 48, 1006–1019 (1954); (in German) *Collection* 20, 649–61 (1955).

43. P. Zuman: The polarography of barbituric acid derivatives IV. The damping of maxima by the action of thiobarbituric acid derivatives. (in Czech) *Chem. listy* 48, 1020–24 (1954); (in German) *Collection* 20, 883–88 (1955).

44. P. Zuman: The influence of adsorption on polarographic anodic curves; the origin of non-turbulent maxima and minima. (in Czech) *Chem. listy* **48**, 1025–30 (1954); (in German) *Collection* **20**, 876–82 (1955).

45. P. Zuman, R. Zumanová, and J. Teisinger: The polarography of some sulfur-containing compounds IV. Anodic waves of 2,3- dimercaptopropanol (BAL). (in Czech) *Chem. listy* **48**, 1499–1505 (1954); (in German) *Collection* **20**, 139–46 (1955).

46. P. Zuman: The influence of the surface active substances on polarographic curves. (in Czech) *Chem. zvesti* **8**, 789–822 (1954).

47. P. Zuman: The role of Hammett's equation in the polarographic solution of structural problems. (in Czech) *Chem. zvesti* **8**, 939–52 (1954).

48. P. Zuman: The constitution and polarographic behaviour of organic substances. (in Czech) *Chem.·zvesti* **8**, 926–38 (1954).

49. P. Zuman, and R. Zumanová: The polarography of some sulfur-containing compounds V. Compounds of 2,3-dimercaptopropanol with heavy metals. (in Czech) *Chem. listy* **49**, 652–67 (1955); (in German) *Collection* **21**, 123–39 (1956).

50. O. Manoušek, and P. Zuman: The polarography of urea and thiourea derivatives V. Anodic depolarisation in the solutions of uracil, its derivatives and 4-methyl-2-thiouracil. (in Czech) *Chem. listy* **49**, 668–78 (1955); (in German) *Collection* **20**, 1340–52 (1955).

51. M. Fedoroňko, and P. Zuman: The polarography of some sulfur-containing compounds VII. Anodic depolarisation in the solution of 2-mercaptobenzthiazole and 2-mercapto-benzimidazole. (in Czech) *Chem. listy* **49**, 1484–93 (1955); (in German) *Collection* **21**, 678–88 (1956).

52. M. Fedoroňko, O. Manoušek, and P. Zuman: The polarography of urea and thiourea derivatives VII. Some substituted ureas, thioureas and isothioureas. (in Czech) *Chem. listy* **49**, 1494–98 (1955); (in German) *Collection* **21**, 672–77 (1956).

53. P. Zuman: Advances of polarography in pharmacy. (a review in Czech) *Čsl. Farmacie* **5**, 44–49 (1956); (in German) *Pharmazie* **11**, 449–56 (1956).

54. P. Zuman, and M. Březina: The polarography in medicine. (a review in German) *Leybold Polarogr. Ber.* **3**, 224–49 (1955).

55. O. Manoušek, and P. Zuman: The polarography of urea and thiourea derivatives VIII. Polarographic determination of thiobarbituric acid derivatives and 4-methy-2-thiouracil in preparations. (in Czech) *Čsl. Farmacie* **4**, 193–95 (1956); (in German) *Pharmazie* **11**, 530–33 (1956).

56. P. Zuman, and R. Zumanová: Polarography of some sulfur-containing compounds XI. Oxidation and certain reactions of 2,3-dimercaptopropanol. (in Czech) *Chem. listy* **50**, 1908–17 (1956); (in German) *Collection* **22**, 929–40 (1957).

57. P. Zuman, J. Sicher, J. Krupička, and M. Svoboda: Polarographic determination of rate of periodate oxidation of epimeric open-chain 1,2-glycols. *Nature* **178**, 1407–08 (1956).

58. P. Zuman, and J. Krupička: Polarographic method for the study of reaction of periodic acid with glycols. (in Czech) *Chem. listy* **51**, 424–32 (1957); (in English) *Collection* **23**, 598–609 (1958).

59. P. Zuman: Polarographic determination of vitamins. (in German) *Acta Chim. Acad. Sci. Hungary* **9**, 279–94 (1956).

60. P. Zuman: A simple modification of the polarographic vessel. (in Czech) *Chem. listy* **51**, 993–95 (1957).

61. P. Zuman, J. Sicher, J. Krupička, and M. Svoboda: Stereochemical studies VII. Oxidation of diastereoisomeric diols of the type RCH(OH)CH(OH)R by periodate. (in Czech) *Chem. listy* **51**, 1068–81 (1957); (in English) *Collection* **23**, 1237–51 (1958).

62. P. Zuman: The role of buffers in analytical chemistry. (in Czech) *Chemie* **9**, 192–83 (1957).

63. R. Zumanová, J. Teisinger, and P. Zuman: Influence of proteins on the polarographic behaviour of metals and their compounds with 2,3-dimercaptopropanol. (in Czech) *Chem. zvesti* **9**, 517–27 (1957).

64. P. Zuman: Two examples of application of polarography to studies of homogeneous chemical reactions (in Polish) Prace Konf. Polarograf., Warszawa 1956, p. 339–48, PWN Warszawa 1957.

65. P. Zuman, and R. Zumanová: Some reactions of 2,3-dimercaptopropanol studied polarographically. (in Polish) Prace Konf. Polarograf., Warszawa 1956, p. 467–78, PWN, Warszawa 1957.

66. P. Zuman, and R. Zumanová: Some reactions of 2,3-dimercaptopropanol (BAL) with heavy metals and some oxidizing agents. *Tetraedron* 1, 289–300 (1957).

67. R. Zahradník, and P. Zuman: Carbamates, thiocarbamates and dithiocarbamates VIII. Kinetics and mechanism of decomposition of dithiocarbamic acids in acid solutions, studied polarographically. (in Czech) *Chem. listy* 52, 231–42 (1958); (in German) *Collection* 24, 1132–45 (1959).

68. P. Zuman, and R. Zahradník: Kinetics and mechanism of the decomposition of dithiocarbamates in acid solutions. A polarographic study. (in German) *Z. physikal. Chem.* (Leipzig) 208, 135–40 (1957).

69. P. Zuman: Polarographic study of biochemically important reactions. (in Czech) *Čs. Farmacie* 7, 84–88 (1958).

70. P. Zuman: Steric effects on the catalytic evolution of hydrogen in ammoniacal solutions of cobalt in the presence of threo- and erythro-phenyl-cystein. (in Czech) *Chem. listy* 52, 1349–50 (1958); (in German) *Collection* 24, 2027–29 (1959).

71. P. Zuman, and V. Černý: Polarography of steroids III. Polarographic reduction of steroids containing an aldehydic group in position 18. (in Czech) *Chem. listy* 52, 1468–73 (1958); (in German) *Collection* 24, 1925–32 (1959). [8]

72. P. Zuman: Polarography of some aromatic compounds: sydnones and azulenes. (in German) *Z. physikal. Chemie* (Leipzig): Sonderheft 1958, 243–267. [38]

73. P. Zuman, and M. Kuik: Influence of sulfur on catalytic waves in ammoniacal solutions of cobaltous salts. (in German) *Naturwiss.* 45, 541 (1958).

74. P. Zuman, and M. Fedoroňko: Polarographic study of alkaline hydrolysis of isothioureas. *Z. Physikal. Chem.* 209, 376–80 (1958).

75. P. Zuman: Discussion to the paper: "Polarographic Reduction of Δ^4-3-Ketosteroids in well Buffered Media" by P. Kabasakalian and J. McGlotten. *J. Electrochem. Soc.* 105, 758–61 (1958).

76. P. Zuman, J. Chodowski, H. Potěšilová, and F. Šantavý: Polarographic study of acid-base reactions of tropylium ions. *Nature* 182, 1535 (1958).

77. P. Zuman: Steric effects in polarography. (in Czech) *Chem. listy* 53, 154–63 (1959); (in German) *Acta Chem. Acad. Sci. Hungar.* 18, 141–54 (1959).

78. E. Banyai, and P. Zuman: Polarographic behaviour of 4-amino-4-methoxyphenylamine (Variamin Blue). (in German) *Collection* 24, 522–25 (1959).

79. P. Zuman, and M. Kuik: Catalysis of hydrogen evolution in the presence of some sulfur derivatives of hydantoin and pyrimidine. (in German) *Collection* 24, 3861–80 (1959).

80. P. Zuman: Polarographic micro-analysis in reaction kinetics. Proc. Internat. Symposium Microchem., Birmingham 1958, Pergamon Press, 294–98 (1960).

81. P. Zuman: Polarography and structure of organic compounds. Proc. Internat. Symposium Microchem., Birmingham 1958, Pergamon Press, 299–304 (1960).

82. O. Manoušek, and P. Zuman: Polarographic determination of pyridoxal and pyridoxal-5-phosphate. (in German) *J. Electroanalyt. Chem.* 1, 324–30 (1960).

83. P. Zuman: A study of the effect of structure on polarographic behaviour of organic substances in Polarographic Institute of the Czechoslovak Academy of Science. (a review in Czech) *Chem. listy* 54, 1244–64 (1960); (in English) *J. Polarograph. Soc.* 7, 66-85 (1961).

84. P. Zuman: Quantitative treatments of substituent effects in polarography I. General equation for the relation between polarographic half-wave potentials and the effects of substituents. *Collection* 25, 3225–44 (1960).

10

85. P. Zuman: Polarography of nonbenzenoid aromatic and related substances II. The course of the reduction of sydnones at the dropping mercury electrode. *Collection* **25**, 3245–51 (1960). [39]

86. P. Zuman: Polarography of nonbenzenoid aromatic and related substances III. The course of the reduction of N,N-polymethylene-bis-sydnones. *Collection* **25**, 3252–64 (1960). [40]

87. P. Zuman: Polarography of nonbenzenoid aromatic and related substances IV. Polar effects of substituents in phenyl sydnones; an application of modified Hammett equation. *Collection* **25**, 3256–70 (1960). [41]

88. G. Horn, and P. Zuman: Polarography of urea and thiourea derivatives XI. A note on polarography of some mercaptopurines. (in German) *Collection* **25**, 3401–03 (1960).

89. P. Zuman, and V. Horák: Polarographic reduction of single carbon-nitrogen bond. Advances in Polarography, Proc. 2nd Internat. Congr. Polarograph., Cambridge 1959, Pergamon Press, 1960, p. 804–11.

90. P. Zuman: Influence of substituents in aliphatic and heterocyclic series. Advances in Polarography, Proc. 2nd Internat. Congr. Polarograph., Cambridge 1959, Pergamon Press, 1960, p. 812–28.

91. P. Zuman: Oscillographic and classical polarography. (in Czech) *Chem. Zvesti* **14**, 869–72 (1960).

92. O. Manoušek, and P. Zuman: Polarographic identification of the reactive form in the hydrolysis of pyridoxal-5-phosphate. *Biochem. Biophys. Acta* **44**, 393–94 (1960).

93. V. Horák, and P. Zuman: Fission of activated carbon-nitrogen and carbon-sulphur bonds. Introductory remarks. *Collection* **26**, 173–75 (1961). [21]

94. P. Zuman, and V. Horák: Fission of activated carbon-nitrogen and carbon-sulphur bonds I. Polarographic reduction of the single C – N bond. *Collection* **26**, 176–92 (1961). [22]

95. P. Zuman: A general equation for the effect of substituents on half-wave potentials. *Contributi teorici e sperimentali di polarografia* **5**, *Ric. Sci.* **30**, 229–59 (1960).

96. P. Zuman: Application of polarography to the study of the structure of organic substances and intermediates. (a review in Czech) *Chem. listy* **55**, 261–81 (1961); (in German) *Z. phys. Chem.* (Leipzig) 3, 161–71 (1963).

97. P. Zuman, J. Chodowski, and F. Šantavý: Polarography of nonbenzenoid aromatic and related substances VII. A polarographic study of the acid-base properties of the tropylium ion. *Collection* **26**, 380–91 (1961). [42]

98. P. Zuman, and O. Manoušek: Polarographic study of the nonenzymatic hydrolysis of pyridoxal-5-phosphate. *Collection* **26**, 2134–43 (1961).

99. P. Zuman, and J. Michl: Role of keto-enol tautomerism in polarographic reduction of some carbonyl compounds. *Nature* **192**, 655–57 (1961). [17]

100. P. Zuman, and D. J. Voaden: Polarography of nonbenzenoid aromatic and related substances VI. Polarographic and spectrophotometric study of hindrance of coplanarity in 3-phenylsydnones. *Tetrahedron* **16**, 130–38 (1961).

101. P. Zuman, and S. Wawzonek: Trends in organic polarography. Progress in Polarography (Ed. by P. Zuman and I. M. Kolthoff), Vol. I, p. 303–17, Interscience, New York 1962.

102. P. Zuman: Current trends in study of the influence of structure on the polarographic behaviour of organic substances. Progress in Polarography (Ed. by P. Zuman and I. M. Kolthoff), Vol. I, p. 319–32, Interscience, New York, 1962.

103. P. Zuman: Important factors in classical polarography. Progress in Polarography (Ed. by P. Zuman and I. M. Kolthoff), Vol. II, p. 583–600, Interscience, New York, 1962.

104. P. Zuman, and M. Březina: Polarographic analysis in pharmacy. Progress in Polarography (Ed. by P. Zuman and I. M. Kolthoff), Vol. II, p. 687–701, Interscience, New York, 1962.

105. P. Zuman, and V. Horak: Fission of activated carbon-nitrogen and carbon-sulphur bonds II. Polarographic reduction of β-aminoketones. *Collection* **27**, 187–98 (1962). [11]

106. P. Zuman: Polarography of benzenoid halogen hydrocarbons and their derivatives. (a review in Czech) *Chem. listy* 56, 219–41 (1962).

107. O. Manoušek, and P. Zuman: Direct polarographic determination of pyridoxal in the presence of pyridoxal-5-phosphate. *Collection* 27, 486–87 (1962). [7]

108. V. Horák, J. Michl, and P. Zuman: A contribution to the studies of elimination reactions of Mannich bases, using polarographic method. *Tetrahedron Letters* 21, 744–45 (1961).

109. V. Horák, and P. Zuman: Synthetic and polarographic studies in the 3-tropinone series. *Tetrahedron Letters* 21, 746–48 (1961).

110. P. Zuman: Application of polarography to organic chemistry. *J. Electroanal. Chem.* 3, 157–68 (1962).

111. P. Zuman: Quantitative treatments of substituent effects in polarography. II. Free energy relationship in monocyclic heterocyclic series. *Collection* 27, 630–47 (1962).

112. P. Zuman: Quantitative treatments of substituent effects in polarography III. Substituents in ortho-position. *Collection* 27, 648–66 (1962).

113. P. Zuman, and J. Chodowski: Polarography of nonbenzenoid aromatic and related substances VIII. Adsorption processes during the electroreduction of the tropylium ion. *Collection* 27, 759–74 (1962). [43]

114. P. Zuman, and H. Zinner: Polarography of tetroses and application of polarography in the study of cleavage of 1,1-bis-alkylsulphonyl- pentose-enes (1). (in German) *Chem. Ber.* 95, 2089–94 (1962).

115. P. Zuman: Quantitative treatments of substituent effects in polarography IV. Linear free energy relationships in quinoid series. *Collection* 27, 2035–57 (1962).

116. P. Zuman: Application of polarography in organic chemistry. (in Hungarian) *Magyar Kemikusok Lapja* 17, 8–11 91962).

117. P. Zuman: Polarography in functional group analysis. Analytical Chemistry 1962 (Proc. Feigl Anniv. Symp., Birmingham), p. 222–27, Elsevier, Amsterdam 1963.

118. P. Zuman, S. Tang: Fission of activated carbon-nitrogen and carbon-sulphur bonds III. Reduction of the $C - S^{(+)}$ bond in methyl butyl phenacylsulphonium perchlorate. *Collection* 28, 829–37 (1963). [24]

119. P. Zuman: Organic polarography, structure and linear free energy relationships. *Review of Polarography* (Japan) 11, 102–116 (1963).

130. S. Tang, and P. Zuman: A new type of maximum on the limiting current of the reduction wave of phenacylsulfonium ions. *Collection* 28, 1524–34 (1963). [25]

121. P. Zuman: Some techniques in organic polarography. *Advances in Analytical Chemistry and Instrumentation* (Ed. C. N. Reilley) Vol. 2, 219–53, Interscience-Wiley, New York (1963).

122. P. Zuman: Structure, linear free energy relationships and polarography. (in Russian) Correlation Equations in Organic Chemistry, Vol. 2, 3–25, Tartu State Univ., Tartu 1964.

123. S. I. Ždanov, and P. Zuman: Polarography of nonbenzenoid aromatic and related compounds IX. Reduction and hydrogen evolution catalysis in solutions of N-troponyl amino acids. *Collection* 29, 960–72 (1964).

124. A. Talvik, P. Zuman, and O. Exner: Studies on the inductive effect III. The kinetics of the reaction of substituted benzoic acids with diphenyldiazomethane. *Collection* 29, 1266–76 (1964).

125. O. Manoušek, and P. Zuman: Polarographic study of the reaction of pyridoxal with hydroxylamine. *Experientia* 20, 301 (1964). [34]

126. O. Manoušek, and P. Zuman: Polarography of pyridoxine and some of its derivatives. *Collection* 29, 1432–57 (1964). [26]

127. O. Manoušek, and P. Zuman: Polarographic reduction of 2-(2-methyl-3-hydroxy-5-hydroxymethyl-4-pyridyl)thiazolidine-4-carboxylic acid. *Collection* 29, 17–22 (1964). [27]

128. M. Fedoroňko, and P. Zuman: Polarography of urea and thiourea derivatives XIII. Kinetics of base catalysed decomposition of S-substituted thioureas in alkaline solution. *Collection* **29**, 2115–23 (1964).

129. P. Čársky, P. Zuman, and V. Horák: Fissions of activated carbon-nitrogen and carbon-sulfur bonds V. Polarographic study of elimination of β-morpholino propiophenone. *Collection* **29**, 3044–56 (1964). [12]

130. P. Zuman, O. Manoušek, and V. Horák: Fission of activated carbon-nitrogen and carbon-sulphur bonds VI. Polarographic reduction of the $C - N^{(+)}$ bond in α-aminonitriles. *Collection* **29**, 2906–12 (1964). [23]

131. P. Zuman, and P. Čársky: Reaction kinetics of a reversible reaction which is coupled with two rapidly established equilibria. *Z. Physikal. Chem.* (Leipzig) **227**, 278–80 (1964).

132. P. Zuman, and O. Exner: Oxime derivatives VIII. Polarographic reduction of O- and N-substituted oximes. *Collection* **30**, 1832–52 (1965). [33]

133. P. Čársky, P. Zuman, and V. Horák: Fission of activated carbon-nitrogen and carbon-sulfur bonds VII. Kinetics of ketol formation from α,β-unsaturated ketones in alkaline media. *Collection* **30**, 4316–36 (1965). [18]

134. P. Zuman: Recent trends in organic polarography. Proc. 3rd Internat. Polarograph. Congress, Southampton 1964, 687–710, MacMillan, London 1966.

135. P. Čársky, V. Horák, and P. Zuman: Polarographic study of the kinetics of elimination reaction of β-morpholino propiophenone. Organic Reaction Mechanisms. *Chem. Soc. Spec. Publ.* 19, 180–81 (1965).

136. O. Manoušek, and P. Zuman: The polarographic and electrolytic reduction of the cyano group in substituted benzonitriles. *Chemical Communs.* 1965, 158.

137. I. Šestáková, P. Zuman, and V. Horák: Fission of activated carbon-nitrogen and carbon-sulphur bonds VIII. Elimination of β-piperidinoethyl phenyl sulphone methoiodide and reaction of phenyl vinyl sulphone with hydroxyl ions. *Collection* 31, 827–34 (1966). [15]

138. N. Kucharczyk, M. Adamovský, V. Horák, and P. Zuman: Slowly established acid-base equilibrium in organic polarography: Tautometric changes between 3-thianaphthenone and 3-hydroxy thianaphthene. *J. Electroanal. Chem.* 10, 503–10 (1965). [16]

139. P. Zuman: Polarography of alicyclic compounds. *Talanta* 12, 1337–79 (1965).

140. P. Zuman: Application of polarography in organic analysis and in the solution of some fundamental problems of organic chemistry. (in German) *Z. anal. Chem.* 216, 151–64 (1966).

141. B. Fleet, and P. Zuman: Quantitative treatment of substituent effects in polarography V. Polarographic reduction of some semicarbazones. *Collection* 32, 2066–88 (1967). [32]

142. L. Suchomelová, J. Zýka, and P. Zuman: Effects of structure of the alkyl group on the polarographic waves corresponding to catalytic hydrogen evolution by organic sulphoxides. *J. Electroanal. Chem.* 12, 194–98 (1966).

143. P. Zuman: Oxidation by chromic oxide. Complete kinetic treatment of the oxidation of secondary alcohols. *Collection* 32, 1610–13 (1967).

144. J. Kargin, O. Manoušek, and P. Zuman: Polarographic behaviour of p-diacetylbenzene. An example of a compound with two electroactive centres. *J. Electroanal. Chem.* 12, 443–61 (1966). [9]

145. P. Zuman: Polarographic behaviour of organic substances bearing two electronegative groups. (in German) *Z. anal. Chem.* 224, 374–83 (1967).

146. I. Šestáková, V. Horák, and P. Zuman: Fission of activated carbon-nitrogen and carbon-sulphur bonds IX. Polarographic study of addition of primary amines to phenyl vinyl ketone. *Collection* 31, 3889–902 (1966). [13]

147. P. Zuman: Physical organic polarography. *Progr. Physical Org. Chem.* (Streitwiesser A. Jr., Taft R. W. Eds.) 5, 81–206 (1967).

148. P. Zuman: Polarography and reaction kinetics. *Adv. Physical Org. Chem.* (R. Gold, Ed.) 5, 1–52 (1967).

149. J. Janata, O. Schmidt, and P. Zuman: Application of analogue computers to the studies of reactions involving electrolytically generated reagent II. Treatment of consecutive and competitive reactions. *Collection* 31, 2344–59 (1966).

150. A. Bashar, A. Townshend, and P. Zuman: Polarographic evidence for the structure of the so-called phenylthiohydantoic acid. *Chem. Commun.* 1967, 901–2.

151. P. Zuman, and D. Barnes: Desorption of 3-phenylpropionaldehyde preceding polarographic reduction. *Nature* 215, 1269–70 (1967). [6]

152. P. Zuman: Preparative methods used in the elucidation of organic electrode processes. *J. Polarograph. Soc.* 13, 53–70 (1967). [1]

153. P. Zuman: Some applications of polarography in organic chemistry. *Mises au Point de Chim. Anal. Org. Pharm. Bromatol.* 17, 219–28 (1968).

154. D. Barnes, R. Belcher, and P. Zuman: The use of submicroanalysis and flame ionization detecting gas-liquid chromatograph in the identification of electrolysis products at the dropping mercury electrode. *Talanta* 14, 1197 (1967).

155. D. J. Halls, A. Townshend, and P. Zuman: Polarography of some sulphur-containing compounds XII. Anodic waves of monoalkyldithiocarbamates and their reaction with heavy metals. *Anal. Chem. Acta* 40, 459–72 (1968).

156. D. J. Halls, A. Townshend, P. Zuman: Polarography of some sulphur-containing compounds. XIII. Anodic waves of dialkyldithiocarbamates. *Anal. Chim. Acta* 41, 51–62 (1968).

157. D. J. Halls, A. Townshend, and P. Zuman: Polarography of some sulphur-containing compounds XIV. Anodic waves of ethylene-bis-dithiocarbamates and their reactions with heavy metals. *Anal. Chim. Acta* 41, 63–74 (1968).

158. D. Barnes, and P. Zuman: Polarographic reduction of cinnamaldehyde; comparison with 3-phenyl-propionaldehyde and phenylpropargylaldehyde. *J. Electroanal. Chem.* 16, 575–82 (1968). [19]

159. O. Manoušek, O. Exner, and P. Zuman: Fission of activated carbon-nitrogen and carbon-sulphur bonds X. Polarographic reduction of substituted methyl phenyl sulphones. *Collection* 33, 3988–99 (1968). [28]

160. O. Manoušek, O. Exner, and P. Zuman: Fission of activated carbon-nitrogen and carbon-sulphur bonds XI. Polarographic reduction of substituted benzene sulphonamides. *Collection* 33, 4000–07 (1968). [29]

161. P. Zuman: Polarographic reduction of aldehydes and ketones I. Course of the electrode process. *Collection* 33, 2548–59 (1968). [2]

162. P. Zuman, and B. Turcsanyi: Polarographic reduction of aldehydes and ketones II. pH-dependence of· polarographic limiting currents of aryl alkyl ketones governed by the rate of simultaneous reactions with several acids producing the electroactive species. *Collection* 33, 3090–96 (1968). [3]

163. P. Zuman, and B. Turcsanyi: Polarographic reduction of aldehydes and ketones III. Effects of alkyl groups in the side chain on the half-wave potentials of deoxybenzoins and some other alkyl aryl ketones. *Collection* 33, 3205–12 (1968). [4]

164. P. Zuman, O. Exner, R. F. Rekker, and W. Th. Nauta: Polarographic reduction of aldehydes and ketones IV. Linear free energy treatment of substituted benzophenones. *Collection* 33, 323–26 (1968). [5]

165. I. Šestáková, J. Pecka, and P. Zuman: Polarographic reduction of aldehydes and ketones V. β-Ketosulphides. *Collection* 33, 3227–34 (1968). [14]

166. P. Zuman: Polarographic methods. *Fast Reactions in Enzymology* (Kustin, Ed.). Academic Press, New York 1969.

167. D. J. Halls, A. Townshend, and P. Zuman: Polarographic determination of some dithiocarbamates and their heavy metal complexes. *Analyst* 93, 219–23 (1968).

14

168. P. Zuman, D. Barnes, and A. Ryvolová-Kejharová: Polarographic reduction of some carbonyl compounds. *Disc. Faraday Soc.* **45**, 202–26 (1968).

169. D. Barnes, and P. Zuman: Polarographic determination of caesium in the presence of other alkali metals based on interaction with the cinnamaldehyde radical anion. *Analyst* **93**, 589–94 (1968).

170. P. Zuman, O. Manoušek, and S. K. Vig: Further examples of *para-* and *ortho*-disubstituted benzene derivatives reduced in a reversible two-electron step into products of limited stability. *J. Electroanal. Chem.* **19**, 147–55 (1968). [10]

171. O. Manoušek, P. Zuman, and O. Exner: Quantitative treatment of substituent effects in polarography VI. Polarographic reduction of benzonitriles *m-* and *p-*substituted with electronegative groups. *Collection* **33**, 3979–87 (1968). [35]

172. P. Zuman, and O. Manoušek: Fission of activated carbon-nitrogen and carbon-sulphur bonds XII. Polarographic reduction of substituted benzonitriles bearing in *para-*position a carbonyl group in acidic media. *Collection* **34**, 1580–94 (1969). [36]

173. P. Čárský, and P. Zuman: The use of simple molecular orbital theory to elucidate the polarographic behaviour of some *para*-substituted benzonitriles. *Collection* **34**, 497–503 (1969). [37]

174. P. Zuman, and A. Ryvolová-Kejharová: Desorption preceding polarographic reduction of some acetophenone derivatives in alkaline media. *Anal. Letters* **1**, 429–35 (1968).

175. B. Kakáč, K. Mnouček, P. Zuman, M. Semonský, V. Zikán, and A. Černý: Substances with antineoplastic activity XXII. Spectral and polarographic properties of β-4-methoxy-benzoyl-β-bromo and β-chloroacrylic acids and related compounds. *Collection* **33**, 1256 -77 (1968).

176. P. Zuman: Polarography in organic chemistry. *Fortschr. Chem. Forschg* **12**, 1–76 (1969).

177. D. Barnes, and P. Zuman: Polarographic reduction of aldehydes and ketones. Part 6. Behaviour of cinnamaldehyde at lower pH values. *Trans. Faraday Soc.* **65**, 1668–80 (1969).

178. D. Barnes, and P. Zuman: Polarographic reduction of aldehydes and ketones. Part 7. Cinnamaldehyde at higher pH-values. *Trans. Faraday Soc.* **65**, 1681–89 (1969).

179. P. Zuman: Polarography of biomolecules. J. Wiley (in press).

180. A. Ryvolová-Kejharová, and P. Zuman: Polarographic reduction of aldehydes and ketones VIII. Polarographic behaviour of chalcone and dihydrochalcone. *J. Electroanal. Chem.* **21**, 197–219 (1969).

181. D. Barnes, and P. Zuman: Effect of pH on concentration of species present in aqueous solutions of carbonyl compounds affected by hydration – dehydration and keto-enol equilibria. *Talanta* **16**, 975-93 (1969).

Part II

Techniques Used in Elucidation of Organic Electrode Processes

1. P. Zuman: Preparative methods used in the elucidation of organic electrode processes. *J. Polarograph. Soc.* 13, 53 (1967).

In the elucidation of organic electrode processes, it is essential to prove the structure of the electrolysis product or stable intermediate. This paper[1] discusses the various techniques used in the elucidation of organic electrode processes; in particular, the results obtained with a dropping mercury and mercury pool electrode are compared. Sources of differences between these two modes of electrolysis are discussed.

Preparative Methods used in the Elucidation of Organic Electrode Processes

by

P. ZUMAN

J. Heyrovský Institute of Polarography, Czechoslovak Academy of Sciences, Prague*

(*Submitted 21st April*, 1967)

For all applications of polarography both for theoretical and for practical purposes it is essential to understand the electrode process involved in the electrolysis with the dropping mercury electrode. How deep the understanding of the electrode process involved should be depends on the purpose for which the polarographic curves are applied.

In most cases electrode processes studied in organic polarography do not consist of a simple electron transfer, but the electrode process proper is accompanied by several steps involving mass transfer, chemical reactions, adsorption and desorption phenomena. Several combinations of these partial steps are possible. A typical example is considered in Table I, showing a two-step reduction process accompanied by chemical reactions and adsorption-desorption phenomena. The actual electrode processes are sometimes simpler and may involve, for example, only one single electrode process proper (in this case reaction IV is a consecutive rather than an interposed chemical reaction). Sometimes, nevertheless, the actual electrode process is more complex and is complicated, for example, by a side-reaction of the intermediate; by a two-step antecedent or consecutive reaction; by production of the original electroactive species by a chemical reaction from the intermediate or product; or by adsorption or desorption of the intermediate etc. The scheme in Table I was chosen to demonstrate the possible types of partial processes involved and their sequence.

* Present Address: Department of Chemistry University of Birmingham.

TABLE I

Scheme of the Electrode Process

Transported particle	
↓ k_1 : Antecedent chemical reaction	(I)
Electroactive particle	
↓ Adsorption; orientation	(II)
[Electroactive particle]$_0$	
↓ n_1i : Interaction with the electrode	
Transition state I	(III)
↓ Interaction with the electrode	
[Primary unstable intermediate]$_0$	
↓ k_2i : Consecutive chemical reaction	(IV)
[Stable intermediate]$_0$	
↓ n_2e : Interaction with the electrode	
Transition state II	(V)
↓ Interaction with the electrode	
[Product]$_0$	
↓ Desorption	(VI)
Product transported from the electrode	

Subscript zero means on the electrode surface.

To identify the partial steps involved and to elucidate the organic electrode process the information is gathered: (i) from the course of polarographic i–E curves and their changes with capillary characteristics and solution composition; (ii) from the results of other electrochemical techniques; (iii) from the results of preparative methods; (iv) from the effects of modifications in the structure of the electroactive substance (Table II).

TABLE II

Elucidation of the Electrode Process

(1)	Polarographic curves and their change with capillary characteristics and solution composition
(2)	Other electrochemical techniques
(3)	Preparative methods
(4)	Structural effects

(i) From the presence and changes of diffusion controlled and kinetic currents and from the shifts of half-wave potentials with solution composition it is possible to identify the electroactive species and to prove the participation of antecedent (Table I, step I), interposed (step IV), parallel (step IV producing the original electroactive particle) or consecutive (step IV, with step V absent) chemical reactions. Adsorption waves, wave-shapes and half-wave potentials can provide us with information on adsorption phenomena (steps II and VI)—the latter two quantities also on the electrode process proper (Scheme I, steps III and V). Finally the presence of two waves in the polarogram of a solution to which a single compound has been added can enable us to identify some intermediates or products provided that they are stable under the conditions under which they were formed (Table III).

TABLE III

Information Obtained from Polarographic Curves

Measured quantity	Information on
i_d, i_k, $E_{\frac{1}{2}}$	Electroactive species
i_k, i_d, $E_{\frac{1}{2}}$	Chemical reactions (antecedent, interposed, parallel, consecutive)
i_a, $E_{\frac{1}{2}}$, $\log{(i_d - i)}/i$	Adsorption
$E_{\frac{1}{2}}$, $\log{(i_d - i)}/i$	Electrode process proper
i_1, i_2 (two waves)	Intermediates and products

(ii) Cyclic polarisation using a hanging mercury drop electrode offers information on stable electrolysis products and the reversibility of the overall electrode process; the analogous technique using cathode-ray polarographs does the same for unstable products and intermediates. Similar information can be obtained from $dE/dt = f(E)$ curves recorded in oscillographic polarography and from commutator techniques. These latter two methods indicate also the presence of adsorption phenomena, whereas chronopotentiometry is useful for the study of consecutive chemical reactions. Significant for the study of adsorption are the changes of the instantaneous current with time (i–t curves) and a.c. polarography, as well as electrocapillary curves, measurements of differential capacity and pulse polarography. Several other new electrochemical techniques has been developed during the last decade, but because the number of systems to which these techniques have been successfully applied is still limited, it is at present difficult to assess their merits and scope of application (Table IV).

TABLE IV

Information Obtained by Means of other Electrochemical Methods

Method used	Information on
Hanging drop, cyclic polarisation	Intermediates, products (stable), reversibility
Single-sweep, cyclic polarisation	Intermediates, products (unstable), reversibility
Oscillographic polarography $dE/dt = f(E)$	Intermediates, products (unstable), reversibility, adsorption
Rectangular voltage polarisation (Commutator)	Intermediates, products (stable), reversibility, adsorption
Chronopotentiometry	Consecutive reactions
A.C. polarography	Adsorption (reversibility)
i — t curves	Adsorption (reversibility)
Electrocapillary curves	Adsorption
Differential capacity	Adsorption
Pulse polarography	Adsorption

(iii) Preparative methods offer us information on both stable and unstable intermediates and products of the electrode process proper. These methods will be dealt with in detail subsequently.

(iv) From a study of the effects of changes in the structure of the electroactive compound on half-wave potentials, generally on the polarographic behaviour and on the behaviour when studied by other electrochemical techniques, it is possible to make assumptions about the mechanism of the electrode process involved and in particular about the structure and stereochemistry of the transition state. Qualitative and quantitative approaches used in general organic chemistry are used for such purposes.

To summarise it can be said that the most important information on accompanying chemical reactions is obtained from kinetic currents, half-wave potentials and chronopotentiometry; on adsorption phenomena from i–t curves, adsorption currents, a.c. polarography and capacity measurements; on products and intermediates from preparative, cyclic and rectangular polarisation techniques and on transition states from structural effects (Table V).

TABLE V

Simplified Summary of the Main Sources of Information

Process	Measured quantity, method
Accompanying chemical	i_k, $E_{\frac{1}{2}}$, chronopotentiometry
Adsorption	$i - t$, i_a, A.C. polarography, differential capacity
Products, intermediates	Preparative, stripping, commutator
Transition state	Structural effects

PREPARATIVE METHODS

In preparative methods the product is formed using a controlled potential electrolysis at a chosen potential*. The electrolysis product can be identified either directly in the solution in which the electrolysis has been carried out or after isolation. When the identification is carried out in the electrolysed solution, both stable and unstable products† can be studied. Isolation is usually restricted to stable products. Only when scavengers are used that trap the reactive product

* According to the potential chosen the species formed is either an intermediate (when the potential corresponds to the first step of a multistep reduction) or the final product (when the potential corresponds to the only step observed or to the most negative reduction step).

formed and transform it in a fast chemical reaction into a less-reactive one, can isolation techniques be adopted also in the study of unstable products (Table VI).

TABLE VI

Classification of Preparative Methods

(1) Identification of the electrolysed product in the solution

    (a) Unstable species (ESR, u.v., electrochemical methods)

    (b) Stable species (ditto + chromatography, pK, colour reactions, identification—m.p.)
    (A)   D.M.E.
    (B)   Mercury pool electrode

(2) Isolation of the electrolysis product

    (a) Stable species

    (b) Trapping of unstable species

The generation of the product can be carried out either continuously, when the same voltage is applied during the whole period of electrolysis, or intermittently, when the voltage is regularly changed and periods during which the electrolysis is carried out are followed by periods when the voltage is chosen so that no electrolysis occurs, or finally interruptedly, when the applied voltage after a chosen period is switched off. When the identification of the product is carried out in the electrolysed solution, the concentration changes of the product are usually followed continuously, but can be recorded after pre-selected time intervals, too.

The method used for identification of the product in the solution during and after electrolysis depends on the stability of the species formed. For unstable species methods enabling identification of the product in dilute solutions and providing us with an electrical signal that is easily continuously recorded are preferred. Because unstable products are often radicals, electron spin resonance (ESR) has proved extremely useful, but u.v. absorption and electrochemical techniques can also be applied.

† A product is considered as stable or unstable according to the relation of the time necessary for its production and its identification. In all cases products with half-lives over 5 minutes are considered as stable, in some instances even those with half-lives over 5 seconds.

For stable species the same methods can also be used, but it is possible to use chromatography—in particular paper, thin-layer or gas chromatography—determination of pK values, chemical colour and other qualitative reactions, as well as identification techniques using preparation of an appropriate derivative and its characterisation. The use of infrared techniques is usually prevented by the presence of smaller or larger amounts of water in the electrolysed solution, the use of mass spectrometry and nuclear magnetic resonance by high dilution and presence of supporting electrolytes.

The identification of the electrolysis product in the electrolysed solution can be carried out either directly in the electrolysis cell or after transferring the solution to another vessel. For less stable products this transfer must take place quickly. Pumping the solution through a closed circuit has proved useful.

The generation of the electrolysis product would be carried out best using the dropping mercury electrode, because then the conditions are most similar to those used in polarography. For various technical reasons this arrangement cannot be used in connection with some of the identification techniques used, mainly due to the small current densities and hence low yields of products formed within the short time-periods comparable with the life-times of the products. Hanging mercury drop and in particular mercury pool electrodes with large surfaces covered with mercury have proved particularly useful.

EXAMPLES OF IDENTIFICATION OF UNSTABLE INTERMEDIATES AND PRODUCTS

The unstable intermediates and products formed in the electrolysis have often the nature of a radical or anion radical. ESR has proved most useful in the identification of the radicals formed, in particular in reduction of various carbonyl and nitro compounds.

For identification purposes the electrolysis can be carried out directly in the microwave cavity. This arrangement can be adopted even for extremely short-lived radicals, but due to the uneven distribution of the generated radicals in the cavity, it is difficult to follow quantitatively concentration changes of the radicals formed. Quantitative measurements are possible when the electrolysis is carried out in an outside cell and the solution containing the generated radical is pumped through the cavity. The shorter the life-time, the faster must be the transfer. Sufficiently stable radicals can be generated mainly in non-aqueous solvents, but some radicals are sufficiently stable to be followed even in water containing media. This is of importance, because the vast majority of polarographic data was obtained in water-containing solutions and it is this data which needs interpretation.

Using external generation—either continuous or interrupted—most detailed information was obtained on the one-electron reduction of nitrobenzene either in dimethylformamide solutions or in aqueous alkaline solutions containing 10% methanol as solvent and an addition of 0·06% triphenylphosphine oxide as a surfactant that prevents the uptake of further electrons.

It is possible to measure the ESR signals corresponding to the radical formed and to follow the time-changes in the anodic waves of the anion radicals and in the more positive one corresponding to phenylhydroxylamine. When moreover the time-changes in the reduction current of the nitro group were followed, and the effect of pH on the cleavage of the radical anion was examined, it was deduced[1] that the anion radical undergoes cleavage by a pH-independent first order reaction of $C_6H_5NO_2^{(-)}$ with $C_6H_5NO_2H$. The stoichiometry of the overall reaction is:

$$4C_6H_5NO_2^{(-)} + 4H_2O \rightarrow 3C_6H_5NO_2 + C_6H_5NHOH + H_2O + OH^-$$

In some nitrocompounds a relatively small change in the structure can result in a profound change in the course of the electrode process. Hence, whereas all three position isomers of chloronitrobenzene yield the expected chloronitrobenzene anion, for ortho-bromo nitrobenzene and all three iodonitrobenzenes in dimethylformanide dehalogenation takes place first.[2,3] The two-electron reduction corresponding to the substitution of halogen by hydrogen is followed by a one-electron uptake, in which the nitrobenzene anion (and not halogeno-nitrobenzene anion) is formed. By adding various amounts of water to the dimethylformamide the contribution of dehalogenation and that of halogenonitrobenzene anion radical formation can be varied.

Not only halogen can be eliminated; the nitro group in the form of nitrite anion can also be eliminated from some polynitrobenzenes[4,5] (with formation of nitrophenol anions), from aliphatic trinitrocompounds[6] (formation of anions of dinitro compounds) and from 9-nitroanthracene.[2]

An example of the application of u.v. spectra is the one-electron electrolysis of camphorquinone that was carried out at pH >7 in the presence of surface active substances.[7] The anion radical formed in this process can be followed by the change of the absorption at 330 mμ, or by the height of its reversible anodic wave. The anion radical formed is stabilised in alkaline media by a second order process. After interruption of electrolysis the camphorquinone concentration does not change with time and the stabilisation reaction is hence not a dismutation process which would produce further camphorquinone even after the interruption of electrolysis. At pH 9 the stabilisation reaction follows first order kinetics and is attributed to protonation of the anion radical.

EXAMPLES OF IDENTIFICATION OF STABLE INTERMEDIATES AND PRODUCTS

For identification of stable species formed during electrolysis in solutions, either the dropping mercury electrode or a mercury pool electrode is usually used as a working electrode.

With a dropping electrode usually a small volume of 0·1–1·0 ml of the electrolysed

solution is used, the concentration of which is usually millimolar or smaller. Currents hence usually do not exceed 30 μA and 50% of the reduced form can be reached within 2–5 hours. During electrolysis a stream of pure nitrogen is passed over the surface of the solution (this is used also in techniques using the pool electrode). Experimentally simple to use is stirring of the solution by means of the falling mercury drops. For this purpose a thin, drawn-out capillary is recommended and its proper position in the centre of the vessel, near the surface of the liquid, is essential.

With a mercury pool electrode large volumes, 5·0 to 100 ml of the solution, are usually electrolysed. The concentration of the electrolysed solution is usually below 0·01M, which with the more conductive solutions gives a current flowing of about 10–300 mA. In this way the electrolysis of 95% of the substance can usually be carried out within 0·5 to 2 hours. Intensive stirring is essential; a magnetic stirrer immersed in the mercury proved useful for this purpose.

In both these techniques the second working electrode can be placed directly in the electrolysed solution, provided that an irreversible electrode process is involved. A mercury pool electrode can be used as the second working electrode when the dropping electrode is used and a graphite rod electrode proved suitable in the combination with the mercury pool electrode. Nevertheless, a separation of the second working electrode by a salt bridge is generally recommended and if reversible systems are studied, such separation is necessary. A separated calomel electrode is usually used in connection with the dropping electrode together with an agar bridge and a sintered glass disc, to prevent diffusion to and from the electrolysis compartment. It should be checked that no substance is adsorbed at the salt bridge and that the surface active substances diffusing from the bridge do not affect the studied polarographic waves. These and other interfering factors can be detected when the solution to be electrolysed is first left in a disconnected electrolysis vessel for a time comparable with the time needed for electrolysis. Any change in polarographic curves during such a period without electrolysis indicates complications to be taken into account and taken care of.

When the dropping mercury electrode is used, it should be checked that the stirring of the solution by the drops of mercury is sufficiently efficient. To prove this a stream of nitrogen is introduced into the solution after the electrolysis has been carried out for some time. The limiting current before and after the introduction of nitrogen should be identical. If there is a difference the capillary position should be adjusted first. When solutions are electrolysed that may react with the metallic mercury the dropped off mercury can be covered by a layer of tetrachloromethane, chloroform or chloro- or bromobenzene.

The second working electrode used in connection with the mercury pool electrode, e.g. a graphite rod, should also be separated by a salt bridge from the electrolysed solution. Because sometimes gaseous or other diffusing products are formed at this electrode, it proved useful to purge nitrogen also through this electrode compartment. A third auxiliary electrode, usually a stem-like calomel electrode, is immersed into the electrolysed solution and placed in the vicinity of the working electrodes, where mercury pool electrodes are used.

The course of electrolysis is conveniently followed by recording polarographic i–E curves after chosen time-intervals. When a dropping mercury electrode is used as a working electrode for electrolysis, the same dropping electrode can also be applied for recording the polarographic curves. In this case the second working separated electrode can be utilised as a reference electrode. When electrolysis is carried out with a mercury pool electrode, an additional dropping electrode is immersed into the electrolysed solution. In this case the third, monitoring auxiliary electrode can be used as the reference electrode for polarography.

The recording of whole i–E curves at selected intervals during electrolysis is pre-

ferred to the continuous recording of limiting currents only. As will be demonstrated on some examples below, changes in the wave heights, wave-shapes and of whole polarographic curves that occur during electrolysis can reveal some important information on the course of the electrolytic process.

When the dropping mercury electrode is used, the potential can often be controlled manually.[8] When ill-separated waves are investigated or when a mercury pool electrode is used, potentiostats are used with a current output of 10–300 mA, a potential range of $+1 \cdot 0$ V to $-2 \cdot 5$ V, a time constant of voltage compensation of the order of seconds, the possibility of working against an iR-drop of 10–100 V and an output potential constant within 2–5 mV.

The methods used for identification of the electrolysis products vary according to the type of electrode used. The procedures used in connection with the dropping mercury and mercury pool electrodes will therefore be discussed separately.

TABLE VII

Factors used in the Identification of Products of Electrolysis using the D.M.E.

(1) $\log i = f(t)$ (linearity, n)

(2) Waves of products

(3) Identification of product [methods given in Table VI, (1) (b)]

(4) Changes in the heights of waves at a more negative potential than that at which the electrolysis was carried out

When the **dropping mercury electrode** is used (Table VII), the logarithm of the wave-height during electrolysis is plotted as a function of time. For many simple electrolytic processes the $\log i = f(t)$ plot is linear. Deviations from linear plot can be caused either by irregularities in the experimental set-up or by a more complex course of the electrode process. When these deviations are reproducible and when under identical experimental conditions no deviations are observed for a suitable model substance (e.g. Cd^{2+}-ions), a more complicated course of the electrode process can be anticipated.

An example of deviations from the linear plot was observed[9] in the reduction of α-furildioxime at pH 6·5 and 9·8. The log $i = f(t)$ plot can be approximated by two linear sections with different slopes. The polarographic reduction waves corresponded to an uptake of 6 electrons, whereas in the controlled potential electrolysis 1,2-bis(2-furyl)ethylenediamine was formed. The formation of this compound would correspond to an eight-electron overall process. This would indicate a mechanism:

$$-C \underline{\quad\quad} C- + 6e \rightarrow \text{Intermediate}$$
$$\overset{\|}{NOH} \quad \overset{\|}{NOH}$$

Intermediate $\overset{k_1}{\rightarrow}$ Electroactive species
Electroactive species $+ 2e \rightarrow -CH-CH-$
$$\underset{NH_2}{|} \quad \underset{NH_2}{|}$$

Reduction of the dioxime into an electroinactive (in the given potential range) intermediate corresponds to the first part of the log $i = f(t)$ plot that is steeper. The rate of the second electrolytic step corresponding to the less steep linear part of the log $i = f(t)$ dependence is governed by the rate of the chemical reaction corresponding to the transformation of the electroinactive intermediate into an electroactive species.

From the slope of the linear parts of the log $i = f(t)$ dependence it is possible to determine the value of the number of electrons consumed in the electrode process, provided that the electrolysis product does not undergo successive chemical reactions.

Recording of the waves of products formed during the controlled potential electrolysis can offer us important information about the course of the electrode process, In the reduction of terephthalic acid dinitrile in 0·02M NaOH anodic waves of cyanide ions are formed[10] according to reaction:

Similarly the reduction of the dehydroascorbic acid phenylhydrazone gave anodic waves of scorbaminic acid[11] and that of 3-phenylsydnone anodic waves of the hydrazine derivatives[12] according to the scheme:

Products of the controlled potential electrolysis using the D.M.E. can be also identified using methods suitable for the small volumes of the solution and the low concentrations of the electrolysis product. For instance the reduction of p-cyanoacetophenone in acid media followed the scheme[13]:

The p-methyleneamino acetophenone formed was identified using the u.v. spectrum of the $-C_6H_4CO-$ grouping and using the ninhydrin reaction for the primary amino group. The pK-value of the hydrazine derivative $R - N(NH_2) CH_2COO^-$ was used[13] for the identification of this compound formed in the sydnone reduction mentioned above.

In some cases important information can be obtained from the polarographic curves recorded during the electrolysis, when changes are observed in wave-heights at other potentials (usually more negative) than those at which the electrolysis was carried out. An example of this type is the polarographic reduction of p-diacetylbenzene.[14]

This compound is reduced at pH 2–5 in one single two-electron reduction wave at potential E_1, corresponding to the formation of a biradical. No wave at more negative potentials that would correspond to the reduction of the carbonyl group in p-$CH_3CH(OH)$-$C_6H_4COCH_3$ was observed (Fig. 1, curve 1). When the electrolysis was carried out at the limiting current of this single wave, a new wave formed at more negative potentials (E_2). This is explained by the scheme:

$$CH_3\overset{\underset{\displaystyle OH^+}{|}}{C}\!-\!\!\underset{}{\bigcirc}\!\!-\!\overset{\underset{\displaystyle OH^+}{|}}{C}\!-\!CH_3 + 2e \overset{E_1}{\rightleftharpoons}$$

$$CH_3\overset{\underset{\displaystyle OH}{|}}{\overset{\bullet}{C}}\!-\!\!\underset{}{\bigcirc}\!\!-\!\overset{\underset{\displaystyle OH}{|}}{\overset{\bullet}{C}}CH_3$$

$$CH_3\overset{\underset{\displaystyle OH}{|}}{\overset{\bullet}{C}}\!-\!\!\underset{}{\bigcirc}\!\!-\!\overset{\underset{\displaystyle OH}{|}}{\overset{\bullet}{C}}CH_3 \overset{k}{\rightarrow}$$

$$CH_3\overset{\underset{\displaystyle OH}{|}}{CH}\!-\!\!\underset{}{\bigcirc}\!\!-\!COCH_3$$

$$CH_3\overset{\underset{\displaystyle OH}{|}}{CH}\!-\!\!\underset{}{\bigcirc}\!\!-\!CO\,CH_3 + e + H^+ \overset{E_2}{\rightarrow}$$

$$CH_3\overset{\underset{\displaystyle OH}{|}}{CH}\!-\!\!\underset{}{\bigcirc}\!\!-\!\overset{\underset{\displaystyle OH}{|}}{\overset{\bullet}{C}}CH_3$$

In the electrolysed solution transformation of the biradical into the acetophenone derivative with a rate constant k takes place too slowly to produce a sufficient amount of the ketol p-CH_3CHOH–$C_6H_4COCH_3$ at the surface of the electrode, so that the wave at potential E_2 is absent before electrolysis. Formation of the ketol in the solution gives rise to the more negative wave (Fig. 1, curves 2–9) during electrolysis.

When the **mercury pool electrode** is used in the electrolysis the methods of treatment of the time-dependence of the limiting current, of using the waves of the reduction products and the identification of the electrolysis products are similar, as in the cases in which the dropping mercury electrode is used. In the identification of the electrolysis product at the mercury pool electrode larger volumes and more concentrated solutions of the product are available so that the choice of the identification techniques is less restricted than with

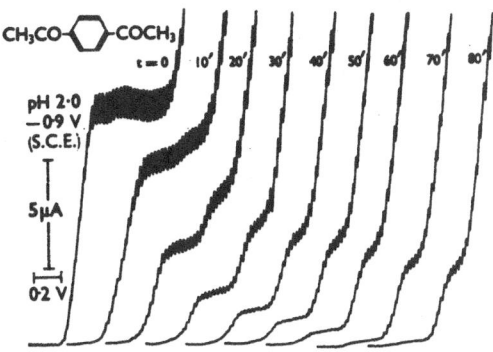

Fig. 1. Change in polarographic curves of p-diacetylbenzene during electrolysis at the potential of the limiting current. $5 \cdot 10^{-3}$M p-diacetylbenzene in a Britton–Robinson buffer, pH 2·0, containing 10% ethanol electrolysed at 0·9 V (S.C.E.); 50 ml of soln., stirred pool electrode with a surface of 20 cm²; polarographic curves recorded starting at $-0\cdot2$ V (S.C.E.).

the D.M.E. Chromatographic methods can be successfully used, as for the identification of ethylbenzoate in the reduction of p-cyanoethylbenzoate[10] or for the identification of acetonitrile in the reduction of diethylamino acetonitrile methoiodide.[15]

For compounds bearing two electroactive groups that give rise to two separated reduction waves, some information can be obtained from the study of the successive changes of polarographic curves recorded in the course of electrolysis. When the electrolysis is carried out at the potential corresponding to the limiting current of the more positive wave, the height of the more negative wave can remain unchanged, can decrease or can increase. The simplest example is when the height of the more negative wave remains unchanged during electrolysis (Fig. 2). Two possibilities exist:

(1) The reduced compound contains one single electroactive group R^1 that is reduced in two successive steps, first to R_I and then to P:

$$R^1 - A + n_1e \underset{E_1}{\rightarrow} R_I - A$$

$$R_I - A + n_2e \underset{E_2}{\rightarrow} P - A \quad (\text{where } E_1 \neq E_2)$$

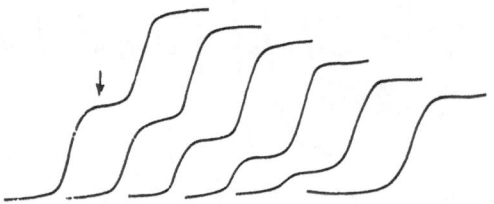

F_{IG}. 2. Change in polarographic curves of a substance showing two waves during electrolysis at the potential of the limiting current of the first wave (schematically).
Arrow indicates potential at which the electrolysis is carried out. The height of the second wave remains unchanged.

(2) The reducible compound contains two electroactive groups R^1 and R^2 which are successively reduced to P^1 and P^2, respectively.

$$R^1 - A - R^2 + n_1e \rightarrow P^1 - A - R^2$$
$$E_1$$

$$P^1 - A - R^2 + n_2e \rightarrow P^1 - A - P^2$$
$$E_2$$

Such a course of the curves as given in Fig. 2 permits the exclusion of a scheme, according to which in the first wave substance $R^1 - A$ with the consumption of n_1 electrons forms a product $P^1 - A$, whereas in the second wave reduction of $R^1 - A$ with consumption of $(n_1 + n_2)$ electrons produces another product $P^2 - A$ and the reduction follows another path, unrelated to that in the first wave :

$$R^1 - A + n_1e \rightarrow P^1 - A$$
$$E_1$$

$$R^1 - A + (n_1 + n_2)e \rightarrow P^2 - A$$
$$E_2$$

An example of the compound containing a single electroactive group which is successively reduced is the reduction of bis-pyridoxyl-disulphide (RSSR) in which pyridoxthiol (RSH) is formed in the first

step and reduced in the second step[16] :
$$RSSR + 2e + 2H^+ \rightarrow 2\ RSH$$
$$RSH + 2e + H^+\quad \rightarrow RH + SH^-$$

An example of the compound bearing two electroactive centres is p-cyanoaceto-phenone. In acid media this compound is first reduced[13] on the nitrilo group and then on the carbonyl group :

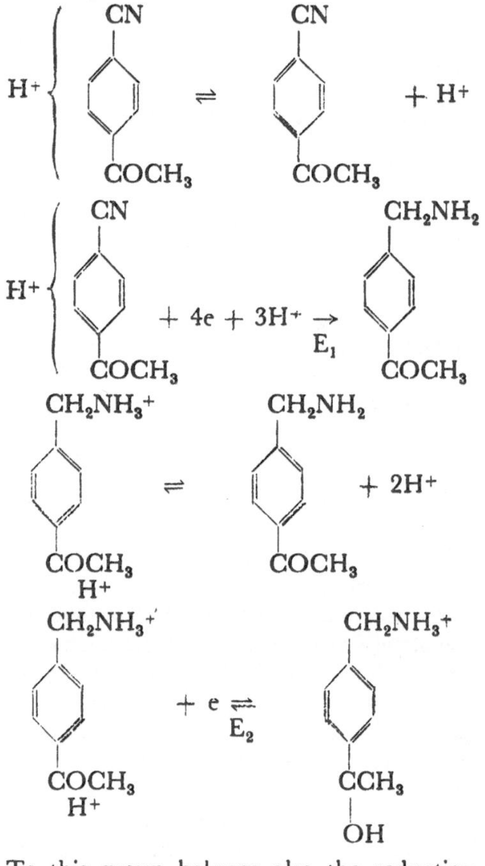

To this group belongs also the reduction of 2-(4-pyridyl)-thiazolidine-4-carboxylic acid[8] in the medium pH-range which takes place as follows :

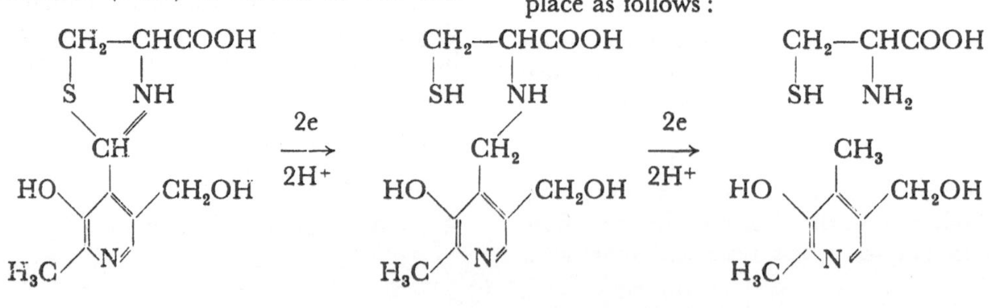

When the height of the second wave decreases during electrolysis carried out at the potential of the limiting current of the first wave, the ratio of the first and second waves i_1/i_2 can remain constant during electrolysis, but can also decrease in the course of the electrolytic process.

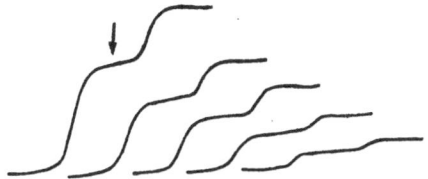

FIG. 3. Change in polarographic curves of a substance showing two waves during electrolysis at the potential of the limiting current of the first wave (schematically).
Arrow indicates potential at which the electrolysis is carried out. The height of the second wave decreases, ratio of the height of the first and second wave remains constant.

The ratio i_1/i_2 remains constant (Fig. 3) when the reduction follows two independent paths :

$$R^1 - A + n_1e \rightarrow P^1 - A \atop E_1$$

$$R^1 - A + (n_1 + n_2)e \rightarrow P^2 - A \atop E_2$$

or when the intermediate is deactivated by a reaction with constant k' corresponding to a half-time of 10–1000 seconds according to scheme :

$$R^1 - A + n_1e \rightarrow R_1 - A \atop E_1$$

$$R_1 - A \xrightarrow[k']{} P^1 - A$$

$$R_1 - A + n_2e \rightarrow P_2 - A \atop E_2$$

The rate of the deactivation reaction with constant k' is too slow to affect polarographic curves, but fast enough to deactivate the intermediate $R_1 - A$ during the electrolytic process. Sometimes an inactive intermediate $P^1 - A - R_1$ is formed that must be first transformed in a fast reaction into the electroactive intermediate $P^1 - A - R^2$:

$$R^1 - A - R_1 + n_1e \rightarrow P^1 - A - R_1 \atop E_1$$

$$P^1 - A - R_1 \xrightarrow[fast]{} P^1 - A - R^2$$

$$P^1 - A - R^2 \xrightarrow{k} P^1 - A - P^2$$

$$P^1 - A - R^2 + n_2e \rightarrow P^1 - A - P^3 \atop E_2$$

Examples of systems, in which a decrease of both waves in the same ratio was observed, are reductions of p-cyanobenzophenone in acid media[13] the detailed course of which has not yet been explained, and those of some azauracil derivatives.[17]

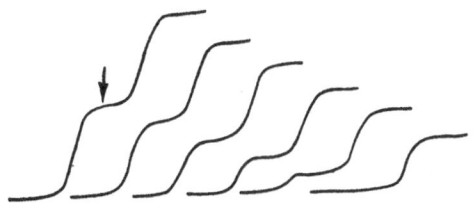

Fig. 4. Change in polarographic curves of a substance showing two waves during electrolysis at the potential of the limiting current of the first wave (schematically).
Arrow indicated potential at which the electrolysis is carried out. The height of the second wave decreases, ratio of the height of the first and second wave decreases.

One possible explanation for systems in which the ratio i_1/i_2 decreases during electrolysis and the height of the first wave decreases more than that of the second wave (Fig. 4) is that the electroactive compound $R^1 - A - R^2$ undergoes parallel to the electroreduction a chemical deactivation. Both electroreduction and chemical reaction produce the same intermediate $P^1 - A - R^2$ by processes of comparable rate :

$$R^1 - A - R^2 + n_1e \rightarrow P^1 - A - R^2 \atop E_1$$

$$R^1 - A - R^2 \xrightarrow{k} P^1 - A - R^2$$

$$P^1 - A - R^2 + n_2e \rightarrow P^1 - A - P^2 \atop E_2$$

26

This type of reaction takes place when the electroactive compound $R^1 - A - R^2$ undergoes a chemical reaction (e.g. hydrolysis) in the medium in which the electrolysis is carried out. It is useful to check for this reaction in the given media, when no voltage is applied. For instance 2-(4-pyridyl)-thiazolidine-4-carboxylic acid undergoes hydrolysis in acid media. When the electrolysis of this compound is carried out in acid solutions the height of the first wave decreases partly due to reductive cleavage, partly due to hydrolysis.

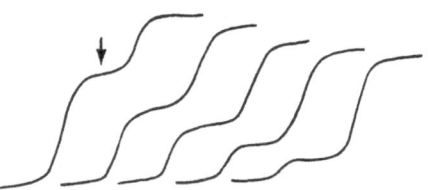

Fig. 5. Change in polarographic curves of a substance showing two waves during electrolysis at the potential of the limiting current of the first wave (schematically).
Arrow indicates potential at which the electrolysis is carried out. The height of the second wave increases.

The increase of the more negative second wave (Fig. 5) indicates that the electro-inactive product of the first reduction step $(P^1 - A - R_1)$ is transformed in a comparable slow process with rate constant k into an electroactive form $(P^1 - A - R^2)$:

$$R^1 - A + R_1 + n_1e \rightarrow P^1 - A - R_1$$
$$E_1$$

$$P^1 - A - R_1 \quad \overset{k}{\underset{slow}{\rightarrow}} P^1 - A - R^2$$

$$P^1 - A - R^2 + n_2e \rightarrow P^1 - A - P^2$$
$$E_2$$

During the drop-time only a part of $P^1 - A - R_1$ can be transformed into $P^1 - A - R^2$, whereas during the time necessary for electrolysis the greater part of $P^1 - A - R_1$ is converted into $P^1 - A - R^2$. Hence the second wave at potential E_2 which is small (or even absent) on the polarographic curve before electrolysis, increases during electrolysis.

Another possibility is that the primary electrolysis product $P^1 - A - R^2$ is transformed by a chemical reaction, into another species $P^1 - A - R^3$, which is reduced at a similar potential as $P^1 - A - R^2$, but by a higher number of electrons :

$$R^1 - A - R^2 + n_1e \rightarrow P^1 - A - R^2$$
$$E_1$$

$$P^1 - A - R^2 + n_2e \rightarrow P^1 - A - P^2$$
$$E_2$$

$$P^1 - A - R^2 \quad \overset{k}{\underset{slow}{\rightarrow}} P^1 - A - R^3$$

$$P^1 - A - R^3 + n_3e \rightarrow P^1 - A - P^3$$
$$E_3$$

where $n_3 > n_2$ and $E_3 \approx E_2$. The transformation of $P^1 - A - R^2$ into $P^1 - A - R^3$ is slow to affect polarographic waves, but can increase the height of the second wave in electrolysis.

The example, in which the second wave increased during electrolysis is the reduction of p-diacetylbenzene (Fig. 1), discussed above. In this case the activating reaction is the transformation of the biradical into p-$CH_3CHOHC_6H_4COCH_3$ which can undergo further reduction.

Determination of yields of products during electrolysis is of importance, because comparison of the theoretical with the experimental yield can offer the proof that the expected product is the only one, that no by-products are formed. In this way the existence of side-reactions or of consecutive reactions which the reduction product undergoes can be detected.

Absence of side-reaction has been proved by determining the yields of reductions of various substituted benzonitriles.[10] In the reduction of p-cyanobenzene sulphonamide the yield of ammonia resulting from the cleavage of the sulfaminic acid, in the reduction of p-cyanophenyl methyl sulphone the yield of sulphur dioxide and in the reduction of p-cyanobenzoic acid the yield of cyanide ions were determined. The yields can be detected by classical methods, e.g. that of ammonia by titrimetry, but it is elegant to use polarography for the determination of yield. In this way yields of sulphur dioxide and cyanide ions were determined.

A special case of that type is the determination of the overall number of protons consumed in the electrode process. For this purpose the reduction is carried out in unbuffered media. Because of hydrogen ions consumed the solution turns alkaline in the course of electrolysis. Equivalents of hydroxyl ions liberated can be then determined titrimetrically. In this way the consumption of one hydrogen ion was confirmed in the reduction of diaryl and aryl alkyl sulphones.[18]

Determination of yields indicated a side-reaction in the reduction of benzaldehyde semicarbazones.[19] A hydration or hydrolysis of the Schiff base formed as an intermediate was considered as an explanation. For p-acetamidobenzaldehyde thiosemicarbazone, in addition to p-acetamido benzylamine p-acetamidobenzyl thiosemicarbazide was also found[20] and the proportion of the latter increased with increasing pH. At pH 6.5, where the wave-height at a dropping mercury electrode corresponded to a four-electron process, the controlled potential electrolysis using a mercury pool electrode yielded 30% of the p-acetamido-benzylsemicarbazide, whereas at pH 8.3 (3.3 electrons) 40% and at 11.5 (2.9 electrons) 75% of this product of the two-electron reduction was found. It is assumed that the protonised form of semicarbazone is reduced in a four-electron step to benzylamine whereas the unprotonised form in a two-electron step to benzylsemicarbazide. The higher yield of the electrolysis at pH 6.5 and 8.3, when compared with the wave-height, indicates that the protonation at the mercury pool electrode occurs to a lesser degree than at the dropping electrode.

Comparison of polarographic curves with the current-voltage curves obtained with stirred mercury pool electrodes is of importance in the use of these electrodes both in identification of the electrolysis products in the solution and in isolation of the electrolysis products.

In most cases the number, the shape and half-wave potentials of current-voltage curves obtained with the dropping mercury and mercury pool electrodes are identical. An example of this type of behaviour is the curves of p-diacetylbenzene. Some cases have been nevertheless reported recently[11] in which differences are observed between the curves recorded with the two types of electrodes. The curve recorded with the pool electrode can be either more negative (as it is in the case of p-cyanoacetophenone) or more positive (some pyridazone derivatives[21] in acid media) than the wave obtained with the dropping mercury electrode. In some cases, e.g. for pyridazone[21] at pH 4.7, the curves recorded with the mercury pool electrode do not resemble at all the curves obtained with the dropping electrode. They differ not only in the shape of the waves and their potentials, but also in the number of waves observed.

When the current-voltage curves are recorded with a rate of scanning comparable with those usually used in classical polarography (i.e. about 400 mV/min.) and with a comparable current density, the curves recorded with a stirred pool electrode are quite reproducible.

TABLE VIII

Factors Used in the Identification of the Products of Electrolysis using a Mercury Pool Electrode

(1) $\log i = f(t)$ (linearity)

(2) Waves of products

(3) Identification of products [methods given in Table VI, (1) (b)]

(4) Changes in the height of the more negative wave

(5) Quantitative yields

(6) $i - E$ curves with a stirred pool electrode

IDENTIFICATION OF ELECTROLYSIS PRODUCTS AFTER ISOLATION

For preparative purposes, when the electrolysis products are to be isolated from the solution, mercury pool electrodes are usually used and the electrolysis is carried out at higher concentrations of the electrolysed substance — 0·01M or even more. Even in this case it proved useful to check the course of electrolysis, by recording polarographic curves, but because at such high concentrations polarographic curves are often not well developed (due to the high iR drop, the high ratio of depolariser to the concentration of the supporting electrolyte, or adsorption phenomena), it is often necessary to dilute the sample for polarographic analysis.

TABLE IX
Identification of the Electrolysis Products after Isolation

Experimental conditions :	(1)	Hg-pool electrode
	(2)	higher concentrations
Isolation methods :	(1)	Classical (extraction, distillation, precipitation)
	(2)	Preparative chromatography
	(3)	Trapping of unstable products
Identification :	(1)	Methods used for stable products in the solution
	(2)	IR, NMR, mass-spectrometry

As isolation methods (Table IX) both classical, such as extraction, distillation and precipitation and the more sophisticated methods, such as various types of preparative chromatography, including ion-exchange, can be used. Presence of high concentrations of neutral salts and buffer components can sometimes present difficulties.

When the electrolysis product or intermediate is unstable, it is necessary to trap it. An electroinactive chemical reagent is added to the electrolysed solution which does not react with the original components of the reaction mixture but which reacts with the unstable intermediate or product and transforms it into a stable compound. This compound is then isolated. In such a way carbon dioxide was introduced into electrolysed dimethylformamide solutions, when the formation of anion and dianion radicals was proved. Carboxy derivatives isolated in this way served as proofs of anion radical formation in the reduction of naphthalene,[22] phenanthrene,[22] diphenylacetylene,[22] benzophenone[23] and some α,β-unsaturated ketones.[24] Also alkyl halides can be used as trapping agents. When the reduction of anthraquinone in acetonitrile solution was carried out in the presence of ethyl bromide[25] a diether was formed and this indicates dianion formation. Similarly, the reduction of benzophenone in a dimethyl formamide solution in the presence of ethyliodide yields diphenylethylcarbinol.[23] This indicates formation of a ketyl anion as intermediate.

Low temperatures can be used in conjunction with the addition of the trapping agent. For example, when the reduction of carbon tetrachloride in acetonitrile solution was carried out[26] at $-20°C$ in the presence of tetramethyl ethylene, 1,1-dichloro-2,2,3,3-tetramethylcyclopropane was identified as one of the products by gas chromatography. This was considered as a proof of the electrolytic formation of dichlorocarbene (CCl_2) according to the scheme :

$$CCl_4 + 2e \rightarrow CCl_3^- + Cl^-$$
$$E_1$$

$$CCl_3^- \rightleftharpoons CCl_2 + Cl^-$$
$$k$$

$$CCl_2 + (CH_3)_2 C = C (CH_3)_2 \rightarrow$$

$$CCl_2 + 2e \rightarrow \text{Products}$$
$$E_2$$

This interpretation is in agreement with the observation that whereas in water-containing solutions carbon tetrachloride is reduced in three successive two-electron steps of equal height, corresponding each to cleavage of one C–Cl bond, in acetonitrile solution carbon tetrachloride gives only two waves of equal height.[26]

The stereospecificity of some electro-organic reductions promises to be a very important tool in the study of the mechanisms of electrode processes. The isolation and identification of the stereoisomers is able to offer information not only about the final products, but also about the structure and stereochemistry of the transition state. This powerful tool has been used so far only in two problems: For α,α'-dibromosuccinic acid, the **threo**form of the free acid, the monoanion and the dianion are reduced to fumaric acid, and so is the **erythro**-epimer in the free acid and dianion-form. The univalent anion is at least partly reduced to the maleic acid. Dialkyldiesters of both **threo**- and **erythro**-dibromosuccinic acids show a behaviour similar to that of the undissociated free acid and are reduced to the dialkylesters of fumaric acid.[27,28]

The second case is the one-electron reduction of $\Delta^{1,4}$-ketosteroids,[29] in which various stereoisomeric pinacols are formed, according to the pH of the electrolysed solution. In acidic solutions the protonised form of the ketosteroid is reduced and a pinacol with hydroxy groups in the α-position is formed. In alkaline media, the unprotonised ketosteroid is reduced and the isomer with hydroxyls in the β-position results.

For the identification of the products isolated after electrolysis all the methods enumerated for the identification of the more stable products in the solution can be used. Furthermore, melting points can be measured and elemental analysis carried out. Infrared spectra, nuclear magnetic resonance and mass-spectrometry are further powerful tools, especially when used in combination.

SOURCES OF DIFFERENCES

In most cases products identified or isolated after a controlled potential electrolysis with a dropping mercury or mercury pool electrode are identical with those formed at the surface of the mercury electrode. Nevertheless, it is important to be aware of the possibility that products identified with a mercury pool electrode can be different from those identified in methods using the dropping mercury electrode.[30] Moreover even when using the dropping

mercury electrode only, it is in some cases possible to find after controlled potential electrolysis in the solution other forms or compounds than those formed at the surface of this electrode during polarographic electrolysis. Some of the reasons for these differences will be discussed now.

The results obtained in the controlled potential electrolysis using the dropping electrode may differ from those obtained when the electrolysis is carried out under polarographic conditions, because of consecutive reactions of products in the bulk of the solution, including the reaction of products with one component of the original reaction mixture. Care should be also taken to avoid the effects of the presence and reduction of traces of oxygen, as well as those caused by the volatility of the components of the original solution, the product or the solvent. The steady flow of nitrogen can cause variations of wave-heights which are best checked when the electrolysis is interrupted and curves recorded for some time with and without nitrogen flowing.

When the controlled potential electrolysis is carried out using a mercury pool electrode the time factor and other effects just mentioned which play a role with the dropping electrode can be involved as well. Moreover, due to the difference in conditions for the electrolytic process, some further factors can affect the results obtained with the pool electrode (Table X). First of these factors is

TABLE X
Sources of Differences and Errors

Source	Causes
Prolonged time	Consecutive reaction of products
Enlarged surface	Surface reactions Adsorption phenomena Change in the rate of accompanying chemical processes
Higher concentrations	Higher order reactions
Buffering, insufficient	Change in pH

the increased surface of the electrode and the prolonged contact of the solution and products (accumulation of products) at the surface of the pool electrode. Surface reactions can take place to a greater extent and even other types of surface reaction can predominate over those occurring with the dropping electrode, the rate of which may be affected by adsorption of the participating species. For electrode processes accompanied by chemical reactions, the rate of the chemical reaction furnishing the electroactive species can be different at the dropping and pool electrodes. The value i/i_d at a given pH measured with a dropping electrode may be different from the value of the ratio i/i_d in the buffer of the same composition and pH measured from the current-voltage curve obtained with a mercury pool electrode. The over-voltage can generally be different at the surface of the pool electrode when compared with the dropping electrode and in particular the current due to hydrogen evolution can affect the i–E curves in different ways.

Furthermore, the higher concentrations of the electroactive species usually used in the mercury pool electrolysis can increase the rate of chemical reactions of higher order. It is understood that the ratio of the concentrations of buffer solutions to the concentration of the electrolysed substance must be higher in controlled potential electrolysis with a pool electrode than with a dropping electrode. But even when well-buffered solutions are used in the electrolysis, it can be questionable, in particular with higher current densities, whether the acidity at the surface of the pool electrode will remain during electrolysis the same as the pH in the bulk of the solution. Finally, especially when working at concentrations below $5 \cdot 10^{-4} M$ and in cases when the electrolysis is carried out at potentials differing little from the potential of the final rise of current of the supporting electrolyte, it is necessary to keep in mind that part of the current is consumed in the charging of the electrode and in the reduction of cations of the supporting electrolyte.

It is of importance to compare results obtained with the dropping mercury electrode with those resulting from electrolysis with a pool electrode. Some examples will be given.

If a chemical reaction interposed between two electrochemical steps proceeds too slowly to affect the electrolysis at the dropping electrode, but with a rate sufficient to affect the electrolysis at the pool electrode, the wave of the intermediate C is not observed on polarographic curve i–E, but the product D is identified after a pool electrolysis:

$$A + n_1e \xrightarrow{E_1} B$$

$$B \xrightarrow{k} C$$

$$C + n_2e \xrightarrow{E_2} D$$

For instance, on polarographic curves fo p-diacetylbenzene at pH 2–5 only one two-electron wave is observed[14]; the pool electrode electrolysis nevertheless produces the alcohol, p-CH_3CHOH-$C_6H_4CHOHCH_3$, which is a product of a four-electron reduction when the reduction is carried out at sufficiently negative potentials (E_2):

A similar example is the reduction of isonicotinic acid amide at pH < 1. Polarographic curves correspond to a two-electron process, but in the controlled potential electrolysis pyridine-4-carbinol is formed, indicating an over-all consumption of four electrons.[30] Formation of pyridine-4-aldehyde as an intermediate in the controlled potential electrolysis was furthermore proved. The system involved can be described schematically:

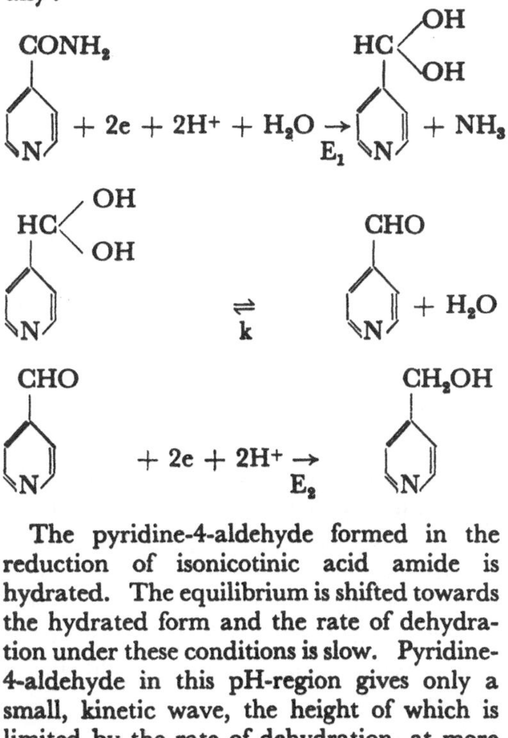

The pyridine-4-aldehyde formed in the reduction of isonicotinic acid amide is hydrated. The equilibrium is shifted towards the hydrated form and the rate of dehydration under these conditions is slow. Pyridine-4-aldehyde in this pH-region gives only a small, kinetic wave, the height of which is limited by the rate of dehydration, at more positive potentials than the reduction of the amide occurs. Therefore polarographic reduction practically stops after the two-electron uptake. During mercury pool electrolysis, the dehydration has time to take place and pyridine-4-aldehyde can be quantitatively reduced to pyridine-4-carbinol. The interesting feature of this system is that potential E_2 is more positive than potential E_1.

Side reactions in which the intermediate C is transformed into an electroinactive form B will affect the electrolysis as well. The process involved can be depicted schematically:

$$A + n_1 e \xrightarrow{E_1} C$$

$$C \xrightarrow{k} B$$

$$C + n_2 e \xrightarrow{E_2} D$$

If the transformation of C by the reaction with rate constant k proceeds relatively slowly, the process does not affect polarographic curves, but during pool electrolysis part of C is inactivated. Polarographic curves of o-benzoylbenzoic acid diethylamide[31] correspond to a two-electron reduction over the whole pH-range. In contrast, during controlled potential electrolysis only 1·65 electrons per molecule were consumed and a dimer was isolated. It is assumed that in this case particle C is a radical and its deactivation occurs through dimerisation, according to the scheme:

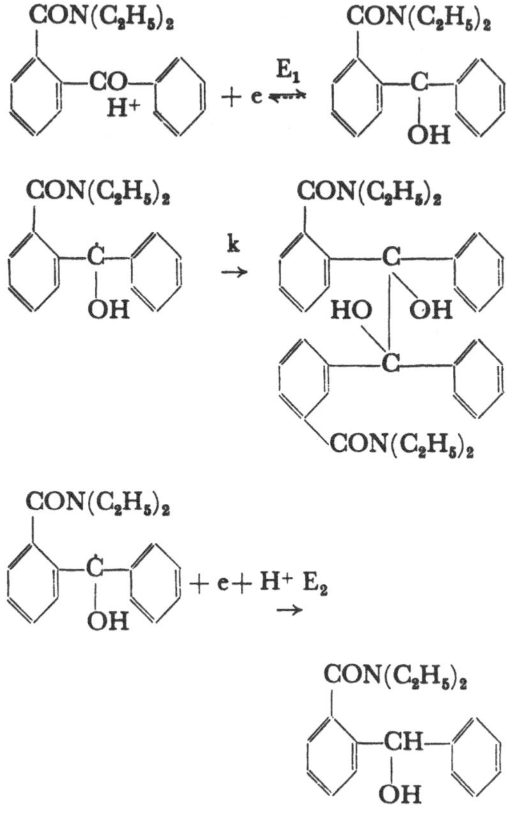

Dimerisation will affect more the large scale electrolysis because of the higher concentrations involved and probably also because of the adsorption of the radicals.

A side reaction also affects the electrolysis of picric acid in acidic media.[32] The complete reduction in which 18 electrons are consumed, takes place only in dilute solution. Above a certain concentration, the value of which depends on pH, the apparent number of consumed electrons decreases. It is assumed that the rate of reaction causing the deactivation of an intermediate C increases with increasing picric acid concentration, but decreases with the increasing concentration of hydrochloric acid. The apparent number of electrons consumed decreases below 17 and therefore the value of the number of electrons (n_1) transferred in the step producing the reactive intermediate C (i.e. $A + n_1e \rightarrow C$) must be smaller than 17.

Antecedent protonation can also affect in a different way the electrolysis on the dropping and on the pool electrode. This can be shown in the case of the reduction of isonicotinic acid at higher pH-values. The protonated form of isonicotinic acid is reduced in a two-electron step and gives pyridine-4-aldehyde.[31] When the dropping mercury electrode is used, the waves of the protonated form of isonicotinic acid can be observed at pH < 8. When the electrolysis is carried out using a pool electrode, it was possible to prove the formation of pyridine-4-aldehyde only at pH < 6. This is attributed to an antecedent protonation

It seems that the rate of the protonation reaction with constant k is not high enough at the surface of the mercury pool to supply a sufficient amount of the protonated form of isonicotinic acid when the pH is higher than about 6.

In some few instances different products were isolated for substances giving principally one wave, when the electrolysis was carried out at various potentials. In the course of such reaction the number of electrons per molecule can be either the same at all potentials at which the electrolysis can be carried out or can vary according to the potential used.

An example of the former type is the reduction of 4,4'-dithiomorpholine which gives a two-electron reduction wave[33] that resembles those obtained in the reduction of organic disulphides. When the electrolysis is carried out at the potential corresponding to the half-wave, morpholine and mercuric sulphide were isolated. When for electrolysis a potential corresponding to the limiting current was chosen, morpholine and elemental sulphur were formed. It is assumed that the primarily formed N-mercaptomorpholine reacts with metallic mercury at more positive potentials, but is cleaved at more negative potentials and forms elemental sulphur. A check on the course of the current-voltage curves using the mercury pool electrode would be of interest.

A different number of electrons is consumed according to the potential chosen in the reduction of 4-cyanopyridine.[34] The yield of 4-picolylamine formed in the electrolysis decreased with increasingly negative potentials, whereas single-sweep i–E curves indicate an increase in cyanide ion formation with increasingly negative potentials. Competitive reductions of the protonised form, producing 4-picolylamine in a four-electron step, and of the unprotonised form in which a two-electron nucleophilic substitution takes place, participate in the overall electrode process :

$$\text{(4-cyanopyridine)} + H^+ \rightleftharpoons \text{(N-protonated 4-cyanopyridinium)}$$

$$\text{(N-protonated 4-cyanopyridinium)} + 4e + 4H^+ \rightarrow \text{(4-aminomethyl-pyridinium, CH}_2\text{NH}_2)$$

$$\text{(4-cyanopyridine)} + 2e + H^+ \rightarrow \text{(pyridine)} + CN^-$$

The reported[35] apparent change with potential in the number of electrons per molecule consumed in the reduction of dimethylglyoxime indicates a catalytic hydrogen evolution.

CONCLUSIONS

Preparative methods in which the product or the intermediate of an electrode process is generated electrochemically are of fundamental importance for the elucidation of the course of organic electrode processes and should be included as a standard procedure in polarographic studies. It should be a rule that the intermediates and products of an electrode process are identified prior to the application of theoretical treatments. It is, nevertheless, surprising how often this fundamental rule is violated.

Even when it is understood, how the preparative techniques can help in the elucidation of the course of electrode processes, their power should not be overestimated. Care should be taken not to mistake the products of secondary reactions for the products sought.

REFERENCES

1. Kastening, B., *Electrochim. Acta*, 1964, 9, 241.
2. Adams, R. N., *J. Electroanal. Chem.*, 1964, 8, 151.
3. Fujinaga, T., Deguchi, Y., and Umemoto, K., *Bull. Chem. Soc. Japan*, 1964, 37, 822.
4. Bernal, I., and Fraenkel, G. K., *J. Am. Chem. Soc.*, 1964, 86, 1671.
5. Chambers, J.Q. III, and Adams, R. N., *J. Electroanal. Chem.*, 1965, 9, 400.
6. Hoffman, A. K., Hodgson, W. G., Maricle, D. L., and Jura, W. H., *J. Am. Chem. Soc.*, 1964, 86, 631.
7. Kastening, B., *Collection Czech. Chem. Commun.*, 1965, 30, 4033.
8. Manošuek, O., and Zuman, P., *Collection Czech. Chem. Commun.*, 1964, 29, 1718.
9. Gelb, R. I., and Meites, L., *J. Phys. Chem.*, 1964, 68, 2599.
10. Manoušek, O., Exner, O., and Zuman, P., in preparation.
11. Manoušek, O., unpublished results.
12. Zuman, P., *Collection Czech. Chem. Commun.*, 1960, 25, 3245.
13. Manoušek, O., and Zuman, P., *Chem. Commun. (London)*, 1965 (No. 8), 158.
14. Kargin, J. H., Manoušek, O., and Zuman, P., *J. Electroanal. Chem.*, 1966, 12, 443.
15. Zuman, P., Manoušek, O., and Horák, V., *Collection Czech. Chem. Commun.*, 1964, 29, 2906.
16. Manoušek, O., and Zuman, P., *Collection Czech. Chem Commun.*, 1964, 29, 1432.
17. Manoušek, O., Krupička, J., Gut, J., and Zuman, P., unpublished results.
18. Bowers, R. C., and Russel, H. D., *Anal. Chem.*, 1960, 32, 405.
19. Fleet, B., and Zuman, P., *Collection Czech. Chem. Commun.*, 1967, 32, 2066.
20. Lund, H., *Acta Chem. Scand.*, 1959, 13, 249.
21. Pflegel, P., Manoušek, O., and Wagner, G., private communication.
22. Wawzonek, S., and Wearring, D., *J. Am. Chem. Soc.*, 1959, 81, 2067.
23. Wawzonek, S., and Gundersen, A., *J. Electrochem. Soc.*, 1960, 107, 537.
24. Wawzonek, S., and Gundersen, A., *J. Electrochem. Soc.*, 1964, 111, 324.
25. Wawzonek, S., Berkey, R., Blaha, E. W., and Runner, M. E., *J. Electrochem. Soc.*, 1956, 103, 456.
26. Wawzonek, S., and Duty, R. C., *J. Electrochem. Soc.*, 1961, 108, 1135.
27. Elving, P. J., Martin, A. J., and Rosenthal, I., *J. Am. Chem. Soc.*, 1955, 77, 5218.
28. Rosenthal, I., and Elving, P. J., *J. Am. Chem. Soc.*, 1951, 73, 1880.
29. Lund, H., *Acta Chem. Scand.*, 1957, 11, 283.
30. Lund, H., paper delivered at the 19th International Congress of Pure and Applied Chemistry, London, 1963.
31. Lund, H., *Abhandl. Deutsch. Akad. Wiss., Kl. Chem., Geol. Biol.*, 1964, 1, 434.
32. Meites, L., and Meites, T., *Anal. Chem.*, 1956, 28, 103.
33. Lund, H., paper read before the 19th International Congress of Pure and Applied Chemistry, London, 1963.
34. Kardos, A. M., Valenta, P., and Volke, J., *J. Electroanal. Chem.*, in press.
35. Spritzer, M., and Meites, L., *Anal. Chim. Acta*, 1962, 26, 58.

Part III

Polarographic Reduction of Some Carbonyl Compounds

2. P. Zuman: Polarographic reduction of aldehydes and ketones. I. Course of electrode process. *Collect. Czechoslov. Chem. Commun.* **33**, 2548 (1968).
3. P. Zuman, and B. Turcsanyi: Polarographic reduction of aldehydes and ketones. II. pH-Dependence of polarographic limiting currents of aryl alkyl ketones governed by the rate of simultaneous reactions with several acids producing the electroactive species. *Collect. Czechoslov. Chem. Commun.* **33**, 3090 (1968).
4. P. Zuman, and B. Turcsanyi: Polarographic reduction of aldehydes and ketones. III. Effects of alkyl groups in the side chain on the half-wave potentials of deoxybenzoins and some other alkyl aryl ketones. *Collect. Czechoslov. Chem. Commun.* **33**, 3205 (1968).
5. P. Zuman, O. Exner, R. F. Rekker, and W. Th. Nauta: Polarographic reduction of aldehydes and ketones. IV. Linear free energy treatment of substituted benzophenones. *Collect. Czechoslov. Chem. Commun.* **33**, 3213 (1968).
6. P. Zuman, and D. Barnes: Desorption of 3-phenylpropionaldehyde preceding polarographic reduction. *Nature* **215**, 1269 (1967).
7. O. Manoušek, and P. Zuman: Direct polarographic determination of pyridoxal in the presence of pyridoxal-5-phosphate. *Collect. Czechoslov. Chem. Commun.* **27**, 486 (1962).
8. P. Zuman, and V. Černý: Polarography of steroids III. Polarographic reduction of steroids containing an aldehydic group in position 18. (in German) *Collect. Czechoslov. Chem. Commun.* **24**, 1925 (1959).
9. J. Kargin, O. Manoušek, and P. Zuman: Polarographic behaviour of *p*-diacetyl-benzene; an example of a compound with two electroactive centres. *J. Electroanal. Chem.* **12**, 443 (1966).
10. P. Zuman, O. Manoušek, and S. K. Vig: Further examples of *para-* and *ortho-* disubstituted benzene derivatives reduced in a reversible two-electron step into products of limited stability. *J. Electroanal. Chem.* **19**, 147 (1968).

Arylalkyl ketones and diaryl ketones are reduced at the dropping mercury electrode in acidic media following the sequence: proton, electron, electron, proton; in the medium pH-range the sequence is electron, proton, electron, proton; and in alkaline media the path is electron, electron, two protons.[2] In the intermediate pH-ranges, the current is governed by the rate of protonation of the carbonyl compound or of the radical anion. In the protonation of the ketone, proton donors other than hydronium ion can participate. Simple pH-dependence can be obtained only in buffers containing a constant concentration of the basic component of the buffer.[3] In substituted acetophenones, steric effects of branched alkyl groups prevent optimum orientation at the electrode surface, whereas in deoxybenzoins, the two phenyl rings fix the ketone and steric effects are negligible.[4] Meta- and *para*-substitution in benzophenones results in shift of half-wave potentials that can be expressed by Hammett equations. Conclusions can be drawn regarding electronic structure of the radical formed for electron-accepting and electron-donating substituents.[5]

Contrary to the assumption that organic compounds must be adsorbed to undergo reduction, it was proved[6] that certain carbonyl compounds are desorbed from the electrode surface in the potential region where the reduction occurs. Different protonation effects allow separation of the pyridoxal and pyridoxal-5-phosphate waves.[7] The pH-independence of the height of the reduction wave of steroids bearing an aldehydic grouping in position 10 was explained by steric hindrance to hydration.[8] Identification of reduction products would be desirable.

In the reduction of *para*-derivatives of the type $R.COC_6H_4CO.R$ in acidic media (which occurs in a two-electron reversible reduction step), a biradical or quinoid product of limited stability is formed.[9,10] The life-time of this product increases in the sequence: $R = H < CH_3 < C_6H_5$. When the acid and base catalysed deactivation of this product, which gives a ketol, is sufficiently fast, two reduction waves, corresponding to two consecutive reductions of the carbonyl groups, are observed.

POLAROGRAPHIC REDUCTION OF ALDEHYDES AND KETONES. I. GENERAL REDUCTION SCHEME

P. ZUMAN*

*J. Heyrovský Institute of Polarography,
Czechoslovak Academy of Sciences, Prague 1*

Received December 27th, 1967

Reduction scheme of polarographic reduction of aryl alkyl ketones is discussed. Schemes are summarised for the reduction of the protonized form in acid media, for the two-electron reduction in medium pH-range and for the one-electron steps in alkaline solutions. Kinetics currents resulting in protonations, antecedent the electrode process or interposed between two electrode processes, which play a role in the transition regions of pH, are discussed. The effect of cations on the reduction waves is interpreted. Studied examples of reductions of various types of carbonyl compounds are summarised.

The course of the polarographic reduction of aldehydes and ketones has been studied rather extensively[1-32]. Nevertheless, in most investigations attention has been paid only to a certain aspect of the problem. Either the study was restricted to a certain — acid or alkaline — pH-range, or the stress was laid on detailed examination of the effects of concentration of the electroactive compounds, cations or solvents, or the investigation was carried out only for a small group of compounds, or attention was limited to a comparison of half-wave potentials.

In the studies that will be discussed in the present series, various compounds have been investigated which have either the carbonyl group attached to a phenyl ring or compounds in which the carbonyl group is conjugated with an unsaturated group. In these studies the behaviour of carbonyl compounds showing a wide variation in structure has been compared and this has made it possible to find a general scheme for the reduction course. According to the structure of the compound some reaction steps can become more important than others, some reaction steps cannot be followed due to interferences (*e.g.* by the supporting electrolyte or overlapping waves), some steps are negligible, or the overall scheme can involve a set of antecedent or con-

* Present address: Department of Chemistry, University of Birmingham, Great Britain.

secutive electrochemical and chemical steps. Nevertheless, the general scheme can be depicted as follows (using Ar for phenyl, substituted benzene ring, polycyclic benzenoid rings and some "neutral" heterocyclic rings, R for alkyl, hydrogen or substituents of the CH_2X-type).

In acid media containing water or proton–donor the sequence of the reduction steps is proton, electron, electron, proton (H^+, e, e, H^+)

$$Ar\overset{(+)}{C}OHR \underset{k_{-1}}{\overset{k_1}{\rightleftharpoons}} ArCOR + H^+, \qquad\qquad pK_1 \qquad (A)$$

$$Ar\overset{(+)}{C}OH^+R + e \underset{E_1}{\rightleftharpoons} Ar\overset{\cdot}{C}OHR, \qquad\qquad i_1 \qquad (B)$$

$$2\,Ar\overset{\cdot}{C}OHR \overset{k_2}{\longrightarrow} dimers, \qquad\qquad\qquad (C_1)$$

$$Ar\overset{\cdot}{C}OHR + Hg \overset{k_3}{\longrightarrow} organometallic\ compound, \qquad\qquad (C_2)$$

$$Ar\overset{\cdot}{C}OHR + solvent \overset{k_4}{\longrightarrow} products, \qquad\qquad\qquad (C_3)$$

$$Ar\overset{\cdot}{C}OHR + e \underset{E_2}{\longrightarrow} Ar\overset{(-)}{C}OHR, \qquad\qquad i_2 \qquad (D)$$

$$ArCHOHR \underset{k_{-5}}{\overset{k_5}{\rightleftharpoons}} Ar\overset{(-)}{C}OHR + H^+. \qquad\qquad pK_5 \qquad (E)$$

Hence in acid media two one-electron waves (i_1, i_2) at potentials E_1 and E_2 are observed (Fig. 1a). The half-wave potential of the wave i_1 is shifted with increasing pH towards more negative potentials (Fig. 1b) at pH > pK, (because the pK_1 value for most aryl ketones studied is smaller than zero, this condition is fulfilled over the acidity range usually studied). Pinacol is the predominating product for electrolysis of most aryl alkyl ketones at the potential of the limiting current of the first one-electron wave i_1 in acid media which corresponds to a reduction of the protonized form $ArCOHR^{(+)}$ as was recognised rather early.[1,3] The half-wave potential of the wave i_1 depends on the concentration of ketone[17,19,22,26] but is practically independent of neutral salt kind and concentration[19,20,22] and of ethanol concentration[22,26]. For some ketones it is assumed that the equilibrium (B) is rapidly established at the electrode surface after the potential is applied and hence the first one-electron uptake is considered to be a reversible process. Correlations between the rate of the electrode process (B) and the structure of the carbonyl compound have not yet been studied.

The half-wave potential of the second one-electron wave i_2 at potential E_2 is practically pH-independent (Fig. 1b), which indicates that there the proton-transfer (E) occurs as a step consecutive to an irreversible electrode process. The sensitivity

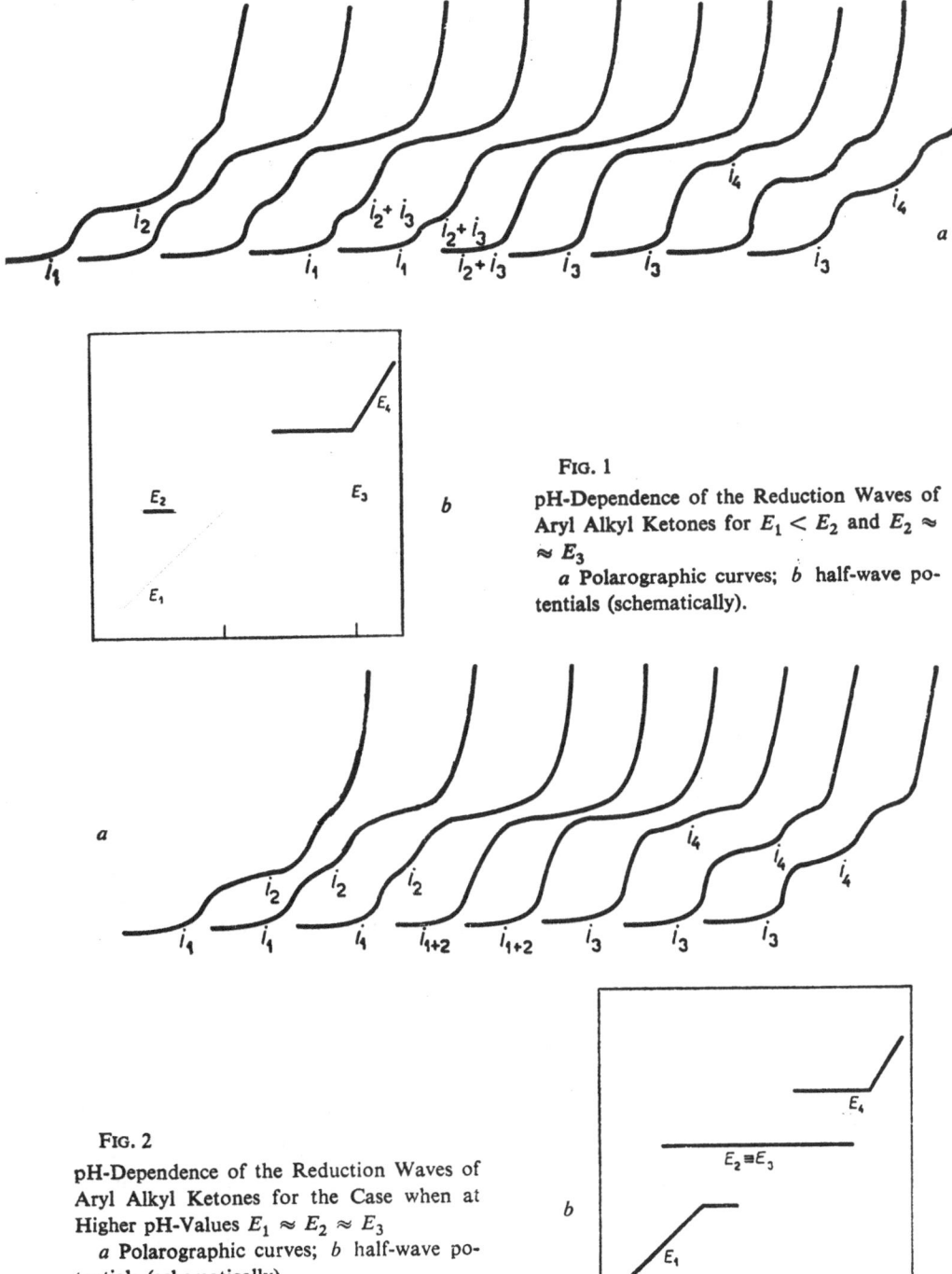

FIG. 1

pH-Dependence of the Reduction Waves of Aryl Alkyl Ketones for $E_1 < E_2$ and $E_2 \approx E_3$

a Polarographic curves; *b* half-wave potentials (schematically).

FIG. 2

pH-Dependence of the Reduction Waves of Aryl Alkyl Ketones for the Case when at Higher pH-Values $E_1 \approx E_2 \approx E_3$

a Polarographic curves; *b* half-wave potentials (schematically).

of the half-wave potential of the wave i_2 to the effects of cations and solvents[19,20,22,26] can be interpreted by the role of the double-layer phenomena and adsorption-desorption processes rather than by adduct formation. Because of the shift of the first wave and the practical pH-independence of the wave i_2, the two waves i_1 and i_2 can merge into one two-electron wave (i_{1+2}) at sufficiently high pH (Fig. 2a). Because the increase in ethanol concentration causes a shift of wave i_2 of benzaldehyde to more negative potentials* whereas the half-wave potential of wave i_1 remains practically unchanged[26], the separation of waves i_1 and i_2 can be observed at high ethanol concentrations even at pH-values when at low ethanol concentration only the combined wave i_{1+2} is observed. The reduction product of the combined two-electron wave (i_{1+2}) is alcohol.[2]

In the medium pH-range in water-containing solutions the sequence is electron, proton, electron, proton (e, H^+, e, H^+) and the scheme can be given as follows:

$$ArCOR + e \xrightarrow[E_3]{} Ar\overset{(-)}{\overset{\bullet}{C}}OR, \qquad\qquad i_3 \qquad (F)$$

$$Ar\overset{\bullet}{C}OHR \underset{k_{-6}}{\overset{k_6}{\rightleftharpoons}} Ar\overset{(-)}{\overset{\bullet}{C}}OR + H^+, \qquad\qquad pK_6 \qquad (G)$$

$$Ar\overset{\bullet}{C}OHR + e \xrightarrow[E_2]{} Ar\overset{(-)}{C}OHR, \qquad\qquad i_2, i_3 \qquad (D)$$

$$ArCHOHR \underset{k_{-5}}{\overset{k_5}{\rightleftharpoons}} Ar\overset{(-)}{C}OHR + H^+. \qquad\qquad pK_5 \qquad (E)$$

Radical $Ar\overset{\bullet}{C}OHR$ can undergo side-reactions $(C_1)-(C_3)$. Because usually the difference between potentials E_2 and E_3 is small (or because E_3 becomes more positive than E_2), one two-electron wave $(i_3$ in Figs 1a and 3a, i_{1+2} in Fig. 2a) is observed in this pH-region. As the value i/c for this wave remains constant, because the coulometrically determined value of the number of transferred electrons is two, and because practically only alcohol was found as the controlled potential electrolysis product, reactions of the radical (C) are probably not fast enough to affect the electrode process.

The half-wave potential of the two-electron wave i_3 (or i_{1+2}) is practically pH-independent (Fig. 1b, 2b). This independence can be explained by assuming that potential determining is the irreversible one-electron step (F). This view is supported by the shape of the wave for which the logarithmic analysis in several instances gave the value corresponding to $n = 1$. It is nevertheless impossible to exclude the explanation that

---

* At low ethanol concentrations it was[4] oppositely reported that with increasing ethanol concentration the half-wave potential of wave i_1 of benzophenone shifts to more negative values, whereas that of wave i_2 remains constant.

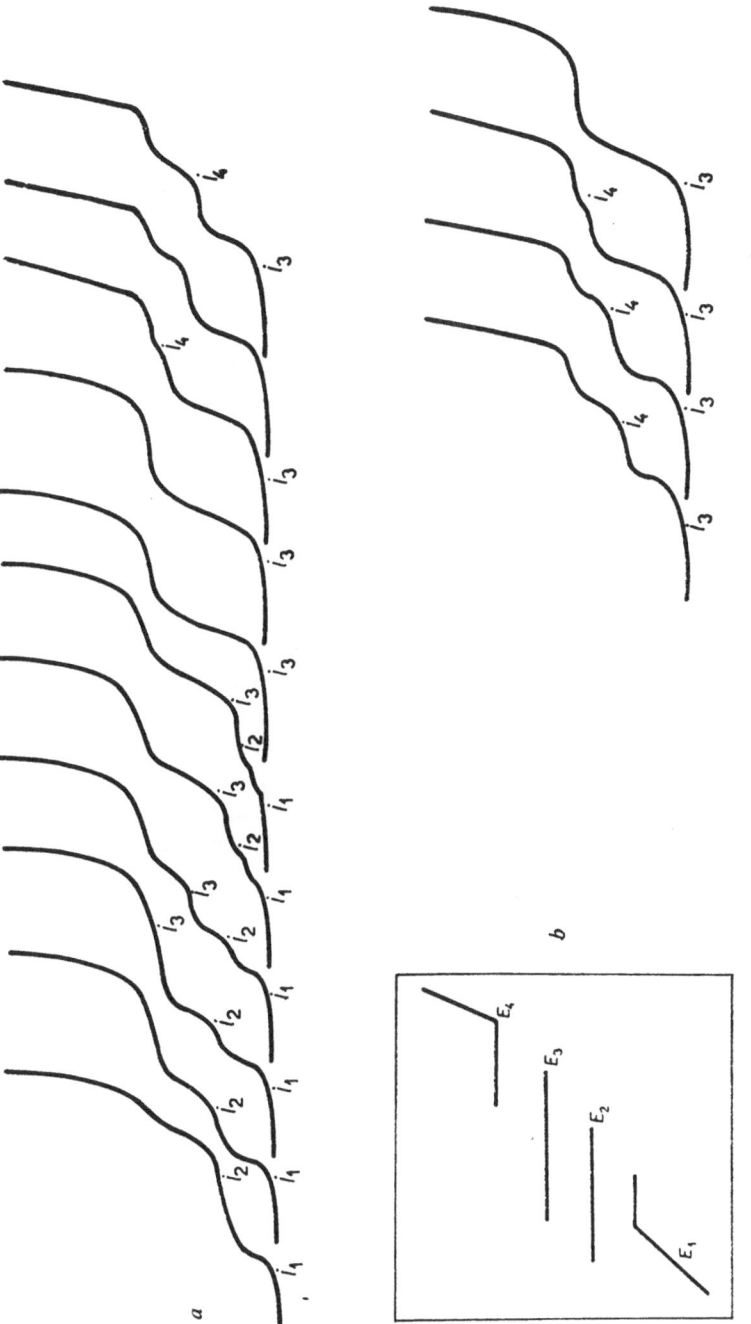

FIG. 3

pH-Dependence of the Reduction Waves of Aryl Alkyl Ketones for $E_1 < E_2 < E_3$ *a* Polarographic curves; *b* half-wave potentials (schematically).

FIG. 4

Effect of Increasing Cation Concentration on the Reduction Waves of Aryl Alkyl Ketones in Alkaline Media (schematically) $E_5 \approx E_3$

the potential of the two-electron wave is pH-independent, because equilibrium (G) is completely shifted to the left hand side $(i.e.$ that $pH \ll pK_6)$.

The scheme (F), (G), (D), (E) takes place in the medium pH-range, where the protonation reaction (A) with constant k_{-1} is already so slow that it cannot produce any considerable amount of the protonized form $Ar\overset{(+)}{C}OHR$. On the other hand the acidity must not be too low, to convert all the radical anion $Ar\overset{(-)}{C}OR$ completely into the radical $Ar\dot{C}OHR$ by keeping the reaction (Ga) with constant k_{-6} fast enough. With most carbonyl compounds this pattern is followed between pH 6 and 10.

In the intermediate pH-range between the acid and medium pH-values $(i.e.$ at pH-values usually between 3 and 7), reduction of the protonated form $Ar\overset{(+)}{C}OHR$ according to (A) and (B) and that of the form $ArCOR$ according to (F) and (G) compete. On polarographic curves this competition is shown in several ways:

Most instructive is the case shown in Fig. 3. This ideal case would take place, if the decrease in the rate of protonation (A) would become of importance in a pH-range in which waves i_1 and i_2 corresponding to processes (B) and (D) are separated, $i.e.$ when $E_1 < E_2$ and when moreover waves i_2 and i_3 are resolved, $i.e.$ when $E_2 < E_3$. With increasing pH waves i_1 and i_2 decrease in the shape of a dissociation curve. Simultaneously a two electron wave i_3 corresponding to processes (F), (G), (D), (E) increases and the total wave-height remains constant. Because $E_2 < E_3$ only one two-electron wave i_3 is observed. In practice the condition $E_2 < E_3$ is not fulfilled and hence this type of dependence is usually not observed.

If the decrease of the rate of protonation (A) plays a rôle in the pH-range in which waves corresponding to processes (B) and (D) are separated, $i.e.$ when $E_1 < E_2$, but waves i_2 and i_3 merge, $i.e.$ when $E_2 \approx E_3$, the waves change with pH as shown in Fig. 1. The first one-electron wave decreases with increasing pH in the form of a dissociation curve and is replaced by a two-electron wave, corresponding to a process (F), (G), (D). This type of behaviour was observed particularly in the reduction of several aryl ketones.

If on the other hand the decrease in the rate of protonation (A) becomes of importance in a pH-range, where both reduction steps (B) and (D) of the protonized form coalesce into one two-electron wave, no decrease in wave-heights is observed (Fig. 2). This is due to the fact that $E_1 \approx E_2$ and $E_2 \approx E_3$. In some instances the change from the process $(A)-(F)$ to the process (F), (G), (D), (E) can be[33] detected from logarithmic analysis, which shows two linear portions the ratio of which changes with pH. The shifts of half-wave potentials allow distinguishing between waves i_{1+2} and i_3: The pH-dependence of half-wave potentials shows two linear portions, the first with a slope of $0.059/n\alpha$, the second pH-independent, which intersect at $pH = pK''$. The reduction of the protonized form (i_{1+2}) can play a considerable rôle at $pH \leqq (pK'' + 1)$. Hence for $pH > (pK'' + 1)$ the reduction of the unprotonized form (F), (G), (D), (E) must predominate. Hence above $pH = (pK'' + 1)$ it would be possible in Fig. 2 to denote the two-electron wave as i_3 (instead i_{1+2}).

Buffer types, composition and concentration affect the decrease of wave i_1 with increasing pH and in reaction (A) therefore other proton donors can participate simultaneously with hydroxonium ions and water[34].

In alkaline media the reduction takes place in the sequence electron, electron, protons (e, e, H^+, H^+) according to the scheme (F), (H), (I):

$$ArCOR + e \xrightarrow{E_3} Ar\overset{(\overset{\cdot}{-})}{C}OR, \qquad i_3 \qquad (F)$$

$$Ar\overset{(\overset{\cdot}{-})}{C}OR + e \xrightarrow{E_4} Ar\overset{(2-)}{C}OR, \qquad i_4 \qquad (H)$$

$$Ar\overset{(2-)}{C}OR + 2H^+ \rightleftharpoons ArCHOHR, \qquad (I)$$

$$Ar\overset{(\overset{\cdot}{-})}{C}OR + Me^+ \underset{k_{-7}}{\overset{k_7}{\rightleftharpoons}} Ar\overset{\cdot}{C}R, \qquad (J)$$
$$\underset{OMe}{}$$

$$\underset{OMe}{Ar\overset{\cdot}{C}R} + e \xrightarrow{E_5} \underset{OMe}{Ar\overset{(-)}{C}R}, \qquad i_3, i_4 \qquad (K)$$

$$\underset{OMe}{Ar\overset{(\overset{\cdot}{-})}{C}R} + 2H^+ \rightleftharpoons ArCHOHR + Me^+. \qquad (L)$$

At sufficiently high pH-values, in a medium of ionic strength less than about 0·1, and for cations for which the equilibrium (J) is less shifted towards the right hand side (e.g.. lithium ions), two one electron waves i_3 and i_4 (Fig. 1−3), appear corresponding to processes (F) and (H). Waves are usually well separated, because the reduction of the radical anion ArCOR occurs at more negative potentials than that of the radical ArĊOHR, i.e. because $E_3 < E_4$. Potential E_4 can be sometimes so negative that wave i_4 can be overlapped by the current of the supporting electrolyte. Similar course of reduction was observed also in nonaqueous media[15].

With increasing concentration of cations in the supporting electrolyte, resulting either from the addition of neutral salt[3,4,9] or from using hydroxides of cations, for which equilibrium* (J) is shifted more to the right hand side[35] (e.g.. for caesium

---

*   Product reduced at potential E_5 results from interaction between radical anion Ar$\overset{(\overset{\cdot}{-})}{C}$OR and cation Me^+. The structure of this product, i.e. whether a covalent bond and between which atoms is formed, whether the structure resembles more a complex, charge-transfer complex, or ion-pair is not understood. In equations $(J)−(L)$ the symbol ArĊR is used only for the sake of convenience.
$$\underset{OMe}{}$$

or tetraalkylammonium ions) the current at potential E_5 increases. The potential E_5 is in all cases more positive than E_4, *i.e.* $E_5 < E_4$, but with respect to potential E_3 two possibilities exist:

If $E_5 \approx E_3$, then with increasing concentration of the cation the more positive wave i_3 increases, the more negative wave i_4 decreases (Fig. 4). If $E_5 > E_3$, a new wave i_5 is formed[35] between the two waves i_3 and i_4 corresponding to processes (F) and (H) (Fig. 5). The wave i_5 corresponds to process (K) and its height is limited by the rate of reaction (J) with rate constant k_7. Nevertheless, in both cases the total wave-height remains unchanged, when a correction is carried out for the changes in viscosity.

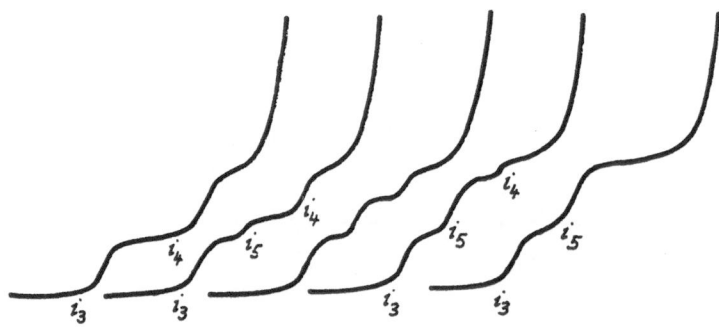

FIG. 5

Effect of Increasing Cation Concentration on the Reduction Waves of Aryl Alkyl Ketones in Alkaline Media (schematically) $E_3 < E_5 < E_4$

The effect of cations usually increases with ionic size and charge. As a further complicating factor the presence of cations can also affect the potential E_4 corresponding to process (H). The double-layer effect causes a shift of wave i_4 which corresponds to the reduction of the radical anion $Ar\overset{(-)}{C}OR$. Because these processes are consecutive to an irreversible electrode process, no information can be obtained from polarography on processes (I) and (L).

In the intermediate pH-range between the medium and alkaline media (*i.e.* usually between pH 9 and 12) the rate of protonation reaction (G) with rate constant k_{-6} is no longer fast enough to transform the radical anion $Ar\overset{(-)}{C}OR$ completely into the radical ArCOHR. In this pH-range both courses — *i.e.* (G), (D), (E) and (H), (I)

can compete. The height of the wave i_3 decreases with increasing pH in the form of a dissociation curve from a height corresponding to a two electron process and reactions (F) and (D) to a height corresponding to a one-electron process (F) (Figs $1a-3a$). Simultaneously the height of the more negative wave i_4 increases until it reaches a one-electron value. The rate of reaction (G) with constant k_{-6} governs the height of the first wave i_3, or more exactly its increase above the value corresponding to a one-electron transfer[12,36].

The half-wave potential of wave i_3 is practically pH-independent, the half-wave potential of wave i_4 is independent up to pH $= pK_6$, then shifts towards more negative potentials are observed (Figs $1b-3b$). Under these conditions reaction (G) is namely a dissociation equilibrium preceding the electrode process (H). As these shifts are observed only in strongly alkaline media, it is not easy to separate them from shifts resulting from specific cation effects.

In the past we recognized relatively early[37] the effect of protonation of carbonyl compounds antecedent to the electroreduction on the limiting currents. This idea was criticized[38,39], but later proved[40] to be correct and essential for the interpretation of the reduction path. The ketones and aldehydes, studied recently, which will be discussed in this series, can be classified into four groups: To the first group belong ketones of the type ArCOR, to the second those ketones which are formed at the surface of the electrode from a more complex compound by electrochemical and chemical processes. Third group consists of aldehydes, in particular unsaturated and those saturated which were of interest as reduction intermediates. Finally the fourth group is composed of compounds which bear two or more carbonyl groups in the molecule.

Among the studied ketones belong the aryl alkyl ketones containing a branched alkyl group. These compounds of the structure $ArCOCHR^1R^2$ were studied[41] as the reduction products of substituted ω-aminoacetophenones $C_6H_5COC(NH_2)R^1R^2$. It has been observed that the alkyl groups exert a pronounced steric effect. In a series of substituted deoxybenzoins of the type $C_6H_5COCHRC_6H_5$ the steric effects were found[42] to be less pronounced than in the previous series. For the sake of comparison the behaviour of acetophenone and other unbranched aryl alkyl ketones was checked over a wide range of conditions. It was observed[34] that the shape of the polarographic dissociation curve depends on the composition of the buffer. A theoretical treatment was developed for the i–pH plots in dependence on buffer composition. The behaviour of the dihydrochalkone $C_6H_5COCH_2CH_2C_6H_5$ has been investigated[43] in connection with the electroreduction of chalkone. The reduction of β-ketosulfides of the type $C_6H_5COCH_2CH_2SR$ follows a path similar to that of other ketones ArCOR, indicating a protonation of the carbonyl group[44], whereas the reduction of β-aminoketones[45] of the type $C_6H_5COCH_2CH_2NR_2$ is affected by the protonation of the aminogroup which is preferred to that of the carbonyl group. This results in a change in the polarographic behaviour.* In both these types the effect of the

---

*   In a similar way a difference in the protonation of $(R')_2C(SR^2)CHO$ (preferably on CO) and of $(R')_2C(NR_2^2)CHO$ (preferably on N) has recently been reported[31]. In this case, however, the C—S or the C—N bond undergoes an electrochemical hydrogenolysis.

substituents on the hetero-atom was examined. Finally for substituted benzophenones (for all three basic types of reduction process) the application of the Hammett equation, the additivity of substituent effect on the same and on different rings, as well as the comparison of contributions from polar substituent effects in *meta-* and *para-*position was investigated[46]. The investigation of the anomalies found for p-cyanobenzophenone lead to the elucidation of the reduction of the cyano group in *para*-substituted acetophenones, benzophenones, and benzaldehydes[47,48].

A study of the polarographic behaviour of the carbonyl compound formed in the reduction of compounds of the type $ArCOCH_2$—X (where $X = NR_2$, $NR_3^{(+)}$, $SR_2^{(+)}$) revealed[49,50] that an enolate is the primary reduction product which must be transformed into the electro-active ketone. The same was assumed[30,31] for aldehydes of the type $(CH_3)_2C(X)CHO$ ($X =$ $= NR_2$, $NR^{(+)}$, SR, OR). A similar postulate was reached for α,β-unsaturated ketones of the type $C_6H_5COCH{=}CHR$, where in the first two-electron step the enolate is formed and this must be transformed into the ketone prior to further reduction. This was proved for phenyl vinyl ketone[51] ($R = H$) and for chalkone[43] ($R = C_6H_5$). The effect of substituents on both benzene rings was compared[52] and the exchange of pyridine rings for one of the benzene rings in chalkone was investigated[53].

In the course of study of the polarographic behaviour of cinnamaldehyde[35,54]. A comparison was made with phenylpropargylaldehyde and 3-phenylpropionaldehyde. The results indicated in agreement with a recent study[55], that in the first step in weakly alkaline media the $C{\equiv}C$ bond is reduced and cinnamaldehyde formed; but whereas the Italian authors[55] assumed the reduction of the aldehydic group and formation of an unsaturated alcohol in the next step, we were able to prove[35,54] that the $C{=}C$ bond is reduced first in a two-electron step and a saturated aldehyde formed, which at more negative potentials is reduced in a consecutive step to a saturated alcohol. The behaviour of 3-phenylpropionaldehyde was compared with that of 2-phenylpropionaldehyde, phenylacetaldehyde and acetaldehyde and the effect of hydration rate and equilibria examined. The effect of hydration on limiting current formally resembles that observed[30,31] in substituted aldehydes of the type $(CH_3)_2C(Z)CHO$ (for $Z = NR_2$, $NH_3^{(+)}$, SR, OR, Cl) even when in the latter case the C—Z bond and not the aldehydic group is reduced.

For cinnamaldehyde no indication of a chemical transformation other then dehydration of the primary two-electron reduction product was found. If the enolate is formed, its conversion into the aldehyde must be fast. The height of the more negative wave corresponding to the reduction of the electrogenerated 3-phenylpropionaldehyde in the solution of cinnamaldehyde is affected by hydration in the same way as when 3-phenylpropionaldehyde was added to the solution.

Finally two groups of carbonyl compounds, the reduction scheme of which varies from the above have been studied recently: The reduction of 3-thianaphthenone[56] was complicated by participation of the slowly established acid–base equilibria involving the CH_2-group. Secondly, the reduction of p-diacetylbenzene[33], terephthalic dialdehyde and similar compounds[57] follow another scheme, due to the stability of the biradical formed in a reversible two-electron process.

Papers belonging to this series will be submitted for publication to this Journal and the Transactions of the Faraday Society.

REFERENCES

1. Baker J. W., Davies W. C., Hemming M. L.: J. Chem. Soc. *1940*, 692.
2. Pasternak R.: Helv. Chim. Acta *31*, 753 (1948).
3. Ashworth M.: This Journal *13*, 229 (1948).
4. Day R. A. Jr., Milliken S. R., Shults W. D.: J. Am. Chem. Soc. *74*, 2741 (1952).
5. Coulson D. M., Crowell W. R.: J. Am. Chem. Soc. *74*, 1290 (1952).
6. Holleck L., Marsen H.: Z. Elektrochem *57*, 301 (1953).
7. Calzolari C., Furlani C.: Ann. Chim. (Roma) *44*, 356 (1954).
8. Prévost Ch., Souchay P.: Chimie Anal. (Paris) *37*, 3 (1955).
9. Elving P. J., Leone J. T.: J. Am. Chem. Soc. *80*, 1021 (1958).
10. Mandell L., Powers R. M., Day R. A. jr.: J. Am. Chem. Soc. *80*, 5284 (1958).
11. Nagata Y.: Rev. Polarog. (Kyoto) *6*, 124 (1958); *7*, 12 (1959).
12. Kastening B., Holleck L.: Z. Elektrochem. *63*, 166 (1959).
13. Powers R. M., Day R. A.: J. Am. Chem. Soc. *81*, 808 (1959).
14. Kemula W., Grabowski Z., Kalinowski M.: Naturwissenschaften *47*, 514 (1960).
15. Wawzonek S., Gundersen A.: J. Electrochem. Soc. *107*, 537 (1960).
16. Suzuki M., Elving P. J.: J. Phys. Chem. *65*, 391 (1961).
17. Majranovskij S. G.: Izv. Akad. Nauk SSSR, Otd. Chim. Nauk *1961*, 2140.
18. Berg H.: Z. Chem. *2*, 237 (1962).
19. Majranovskij S. G.: J. Electroanal. Chem. *4*, 166 (1962).
20. Stradiš J. P.: Electrochim. Acta *9*, 711 (1964).
21. Grabowski Z. R., Kemula W.: Abhandl. Deutsch. Akad. Wiss. Berlin No. *1*, 377 (1964).
22. Majranovskij S. G., Pavlov V. N.: Ž. Fiz. Chim. *38*, 1804 (1964).
23. Majranovskij S. G., Pavlov V. N.: Elektrochimija *1*, 226 (1965).
24. Stradiš J. P., Terauds V. V.: Latvijas PSR Zinatnu Akad. Vestis, Kimi. Ser. *1965*, 43.
25. Laviron É.: This Journal *30*, 4219 (1965).
26. Laviron É., Lucy J. C.: Bull. Soc. Chim. France *1966*, 2202.
27. Vodzinskij Ju. V., Koršunov I. A.: Učen. Zapiski Gorkov. Univ., Ser. Chim. *32*, 25 (1958).
28. Savéant J. M.: Bull. Soc. Chim. France *1967*, 471.
29. Savéant J. M.: Bull. Soc. Chim. France *1967*, 481.
30. Savéant J. M.: Bull. Soc. Chim. France *1967*, 486.
31. Savéant J. M.: Bull. Soc. Chim. France *1967*, 493.
32. Capobianco G., Vianello E., Giacometti G.: Gazz. Chim. Ital *97*, 243 (1967).
33. Kargin Yu., Mânoušek O., Zuman P.: J. Electroanal. Chem. *12*, 443 (1966).
34. Zuman P., Turcsányi M.: This Journal, in the press.
35. Barnes D., Zuman P.: Unpublished results.
36. Ryvolová A.: This Journal *25*, 420 (1960).
37. Zuman P., Tenygl J., Březina M.: This Journal *19*, 46 (1954); Chem. listy *47*, 1152 (1953).
38. Grabowski Z. R.: *Prace Konf. Polarograf., Warszawa 1956*, p. 91. PWN, Warsaw 1957,
39. Kabasakalian P., McGlotten J.: J. Electrochem. Soc. *105*, 261, 760 (1958).
40. Zuman P.: J. Electrochem. Soc. *105*, 758 (1958).
41. Zuman P.: Unpublished results.
42. Zuman P., Turcsányi M., Mills A. K.: This Journal, in the press.
43. Ryvolová-Kejharová A., Zuman P.: This Journal, in the press.
44. Šestáková I., Pecka J., Zuman P.: This Journal, in the press.
45. Zuman P., Horák V.: This Journal *27*, 187 (1962).
46. Zuman P., Nauta J. P., Exner O.: This Journal, in the press.
47. Manoušek O., Zuman P.: Chem. Commun. *1965*, 158.
48. Zuman P., Manoušek O., Exner O.: This Journal, in the press.

49. Zuman P., Horák V.: This Journal *26*, 176 (1961).
50. Zuman P., Tang S.: This Journal *28*, 829 (1963).
51. Zuman P., Michl J.: Nature *192*, 655 (1961).
52. Wagner M., Zuman P.: This Journal, in the press.
53. Laviron E., Zuman P.: Unpublished results.
54. Barnes D., Zuman P.: J. Electroanal. Chem. *16*, 576 (1968).
55. Capobianco G., Vianello E., Giacometti G.: Gazz. Chim. Ital. *97*, 243 (1967).
56. Kucharczyk N., Adamovský M., Horák V., Zuman P.: J. Electroanal. Chem. *10*, 503 (1965).
57. Zuman P., Manoušek O., Vig S. K.: J. Electroanal. Chem., in the press.

POLAROGRAPHIC REDUCTION OF ALDEHYDES AND KETONES. II.*

pH-DEPENDENCE OF POLAROGRAPHIC LIMITING CURRENTS OF ARYL ALKYL KETONES GOVERNED BY THE RATE OF SIMULTANEOUS REACTIONS WITH SEVERAL ACIDS PRODUCING THE ELECTROACTIVE SPECIES

P.Zuman** and B.Turcsanyi***

J. Heyrovský Institute of Polarography,
Czechoslovak Academy of Sciences, Prague 1

Received December 27th, 1967

For reactions in which the electroactive protonated form is generated at the electrode surface not only by reaction with hydrogen ions, but also with other proton-donors, the pH-dependence of limiting currents studied in buffers with a constant concentration of the acid buffer component gives a distorted dissociation curve. The dependence has the shape of a theoretical dissociation curve only, when the concentration of the base buffer component is kept constant. The inflexion point of such curves (pK'') depends on the concentration of the base buffer component $[S^-]$ and the curve is shifted towards lower pH-values with decreasing base component concentration. From dependence of $1/(K'')^2$ on $[S^-]$, values of rate constants with the acid buffer component (k_{SH}) and with hydrogen ions ($k_{H_3O^+}$) can be evaluated. Derived equations are used for the study of the one-electron wave of the protonated form of aryl alkyl ketones, in particular some deoxybenzoins.

In the original treatment[1-5] of acid-base reactions preceding the electrode process proper (C) only proton-transfer reactions involving hydrogen ions (A) were considered. The possibility of participation of acids SH other than hydrogen ion in simultaneous reactions was soon recognised, however[5-9]. Equation (1) was derived[7,8] for the mean limiting current (i) using the exact theory[3] for systems involving reactions $(A)-(C)$:

$$A + H_3O^{(+)} \underset{k^-_{H_3O^+}}{\overset{k_{H_3O^+}}{\rightleftharpoons}} AH^{(+)} + H_2O \qquad (A)$$

$$A + SH \underset{k^-_{SH}}{\overset{k_{SH}}{\rightleftharpoons}} AH^{(+)} + S^{(-)} \qquad (B)$$

---

*   Part I: This Journal *33*, 2548 (1968).
**   Present address: Department of Chemistry, University of Birmingham, Great Britain.
*** Present address: Central Research Institute for Chemistry the Hungarian Academy of Sciences, Budapest, Hungary.

$$AH^{(+)} + e \xrightarrow{\quad E \quad} \text{Products} \tag{C}$$

$$\frac{i}{i_d - i} = 0{\cdot}886[H_3O^+]\left(\frac{t_1}{K_a}\right)^{\frac{1}{2}}\left(k_{H_3O^+} + \frac{k_{H_2O}[H_2O]}{[H_3O^+]} + \frac{k_{SH}[SH]}{[H_3O^+] + K_{SH}}\right), \tag{1}$$

where i_d is the mean diffusion current corresponding to the reduction of the species $AH^{(+)}$, i the mean kinetically controlled current of the same species, and t_1 the drop-time at the potential corresponding to the limiting current i, symbol K_a stands for the acid dissociation constant of the electroactive species $AH^{(+)}$, symbol k_{H_2O} for the rate constant k_{SH} of reaction (B) for $SH \equiv H_2O$, $[SH]$ for the acid buffer component concentration and k_{SH} for the acid dissociation constant of the buffer acid. A more complex equation was derived for α-ketoacids, where a general acid-base catalysed dehydration was taken into account[10] in addition to the proton-transfer reaction.

To determine separately the values of $k_{H_3O^+}$ and k_{SH} in buffered solutions containing only one acid SH and one base $S^{(-)}$, the dependence of the limiting current i was evaluated[7-14] by varying the buffer concentration of a constant ratio of $[SH]:[S^-]$, i.e. at a given concentration of hydrogen ions. In addition to a quantitative treatment of the wave-heights at a selected pH-value, qualitative handling has indicated[15-20] that for a number of structurally different substances the height of the kinetically controlled limiting current depends on the kind of buffer and its concentration[15-20]. A special example of this type is reaction in which a strong adsorption of proton donors at the electrode surface[13,14,19,21] can affect the rate of the protonation reaction.

The most common type of dependence followed in polarographic studies of organic compounds for their characterization and for the elucidation of the course of the electrode process, is the dependence of the mean limiting current i on pH (for reactions in aqueous and water-containing systems). The shape of the plot of this dependence for systems in which only reaction (A) and (C) is involved was derived in the early stages of the development of the theory[1-3] and its application has since proved useful for numerous compounds. For systems involving both reactions (A) and (B), even when it might be apparent from the form of equation (1), the expression for the dependence of limiting current on pH for varying buffer compositions was neither explicitly derived nor experimentally verified. In this paper such a treatment is given for a system of reactions $(A)-(C)$ involving a single acid SH. This corresponds to the behaviour observed for the reduction of a protonated form in simple buffers, consisting only of one acid and one conjugate base. The derived equations were applied to the change in limiting currents of the one-electron wave of aryl-alkyl ketones with pH in acetate buffers.

THEORETICAL

In buffers consisting of a monobasic acid and the conjugate base, the pH can be changed in three ways, according to method of preparation: Either the analytical concentration of the buffer substance (*i.e.* $[S^-] + [SH]$) is kept constant and the ratio $[S^-]/[SH]$ changed, or the acid form concentration $[SH]$ is kept constant and that of the base form $[S^-]$ changed, or, *vice versa*; the base form concentration $[S^-]$ is kept constant and that of the acid form $[SH]$ changed. Since in the first alternative ($[S^-] + [SH] = $ const.) three quantities *i.e.* $[S^-]$, $[SH]$ and pH, are changed simultaneously, the separation of rate constants $k_{H_3O^+}$ and k_{SH} is not straightforward. Hence this type of buffer composition was not taken into consideration here and the treatment was restricted to buffer solutions in which the concentration of either the acid or the base form was kept constant.

Acid component constant. When the concentration of the acid buffer component $[SH]$ is kept constant and the pH is changed by varying the base buffer component concentration $[S^-]$, the dependence of the ratio of kknetic and diffusion mean currents (i/i_d) follows equation (2):

$$i/i_d = \frac{0{\cdot}88\sqrt{(t_1/K_a)^{\frac{1}{2}}}\,[H_3O^+]\,(k_{H_3O^+} + k_{SH}[SH]/[H_3O^+])^{\frac{1}{2}}}{1 + 0{\cdot}88(t_1/K_a)^{\frac{1}{2}}\,[H_3O^+]\,(k_{H_3O^+} + k_{SH}[SH]/[H_3O^+])^{\frac{1}{2}}}, \qquad (2)$$

where the value of $k_{H_2O}/[H_3O^+]$ was negligible compared to the other two terms in the bracket. For $k_{H_3O^+} \gg k_{SH}[SH]/[H_3O^+]$, the shape of the i/i_d–pH plot does not depend on kind and concentration of the buffer component and has a form corresponding to a simple dissociation curve derived for a system involving only reactions (A) and (C) (Fig. 1, curve 1).

For $k_{H_3O^+}[H_3O^+] \ll k_{SH}[SH]$, the i/i_d–pH plot is independent of the value of $k_{H_3O^+}$ and follows equation (3) which corresponds to a dissociation curve with half of the slope of that observed for a dissociation curve for a system involving only reactions (A) and (C). Because the Brönsted relation is generally valid, the value

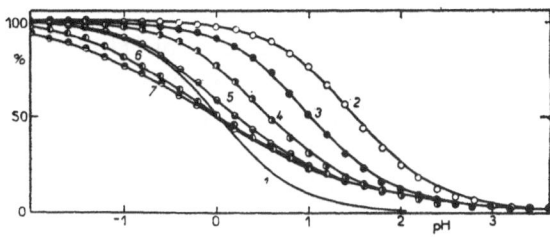

FIG. 1

pH-Dependence of the Ratio i/i_d for Various Values of $k_{SH}[SH]/k_{H_3O^+}$ in Buffers with Acid Component Concentration Constant

Theoretical curves for: *1* $k_{H_3O^+} \gg k_{SH}[SH]/[H_3O^+]$; ratio $k_{SH}[SH]/k_{H_3O^+}$ is: *2* 10^{-3}; *3* 10^{-2}; *4* 10^{-1}; *5* 1; *6* 10; *7* 100.

of $k_{H_3O^+}$ is greater than any value of the specific rate constant k_{SH} for the acids usually used as buffers in aqueous media, but the condition $k_{H_3O^+}[H_3O^+] \ll k_{SH}[SH]$ has not been fulfilled in practice.

$$i/i_d = \frac{0.88\,(t_1/K_a)^{\frac{1}{2}}\,([H_3O^+]\,k_{SH}[SH])^{\frac{1}{2}}}{1 + 0.88(t_1/K_a)^{\frac{1}{2}}\,([H_3O^+]\,k_{SH}[SH])^{\frac{1}{2}}}\,. \tag{3}$$

Because for a given system the change in the value of i/i_d with pH can be observed only over a given pH-range, the shape of the i/i_d-pH plot depends on the ratio $k_{SH}[SH]/k_{H_3O^+}$. Theoretical curves for several values of this ratio are given in Fig. 1.

Base component constant. When the concentration of the base buffer component $[S^-]$ is kept constant and the pH-value changed by varying the acid buffer component $[SH]$, the ratio i/i_d is given by equation (4):

$$i/i_d = \frac{0.88(t_1/K_a)^{\frac{1}{2}}\,[H_3O^+]\,(k_{H_3O^+} + k_{SH}[S^-]/K_{SH})^{\frac{1}{2}}}{1 + 0.88(t_1/K_a)^{\frac{1}{2}}\,[H_3O^+]\,(k_{H_3O^+} + k_{SH}[S^-]/K_{tH})^{\frac{1}{2}}} \tag{4}$$

For a given concentration of the base buffer component $[S^-]$, the value of $k_{SH}[S^-]/K_{SH}$ is pH-independent and its sum with $k_{H_3O^+}$ is constant. Hence for each base buffer component concentration $[S^-]$, the plot of i/i_d against pH follows a simple dissociation curve, that corresponds to a system involving only reactions (A) and (C) (Fig. 1, curve 1). The position of this dissociation curve is, nevertheless, shifted on the pH-scale towards higher pH-values, as compared with a system with the same values of K_a and $k_{H_3O^+}$, but in which reaction (B) does not participate. The greater is the concentration of the base buffer component $[S^-]$, the higher is the pH-value at which the dissociation curve is observed and the greater the shift between the curves without and with participation of reaction (B).

To express this shift the concept of a polarographic dissociation constant (K') for the case in which only reaction (A) is involved (or when $[S^-]$ approaches zero) is introduced and its value is put equal to the value of $[H_3O^+]$ at which $i = i_d/2$. For systems in which the value of $k_{SH}[S^-]/K_{SH}$ is not negligible compared to $k_{H_3O^+}$ and in which reaction (B) ought to be taken into consideration, the value of $[H_3O^+]$ at which $i = i_d/2$ is denoted as K''. The change in the value of K'' with concentration of the base buffer component $[S^-]$ is given by equation (5):

$$1/(K'')^2 = 1/(K')^2 + 0.77(t_1/K_a)\,.\,(k_{SH}[S^-]/K_{SH})\,. \tag{5}$$

The value of pK'' is determined from the pH-value using the $\bar{z}/\bar{z}_d$–pH plots for $i = i/2$ for various concentrations $[S^-]$. The plot of $1/(K'')^2$ against $[S^-]$ should be linear. From the slope of this plot, which is equal to $0.77\,t_1\,k_{SH}/(K_{SH}K_a)$ the value of k_{SH}/K_a can be evaluated and from the intercept, the value of $1/(K')^2$ determined.

DISCUSSION

The equations derived above were used for the evaluation of changes of limiting current with pH for some aryl alkyl ketones. To study substituent effects in alkyl substituted deoxybenzoins[22] their polarographic behaviour was compared with that of aceto-phenone and some other aryl alkyl ketones in order to verify their reduction scheme. The decrease of the one-electron wave of acetophenone in acidic media, which corresponds to the reduction of the protonated carbonyl group[23] (equation (A) and (B), Part I), was shown to depend on the kind and concentration of the buffer. The same effect was observed also for deoxybenzoin derivatives (Part III of this Series) which moreover gave waves that were better developed in the pH-region where the decrease of the limiting current occured and hence were measurable with a greater accuracy (Fig. 2).

The effect of buffer composition was interpreted by participation of acids other than the hydrogen ion in the protonation reaction, according to the reaction (B). To separate specific rate constants it proved to be useful in homogeneous general acid catalysed reactions to kept the acid buffer component concentration constant. When the same procedure was applied to this electrode process accompanied by a simultaneous reaction with acid components and when accordingly acetate buffers were prepared containing identical concentrations of acetic acid and varying con-centrations of sodium acetate, the i/i_c-plots were drawn out, in agreement with

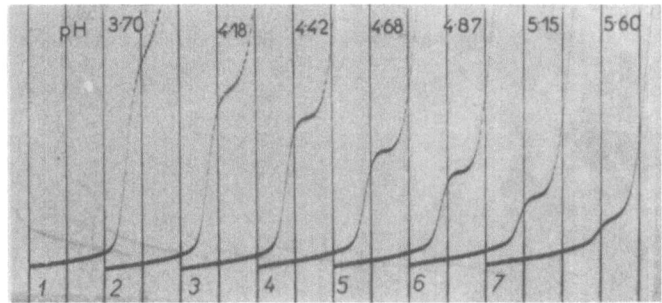

FIG. 2

pH-Dependence of Polarographic Waves of n-Propyldeoxybenzoin

$5 . 10^{-4}$M Propyldeoxybenzoin, acetate buffers, sodium acetate concentration constant (0·05M), 50% ethanol, pH-values of aqueous buffers: 1 3·70; 2 4·18; 3 4·42; 4 4·68; 5 4·87; 6 5·15; 7 5·60. Curves starting at −0·8 V, s.c.e., 200 mV/absc., $h = 39$ cm, $t_1 = 2·5$ s, $m = 2·2$ mg/s, full scale sensitivity 1·4 μA.

54

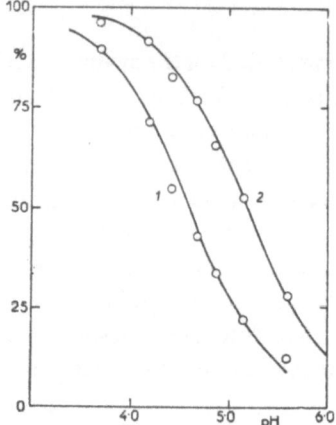

FIG. 3

pH-Dependence of the Ratio i/i_d for the One-Electron Wave of n-Propyldeoxybenzoin in Acetate Buffers with Sodium Acetate Concentration Constant

Curve *1* 0·05M sodium acetate; *2* 0·5M sodium acetate. Theoretical curves for pK'' 4·60 *1* and 5·18 *2*.

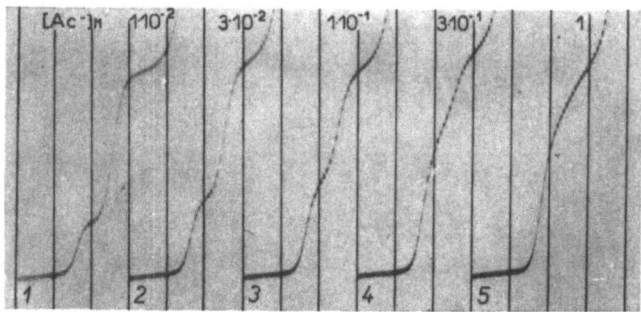

FIG. 4

Effect of Buffer Concentration on Reduction Waves of Acetophenone

Acetate buffer pH 5·68, 4 . 10^{-4}M acetophenone, ionic strength and sodium ion concentration kept constant by addition of sodium chloride. Concentration of the acetate: *1* 0·01M; *2* 0·03M; *3* 0·1M; *4* 0·3M; *5* 1·0M. Curves starting at −0·8 V, s.c.e., 200 mV/absc., $h = 48$ cm, full scale sensitivity 2 µA.

equation (2). To exclude the possibility of specific ion or ionic strength effects, the neutral salt added for the control of the ionic strength was chosen to contain the same cation (Na^+) and an anion of the same unit charge as the base component of the buffer used.

When on the other hand buffers were prepared containing constant concentration of acetate and varying concentration of acetic acid, the plot had the shape of a simple dissociation curve of a monobasic acid, in accordance with equation (4). The curves (Fig. 3) were shifted towards higher pH-values with increasing concentration of acetate.

The dependence of $1/(K'')^2$ on acetate concentration was found to be linear in accordance with equation (5). From the intercept, $K' = 3 . 10^{-5}$ for propyl deoxy-benzoin and $K' = 1\cdot8 . 10^{-4}$ for sec-butyldeoxybenzoin were computed. Because the $1/(K'')^2$ — scale is not logarithmic, the accuracy of these values should be considered rather low. The slope of the linear plot gave $k_{CH_3COOH}/K_a = 5 . 10^5$ for the propyl derivative and $6 . 10^4$ was obtained for sec-butyl deoxybenzoin. The accuracy of these values is better than of those for K'. When compared with corresponding values calculated at one pH-value for different buffer concentrations (Fig. 4*) using equation (1), values for K' and k_{CH_3COOH}/K_a of the same order of magnitude were obtained.

It was indicated[9] that participation of reaction (B) can be an explanation for abnormally and unreasonably high values of rate constants computed for recombination reactions from polarographic data. The values of dissociations constants K_a for the studied compounds are unknown[24], but as a first approximation it can be assumed that they are of the order of 10^6. This would give a just acceptable value of the order 10^{11} ($l\ mol^{-1}\ s^{-1}$) for the constants k_{CH_3COOH}, but a non-plausible value of the order $10^{13} - 10^{14}$ for the rate constant $k_{H_3O^+}$. Therefore, even when the value of $k_{H_3O^+}$ obtained by the above treatment using equation (5) or at constant pH and variable buffer concentration using equation (1) is smaller than the apparent rate constant $[k_{H_3O^+} + k_{SH}[SH]/([H_3O^+] + K_{SH})]$ obtained with a finite concentration of the buffer, its value is still too large. The reaction involved is a surface reaction[13,14] and either adsorption[13,14,19] or other double-layer effects[21] must be taken into consideration in the evaluation of the homogeneous rate constant $k_{H_3O^+}$ in addition to the participation of reactions of the type (B).

56

REFERENCES

1. Brdička R., Wiesner K.: This Journal *12*, 39, 139 (1947).
2. Koutecký J., Brdička R.: This Journal *12*, 337 (1947).
3. Koutecký J.: This Journal *18*, 597 (1953).
4. Henke K. H., Hans W.: Z. Elektrochem. *57*, 591 (1953).
5. Hans W., Henke K. H.: Z. Elektrochem. *57*, 595 (1953).
6. Hanuš V.: *Proc.. 1st Internat. Polarograph. Congr. Prague 1951*, Part I, p. 881. Published by Přírodovědecké nakladatelství, Prague 1951.
7. Wiesner K., Wheatley M., Los J. M.: J. Am. Chem. Soc. *76*, 4858 (1954).
8. Green J. H., Walkley A.: Australian J. Chem. *8*, 51 (1955).
9. Bartel E. T., Grabowski Z. R., Kemula W., Turnowska-Rubaszewka W.: Roczniki Chem. *31*, 13 (1957).
10. Becker M., Strehlow H.: Z. Elektrochem. *64*, 818 (1960).
11. Bartel E. T., Grabowski Z. R., Kemula W.: Roczniki Chem. *34*, 345 (1960).
12. Kemula W., Grabowski Z. R., Bartel E. T.: Roczniki Chem. *33*, 1125 (1959).
13. Majranovskij S. G., Liščeta L. I.: This Journal *25*, 3025 (1960); Izv. Akad. Nauk. U.S.S.R. *1962*, 1984.
14. Majranovski S. G.: J. Electroanal. Chem. *3*, 166 (1962).
15. Wheatley M. S.: Experientia *12*, 339 (1956).
16. Zahradník R., Svátek E., Chvapil M.: This Journal *24*, 347 (1959); Chem. listy *51*, 2232 (1957).
17. Hrdý O.: This Journal *24*, 1180 (1959); Chem. listy *52*, 1058 (1958).
18. Zuman P.: J. Electrochem. Soc. *105*, 758 (1958).
19. Volková V.: Nature *184*, 743 (1960).
20. Manoušek O., Zuman P.: This Journal *29*, 1432 (1964).
21. Koryta J.: Rev. Polarog. (Kyoto) *13*, 1 (1965).
22. Zuman P., Turcsanyi B., Mills A. K.: This Journal, in the press.
23. Zuman P.: This Journal *33*, 3205 (1968).
24. Arnett E. M.: Progr. Phys. Org. Chem. *1*, 223 (1963).

Translated by the author (P. Z.).

POLAROGRAPHIC REDUCTION
OF ALDEHYDES AND KETONES. III.*

EFFECTS OF ALKYL GROUPS IN THE SIDE CHAIN
ON THE HALF-WAVE POTENTIALS OF DEOXYBENZOINS
AND SOME OTHER ALKYL ARYL KETONES

P.Zuman**, B.Turcsanyi*** and A.K.Mills****

J. Heyrovský Institute of Polarography,
Czechoslovak Academy of Sciences, Prague 1

Received December 27th, 1967

The polarographic waves of deoxybenzoins are in principle similar to the reduction waves of acetophenone. In acidic media the protonised form is reduced and the rate of the protonation (in which also acids other than hydroxonium ion participate) limits the wave-height. Even when this reaction is a "surface reaction", the wave height is governed primarily by changes in chemical reactivity in the protonation reaction rather than by structural changes in adsorptivity. Alkyl groups R in deoxybenzoins of the type $C_6H_5COCHRC_6H_5$ exert predominantly a polar effect on half-wave potentials.

Studies of some ω-aminoacetophenones[1] of the type $C_6H_5COCR'R^2NH_2$ and branched alkyl phenyl ketones[1] of the type $C_6H_5COCR'R^2R^3$ indicated that branching of the side-chain at the carbon in the α-position to the carbonyl group results in considerable steric effects. These were shown by deviations from linear $E_{1/2}-\sigma^*$ plots[2]: The half-wave potentials for the compounds with α-branched chains were shifted to potentials more negative than those expected if only polar effects were operating,

$$C_6H_5COCHC_6H_5$$
$$|$$
$$R$$

I, R = CH$_3$	*IV*, R = i-C$_3$H$_7$	*VII*, R = sec-C$_4$H$_9$
II, R = C$_2$H$_5$	*V*, R = n-C$_4$H$_9$	*VIII*, R = tert-C$_4$H$_9$
III, R = n-C$_3$H$_7$	*VI*, R = i-C$_4$H$_9$	*IX*, R = n-C$_5$H$_{11}$

* Part II: This Journal *33*, 3090 (1968).
** Present address: Department of Chemistry, University of Birmingham, Great Britain.
*** Present address: Central Research Institute for Chemistry Hungarian Academy of Sciences, Budapest, Hungary.
**** Present address: Research Laboratory, Arthur Guinness, Son & Co., Dublin, Ireland.

as in straight-chain compounds. Because in both the series mentioned above only a limited number of derivatives were available, the effect of alkyl groups on the reduction of a carbonyl group was investigated in the deoxybenzoin series, prepared by one of us[3].

The polarographic reduction of substituted deoxybenzoins has not been studied previously; even for the parent compound only the half-wave potential in one or two supporting electrolytes has been quoted[4-8] and it has been mentioned that the polarographic behaviour of deoxybenzoin resembles that of acetophenone[5].

In this paper the polarographic behaviour of compounds $I-IX$ was studied. For the study of structural effects, the half-wave potentials of the pH-independent two-electron waves of pH 9·3 proved best.

EXPERIMENTAL

Apparatus. The polarographic curves were recorded on a Heyrovský polarograph, Mark V/301 (Zbrojovka, Brno) with photographic recording, using a galvanometer of sensitivity $2 . 10^{-9}$ A/min. A Kalousek vessel with a separated saturated calomel electrode (s.c.e.) was used. The capillary electrode used had the following constants: outflow velocity $m = 2·2$ mg/s and drop-time $t_1 = 2·5$ s in 0·1M-KCl at the potential of the s.c.e. at mercury pressure $h = 39$ cm. All half-wave potentials were measured using two techniques: Firstly, the half-wave potentials were determined from polarographic curves recorded from positive to negative potentials and then with the reverse scan from negative to positive potentials. The mean values of these measurements were evaluated using the half-wave potential of thallous ions as internal standard ($E_{\frac{1}{2}} = 0·445$ V *versus* s.c.e.) and expressed *versus* the s.c.e. Secondly, the three electrode technique[9,10] as modified by Majranovskij[11] was used for potentiometric measurement of the potential at the half-wave. The pH-values of buffer solutions used were checked using a pH-meter Mark PHM 4 (Radiometer-København) with a glass electrode type G 200 B.

Substances and solutions. The preparation of alkyl substituted deoxybenzoins $I-IX$ has been described elsewhere[3]. Chemicals for preparation of buffers were Analar Grade. 0·01M ethanolic stock solutions of substances $I-IX$ were prepared by dissolving a weighed amount of the substance in 96% ethanol at 25°C.

Techniques. 5 ml of the buffer solution was added to 4·8 ml of ethanol and from the solution the atmospheric oxygen was removed using a stream of pure nitrogen. 0·2 ml of a 0·01M stock solution of the deoxybenzoin were added and the purging continued for a further 30 seconds. Polarographic curves were then recorded.

RESULTS AND DISCUSSION

Polarographic Behaviour of Deoxybenzoins $I-IX$

Substituted deoxybenzoins $I-IX$ follow in principle the reduction scheme described in Part I of this Series[12]. The waves of compounds $I-IX$ in solutions of strong mineral acids nevertheless were indistinct in the presence of 50 per cent ethanol (needed due to their limited solubility) and unsuitable for quantitative treatment. The solubility of the methyl derivative I only was sufficiently high for measurements to be made in solutions containing 20 per cent ethanol. Under these conditions in solutions of hydrochloric acid, methyl derivative I gave measurable waves (i_1). With increasing ethanol concentration the wave i_1 is shifted towards more negative potentials. This explains the behaviour observed for solutions containing 50 per cent ethanol mentioned above.

At pH < 5 in acetate buffers containing 20 per cent ethanol, the methyl derivative I gave two waves $(i_1$ and $i_2)$ of approximately the same height. At pH > 5 in acetate or citrate buffers, the two waves coalesce and form a two-electron wave $(i_1 + i_2)$ at potential $E_1 \approx E_2$. With increasing pH, the height of the wave $(i_1 + i_2)$ decreases and a new wave i_3 appears at more negative potentials E_3. Hence for compound I in the presence of 20 per cent ethanol at pH > 5 the following condition is valid: $|E_1| \approx |E_2| < |E_3|$. Hence the pH-dependence is similar to that depicted in Fig. 3, part I of this Series[12], with the exception that at pH > 5 the half-wave potentials E_1 and E_2 differ so little that waves i_1 and i_2 coalesce.

In acetate buffers containing 50 per cent ethanol all the deoxybenzoins $I-IX$ gave measurable waves of the protonized form. The decrease of the limiting current with increasing* pH depends on the composition of the buffer in the way predicted by equations $(2)-(4)$ derived in Part II of this Series[13]. In the protonation of deoxybenzoins at the surface of the dropping mercury electrodes acid buffer components can participate in addition to hydrogen ions, according to the scheme $(A)-(C)$

$$C_6H_5COCHRC_6H_5 + H_3O \underset{k^-_{H^+}}{\overset{k_{H^+}}{\rightleftharpoons}} \underset{\underset{OH^{(+)}}{\|}}{C_6H_5CCHRC_6H_5} \quad (A)$$

$$C_6H_5COCHRC_6H_5 + HS \underset{k_{SH}^-}{\overset{k_{SH}}{\rightleftharpoons}} \underset{\underset{OH^{(+)}}{\|}}{C_6H_5CCHRC_6H_5} \quad (B)$$

$$\underset{\underset{OH^{(+)}}{\|}}{C_6H_5CCHRC_6H_5} + e \overset{E_1}{\longrightarrow} \underset{\underset{OH}{\mid}}{C_6H_5\dot{C}CHRC_6H_5} \quad (C)$$

where HS is the acid buffer component. The radical formed in reaction (C) can undergo dimerization, interaction with mercury or components of the solution or be further reduced.

Both the value of pK'' (defined[13] in Part II) and the shape of the limiting current depend on the nature of the alkyl group R (Fig. 1). The substances with straight chain

---

* The values of "pH" in 50 per cent ethanol were measured with a glass electrode. To prove that the presence of ethanol affects only the pH value of the acid buffer component, half-wave potentials of naphthoquinone were measured in buffers used in the presence of 50 per cent ethanol. The plot of half-wave potentials against "pH" was linear.

alkyl group *I, II, III, V* and *IX* show a similar value of pK'' and hence their wave-heights at pH 4·6 differ only little (Fig. 1). The pK''-values for branched chain alkyl derivatives *IV, VI, VII* and *VIII* are considerably smaller, as is indicated by the smaller wave heights at pH 4·6 (Fig. 1). This effect is most marked in the case of the tert-butyl derivative *VIII*.

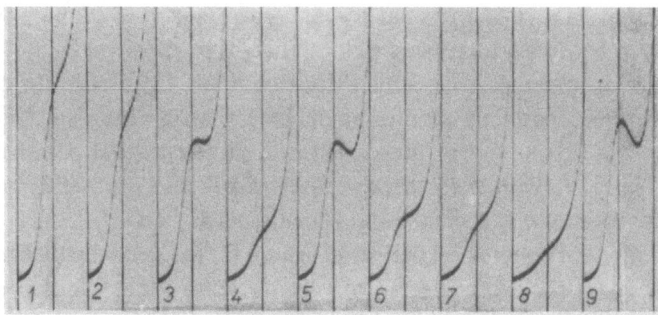

Fig. 1

Comparison of Wave Heights of Alkyl Substituted Deoxybenzoins ($C_6H_5COCHRC_6H_5$)

$2 . 10^{-4}$M deoxybenzoin, 0·05M acetate buffer pH 4·56, 50% ethanol, $R = 1$ CH_3; 2 C_2H_5; 3 C_3H_7; 4 i-C_3H_7; 5 C_4H_9; 6 i-C_4H_9; 7 s-C_4H_9; 8 t-C_4H_9; 9 C_5H_{11}. Curves starting at $-1·2$ V, s.c.e., 200 mV/absc. $h = 38$ cm, $t_1 = 2·6$ s, $m = 2·1$ mg/s, full scale sensitivity $= 1$ μA.

This change in the value of pK'' can be a result of polar and steric effects of the alkyl groups on rate and dissociation constants of the protonation reaction. Because the values of the dissociation constants of the reaction are not accessible, it was not possible to calculate the values of the rate constants using the treatment shown in Part II of this Series[13]. As it was possible to assume that linear free energy relationships will be valid not only for dissociation constants but also for rate constants and their combination, the wave-heights at a certain pH (4·6) were plotted[2,14] against Taft polar substituent constants σ^*. The limiting currents fitted the linear relationship reasonably well (Fig. 2) with the exception of the isobutyl derivative *VI*. Even when the approximative character of such a treatment is well understood, nevertheless it allows one to deduce that the predominant factor affecting the reactivity in the antecedent proton transfer reaction is the polar rather than steric effect of the alkyl groups.

The antecedent protonation is a surface reaction. This was indicated in Part II of this' Series[13] and will be discussed below. One of the assumptions made in the interpretation of surface reactions is that the electroactive species is adsorbed at the surface of the electrode (ref. in Part I of this Series[12]). Even when quantitative data on the correlation between adsorptivity and structure are rather scarce, it is usually accepted that the longer and the bulkier is the side-chain the more strongly is the substance adsorbed at the mercury surface. Increase in adsorption, *i.e.* increase

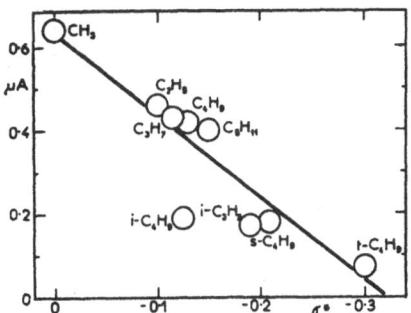

FIG. 2

Dependence of the Limiting Currents (i_1) of Alkyl Substituted Deoxybenzoins at pH 4·56 on Taft Polar Substituent Constants σ^*

in surface concentration, would be expected to lead to an increase in the rate of the rate of the protonation reaction, *i.e.* to an increase in the wave height at a chosen pH-value. Inspection of Figs 1 and 2 indicates oppositely, that the bulkier the alkyl groups the slower the reaction which limits the rate of the wave. Hence it can be concluded that the substitution of alkyl groups in deoxybenzoins affects the chemical reactivity rather than adsorptivity of those compounds.

The values of estimated rate constants of the recombination reaction[13] indicated that the reaction accompanying the electrode process proper is a surface reaction. This is further substantiated by the shape of the waves, in particular of compounds *III*, *V* and *IX* with a longer straight chain. For these compounds, when the limiting current was between about 70 and 30 per cent of the diffusion controlled limiting current, a decrease (trough) was observed on the limiting currents. Such a decrease in the value of the limiting current with increasingly negative potentials is assumed to be a proof of the surface character of the reaction involved.

The behaviour at pH > 7 showed the regular pattern discussed[12] in Part I. The height and the half-wave potential of the two-electron wave observed between pH 7 and 10 were practically pH-independent. The half-wave potentials showed small but significant changes, when the kind and concentration of the cation in the buffered supporting electrolyte were changed.

The controlled potential electrolysis of α-methyldeoxybenzoin (*I*) at pH 8 yielded[15] *erythro*-1,2-diphenyl-1-propanol. A sequence: electron, electron, hydrogen ions was assumed and was suggested[15] that either the rate controlled protonation of the dianion or the transfer of the second electron occurs in a stereospecific manner.

Substituent Effects on Half-Wave Potentials

For the study of substituent effects 0·05M borax was chosen as a supporting electrolyte for measurements of the half-wave potentials, where the electrode process follows the scheme $(D)-(G)$

$$C_6H_5COCHRC_6H_5 + e \xrightarrow{E_3} C_6H_5\overset{(\doteq)}{C}OCHRC_6H_5 \qquad (D)$$

$$C_6H_5\overset{(\doteq)}{C}OCHRC_6H_5 + H^+ \rightleftharpoons \underset{OH}{C_6H_5\overset{\cdot}{C}CHRC_6H_5} \qquad (E)$$

$$\underset{OH}{C_6H_5\overset{\cdot}{C}CHRC_6H_5} + e \xrightarrow{E_2} \underset{OH}{C_6H_5\overset{(-)}{C}CHRC_6H_5} \qquad (F)$$

$$\underset{OH}{C_6H_5\overset{(-)}{C}CHRC_6H_5} + H^+ \rightleftharpoons \underset{OH}{C_6H_5CHCHRC_6H_5} \qquad (G)$$

Because the half-wave potential of the two-electron wave is pH-independent, it may be deduced that either reaction (D) is activation controlled and potential determining or the reaction (D) will be assumed to be reversible and equilibrium (E) shifted completely to the right hand side, so that reaction (F) becomes potential determining, as it is in the second wave in more acidic media. Because potential E_2 is either comparable with E_3 or even more positive, only one two-electron wave is observed.

The half-wave potentials measured from polarographic curves by comparison with thallium (Fig. 3) and by a three electrode method are summarized in Table I. Comparison of the values indicates that the reproducibility of the values measured

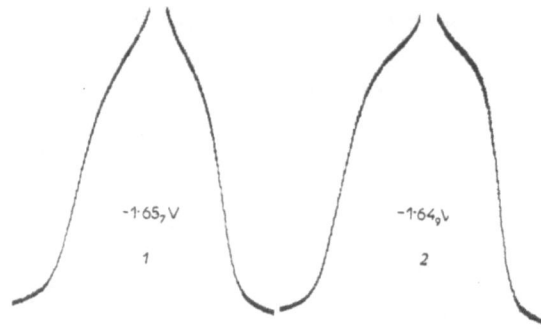

-1·65₇V   -1·64₉V

1   2

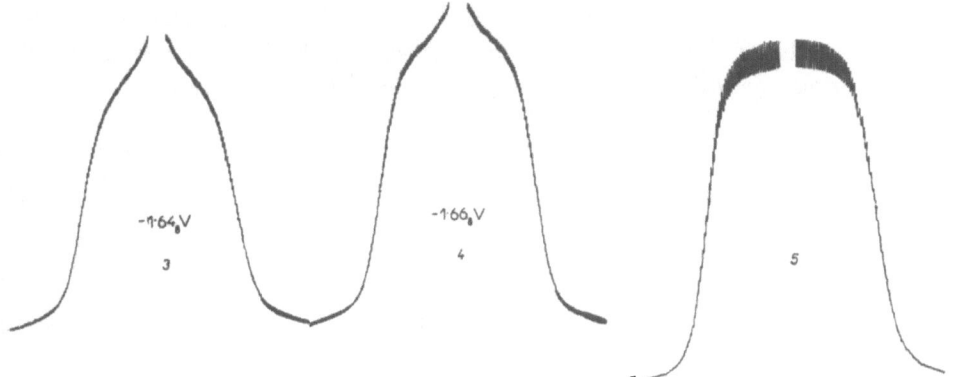

FIG. 3

Measurements of Half-Wave Potential of Deoxybenzoins ($C_6H_5COCHRC_6H_5$) from Polarographic Curves by Comparison with a Thallium Wave

0·05M borax pH 9·3, 50% ethanol. $2·10^{-4}$M deoxybenzoin, $R = 1$ C_5H_{11}; 2 C_2H_5; 3 C_3H_7;
4 i-C_3H_7. Curve 5: $1·10^{-4}$M-$TlNO_3$. Curves 1 and 2 starting at $-1·45$ V, 3 and 4 at $-1·40$ V,
5 at $-0·2$ V, s.c.e., 50 mV/absc., $h = 65$ cm. $t_1 = 2·3$ s, $m = 2·4$ mg/s, full scale sensitivity
2 µA.

TABLE I

Half-Wave Potentials of Alkyl Substituted Deoxybenzoins ($C_6H_5COCHRC_6H_5$)

0·05M Borax pH 9·26, 20°C, $2 . 10^{-4}$M depolarizer. A Comparison with Tl^+, B three electrode
method.

No	Compound, R	$-E_{1/2}$, V s.c.e.							σ^*	$\dfrac{RT}{\alpha nF}$
		A				B		mean		
I	CH_3	$1·63_1$	$1·63_5$			$1·63_4$	$1·64_0$	$1·63_5$	0	$0·071$
II	C_2H_5	$1·64_4$	$1·64_3$	$1·64_9$	$1·64_6$	$1·64_4$	$1·64_9$	$1·64_6$	$-0·10$	$0·068$
III	C_3H_7	$1·66_2$	$1·65_4$	$1·64_8$	$1·65_9$	—	—	$1·65_6$	$-0·11_5$	—
IV	i-C_3H_7	$1·66_5$	$1·66_1$	$1·66_8$	$1·66_2$	$1·65_8$	$1·66_1$	$1·66_2$	$-0·19$	$0·067$
V	C_4H_9	$1·66_0$	$1·66_4$	$1·66_3$	$1·66_0$	$1·65_7$	$1·65_7$	$1·66_0$	$-0·13$	$0·050$
VI	i-C_4H_9	$1·65_4$	$1·66_2$	$1·65_4$	$1·66_2$	$1·65_7$	$1·65_8$	$1·65_8$	$-0·12_6$	$0·058$
VII	s-C_4H_9	$1·66_1$	$1·66_4$			—	—	$1·66_3$	$-0·21$	—
VIII	t-C_4H_9	$1·68_5$	$1·69_2$	$1·69_6$	$1·69_6$	$1·69_5$	$1·69_6$	$1·69_3$	$-0·30$	$0·051$
IX	C_5H_{11}	$1·65_7$	$1·66_0$			$1·65_9$	$1·65_8$	$1·65_9$	$-0·15$	$0·056$

with the three-electrode method is slightly better than of those obtained from polarographic curves (the latter values were nevertheless gathered over a period of several
months).

The values of $RT/\alpha nF$ (Table I) show a decrease with increasing negative value of σ^*. Owing to the low reproducibility of values of $RT/\alpha nF$ this relation is to be considered rather as a trend than as a dependence. The half-wave potentials show a good correlation with Taft[14,15] polar substituent constants σ^* (Fig. 4). Because

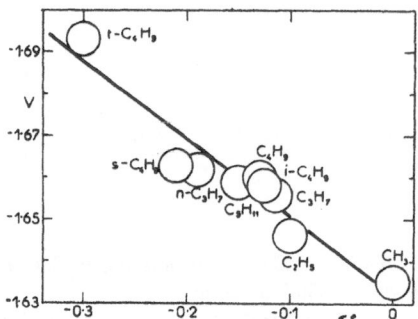

Fig. 4

Dependence on Half-Wave Potentials ($E_{1/2}$) of Alkyl Substituted Deoxybenzoins at pH 9·3 on Taft Polar Substituent Constant σ^*
Line corresponds to a rate constant $\varrho^* = 0·18$ V.

of the above mentioned change in the value of αn with structure, it should be kept in mind that the value of the reaction constant $\varrho^* = 0·18$ V for the relationship shown in Fig. 4 reflects not only the effect of structure on the value of the rate constant k_e^f at the potential of the half-wave, but also the structural changes in the value of the transfer coefficient α.

The linear relationship shown in Fig. 4 indicates that in the studied series the polar effect of the alkyl groups predominates over their steric effects. In compounds of the type $C_6H_5COCR^1R^2R^3$ (with R standing for alkyl groups) steric effects of branched alkyl groups were observed[1]. The effects of groups R^1, R^2 and R^3 were not additive and this also indicates the participation of steric effects (small non-representative groups of compounds were nevertheless compared and hence complications arrousing from accidental choice of substituents cannot be excluded). It is assumed that these steric effects results from bulky groups preventing the optimum orientation of the ketone $C_6H_5COCR^1R^2R^3$ at the surface. Hence in the case of deoxybenzoins it is assumed that both phenyl rings are orientated (probably flat) to the surface of the electrode in such a way that the orientation of the carbonyl group is fixed and unaffected by the presence of bulky alkyl groups that are forced into a position away from the electrode surface. If the adsorbed phenyl rings are in *trans*-position and if the transfer of the second proton (on carbon) takes place before desorption, the reduction yields *erythro*-isomer, as found[15].

REFERENCES

1. Zuman P.: Unpublished results.
2. Zuman P.: *Substituent Effects in Organic Polarography*. Plenum Press, New York 1967.
3. Mills A. K., Wilder Smith A. E.: Helv. Chim. Acta *43*, 1915 (1960).
4. Adkins H., Cox F. W.: J. Am. Chem. Soc. *60*, 1151 (1938).
5. Pasternak R.: Helv. Chim. Acta *31*, 753 (1948).
6. Ashworth M.: This Journal *13*, 229 (1948).
7. Calzolari C.: Ann. Triestini *1954*, No. 1, p 16.
8. Calzolari C., Furlani C.: Ann. Chim. *44*, 356 (1954); Bull. Sci. Fac. Chim. Ind. (Bologna) *12*, 14 (1954).
9. Gardner H. J.: Nature *167*, 158 (1951).
10. Vlček A. A.: This Journal *19*, 862 (1954); Chem. listy *48*, 189 (1954).
11. Majranovskij S. G.: Private communication.
12. Zuman P.: This Journal *33*, 2548 (1968).
13. Zuman P., Turczanyi B.: This Journal *33*, 3090 (1968).
14. Taft R. W. jr. in the book: *Steric Effects in Organic Chemistry* (M. S. Newman, Ed.), Wiley, New York 1956.
15. Mandell L., Powers R. M., Day R. A. jr.: J. Am. Chem. Soc. *80*, 5284 (1958).

Translated by the author (P. Z.).

POLAROGRAPHIC REDUCTION
OF ALDEHYDES AND KETONES. IV.*
LINEAR FREE ENERGY TREATMENT
OF SUBSTITUTED BENZOPHENONES

P.Zuman**, O.Exner, R.F.Rekker*** and W.Th.Nauta***

J. Heyrovský Institute of Polarography,
Czechoslovak Academy of Sciences, Prague 1

Received December 27th, 1967

Reduction of thirty nine *meta*- and *para*-substituted benzophenones at the dropping mercury electrode was investigated. Substituent effects on half-wave potentials were determined for the one-electron wave of the protonized form in acidic media, for the two-electron wave at pH 9·3 and for the wave in sodium hydroxide solutions that corresponds predominantly to a one-electron reduction. For correlations the Hammett equation and an equation relating *meta* with corresponding *para* values derived earlier were used. The deviations from the latter equation as well as the use of constants σ_{p-x}^- in the Hammett plot has enabled some conclusions to be drawn about the electronic structure of the radicals involved. Even when the correlations of the half-wave potentials with Hammett substituent constants show only a limited precision, they allow the conclusion that half-wave potentials are controlled by substituent I and M effects transmitted through the benzene nucleus.

Polarographic literature on the reduction of benzophenones is numerous[1-24], but few papers treated the more fundamental problems of the electrode process. The few papers[6,20-22] in which the effect of substituents on benzene rings on half-wave potentials was studied were restricted in the number and range of compounds studied[6] or the half-wave potentials were measured under conditions where the scheme of the electrode process was just on the transition between two mechanisms[20]. Moreover, some of the half-wave potentials used for structural correlations were measured in the present of an excessive amount of a surface active agent[20]. The most extensive recent studies were restricted to mono- and polysubstituted methyl[22] and methoxy[21] benzophenones.

The reduction of benzophenones follows three main paths, depending on the pH of the buffer solution used, as was discussed in some detail in Part I of this series[25]. In acid media two one-electron waves are observed and the sequence is: proton,

* Part III: This Journal *33*, 3205 (1968).
** Present address: Department of Chemistry, University of Birmingham, Great Britain.
*** Present address: Department of Chemistry, Vrije Universiteit Amsterdam, and Research Department Brocades, Amsterdam, The Netherlands.

electron, electron, proton, the first step of which is described by equations (A) and (B):

$$C_6H_5\overset{(+)}{C}OHC_6H_5 \underset{k_{-1}}{\overset{k_1}{\rightleftharpoons}} C_6H_5COC_6H_5 + H^+ \qquad (A)$$

$$C_6H_5\overset{(+)}{C}OHC_6H_5 + e \overset{E_1}{\dashleftarrow\dashrightarrow} C_6H_5\overset{\cdot}{C}C_6H_5 \qquad (B)$$
$$\underset{OH}{|}$$

The radical formed in the first one-electron step can dimerize, react with mercury or with solvent or is further reduced at a more negative potential E_2 (equation (E)).

In the medium pH-range, the protonation reaction (A) with the rate constant k_{-1} is no longer fast enough to convert at the surface of the electrode benzophenone into the conjugate acid. A two-electron wave is observed having a pH-independent half-wave potential, which corresponds to a sequence: electron, proton, electron, proton, corresponding to scheme $(C)-(F)$:

$$C_6H_5COC_6H_5 + e \overset{E_3}{\longrightarrow} C_6H_5\overset{(\dot{-})}{C}OC_6H_5 \qquad (C)$$

$$C_6H_5\overset{\cdot}{C}C_6H_5 \underset{k_{-6}}{\overset{k_6}{\rightleftharpoons}} C_6H_5\overset{(\dot{-})}{C}OC_6H_5 + H^+ \qquad (D)$$
$$\underset{OH}{|}$$

$$C_6H_5\overset{\cdot}{C}C_6H_5 + e \overset{E_2}{\longrightarrow} C_6H_5\overset{(-)}{C}C_6H_5 \qquad (E)$$
$$\underset{OH}{|} \qquad\qquad \underset{OH}{|}$$

$$C_6H_5CHC_6H_5 \rightleftharpoons C_6H_5\overset{(-)}{C}C_6H_5 + H^+ \qquad (F)$$
$$\underset{OH}{|} \qquad\qquad \underset{OH}{|}$$

To observe a single two-electron wave two conditions must be fulfilled in this pH-range: Firstly, the potential E_2 of the second reduction step (E) corresponding to the reduction of the radical $C_6H_5\overset{\cdot}{C}(OH)C_6H_5$, must be comparable to or even more positive than the potential E_3 at which the first electron is transferred onto the unprotonized benzophenone. Secondly, the protonation of the radical anion must be sufficiently fast to convert all the radical anion into the radical (Eq. (D)).

Finally, in alkaline media, in solutions containing a total concentration of cations smaller than about 0.1M and containing cations of smaller ionic radii, the reduction of benzophenone takes place again in two one-electron waves. The sequence is electron, proton, proton and the first step is described by equation (C). The whole system described by equations (F), (H) and (I) in Part I (ref. [25]) is realized, when the rate of protonation of the radical anion is too slow. This radical anion $R\overset{(-)}{C}OR$ can also undergo competitive reactions.

In this contribution substituted benzophenones were chosen primarily as a model system for investigation of substituent effects in a benzene ring. Therefore a broader scope in the structural variations of the studied substances $I-XXXIX$ was chosen than in previous studies. The selected series of compounds included substances bearing various types of substituents in positions *meta-* and *para-* to the carbonyl grouping, and some representative polysubstituted derivatives. The effects of substituents on polarographic half-wave potentials were examined in 0·1M-HCl for the wave which corresponds to the one-electron reduction of the protonized form, according to equations (A) and (B), in borate buffer pH 9,3 in which the two-electron wave corresponds to the scheme $(C)-(F)$, and in 0·1M-NaOH in which the polarographic wave corresponds predominantly to a one-electron process described by equation (C). Our goal was on one hand to verify the possibility of applying the Hammett equation to the treatment of half-wave potentials[26] for a broader range of experimental data and to evaluate the precision of the $E_{1/2}-\sigma$ plots; on the other hand to investigate the possibility of application of equation (1) on polarographic data. The latter equation was earlier derived[27] for homogeneous equilibrium and rate constants in the form (2):

$$(E_{1/2})_{p-x} - E^0_{1/2} = \lambda[(E_{1/2})_{m-x} - E^0_{1/2}] \tag{1}$$

$$\log k_{p-x} - \log k^0 = \lambda(\log k_{m-x} - \log k^0) . \tag{2}$$

Validity of equations (1) and (2) can be expected only for substituents which do not have a free electron pair in the α-position. These equations correlate the effects of these substituents on the *meta-* and *para*-position. Finally, attention has been paid to the additivity of substituent effects[28].

EXPERIMENTAL

Apparatus

The polarographic curves were recorded using a pen-recording polarograph Mark LP 60 (Laboratorní přístroje, Prague). A Kalousek vessel with a separated saturated calomel electrode (s.c.e.) was used, whereas for a three-electrode measurement an adapted Kalousek vessel[29,30] was used. Several capillaries with drop-times between 3 and 4 seconds and out-flow velocities between 1·5 and 2·5 mg/s at the potential of s.c.e. in 1M-KCl were used. The pH-values of the buffer solutions were checked using a pH-meter Mark PHM 4 (Radiometer-København) with a glass electrode type G 200B. For potentiometric measurements of the potential at the half-wave, a compensator Mark QTK (Metra, Blansko) was used.

Substances

The substances used were either commercially available compounds, or were prepared according to the literature references, or by application of commonly used methods, in particular of the

Friedel-Crafts reaction. Their melting points (corr.) were in good agreement with the reported values. No attempts have been made to find optimum reaction conditions which would allow highest yields to be obtained. The synthesis of alkyl derivatives *IV—XII* has already been described[31], that of methylsulphonyl derivatives *XXXVIII* and *XXXIX* was reported in another paper[32].

Special attention has been paid to the synthesis of 4-*cyanobenzophenone* (XXIV). The anomalous polarographic behaviour of this compound[33,34] made it necessary to exclude the possibility of doubt about the identity of compounds described in the literature (*e.g.* ref.[35,36]).

3-*Bromomethylbenzophenone*: 3-Methylbenzophenone (9·8 g) was heated to 150°C. Contamination by atmospheric moisture was prevented. During 30 minutes bromine (2·6 ml) was added drop-wise, and the solution was heated for another 10 minutes. After cooling, ethanol (2 ml) was added and the reaction mixture was stored over night in a refrigerator. Crystals were separated and washed with cold ethanol. Yield 8·4 g (61%), m.p. 73°C (methanol). For $C_{14}H_{11}BrO$ (275·2) calculated: 61·10% C, 4·03% H, 29·04% Br; found 60·78% C, 3·96% H, 29·70% Br.

3-*Phenoxymethylbenzophenone* (XVI): To a solution of sodium (0·26 g) in methanol (10 ml) phenol (1·05 g) and 3-bromomethylbenzophenone (2·75 g) were added and the mixture was refluxed for 3 hours. Water (20 ml) was added to the mixture which deposited 2·66 g (92%) of the crude product on cooling; m.p. 53°C (hexane). For $C_{20}H_{16}O_2$ (288·3) calculated: 83·31% C, 5·59% H; found: 83·50% C, 5·64% H.

4-*Phenoxymethylbenzophenone* (XVII) was prepared from 4-bromomethylbenzophenone[37] in a way analogous to compound *XVI*. Yield 2·70 g (94%), m.p. 77°C (cyclohexane). For $C_{20}H_{16}O_2$ (288·3) calculated: 83·31% C, 5·59% H; found 83·65% C, 5·77% H.

3-*Cyanomethylbenzophenone* (XIV): The solution of 3-bromomethylbenzophenone (2·75 g) and potassium cyanide (0·8 g) in ethanol (30 ml) and water (3 ml) was refluxed for 2 hours. The greater part of ethanol was then distilled off under reduced pressure, the reaction mixture was diluted with water and extracted with ether. The extract was dried over sodium sulphate and ether was distilled off. The oily residue (1·84 g; 84% of the crude product) formed crystals on prolonged storage in a desiccator over phosphorous pentoxide at −20°C, m.p. 63°C (benzene and hexane mixture at −10°C). For $C_{15}H_{11}NO$ (221·3) calculated: 81·43% C, 5·01% H, 6·33% N; found: 81·55% C; 5·08% H, 6·58% N.

4-*Cyanomethylbenzophenone* (XV) was prepared from 4-bromomethylbenzophenone[37] analogously to compound *XIV*. Yield 1·80 g (82%) of a crude oily product. Crystallization was achieved after distillation of a small aliquot at 0·5 mm Hg, m.p. 67°C (diluted ethanol). For $C_{15}H_{11}NO$ (221·3) calculated: 81·43% C, 5·01% H, 6·33% N; found: 81·28% C, 4·94% H, 6·48% N.

Methyl benzophenone-3-carboxylate (XIX) was prepared from the corresponding acid[38] (*XVIII*) *via* the chloride which was not isolated, m.p. 65°C (methanol); literature[39] gives 62°C. Isomeric *methyl benzophenone-4-carboxylate* (XX) was prepared analogously according to the literature[40], m.p. 109°C (methanol), literature[40] records 106—107°C.

Benzophenone-3-carboxamide (XXI). Benzophenone-3-carboxylic acid[37] (2.26 g) was brought to boiling with an excess of thionyl chloride and 0·1 ml of dimethylformamide to achieve solution. The unreacted thionyl chloride was removed by distillation and the crude acid chloride was allowed to react overnight with aqueous ammonia. The crude amide was obtained in a yield of 1·80 g (80%), m.p. 101°C (diluted ethanol). For $C_{14}H_{11}NO_2$ (225·2) calculated: 74·65% C, 4·92% H, 6·22% N; found: 74·94% C, 4·95% H, 6·35% N. This amide was also obtained from 3-cyano-benzophenone (*XXIII*) by treatment with hydrogen peroxide in alkaline media in diluted acetone; yield 90%, m.p. 101°C.

Benzophenone-4-carboxamide (XXII) was prepared by both routes used for compound *XXI* with yields of 85% and 89%. The two products were identical, m.p. 169°C (ethanol). Literature[36] quotes the same melting for a product obtained by cleavage of a complex derivative. For $C_{14}H_{11}NO_2$ (225·2) calculated: 74·65% C, 4·92% H, 6·22% N; found: 74·73% C, 5·07% H, 6·50% N.

3-Cyanobenzophenone (XXIII). 3-Cyanobenzoic acid (2·94 g) was converted to the corresponding acid chloride by treatment with an excess of thionyl chloride and 0·1 ml of dimethylformamide. After removal of the unreacted thionyl chloride by distillation, the crude residue was heated with aluminium chloride (3 g) in benzene (20 ml) for 1 hour at 70°C. The reaction mixture was continuously purged of hydrogen chloride by a slow stream of dry air. Dilute hydrochloric acid was then added and the aqueous layer was extracted repeatedly with small portions of benzene. The collected benzene extracts were distilled under reduced pressure to give the product (2·94 g, 71%), m.p. 90°C (cyclohexane). Literature[41] gives m.p. 92°C for a product prepared by a Sandmayer reaction. For $C_{14}H_9NO$ (207·2) calculated: 81·14% C, 4·38% H, 6·76% N; found: 81·19% C, 4·32% H, 6·68% N.

4-Cyanobenzophenone (XXIV) was prepared from 4-cyanobenzoic acid following the procedure as that used for the preparation of the 3-cyano derivative *XXIII*. Yield 3·43 g (83%), m.p. 111°C (cyclohexane). For $C_{14}H_9NO$ (207·2) calculated: 81·14% C, 4·38% H, 6·76% N; found: 81·30% C, 4·44% H, 6·65% N. For purposes of comparison another sample was prepared. Benzophenone-4-carboxamide (*XXII*) (2·45 g) and phosphor pentoxide (2·5 g) were heated together to 150°C for 15 minutes. The reaction mixture was boiled with benzene, and the benzene solution was evaporated to dryness; yield 1·47 g (71%), m.p. 112°C (cyclohexane). The two specimens were identical. The literature[35,36] gives m.p. 107° and 110°C for products prepared by other methods.

Procedures

0·01M ethanolic stock solutions were prepared by dissolving a weighed amount of the substance in 96% ethanol at 20°C. The supporting electrolytes used were prepared from reagent grade chemicals.

For the preliminary investigation of the substituted benzophenones the atmospheric oxygen was removed from 5 ml of the buffer solution diluted with 4·8 ml of ethanol in polarographic vessel by a stream of pure nitrogen, the final concentration of ethanol being thus 46% by weight. Then 0·2 ml of the 0·01M solution of the substance to be studied was added and nitrogen purging continued for some further 30 seconds and the polarographic curve recorded.

For determination of half-wave potentials the three-electrode system was used[42,43]. The measurements were carried out in a modified Kalousek vessel, containing an immersed stem-like saturated calomel electrode in addition to the dropping electrode and working reference electrode[29,30]. The half-cell E.M.F. of the saturated calomel electrode used was controlled against a standard calomel electrode using the compensation method and by comparison with the half-wave potentials of thallium ions in 1M-KCl, put equal to $-0·48$ V, S.C.E. This value was reproducible to ± 1 mV.

Directly on the polarographic curve recorded with the pen-recording polarograph LP 60 the wave-height was measured graphically and the point corresponding to the half-wave was determined. The voltage corresponding to this point was applied and the potential of the dropping electrode of this voltage was measured using a compensation method and a QTK compensator (Metra, Blansko). The instrument enables the value of potentials to be read with an accuracy of 0·1 mV. The accuracy of the measurement of half-wave potentials depends on the accuracy of the value of the e.m.f. of the saturated calomel electrode half-cell and is ± 2 mV. The reproducibility of measured half-wave potentials of well-developed, sufficiently positive waves is $\pm 1-2$ mV, but for very negative, ill-developed waves this had dropped to $\pm 5-8$ mV. This decrease in accuracy is caused by the uncertainity in the determination of the wave-height and the position of the half-wave.

RESULTS AND DISCUSSION

The half-wave potentials measured with reproducibility of about $2-3$ mV together with the values of αn and the limiting current measured at $2 . 10^{-4}$M are given in Table I. The values of half-wave potentials were correlated with the Hammett total substituent constants σ_X, using the modified Hammett equation[26] (3):

$$\Delta E_{1/2} = \varrho_{\pi,\text{co}} \sum \sigma_X .$$

(3)

Correlations given in Figs $1-3$ show a reasonable precision, typical of similar cases[26,44]. Characteristic for the treatment of polarographic half-wave potentials in correlations with substituent constants is the necessity of applying for electron attracting substituents the values of constants σ^-_{p-X}. They express increased conjugation interaction between substituent and the electroactive group in which the substituent plays the rôle of an acceptor. Also significant is the deviation for the *para*-amino substituted compound (Fig. 2,3). Because this deviation is observed in the same direction and is of the same magnitude in borax and sodium hydroxide solutions the participation of the effect of the protonized ammonium group can be ruled out. This deviation indicates for the electrode reaction a diminished conjugation in the function of a donor in comparison with homogeneous reactions. Deviations of the same type were observed also for other *para*-aminosubstituted electroactive compounds[44].

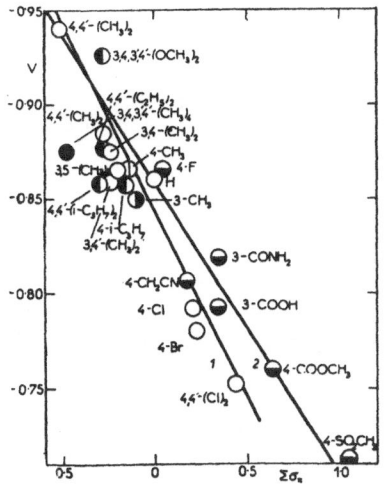

FIG. 1

Dependence of Half-Wave Potentials of Substituted Benzophenones $(E_{1/2})$ on the Sum of Hammett Substituent Constants $\sum \sigma_X$ in 0.1M-HCl

○ $1.25 < \alpha n < 1.52$; $0.30 < I < 0.42$; ◒ αn not determined; ● $\alpha n < 1.25$; ◕ $\alpha n < 1.52$; ◐ $I < 0.42$; ● $I < 0.30$. Not plotted: values for hydroxy and amino derivatives. Line 1: Only points ○ were considered, $\varrho = 0.197$ V; line 2: All points were considered, $\varrho = 0.154$ V. Substituent constants used for 4-COOCH$_3$ and 4-SO$_2$CH$_3$ were σ^-_{p-X} values.

TABLE I

Half-Wave Potentials, ($E_{1/2}$, V vs. s.c.e.), Transition Coefficients (αn) and Limiting Currents (I, μA for $2 \cdot 10^{-4}$M) of Substituted Benzophenones (in 46% ethanol by weight)

No	Substituent	σ	0·1M-HCl			Borax pH 9·3			0·1M-NaOH		
			$-E_{1/2}$, V	αn	I, μA	$-E_{1/2}$, V	αn	I, μA	$-E_{1/2}$, V	αn	I, μA
I	H	0	0·86	1·36	0·36	1·404	0·98	0·55	1·497	1·18	0·41
II	3-CH$_3$	−0·10	0.85_0	1·64	0·34	1·416	1·12	0·55	1·507	1·24	0·44
III	4-CH$_3$	−0·14	0.86_6	1·37	0·55	1·428	1·12	0·57	1·531	1·24	0·39
IV	4-i-C$_3$H$_7$	(−0·15)^a	0.85_7	1·07	0·57	1·436	1·06	0·50	1·525	1·24	0.31_5
V	3,5-(CH$_3$)$_2$	−0·20	0.86_5	1·52	0·38	1·409	1·02	0·52	1·515	1·24	0.38_5
VI	4,4'-(CH$_3$)$_2$	−0·28	0.88_5	1·37	0·41	1·470	1·00	0·54	1·567	1·40	0·36
VII	3,4'-(CH$_3$)$_2$	−0·24	0.85_9	1·20	0·55	1·439	1·02	0·50	1·547	1·24	0·39
VIII	2,4-(CH$_3$)$_2$	−0·24	0.87_5	1·35	0·34	1·459	1·02	0·41	1·551	1·12	0·30
IX	4,4'-(C$_2$H$_5$)$_2$	−0·28	0.87_7	1·00	0·48	1·465	1·24	0·40	1·559	1·34	0·27
X	4,4'-(i-C$_3$H$_7$)$_2$	(−0·30)	0.85_8	0·99	0·29	1·472	1·06	0·42	1·560	1·18	0·27
XI	4,4'-(t-C$_4$H$_9$)$_2$	−0·38	—	—	—	1·477	1·50	0.24^b	1·553	1·50	0.17^b
XII	3,4,3',4'-(CH$_3$)$_4$	−0·48	0.87_5	1·52	0·28	1·504	1·02	0·44	1·602	1·32	0·31
XIII	4-C$_6$H$_5$	0·03	—	—	—	1·351	0·72	0·49	1·433	0·87	0·38
XIV	3-CH$_2$CN	0·16	—	—	—	1·317	1·02	0·54	1·417	0·97	0·48
XV	4-CH$_2$CN	0·18	0.80_7	—	—	1·327	0·77	0·58	1·425	1·02	0·51
XVI	3-CH$_2$OC$_6$H$_5$	0·05	—	—	—	1·354	0·89	0·51	1·433	0·83	0·39
XVII	4-CH$_2$OC$_6$H$_5$	0·06	—	—	—	1·298	1·73	0·50	1·385	1·12	0·41
XVIII	3-COO$^{(-)}$	0·35^c; −0·1^a	$0.79_3{}^c$	—	—	1.418^d	1·32	0·43	1.510^d	1·32	0·33
XIX	3-COOCH$_3$	0·36	—	—	—	1·334	0·93	0·50	1·421	1·32	0·30
XX	4-COOCH$_3$	0·64^e	0·76	—	—	1·222	1·40	0·50	1·264	1·24	0·40
XXI	3-CONH$_2$	(0·35)^a	0·82	—	—	1·332	1·00	0·50	1·409	1·15	0·32
XXII	4-CONH$_2$	0·63^e	—	—	—	1·251	1·18	0·49	1·296	1·60	0·52
XXIII	3-CN	0·61	—	—	—	1·277	1·07	0·50	1·310	1·24	0·31
XXIV	4-CN	1·00^e	—	—	—	1·182	1·12	0·47	1·222	1·60	0·55
XXV	3-F	0·36	—	—	—	1·346	1·12	0·51	1·430	1·32	0·35
XXVI	4-F	0·05	0.86_6	—	—	1.410_5	1·32	0·54	1.502_5	1·32	0·37
XXVII	3-Cl	0·39	—	—	—	1·332	1·02	0·53	1·421	1·18	0·38
XXVIII	4-Cl	0·22	0.79_2	1·38	0·42	1·356	0·97	0·58	1·440	1·24	0·40

No	Substituent										
XXIX	3-Br	0·39	—	—	—	1·312	0·81	0·49	1·406	1·02	0·36
XXX	4-Br	0·23	0·78$_0$	1·38	0·42	1·338	0·97	0·50	1·441	0·57	0·41
XXXI	4,4'-(Cl)$_2$	0·44	0·75	1·25	0·38	1·315	1·02	0·59	1·387$_5$	1·32	0·35
XXXII	4-NH$_2$	−0·66[f]	0·86[g]	—	—	1·515[f]	1·06	0·42	1·608[f]	1·12	0·39
XXXIII	4-OH	−0·33[h]	0·96[h]	1·46	0·27	1·591[j]	1·06	0·50	1·735[j]	1·40	0·83
XXXIV	2,4-(OH)$_2$	—	—	—	—	1·698[j]	1·02	0·73	1·728	1·60	0·78
XXXV	4,4'-(OH)$_2$	−0·66[h]	1·03[h]	1·59	0·36	1·705[j]	0·92	0·72	—	—	—
XXXVI	4,4'-(OCH$_3$)$_2$	−0·52	0·94	1·31	0·30	1·53$_0$	1·12	0·53	1·616	1·32	0·42
XXXVII	3,3',4'-(OCH$_3$)$_4$	−0·28	0·92$_6$	1·13	0·43	1·479	1·18	0·44	1·578	1·24	0·40
XXXVIII	3-SO$_2$CH$_3$	0·64	—	—	—	1·212	1·12	0·59	1·313	1·40	0·41
XXXIX	4-SO$_2$CH$_3$	−1·05[e]	0·72$_3$	—	—	1·155	1·24	0·58	1·223	1·40	0·54

[a] Approximate value; [b] values can be affected by low solubility; [c] for COOH; [d] for $COO^{(-)}$; [e] σ_x^--values; [f] for NH$_2$; [g] for $NH_3^{(+)}$; [h] for OH; [j] for $O^{(-)}$.

TABLE II

Polarographic Data on Some Substituted Benzophenones in Sodium Hydroxide Solutions

No	Substituent	0·01M-NaOH			0·1M-NaOH			1·0M-NaOH		
		−E$_{1/2}$, V	αn	I, μA[a]	−E$_{1/2}$, V	αn	I, μA[a]	−E$_{1/2}$, V	αn	I, μA[a]
XV	4-CH$_2$CN	1·434	0·90	0·49	1·425	1·02	0·48	1·425	1·31	0·43
XX	4-COOCH$_3$	1·262	1·24	0·44	1·264	1·42	0·31	1·405[b]	1·86[b]	0·50[b]
XXI	3-CONH$_2$	1·412[c]	1·18	0·37[c]	1·409[c]	1·31	0·34[c]	1·393[c]	1·60	0·28[c]
XXII	4-CONH$_2$	1·316	1·31	0·45	1·298	1·60	0·54	1·273	1·60	0·49
XXIII	3-CN	1·28	1·31	0·28	1·317	1·49	0·29	1·305[d]	2·48[d]	0·23[d]
XXIV	4-CN	1·229	1·40	0·49	1·218	1·60	0·45	1·215	1·31	0·59
XXXIII	4-OH	1·725	1·31	0·38	1·727	1·72	0·38	1·728	1·86	0·31
XXXI	4,4'-(Cl)$_2$	1·395	1·31	0·27	1·396	1·49	0·37	1·389	1·72	0·32
XXXIX	4-CH$_3$SO$_2$	1·235	1·40	0·41	1·225	1·60	0·50	1·215	2·03	0·49

[a] Limiting current for $2 \cdot 10^{-4}$ M benzophenone; [b] values affected by hydrolysis; [c] 0·015% of Dowex added as surfactant; [d] 0·03% Dowex added as surfactant.

Because the measurement of polarographic half-wave potentials is as reproducible as it is approximately shown by the size of points in Figs 1–3, it is necessary to look for other reasons for the deviations from the linear $E_{1/2}-\sigma$ plots. Two reasons for the scattering of points and for the above mentioned special deviations can be pointed out: the complexity of the reaction mechanism and the character of the electrode process proper. The importance of the course of the electrode process becomes obvious by comparison of the individual correlations at various pH.

The simplest relation between the half-wave potential and the substituent effect was observed for values obtained in borax media. In acidic solutions the protonation and, in principle also, the dimerization reaction affect the half-wave potential. Also the value an varies in this media more than in the other two. In alkaline media it was necessary to choose such a concentration of sodium hydroxide so as to make the protonation in reaction (D) with rate constant k_{-6} negligible. But with increasing concentration of sodium hydroxide the reaction with sodium ions (Part I, ref.[25]) begins to contribute to the increase in the height of the more positive wave. The com-

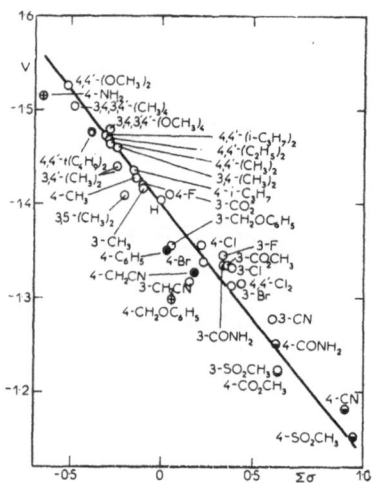

FIG. 2

Dependence of Half-Wave Potentials of Substituted Benzophenones $(E_{1/2})$ on the Sum of Hammett Substituent Constants σ_X in a Borate Buffer pH 9·3

○ $0·8 < an < 1·32$; ● $0·8 > \alpha > 1·32$; ◗ σ_{p-X}^- used; ⊕ deviates. Line corresponds to $\varrho = 0·25$ V. Value for 4-OH left out.

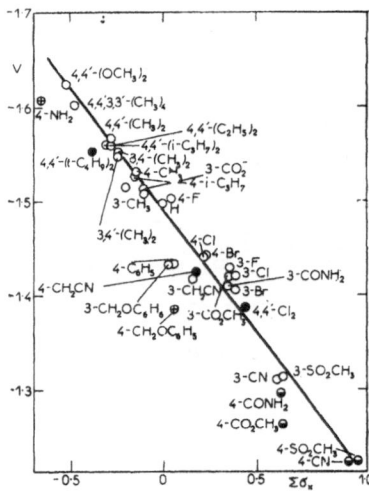

FIG. 3

Dependence of Half-Wave Potentials of Substituted Benzophenones $(E_{1/2})$ on the Sum of Hammett Substituent Constants σ_X in 0·1M-NaOH

$\varrho = 0·26$ V. Value for 4-O$^{(-)}$ left out.

petition of the proton transfer reaction and of the interaction with alkali metal ions results in the height of the wave in 0·1M-NaOH which changes with the substituent. The wave is higher than it would correspond to a simple one-electron process. Also the irregular dependence of the limiting current on sodium hydroxide concentration (Table II) recorded when the ionic strength was not kept constant, indicates the competition of the two above mentioned processes.

Under such conditions it might be surprising that the half-wave potentials in acidic and alkaline media give at all a linear $E_{1/2}$–σ plot. It should nevertheless be kept in mind that not only the rate of the electrode process proper, but also the transfer coefficient α, the rate of the protonation and dimerisation reactions affecting the half-wave potential in acid media and the rate sof reactions of the radical anion with hydrogen and alkali ions in alkaline media depend on the structure in a similar way.

The values of the product αn given in Tables I and II are determined with such a big relative error that they cannot be used for structural correlations. Deviating value for compound *XVII* shows a considerably higher value of αn in borax, but in sodium hydroxide media, where the same type of deviation was observed for half-wave potentials, the value of αn was in the expected. range. The smaller values of αn in borax for *XIII* and *XV* and in hydroxide media for *XIII* and *XXX* as well as large values for *XXII* and *XXIV* remain unexplained. Values of αn increase in most instances with increase in sodium hydroxide concentration (Table II). The large values for limiting currents of hydroxy derivatives *XXXIII*—*XXXV* indicate a change in mechanism. The half-wave potentials of these compounds clearly deviate from $E_{1/2}$–σ plots.

The application of substituent constants σ^-_{p-x} for acceptor groups has proved[26,44] necessary in numerous applications of Hammett equation for half-wave potentials. In particular in such cases in which the substituent and the reaction centre are of the same type and therefore is little reason to expect a conjugation interaction between these two groups. The reason for the application of constants σ^-_{p-x} must be therefore a conjugation in the transition state. In the studied system, in which the course of the reaction is known in some detail, it is possible to offer a more detailed interpretation. If it is assumed that the structure of the transition state of the reaction (C) resembles more the product that the initial compound, it is possible to consider the rôle of substitution on the electron density distribution in the resulting radical anion. Its structure can be expressed by a series of mesomeric formula $a - d$. Of these the greatest weight can be attributed to the formula a, in which the unpaired electron is situated on carbon an the negative charge on oxygen. If Y is an acceptor group, the relative weight of structure d and c is increased; localisation of the unpaired electron also on the oxygen atom increases the stability of the radical and results in a shift of the half-wave potential to more positive values. Delocalization of the free electron pair expressed by formulas $c \rightleftarrows d$ corresponds to a model system, used for definition of constants σ_{p-x} for acceptor groups[45]. If Y is a donor grouping, it can contribute to the delocalization of the unpaired electron in the way indicated by structure b; this delocalization is nevertheless considerably weaker than the conjugation of such groups with an electron pair[46] and is overweighted by the suppression of contributions of structure c. Therefore the effect of electron donating groups, which diminish the weight of structures c and d, can be adequately expressed by the common substituent constants σ^-_{p-X}. All these arguments can be analogously applied to the discussion of the stability of the radical formed in acid media according to the equation (B), as this radical differs only by the presence of one proton.

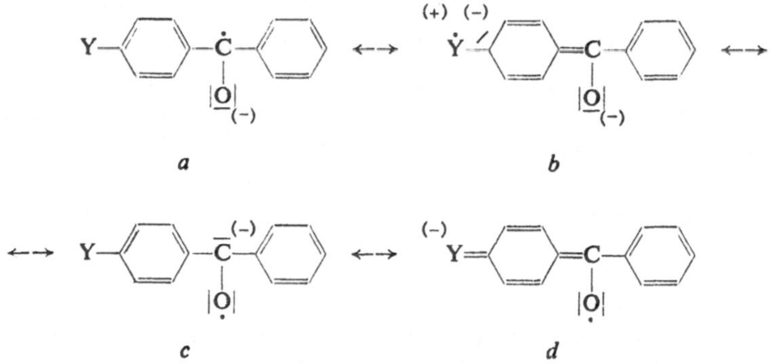

The above discussion corresponds qualitatively to the deductions of Staab[47] concerning the effect of the substituent NO_2 on the stability of diphenylpicrylhydrazyl. The stability of similar radicals is discussed semiquantitatively in the terminology of constant σ_X by Walter[46]. In this contribution[46], nevertheless, only the substitution on the opposite side of the molecule is treated and there are only correlations with constants σ^+ involved. Application of constants σ^-_{p-X} for reactions involving radicals has not yet been described, whereas the application of constants σ^+_X is rather frequent[46,48].

The observed linear relations of half-wave potentials and substituent constants (Figs 1–3), even when possibly composed of several factors (in particular for the half-wave potentials measured in hydrochloric acid) indicate, that substituents affect the reducibility of the carbonyl group by inductive and mesomeric effects *via* the benzene ring. The reaction constants ϱ_π have a sign expected for a nucleophilic attack by an electron and their absolute values are of the generally found order of magnitude[26,44]. Practically identical values found in borax $(\varrho_{\pi,CO} = 0.25 \text{ V})$ and in sodium hydroxide $(\varrho_{\pi,CO} = 0.26 \text{ V})$ solutions indicates that the potential determining step might be the same, for instance reaction (C). The value found in the hydrochloric acid medium $(\varrho_{\pi,CO} = 0.15 \text{ V})$ indicates another type of process.

In agreement with the relation[26,44] showing that the values of reaction constants $\varrho_{\pi,R}$ in the benzenoid series are greater the more negative is the half-wave potential of the unsubstituted parent compound, the sequence of the values of reaction constants parallelled the sequence of half-wave potentials of unsubstituted benzophenone which was -1.497 V in 0.1M-NaOH $(\varrho_{\pi,CO} = 0.26 \text{ V})$, -1.404 in borax $(\varrho_{\pi,CO} = 0.25 \text{ V})$ and -0.86 V in 0.1M-HCl $(\varrho_{\pi,CO} = 0.15 \text{ V})$.

For an understanding of organic electrode processes perhaps the observation of deviations from linear $E_{1/2}-\sigma$ plots is even of greater importance than the existence of a linear relationship. In the present case the observation that the value of the half-wave potential of 4-cyanobenzophenone (not shown on Fig. 1) deviates in acidic media from the linear $E_{1/2}-\sigma$ plot, led to a detection of the reducibility of the cyano group in the protonated from. This observation led to the discovery that such a process is not restricted to 4-cyanobenzophenone, but can be proved for other *para*-substituted benzonitriles as well[49]. It is hoped to study the reduction course of 4-phenoxymethylbenzophenone (*XVII*), showing a deviation at higher pH-values (Figs 2, 3) in more detail.

Equation (2) was derived[27] using extensive material concerning equilibria and rates. Its validity is restricted to these substituents which do not have a free electron pair in the α-position, and moreover only to those reactions, in which direct conjugation of the substituent with the electroactive centre is excluded. This range of validity is hence similar to that of the original Hammett equation (without the dual constant), even when the two equations are mutually independent. One of the goals of this paper was to find out if equation (2) can be used for half-wave potentials in the form (1) with the same precision as the Hammett equation, or even greater, as some specific effects can be cancelled out when data for the same substituent in *meta*- and *para*-position are correlated. A reliable verification of equation (1) using the

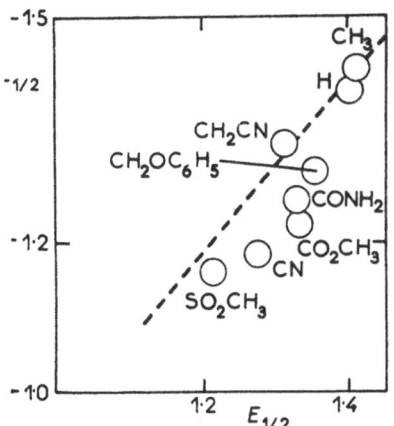
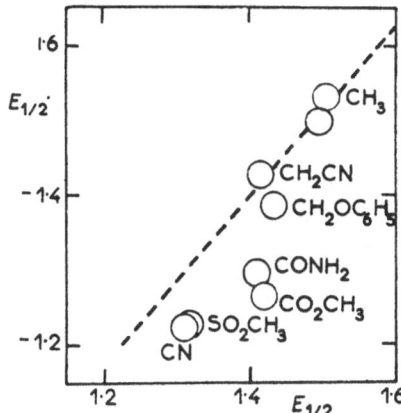

FIG. 4

Dependence of Half-Wave Potentials of *para*-Substituted Benzophenones on Half-Wave Potentials of Benzophenones Bearing the Same Substituent in *meta*-Position

Half-wave potentials measured: A in a borate buffer pH 9·3; B in 0·1M-NaOH. Line corresponds to a slope $\lambda = 1·14$.

data reported in the literature has not been possible, since in none of the reaction series investigated[44] a sufficient number of derivatives bearing the suitable substituents was studied.

For the studied series it was necessary to use constants σ_p^- (Figs 1–3). This, together with the suggested theoretical interpretation, indicate that this series is beyond the validity of equation (1), because the electrode process is affected by a direct conjugation of the substituent and the electroactive grouping in the transition state (resembling formula d). Equation (1) can therefore be applied only to substituents without a free electron pair in the α-position which moreover cannot act as acceptors. Among those substituents only H, CH_3, CH_2CN and $CH_2OC_6H_5$ were studied in this paper. The expected relation corresponding to equation (1) with $\lambda = 1\cdot14$ is found for the first three substituents with reasonable precission (Fig. 4); the last one shows a similar deviation as for the correlation using the Hammett equation (3). More conclusive verification of validity of equation (1) is prevented by the small distance between the experimental points, which cannot be increased by a selection of further substituents of the same type. Significant deviations of the points corresponding to substituents $COOCH_3$, CN, $CONH_2$ and SO_2CH_3 in the same direction indicate their participation in conjugative interaction with the reaction centre in transition state even more clearly than corresponding Hammett plots.

Finally it is possible to handle our experimental data from the point of view of the additivity of structural effects on half-wave potentials[28,44]. For the conclusion whether these effects are additive or not, the level of precision of the measured data is of importance. For data precise to 0·01 V the shifts caused by individual substituents can be considered as additive[21,22]. With our more precise data it is possible to deduce that even when for a number of substituent combinations equation (3) is valid (Fig. 2), some, in particular 3,5-dimethyl, 3,4'-dimethyl and 4,4'-di-tert-butyl, derivatives show deviations (Fig. 2, 3) in one direction, even when only just significant. In acidic media the greatest deviations were observed for the 3,4; 3',4'-tetramethyl (XII) and 4,4'-dichloro (XXXI) derivatives (Fig. 1); hence bulkiness of the substituent group can be of some importance.

For experimental help we thank to Mrs H. Bilyková and Miss M. Korneenková. Elemental analyses were carried out in the Analytical Department, Institute of Organic Chemistry and Biochemistry, Czechoslovak Academy of Sciences, Prague, under the supervision of Dr J. Horáček.

LITERATURE

1. Pasternak R.: Helv. Chim. Acta 31, 753 (1948).
2. Ashworth M.: This Journal 13, 229 (1948).
3. Calzolari C.: Anuali Triestini 29, 21 (1951).
4. Calzolari C., Furlani C.: Ann. Chim. (Rome) 44, 356 (1954).
5. Sartori C., Furlani C.: Ann. Chim. (Rome) 44, 95 (1954).
6. Sartori C., Silverstroni P., Calzolari C.: Ric. Sci. 24, 1471 (1954).

7. Suzuki M.: Mem. Coll. Agr. Kyoto Univ. 67, 1 (1954).
8. Day R. A. jr., Kirkland J. J.: J. Am. Chem. Soc. 72, 2766 (1960).
9. Day R. A. jr., Milliken S. R., Shults W. D.: J. Am. Chem. Soc. 74, 2741 (1952).
10. Day R. A. jr., Biggers R. E.: J. Am. Chem. Soc. 75, 738 (1953).
11. Elving P. J., Leone J. T.: J. Am. Chem. Soc. 80, 1021 (1958).
12. Suzuki M., Elving P. J.: J. Phys. Chem. 65, 391 (1961).
13. Vodzinskij Ju. V., Koršunov I. A.: Učenyje Zapiski Gork. Univ. Ser. Chim. 32, 25 (1958).
14. Given P. H., Poever M. E.: J. Chem. Soc. 1960, 385.
15. Given P. H.: Abhandl Deut. Akad. Wiss. Berlin 1964, No. 1, 481.
16. Wawzonek S., Gundersen A.: J. Electrochem. Soc. 107, 535 (1960).
17. Majranovskij S. G., Pavlov V. N.: Ž. Fiz. Chim. 38, 1804 (1964).
18. Majranovskij S. G.: Izv. Akad. Nauk SSSR, Otd. Chim. Nauk 1961, 2140.
19. Vettig K.: Elektrochimija 1967, 269.
20. Brockman R. W., Pearson D. E.: J. Am. Chem. Soc. 74, 4128 (1952).
21. Pfister-Guillouzo G., Bonastre J.: Bull. Soc. Chim. France 1965, 1993.
22. Pfister G., Bonastre J.: Bull. Soc. Chim. France 1966, 2053.
23. Kalvoda R., Budnikov G.: This Journal 28, 838 (1963).
24. Kalvoda R., Budnikov G.: Abhandl. Deut. Akad. Wiss. Berlin 1964, No. 1, 460.
25. Zuman P.: This Journal 33, 2548 (1968).
26. Zuman P.: This Journal 25, 3225 (1960).
27. Exner O.: This Journal 31, 65 (1966).
28. Exner O.: This Journal 25, 1044 (1960).
29. Majranovskij S. G., Titov F. S.: Ž. Anal. Chim. 15, 121 (1960).
30. Manoušek O.: Unpublished results.
31. Rekker R. F., Nauta W. T.: Rec. Trav. Chim. 80, 747 (1961).
32. Manoušek O., Exner O., Zuman P.: This Journal, in the press.
33. Manoušek O., Zuman P.: Chem. Commun. 1965, 158.
34. Zuman P., Manoušek O.: This Journal, in the press.
35. Tadros W., Akhnookh Y., Aziz G.: J. Chem. Soc. 1953, 186.
36. Szmant H. H., Yoncoskie R.: J. Org. Chem. 21, 78 (1956).
37. Bourcet P.: Bull. Soc. Chim. France [3] 15, 945 (1896).
38. White W. N., Schlitt R., Gwynn D.: J. Org. Chem. 26, 3613 (1961).
39. Senff P.: Ann. 220, 225 (1883).
40. Smith M. E.: J. Am. Chem. Soc. 43, 1920 (1921).
41. Novák L., Protiva M.: This Journal 24, 3966 (1959).
42. Gardner H. J.: Nature 167, 158 (1951).
43. Vlček A. A.: This Journal 19, 862 (1954); Chem. listy 48, 189 (1954).
44. Zuman P.: Substituent Effects in Organic Polarography. Plenum Press, New York 1967.
45. Exner O.: Chem. listy 53, 1302 (1959).
46. Walter R. I.: J. Am. Chem. Soc. 82, 1923 (1966).
47. Staab H. A.: Einführung in die Theoretische Organische Chemie, p. 457. Verlag Chemie, Weinheim 1959.
48. Russel G. A.: J. Org. Chem. 23, 1407 (1958).
49. Zuman P., Manoušek O.: This Journal, in the press.

Translated by the author (P. Z.).

Desorption of 3-Phenylpropionaldehyde preceding Polarographic Reduction

IN the interpretation of polarographic measurements, it is usually assumed that organic compounds are only reduced when they are adsorbed onto the surface of the mercury electrode[1,2]. Unfortunately, the term "adsorption" is used for a broad range of interactions between the solid and liquid phase.

Two of the simplest electrochemical methods which are assumed to indicate the potential range over which adsorption occurs, are the change in capacity current on d.c. polarographic curves[3] and the presence of tensammetric peaks[4] in a.c. polarography. In many previously reported cases, there have been changes in the double layer, associated with adsorption/desorption effects, at potentials so near to the potential corresponding to a faradaic process that it was difficult to separate the components affecting the a.c. polarographic curves.

3-Phenylpropionaldehyde exhibits good resolution between the adsorption/desorption phenomena and the faradaic reduction process.

The d.c. polarogram comparing the capacity current of the pure buffer and the buffer in the presence of aldehyde (Fig. 1) shows a well defined region between -0.2 V (Fig. 1, potential A) and -1.4 V (potential D) where the capacity current in the presence of the aldehyde is below that of the pure buffer. The reduction wave occurs at a more negative potential (-1.64 V) than the desorption potential (-1.4 V).

The a.c. polarographic curves of the same solutions Fig. 2) show a decrease of the a.c. current below the value of the pure supporting electrolyte between -0.1 V and -1.42 V (potential D, Fig. 2). The reduction peak (R) at -1.6 V is therefore preceded by a potential region in which the curves of the pure buffer and aldehydic solution overlap. The nature of the indistinct peaks of -0.8 V and -1.2 V is being investigated.

Thus both methods suggest that 3-phenylpropionaldehyde is adsorbed at about -0.1 V and is desorbed at about -1.4 V. The reduction at -1.65 V must therefore occur at a potential at which the compound is usually considered to be desorbed. Such behaviour might be more common than is usually assumed—for example, benzophenone in 60 per cent ethanol also shows no decrease in the a.c. polarographic curve before the reduction peak (Fig. 2b in ref. 5).

3-Phenylpropionaldehyde is therefore a suitable model for the study of adsorption/desorption phenomena,

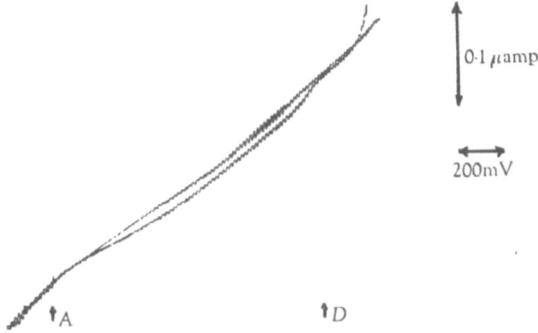

Fig. 1. D.c. polarographic curves of 3-phenylpropionaldehyde. (1) Borate buffer, pH 9·2; (2) 1·5 × 10⁻⁴ molar 3-phenylpropionaldehyde in borate buffer, pH 9·2. Curves start at 0·0 V.

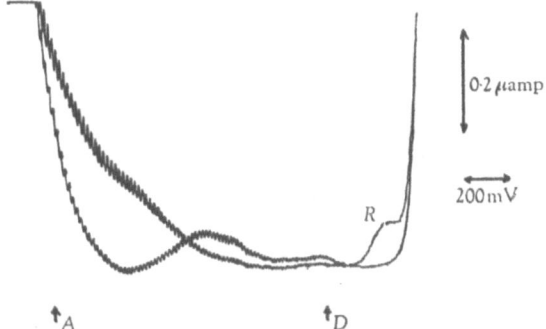

Fig. 2. A.c. polarographic curves of 3-phenylpropionaldehyde. (1) Borate buffer, pH 9·2; (2) 1·5 × 10⁻⁴ molar 3-phenylpropionaldehyde in borate buffer, pH 9·2. Curves start at 0·0 V.

because it is a relatively simple molecule and is sufficiently soluble, and particularly because it shows a clear separation of the capacity and faradaic processes.

We thank Professor R. Belcher for his interest and support, and the Scientific Research Council (P. Z.) and Lyons and Co., Ltd. (D. B.), for grants.

P. Zuman*
D. Barnes

Department of Chemistry,
University of Birmingham.

Received July 10; revised August 9, 1967.

* On leave from Jaroslav Heyrovsky Institute of Polarography, Czechoslovak Academy of Sciences, Prague.

¹ Majranovskii, S. G., *Catalytic and Kinetic Waves in Polarography* (Nauka, Moscow, 1966) (in Russian).
² Tedoradze, C. A., *Usp. Electrochim. Org. Soed.*, 23 (Nauka, Moscow, 1966).
³ Valenta, P., *Chem. Zvesti*, **8**, 767 (1954).
⁴ Breyer, B., and Bauer, H. H., *Alternating Current Polarography and Tensammetry* (Interscience, New York, 1963).
⁵ Vettig, K., *Elektrokhimiya*, **3**, 269 (1967).

DIRECT POLAROGRAPHIC DETERMINATION OF PYRIDOXAL
IN THE PRESENCE OF PYRIDOXAL-5-PHOSPHATE

O. Manoušek and P. Zuman

Central Research Institute for Food Industry and Polarographic Institute, Czechoslovak Academy of Science, Prague

Received June 16th, 1961

The determination of pyridoxal in the presence of pyridoxal-5-phosphate may be[1] based on the difference between the total wave-height obtained at pH about 9, where both substances give a diffusion-controlled wave and that obtained in acid media, where pyridoxal alone shows a kinetic current[2]. The mixture of pyridoxal and pyridoxal-5-phosphate alternatively can be converted to oximes and their four-electron waves at pH 8,5 can be resolved using derivative curves[1].

A third possibility is offered by the difference of the half-wave potentials of polarographic waves for pyridoxal and its phosphate in alkaline solutions[3]. In 0·1M-NaOH the half-wave potential of the tribasic anion of pyridoxal-5-phosphate is some 350 mV more negative than that of the monobasic anion of pyridoxal (Fig. 1). At higher pH-values the separation of the half-wave potentials decreases owing to the shift of the half-wave potentials of pyridoxal towards more negative values, whilst the half-wave potentials for pyridoxal-5-phosphate are pH-independent. The limiting current of the latter at pH 13 is about 20% lower than that of pyridoxal, owing to the difference in diffusion coefficients.

0·1 to 0·15M-NaOH can be used for determination of pyridoxal in the presence of a twenty fold excess of pyridoxal-5-phosphate. The simultaneous determination of both forms is possible only when the concentrations are comparable (Fig. 1).

The method based on the reduction of the corresponding oximes[1] is less accurate, because of the small difference in half-wave potentials. That based on measurement of wave heights in acid media[1] necessitates a careful temperature control, as a kinetic current is involved. Nevertheless, it has been succesfully used for folowing the continuous changes of pyridoxal concentration during the hydrolysis of pyridoxal-5-phosphate in acid solution directly in the reaction mixture[4].

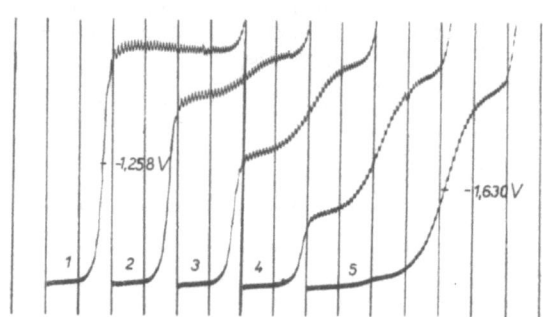

Fig. 1

Mixtures of Pyridoxal and Pyridoxal-5-phosphate in 0·1M-NaOH

Concentrations of pyridoxal-5-phosphate: *1* 0; *2* 1·25 . 10^{-4}M; *3* 2·5 . 10^{-4}M; *4* 3·75 . 10^{-4}M; *5* 4·9 . 10^{-4}M; concentration of pyridoxal: *1* 5·0 . 10^{-4}M; *2* 3·75 . 10^{-4}M; *3* 2·5 . 10^{-4}M; *4* 1·25 . 10^{-4}M; *5* 1 . 10^{-5}M. Half-wave potentials given against the s.c.e., 159 mV/absc., full-scale sensitivity 4·06 μA.

The method suggested in this contribution has an accuracy of $\pm 2-3\%$ and was successfully applied to the study of enzymatic hydrolysis of pyridoxal-5-phosphate[5]. The method proved useful for determining the purity of specimens of pyridoxal-5-phosphate, when about 1·5% of pyridoxal were found in preparations of this compound made by L. Light Ltd. (Colnbrook, England) and by Fluka AG. (Buchs, Switzerland).

References

1. Manoušek O., Zuman P.: J. Electroanal. Chem. *1*, 324 (1959—60).
2. Volke J.: Z. physik. Chem. (Leipzig) Sonderheft *1958*, 268.
3. Manoušek O., Zuman P.: This Journal, in the press.
4. Zuman P., Manoušek O.: This Journal, *26*, 2134 (1961).
5. Manoušek O., Kučerová Z.: Unpublished results.

Translated by the Author (P. Z.)

POLAROGRAPHIE DER STEROIDE III.*
POLAROGRAPHISCHE REDUKTION DER STEROIDE MIT ALDEHYD-GRUPPE IN STELLUNG 18**

P. Zuman und V. Černý

Polarographisches Institut und Chemisches Institut, Tschechoslowakische Akademie der Wissenschaften, Praha

Eingegangen am 13. Dezember 1957

Die polarographische Reduktion der Steroide mit Aldehydgruppe in Stellung 18 und mit verzweigter Kette in Stellung 17 unterscheidet sich von der Reduktion anderer Aldehyde. Die Höhe ihrer Stufen sowie ihre Halbstufenpotentiale verändern sich praktisch nicht mit dem pH-Wert, was durch eine sterische Hinderung der Hydratation erklärt wurde. Die umfangreiche axiale 3α-Aminogruppe des Holarrhidins verhindert dessen Adsorption an der Tropfelektrode. Holarrhidin weist keinen Kapazitätseffekt auf und hat eine wesentlich geringere Wirksamkeit auf die Maximaunterdrückung als Verbindungen mit 3β-Dimethylaminogruppe.

Im Rahmen eines Studiums der Konstitution des Conessins[1-4], Holarrhimins[1-4] und Holarrhidins wurde auch den Verbindungen dieser Gruppe Aufmerksamkeit gewidmet, die am Kohlenstoff 18 des Steroidkerns eine Aldehydgruppe tragen, und die aus dem entsprechenden Alkohol durch Oxydation mit Chromtrioxyd dargestellt wurden. Durch ihre leichte Reduzierbarkeit und die Unabhängigkeit ihres Reduktionsverlaufs vom pH-Wert unterscheiden sich diese Stoffe von den meisten bisher untersuchten Aldehyden, deren Aldehydgruppe nicht mit einer Doppelbindung konjugiert ist.

Experimenteller Teil

Es wurde mit dem Polarographen der Firma Nejedlý, Type VIII, mit Galvanometer von der Empfindlichkeit $2 . 10^{-9}$ A/cm gearbeitet. Verwendet wurde ein abgeändertes Gefäß[6] mit getrennter Elektrode für kleine Volumen der Untersuchungslösung. Die Dämpfung der Sauerstoffmaxima wurde in einem einfachen Gefäß mit Bodenquecksilber als Bezugselektrode verfolgt. Die benutzte Kapillare besaß eine Ausströmungsgeschwindigkeit von 2,1 mg/s und eine Tropfzeit von 2,9 s bei 60 cm Quecksilberniveauhöhe und beim Potential 0 V (SKE). Die oszillographischen Messungen wurden mit dem Oszilloskop der Firma Křižík durchgeführt. Die potentiometrischen Kurven wurden durch Titration unter Verwendung des pH-Meters der Firma Radiometer mit der Glaselektrode erhalten.

Dehydro-5,6-dihydrotetramethylholarrhimin (*I*), sein Oxim (*II*), 5,6-Dihydrotetramethylholarrhimin (*III*), Dehydrotetramethylholarrhimin (*IV*) und Dehydrotetramethylholarrhidin (*V*) wurden von uns dargestellt. Das zu Vergleichszwecken benutzte Strophantidin wurde von F. Šantavý (Olomouc) isoliert.

Die Vorratslösungen für die Polarographie wurden in der Konzentration 0,005M durch Auflösen der betreffenden Verbindung in 0,1M-H_2SO_4 hergestellt. Die 0,005M Strophantidinlösung wurde durch Lösen der Verbindung in Äthanol bereitet.

---

* II. Mitteilung: Chem. listy *48*, 194 (1954); diese Zeitschrift *19*, 894 (1954).
** Diese Arbeit erschien bereits in tschechischer Sprache in Chem. listy *52*, 1468 (1958).

I, R = HC=O

II, R = HC=NOH

III, R = HCHOH

IV

V

Zur Ermittlung der Dissoziationskonstanten wurden die 0,005M Lösungen der Stoffe *I* bis *V* in 70%igem Äthanol nach Beseitigung des Kohlendioxyds in Stickstoffatmosphäre mit 0,05M-HCl titriert. Bei dem Oxim (*II*) war es im Hinblick auf die Unlöslichkeit der freien Base notwendig, die Lösung zuerst anzusäuern und dann mit 0,05N-NaOH zurückzutitrieren.

Die Pufferlösungen nach Britton und Robinson sowie die Phosphat- und Boratpuffer wurden ebenso wie die Lithiumhydroxydlösungen aus p. a. Chemikalien dargestellt. Das Äthanol wurde nach dem in der Literatur beschriebenen Verfahren[7] von aldehydischen Stoffen befreit. Dem Grundelektrolyten (1 bis 4 ml) wurde die Vorratslösung zugegeben und nach Entlüftung mit Stickstoff wurde die Lösung polarographiert.

Ergebnisse

Die aldehydischen Substanzen *I*, *IV* und *V* werden in Pufferlösung zwischen pH 7 bis 10,5 in einer irreversiblen Reduktionsstufe mit Halbstufenpotential um −1,65 V (SKE) reduziert (Abb. 1). Im angeführten Bereich ändert sich praktisch weder das Halbstufenpotential noch die Höhe der Stufe mit dem pH-Wert (Abb. 2). Aus der Abhängigkeit von der Behälterhöhe wurde er-

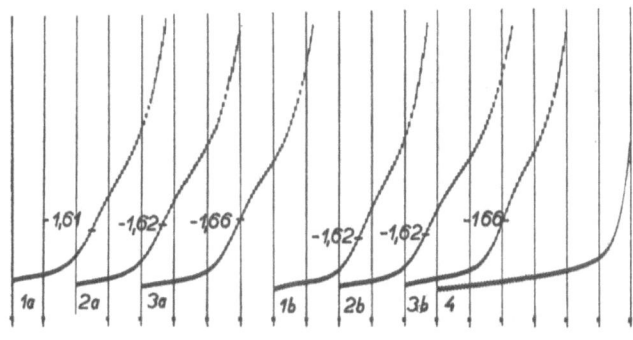

Abb. 1

Reduktionsstufen der Steroide mit Aldehydgruppe in Stellung 18

Pufferlösung nach Britton und Robinson *a* pH 8,0; *b* pH 9,9. Konzentration der Steroide $1 \cdot 10^{-4}$M. *1* Dehydrotetramethylholarrhimin (*IV*); *2* Dehydrotetramethylholarrhidin (*V*); *3* Dehydro-5,6-dihydrotetramethylholarrhimin (*I*); *4* Strophantidin. Von −1,35 V angefangen, SKE, 100 mV pro Absz., $h = 60$ cm, Empf. 1:10.

wiesen, daß die Höhe der Stufe durch die Diffusion des Depolarisators begrenzt ist. Bei niedrigeren pH-Werten wird die Stufe von dem der Wasserstoffabscheidung aus dem Grundelektrolyten angehörenden Strom verdeckt. Bei pH größer als etwa 10,5 fällt in wäßrigen Lösungen die freie Base aus der Lösung aus. Bei kleinerem pH als 10,5 können aus den Verbindungen *I* und *V* auch 1 . 10⁻³M wäßrige Lösungen hergestellt werden, die Stufe ist jedoch bei kleineren Konzentrationen als 1,5 . 10⁻⁴M besser ausgebildet. Die untersuchten steroiden stickstoffhaltigen Verbindungen setzen nämlich die Wasserstoffüberspannung herab und bei größerer Depolarisatorkonzentration verschmilzt die Reduktionsstufe der Steroide mit der katalytischen Stufe des Wasserstoffes.

Eine Äthanolzugabe bei niedrigeren pH-Werten hat eine schlecht ausgebildete Stufe zur Folge. In 0,1M-LiOH mit 50% Äthanolgehalt wurde bei den Verbindungen *I* und *IV* ein Stromanstieg in Form einer Stufe bei −1,5 V

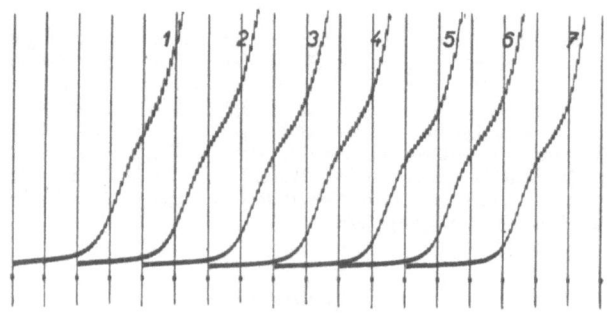

Abb. 2.

Abhängigkeit der Reduktionsstufen vom pH-Wert

Pufferlösungen nach Britton und Robinson, pH: *1* 7,5; *2* 8; *3* 8,5; *4* 9; *5* 9,5; *6* 10; *7* 10,3. 3,1 . 10⁻⁴M Aldehyd *I*. Von −1,3 V angefangen, SKE, 100 mV/Absz., $h = 60$ cm, Empf. 1 : 20.

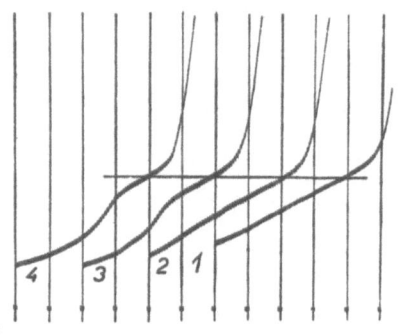

Abb. 3

Kapazitätseffekte

0,1M-LiOH, 50% Äthanol, 2,5 . 10⁻⁴M Aldehyd: *1* der leere Grundelektrolyt; *2* Verbindung *V*; *3* Verbindung *IV*; *4* Verbindung *I*. Von −1,3 V angefangen; SKE, 100 mV pro Absz., $h = 60$ cm, Empf. 1 : 7.

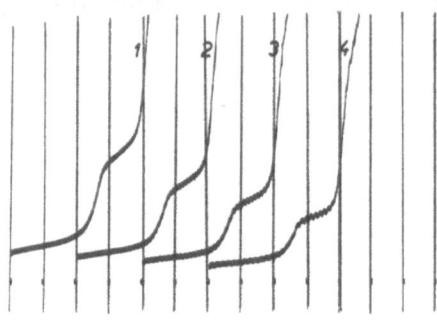

Abb. 4

Abhängigkeit der Kapazitätserscheinungen von der Behälterhöhe

0,5N-LiOH. 50% Äthanol und 2,5 . 10⁻⁴M Dehydro-5,6-dihydrotetramethylholarrhimin. Be; hälterhöhen $h = 1$ 60 cm; *2* 50 cm; *3* 40 cm; *4* 30 cm. Von −1,3 V angefangen, SKE, 100 mV/Absz , Empf. 1 : 7.

beobachtet (Abb. 3, 4). Es handelt sich jedoch nicht um eine Erhöhung des Faradayschen Stromes bei negativeren Potentialen als −1,5 V, sondern um eine Verminderung des Kapazitätsstromes bei positiveren Potentialen als −1,5 V infolge von Adsorption (sog. Kapazitätserscheinung[8]), wie die Abb. 3 bestätigt, aus der ersichtlich ist, daß sich der Strom bei −1,57 V nach Zugabe des Stoffes nicht erhöht hat. Der Stromanstieg bei −1,5 V entspricht folglich der Desorption des Steroids, wonach der Strom wieder auf den Wert des Kapazitätsstroms zurückkehrt. In Übereinstimmung mit der Voraussetzung, daß es sich um einen Kapazitätsstrom handelt, ist auch die lineare Abhängigkeit dieses Stromes von der Behälterhöhe (Abb. 4). Bei der Verbindung V wurde keine ähnliche Kapazitätserscheinung beobachtet.

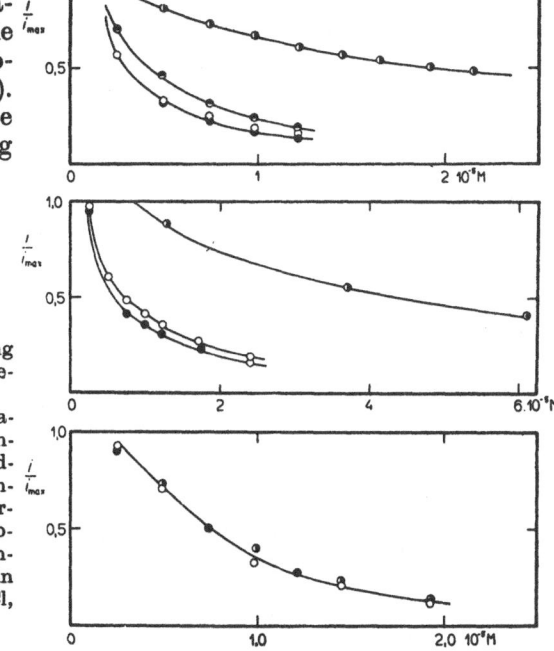

Abb. 5

Einfluß der Steroide auf die Dämpfung des Sauerstoffmaximums in verschiedenen Milieus

Abhängigkeit des Verhältnisses der Maximumhöhe bei gegebener Steroidkonzentration (i) zum Strom in der steroidfreien Lösung (i_{max}) von der Steroidkonzentration. ○ Dehydrotetramethylholarrhimin (IV). ● Dehydro-5,6-dihydrotetramethylholarrhimin (I), ◐ Strophantidin, ◑ Dehydrotetramethylholarrhidin (V). Milieu: a 0,001m-HCl, b 0,001m-KCl, c 0,001m-NaOH.

Der Unterschied zwischen dem Verhalten der Stoffe IV und V (Abb. 3), die sich nur durch die Lage der tertiären Aminogruppe voneinander unterscheiden, zeugt von der verschiedenen Adsorbierbarkeit beider Verbindungen. Die Polarographie ermöglicht es, die Adsorbierbarkeit und die Oberflächenaktivität an Hand der Maximadämpfung zu verfolgen[9]. Es wurde festgestellt, daß die Stoffe I und IV ein weitaus größeres Maximadämpfungsvermögen besitzen als die Verbindung V, wie aus dem Vergleich der Kurven in Abb. 5 hervorgeht. Zum Vergleich wurde auch die Dämpfungsfähigkeit des Strophantidins verfolgt. Die Ergebnisse der Dämpfungsfähigkeit auf das Sauerstoffmaximum sind in Tabelle I zusammengefaßt.

Die Dämpfungsfähigkeit des Aldehyds IV ist also mehr als sechsmal größer als die des Aldehyds V. Die Dämpfungswirkungen, besonders in saurem Medium, sind außerordentlich groß, ähnlich den bei den Gallensäuren[10,11] beobachteten Wirkungen. Die gleiche Dämpfungswirkung aller drei Stoffe in alkalischem Milieu (Abb. 3, Tab. I) kann am ehesten dadurch erklärt werden, daß die Dämpfung am positiven Sauerstoffmaximum verfolgt wurde, wogegen sich die Unterschiede in der Adsorption bei den Kapazitätseffekten an der negativ geladenen Zwischenphase äußerten. Es gelang uns vorläufig nicht, ein System mit ausgeprägtem Maximum erster Art beim Potential −1,3 V im Milieu von Hydroxyden aufzufinden. Mit dem Anstieg der Oberflächenaktivität in alkalischem Milieu ist das starke Schäumen der Lösung im Einklang, ähnlich wie es bei Steroiden mit Aldehydgruppe[12] in Stellung 10 und bei den Δ⁴-3-Ketosteroiden[13] beobachtet wurde. Die Löslichkeit des Stoffes *IV* in äthanolischer Lösung ist wesentlich geringer als bei dem Aldehyd *V*.

Tabelle I

Die Verdünnung, die zur Unterdrückung des Sauerstoffmaximums auf 50% in verschiedenen Milieus erforderlich ist

Verbindung	0,001M-HCl	0,001M-KCl	0,001M-NaOH
I	3 400 000	140 000	1300 000
IV	3 400 000	140 000	1300 000
V	520 000	23 000	1300 000
Strophantidin	2 400 000	80 000	—

Tabelle II

Titrationsexponenten in 70%igem Äthanol

Verbindung	pK_{a_1}	pK_{a_2}
I	8,75	9,7
II	8,4[a]	9,1[a]
III	8,75	9,5
V	8,3[a]	9,3[a]

[a] Angenäherte Werte.

Setzt man den Aldehyd *I* einer 0,02M Semicarbazidlösung in 0,1M-HCl zu, so erfolgt nur eine Verschiebung des Endstromanstiegs im Grundelektrolyten zu positiveren Potentialen (am wahrscheinlichsten infolge katalysierter Wasserstoffabscheidung), es kommt jedoch nicht zur Bildung der Reduktionsstufe des Semicarbazons[14]. Ähnlich wurde auch in Methylamin-Methylaminhydrochlorid-Lösungen[15-17] von pH 9,7 nur die Reduktionsstufe des Aldehyds beobachtet und keine weitere Reduktionsstufe, die dem Aldimin angehören würde. Nach Kondensation des Aldehyds *I* mit Girardschem Reagenz T[18-20] wurden in Lösungen von etwa pH 10 Reduktionsstufen bei −1,78 V aufgefunden.

In Lösungen des Oxims *II* wurden im gesamten pH-Bereich keine Reduktionsstufen beobachtet. Bei pH 10,3 konnte bei 25°C keine Hydrolyse festgestellt werden, die durch die Stufe des freien Aldehyds zum Ausdruck käme, und zwar nicht einmal nach 12 Stunden. Der Alkohol *III* wird erwartungsgemäß nicht an der Quecksilbertropfelektrode reduziert.

Die potentiometrischen Titrationskurven hatten in allen Fällen den der Titration einer zweibasischen Base entsprechenden theoretischen Verlauf. Die aufgefundenen Werte der sauren Dissoziationskonstanten (in Anwesenheit von 70% Äthanol und bei genügend kleiner Konzentration der Stoffe, um die Ionenstärke als Null betrachten zu können) sind in Tabelle II zusammengefaßt.

Diskussion

Vergleicht man das beschriebene Verhalten mit dem Verhalten der eine Aldehydgruppe am Kohlenstoff 10 des Steroidkerns tragenden Steroiden[12], so ist vor allem die leichte Reduzierbarkeit der Aldehyde *I*, *IV* und *V* überraschend. Im Hinblick darauf, daß die Stufenhöhe pH-unabhängig ist, kann angenommen werden, daß die Aldehydgruppe nicht in merklichem Maß hydratisiert ist. Für eine sterische Hinderung der Solvatation (wodurch auch die

Anomalien einiger chemischer Reaktionen erklärt werden[21]) spricht auch das abweichende Verhalten dieser Stoffe vom Verhalten der Steroide[12] mit Aldehydgruppe in Stellung 10, weiter von den Terpenaldehyden[22], wo die Reduktion bei $-2,3$ V verläuft, und von den übrigen gesättigten Aldehyden. Für die Elektronenaufnahme ist die sterische Anordnung offensichtlich kein Hindernis. Durch sterische Hinderungen läßt sich auch erklären, warum keine Kondensation mit Methylamin und mit Semicarbazid bei milden Bedingungen stattfindet.* Auch die schwierige Hydrolyse des Oxims ist am ehesten sterisch beeinflußt, ähnlich wie auch die Hydrolyse des Esters der entsprechenden Carbonsäure[1] erschwert ist.

Die schwierigere Reduzierbarkeit des Betainylhydrazons im Vergleich zum entsprechenden Aldehyd, die durch ein negativeres Halbstufenpotential zum Ausdruck kommt, und die Unreduzierbarkeit des Oxims (und gegebenenfalls auch des Aldimins und Semicarbazons) im erreichbaren Potentialbereich steht im Widerspruch zu den bisherigen Erfahrungen, die für eine leichtere Reduzierbarkeit der Bindung C=N als der Bindung C=O sprechen[15,23]. Dies ließe sich so erklären, daß bisher mit der Reduktion der Bindung C=N immer nur die Reduktion der hydratisierten Carbonylgruppe, die schwieriger reduzierbar ist, verglichen werden konnte. Die geringeren Veränderungen der Halbstufenpotentiale mit der Konstitution sind bei den irreversiblen gedehnten Stufen bedeutungslos.

Die erste Dissoziationskonstante gehört am wahrscheinlichsten der Aminogruppe in Lage 3 an, denn bei den Verbindungen *I* und *III* verändert sie sich nicht, und die Abweichung beim Stoff *V* (Tab. II) ist in Übereinstimmung mit der Annahme, daß andere sterische Bedingungen zur Geltung kommen. Die niedrigeren Werte der Dissoziationskonstanten der tertiären Aminogruppen, als in wäßrigen Lösungen zu erwarten wäre, sind am ehesten durch die Gegenwart des Äthanols verursacht, das bei den Aminen eine Verschiebung von pK_a zu niedrigeren Werten bewirkt[24].

Aus dem geringen Unterschied in den zweiten Dissoziationskonstanten, die am wahrscheinlichsten der Aminogruppe in der Seitenkette beim Aldehyd *I* und beim Alkohol *III* entspricht, kann geschlossen werden, daß beim Aldehyd *I* kein Ringschluß unter Bildung eines inneren Aldimins eintritt. In Übereinstimmung mit dieser Schlußfolgerung ist neben den chemischen Erkenntnissen auch die polarographische Reduzierbarkeit, die bei einer cyclisch geschlossen Verbindung nicht vorausgesetzt werden könnte.

Beachtenswert ist der starke sterische Effekt der Aminogruppe in Stellung 3 bei den Derivaten des Holarrhimins *I* und *IV* (zum Unterschied von den Derivaten des Holarrhidins *V*) auf die Kapazitätserscheinungen und auf das Maximadämpfungsvermögen. Derartige Unterschiede im Verhalten von zwei Stereoisomeren sind bisher nicht in der Polarographie beschrieben worden.

Es ist bekannt, daß bei einer Reihe von chemischen Reaktionen das Steroidmolekül dem Angriff des Reagenz von der hinteren Seite des Moleküls zugänglicher ist. Für dieses Verhalten schlug Fieser[25] den Termin "the rule of the rear" vor. Diese Regel kann sich auch bei solchen Eigenschaften des Moleküls geltend machen, wie es seine Adsorbierbarkeit ist. Dies folgt aus dem Verlauf

---

* Die Stufen, die der Reduktion einer Verbindung mit der Bindung C=N— entsprechen würden, sind entweder deshalb nicht sichtbar, daß es überhaupt nicht zur Bildung einer solchen Form in der Lösung kommt, oder deshalb, daß die Reduktion zu negativ verläuft, so daß sie auf den Kurven nicht sichtbar ist.

von zahlreichen Fällen der katalytischen Hydrierung steroider Verbindungen, die durch Adsorption bedingt sind und die genannte Regel befolgen. Im Einklang mit der leichteren Adsorbierbarkeit von der hinteren Seite des Moleküls ist auch die leichtere Adsorbierbarkeit des Holarrhimins als die des Holarrhidins, die aus den polarographischen Messungen hervorgeht. Beim Holarrhidin wirkt sich nämlich der Einfluß der umfangreichen axialen 3α-Dimethylaminogruppe aus, die die Adsorption von hinten verhindert.

Ein ähnlicher Einfluß der Dimethylaminogruppe in Stellung 3 wurde auch bei den 3-Dimethylamino-Δ^5-cholestenen beobachtet[26,27]. Die damaligen Ergebnisse der katalytischen Hydrierung lassen sich in der Weise erklären, daß bei der katalytischen Hydrierung des 3α-Dimethylamino-Δ^5-cholestens ein Koprostanderivat aus dem Grunde entstand, daß die Adsorption am Katalysator von der vorderen Seite des Moleküls stattfand. Bei der Hydrogenierung des 3β-Dimethylamino-Δ^5-cholestens, wo die Dimethylaminogruppe die Adsorption von der hinteren Molekülseite nicht hinderte, entstand das Cholestanderivat.

Literatur

1. Černý V., Šorm F.: Chem. listy *49*, 909 (1955); diese Zeitschrift *20*, 1473 (1955).
2. Lábler L., Černý V., Šorm F.: Chem. listy *49*, 1389 (1955); diese Zeitschrift *20*, 1484 (1955).
3. Lábler L., Černý V., Šorm F.: Chem. & Ind. (London) *1955*, 1119.
4. Černý V., Lábler L., Šorm F.: Chem. listy *50*, 1126 (1956); diese Zeitschrift *22*, 76 (1957).
5. Černý V., Lábler L., Šorm F.: Chem. listy *50*. 1126 (1956); diese Zeitschrift *22*, 76 (1956).
6. Zuman P.: Chem. listy *51*, 993 (1957).
7. Weissberger A., Proskauer E. S., Riddick J. A., Toops E. E. Jr.: *Organic Solvents*. Interscience, New York 1955.
8. Heyrovský J., Šorm F., Forejt J.: diese Zeitschrift *12*, 11· (1947).
9. Vavruch J.: *Polarografická maxima v theorii i praxi*. Verlag SPICH, Praha 1949.
10. Zambotti V., Rocco L.: Arch. sci. biol. *26*, 384 (1940); Chem. Zentr. *1941*, II, 1656.
11. Zambotti V.: Boll. soc. ital. biol. sper. *15*, 403 (1940); Chem. Abstr. *35*, 4262 (1941).
12. Zuman P., Šantavý F.: Chem. listy *46*, 393 (1952); diese Zeitschrift *18*, 28 (1953).
13. Zuman P., Tenygl J., Březina M.: Chem. listy *47*, 1152 (1953); diese Zeitschrift *19*, 46 (1954).
14. Souchay P., Graizon M.: Chim. anal. *36*, 85 (1954).
15. Zuman P.: Chem. listy *45*, 65 (1952), diese Zeitschrift *15*, 839 (1950).
16. Zuman P., Březina M.: Chem. listy *46*, 599 (1952).
17. Březina M., Zuman P.: Chem. listy *47*, 975 (1953).
18. Eisenbrand J., Picher H.: Z. physiol. Chem. *260*, 83 (1939).
19. Wolfe J. K., Hershberg E. B., Fieser L. F.: J. Biol. Chem. *136*, 653 (1940).
20. Březina M., Volke J., Volková V.: Chem. listy *48*, 194 (1954); diese Zeitschrift *19*, 894 (1954).
21. Leffler J. E.: J. Org. Chem. *20*, 1202 (1955).
22. Schwabe K., Ohloff G., Berg H.: Z. Elektrochem. *57*, 293 (1953).
23. Zuman P.: Chem. listy *48*, 94 (1954).
24. Braude E. A., Nachod F. C.: *Determination of Organic Structures by Physical Methods*. Academic Press, New York 1955.
25. Fieser L. F., Fieser M.: Experientia *4*, 285 (1948).
26. Šorm F., Lábler L., Černý V.: Chem. listy *47*, 418 (1953); diese Zeitschrift *18*, 842 (1953).
27. Lábler L., Černý V., Šorm F.: Chem. listy *48*, 1058 (1954); diese Zeitschrift *19*, 1249 (1954).

Übersetzt von H. Bažantová.

POLAROGRAPHIC BEHAVIOUR OF p-DIACETYLBENZENE; AN EXAMPLE OF A COMPOUND WITH TWO ELECTROACTIVE CENTERS

YU. KARGIN*, O. MANOUŠEK AND P. ZUMAN

J. Heyrovský Institute of Polarography, Czechoslovak Academy of Sciences, Prague (C.S.S.R.)

(Received June 20th, 1966)

Dedicated to Professor Dr. M. VON STACKELBERG on his 70th birthday

Organic compounds bearing two electroactive centers (R^1, R^2) in their molecule (R^1–A–R^2) can be divided, according to their polarographic behaviour, principally into three groups.

To the *first* group belong compounds in which the two reactive centres are well separated and screened off. Each of these centers, R^1 and R^2, shows a polarographic behaviour which resembles that of a molecule bearing on the same molecular frame, A, only one electroactive group. The polarographic curve in such cases resembles that of a mixture of R^1-A-H and H-A-R^2. Examples of this type of behaviour are surprisingly rare and even with molecules containing groups R^1 and R^2 well separated by a saturated chain, a certain degree of mutual interaction, shown by a shift of the half-wave potential, is usually observed.

To the *second* group belong compounds in which the electrode process for the first step in the electrolysis of R^1–A–R^2 follows principally the same path as that observed for the electrolysis of R^1 in the molecule R^1-A-H, but which occurs at a somewhat different potential. This means that the number of electrons transferred, the electrolysis product, and the shape of the waves, are the same for the reduction of R^1 in R^1–A–R^2 as in R^1–A–H. On the other hand, the half-wave potential of the first wave in R^1–A–R^2 is shifted compared with the half-wave potential of the single wave observed for R^1–A–H. The magnitude of this shift depends on the nature of group R^2, which during the reduction in the first step plays the role of a substituent. Most examples found in organic polarography of compounds with two electroactive centers belong to this group.

Finally, to the *third* group belong compounds R^1–A–R^2, the behaviour of which is fundamentally different from both the behaviour of R^1–A–H and H–A–R^2 and which indicates a strong mutual interaction of groups R^1 and R^2 usually observed when these groups are in adjacent positions or situated on a conjugated system. A typical example of this type is the polarographic behaviour[1] of α,β-unsaturated ketones of type $C_6H_5CH=CHCOPy$ (Py = pyridine ring). In this case the reduction in the first step produces neither $C_6H_5CH_2CH_2COPy$ nor $C_6H_5CH=CHCHOHPy$.

In the course of a study of benzene derivatives bearing on the benzene ring two electronegative groups, such as benzonitriles[2,3] of type $CN-C_6H_4-X$ (where X = $COCH_3$, CHO, COO⁻, $COOC_2H_5$, or CN), or substituted methylphenylsulphones[4] of

---

* On leave from the Institute of Organic and Physical Chemistry "A. E. Arbuzov", Kazan, Academy of Sciences, U.S.S.R.

type $CH_3SO_2-C_6H_4-X$ (where $X = CN$, COO^-, $COOC_2H_5$, SO_2NH_2, or SO_2CH_3), it was decided to study in some detail the polarographic behaviour of p-diacetylbenzene. It could be anticipated that this compound, with the two electroactive groupings well separated, belongs to the second group mentioned above and that either a simultaneous or a successive attack of the two carbonyl groupings occurs. Nevertheless, the observation that the half-wave potential of the first wave is by some 0.3–0.4 V more positive than predicted by the modified[5] Hammett equation for an acetophenone substituted in the *para* position by an acetyl group, already indicates that the reduction mechanism of p-diacetylbenzene differs from that of the remaining substituted acetophenones. This idea was further substantiated by the fact that the reduction current observed with p-diacetylbenzene in the potential region corresponding to the reduction of acetophenone, is considerably smaller than expected for an equimolar solution of a substituted acetophenone. This indicated a strong mutual interaction of the two acetyl groups *via* the benzene nucleus.

The polarographic behaviour of p-diacetylbenzene has been mentioned recently in a paper by DAY and co-workers[6], but the results were not reported in sufficient detail to allow a more complete scheme of the electrode process to be deduced. A more detailed study reported here has shown that the behaviour of p-diacetylbenzene can be classified either as belonging to the third or to the second group of compounds bearing two electroactive groups, according to the acidity of the media. Whereas in more acidic media than about pH 0 and at pH-values greater than about 7, a successive reduction of the two carbonyl groups takes place, in the pH range 2–5 the system shows no analogy with that of simple acetophenone. Hence, in the more acidic and more alkaline media the behaviour can be classified as belonging to group 2; in the medium pH-range characteristics of group 3 are shown. The symmetry of the system is assumed to be responsible for this unusual behaviour and hence the study was extended to some other systems bearing two carbonyl groups in *para*- and *ortho*-positions on an aromatic system.

EXPERIMENTAL

A photographic recording polarograph LP 55 and in some cases pen-recording polarograph LP 60 (both Laboratorní přístroje, Praha) were used. Polarographic electrolysis was carried out using a Kalousek vessel with a saturated calomel electrode as reference electrode and a mercury dropping electrode with constants $m = 2.8$ mg/sec, $t_1 = 2.2$ sec at $h = 78$ cm in 1 M KCl at 0.0 V. p-Diacetylbenzene (Light, Colnbrook, England) after recrystallization from ethanol had m.p. 113° (lit.[7], 113°). p-Aldehydobenzoic acid and its methylate (m.p. 59–61°) were kindly donated by Dr. J. FRANC and Dr. J. ARIENT (Research Institute for Organic Syntheses, Pardubice, Rybitví, Czechoslovakia). Acetophenone was freshly redistilled.

0.01 M ethanolic stock solutions were prepared from the electroactive substance. These stock solutions were freshly prepared each week and were added to the supporting electrolyte to ensure a $2 \cdot 10^{-4}$ M solution. Because polarographic curves recorded in solutions containing 10% of ethanol have shown adsorption effects, all experiments were carried out in the presence of 30% ethanol in the supporting electrolyte.

Britton–Robinson, acetate, and phosphate buffers, and solutions of sulphuric acid were prepared from reagent-grade chemicals.

The dependence of wave-heights and of half-wave potentials of p-diacetyl-benzene on pH and concentration were measured directly from the polarogram and compared with the waves for acetophenone. For p-aldehydobenzoic acid and its methylate, only the pH-dependence of the wave-heights and the absence of any more negative wave (following the two-electron reduction of the aldehydic group) was checked.

The logarithmic analysis of the p-diacetylbenzene waves at various pH-values was made using the curves recorded with the LP 60 polarograph, using a slow rate of scanning (100 mV/min). The effect of concentration and drop-time on the half-wave potentials was measured by comparison with the half-wave potentials of cadmium. The effect of drop-time was investigated using an electrode with a mechanically controlled drop-time.

The single-sweep method was carried out using an oscillographic polarograph OP 1 produced by the Scientific laboratory of the Soviet Academy of Sciences. This instrument enables single-sweep curves to be recorded with a scanning rate adjustable from 0.1–50 V/sec.

The rectangular voltage polarization was followed using a commutator constructed according to the principle suggested by KALOUSEK[8,9]. In the scheme used, the auxiliary potential was always kept constant at the potential corresponding to the limiting current of the first wave. In all cases, the capacity current using the commutator circuitry was recorded and subtracted from the current–voltage curves recorded in the presence of the electroactive component. The effect of pH on the anodic wave on the commutated curve in the presence of p-diacetylbenzene was studied in the range between 5 N H_2SO_4 and pH 8. No anodic waves were observed over this range in solutions of p-aldehydobenzoic acid or its methylate.

Preliminary chronopotentiometric examinations were made using the apparatus built by O. FISCHER (Faculty of Science, The University, Brno). After the electrolysis had been carried out for a time, τ, the direction of the current was switched and the transition time determined.

The pH-values of the polarographed solutions were determined using a glass electrode, G 200 B, in connection with pHM-4 meter (Radiometer, Copenhagen).

Coulometric measurements were carried out using a mercury electrode dropping into a small volume of the electrolysed solution. The limiting current was recorded after selected time intervals and the number of electrons consumed was determined from the slope of a plot of the logarithm of this current against time.

A potentiostat designed by L. NĚMEC and I. HOLUB in this Institute was used for preparative electrolysis. It has a current capacity up to 600 mA, a response time of about 1 sec, a range of available controlled potentials of ± 2.0 V, and a potential control to 3 mV.

The reactions of p-diacetylbenzene with hydroxylamine, sodium borohydride, hydrogen catalysed by platinum asbestos, and univalent cobalt–dipyridyl complex took place directly in the electrolysis cell. The polarographic curves were recorded during the various stages of each reaction. The equilibrium concentrations in the reaction of p-diacetylbenzene with hydroxylamine were measured after the solution had been kept in the dark in closed vessels for 3–5 days. The solutions were not heated because of the volatility of the components of the reaction mixture.

RESULTS

Polarographic behaviour of p-diacetylbenzene

Three waves, i_1, i_2 and i_3 can be observed on polarographic curves of p-diacetyl-benzene in the pH-range studied. Whereas the height of the most positive wave, i_1, remains practically constant over the whole pH-range, the wave-heights of waves i_2 and i_3 are strongly pH-dependent (Figs. 1 and 2). Wave i_2 is observed in solutions of strong acids and in buffers of $3.0 > $ pH $ > 5.5$; wave i_3 occurs at pH $ > 6$.

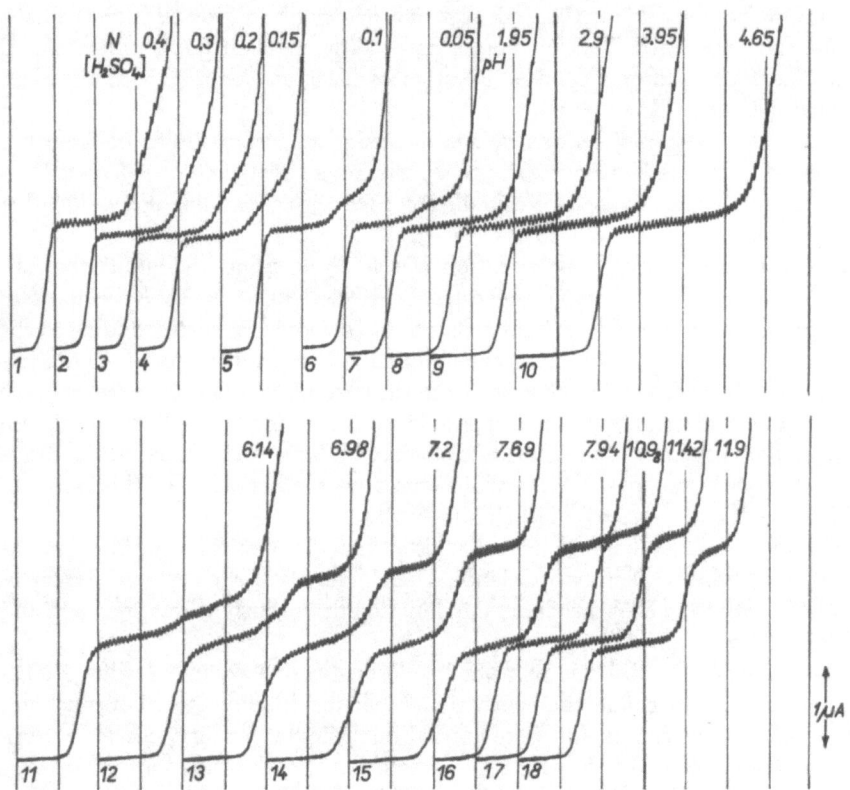

Fig. 1. pH-dependence of waves of $2 \cdot 10^{-4}$ M p-diacetylbenzene; 30% ethanol; pH given on the polarogram; (1)–(6) H$_2$SO$_4$, (7)–(18) Britton–Robinson buffers. Curves starting at: (1)–(10), -0.4 V; (11)–(15), -0.6 V; (16)–(18), -0.8 V; 200 mV/absc., S.C.E.

It has been shown that the half-wave potentials of wave i_2, their pH-dependence, the shape of the wave and the pH-dependence of the wave-height expressed as $i_2/(i_2+i_3)$ as well as the pK'-value (*i.e.* the value at which $i_2 = \frac{1}{2}(i_2+i_3)$), are analogous and similar to these characteristics of the wave of the protonized form of acetophenone. On the other hand, the value of half-wave potential of wave i_3, the almost complete pH-independence of this half-wave potential and the increase of the wave-height expressed as $i_3/(i_2+i_3)$, was found analogous to the two-electron reduction wave of unprotonized acetophenone. Hence, it was concluded that wave i_2 corresponds to the reduction of the protonized form, p-CH$_3$CHOHC$_6$H$_4$COH$^+$CH$_3$, and wave i_3 to the reduction of the unprotonized ketol, p-CH$_3$CH(OH)C$_6$H$_4$COCH$_3$. It can, therefore, be concluded that the ketol [p-(1-hydroxyethyl)acetophenone] can be formed in reduction wave i_1.

The pH-dependence of the total wave height ($i_2 + i_3$) shows an increase both in acid and in alkaline media. In acid media, wave i_2 increases, but the shape of this increase was difficult to estimate due to the ill-defined shape of the wave under these conditions. The increase of i_3 at higher pH-values (Fig. 2) gives the shape of a dis-

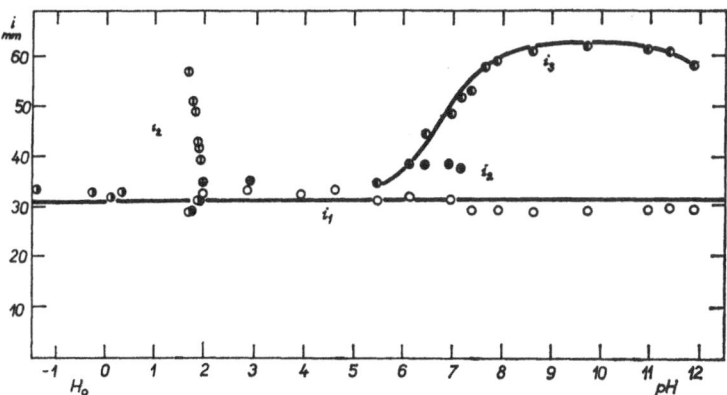

Fig. 2. pH-dependence of limiting currents of $2 \cdot 10^{-4}$ M p-diacetylbenzene; 30% ethanol; halved circles at pH-values below 2, sulphuric acid; all others, Britton-Robinson buffers. 10 mm on current scale correspond to 0.48 μA.

sociation curve with pK_2' of about 7.2. The kinetic character of the wave-height ($i_2 + i_3$) indicates that the formation of p-(1-hydroxyethyl)acetophenone is not a straightforward electrochemical process but that chemical reactions are involved the rates of which are comparable to the rate of growth of the mercury drop. The rate of the overall process is pH-dependent, being slowest at pH 3–5. The rate of the production of p-(1-hydroxyethyl)acetophenone increases both on increasing and decreasing the acidity.

Wave i_1 corresponds to a two-electron reduction, as was confirmed by controlled-potential electrolysis and coulometry. Even when no separation of the two waves was apparent on the recorded curves, it was possible to show, using logarithmic analysis, the presence of two forms. The logarithmic analysis (Fig. 3) indicates the formation of two waves, the height of the more positive decreasing with increasing pH, the height of the more negative increasing with increasing pH. The pH-dependence has the form of a dissociation curve with pK_2' = 7.2. The pH-dependence of the half-wave potentials of the first wave, i_1, (Fig. 4) shows a break at a pH value of about 8 (pK_2''). Below this value, a proton-transfer precedes the electrode process proper; above pH 8 the half-wave potentials are little pH-dependent. The pH-independent section, corresponding to the reduction of the protonized form, merges here with the reduction of the unprotonized form. The small difference at a pH-value of about 8 between the half-wave potentials, corresponding to the reduction of the protonized and unprotonized forms, explains the observed lack of separation of the two waves.

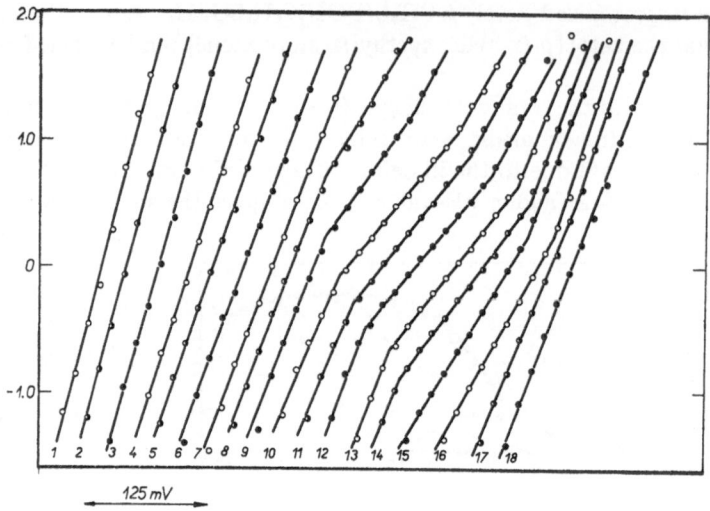

Fig. 3. Logarithmic analysis. Dependence of log $i/(i_d - i)$ on potential for different pH-values; $2 \cdot 10^{-4}$ M p-diacetylbenzene; 30% ethanol. (1), 5 N H_2SO_4; (2), 1 N H_2SO_4; all others Britton–Robinson buffers of pH: (3), 1.95; (4), 2.8 (5), 3.8; (6), 4.8; (7), 5.5; (8), 6.1; (9), 6.5; (10), 7.0; (11), 7.2; (12), 7.4; (13), 7.7; (14), 7.9; (15), 8.65; (16), 9.15; (17), 9.8; (18) 10.35.

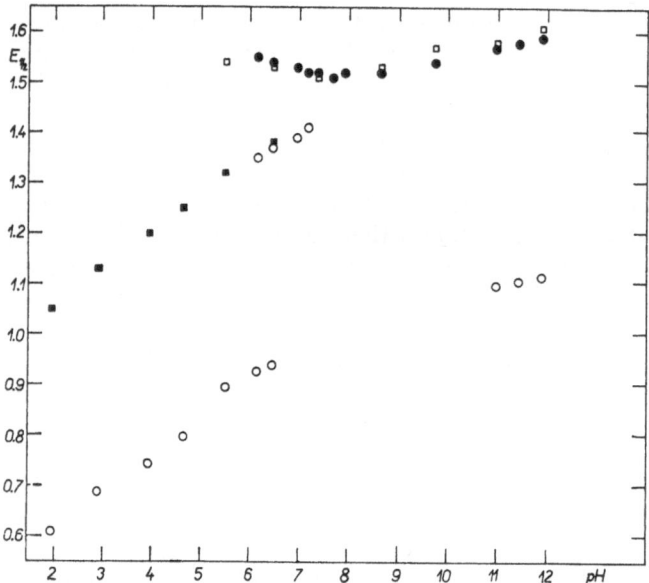

Fig. 4. pH-dependence of half-wave potentials; $2 \cdot 10^{-4}$ M p-diacetylbenzene in Britton–Robinson buffers, 10% ethanol.

The half-wave potentials were practically independent of the drop-time (from $t_1 = 0.24$ sec to $t_1 = 2.84$ sec) in 5 N H_2SO_4. With increasing concentration of p-diacetylbenzene in acetate buffer pH 4.7, the half-wave potentials changed as follows:

p-diacetylbenzene $(mole/l \cdot 10^{-4})$	$E_{\frac{1}{2}}$ (V)
1	−0.818
2	−0.823
5	−0.824
10	−0.829

Single-sweep investigations

The peak current of the $i–E$ curve in the single-sweep method, corresponding to wave i_1 has been shown to remain proportional to the square-root of the scanning rate, $v^{\frac{1}{2}}$. Hence the current of the first wave is diffusion-controlled even for time-intervals considerably shorter than in classical polarography.

$i–E$ curves obtained with the single-sweep technique revealed that, in the pH-range 2–5, the electrode process is reversible. When scanning from negative to positive potentials, an anodic peak was observed, the peak-potential fulfilling the condition for a two-electron reversible system. In order to obtain conditions comparable with those of classical polarography, the quantitative treatment was restricted to the commutator method.

Commutator and controlled-potential electrolysis

Using rectangular voltage polarization in the commutator method according to KALOUSEK[8,9], and choosing the auxiliary potential at the limiting current of the first wave, i_1, we observed that an anodic wave is formed. The half-wave potential of this wave was practically identical with that of the cathodic wave in that particular media, showing the reversibility of the process. No separation of two waves, indicated in Fig. 3 of ref. 6 has been observed.

The height of the anodic wave, i_A, decreased at pH > 5.5 in the form of a dissociation curve with a $pK_2' \doteq 7.3$, and at pH < 0 with a $pK_1' \doteq -2.3$. (Figs. 5 and 6.) This behaviour indicates that the reduction of the form (having one proton less than the form predominantly reducible at pH 2–5) which predominates at pH > 7, is reduced irreversibly. The coincidence of the increase of the fraction of wave i_1, of the sum of the waves $(i_2 + i_3)$ and of the anodic wave, i_A, indicates that the product of the irreversible reduction is p-(1-hydroxyethyl)acetophenone which is then further reduced in the more negative waves, $(i_2 + i_3)$.

The decrease of the anodic wave in strongly acid media indicates the deactivation of the product of the reversible reduction process. The height of the anodic wave i_A, increases with decreasing ethanol concentration (Fig. 7). This may be due to the increase in the rate of the deactivating process with increasing ethanol concentration, but a decrease in adsorbability of the reduction product with increasing ethanol concentration cannot be ruled out. Moreover, the effect of such adsorption is demonstrated by the fact that the anodic wave, i_A, obtained with the commutator is higher than wave i_1.

When controlled-potential electrolysis with a mercury pool cathode was carried out at the potential of the limiting current of the first wave, i_1, for one or two hours, only p-(1-hydroxyethyl)acetophenone—identified from polarographic curves—and not the product of the reversible reduction was detected among the electrolysis products. The wave of p-(1-hydroxyethyl)acetophenone, absent on the curve of

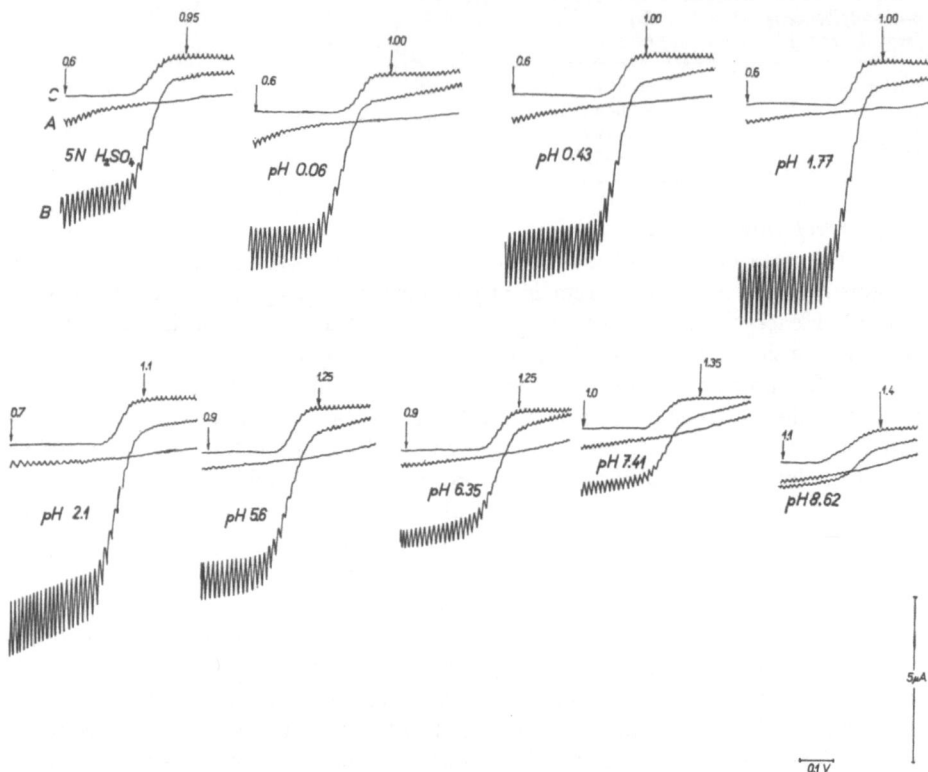

Fig. 5. Application of the commutator method according to KALOUSEK. pH-dependence of the wave of the product. 2.5×10^{-4} M p-diacetylbenzene, Britton–Robinson buffers, 15% ethanol; frequency 6 c/sec; values of pH, potential at which the recording started, and auxiliary potential are denoted in the polarogram; (A), residual current obtained with the commutator; (B), p-diacetylbenzene added, curve recorded with the commutator; (C), curve recorded without commutator; mercurous sulphate electrode.

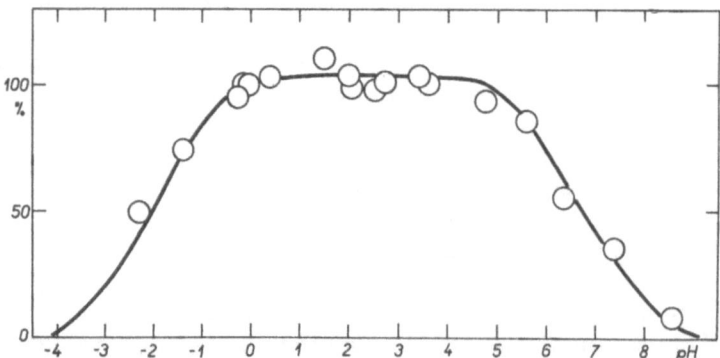

Fig. 6. pH-dependence of the anodic limiting current of the reduction product of p-diacetylbenzene. Limiting current measured with the commutator, expressed as % of the maximum limiting current. Sulphuric acid and Britton–Robinson buffers, 15% ethanol.

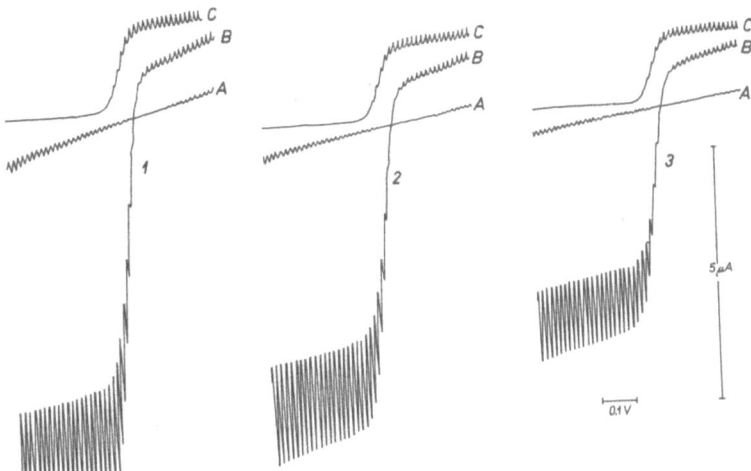

Fig. 7. Effect of ethanol concn. on the curves obtained with the commutator method: 2.5×10^{-4} M p-diacetylbenzene in 0.075 N H_2SO_4. Ethanol concn.: (1), 5; (2), 15; (3), 25%. (A) and (B) recorded with the commutator, (C) without; (A) supporting electrolyte; (B) and (C) p-diacetylbenzene added. Commutator frequency 6 c/sec; curves recorded starting at -0.4 V; auxiliary potential -0.85 V; mercurous sulphate electrode.

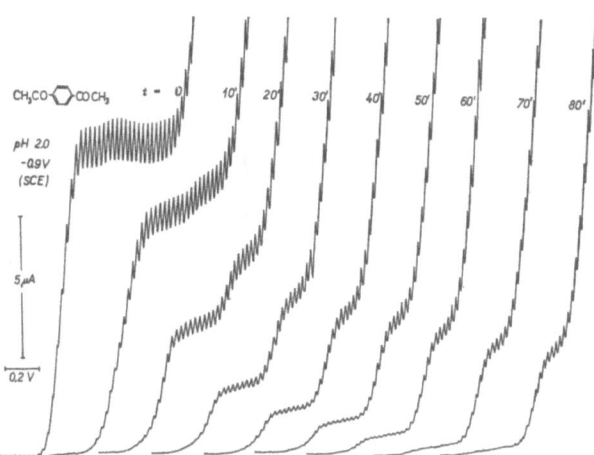

Fig. 8. Record of the course of the electrolysis on a mercury pool electrode; $5 \cdot 10^{-3}$ M p-diacetylbenzene in a Britton–Robinson buffer, pH 2.0, containing 10% ethanol electrolysed at 0.9 V (S.C.E.); 50 ml of soln., stirred pool electrode with a surface of 20 cm^2; polarographic curves recorded starting at -0.2 V (S.C.E.).

p-diacetylbenzene at pH 2.0, increased during electrolysis at a mercury pool (Fig. 8). The absence of the anodic wave indicated that in each interval the primary electrolysis product formed had already been cleaved into p-(1-hydroxyethyl)acetophenone. Even when the electrolysis was carried out using a pool electrode with a larger current for only 10 sec, it was impossible to detect any anodic wave. A comparison with results obtained with the commutator method indicate that the half-life of the primary product between pH 0 and 7 is between 0.5 and 10 sec. Preliminary results of chronopotentiometric measurements[13] indicate a half-life of the order of 10^1 sec at pH 2–3.

Chemical reactions

The limiting current of the first wave at pH 4.7, recorded after establishment of the equilibrium following the addition of hydroxylamine to a buffered solution of p-diacetylbenzene, increased with increasing hydroxylamine concentration from a value corresponding to a two-electron transfer to a value corresponding almost to an eight-electron process (Fig. 9). This indicates that both carbonyl groups react with

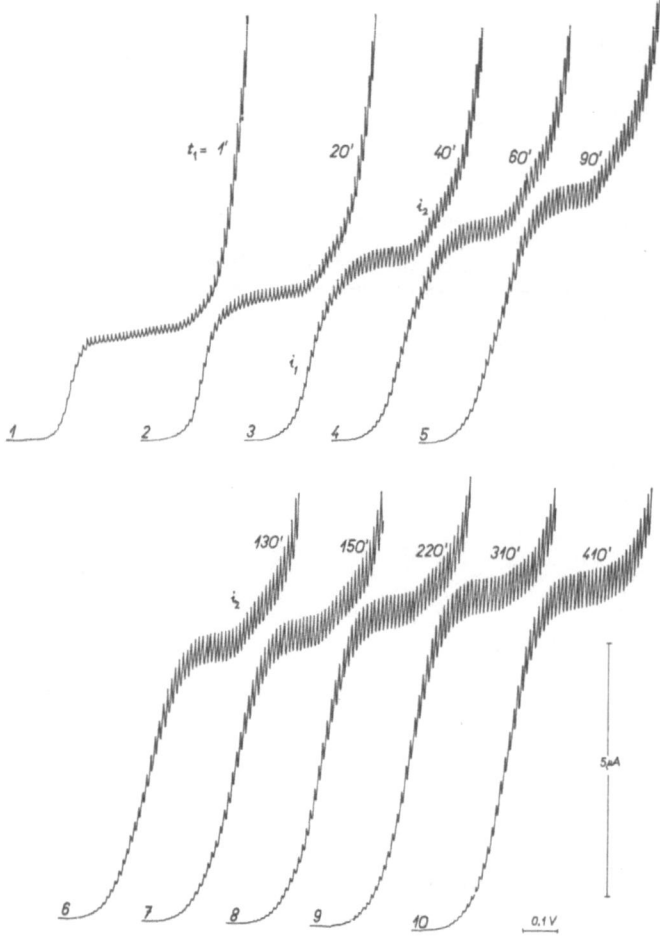

Fig. 9. Reaction of p-diacetylbenzene with hydroxylamine; 2.4×10^{-4} M p-diacetylbenzene, 5.7×10^{-3} M hydroxylamine sulphate, acetate buffer pH 4.7, 30% ethanol; reaction temp. $t = 50°$; polarographic curves recorded starting at -0.7 V (S.C.E.).

hydroxylamine. The two equilibrium constants are greater than 10^4 and the reaction proceeds practically quantitatively.

The half-wave potentials of the first wave of p-diacetylbenzene and of its monoxime and dioxime, differ too little to allow a separation of two or three waves, respectively. The appearance of the wave i_2 at more negative potentials (Fig. 9) was ascribed to the reduction of the carbonyl group of p-diacetylbenzene monoxime in the reduction step B. The occurrence of such a wave indicates that the reaction of p-diacetylbenzene with hydroxylamine takes place in two consecutive steps rather than in a simultaneous attack by two hydroxylamine molecules on both carbonyl groups.

$$\text{[ring: COCH}_3\text{ / C(CH}_3\text{)=NOH]} + 4e + 4H^+ \longrightarrow \text{[ring: COCH}_3\text{ / CHCH}_3\text{–NH}_3^+] + OH^- \quad A$$

$$\text{[ring: COCH}_3\text{ / CHCH}_3\text{–NH}_3^+] + 2e + 2H^+ \longrightarrow \text{[ring: CHCH}_3\text{–OH / CHCH}_3\text{–NH}_3^+] \quad B$$

Scheme A and B.

In agreement with this assumption is the fact that the height of the more negative wave (after the establishment of equilibrium) changed with increasing hydroxylamine concentration in the way expected for an intermediate, *i.e.*, first increased, when the monoxime was formed; then decreased, when all the p-diacetylbenzene had been transformed into the dioxime. Similarly, the more negative wave corresponding to the monoxime first increased and then decreased with time when the curves were recorded at a given hydroxylamine concentration before equilibrium had been established (Fig. 9).

None of the reducing agents used, *i.e.*, sodium borohydride, univalent cobalt–dipyridyl complex, or hydrogen in the presence of platinum asbestos catalyst, gave the product exhibiting the anodic wave.

Comparison with reduction processes of related compounds

If the reduction of the monoxime and dioxime follow a similar path to that found for p-diacetylbenzene, a two-electron process would also be observed with the oximes. The experimental evidence described in the preceding paragraph shows, on the contrary, that the reduction of p-diacetylbenzene monoxime occurs in one four-electron and one two-electron step, whereas that of p-diacetylbenzene dioxime is a single eight-electron step. Hence, a process different from that observed for p-diacetylbenzene occurs in solutions of both monoxime and dioxime. This was confirmed by the commutator method, when no anodic currents for the electrolysis products of the oximes were observed. The two oxime groups or one oxime and one carbonyl group do not show the type of mutual interaction found for p-diacetylbenzene. The results of experiments with p-aldehydobenzoic acid were similarly negative; the behaviour of this compound was found to be similar to that described in the literature[10,11]. Only one two-electron reduction wave of the acid or anion form, respectively, was observed; no additional reduction process occurred at more negative potentials in the potential range available. The commutator indicated no reversibility over the whole pH-range. Hence the increase in the a.c. current observed[11] at pH 3–5 corresponds rather to a change in adsorbability than in reversibility of the electrode process. Similarly, for p-aldehydobenzoic acid methylate the two-electron step was not followed by a further reduction and the irreversible process yielded no anodic wave with a commutator. Whereas for the free acid in acidic media the two-electron wave was split into two one-electron steps, only one two-electron wave was observed for the methyl ester, even at low pH-values.

On the other hand, preliminary results obtained with terephthaldialdehyde indicated a resemblance to p-diacetylbenzene. An anodic wave was observed at pH 2–5, using the commutator, pointing to reversibility. The decrease of the anodic wave both in more acidic and more alkaline solutions was accompanied by a simultaneous increase of a more negative cathodic wave.

DISCUSSION

The polarographic behaviour of p-diacetylbenzene indicates that, depending on the acidity, the two acetyl groups can exert either, (i) a moderate interaction, the reduction process giving rise to p-(1-hydroxymethyl)acetophenone with one carbonyl group reduced, or (ii) a strong mutual interaction (at pH 0–7) which results in a behaviour fundamentally different from that of acetophenone.

The mutual interaction is exhibited by a two-electron reduction step that is not followed by further reduction of an acetophenone derivative. This reduction process occurs, moreover, at potentials by some 0.3 V more positive than would be expected[5] for a p-acetyl-substituted acetophenone. And finally, the two-electron reduction process is reversible. The product of this reversible reduction is sufficiently stable to give an anodic wave with the commutator and an anodic peak with the single-sweep method.

The form that is electroactive at pH 0–7 at the potentials of the first wave, i_1, gives a product that shows anodic waves with a commutator but no cathodic wave in the potential region of waves i_2 and i_3 in which the substituted acetophenones are reduced. The pH-dependence of the half-wave potentials of wave i_1 at pH-values below 6 (Fig. 4) indicates that the electroactive form accepts protons. Information regarding acid–base equilibria in the bulk of the solution are of little value in elucidating the structure of the electroactive particle at the electrode surface and it was first important, therefore, to ascribe a proper degree of protonation to the form electroactive in wave i_1. It was necessary to decide whether the mono- or diprotonated form of p-diacetylbenzene is reduced in the first, reversible step.

It is difficult to imagine the formation of a relatively stable, easily oxidisable product of a two-election reduction of the monoprotonated p-diacetylbenzene in aqueous media. For this species, an attack on the protonated carbonyl seems to have a higher probability than an attack on the unprotonated carbonyl. Products of a two-electron reduction of a carbonyl group are known to be reactive towards the solvent. It seems rather improbable, therefore, that such a species would be able to participate in a reversible electrode process and also possess the properties found for the reduced form of p-diacetylbenzene. Hence it is assumed that at pH 2–5 the predominant form at the surface of the electrode that undergoes the reversible two-electron process is the diprotonated form of *p*-diacetylbenzene. This assumption leads to the conclusion that the reduced form, formed by the reversible electrode process in this pH-range, is the bi-radical (I). The stability of such a bi-radical is favoured by the symmetry of its structure and by the possible resonance stabilisation involving a quinonedimethane structure. Because the radicals derived from acetophenone are rather unstable in aqueous solution, it is necessary to assume that the observed stabilisation results from the mutual interaction of the two groups in the bi-radical; this seems plausible.

The observed system corresponds in the potential-determining step, to a simple electron transfer and not a group transfer, as takes place in most organic reductions.

At pH-values below 2 and above 5, the reversibility of the system is perturbed. In the system described by Scheme I the reasons for deviations from reversibility in acidic media are discussed first.

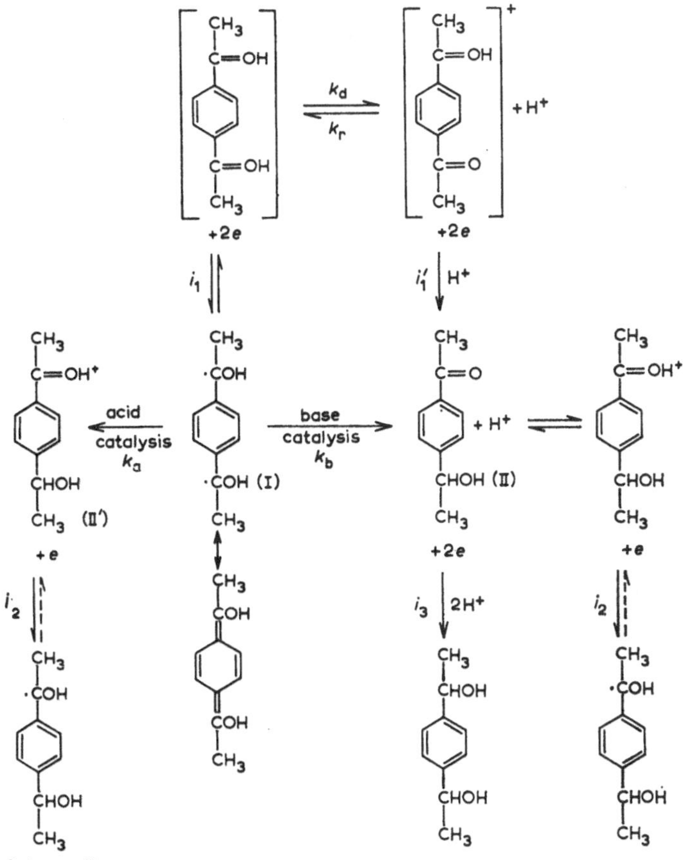

Scheme I

The decrease of the anodic wave, i_A, obtained by the commutator method at pH-values below 2 cannot be caused by a change in the rate of a chemical reaction preceding the electrode process proper. This possibility was excluded by (i) the independence of pH of the limiting current i_1, in this region, (ii) the absence of a change in the slope of the pH-dependence of the half-wave potentials of wave i_1 and (iii) the strictly linear plots of the logarithmic analysis of wave i_1 in this pH-region.

The decrease of the reversible component can originate in a reaction interposed between the acceptance of the first and second electron or between the transfer of the second electron and the following electron uptake. There is no information available on this reaction, it is known only that p-(1-hydroxyethyl)acetophenone is formed as the final product in the first two-electron step. One of the possibilities is that the rearrangement of the bi-radical with rate constant k_a is acid-catalysed (Scheme I) and yields rapidly in acidic media the acetophenone derivative, II'. A reaction course following the other type, $i.e.$, the protonation of the mono-radical formed as intermediate in the first two-electron step, shown in Scheme II, cannot be excluded.

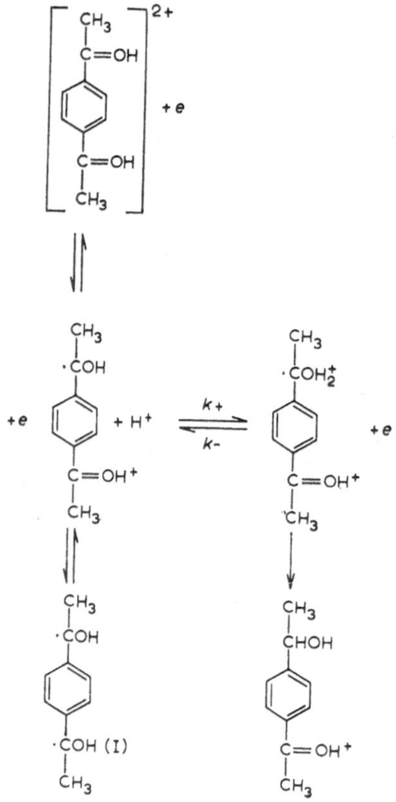

$$
\left[\begin{array}{c} CH_3 \\ | \\ C=OH \\ \bigcirc \\ C=OH \\ | \\ CH_3 \end{array} \right]^{2+} \quad +e
$$

$$+e \quad
\begin{array}{c} CH_3 \\ | \\ \cdot COH \\ \bigcirc \\ C=OH^+ \\ | \\ CH_3 \end{array} + H^+ \quad
\underset{k-}{\overset{k+}{\rightleftharpoons}} \quad
\begin{array}{c} CH_3 \\ | \\ \cdot COH_2^+ \\ \bigcirc \\ C=OH^+ \\ | \\ CH_3 \end{array} \quad +e
$$

$$
\begin{array}{c} CH_3 \\ | \\ \cdot COH \\ \bigcirc \\ \cdot COH \text{ (I)} \\ | \\ CH_3 \end{array} \qquad
\begin{array}{c} CH_3 \\ | \\ CHOH \\ \bigcirc \\ C=OH^+ \\ | \\ CH_3 \end{array}
$$

Scheme II

The almost complete independence of the half-wave potentials of the drop-time, and of the concentration of the electroactive species, indicate that the reaction deactivating the bi-radical I is slow when compared with the drop-time. The reaction in which the bi-radical I is deactivated cannot be a reaction that is second-order in the bi-radical I, such as dimerisation. These observations are in accordance with the half-lives estimated from our measurements, and those determined[13] using chronopotentiometry. Reactions with half-lives of the order of seconds cannot affect the half-wave potentials.

The next question to be considered is the reason for deviations from reversibility at pH-values above 5. Experimental evidence shows that they also are caused by a chemical reaction accompanying the electrode process proper. There are in principle two possibilities: either this chemical reaction involves the oxidised form (*i.e.*, *p*-diacetylbenzene), or the product of the first reversible electrode process (*i.e.*, the bi-radical I). The chemical reaction that causes the decrease of the anodic wave i_A of the electrolysis product gives rise also to the increase in the height of the more negative cathodic waves, i_2 and i_3. The inflexion point of the decreasing pH-dependence of the anodic wave, i_A ($pK_2' \doteq 7.3$) is practically equal to the inflexion point of the increasing pH-dependence of the sum ($i_2 + i_3$), *i.e.*, $pK_2' = 7.2 \pm 0.2$. The kinetic character of the sum ($i_2 + i_3$) at pH 6 indicates that under these conditions the reaction is fast and its rate comparable with the drop-time.

Waves i_2 and i_3 were shown to correspond to *p*-(1-hydroxyethyl)acetophenone (II). This was proved by a comparison of their half-wave potentials, their pH-dependence (Fig. 4) and the dependence of the ratio i_2/i_3 on pH, with the corresponding characteristics for acetophenone.

Hence, the final product of the chemical reaction that governs the height of the anodic wave, i_A, and the heights of cathodic waves, i_2 and i_3, is the acetophenone derivative II. When the second alternative is dealt with first, it can be assumed that this chemical reaction is a deactivation of the bi-radical I to form compound II in a similar way as was considered in acid solutions, but through a base-catalysed reaction with rate constant, k_b (Scheme I).

Two possibilities can be suggested for the other alternative (*i.e.*, the chemical reaction affecting *p*-diacetylbenzene itself)—acid–base equilibria and hydration–dehydration equilibria.

The fundamental assumption of the interpretation involving acid–base equilibria is that the monoprotonated form of *p*-diacetylbenzene is irreversibly reduced in two main steps. In the first step, coinciding with wave i_1, the protonated carbonyl group is attacked with consumption of two electrons and compound II is formed. In a consecutive reaction, this acetophenone derivative II is reduced in waves, i_2 and i_3, in a similar manner as most other substituted acetophenones.

Supporting this interpretation is the evidence that a protonated carbonyl group is usually reduced at more positive potentials than the unprotonated group. Furthermore, the change in the plots for logarithmic analyses can be interpreted as a change of poorly separated waves of the conjugate acid (diprotonated) and conjugate base (monoprotonated) forms with pH. This is equivalent to a conclusion that in the monoprotonated form only a limited interaction between the two acetyl groups occurs and that CH_3CO is substituent X in p-X-C_6H_4-COH^+CH_3. The inflexion point of the pH-dependence of the two parts of wave i_1 and also the change in slope of the plot of the pH-dependence of the half-wave potentials of wave i_1 again occur in the region $pK_2' = 7.0 \pm 0.2$. If this is not a coincidence, it indicates a relation between the chemical process affecting wave i_1 on one side, and the reactions that govern the anodic wave, i_A, and the height of cathodic waves ($i_2 + i_3$), on the other side.

There are two points that can be used as an argument against this interpretation. First, the half-wave potential of wave i_1 at pH-values above 7 is still more positive than predicted[5] from the effect of a p-CH$_3$CO substituent. Secondly, wave i_1, which according to the present scheme would, at pH-values above 7, correspond to the reduction of a protonated carbonyl, does not show the separation of two one-electron waves typical for the reduction of the protonated carbonyl group in aryl-alkyl-ketones. It cannot be excluded that at these higher pH-values the two one-electron steps have already merged. Moreover, some other alkyl–aryl-ketones with electronegative substituents show single two-electron steps even in acid media.

None of these arguments makes this interpretation unreasonable and at present it is impossible to decide between the two possibilities considered in Scheme I. The changes in waves i_1, i_2, i_3 and i_A can be interpreted by assuming that bi-radical I is deactivated in acidic media by an acid-catalysed reaction with rate constant k_a, and at higher pH-values by a base-catalysed reaction with constant k_b. Alternatively, the change in alkaline media can be interpreted by a change in the equilibrium between the diprotonated and monoprotonated forms of p-diacetylbenzene. The decrease of the anodic wave, i_A, would then be due to a decrease in the rate of the recombination reaction with constant k_r, as in the splitting of wave i_1. The increase of cathodic waves i_2 and i_3 would, in this case, result from the increase in the rate of the dissociation reaction with constant k_d.

As mentioned on p. 457 there is still another possible explanation for the changes in the wave-heights with pH (in particular in alkaline media), based on the effects of hydration.

Because with other aldehydes and ketones, certain changes in the polarographic curves are explained by a dehydration reaction[12], the possible role of dehydration was considered even in the interpretation of the polarographic behaviour of p-diacetyl-benzene. So far, it has been assumed[12] that the hydrated form of a carbonyl compound is electro-inactive and only the dehydrated form undergoes reduction.

Three possible forms, III–V, which in principle can exist in solutions of p-diacetylbenzene, are given in Scheme III. It cannot be assumed that the dihydrated form, III, undergoes reduction in the potential range of the first wave, i_1. The possibility that either form IV or form V is the electroactive species undergoing the reversible reduction is next discussed.

If form V is the reactive form undergoing reduction, it is necessary to assume that hydration with rate constant k_{-2} is acid- and base-catalysed and that above pH 7 or below pH 0, form IV is reduced. This would mean that form IV would be more easily reducible at higher pH-values than form V, which seems little probable. Furthermore, it would be necessary to assume that with the hydrated form of p-(1-hydroxy-ethyl)acetophenone, VI, dehydration with rate constant k_3 takes place.

The total height of the more negative waves $(i_2 + i_3)$ would then be limited by the rates of hydration of form V with constants k_{-1} and k_{-2}, rather than that of de-hydration of compound VI with rate constant k_3, since the shape of the pH-dependence of waves $(i_2 + i_3)$ corresponds to changes in the current, i_1, and in i_A. Only if the change due to the acid–base controlled dehydration of form VI appears in the same pH-range as the changes in waves i_1 and i_A, is it possible to assume that waves $(i_2 + i_3)$ are limited by the rate of the dehydration reaction with rate constant k_3.

Nevertheless, it is usually supposed that the hydrated form predominates in the medium pH range, and dehydration occurs in acidic and basic media and not conversely, as would have to be assumed when considering Scheme III. This scheme cannot, therefore, be excluded from consideration on the basis of the experimental material available at present, but is seems less probable than the other possibilities.

The last alternative is the assumption that the monohydrated form, IV, undergoes the reversible reduction. The main argument against this interpretation is that it is difficult to find reasons why a system like IV should be reduced reversibly. For a

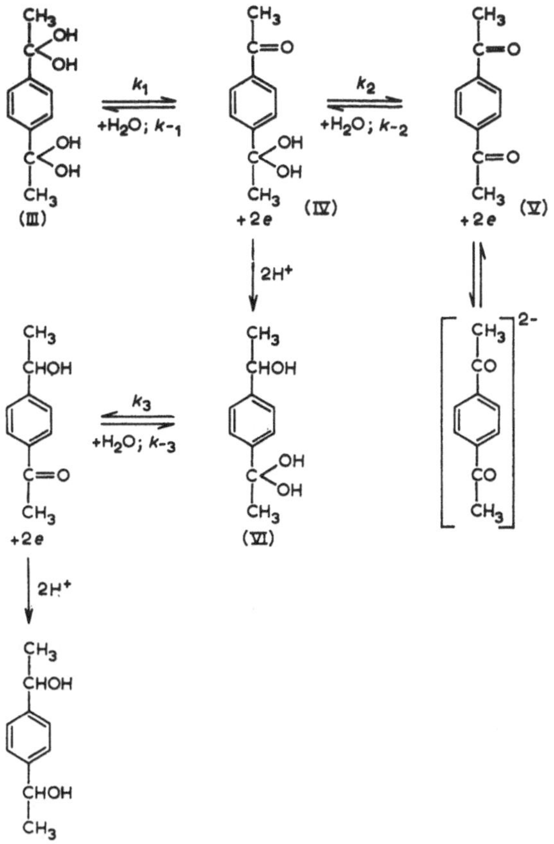

Scheme III

compound of this type it can hardly be assumed that the nucleophilic attack of electrons would result in the formation of a reversible oxidisable species. Substances of type IV would probably be attacked by both electrons on the same, unhydrated carbonyl group. Furthermore, it would be even more difficult to explain the positive potential of species IV than in previous cases, as for the substituent, $C(OH)_2CH_3$, even a smaller shift of the half-wave potential (compared with acetophenone) than for the acetyl group, can be expected.

The reaction with hydroxylamine indicates that the reactivity of the two carbonyl groups for hydroxylamine at pH 4.7 does not differ fundamentally. This indicates a reaction of a symmetrical rather than an unsymmetrical (like IV) molecule. The proof is nevertheless not convincing since the reaction with hydroxylamine is considerably slower than would be expected for a dehydration process.

Although unequivocal proofs cannot be given at present, it seems more probable that the changes in the limiting currents are explained by one of the reactions considered in Scheme I than by equilibria between hydrated and non-hydrated forms.

CONCLUSIONS

A more detailed knowledge of the reversible system of p-diacetylbenzene and the product of its two-electron reduction has been obtained; Scheme I is in best accord with the experimental data. The extension to other similar systems is studied. Some contributions to the techniques used in the elucidation of organic electrode processes are described, such as the observation of the formation of a new wave during electrolysis, the use of logarithmic analysis for the detection of a preceding chemical reaction, and the study of the pH-dependence of the yield of the electrolysis product using the commutator method.

ACKNOWLEDGEMENT

Thanks are expressed to Dr. O. FISCHER for assistance in some preliminary chronopotentiometric measurements, to Dr. R. KALVODA for assistance in oscillographic measurements and to Drs. A. A. VLČEK and A. RUSINA for suggestions concerning the inorganic reducing agents. We are also grateful to Dr. J. SHORTER (University, Hull) and Dr. J. FRANC and Dr. J. ARIENT (Pardubice) for their gifts of chemicals, and to Miss H. JÄNCHENOVÁ for technical assistance.

SUMMARY

p-Diacetylbenzene is reduced at pH 2–5 in the diprotonised form in a reversible two-electron step to a bi-radical with a half-life of the order of seconds. Reversibility was confirmed by single-sweep and commutator methods. At pH-values below 2 the reversibility is perturbed by transformation of the bi-radical into p-(1-hydroxyethyl)acetophenone in an acid-catalysed reaction. The same product results also at pH-values above 5, where the reversibility is affected either by a base-catalysed deactivation of the bi-radical or by the formation of a monoprotonated form, reduced irreversibly in two successive steps.

ZUSAMMENFASSUNG

p-Diacetylbenzol wird zwischen pH 2 und 5 in der diprotonierten Form unter Aufnahme von 2 Elektronen zu einem Biradikal reversibel reduziert. Die Lebensdauer des Biradikals in den wässrig–äthanolischen Lösungen liegt in der Grössenordnung von Sekunden. Die Reversibilität wurde mit Hilfe der Impulspolarographie und des Umschalters nach Kalousek bewiesen. Bei pH unter 2 wird die Reversibilität durch eine sauer katalysierte Umsetzung des Biradikals in p-(1-Hydroxyäthyl)acetophenone beeinträchtigt. Dasselbe Produkt entsteht auch bei pH >5, wo die Reversibilität entweder durch eine basisch katalysierte Umsetzung des Biradikals oder durch die Bildung einer monoprotonierten irreversibel zweistufig reduzierbaren Form beeinflusst wird.

REFERENCES

1 E. LAVIRON AND P. ZUMAN, unpublished results.
2 O. MANOUŠEK AND P. ZUMAN, *Chem. Commun. (London)*, (1956) 158.
3 P. ZUMAN AND O. MANOUŠEK, *Collection Czech. Chem. Commun.*, in the press.
4 O. MANOUŠEK, O. EXNER AND P. ZUMAN, *Collection Czech. Chem. Commun.*, in the press.
5 P. ZUMAN, *Collection Czech. Chem. Commun.*, 25 (1960) 3225.
6 R. H. PHILIP, JR., R. L. FLURRY AND R. A. DAY, JR., *J. Electrochem. Soc.*, 111 (1964) 328.
7 W. F. BEECH, *J. Chem. Soc.*, (1954) 1297.
8 M. KALOUSEK, *Collection Czech. Chem. Commun.*, 13 (1948) 105.
9 M. RÁLEK AND L. NOVÁK, *Collection Czech. Chem. Commun.*, 21 (1956) 248.
10 L. HOLLECK AND H. MARSEN, *Z. Elektrochem.*, 57 (1953) 944.
11 M. NISHIYAMA, M. MARUYAMA AND H. HAMAGUCHI, *Bull. Chem. Soc. Japan*, 37 (1964) 616.
12 J. HEYROVSKÝ AND J. KŮTA, *Principles of Polarography*, Academic Press, London, 1965, p. 358.
13 O. FISCHER, private communication.

FURTHER EXAMPLES OF *para*- AND *ortho*-DISUBSTITUTED BENZENE DERIVATIVES REDUCED IN A REVERSIBLE TWO-ELECTRON STEP INTO PRODUCTS OF LIMITED STABILITY

P. ZUMAN*, O. MANOUŠEK AND S. K. VIG**

J. Heyrovský Institute of Polarography, Czechoslovak Academy of Sciences, Prague (Czechoslovakia)

(Received February 24th, 1968)

In the course of the study of benzene derivatives bearing two electronegative groups on the ring, it was possible to demonstrate[1] that the behaviour of p-diacetylbenzene is different from that of most of the other substances studied. The change in the wave-height of the more negative wave in the region where reduction of the $C_6H_5COCH_2$–grouping takes place, and the change of the anodic wave of the electrolysis product obtained by rectangular voltage polarization, indicated formation of an intermediate of biradical character[†] with limited stability. In principle, three schemes were considered, involving hydration, protonation of the parent compound, and acid–base catalysed cleavage of the biradical intermediate. A chronopotentiometric investigation[2] confirmed that Scheme I gives the best explanation for all the experimental results. pH-Dependence of experimentally found rate constants, k_a, k_b, obtained by chronopotentiometry, is analogous to the dependence of the height of the second reduction step of p-diacetylbenzene and of the anodic wave of the primary electrolysis product obtained with the commutator. The reaction limiting the height of the second reduction step and of the anodic wave of the product, is hence the acid–base catalysed cleavage of the biradical. The specific rate constants of this reaction obey a Brønsted relation[2].

For a more detailed understanding of the mechanism of the chemical reaction responsible for the transformation of the primary electrode product, it was of importance to determine the limits of applicability of the suggested scheme. It was necessary to show how large changes may be realised in the structure of the molecule without changing the course of the electrode process. Furthermore, it was of interest to compare, at least qualitatively, how changes in structure of the parent compound (within the scope of "permitted" changes) affect the stability and reactivity of the primary product.

The result of some investigations along these lines with compounds bearing a carbonyl, nitro- or cyano-group in *para*- or *ortho*-positions are discussed in this paper. The main technique used for the characterization of the stability of the biradical intermediate in this investigation was a rectangular voltage polarization in the commutator method by KALOUSEK[3].

---

* Present address: Department of Chemistry, University of Birmingham, Great Britain.
** Present address: St. Stephen's College, Delhi, India.
† In the text, the term biradical is used for the primary product of the two-electron reduction, even though we have no proof for its existence in the biradical or quinoid structure shown in Scheme I.

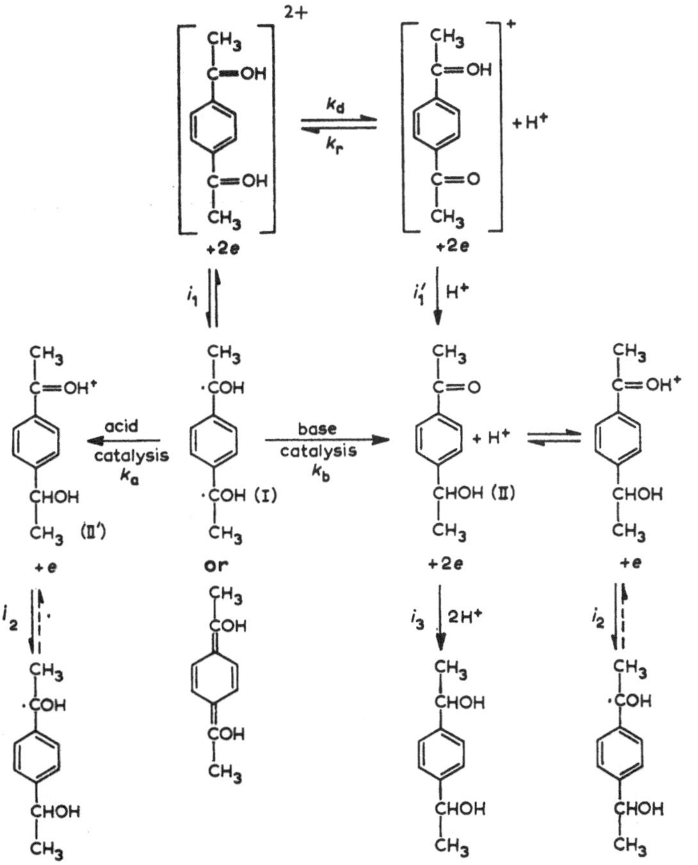

Scheme I

EXPERIMENTAL

A pen-recording polarograph LP 60 (Laboratorní přístroje, Praha) was used in most experiments. Polarographic electrolysis was carried out using a Kalousek vessel with a saturated calomel electrode as reference electrode and a dropping mercury electrode with constants, $m = 2.1$ mg/sec, $t_1 = 2.8$ sec at $h = 66$ cm in $1\ M$ KCl at 0.0 V. o-Phthaldialdehyde, terephthaldialdehyde, o-phthaldinitrile and p-dinitrobenzene were commercial products; terephthalophenone, 4,4'-diacetyldiphenylmethane, terephthaldinitrile and isophthaldinitrile were kindly donated by Dr. O. EXNER.

0.01 M Ethanolic stock solutions were prepared from the electroactive substance. Freshly prepared stock solutions were added to the supporting electrolyte to ensure a $2 \cdot 10^{-4}\ M$ solution. The final solution usually contained about 5% ethanol; only with terephthalophenone was it necessary to use solutions containing 50% ethanol.

The rectangular voltage polarization was followed using a commutator constructed according to the principle suggested by KALOUSEK[3,4]. In the scheme used, the auxiliary potential was always kept constant at the potential corresponding to the limiting current, usually of the first wave. Capacity current was recorded in the given supporting electrolyte and subtracted from the current–voltage curve recorded in the presence of the electroactive component.

Cyclic voltametric i–E curves were recorded using a hanging mercury drop electrode in conjunction with the polarograph OH 102 (Radelkis) using a scan rate of 50 mV/sec.

RESULTS

Carbonyl compounds

Terephthalophenone (I), 4,4′-diacetyldiphenylmethane (II), terephthalalde-hyde (III) and *o*-phthalaldehyde (IV) were investigated:

(I) (II) (III) (IV) (V)

The polarographic behaviour of terephthalophenone (I) is analogous to that of *p*-diacetylbenzene (V)[1]. In acidic media, the first two-electron wave corresponds to a reversible process; this was shown with the commutator by the formation of an anodic wave on the curve obtained (Fig. 1) at the same potential as that of the original cathodic wave. In comparison with *p*-diacetylbenzene, the decrease of height of the anodic wave with increase in pH is shifted with (I) towards higher pH-values. With increasing acidity, no decrease in the height of the anodic wave obtained with

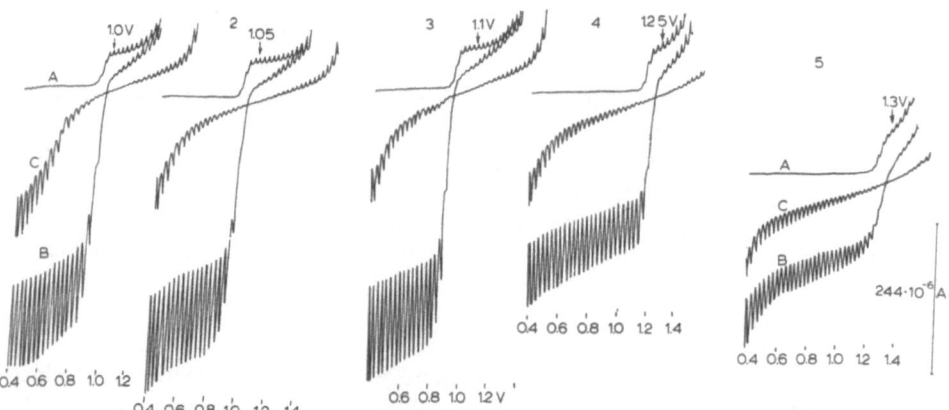

Fig. 1. Polarographic curves of terephthalophenone using Kalousek commutator method. $2 \cdot 10^{-4}$ M depolarizer, 50% ethanol. (A), original i–E curve; (B), curve obtained with commutator; (C), curve obtained with commutator for the supporting electrolyte: (1), 2.5 M; (2), 0.5 M; (3), 0.05 M H$_2$SO$_4$; (4), acetate buffer pH 4.9; (5), phosphate buffer pH 6.9. Mercurous sulphate reference electrode; arrows on original i–E curves indicate value of the auxiliary potential; frequency of commutation 6 cycles/sec.

the commutator was observed and the wave shows practically the same wave-height in 0.1 N and 5 N H$_2$SO$_4$ (Fig. 1). Both the increase in size of the pH-region in which the anodic wave is observed, and the extension of the range in which the height of the wave reaches its limiting value, indicates greater stability of the biradical formed from terephthalophenone (I) than that from p-diacetylbenzene. This was also confirmed by polarization with triangular sweeps. Whereas for p-diacetylbenzene (V) it was possible to obtain anodic peaks only when using oscilloscopic techniques and scanning rates of the order 1 V/sec, for terephthalophenone (I) it was possible to obtain a reversible anodic peak with a hanging mercury drop electrode with a scanning rate of 0.05 V/sec (Fig. 2).

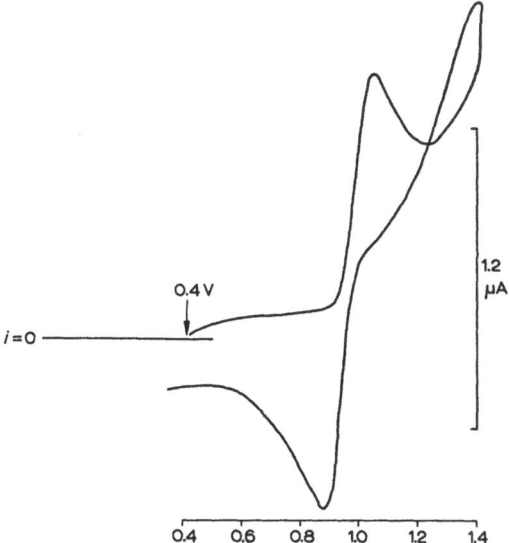

Fig. 2. Current–voltage curve of terephthalophenone using hanging mercury drop electrode. $2 \cdot 10^{-4}$ M depolarizer, 0.25 M H$_2$SO$_4$, 50% ethanol, triangular voltage sweep, 50 mV/sec, hanging mercury drop "Metrohm", polarograph OH 102.

As expected, 4,4'-diacetyldiphenylmethane (II) does not give a reduction product of limited stability in acidic media, which would give an anodic wave using the commutator method.

Terephthaldialdehyde (III) at pH < 3 behaves similarly to p-diacetylbenzene. A two-electron reversible cathodic wave was followed by a more negative wave, the height of which approached that of a one-electron process. Simultaneously, an anodic wave was observed on the curve obtained with the commutator, the height of which changed with pH in a bell-shaped curve (Fig. 3). The discontinuity at that part of the curve corresponding to a change from solutions of a strong acid to buffered solutions, indicates that the reaction causing the cleavage of the biradical is generally acid–base catalysed. The range in which the biradical of (III) gives an anodic wave, is more narrow than that observed for (V) and does not show a pH-independent section. This indicates that the biradical formed in the reduction of terephthalaldehyde (III) has a shorter life-time than that formed from p-diacetylbenzene (V).

The waves of o-phthalaldehyde (IV) are strongly affected by hydration and do not show any measurable anodic wave with the commutator.

Nitro compounds

p-Dinitrobenzene is known[5] to give a reversible wave in water-containing dimethylformamide media. We were able to show that in aqueous acidic media a product of limited stability is also formed, which gives an anodic wave when the commutator is used. The anodic wave is small and observed over a narrow pH-range, but the shape of the pH-dependence shows the general pattern, and the abrupt change accompanying transition from strong acid solution to buffer solutions indicates acid–base catalysis of the cleavage reaction. The product formed in this process is hence considerably less stable than that obtained with terephthalaldehyde (III).

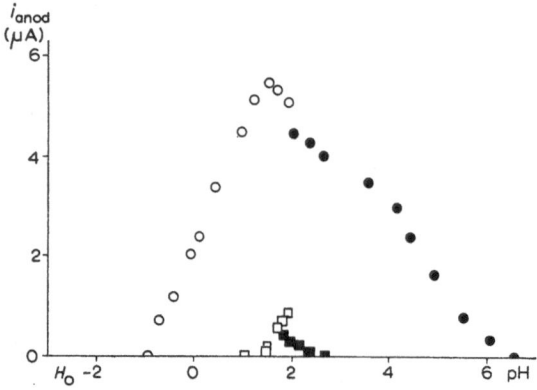

Fig. 3. pH-Dependence of heights of anodic waves obtained with Kalousek commutator method for terephthalaldehyde and p-dinitrobenzene. $2 \cdot 10^{-4} M$ depolarizer, 5% ethanol, auxiliary potential at limiting current of original reduction wave. (Circles), terephthaldialdehyde; (squares), p-dinitrobenzene; (white points), sulphuric acid solns.; (black points), Britton–Robinson buffers.

Dinitriles

o-Phthaldinitrile (VI), isophthaldinitrile (VII) and terephthaldinitrile (VIII) were shown to behave differently from p-diacetylbenzene (V). At lower pH-values, no anodic wave is observed for these substances using the commutator method.

(VI)   (VII)   (VIII)

At high pH-values, for compounds (VI) and (VIII) the formation of the anodic cyanide wave was observed, and is discussed elsewhere[6]. At pH > 5, in addition to this cyanide wave, a small anodic wave was observed on the curves obtained with the commutator for o- (VI) and terephthaldinitrile (VIII). The height of this anodic wave, indicating a reversible process, was higher for the *ortho*-derivative (VI) than for the *para*-derivative (VIII). No such effect was observed for isophthaldinitrile (VII) (Fig. 4).

The height of the anodic waves obtained with the commutator for dinitriles (VI) and (VIII) increases with increasing pH (Fig. 5) and reaches its limiting value at about pH 8. At higher pH-values, up to pH 12, the height of this wave remains unchanged, provided hydrolysis does not take place.

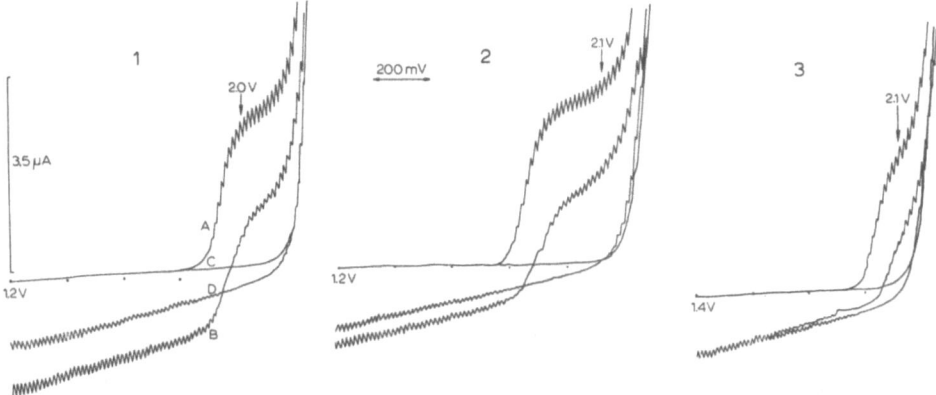

Fig. 4. Comparison of anodic waves of benzodinitriles obtained with Kalousek commutator method. Britton–Robinson buffer pH 9.9. (A) and (C), original *i–E* curves; (B) and (D), curves obtained with commutator; (A) and (B), $5 \cdot 10^{-4} M$ depolarizer; (C) and (D), supporting electrolyte. (1), *o*-phthaldinitrile; (2), terephthaldinitrile; (3), isophthaldinitrile. Mercurous sulphate electrode; arrows on original *i–E* curves indicate value of auxiliary potential; frequency of commutation 6 cycles/sec.

The potential of the anodic wave on the curve obtained with the commutator is identical with the half-wave potential of the original reduction wave. Therefore, only one anodic–cathodic wave is observed on the curve obtained with rectangular voltage polarization (Fig. 5). The ratio of the height of the anodic wave obtained with the commutator and the original cathodic wave shows no dependence on dinitrile concentration below $1 \cdot 10^{-3} M$.

The height of the anodic wave corresponding to an oxidation of the product of electrolysis formed at the potential chosen on the auxiliary potentiometer, increased with increasing negative auxiliary potential. The height reaches a limiting value when the auxiliary potential reaches the value corresponding to the limiting current of the original d.c. curve. The ratio of the height of the anodic wave obtained with the commutator, $(i_a)_{com}$, and the original cathodic wave, $(i_d)_{orig}$, increases with decreasing temperature. This indicates that the oxidizable species is more stable at lower temperatures.

T (°C)	6°	10°	15°	20°	25°
$\dfrac{(i_a)_{com}}{(i_d)_{orig}}$	0.28	0.27₅	0.24	0.23	0.21

The ratio of the two wave-heights also increases with increasing ethanol concentration:

$[C_2H_5OH]$ $(\%)$	5	10	15	20
$(i_a)_{com}$ (μA)	0.55	0.82	1.00	0.98
$(i_d)_{orig}$ (μA)	3.37	3.12	2.92	2.42
$(i_a)_{com}/(i_d)_{orig}$	0.16	0.26	0.34	0.40

With increasing ethanol concentration, not only does the ratio of wave-heights increase (indicating that the intermediate is more stable at higher ethanol concentrations), but the absolute value of the height of the original cathodic wave, $(i_d)_{orig}$, decreases, whereas that of the anodic wave increases.

Controlled-potential electrolysis at the potential of the limiting current of nitriles (VI)–(VIII) gave only cyanide ions. No reversible anodic wave was observed after prolonged electrolysis.

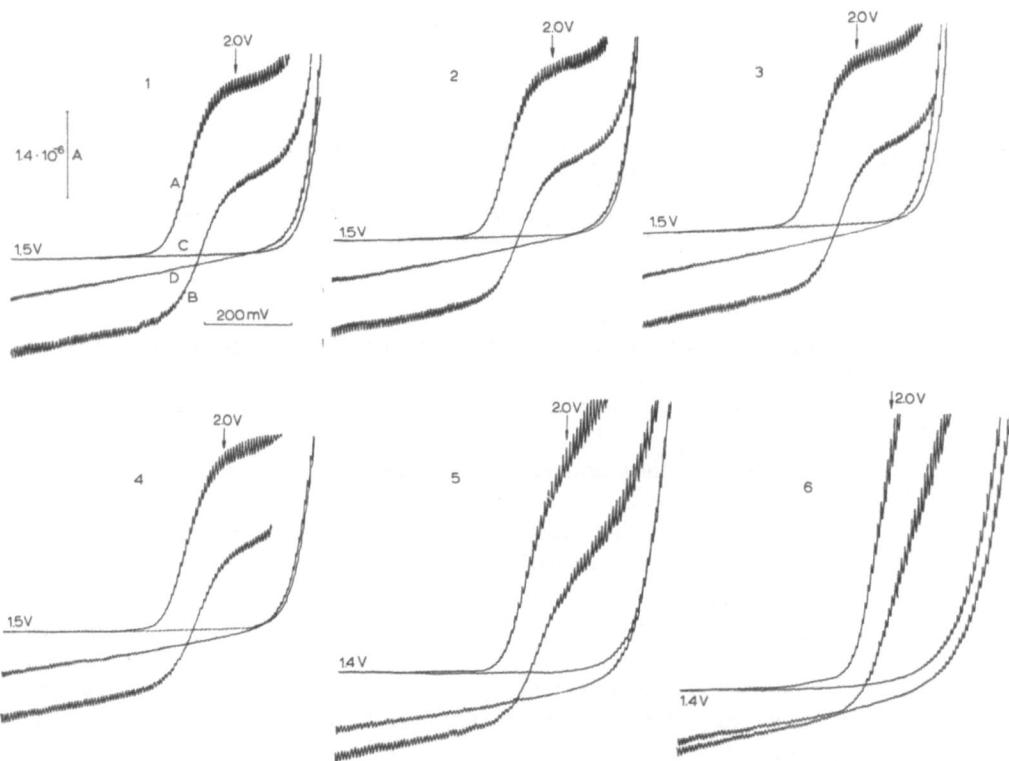

Fig. 5. pH-Dependence of anodic waves obtained with Kalousek commutator method for o-phthaldinitrile. $5 \cdot 10^{-4}$ M depolarizer, Britton–Robinson buffers, pH: (1), 12.7; (2), 11.2; (3), 10.0; (4), 8.9; (5), 7.9; (6), 7.1. (A) and (C), original i–E curves; (B) and (D), curves obtained with commutator; (A) and (B), solns. o-phthaldinitrile; (C) and (D), supporting electrolyte. (1)–(4), starting at -1.5 V; (5) and (6), at -1.4 V; mercurous sulphate electrode, arrows on original i–E curves indicate value of auxiliary potential; frequency of commutation 6 cycles/sec; regulated drop-time, $t_1 = 2.2$ sec; $m = 2.8$ mg/sec.

DISCUSSION

The type of behaviour corresponding to Scheme I, in which the biprotonated form is reversibly reduced in a two-electron step to a biradical, has so far been found only for p-dicarbonyl compounds, (I), (III) and (V). Neither is the scheme followed when one or both of the carbonyl groups are transformed into ketimino groups in semicarbazones, nor for systems in which one or both of the carbonyl groups are exchanged for cyano groups. Both carbonyl groups need to be ketonic or aldehydic groups, since it was found[1] that the reduction according to Scheme I does not take place when one or both of the carbonyl groups are a part of a carboxyl grouping*.

The stabilization of the biradical is attributed to the contribution of the canonical form with quinoid structure. This is supported by the fact that in 4,4'-diacetyldiphenylmethane, where the conjugation interaction of the two acetyl groups in the aromatic system is blocked and where a formation of a two-electron reduction product with a quinoid structure is impossible, the formation of the product giving a reversible anodic wave was not observed. The life-time and the pH-range in which the biradical is stable increases in compounds of the type, $RCO-C_6H_4-COR$ in the sequence, $R = H < CH_3 < C_6H_5$. This is in accordance with the expected polar and conjugation effects of these groups. For aldehydes, hydration can be a complicating factor.

Reduction of o-phthaldinitrile (VI) and terephthaldinitrile (VIII) differ from that of isophthaldinitrile (VII). This is in accordance with the observation[7] of the reduction of these compounds in aprotic media, where compounds (VI) and (VIII) give stable radical anions and two one-electron reduction steps, whereas (VII) gives only one one-electron step and a low yield of a radical anion, which is attributed to its cleavage in side reactions.

As the half-wave potentials of dinitriles (VI) and (VIII) are pH-independent, it can be deduced that no proton transfer occurs, prior to the first electron uptake. It can thus be assumed that a radical anion is formed in the first electron uptake and that this is oxidised in the anodic wave. Even when the formation of the radical dianion cannot be excluded, the sufficient stability of the observed species to give the anodic wave with the commutator suggests that the dianion formation is less probable in view of the findings on reactivity of dianions reported by REINMUTH[7]. The reaction causing the cleavage of the oxidizable species is independent of pH and dinitrile concentration, but its rate decreases with decreasing temperature and increasing ethanol content. It seems that the radical anion, which is reported for compound (VI) to be stable in dimethylformamide media for several hours, has a life-time of the order of seconds in aqueous media. The stability of this radical anion is greater for compound (VI) than for the *para*-derivative (VIII), but undergoes consecutive chemical and electrochemical reactions to produce benzonitrile and cyanide ions. The process seems to be analogous to that observed[7] at very negative potentials (-2.8 V) in dimethylformamide solutions. Whereas in aqueous solutions benzonitrile is the final product (at -1.8 V), in dimethylformamide benzonitrile is further reduced.

The products of reductions of o- and p-dinitriles in aqueous media, therefore

---

* Experimental evidence for the p-dinitro compound is insufficient to enable us to draw any conclusions as to the course of the electrode process concerned.

show limited stability, but the process is different from that observed[1] for *para*-dicarbonyl compounds.

SUMMARY

Only those benzene derivatives (bearing two electronegative groups) with two carbonyl groups in the *para*-position were found to be reduced in acid media in one two-electron reversible step to form a biradical or a quinoid product of limited stability. The life-time of the primary product of $RCOC_6H_4COR$ increased in the sequence: $R=H < CH_3 < C_6H_5$. Exchange of one of the carbonyl groups for an oxime- or carboxy-group prevents this type of process. *p*-Dinitrobenzene was shown by rectangular voltage polarization to give a rather unstable reduction product in acid media. *Ortho*- and *para*-benzenedinitriles give radical anions of limited stability in aqueous alkaline media.

REFERENCES

1 Yu. KARGIN, O. MANOUŠEK AND P. ZUMAN, *J. Electroanal. Chem.*, 12 (1966) 443.
2 O. FISCHER, L. KIŠOVÁ AND J. ŠTĚPÁNEK, *J. Electroanal. Chem.*, 17 (1968) 233.
3 M. KALOUSEK, *Collection Czech. Chem. Commun.*, 13 (1948) 105.
4 M. RÁLEK AND L. NOVÁK, *Collection Czech. Chem. Commun.*, 21 (1956) 248.
5 J. Q. CHAMBERS III AND R. N. ADAMS, *J. Electroanal. Chem.*, 9 (1965) 400.
6 O. MANOUŠEK, P. ZUMAN AND O. EXNER, *Collection Czech. Chem. Commun.*, in the press.
7 P. H. RIEGER, I. BERNAL, W. H. REINMUTH AND G. K. FRAENKEL, *J. Am. Chem. Soc.*, 85 (1963) 683.

Part IV

Polarographic Reduction and Some Reactions of β-Substituted Carbonyl Compounds

11. P. Zuman, and V. Horák: Fission of activated carbon-nitrogen and carbon-sulphur bonds. II. Polarographic reduction of β-amino ketones. *Collect. Czechoslov. Chem. Commun.* **27**, 187 (1962).
12. P. Čársky, P. Zuman, and V. Horák: Fission of activated carbon-nitrogen and carbon-sulphur bonds. V. Polarographic study of elimination of β-morpholino propiophenone. *Collect. Czechoslov. Chem. Commun.* **29**, 3044 (1964).
13. I. Šestáková, V. Horák, and P. Zuman: Fission of activated carbon-nitrogen and carbon-sulphur bonds. IX. Polarographic study of addition of primary amines to phenyl vinyl ketone. *Collect. Czechoslov. Chem. Commun.* **31**, 3889 (1966).
14. I. Šestáková, J. Pecka, and P. Zuman: Polarographic reduction of aldehydes and ketones. V. β-Ketosulphides. *Collect. Czechoslov. Chem. Commun.* **33**, 3227 (1968).
15. I. Šestáková, P. Zuman, and V. Horák: Fission of activated carbon-nitrogen and carbon-sulphur bonds. VII. Elimination of β-piperidinoethyl phenyl sulphone methoiodide and reaction of phenyl vinyl sulphone with hydroxyl ions. *Collect. Czechoslov. Chem. Commun.* **31**, 827 (1966).
16. N. Kucharczyk, M. Adamovský, V. Horák, and P. Zuman: Slowly established acid-base equilibrium in organic polarography: Tautometric changes between 3-thianaphthenone and 3-hydroxythianaphthene. *J. Electroanal. Chem.* **10**, 503 (1965).

In β-aminoketones, protonation takes place preferably on the amino group, and the carbonyl group is reduced without pre-protonation.[11] Elimination of Mannich bases of the β-amino-ketone type is complicated by two acid-base equilibria: protonation of the Mannich base and of the amine formed. Only unprotonated Mannich base reacts with the solvent to eliminate amine.[12] Additions of alkylamines to phenyl vinyl ketone show a pronounced steric effect indicating the possibility of a cyclic transition state.[13]

In β-ketosulphides the protonation occurs preferably at the carbonyl group, and therefore the reduction[14] resembles that of simple arylalkyl ketones rather than of β-aminoketones. For comparison, β-amino sulphones were investigated[15] and the addition of hydroxyl ions to phenyl vinyl sulphone followed.

Formation of carbanion-enolate from carbonyl compounds of the type $R.COCH_2.X$ (where X = COR, COOR, CN or SR in an aromatic system) can be followed from polarographic curves. For 3-thianaphthenone[16] and ethyl benzoylacetate, the acid-base equilibrium is relatively slow and only dissociation constants can be determined. For R = CN, or $COCH_3$, the rate constant for the recombination of the anion and hydrogen ion can be calculated.

119

FISSION OF ACTIVATED CARBON–NITROGEN
AND CARBON–SULPHUR BONDS. II.*
POLAROGRAPHIC REDUCTION OF β-AMINO KETONES

P. Zuman and V. Horák

*Polarographic Institute, Czechoslovak Academy of Science and Department of Organic Chemistry,
Charles University, Prague*

Received April 14th, 1961

Polarographic reduction of β-amino ketones proceeds on the carbonyl group. The protonized form is reduced with an uptake of two electrons to the corresponding alcohol. The influence of structure in the ketones of the type $R-COCH_2CH_2NR^1R^2R^{3(+)}$ is discussed. At higher pH-values an elimination reaction occurs. The α,β-unsaturated ketone formed in this reaction is manifested by a separate wave on polarographic curves.

In the preceding part of this series[1] it was shown that in α-amino ketones and related substances the reductive cleavage of the $C-N^{(+)}$ bond is preferred to the reduction of the C=O bond. In this part the polarographic behaviour of some β-amino ketones containing the grouping $CO-CH_2CH_2-NR^1R^2R^{3(+)}$ was studied. To detect the course of the polarographic reduction as well as the effect of the substituent $-NR^1R^2R^{3(+)}$ (or $CH_2CH_2NR^1R^2R^{3(+)}$ respectively) simple model substances were chosen. β-Piperidinopropiophenone represented substances in which the carbonyl group is activated by an adjacent phenyl group. β-Piperidino ethyl methyl ketone is a typical example of aliphatic ketones and finally 2-piperidinomethylcyclopentanone represents cyclanone derivatives. The first two substances were studied both in the form of methoiodides and as hydrochlorides.

Experimental

Polarographic equipment, pH-meter, controlled potential electrolysis, as well as the techniques of measurement were the same as in the preceding paper[1]. The capillary used had the following constants: out-flow velocity $m = 2\cdot1$ mg/s and drop-time $t_1 = 3\cdot7$ s at the potential of s. c. e. at mercury pressure $h = 65$ cm. All the values of half-wave potentials are expressed versus the saturated calomel electrode (s. c. e.) and measured using Tl^+ ($-0\cdot445$ V) as an internal standard.

β-*Piperidinopropiophenone Methoiodide* (I). Whereas in methanolic solution the reaction of β-piperidinopropiophenone with methyliodide occurred under elimination and resulted mainly in the formation of a methoiodide of N-methylpiperidine, when performed in anhydrous ether

---

* Part I.: This Journal 26, 176 (1961).

in the presence of excess of methyliodide the reaction yielded β-piperidinopropiophenone (*I*) as the main product. The purification of *I* is difficult. Best results were obtained by repeated crystal-lization from acetone, where the substance is dissolved after addition of a small amount of water. *I* is separated in crystalline from after addition of a minimum amount of ether. M. p. 148—155° (decomp.). For $C_{15}H_{22}INO$ (359·24) calculated: 50·6% C, 6·22% H, 35·62% I, 3·93% N; found: 50·56% C, 6·45% H, 35·3% I, 3·81% N.

Compound *I* is unstable in aqueous solutions. After a few minutes of boiling it decomposes totally under formation of N-methylpiperidine and phenyl vinyl ketone. This substance was isolated from the steam distillate and identified as diphenylpyrazoline m. p. 152° (lit.[2] 152°).

β-Piperidinopropiophenone[3] (*II*) (m. p. 193°), β-piperidino ethyl methyl ketone methoiodide[4] (*III*) and hydrochloride[4] (*IV*) and 2-piperidinomethylcyclopentanone hydrochloride[5] (*V*) were prepared by known methods. Priopiophenone (*VI*) (B. D. H.) was freshly distilled.

$$R^1-COCH_2CH_2\overset{(+)}{N}\langle\rangle \overset{(-)}{X}$$

$$\overset{|}{R^2}$$

I, $R^1 = C_6H_5$, $R^2 = CH_3$, X = I
II, $R^1 = C_6H_5$, $R^2 = H$, X = Cl(I)
III, $R^1 = CH_3$, $R^2 = CH_3$, X = I
IV, $R^1 = CH_3$, $R^2 = H$, X = Cl

V

0·01M stock solutions were prepared by dissolving a weighed amount of the substances in cold water. For β-piperidinopropiophenone methoiodide (*I*) and the other methoiodide *III* fresh stock solution were prepared daily and acidified with 0·01M-HCl, since in unacidified aqueous solutions an elimination reaction occurred. The aqueous stock solutions of the hydrochlorides *II*, *IV* and *V* were more stable and it was sufficient to prepare fresh solutions every third or fourth day. The buffer solutions used (Britton-Robinson universal buffer, acetate and phosphate buffers) were prepared from analytical reagent grade chemicals.

Results and Discussion

β-Piperidinopropiophenone Hydrochloride (*II*)

Substance *II* gives in the whole pH-range studied one two-electron wave (Fig. 1) corresponding to the reduction of the carbonyl group. At concentrations lower than about $1 . 10^{-4}$M an ill-separated adsorption pre-wave appears at more positive poten-tials. The total height of the limiting current is diffusion controlled: it is linearly pro-portional to the concentration of the depolarizer as well as to the square root of height of the mercury head *h*. The half-wave potential is shifted towards more negative values with increasing pH-values according to Fig. 2.

In strongly acidic solution (pH 2 and lower) another indistinct wave is observed at more negative potentials, probably due to the catalytic evolution of hydrogen. At pH higher than about 10 two processes occur: a new double-wave appears (curve *11*, Fig. 1) and the total height decreases. The more positive waves increases at the expenses of the original wave with time. One of these waves corresponds to the product of the elimination reaction, *i. e.* to the phenyl vinyl ketone together with

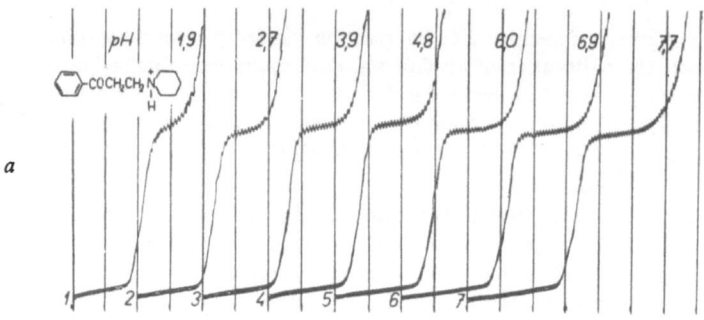

a

Fig. 1

Dependence of Wave for β-Piperidinopropio-
phenone Hydrochloride (*II*) on pH
Britton-Robinson buffers, $2 \cdot 10^{-4}$M depo-
larizer. pH-values given on the polarogram.
Curves *1—7* starting at $-0 \cdot 6$ V, *8—11* at
$-0 \cdot 8$ V, 200 mV/absc., s. c. e., sens. 1 : 15.

b

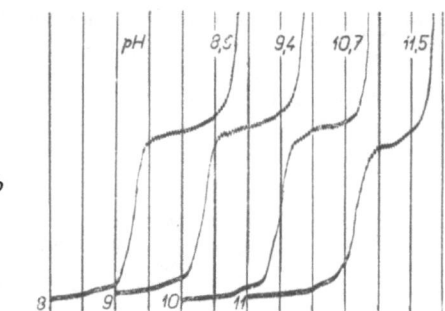

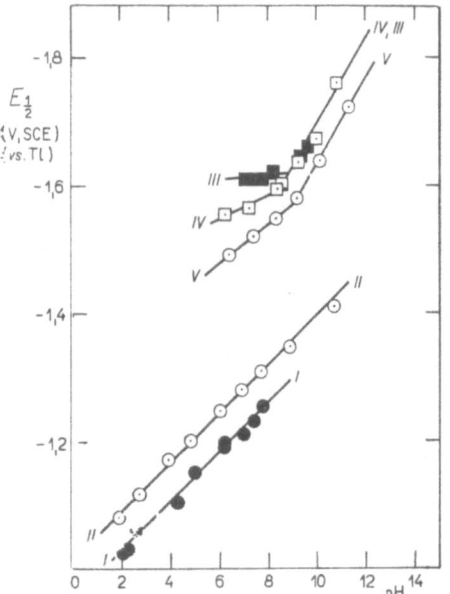

Fig. 2

Dependence of the Half-Wave Potentials of
β-Amino Ketones on pH in Britton-Robinson Buffers

Substances *I — V* as given in the Experimental.

a wave at more negative potentials than that
for the original amino ketone. When the
change of the curves with time was followed at a given pH, practically no decrease
of the total wave-height was observed (actually a slight increase was recorded due to
the higher electron uptake by the phenyl
vinyl ketone). The decrease of the total
wave height is thus not due to the elimination reaction. When the change of polarographic curves was followed at 0°C where
the elimination reaction is so slow that no
waves for phenyl vinyl ketone could be
observed on polarographic curves, the decrease of the total wave-height has been
observed again. This decrease of the total
wave-height, which possesses a form of a
part of a dissociation curve, was thus attributed to the decrease of rate of protonisation of *II*. In the two electron-wave thus the

protonized form of *II* is reduced, both present in the solution and formed by recombination at the dropping electrode. Due to the complications caused by the elimination reaction, a quantitative treatment of the recombination reaction was abandoned.

β-Piperidinopropiophenone Methoiodide (*I*)

In acid solutions at pH lower than about 5 one single two electron wave, similar to that for *I*, was observed (Fig. 3). At higher pH-values two new waves A, B grow at the expense of the original wave for *I* with time (Fig. 4). Both these waves — one at more positive (A) and one at more negative potentials (B) than that of *I* — correspond to the phenyl vinyl ketone formed in the elimination reaction.

The waves A and B were identified by preparation of the elimination product at pH 9. The product was transferred into four different buffer solutions. The waves obtained in this way were found identical in form and half-wave potentials with that of synthetic phenyl vinyl ketone in corresponding buffers.

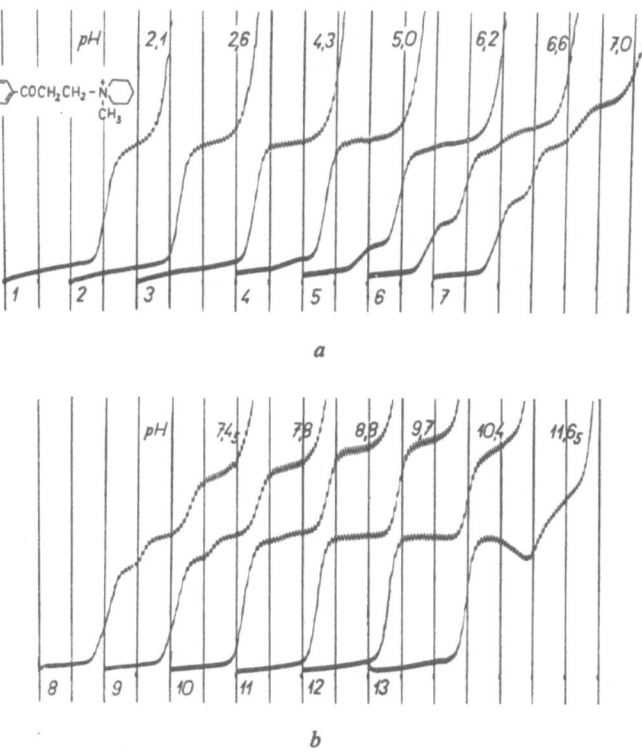

Fig. 3

Dependence of Waves for β-Piperidinopropiophenone Methoiodide (*I*) on pH

Britton-Robinson buffers, 2 . 10^{-4}M depolarizer. pH-values given on the polarogram. The curves were recorded two minutes after mixing the stock solution with the deaerated buffer solution. Curves *1—3* starting at −0·4 V, curves *4—13* at −0·6 V, 200 mV/absc., S.C.E., sens. 1 : 20.

When recorded after a constant time-period after mixing the stock solution of *I* with the appropriate buffer, curves given in Fig. 3 were obtained. The curves in Fig. 3 show the participation of the elimination reaction, the rate of which increases with increasing pH-value. The dependence of the yield of the elimination reaction on pH-value, given by the increase of the first wave of phenyl vinyl ketone or by the decrease of the wave for *I*, at pH about 7 is S-shaped.

The product of the elimination reaction, *phenyl vinyl ketone* is reduced at the dropping mercury in one or (at pH higher than about 6·0) in two two-electron steps. This increase in the electron-uptake offers an explanation for the increase in the total limiting current observed in Fig. 3. The height of the more positive wave of *phenyl vinyl ketone*, accompanied by an adsorption pre-wave, is practically pH-independent. The more negative wave cannot be observed at lower pH-values than about 6·0. With increasing pH its height increases in an S-shaped form. The independence of the mercury pressure shows that the current is determined by the rate of chemical reaction. The half-wave potentials of this more negative wave are similar to those for propiophenone. The decrease of the height of this wave at pH above 10 follows the same pattern as the decrease of the wave for propiophenone. This behaviour can be explained by a chemical change of the product of the first electrode process preceding the second one. The possible mechanisms are discussed elsewhere[8].

Similar changes of a product of an electrode reaction, occuring before or between two successive electron-transfers has been observed recently for other phenacyl derivatives, like benzil[6,7], α-amino ketones[1,8], triethylaminoacrolein[1], phenacyl alcohols and their derivatives[9], phenacyl sulfides[9] etc.

The dependence of the half-wave potential of β-*piperidinopropiophenone* methoiodide (*I*) on pH is given in Fig. 2 for the pH-range, where the elimination did not proceed completely. At 0°C this range could be extended over more than one pH-unit towards higher pH-values.

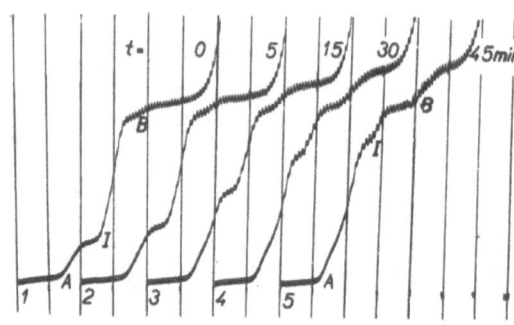

Fig. 4

Time-Dependence of the Wave for β-Piperidinopropiophenone at pH 6·23

Britton-Robinson buffer, 2 . 10^{-4}M depolarizer, time at which the recording was started after preparation at the reaction mixture is given on the polarogram. Waves for phenyl vinyl ketone are formed. Curves starting at −0·6 V, 200 mV/absc., 200 mV/min., s. c. e. sens. 1 : 20.

β-Piperidino Ethyl Methyl Ketone Methoiodide (*III*)
and Hydrochloride (*IV*)

Hydrochloride *IV* is reduced in one two-electron wave in the pH-range from pH 6 to 10 (Fig. 5). At lower pH-values this wave is obscured by hydrogen evolution from the buffer solution. At pH above 10·5 the height of this wave decreases in the form of a dissociation curve. The decreasing wave is shifted towards more negative potentials and therefore ill-developed. The two-electron wave is diffusion controlled, the small wave at pH 11 possesses a kinetic character. Polarographic reduction is thus preceded by a recombination reaction and the reduced form is more protonated than the form predominating in the solution. The shifts of half-wave potentials are given in Fig. 2.

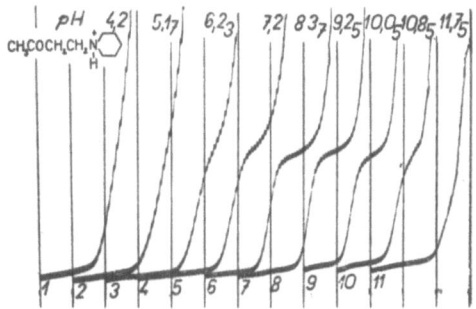

Fig. 5
Dependence of Waves for β-Piperidino Ethyl Methyl Ketone Hydrochloride (*IV*) on pH Britton-Robinson buffers, 2 . 10^{-4}M depolarizer. pH-values given on the polarogram. Curves *1 — 4* starting at −1·0 V, *5* and *6* at −1·2 V, *7—9* at −1·4 V, 200 mV/absc., s. c. e., sens. 1 : 20.

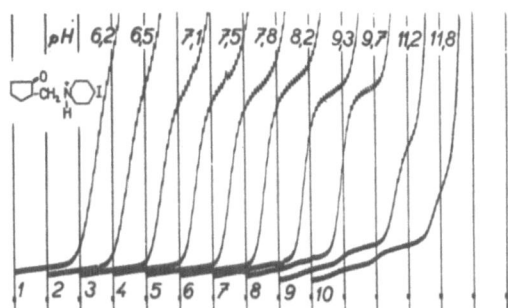

Fig. 6
Dependence of Waves for β-Piperidino Ethyl Methyl Ketone Methoiodide (*III*) on pH Britton-Robinson buffers, 1 . 10^{-4}M depolarizer. pH-values given on the polarogram. The curves were recorded two minutes after mixing the stock solution with the deaerated buffer solution. Curves starting at −1·0 V, 200 mV/absc., s. c. e., *h* = 80 cm, sens. 1 : 10.

The curves of the methoiodide *III* are complicated by the homogeneous elimination reaction (Fig. 6). The two electron wave for *III* can be observed only between pH 7·0 and 10·0 shortly after preparation of the solution. At lower pH-values this wave is obscured by the hydrogen evolution. In this whole pH-range the curves are complicated by the elimination reaction. The methyl vinyl ketone, formed in this reaction, is reduced at potentials some 300 mV more positive than piperidino methyl ethyl ketone. The wave *V* for methyl vinyl ketone increases in time at the expense of the wave for *III*. When registered in a given time after mixing the fresh stock solution of *III* with the appropriate buffer, the wave for methyl vinyl ketone increases with pH in the form of a dissociation curve with $pK \doteq 9$.

The methyl vinyl ketone is reduced[10] in two one-electron waves at lower pH values and in one two-electron-wave at pH above 7·0. Contrary to phenyl vinyl ketone the corresponding saturated ketone is non-reducible in the potential range available. The observed one-step reduction can alone be no proof either for the reduction of the C=C or for the reduction of the C=O bond. – At pH above 10·5 the rate of the elimination reaction is so high that after addition of *III* to a buffer only waves *V* of methyl vinyl ketone are registered (Fig. 6, curves *9, 10*). The shift of the half-wave potentials for *III* is given in Fig. 2.

2-Piperidinomethylcyclopentanone Hydrochloride (*V*)

The polarographic curves of *V* resemble closely those for *IV*. Two-electron diffusion controlled waves are observed at pH above 6 (at lower pH-values the hydrogen evolution interferes with the development of the waves) (Fig. 7). At pH above about 10 the limiting current decreases in the form of a dissociation curve and becomes rate controlled. At pH above 9 the elimination reaction causes an increase of a wave at more positive potentials (Fig. 7), corresponding to the reduction of 2-methylene-

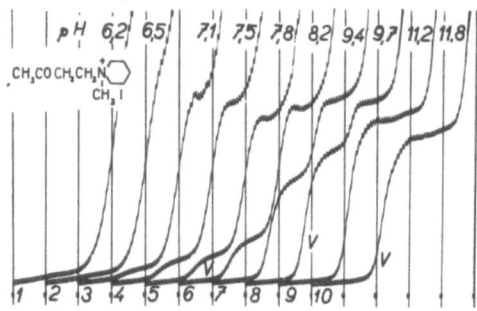

Fig. 7
Dependence of Waves for 2-Piperidinomethylcyclopentanone Hydrochloride (*V*) on pH
Britton-Robinson buffers, $1 . 10^{-4}$M depolarizer, at 20°C. pH-values given on the polarogram. Curves were recorded two minutes after mixing the stock solution with deaerated buffer solution. Curves starting at −1·0 V, 200 mV/absc., s. c. e., $h = 80$ cm, sens. 1 : 15.

cyclopentanone. This wave increases with time at the expense of the wave for V (Fig. 8). At such pH-values, where the rate of the recombination reaction is not high enough to form polarographically active substances sufficiently rapidly (cf. Fig. 7, curve *10*), the unsaturated ketone is formed from the polarographically inactive form and an increase of a more positive wave was observed. — The shift of half-wave potentials of V is given in Fig. 2.

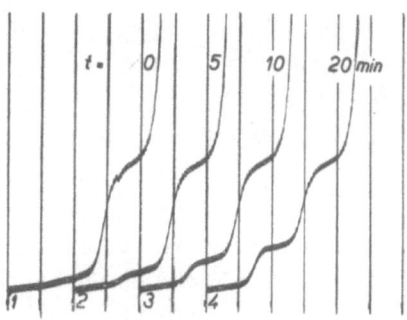

Fig. 8

Time-Dependence of the Wave for 2-Piperidinomethylcyclopentanone Hydrochloride (*V*) in Borate Buffer at pH 9·3

0·05M Borax, $2 . 10^{-4}$M depolarizer, 20°C. Time, at which the recording was started after preparation of the reaction mixtures is given on the polarogram. Curves starting at $-1·0$ V, 200 mV/absc., 200 mV/min, s. c. e., sens. 1 : 20.

Conclusions

Course of the Polarographic Reduction. No systematic discussion of the polarographic behaviour of β-amino ketones could be found in the literature. Only the reducibility of falicain[12] (*p*-propoxy-β-piperidinopropiophenone), kynurenin[13], lobelin[14−16] and lobelanine[17] was recognized.

Contrary to α-amino ketones[1], where the reduction of the C−N bond is facilitated by the presence of the carbonyl group, the reduction of β-amino ketones proceeds by a two-electron uptake on the carbonyl group. The reducibility of the C=O group is influenced by the presence of an amino group or ammonium group respectively.

Principally the reduction of the β-amino ketone is shifted towards more positive potentials relatively to the parent ketone. This shift, due (beside other factors) to the strong polar effect of the ammonium group, is not so marked for phenacyl derivatives (cf. Table I, values for *I, II* and *VI*), but for the saturated ketones *III − V* this shift is considerable, since parent ketones are not reduced[18] until $-2·4$ V.

However, this comparison is only a very rough one, since the reduction of aminoketones follows another pattern than that of the parent ketones. Acetophenone and its homologues[19−25] as well as the derivatives substituted on phenyl ring[26−29] are reduced in the protonized form principally in one one-electron wave (the second one-electron process being obscured by the hydrogen evolution).* With increasing pH-value the height of this wave decreases and a new wave increases at more negative potentials. This more negative wave corresponds to a two-electron reduction step to alcohol.

The height of the more positive one-electron wave of aryl ketones at lower pH-values is rate controlled, proton transfer being the rate determining step. The reduction process results in the formation of a radical, followed at least partially by dimerisation with the formation of pinacol. The effect of gelatin, of buffer composition and the large difference between the measured pK'-value (about $pK' \doteq 6.5$) and the equilibrium constant (pK below 0.0) show that the proton transfer can proceed both by a homogeneous general catalyzed reaction[31-35] and by a heterogeneous reaction[30,36-39].

In β-amino ketones the formation of a radical cannot be detected from polarographic curves and one two-electron wave is observed. Such a course of the reduction was also observed for acetophenones[40-43] bearing the amino group on the phenyl ring.** This behaviour can be caused by lowering the stability of the radical against the uptake of a further electron, or by a change in rate of the dimerization reaction,

Table I

Half-Wave Potentials ($E_{1/2}$, vs. s.c.e.) in Britton-Robinson Buffers at 20°C

Compounds					
I	pH	2.2_5	6.2	—	—
	$E_{1/2}$	-1.02_8[a]	-1.19_5[b]	—	—
II	pH	2.7	6.0	8.9	10.7
	$E_{1/2}$	-1.16	-1.24_7	-1.34_8	-1.41_0
III	pH	—	7.1	7.8	—
	$E_{1/2}$	—	-1.61	-1.61	—
IV	pH	—	7.3	8.5	10.2
	$E_{1/2}$	—	-1.56_2	-1.59_9	-1.69_3
V	pH	—	8.3	9.2	10.1
	$E_{1/2}$	—	-1.54_9	-1.57_9	-1.63_r
VI	pH	2.5	6.4_5	9.5_5	12.0_5
	$E_{1/2}$	-1.06^c	-1.33^c	—	—
			-1.50^d	-1.51^d	-1.57^d

^{a} At 0°−1·03₈ V, ^{b} At 0° −1·20₃ V; ^{c} First Wave. ^{d} The more negative wave.

* The height of this wave, dependent on buffer composition (not just on pH), increases with gelatin concentration as do some other rate controlled currents[30]. The controversy concerning the number of waves for acetophenone (mentioned by Elving and Leone[24]) and the inability of some authors to distinguish the decrease of the more positive one-electron wave may be explained by the presence of different solvents and also different concentrations of surface active agents.

** Such changes in reduction mechanism were observed also for some derivatives of benzaldehyde[34,44].

removing the radical formed from the electrode surface. Finally such a reduction course can be caused by a preferential orientation at the electrode surface, enabling a simultaneous uptake of the two electrons.

The role of the ammonium group in the first two mechanisms would be mainly a polar effect (both the internal inductive effect and the direct field effect).*

The ammonium group can furthermore be adsorbed in the negative double layer, enabling thus an advantageous position of the β-amino ketone at the electrode. In agreement with this conclusion are the electrocapillary curves demonstrating the greater adsorbability of *I* as compared to propiophenone.

The decrease of the wave-height with increasing pH at pH above 10 for hydro-chlorides *II*, *IV* and *V* (Fig. 7) demonstrates that the reducible form is protonized. For methoiodides *I* and *III* this protonization could not be proved as the elimination reaction occurred at substantially lower pH-values. The proton transfer which can occur either on an amino or on a carbonyl group (for hydrochlorides and on carbonyl group for methoiodides), proceeding both by homogeneous and heterogeneous[36-39] paths, occurs at pH-values substantially higher than those for unsubstituted ketones. It is believed that the rate of protonation is increased due to the effect of ammonium ions in the double layer in a similar way as with other proton transfers[30].

The scheme for the reduction path of methoiodides *I* and *III* can thus be summarized as follows (as the sequence of the protonisation relatively to the electron transfer cannot be distinguished):

$$R-COCH_2CH_2\overset{(+)}{N}R^1R^2R^3 + 2e + 2H^{(+)} \rightarrow RCH(OH)CH_2CH_2\overset{(+)}{N}R^1R^2R^3$$

For hydrochlorides the uptake of one proton can be separated from the electrode process proper (based on the decrease of the limiting current at higher pH-values), and at pH above pK^+_{NH} reduction path is as follows:

$$R-COCH_2CH_2NR^1R^2 + H^{(+)} \underset{\text{not very fast}}{\rightleftarrows} [\underbrace{R-COCH_2CH_2NR^1R^2}]^{(+)}_{H}$$

$$[\underbrace{R-COCH_2CH_2NR^1R^2}_{H}]^{(+)} + 2e + H^{(+)} \rightarrow R-\underset{OH}{CHCH_2CH_2NR^1R^2}$$

An explanation based on hydration and dehydration rates of the carbonyl groups seems to be less plausible.

Effect of Structure. Similar restriction of comparison as for the effects of intro-duction of the amino group (p. 194) is valid also for comparing the effect of R in $R.CO.CH_2CH_2N\overset{+}{<}{}^{(+)}$. The compounds *I* and *II* bearing a phenyl ring are reduced

---

* The possibility of mesomeric effects caused by the formation of quinoid systems which can play a role with *p*-aminoacetophenone or substituted benzaldehydes, are excluded here.

130

at more positive potentials than *III — V*, bearing the ketogroup on a saturated system. As presumed, the ring closure in *V* has only little effect as can be seen when *V* is compared with *III* and *IV*.

Finally we can compare the methoiodides *I* and *III* with the corresponding hydrochlorides *II* and *IV*. In piperidinopropiophenone the exchange of $N^{(+)}R^1R^2CH_3$ for $N^{(+)}R^1R^2H$ causes a shift of half-wave potentials (Table I, Fig. 2). The difference between the half-wave potentials for methoiodide *III* and for hydrochloride *IV* is not significant (Fig. 2, Table I).

Elimination. In all cases studied an elimination reaction proceeds in buffered solutions of the studied Mannich bases and their methoiodides. This reaction can be followed on polarographic curves from the increase of the wave of the corresponding unsaturated ketone.

The pH-dependence of the extent of the elimination reaction can be explained by a proton transfer involved. The mechanism of the elimination reaction is being further studied. The p*K*-value of the acid-base reaction (and thus also the tendency towards elimination reaction) depends on structure. For derivatives of propiophenone and ethyl methyl ketone the elimination reaction proceeds for methoiodides *I* and *III* at a substantially lower pH values than for corresponding hydrochlorides *II* and *IV*. The pH-values, above which the elimination reaction was observed under conditions used, were as follows: *I* pH > 5, *III* pH > 7; *V* pH > 9; *II* pH > 10; *IV* pH > 11. The acid properties of the amino ketone and the structure of the unsaturated ketone formed seems to be the deciding factors.

References

1. Zuman P., Horák V.: This Journal *26*, 173 (1961).
2. Schäfer H., Tollens B.: Ber. *39*, 2186 (1906).
3. Blicke F.: Org. Reactions *1*, 329 (1942).
4. Wilds A. L., Werth R. G.: J. Org. Chem. *17*, 1149 (1952).
5. Mannich C., Schaler P.: Arch. Pharm. *276*, 575 (1938).
6. Pasternak R.: Helv. Chim. Acta *31*, 753 (1948).
7. Prévost Ch., Souchay P., Malen Ch.: Bull. soc. chim. France *1953*, 78.
8. Zuman P., Michl J.: Nature *192*, 655 (1961).
9. Lund H.: Acta Chem. Scand. *14*, 1927 (1960).
10. Buděšínský B., Mňouček K., Jančík F., Kraus E.: This Journal *23*, 434 (1958); Chem. listy *51*, 1819 (1957).
11. Erskine R. L., Waight E. S.: J. Chem. Soc. *1960*, 3425.
12. Bräuniger H., Raudonat H. W.: Arch. Pharm. *287*, 109 (1954).
13. Šantavý F.: Biol. listy *27*, 60 (1946); Experientia *5*, 70 (1949).
14. Nyman P.: Svensk Farm. Tidskr. *47*, 401, 429 (1943).
15. Nyman P., Reimers F.: Dansk Tidskr. Farm. *15*, 292 (1941).
16. Šantavý F.: Čas. čes. lékárn. *56*, 1 (1943).
17. Šantavý F.: Čas. čes. lékárn. *57*, 109 (1944).
18. Stackelberg M. v., Stracke W.: Z. Elektrochem. *53*, 118 (1949).
19. Baker J. W., Davies W. C., Hemming M. L.: J. Chem. Soc. *1940*, 692.
20. Davies W. C., Evans D. P.: J. Chem. Soc. *1939*, 546.

21. Calzolari C., Furlani C.: Ann. chim. *44*, 356 (1954); Boll. Sci. fac. chim. ind. (Bologna) *12*, 14 (1954).
22. Calzolari C.: Ann. Triestini *1954*, No. 1, 16.
23. Koršunov I. A., Kirillova A. S., Kuznecova Z. B.: Ž. fiz. chim. *24*, 551 (1950).
24. Elving P. J., Leone J. T.: J. Am. Chem. Soc. *80*, 1021 (1958).
25. Mandell L., Powers R. M., Day A. Jr.: J. Am. Chem. Soc. *80*, 5284 (1958).
26. Schultz F. S.: Iowa State College J. Sci. *26*, 280 (1952).
27. Sartori G., Silvestroni P., Calzolari C.: Ricerca sci. *24*, 1471 (1954).
28. Giacometti C., Galdato E.: Ricerca sci. *26*, suppl. 287 (1956).
29. Elving P. J., Leone J. T.: J. Am. Chem. Soc. *82*, 5076 (1960).
30. Volková V.: Nature *185*, 743 (1960).
31. Hanuš V.: *Proc. Ist Intern. Polarograph. Congr. Prague* 1951, Part I, p. 811. Published by Přírodovědecké vydavatelství, Prague 1951.
32. Hans W., Henke K. H.: Z. Elektrochem. *57*, 595 (1953).
33. Wiesner K., Wheatley M., Los J. M.: J. Am. Chem. Soc. *76*, 4858 (1954).
34. Bartel E. T., Grabowski Z. R., Kemula W., Turnowska-Rubaszewska W.: Roczniki Chem. *31*, 13 (1957).
35. Becker M., Strehlow H.: Z. Elektrochem. *64*, 42 (1960).
36. Koryta J.: Z. Elektrochem. *64*, 23 (1960).
37. Grabowski Z. R., Bartel E. T.: Roczniki Chem. *34*, 611 (1960).
38. Strehlow H.: Z. Elektrochem. *64*, 45 (1960).
39. Majranovskij S. G., Liščeta L. I.: This Journal *25*, 3025 (1960).
40. Giacometti G., Del Marco A.: Rend. Acad. Nazl. Lincei, Cl. Sci. fis. mat. nat. [8] *14*, 511 (1953).
41. Knobloch E., Svátek E.: This Journal *20*, 1113 (1955); Chem. listy *49*, 37 (1955).
42. Biester F., Wolff J. H.: Arzneimittel-Forsch. *3*, 481 (1953).
43. Isshiki T., Kajima T., Noguchi S.: Pharm. Bull. Japan *2*, 253 (1945).
44. Holubek J., Volke J.: This Journal *24*, 1436 (1959); Chem. listy *52*, 589 (1958) and reference therein.

FISSION OF ACTIVATED CARBON–NITROGEN AND CARBON–SULPHUR BONDS. V.*
POLAROGRAPHIC STUDY OF ELIMINATION OF β-MORPHOLINO PROPIOPHENONE

P. Čársky, P. Zuman and V. Horák

J. Heyrovský Institute of Polarography, Czechoslovak Academy of Sciences and Department of Organic Chemistry, Charles University, Prague

Received March 3rd, 1964

Mannich bases of the β-amino ketone type undergo elimination only in the unprotonated free base form. With stronger bases like β-piperidino propiophenone the elimination reaction is complicated by the consecutive reaction of the resulting vinyl ketone with hydroxyl ions. However, with weaker bases like β-morpholino propiophenone the simple elimination process can be studied as the vinyl ketone formed is stable in the lower pH-range in which now the reaction proceeds. Owing to the competitive effects of fastly established equilibria of protonation of the Mannich base and of the amine formed the reaction corresponds to reversible or irreversible kinetics according to pH-value. The values $k_1 = (5 \cdot 5 \pm 1) \cdot 10^{-5}\ \mathrm{s}^{-1}$ for elimination, $k_{-1} = (20 \pm 5)\ 1\ \mathrm{mol}^{-1}\ \mathrm{s}^{-1}$ for addition reaction and $K = k_{-1}/k_1 = (4 \pm 1) \cdot 10^5$ for the equilibrium constant have been found.

As stated in the introductory paper of this series[1], β-amino ketones are the most important type of so called Mannich bases, used often as intermediates in organic synthesis. Alkylation properties of these substances are frequently applied for introducing a β-acylethyl group in the active position of an organic molecule or more generally for reactions with certain nucleophilic reagents. The elimination-addition mechanism is usually postulated as a working hypothesis for the alkylation process. The unequivocal verification of this hypothesis would require detailed study of both fundamental processes, *i. e.* the cleavage of the β-amino ketone giving the vinyl ketone and the addition of a nucleophilic reagent on this unsaturated ketone, together with an analysis of the additional consecutive and side reactions that can occur between particular components of the reaction mixture. In this paper, however, our interest is restricted to the elimination process and the corresponding addition reaction only.

Preliminary kinetic studies of the elimination reaction were carried out with β-piperidino propiophenone[2], where it was shown that the initial reaction rate

---

* Part IV. (erroneously denoted as Part III.): This Journal *28*, 1614 (1963).

increases with increasing pH-value in a manner corresponding to the form of a dissociation curve having an inflexion point that coincides with the pK_a-value of the Mannich base $(pK_a = 9\cdot3)$. From this observation it was deduced that only the unprotonized form of the Mannich base undergoes the elimination reaction[2,3].

An exact kinetic treatment was made difficult by the fact that at pH only slightly above 9 the concentration change of phenyl vinyl ketone did not follow simple kinetic equations, but showed features which suggested that a consecutive reaction of the phenyl vinyl ketone was taking place. This reaction was, therefore, studied in greater detail[4] using buffered solutions of phenyl vinyl ketone. The results obtained revealed that the rate of disappearance of phenyl vinyl ketone was first order with respect to the concentration of both phenyl vinyl ketone and hydroxyl ions. This dependence ruled out the possibility that polymerization was occuring at concentrations used and rendred formation of β-hydroxyketones probable[4]. The rate of this competitive reaction increases rapidly with rising pH of the reaction mixture and could have considerable importance in relation to the observed rates of the elimination reactions at pH greater than about 9.

Due to the shape of the pH-dependence of the rate of elimination it was practically possible to follow the elimination only at $pH > (pK_a - 1)$. Hence in the case of Mannich bases with pK 9 to 11 no simple kinetics could have been expected and it was necessary to seek a Mannich base with pK in the range 6 to 8.

The choice of the appropriate model substance was based on deductions from linear free energy relationships. The validity of the Taft equation[5] has been proved for addition of various nucleophilic reagents to phenyl vinyl ketone[6] and hence it was assumed to be valid also for the reverse reaction. However, no data were available for the pK_a values of Mannich bases. It was hence necessary to make a further assumption that substitution of hydrogen atom in a secondary amine HNR^1R^2 by a β-benzoyl ethyl group $C_6H_5COCH_2CH_2$ would affect the pK_a-value relatively in the same manner,* irrespective of the nature of alkyls R^1 and R^2. On this basis morpholine $(pK_a\ 8\cdot3)$ was indicated as a potentially suitable base and β-morpholino propiophenone (for which $pK_a = 6\cdot6$ was subsequently determined titrimetrically) was chosen for the present study of the kinetics of the elimination reaction.

Experimental

Apparatus and equipment. An LP 55 polarograph (Laboratorní přístroje) with photographic recording and a sensitivity $3\cdot5.10^{-9}$ A/mm/m was used. The capillary characteristics were: drop-time $t_1 = 4\cdot1$ s at $h = 53$ cm in 0·1N-KCl, outflow velocity $m = 2\cdot3$ mg/s. Reaction vessel was a water-jacketed cell[7] with a saturated calomel electrode, separated by an agar bridge and sintered glass disc. The temperature was kept at $(25 \pm 0\cdot1)°C$ using Ultrathermostat Hoeppler. For titrimetric determination of the pK-value of β-morpholino propiophenone and for

---

* In the compounds studied this assumption proved valid, introduction of $C_6H_5COCH_2CH_2$ resulted in the shift of pK_a of about $1\cdot5-1\cdot7$ pH-units towards lower pK_a-values.

measurements of pH-values of the buffers a glass electrode G 200B was used in connection with Radiometer PHM 4 pH-meter.

Substances. β-Morpholino propiophenone hydrobromide was prepared according to the procedure described by Williams and Day[8] from acetophenone, paraformaldehyde and morpholine hydrobromide — m. p. 189—190°C (ethanol) (reported[8] 189·5—190·5°C). Morpholine was purchased from Light (Colnbrook), all chemicals for the preparation of buffers, supporting electrolytes and reaction mixtures were reagent grade.

Solutions. A stock solution of 0·02M β-morpholino propiophenone hydrobromide was prepared in 0·01M-HCl and was found stable for several months. Aqueous solutions of phenyl vinyl ketone were prepared from a solution of β-piperidino propiophenone methoiodide by steam distillation. The approximate concentration of the resulting aqueous solution was determined by comparing the height of the first, two-electron wave[9] of phenyl vinyl ketone with the height of the two-electron wave of β-morpholino propiophenone in a solution of known concentration. The aqueous solution of phenyl vinyl ketone was kept at 2°C and the stock solution stood not longer than for three weeks. After this time cleavage occured, as demonstrated by a change in the ratio of the two waves of phenyl vinyl ketone, the second two-electron wave increasing in relative magnitude. In the buffers used the concentrations of the dihydrogen phosphate and borate respectively were kept constant, whilst the ionic strength was controlled by addition of sodium chloride. The compositions of the buffers used in the final measurements are given in Table I.

Table I

Phosphate and Borate Buffers

($\mu = 0.31$)

Phosphate: $5 . 10^{-2}$M-$H_2PO_4^-$			Borate: $3 . 10^{-2}$M-$H_4BO_4^-$; $2.95 . 10^{-1}$M-NaCl	
$[HPO_4^{2-}]$ g mol 1^{-1}	$[NaCl]$ g mol $1^{-1} . 10$	pH	$[H_3BO_3]$ g mol 1^{-1}	pH
$5 . 10^{-3}$	2·75	5·7	$1.5 . 10^{-1}$	8·2
$1 . 10^{-2}$	2·7	6·0	$3 . 10^{-2}$	9·1
$3 . 10^{-2}$	2·1	6·5	$4 . 10^{-3}$	10·1
$5 . 10^{-2}$	1·5	6·7		
$1 . 10^{-1}$	0	7·0		

Procedure. For elimination reactions 0·2 ml of the stock solution of 0·02M β-morpholino propiophenone hydrobromide was added to 19·8 ml of deaerated buffer at 25°C. After a short introduction of nitrogen the recording of the polarographic wave of phenyl vinyl ketone was started.

The application of the polarographic method for the study of the elimination of β-morpholino propiophenone is based on the separation of the waves of the reactant and of the phenyl vinyl ketone formed. Phenyl vinyl ketone in the pH-range studied, *i. e.* at pH > 5, is reduced[9] at the dropping mercury electrode in two two-electron steps. β-morpholino propiophenone behaves similarly to β-piperidino propiophenone[10] and shows but one two-electron wave. Over the whole pH-range studied the first wave of phenyl vinyl ketone occurs at more positive potentials than the wave for β-morpholino propiophenone. These well separated two waves allow measurements of the height of the first wave of phenyl vinyl ketone to be made with a reproducibility of about 3 per cent. The second wave of phenyl vinyl ketone is observed in phosphate buffer at more negative potentials than the wave of β-morpholino propiophenone whereas in borate buffers both waves coalesce.

The course of the reaction was followed either by recording whole i–E curves after chosen time-intervals (Fig. 1) or by recording the time-change of the current at the potential corresponding to the limiting current of the first wave of phenyl vinyl ketone (Fig. 2). In the latter case the residual current at the selected potential was recorded before addition of the Mannich base; after the particular run was completed, an i–E curve was recorded to show that an appropriate

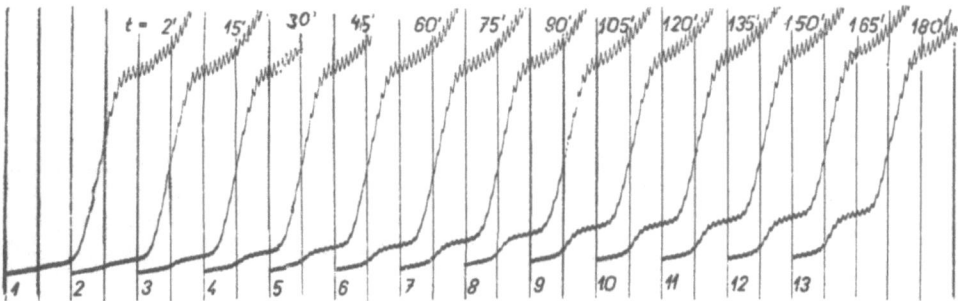

Fig. 1

Elimination of β-Morpholino Propiophenone

$2 . 10^{-4}$M β-morpholino propiophenone, phosphate buffer pH 6·5, $\mu = 0.31$. Recording of curves started after periods given on the polarogram at −0·6 V, s.c.e., 200 mV/absc., 30 s/absc. $h = 53$ cm, full scale sensitivity 3·5 μA.

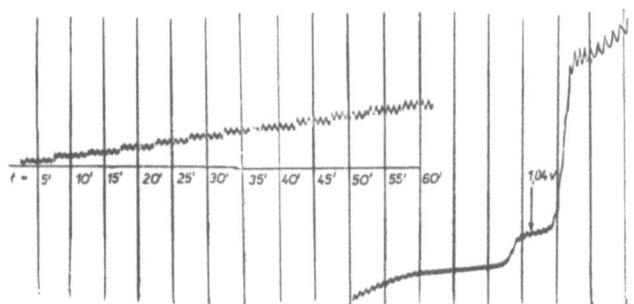

Fig. 2

Time-Change of Limiting Current of Phenyl Vinyl Ketone Formed by Elimination of β-Morpholino Propiophenone

$2 . 10^{-4}$M β-morpholino propiophenone, phosphate buffer pH 7·1, $\mu = 0.31$. Limiting current of phenyl vinyl ketone recorded at −1·04 V after intervals given on the polarogram, s.c.e., full scale sensitivity 0·7 μA. After 65 min. i–E curve recorded, starting at 0 V, s.c.e., 200 mV/abs., 30 s/absc., full scale sensitivity 2·45 μA.

potential had been applied. The first kind of measurement was used for the study of the reaction course and for choice of the most suitable reaction conditions, the second for determination of rate constants.

For addition reactions $1 . 10^{-4}$M phenyl vinyl ketone was added to a borate buffer and after deaeration $2 . 10^{-4}$M morpholine was introduced (with greater excess of morpholine the reaction was too fast for precise computation of the rate constant). The change in height of phenyl vinyl ketone wave was followed by the second method mentioned above.

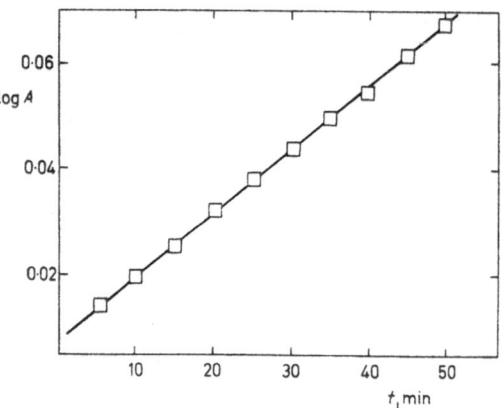

Fig. 3

Elimination of β-Morpholino Propiophenone Under Conditions for Irreversible Reaction

$A = 1.25h_0/(1.25h_0 - h_i)$, where h_0 is the extrapolated height of wave of β-morpholino propiophenone at $t = 0$, h_i the height of the first wave of phenyl vinyl ketone at $t = t_i . 2 . 10^{-4}$M β-morpholino propiophenone, phosphate buffer pH 6·5, $\mu = 0.31$.

Determination of rate constants. The rates of the elimination and addition reactions are pH-dependent, thus at a given pH-value the effective rate constant (denoted as k_i') was determined. The elimination reaction followed irreversible kinetics at lower pH-values changing to a reversible system at higher pH-values (*cf.* p. 3053).

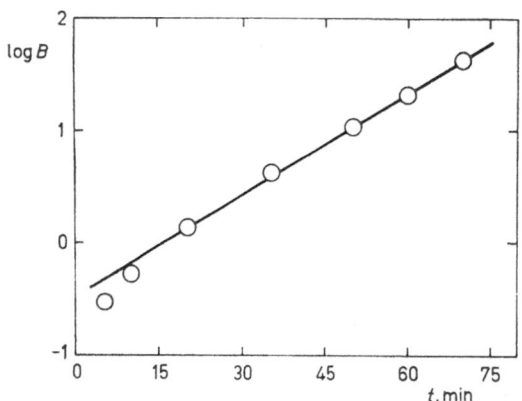

Fig. 4

Elimination of β-Morpholino Propiophenone Under Conditions for Reversible Reaction

$B = [h_0h_i + h_i(h_0 - h_e)]/[h_0(h_e - h_i)]$, symbols h_0 and h_i possess the same meaning as in Fig. 3, h_e is the height of the first wave of phenyl vinyl ketone after establishment of equilibrium ($t = \infty$). $2 . 10^{-4}$M β-morpholino propiophenone, borate buffer pH 9·1, $\mu = 0.31$.

For irreversible reactions the effective rate constant k_1' was determined using equation $k_1't = 2.3 \log 1.25h_0'/(1.25h_0' - h_i)$, where h_0' is the extrapolated height of the β-morpholino propiophenone wave at $t = 0$ and h_i the height of the first wave of phenyl vinyl ketone at $t = t_i$. The numerical factor 1·25 arises in the conversion from concentration to wave-heights, since for equimolar concentrations the ratio of the wave-heights of phenyl vinyl ketone and β-morpholino propiophenone is about 1·25. The values for k_1' were usually computed graphically by plotting $\log 1 \cdot 25 h_0'/(1 \cdot 25 h_0' - h_1)$ or $\log (1 \cdot 25 h_0' - h_1)$ against t and determining the slope (Fig. 3).

In runs, for which the reaction required an equation for a reversible process, the effective rate constant k_1' was calculated using expression $k_1' t (2h_0' - h_e)/h_e = 2 \cdot 3 \log \{[h_0' h_i + h_1(h_0' - h_e)]/ /h_0'(h_e - h_i)\}$, where h_e stands for the height of first wave of phenyl vinyl ketone after establishment of equilibrium $(t = \infty)$. Plotting expression $\log \{[h_0' h_i + h_1(h_0' - h_e)]/h_0'(h_e - h_i)\}$ against time (Fig. 4) enabled us to determine the value of the effective rate constant k_1' from the slope of the linear part. The rate constants found are given in Table II.

Table II
Values of Measured Rate Constants (k_1') for the Elimination of β-Morpholino Propiophenone $(\mu = 0 \cdot 31, 25°C)$
Phosphate Buffer[a]. Equation (2).

pH	5·7	6·0	6·5	6·7₅	7·1	7·3
$k_1'^{[b]}$	0·53	0·82	2·45	3·4	3·9—4·1₅	4·6

Borate Buffer[a], Equation (3)

pH	8·2	9·1	10·1
$k_1'^{[b]}$	5·5	4·8₅—5·4₅	5·6[c]

[a] Composition cf. Table I.; [b] $k_1' = k_1/(1 + [H^+]/K_{MBH^+})$; [c] in the initial phase (to 60 minutes) of reaction when the reaction of phenyl vinyl ketone with hydroxyl ions does not interfere.

The addition reaction was treated under the assumption (p. 3052) that it can be regarded as virtually irreversible. The value of the effective rate constant of the addition reaction k_{-1}' was determined from the general form of second order reaction expression $k_{-1}' t = [2 \cdot 3/(h_N - h_0)] \log [h_0(h_N - h_0 + h_i)/h_i h_N]$ where h_0 is the extrapolated height of the first phenyl vinyl ketone wave for $t = 0$ and h_N the concentration of morpholine for $t = 0$ (expressed as phenyl vinyl ketone wave-height). Under our reaction conditions used, viz. that the initial concentration of morpholine was just double of that of phenyl vinyl ketone, the equation reduces to $k_{-1}' t = 2 \cdot 3/h_0 \log[(h_0 + h_i)/2h_i]$. The evaluation was carried out by plotting the right hand term in one of the equations for $k_{-1}' t$ against t.

Determination of equilibrium constant. To study the established equilibrium reaction mixture was prepared from phenyl vinyl ketone and morpholine and equilibrium constant of reaction (B) was computed from the initial and equilibrium wave-heights of phenyl vinyl ketone using equation $K = (h_0 - h_e)/h_e [NR_2H]_e$, where $[NR_2H]_e$ is equilibrium concentration of unprotonised morpholine. On the other hand when the elimination of Mannich base was carried out, the equation $K = (1 \cdot 25)^2 h_e' h_0/[(h_e)^2 \cdot 2 \cdot 10^{-4}]$, where h_e' is the height of β-morpholino propiophenone wave at equilibrium, was used for the determination of values of K.

Dissociation constant of β-morpholino propiophenone. The determination of pK_{MBH^+}-value* (cf. reaction (A)) of β-morpholino propiophenone cannot be carried out by the usual potentiometric titration in which sodium hydroxide is added, since the elimination reaction produces

---

* Indices MBH[+] stay for protonized form of Mannich base, MB for Mannich base and NH[+] for protonized morpholine.

morpholine, a stronger base and the pH-value changes with time after each addition of sodium hydroxide (Fig. 5). However, the drift of pH after each addition of hydroxide was plotted against time and extrapolation to $t = 0$ enabled us to plot a titration curve for β-morpholino propiophenone with a theoretical shape corresponding to pK 6·6 (Fig. 6).

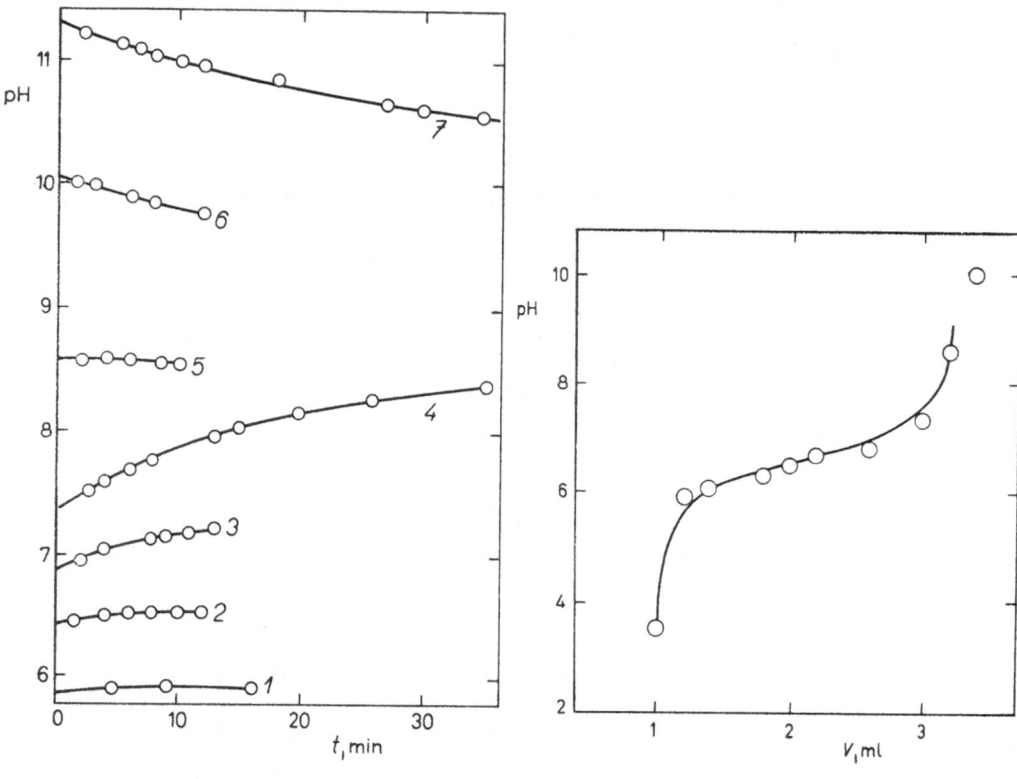

Fig. 5

Change of pH Values During Elimination of β-Morpholino Propiophenone in Unbuffered Media

To 10 ml of 2 . 10^{-2}M β-morpholino propiophenone hydrobromide 0·05M-NaOH was added: 1 1·2 ml; 2 1·8 ml; 3 2·6 ml; 4 3·0 ml; 5 3·2 ml; 6 3·4 ml; 7 4·0 ml. Nitrogen atmosphere, glass electrode.

Fig. 6

Extrapolated Potentiometric Titration Curve of β-Morpholino Propiophenone Hydrobromide

pH-values in solutions of 2 . 10^{-2}M β-morpholino propiophenone hydrobromide after addition of V ml of 0·05M-NaOH extrapolated to $t = 0$. Glass electrode, nitrogen atmosphere.

Results and Discussion

General equations. The course of the pH-dependence shown by the elimination reaction rates of β-morpholino and β-piperidino propiophenone is analogous (Fig. 7). The plot given in Fig. 7 confirmed the estimate that reaction of the morpholino derivative would occur at lower pH-values than that of the corresponding piperidino propiophenone. This fact together with the value of pK_a for β-morpholino propiophenone (pK_a 6·6) proves that both assumptions made in the application of the linear free energy relationships to the choice of a suitable model substance are valid. Hence the elimination reaction of β-morpholino propiophenone in the whole pH-range, where its rate changes with pH, is unaffected by the consecutive reaction of the phenyl vinyl ketone produced, a complication[4], affecting similar elimination reactions[4] at pH greater than about 9.

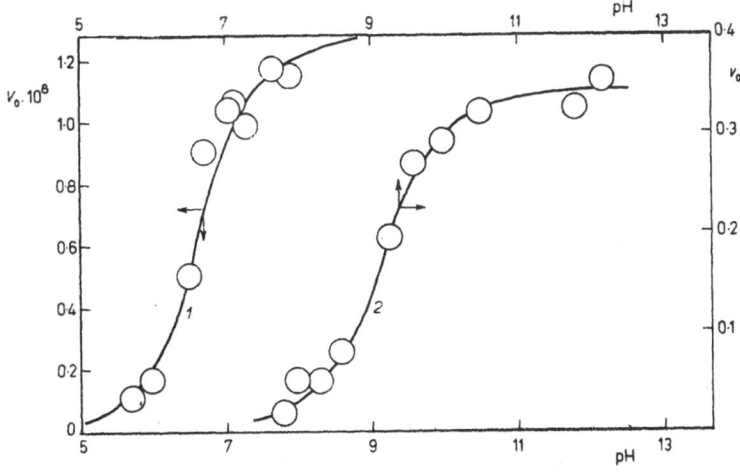

Fig. 7

Initial Velocities (V_0, v_0 — relative values on different scales) of the Elimination of β-Morpholino Propiophenone (1) and β-Piperidino Propiophenone (2)

1 Left and bott m axis, 2 right and top axis. $2 . 10^{-4}$M Mannich base.

Hence the concentration change of phenyl vinyl ketone with time in solutions of β-morpholino propiophenone allowed us to study directly the kinetics of the elimination process. The results were surprising in that in the low pH-range irreversible reaction kinetics were observed changing to reversible kinetics when the pH was increased above 7·3. In the explanation of this rather unusual behaviour together with the dependence of the observed rate constants on pH (Fig. 7) the following scheme is proposed:

$$ArCOCH_2CH_2NR_2H^{(+)} \xrightleftharpoons{K_{MBH}^{+} \text{ (fast)}} ArCOCH_2CH_2NR_2 + H^{(+)} \qquad (A)$$

$$ArCOCH_2CH_2NR_2 \xrightleftharpoons[k_{-1}]{k_1} ArCOCH{=}CH_2 + NR_2H \qquad (B)$$

$$NR_2H_2^{(+)} \xrightleftharpoons{K_{NH}^{+} \text{ (fast)}} NR_2H + H^{(+)} \qquad (C)$$

$$ArCOCH=CH_2 + OH^{(-)} \rightleftharpoons ArCO\overset{(-)}{C}HCH_2OH \tag{D}$$

$$ArCO\overset{(-)}{C}HCH_2OH \underset{(+H)^+}{\overset{\text{several steps}}{\rightleftharpoons}} ArCOCH_3 + HCHO \tag{E}$$

where $-NR_2$ equals to $-N(CH_2CH_2)_2O$.

Elimination reaction. In discussing the relevance of this reaction scheme to the observed kinetics, reactions occuring in media with pH < 9 were considered. Under these conditions namely the rate of reaction (D) is negligibly small as demonstrated by an independent study of the base catalysed hydration of the phenyl vinyl ketone[4] and reactions (D) and (E) need not be considered. A further restriction is imposed by the fact that the elimination reactions of Mannich bases occur with measurable velocities only at pH > ($pK_{MBH^+} - 1$). Hence for the chosen compound, β-morpholino propiophenone (for which $pK_{MBH^+} = 6\cdot6$ and $pK_{NH^+} = 8\cdot3$) the study is limited to the range pH $5\cdot6-9$, where the kinetics of the whole scheme reduce to those of the linked equilibria represented by reactions (A)−(C). In terms of the constants for these reactions the rate (v) of formation of phenyl vinyl ketone (PVK) is given by equation (1):

$$v = \frac{d\,[PVK]}{dt} = \frac{k_1 S_{MB}}{1 + [H^+]/K_{MBH^+}} - \frac{k_{-1} S_N [PVK]}{1 + [H^+]/K_{NH^+}} \tag{1}$$

where $S_{MB} = [ArCOCH_2CH_2NR_2H^{(+)}] + [ArCOCH_2CH_2NR_2]$ is analytical concentration of Mannich base and $S_N = [NR_2H_2^{(+)}] + [NR_2H]$ is analytical concentration of morpholine.

It is now convenient to discuss separately the situations arising in the critical pH-ranges $5\cdot6-7\cdot3$, $7\cdot3-7\cdot6$ and $7\cdot6-9\cdot0$, where the acid-base equilibria of reactions (A) and (C) exert kinetically opposed effects.

pH *Range* $5\cdot6-7\cdot3$. This range is related to equilibrium constants of reactions (A) and (C) and corresponds with ($pK_{MBH^+} - 1$) to ($pK_{NH^+} - 1$). In the elimination reactions the initial concentrations of phenyl vinyl ketone and morpholine were zero. Hence in the second right hand term of equation (1) [PVK] is very small and at the same time $[H^+] \gg K_{NH^+}$ (i. e. the ratio $[H^+]/K_{NH^+}$ varies from $10^{2\cdot7}$ to 10 over the pH range presently considered). Consequently this term may be ignored and the rate expression reduces to equation (2):

$$\frac{d\,[PVK]}{dt} = \frac{k_1 S_{MB}}{1 + [H^+]/K_{MBH^+}}. \tag{2}$$

This equation agrees with the irreversible first order kinetics of an elimination process preceded by equilibrium (A) established by our rate measurements in this pH range.

pH *Range* $7\cdot3-7\cdot6$. This range corresponds with that defined by ($pK_{NH^+} - 1$) and ($pK_{MBH^+} + 1$). In this range the both terms of equation (1) control the kinetics which are consequently those appropriate to a reversible reaction as found experimentally.

pH *Range* 7·6−9·0. This range is set by the lower limit $(pK_{MBH^+} + 1)$ and the complexity introduced by the outset of reactions (D) and (E). We may now apply similar arguments to those used in the lowest pH range, but relating to the equilibrium constant of reaction (A) K_{MBH^+} and the first term of equation (1). Thus $[H^+] \ll K_{MBH^+}$ and the ratio $[H^+]/K_{MBH^+}$ varies from 10^{-1} to $10^{-2·4}$ over the range. This approximation allows equation (1) to be written as (3):

$$\frac{d[PVK]}{dt} = k_1 S_{MB} - \frac{k_{-1} S_N [PVK]}{1 + [H^+]/K_{NH^+}}. \qquad (3)$$

This simplification is equivalent to ignoring the effect of reaction (A) on the concentration of the reactive form of the Mannich base at these relatively high pH values. The kinetics are reversible and involve reaction (B), with its reverse rate modified by the following acid-base equilibrium of reaction (C).

Summarizing this argument we see that, when the pH < 7·3 reaction (C) is unimportant, and the first order rate constant found experimentally corresponds to the constant term on the right hand side of equation (2). Between pH 7·3 and 7·6 the reaction is reversible and the rate is described by equation (1). Above pH 7·6 to the experimental limit 9·0 the reaction is also reversible but since reaction (A) has become unimportant the kinetics observed followed (3) instead of (2).

Using equations $(1)-(3)$ values of effective rate constants $k_1' = k_1/(1 + [H^+]/K_{MBH^+})$ were computed for particular pH-ranges, at pH < 7·3 using the equations for irreversible, and at pH > 7·3 those for reversible systems (Table II). The experimental points given in Fig. 8 fit reasonably well upon the theoretical curve also plotted. The titration curve of morpholine (pK_{NH^+}), using an arbitrary scale, is included in Fig. 6 to demonstrate the pH-region in which the effect of addition reaction with k_{-1} cannot be neglected.

Hence scheme $(A)-(C)$ corresponds both to the observed pH-dependence of the elimination rate and to the observed change from irreversible to reversible reaction. The above treatment allows us to propose an approximate average value $(5·5 \pm 1)$. $. 10^{-5} s^{-1}$ for k_1, the rate constant for the unimolar decomposition of β-morpholino propiophenone.

Addition. The addition reaction, with the rate constant k_{-1} could be followed only at pH > 7·3 [*i.e.* at pH > $(pK_{NH^+} - 1)$] since at lower pH-values the concentration of the reactive unprotonized form of morpholine is negligibly small. Addition reactions were studied under conditions, where initial concentration of Mannich base was nil, $(S_N)_0 = [NR_2H_2^{(+)}]_0 + [NR_2H]_0$, and morpholine was present in excess over phenyl vinyl ketone.

At pH 9 to 7·6 it was then possible to neglect first right hand term in equation (3)

142

and the rate of addition (v') followed equation:

$$v' = \frac{k_{-1} S_N [PVK]}{1 + [H^+]/K_{NH^+}}.$$ (4)

In a borate buffer of pH 9 and for $(S_N)_0 = [PVK]_0/2$ ($\mu = 0.31$) value $k_{-1} = (20 \pm 5)$ 1 mol^{-1} s^{-1} has been found from $k'_{-1} = k_{-1}/(1 + [H^+]/K_{NH^+})$. The low reproducibility is due to the inaccuracy in determination of initial concentration of phenyl vinyl ketone for this fast reaction with $\tau_{1/2} = 180$ s.

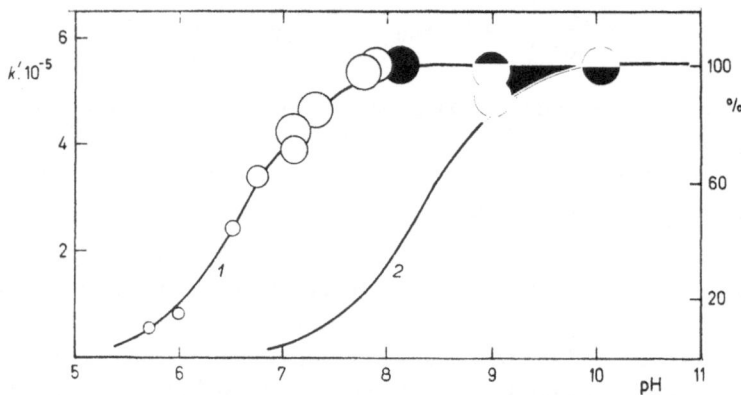

Fig. 8

Dependence of Effective Rate Constant of Elimination of β-Morpholino Propiophenone on pH

$2 . 10^{-4}$M β-morpholino propiophenone; ○ experimental values obtained in phosphate buffers, rate constants computed using equation (2) for irreversible reactions; ● experimental values obtained in borate buffers, rate constants computed using equation (3) for reversible reactions; 1 theoretical curve computed from equation (1), where $k' = k/(1 + [H^+]/K)$ for p$K_{MBH^+} = 6.6$ and $k_1 = 5.5 . 10^{-5}$ s^{-1}; 2 titration curve of morpholine for p$K_{NH^+} = 8.3$ in the same scale, setting numerical value for k_1 equal to 100% of dissociation.

This value is in good agreement with the value $k_{-1} = 17$ 1 mol^{-1} s^{-1} found by Hrubcová[6] who worked in excess of morpholine in buffers consisting of morpholine and morpholine hydrochloride. This author[6] also has proved that observed pH-dependence of the effective rate constants of addition k_{-1} fits well equation (4).

Equilibrium constant. Values of equilibrium constants were determined from measurement of equilibrium wave-heights (Table III). They were in reasonably good agreement with value $K = k_{-1}/k_1 = (4 \pm 1) . 10^5$ obtained from kinetic data. Lower accuracy of the values of equilibrium constants when measured from the wave-heights after completion of the addition reaction is due to inaccuracy in determination of the initial concentration of the phenyl vinyl ketone. The agreement of data obtained from dynamic and equilibrium measurements is another proof of the validity of scheme $(A)-(C)$.

Table III

Equilibrium Constants

Initial Concentrations g mol $l^{-1} . 10^4$			Reaction[b]	Borate Buffer pH	K^c
[PVK][a]	$(S_N)_0$	$(S_{MB})_0$			
5	3	0	ad	8·2	$3\cdot75 . 10^5$
5	20	0	ad	8·2	$4\cdot7 . 10^5$
5	4	0	ad	10·1	$7\cdot7 . 10^5$
5	6	0	ad	10·1	$5\cdot1 . 10^5$
5	8	0	ad	10·1	$8\cdot0 . 10^5$
0	0	2	el	8·2	$(4 \pm 2) . 10^4$
0	0	2	el	9·1	$(6 \pm 1) . 10^4$
0	0	2	el	10·1	$(4 \pm 1) . 10^4$

[a] Phenyl vinyl ketone; [b] ad addition reaction; el elimination reaction; [c] equilibrium constant of reaction (2) defined as $K = [Ar\ COCH_2CH_2NR_2]/[Ar\ COCH = CH_2]\ [NR_2H]$.

Reaction mechanism. The kinetic behaviour observed allows us to eliminate bimolecular mechanisms for the elimination reaction. Hence E1 or a cyclic mechanism can be considered for elimination of Mannich bases of the type of β-morpholino propiophenone. The facts that methoiodides[2] and protonized forms of Mannich bases[2] do not react by this mechanism show the importance of the unshared electron pair on nitrogen. Furthermore the fact that the β-ketosulfides follow another type of kinetics[6] further stress the importance of nitrogen atom in this reaction. These findings together with the observation[6] that addition of primary amines to phenyl vinyl ketone is affected more by steric than by polar effects of substituents seem to indicate a cyclic mechanism, involving either the hydrogen on the α-carbon or the enol form. Work on effects of steric factors and on the transmition of polar effects through this system, aimed for mechanistic studies, is in progress.

References

1. Horák V., Zuman P.: This Journal *26*, 173 (1961).
2. Horák V., Michl J., Zuman P.: Tetrahedron Letters *21*, 744 (1961).
3. Wagner M.: *Thesis.* Charles University, Prague 1962.
4. Čársky P.: *Thesis.* Charles University, Prague 1964.
5. Taft R. W. Jr. in the book: *Steric Effects in Organic Chemistry* (Ed. M. S. Newman). Wiley, New York 1956.
6. Hrubcová I.: *Thesis.* Charles University, Prague 1964.
7. Zuman P., Krupička J.: This Journal *23*, 598 (1958); Chem. listy *51*, 424 (1957).
8. Williams A. L., Day A. R.: J. Am. Chem. Soc. *74*, 3875 (1952).
9. Zuman P., Michl J.: Nature *192*, 655 (1961).
10. Zuman P., Horák V.: This Journal *27*, 187 (1962).

Translated by the author (P. Z.).

FISSION OF ACTIVATED CARBON–NITROGEN AND CARBON–SULFUR BONDS. IX.*

POLAROGRAPHIC STUDY OF ADDITION OF PRIMARY AMINES TO PHENYL VINYL KETONE

I.Šestáková, V.Horák and P.Zuman

J. Heyrovský Institute of Polarography,
Czechoslovak Academy of Sciences
and Department of Organic Chemistry,
Charles University, Prague

Received February 17th, 1966

Free base form of primary amines is added in a reversible reaction to phenyl vinyl ketone. Kinetics verified the scheme $(B)-(D)$. Alkyl groups R in RNH_2 exert considerable steric effects on this reaction. The nucleophilic reactivity of amines in this reaction parallels the reactivity in other nucleophilic reactions of amines.

In the course of the study of the elimination reactions of Mannich bases it has been proved[1] that under certain conditions this reaction behaves as reversible. Hence it appeared important to obtain more informations on the reverse reaction, *i.e.* on the addition of the leaving amine with the α,β-unsaturated ketone formed in the course of the elimination process.

Contrary to additions to isolated double bonds that have been extensively studied and the mechanism of which has been examined from various points of view[2-4], relatively little attention has been paid to the additions to —CO—CH=CH— and similar systems with the exception of the recent work by Patai and Rappoport[4-10] who nevertheless concentrated their efforts mainly to the reactions of α,β-unsaturated nitriles.

Most common type of the addition reactions belonging to this group is the Michael reaction[11]. In spite of its wide scope, only few kinetic studies were devoted to this problem. It has been supposed[12] that the rate determining step is the nucleophilic attack of the conjugate base on the positively polarised β-carbon of the α,β-unsaturated compound. In the agreement with this assumption is the first order kinetics with respect to α,β-unsaturated compound and first order with respect to the conjugate base of the nucleophilic agent that has been proved for addition of cyanide to α-cyanoacrylate[13], of anion of barbituric acid[14] and of trinitromethane[15] to β-nitrostyrene, of the ethanolamine[16], methoxide ion[16] and anions of acetylacetone[16], malonic esters[17] and thiomalonic esters[17], of aminoacids[18,19], and aminothiols[19] to acrylonitrile.

---

*   Part VIII: This Journal *31*, 827 (1966).

None of these studies treats hence the addition to α,β-unsaturated carbonyl compounds, even when this reaction belongs to the common operations in organic synthesis. In all cases the presence of conjugate base species of the nucleophilic reagent was essential. Basing on the yields of reactions of α,β-unsaturated carbonyl compounds with amines[20-23], β-diketones[24], ketones with $CH_2COC_6H_5$ grouping[25-28] and mercaptans[29] some hypotheses concerning the reaction mechanism have been formulated.

Because the product of elimination reactions in most of our studies[1,30] was phenyl vinyl ketone, this compound has been chosen for the studies on addition reactions. In preliminary investigations it has been shown[31] that reactivity increases principally with increasing base strength and that an approximately linear relation or at least a trend exists between the value of $\log k_1^{bim}$ and pK_a (where k_1^{bim} is the rate constant for the second order irreversible addition reaction and K_a is the acid dissociation constant of the nucleophilic reagent). It has been proved[31] that polarography can be applied for the study of addition reactions of rhodanides, cyanides, ammonia, primary and secondary amines, α-branched mercaptans and acetylacetone. For some strong nucleophilic reagents, such as benzylmercaptan or aminoethylmercaptan the reaction was too fast to be measured. In this paper the time-changes of polarographic waves of phenyl vinyl ketone were used for the study of some structural effects on reactivity of primary amines, reacting according to the reaction scheme (A).

$$C_6H_5COCH=CH_2 + HNHR \; \underset{k_{-1}}{\overset{k_1}{\rightleftarrows}} \; C_6H_5COCH_2CH_2NHR. \qquad (A)$$

EXPERIMENTAL

Apparatus and equipment. Polarographic curves were recorded either photographically using a Heyrovský type polarograph (Nejedlý, Mark VII) or a pen recording system using EZ-2 recorder in connection with a polarograph LP 55 (Laboratorní přístroje, Prague). Capillary used had the following characteristics: $t_1 = 3.5$ s at $h = 40$ cm in $0.1N$-KCl, out-flow velocity $m = 3.05$ mg s^{-1}. The reaction was carried out in a water jacketed cell[32], separated by an agar bridge and sintered glas disc. The temperature was kept constant at $(25 \pm 0.1)°C$ using an Ultrathermostat Hoeppler and Universal Thermostat Wobser U 8 (Dresden). pH-Values of buffer solutions were measured with a glass electrode G 200 B in connection with an Autojonometr 21d (Radiometer, Copenhagen).

Substances. Phenyl vinyl ketone was prepared in the same way as in Part V of this series[1]. Primary amines used (R—NH_2, R = CH_3 (I), C_2H_5 (II), n-C_4H_9 (III), n-C_5H_{11} (IV), n-C_6H_{13} (V), iso-C_3H_7 (VI), tert-C_4H_9 (VII), cyclohexyl (VIII), CH_2CH_2OH (IX), $CH_2C_6H_5$ (X) $CH(C_6H_5)_2$ (XI), $CH_2CH=CH_2$ (XII), H (XIII) were commercial products, purified by distillation or crystallisation of hydrochlorides. Physical constants were in agreement with the data of literature, purity was checked polarographically and by gas chromatography. Chemicals for preparation of buffer solutions were reagent grade.

Solutions. Solution of phenyl vinyl ketone resulting in steam distillation was approximately $1 \cdot 10^{-3}$M and was used as a stock solution. This solution when kept in refrigerator in absence of light remained unchanged for 2—3 weeks as proved by comparison of polarographic waves. Hence a fresh stock solution was prepared every one or two weeks. Purity was checked polarographically, concentration of the stock solution was estimated using propiophenone as a standard substance under assumption that the diffusion coefficients of these two substances are practically identical.

Stock solutions of amines were prepared either by addition of a standard hydrochloric acid to a solution of amine or by addition of a standard sodium hydroxide solution to a solution of amine hydrochloride. The ratio of free base to hydrochloride in reaction mixture was changed from 1 : 10 to 10 : 1 in such a way that the concentration of the free base was always at least ten times greater than the concentration of the phenyl vinyl ketone. Aqueous solutions were used as reaction mixtures for water–soluble amines (*I, II, IV* and *XIII*) and contained a maximum of 1·5% ethanol for the other substances, with the exception of amine *XI*, where the use of 2·5 to 20% was necessary.

Technique. Phenyl vinyl ketone is reduced polarographically principally in two two-electron steps[33]. After the addition of an amine both these waves decrease with time (Fig. 1). At potentials between the two waves of phenyl vinyl ketone a new wave increases with time. It has been proved that this wave belongs to the two-electron reduction of the Mannich base[34] formed in this reaction. The sum of heights of this wave and of the more positive wave of phenyl vinyl ketone remains practically constant. The small differences are due to the complications in the phenyl vinyl ketone reduction[33]. The unchanged total height can be taken as a proof that no side reaction occurs. In some cases the wave of the Mannich base is less well separated from the more negative wave of phenyl vinyl ketone than given in Fig. 1.

The decrease of the more positive wave of phenyl vinyl ketone was used for the study of kinetics of the addition reaction. The wave-height was recorded either continuously at the potential of limiting current of the first wave (Fig. 2), or recorded after pre-selected time intervals.

Procedure. Supporting electrolyte was deaerated and thermally equilibrated. Nucleophilic reagent was added, nitrogen purged for another 3 minutes, stock solution of phenyl vinyl ketone was added so that its final concentration was $1 . 10^{-4}$M ($5 . 10^{-5}$M in some few cases) so that the

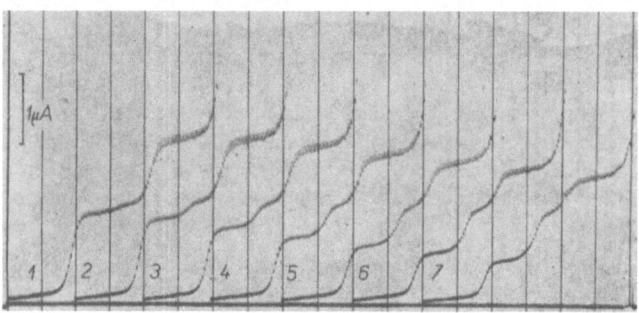

FIG. 1

Reaction of Phenyl Vinyl Ketone with Ammonia

$2 . 10^{-4}$M phenyl vinyl ketone, 0·01M-NH_4Cl, 0·01M-NH_3, 0·09M-NaCl, curves recorded after: *1* 0; *2* 10; *3* 20; *4* 30; *5* 40; *6* 50; *7* 60 minutes. Curves recorded from -0.7 V, s.c.e., 200 mV/absc., $h = 45$ cm, $t_1 = 3.22$ s, $m = 2.06$ mg s^{-1}, full scale sensitivity 4·4 µA.

final volume was 20 ml, solution was mixed and deaerated by nitrogen for another 10—15 seconds and limiting current was recorded.

Conditions for kinetic measurements. In the study of the effects of medium on reaction rate, three types of measurement were carried out:

1. pH-values were kept constant by constant ratio of concentrations of amine and its hydrochloride; ionic strength was kept constant by addition of 2M-NaCl; concentration of the nucleophilic reagent was changed by changing the buffer capacity.

2. pH-values and concentration of the nucleophilic reagent were kept constant by using buffer of the same composition; ionic strength was changed by additions of 2M-NaCl.

3. At a constant ionic strength concentration of the nucleophilic reagent was kept constant, pH was changed by adding to a chosen concentration of free amine varying concentrations of its hydrochloride.

Computation of rate constants. In this study reaction (A) was followed only under conditions that at $t = 0$ the concentration of Mannich base was nil. In all cases was $[NH_2R] \gg \gg [C_6H_5COCH = CH_2]$ and hence reaction with rate constant k_1 followed the first order kinetics. According to whether $k_1 [NH_2R] \approx k_{-1}$ or $k_1 [NH_2R] \gg k_{-1}$ the reaction was treated as reversible or irreversible.

1. When the value of the product $k_1 [NH_2R]$ was comparable with that of k_{-1}, the reaction was reversible and the change of phenyl vinyl ketone wave corresponded to equation (*1*)

$$k't = 2 \cdot 3 \log (h_0 - h_e)/(h - h_e), \tag{1}$$

where

$$k' = k_1 [NH_2R] + k_{-1}. \tag{2}$$

Here h is the wave-height of phenyl vinyl ketone, namely h_0 at $t = 0$, h_e at $t = \infty$ and h at $t = t_1$, k_1 and k_{-1} rate constants of reactions (A), k' the formal rate constant.

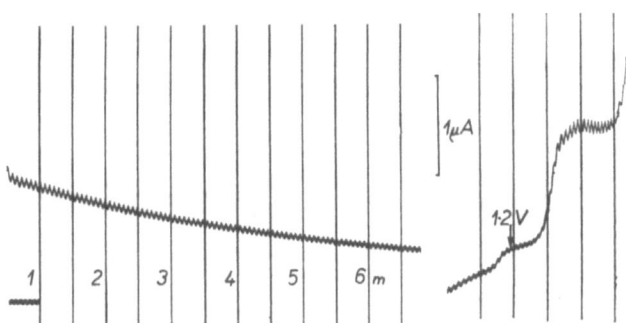

FIG. 2

Reaction of Phenyl Vinyl Ketone with Ammonia

Time-dependence of the height of the first wave of phenyl vinyl ketone at constant potential −1·2 V; 2 . 10^{-4}M phenyl vinyl ketone, 0·1M-NH$_3$, 0·02M-NH$_4$Cl, 0·5M-NaCl; curve started 30 seconds after preparation of the reaction mixture, abscissae recorded each 30 seconds. The curve on the right-hand side recorded after 8 minutes starting from −0·8 V, s.c.e., 200 mV/absc., $h = 40$ cm, $t_1 = 3 \cdot 5$ s, $m = 3 \cdot 05$ mg s^{-1}, full scale sensitivity 3·1 μA.

Rate constants were computed by plotting log $(h - h_e)$ against time t (Fig. 3). The slope gave the value of the formal rate constant k'. Values of formal rate constants k' were plotted against $[NH_2R]$, k_1 determined from the slope of this linear dependence, k_{-1} from the intercept (when present) (Fig. 4). When $[NH_2R]$ was kept constant the value of k_1 was determined using relation $k_1 = k'(h_0 - h_e)/[RNH_2] \cdot h_0$.

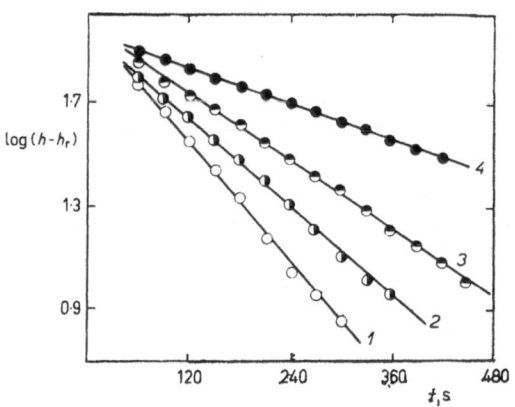

FIG. 3

Reaction of Phenyl Vinyl Ketone with Benzhydrylamine
$1 \cdot 10^{-4}$M phenyl vinyl ketone, equal concentrations of the free amine and its hydrochloride: 1 $6 \cdot 10^{-3}$M; 2 $4 \cdot 7 \cdot 10^{-3}$M; 3 $3 \cdot 5 \cdot 10^{-3}$M; 4 $2 \cdot 5 \cdot 10^{-3}$M. For evaluation of equation (1) log $(h - h_r)$ plotted against time (t).

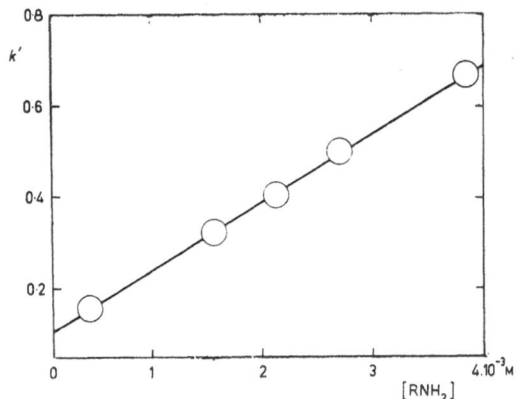

FIG. 4

Reaction of Phenyl Vinyl Ketone with Benzylamine
Dependence of the formal rate constant k' (in reciprocal minutes) on benzylamine concentration $[RNH_2]$.

2. When $k_1[NH_2R] \gg k_{-1}$ the reaction behaved actually as irreversible and change of the phenyl vinyl ketone wave corresponded to equation (3)

$$k't = 2 \cdot 3 \log h/h_0 , \qquad (3)$$

with the same symbols as in (1).

Rate constants were computed by plotting $\log h$ against t. This plot was linear for most reactions studied at conversions lower that about 50% (Fig. 5). Value of k_1 was determined from relation $k' = k_1[NH_2R]$, from the slope of the linear dependence of formal rate constant k' on $[NH_2R]$.

Computation of approximate values of equilibrium constants was carried out using equation (4)

$$K = \frac{h_{MB}}{h_e \cdot [RNH_2]} \qquad (4)$$

where h_{MB} is the height of the Mannich base wave at equilibrium. The reaction mixture was prepared in most cases in the ratio $RNH_2 : RNH_3Cl = 1 : 1$ at $\mu = 0 \cdot 5$. The concentration of the free amine was changed between $5 \cdot 10^{-4}$M and $2 \cdot 10^{-3}$M, that of phenyl vinyl ketone was kept at $5 \cdot 10^{-5}$M. The reaction mixture was left for 24—48 hours before the wave-heights were measured. The concentration of the free amine in equation (4) was determined titrimetrically after the measurement of polarographic waves has been carried out. As the value of K has been computed only for few values of $[RNH_2]$, resulting values of K are approximate and valid only for the ionic strength used.

Measurements of half-wave potentials. Half-wave potentials recorded at 100 mV/absc. were measured against the half-wave potential of thallous ions ($-0 \cdot 445$ V, SCE) and expressed against saturated calomel electrode.

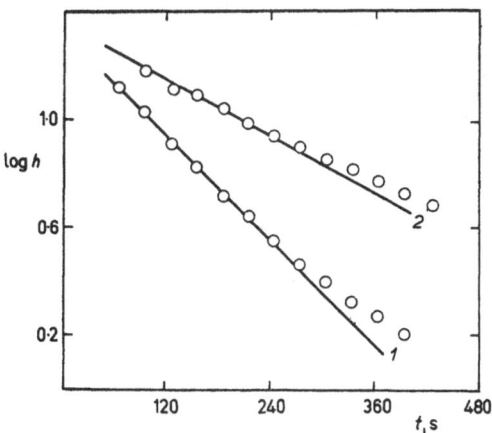

FIG. 5

Reaction of Phenyl Vinyl Ketone with Benzylamine

$5 \cdot 10^{-5}$M phenyl vinyl ketone, equal concentrations of the free amine and its hydrochloride: 1 0·0028M; 2 0·0011M. For evaluation of equation (3) $\log h$ was plotted against time (t).

RESULTS AND DISCUSSION

REACTION SCHEME

It has been shown[1,35] that the system of elimination-addition reactions of Mannich bases can be described by the scheme $(B)-(D)$.

$$ArCOCH_2CH_2NRH_2^{(+)} \underset{fast}{\overset{K_{MBH^+}}{\rightleftharpoons}} ArCOCH_2CH_2NRH + H^{(+)} \qquad (B)$$

$$ArCOCH_2CH_2NRH \underset{k_1}{\overset{k_{-1}}{\rightleftharpoons}} ArCOCH=CH_2 + NRH_2 \qquad (C)$$

$$NRH_3^{(+)} \underset{fast}{\overset{K_{NH^+}}{\rightleftharpoons}} NRH_2 + H^{(+)}. \qquad (D)$$

The analytical concentration of Mannich base (S_{MB}) was defined as $S_{MB} =$ $= [ArCOCH_2CH_2NRH_2^{(+)}] + [ArCOCH_2CH_2NRH]$ and that of amine $S_N =$ $= [NRH_2] + [NRH_3^{(+)}]$. Then the reaction rate defined as the time-change of the unsaturated ketone concentration $[Ke]$ equation (5) was derived:

$$-\frac{d[Ke]}{dt} = k_1[Ke] \frac{S_N}{1 + [H^+]/K_{NH^+}} - k_{-1} \frac{S_{MB}}{1 + [H^+]/K_{MBH^+}}. \qquad (5)$$

For $(S_{MB})_0 = 0$, for $t = 0$, for $pK_{MBH^+} < pK_{NH^+}$ and for $(S_N)_0 \gg (Ke)_0$ if $k_{-1} \ll$ $\ll k_1 S_N$, then for $pH > (pK_{NH^+} + 1)$ it is possible to neglect the first right-hand term and $[H^+]/K_{NH^+}$ against 1 and the equation (5) is transformed into (6):

$$-d[Ke]/dt = k_1 S_N \cdot [Ke]. \qquad (6)$$

Since the amine concentration S_N remains unchanged during the course of the reaction, the reaction under these conditions behaves as an irreversible process and follows the first order kinetics.

For pH comparable with the value of pK_{NH^+} (the other conditions remaining constant) equation (5) can be used in the form (7):

$$-\frac{d[Ke]}{dt} = k_1 \frac{S_N}{1 + [H^+]/K_{NH^+}} [Ke]. \qquad (7)$$

The kinetics correspond again to an irreversible process with the formal rate constant $k' = k_1 S_N/(1 + [H^+]/K_{NH^+})$. At constant analytical amine concentration S_N the formal rate constant k' depends on pH in the form of a dissociation curve with $pK = pK_{NH^+}$.

Other conditions being fulfilled but for k_{-1} comparable with $k_1 S_N$ at $pH > (pK_{NH^+} + 1)$ equation (5) can be used in the form (8), in which $[H^+]/K_{MBH^+}$ can be neglected against 1:

$$- d[Ke]/dt = k_1 S_N[Ke] - k_{-1} S_{MB} \tag{8}$$

The kinetics correspond to a reversible process, with both rate constants pH-independent.

For pH comparable with pK_{NH^+} but greater than pK_{MBH^+} equation (5) reduces to (9):

$$- \frac{d[Ke]}{dt} = k_1 \frac{S_N}{1 + [H^+]/K_{NH^+}} [Ke] - k_{-1} S_{MB} \tag{9}$$

TABLE I

Determined Values of Rate and Equilibrium Constants for the Reaction of Phenyl Vinyl Ketone with Primary Amines RNH_2

Amines used as buffers, $t = 25°C$, $\mu = 0.5$.

Amine	R	$k_1^a \cdot 10$ l mol^{-1} s^{-1}	% Conversion	$k_1^b \cdot 10$ l mol^{-1} s^{-1}	% Conversion	$K_r^c \cdot 10^{-3}$ l mol^{-1}	$k_{-1}^d \cdot 10^5$ s^{-1}
I	CH_3	63·1 ± 2·4	66	56·8 ± 2·2	90	3·5	170
II	C_2H_5	29·9 ± 0·9 21·5 ± 1·7	64	27·6 ± 1·9	86	–	–
III	n-C_4H_9	19·8 ± 1·2	37	–	–	7·0	28
IV	n-C_5H_{11}	27·7 ± 1·2	62	–	–	–	–
V	n-C_6H_{13}	33·5 ± 4·3 42·9 ± 2·7	52	48·4	88	–	–
VI	iso-C_3H_7	9·9 ± 0·5	53	13·5 11·6 ± 0·4	82	3·9	31
VII	t-C_4H_9	2·21 ± 0·07	41	2·47 ± 0·19	70	2·86	7·7
VIII	cyclo-C_6H_{11}	23·0 ± 1·7	70	26·7	95	1·77	13
IX	CH_2CH_2OH	37·8 ± 1·9	94	–	–	–	–
X	$CH_2C_6H_5$	22·9 ± 1·9 24·5 ± 2·7	67	24·8	90	1·8	13
XI	$CH(C_6H_5)_2$	10·53 ± 0·9	53	13·7 11·3 ± 0·5	76	–	–
XII	$CH_2CH=CH_2$	19·8 ± 1·8	64	–	–	–	–
XIII	H	0·29 ± 0·1	45	–	–	0·0014	–

[a] Rate constants of the addition reaction computed for an irreversible reaction using equation (3); [b] rate constants computed for a reversible reaction using equation (1); values without given mean deviation determined graphically; [c] equilibrium constants; [d] rate constants for the elimination reaction computed from K_r and k_1.

Finally when k_{-1} is comparable with $k_1 S_N$ and pH with pK_{MBH^+} equation (5) must be used in its full extent. Usually at pH much smaller than pK_{NH^+} the reaction is too slow to be measured.

Under experimental conditions used in this study when the amine had simultaneously the function of a buffer, it was necessary to restrict ourselves to pH $< (pK_{NH^+} + 1)$. Hence conditions for applications of equations (6) and (8) were not met.

Hence the verification of the scheme $(B) - (D)$ was carried out using equations (7) and (9). At pH $= pK_{NH^+}$ the participation of the elimination reaction depends on the structure of the Mannich base formed. The degree of reversibility is best expressed by the % of conversion to which equation (7) can be applied. For substances VIII (70%) and IX (94%) the reaction can be considered as virtually irreversible. For substances I, II, IV, X and XII equation (7) expressed the concentration change to conversions between 60 and 70% whereas for III, V, VI, VII, XI and XIII between 30 and 60%. Hence all substances with the exception of VIII and IX behaved under conditions used as virtually reversible reactions and equation (9) was used for verification of the scheme $(B) - (D)$ as well as for calculations of the rate constants (Table I). With increasing pH the reverse reaction (elimination) played less role. Accordingly, the conversion to which equation (7) can be used increased with increasing pH.

Further possibility to verify equation (7) and (9) was the effect of pH-changes. These were carried out both at constant analytical amine concentration where the pH-dependence of the formal rate constant for the addition process possessed a shape of a dissociation curve with $pK = pK_{NH^+}$ and at constant concentration of the

TABLE II

Rate Constants for the Reaction of Phenyl Vinyl Ketone with Amines Determined at Various pH- Values

Amines used as buffers, $t = 25°C$, $\mu = 0.5$.

Substance	pH	$\dfrac{k_1 \cdot 10}{1\ \mathrm{mol}^{-1}\,\mathrm{s}^{-1}}$
$C_2H_5NH_2$	9·63	31·3
	10·33	29·9
	11·03	28·4
$C_6H_5CH_2NH_2$	9·0	25·5
	9·4	24·4
	10·4	25·2
$(CH_3)_2CHNH_2$	9·3	11·40
	10·63	11·51
	11·33	11·46

free amine form, where the formal rate constant for the addition process was virtually pH independent (Table II). As $pK_{NH^+} > (pK_{MBH^+} + 1)$ in all cases it was not necessary to consider the protonation of the Mannich base and to use equation (5).

The ionic strength exerted a small effect on the value of rate constant as demonstrated for the reaction of phenyl vinyl ketone with benzylamine (Table III). This is in accordance with the behaviour expected for the reaction of two uncharged particles in the addition reaction (C).

TABLE III

Effect of Ionic Strength on the Reaction of Phenyl Vinyl Ketone with Benzylamine
$5 . 10^{-5}$M-$C_6H_5COCH{=}CH_2$; $2\cdot3 . 10^{-3}$M-$C_6H_5CH_2NH_2$; $2\cdot3 . 10^{-3}$M-$C_6H_5CH_2NH_2.HCl$
sodium chloride added, $t = 25°C$.

μ	0·1	0·2	0·5	0·8	1·0
$k_1 . 10$ $1\,mol^{-1}\,s^{-1}$	24·4	28·4	24·4	21·1	31·8

STRUCTURAL EFFECTS

For structural correlations most reliable values were selected and are given in Table IV. Whereas in the wide range of nucleophilic reagents with pK ranging from 4 to 11 an increasing reactivity towards addition with increasing basicity of the nucleophilic reagent was observed, no such correlation has been found in the presently studied groups of primary amines. The values of $\log(k/k_{CH_3})$ have shown no distinct correlation either with Taft [36] σ^* or with the pK_a-values. Hence polar effects are not responsible alone for the observed changes in the reactivity in the addition reaction. On the other hand, for additions of amino acids to acrylonitrile[18,19] a linear correlation between the rate and dissociation constant has been found.

But also the correlation of $\log(k/k_{CH_3})$ with Taft steric substituent constants[36] E_s is rather poor and the equation $\log(k/k_{CH_3}) = 0\cdot506\,E_s - 0\cdot208$ shows a correlation coefficient $r = 0\cdot78$ and $s = 0\cdot71$. The correlation was somewhat improved $(r = 0\cdot85, s = 0\cdot59)$ when the steric constants were corrected[37] for the number of hydrogen atoms in α-position $[E_s^0 = E_s + 0\cdot306\,(n-3)]$ and equation $\log(k/k_{CH_3}) = 0\cdot40\,E_s^0 - 0\cdot10$ was employed but the improvement is not statistically singificant. Neverthless, it seems to be questionable, whether the steric effects alone can be assumed to be responsible for the observed changes in reactivity and the polar effects considered in the given series as almost constant.

When considering both steric and polar effects a correlation (Fig. 6) can be found for the equation $\log(k/k_{CH_3}) = 0\cdot86\,\sigma^* + 0\cdot58\,E_s - 0\cdot124$ $(r = 0\cdot90$ and $s = 0\cdot49)$

or for the equation $\log (k/k_{CH_3}) = 0.62\sigma^* + 0.42 E_s^0 - 0.054$ $(r = 0.92,\ s = 0.45)$. None of these last two correlations shows a significant improvement when compared with the equations considering steric effects only. Hence present experimental data show a contribution of steric effects but do not allow to decide whether in the addition reaction of primary amines the polar effects contribute considerably to the observed changes in the reactivity when compared with steric effects. The significant contribution of steric effects would be in agreement with a synchronous cyclic

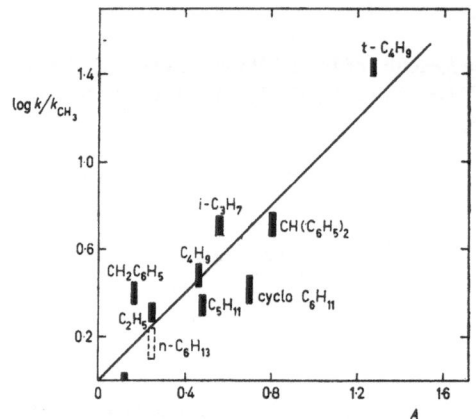

FIG. 6

Dependence of $\log (k/k_{CH_3})$ for the Reaction of Phenyl Vinyl Ketone on the Expression $A = 0.84\ \sigma^* + 0.58\ E_S - 0.124$

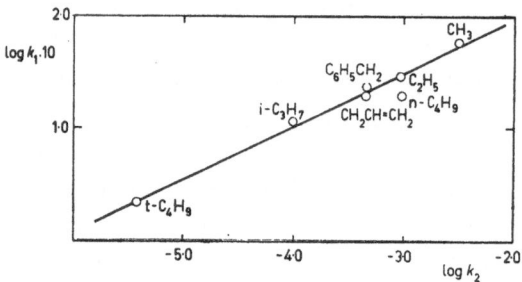

FIG. 7

Correlation of the Reactivity of Primary Amines towards Phenyl Vinyl Ketone and 2·4-Dinitro-chlorobenzene

k_1 Second order rate constant $(1\ \text{mol}^{-1}\ \text{s}^{-1})$ for reaction of amines with phenyl vinyl ketone (this paper) in aqueous ethanol at 25°C; k_2 second order rate constant for reaction of amines with 2,4-dinitrochlorobenzene (according to ref.[38]) in ethanol at 25°C.

mechanism, considered for the elimination addition reaction of type (C) of Mannich bases.

It has been also shown by Hall[38] that whereas dissociation constants of amines can be correlated[39] with polar effects using the Taft equation[36], the rate constants of nucleophilic reactions[38] of amines show a poor correlation with polar substituent constants σ^*. On the other hand it has been shown[38] that the reactivity of amines changes with the structure of amine in reactions of amines with various substrates in an analogous way. It was possible to express the relationship between the nucleophilic reactivities against various substrates quantitatively, analogously as in the Swain-Scott equation[40]. The "nucleophilic constant" n was defined[40], using reactions with 2,3-epoxy-1-propanol and epichlorohydrin as reference reactions.

TABLE IV

Selected Values of Rate Constants for the Reaction of Phenyl Vinyl Ketone with Primary Amines (RNH$_2$) for Structural Correlations

Amine	pK_a	σ^*	E_s (E_s^0)	$k_1 \cdot 10$ l mol^{-1} s^{-1}	$\log k_1 \cdot 10$	$\log (k/k_{CH3})$	pH=10.2 $-E_{\frac{1}{2}}$ V	$dE_{\frac{1}{2}}/dpH$ (αn)
I	10·62	0	0 (0)	59·8 ± 3·5	1·776 ±0·026	0±0·052	1·38	0·08 (1·02)
II	10·63	−0·10	−0·07 (−0·376)	29·4 ± 0·8	1·468 ±0·012	−0·308 ±0·038	1·42	0·08 ·—
III	10·59	−0·13	−0·39 (−0·696)	19·8 ± 1·2	1·296 ±0·026	−0·480 ±0·052	1·355	0·07 (1·09)
IV	10·64	−0·15	−0·40 (−0·706)	27·7 ± 1·2	1·442 ±0·019	−0·334 ±0·045	—	—
V	10·64	−0·15	—	40·03 ± 4·8	1·605 ±0·052	−0·171 ±0·078	1·34	0·02 —
VI	10·63	−0·19	−0·47 (−1·082)	11·6 ± 0·4	1·065 ±0·015	−0·711 ±0·041	1·385	— (1·14)
VII	10·45	−0·30	−1·54 (−2·458)	2·21 ± 0·07	0·344 ±0·014	−1·432 ±0·040	1·405	— (1·02)
VIII	10·64	−0·15	−0·79 (−1·402)	23 ± 1·7	1·362 ±0·033	−0·414 ±0·059	1·327	0·075) (1·18)
IX	9·50	+0·205	—	37·8 ± 1·9	1·577 ±0·022	—	—	—
X	9·40	+0·215	−0·38 (−0·686)	23·4 ± 1·5	1·369 ±0·028	−0·407 ±0·054	1·355	0·07 (1·02)
XI	—	+0·405	−1·76 (−2·372)	11·3 ± 0·5	1·053 ±0·024	−0·723 ±0·050	—	—
XII	9·49	+0·15	—	19·8 ± 1·8	1·298 ±0·020	—	1·37	0·08 —
XIII	9·21	+0·49	+1·24 —	0·29 ± 0·01	−0·538 ±0·016	—	1·42	0·08 —

In a similar way it was possible to correlate our measured rate constants with the logarithms of rate constants[38] of the reaction of amines with 2,4-dinitrochloro-benzene (Fig. 7). Hence it can be concluded that even in the presently studied re-action of amines with α,β-unsaturated ketones the changes in reactivity of the amine with the structure are not governed solely by the type of the polar effects shown in ester hydrolysis[36]. The observed changes in nucleophilic reactivity are rather affected by structural changes in a similar way as in other nucleophilic reactions of amines.

The half-wave potentials (Table IV) have shown practically no correlations either with σ^* or with E_S. Similarly no simple correlations have been found between the logarithms of equilibrium constants (Table I) and σ^*, E_S, k_1 or $E_{1/2}$. Hence both equilibria and half-wave potentials seem to be affected by polar, mesomeric and steric effects of the substituents and the present material is too limited to allow the separa-tion of these effects.

Our thanks are expressed to Dr O. Exner for his assistance in the statistical treatment of structural correlations and to Mr J. Pecka for assistance in the purification of amines.

REFERENCES

1. Čársky P., Zuman P., Horák V.: This Journal 29, 3044 (1964).
2. Bunnett J. F.: Angew. Chem. 74, 731 (1962).
3. Banthorpe D. V.: *Elimination Reactions.* Elsevier, Amsterdam 1963.
4. Patai S., Rappoport Z.: J. Chem. Soc. 1962, 377, 383, 392, 396.
5. Rappoport Z.: J. Chem. Soc. 1963, 4498.
6. Rappoport Z., Degani Ch., Patai S.: J. Chem. Soc. 1963, 4513.
7. Rappoport Z., Grenzaid P., Horowitz A.: J. Chem. Soc. 1964, 1334.
8. Rappoport Z., Horowitz A.: J. Chem. Soc. 1964, 1348.
9. Rappoport Z., Gertler S.: J. Chem. Soc. 1964, 1360.
10. Patai S., Weinstein S., Rappoport Z.: J. Chem. Soc. 1962, 1741.
11. Bergmann E., Ginsburg D., Pappo R.: *Organic Reactions*, Vol. X, p. 179. Wiley, New York 1959.
12. Ingold C. K.: *Otázky struktury a mechanismu v organické chemii*, p. 624. Published by Nakla-datelství ČSAV, Prague 1957.
13. Jones W. J.: J. Chem. Soc. 105, 1547 (1914).
14. Kamlet M. J., Glover D. J.: J. Am. Chem. Soc. 78, 4556 (1956).
15. Hine J., Kaplan L. A.: J. Am. Chem. Soc. 82, 2915 (1960).
16. Ogata Y., Okano M., Furuya Y., Tabushi J.: J. Am. Chem. Soc. 78, 5426 (1956).
17. Schmidt U., Kubitzek H.: Chem. Ber. 93, 866 (1960).
18. Friedman M., Wall J. S.: J. Am. Chem. Soc. 86, 3735 (1964).
19. Friedman M., Cavins J. F., Wall J. S.: J. Am. Chem. Soc. 87, 3672 (1965).
20. Cromwell N. H., Cram D. J.: J. Am. Chem. Soc. 65, 301 (1943).
21. Cromwell N. H.: Chem. Rev. 38, 83 (1946).
22. Rivière H.: Bull. Soc. Chim. France 1962, 1389.
23. Inokawa H.: Bull. Chem. Soc. Japan 37, 568 (1964).
24. Lacey R. N.: J. Chem. Soc. 1960, 1625.
25. Colonge J., Dreux J., Chapurlat R.: Compt. Rend. 251, 252 (1960).
26. Chapurlat R., Dreux J.: Compt. Rend. 253, 2361 (1960).

27. Chapurlat R., Dreux J.: Bull. Soc. Chim. France *1962*, 349.
28. Bertocchio R., Dreux J.: Bull. Soc. Chim. France *1962*, 1809.
29. Sirotanović K., Bajlon-Roćén M., Galović D.: Bull. Soc. Chim. Beograd *25—26*, 506 (1960 to 1961).
30. Horák V., Michl J., Zuman P.: Tetrahedron Letters *21*, 744 (1961).
31. Šestáková-Hrubcová I.: *Thesis*. Charles University, Prague 1964.
32. Zuman P., Krupička J.: This Journal *23*, 598 (1958); Chem. listy *51*, 424 (1957).
33. Zuman P., Michl J.: Nature *192*, 655 (1961).
34. Zuman P., Horák V.: This Journal *27*, 187 (1962).
35. Zuman P., Čárský P.: Z. Physik. Chem. (Leipzig) *227*, 278 (1964).
36. Taft R. W. jr., in the book: *Steric Effects in Organic Chemistry* (M. S. Newman, Ed.). Wiley, New York 1956.
37. Hancock K., Meyers E. A., Yager B. J.: J. Am. Chem. Soc. *83*, 4211 (1961).
38. Hall H. K. jr.: J. Org. Chem. *29*, 3539 (1964).
39. Hall H. K. jr.: J. Am. Chem. Soc. *79*, 4544 (1957).
40. Swain C. G., Scott C. B.: J. Am. Chem. Soc. *75*, 141 (1953).

Translated by the author (P. Z.).

POLAROGRAPHIC REDUCTION
OF ALDEHYDES AND KETONES. V.*

REDUCTION COURSE OF SOME β-KETOSULPHIDES
AND EFFECTS OF ALKYL GROUPS ON SULPHUR

[a]I.Šestáková, [b]J.Pecka and [a]P.Zuman**

[a] *J. Heyrovský Institute of Polarography,*
Czechoslovak Academy of Sciences, and
[b] *Department of Organic Chemistry,*
Charles University, Prague

Received December 27th, 1967

β-Ketosulphides of the type $C_6H_5COCH_2CH(C_6H_5)SR$ are reduced at the carbonyl group. The protonized carbonyl group is reduced in two one-electron steps, the unprotonized form in one two-electron process, which differs from the reduction of corresponding β-aminoketones. The effects of alkyl groups at sulphur and nitrogen on the reduction of the carbonyl group are compared.

In the course of studying the reactions of α,β-unsaturated ketones with various nucleophilic reagents and, on the other hand, the elimination of the β-ketosulphides formed[1] in such reactions, it was of importance to elucidate the course of the electroreduction these β-ketosulphides. Even when the reduction of the carbonyl groups seemed to be probable, it was necessary to prove that reduction of the C—S bond is not involved. This was of particular importance due to the fact that in α-ketosulphides[2] and in phenacylsulphonium salts[3,4] the reduction of the C—S bond has been reported. In connection with studies of substituent effects on half-wave potentials[5] the polar effects of substituents on the sulphur atom in β-ketosulphides were examined.

EXPERIMENTAL

Equipment: Polarographic curves were recorded using a polarograph LP 60 (Laboratorní přístroje, Prague) in a Kalousek vessel with a separated calomel reference electrode. The capillary used possessed the following characteristics: Drop-time $t_1 = 3.5$ s, out-flow velocity $m = 3.05$ mg s^{-1} at 0 V in 0·1M-KCl at mercury head $h = 40$ cm. The half-wave potentials were measured using a three-electrode system, with a QTK-compensator (Metra Blansko) which for well developed

*   Part IV: This Journal *33*, 3213 (1968).
**   Present address: Department of Chemistry, University of Birmingham, Great Britain.

waves was reproducible to ± 0.003 V against the saturated calomel electrode (s.c.e.). pH of the buffer solutions was measured with a glass electrode type G 200 B using a pH-meter PHM 4c (Radiometer Copenhagen). The rectangular voltage polarisation was carried out with the commutator method developed by Kalousek, using a device built in the Polarographic Institute[6]. The auxiliary voltage was chosen in the potential range of the limiting current of the β-ketosulphide and the current–voltage curve corresponding to the electrolysis product was recorded.

TABLE I

Properties of Studied β-Ketosulphides
($C_6H_5COCH_2CHSR$)
$|$
$\overset{|}{C_6H_5}$

No	R	M.p., °C found (reported)	Ref.	Calculated/Found		
				% C	% H	% S
I	CH_3	56·5—57·5 (47—48)	8	75·07 74·96	6·42 6·29	12·51 12·50
II	C_2H_5	66·5—67·5 (66—67)	8	—	—	—
III	$n\text{-}C_3H_7$	48—49·5 (36—40)	8	75·46 76·08	7·04 7·08	11·28 11·27
IV	$i\text{-}C_3H_7$	81—82 (—)	—	75·82 76·08	7·22 7·08	10·91 11·27
V	$n\text{-}C_4H_9$	41·5—42·5 (—)	—	76·37 76·46	7·60 7·43	10·63 10·74
VI	$s\text{-}C_4H_9$	46—47 (—)	—	76·61 76·46	7·59 7·43	10·71 10·74
VII	$t\text{-}C_4H_9$	81 (80—81)	7	—	—	—
VIII	C_6H_5	118—119 (119—120)	9	—	—	—
IX	$CH_2C_6H_5$	70—71 (71)	10	—	—	—
X	cyclo C_6H_{11}	60—605 (—)	—	77·87 77·73	7·52 7·45	9·76 9·88

Substances. β-Ketosulphides were prepared by a base catalysed addition of the corresponding mercaptan to chalcone[7-10]. Constants and analyses ware given in Table I. Chemical used in the preparation of buffer solutions and other supporting electrolytes were Analar Grade.

Procedures. 0·01M stock solution of the β-ketosulphides were prepared in ethanol. Ethanol was added to Britton-Robinson buffers and sulphuric acid solutions so that its final concentration was 40%. Apparent pH-values were measured using the glass electrode in buffer solutions after the addition of ethanol. Sodium perchlorate solution was added so that the final concentration was 0·18M. The stock solution was added to a deaerated buffer solution to ensure a $2 \cdot 10^{-4}$M final solution, containing 42% ethanol. Curves were recorded shortly after mixing and after 10 minutes to check the absence of a homogeneous chemical reaction occurring in the bulk of the solution.

RESULTS AND DISCUSSION

It was possible to study the polarographic behaviour of β-ketosulphides from pH 0 to 11. At pH higher than about 10 the polarographic curves are complicated by the waves of chalcone and mercaptan, which are formed by elimination[1], the rate of which increases with increasing pH.

In acid media polarographic curves show only a one one-electron wave (i_1) the height of which remains practically pH-independent (Fig. 1, 2). At pH above about 4 this wave is accompanied by another one-electron wave (i_2) at more negative poten-

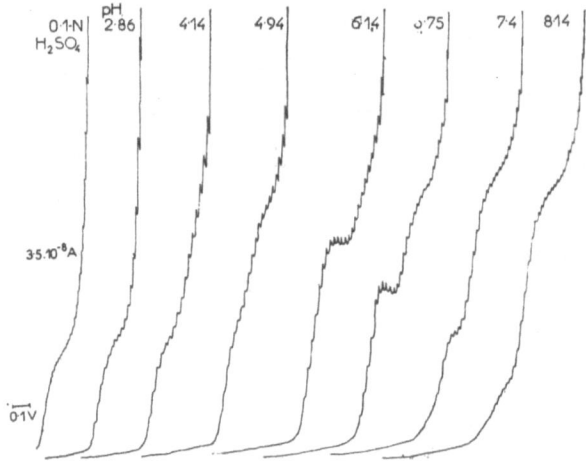

FIG. 1

pH-Dependence of Polarographic Waves of α-Phenacyldibenzylsulphide $(C_6H_5COCH_2CH$. $.(C_6H_5)SCH_2C_6H_5$

$2 . 10^{-4}$M α-Phenacyldibenzylsulphide, sulphuric acid and Britton-Robinson buffers, pH given on the polarogram. Curves starting at -0.8 V, s.c.e., $m = 3.05$ mg/s, $t_1 = 3.5$ s, $h = 40$ cm.

tials. Because the half-wave potential of wave i_1 is shifted towards more negative values with increasing pH, whereas the half-wave potential of wave i_2 remains practically constant (Fig. 3), at pH above 5 these two waves merge. On polarographic curves a single two-electron wave $i_{(1+2)}$ is observed the half-wave potential of which is practically pH-independent. The height of the wave $i_{(1+2)}$ with increasing pH decreases in the form of a dissociation curve (Fig. 2). Simultaneously, another, more negative wave i_3 is formed and its height increases with increasing pH, until between pH 8·5 and 9·5 it reaches a value corresponding to a two-electron process. At pH greater than about 10 the height of the wave i_3 decreases with increasing pH value. The elimination process $(k = 5 . 10^{-3} \, \text{l} \, \text{mol}^{-1} \, \text{s}^{-1})$ prevents an examination of the shape of this decrease at pH about 11 (Fig. 2). The half-wave potential of wave i_3 is practically pH-independent (Fig. 3). At pH > 9 wave i_3 is accompanied by an adsorption pre-wave (i_a) at a potential similar to that of the wave $i_{(1+2)}$.

When waves i_1 and i_3 reach their limiting values the current is diffusion controlled and is linearly proportional to the square root of mercury head. When the height of the wave $i_{(1+2)}$ is small compared with the value corresponding to a diffusion controlled two-electron reduction, it is independent of mercury pressure. Therefore the decrease in current $i_{(1+2)}$ with increasing pH is caused by the decrease in the rate of establishment of an acid–base equilibria antecedent the electrode process proper.

The reduction processes involved are irreversible. This has been proved by the

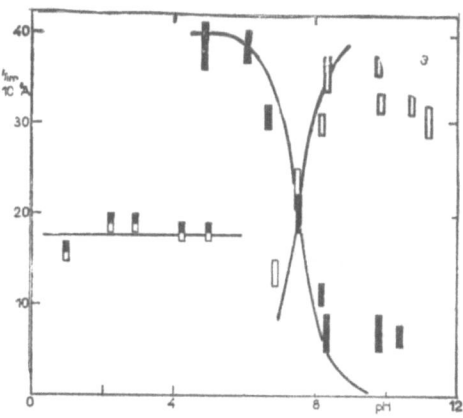

FIG. 2

pH-Dependence of Wave-Heights (i) of α-Phenacyldibenzylsulphide

1 Wave i_1; 2 the sum of waves $i_1 + i_2$; 3 wave i_3.

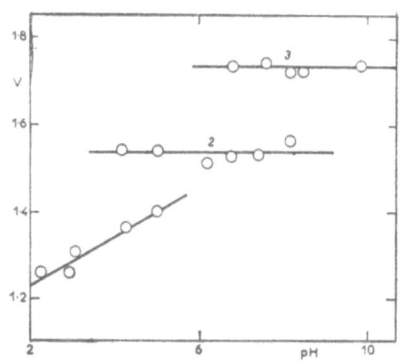

FIG. 3

pH-Dependence of Half-Wave Potentials $(E_{\frac{1}{2}})$ of α-Phenacyldibenzylsulphide

1 Wave i_1; 2 the sum of waves $i_1 + i_2$; 3 wave i_3. Half-wave potentials against s.c.e.

rectangular voltage polarisation, using the commutator according to Kalousek[11]. The check of irreversibility was carried out for the n-butyl derivative V and for the benzyl derivative IX. The auxiliary voltage was chosen for the wave i_1 at $-1\cdot1$ V and $-1\cdot2$ V (S.C.E.) in $0\cdot1$N-H_2SO_4, pH $2\cdot2$ and $2\cdot9$, for the wave $i_{(1+2)}$ at $-1\cdot6$ V (S.C.E.) at pH $6\cdot25$ and for wave i_3 at $-1\cdot8$ V (S.C.E.) at pH $8\cdot0$. In none of these solutions was it possible to prove the formation of mercaptan by the presence of anodic wave.

The absence of mercaptan among the electrolysis products indicates that only the carbonyl group undergoes reduction. Moreover, because the observed behaviour resembles that of aryl alkyl ketones, it is possible to suggest the scheme $A - G$, which is in agreement with the experimental data.

$$\text{pH} < 7 \begin{cases} \text{ArCOCH}_2\text{CH(Ar)SR} + \text{H}^+ \rightleftharpoons \overset{(+)}{\text{ArCOCH}_2\text{CH(Ar)SR}} & (A) \\ \qquad\qquad\qquad\qquad\qquad\quad \overset{|}{\text{H}} \\[4pt] \overset{(+)}{\text{ArCOCH}_2\text{CH(Ar)SR}} - e \xrightarrow{E_1} \text{Ar}\overset{\cdot}{\text{C}}\text{OCH}_2\text{CH(Ar)SR} & (B) \\ \overset{|}{\text{H}} \qquad\qquad\qquad\qquad\qquad\qquad\qquad \overset{|}{\text{H}} \\[4pt] \text{Ar}\overset{\cdot}{\text{C}}\text{OCH}_2\text{CH(Ar)SR} + e \xrightarrow{E_2} \overset{(-)}{\text{ArCOHCH}_2\text{CH(Ar)SR}} & (C) \\ \overset{|}{\text{H}} \\[4pt] \overset{(-)}{\text{ArCOHCH}_2\text{CH(Ar)SR}} + \text{H}^+ \rightleftharpoons \text{ArCHOHCH}_2\text{CH(Ar)SR} & (D) \end{cases}$$

pH > 7 (D) bracket applies

$$\text{pH} > 10 \begin{cases} \text{ArCOCH}_2\text{CH(Ar)SR} + e \xrightarrow{E_3} \overset{(-)}{\text{ArCOCH}_2\text{CH(Ar)SR}} & (E) \\ \overset{(-)}{\text{ArCOCH}_2\text{CH(Ar)SR}} + \text{H}^+ \rightleftharpoons \text{Ar}\overset{\cdot}{\text{C}}\text{CH}_2\text{CH(Ar)SR} & (F) \\ \qquad\qquad\qquad\qquad\qquad\qquad\qquad\quad \overset{|}{\text{OH}} \\[4pt] \overset{(-)}{\text{ArCOCH}_2\text{CH(Ar)SR}} + e \xrightarrow{E_4} \overset{(2-)}{\text{ArCOCH}_2\text{CH(Ar)SR}} & (G) \end{cases}$$

Wave i_1 occurs at potential E_1; its half-wave potential is shifted with pH due to the equilibrium (A), but its wave-height remains pH-independent, because in acid media the establishment of the equilibrium (A) is fast. Wave i_2 occurs at potentials E_2 and corresponds to reaction (C). Its half-wave potential is pH-independent, because the proton-transfer (D) takes place as a consecutive reaction. At pH > 5 potentials E_1 and E_2 become so close that only one wave $i_{(1+2)}$ is observed, this corresponding to the set of reactions $(A)-(D)$.

The decrease in the rate of establishment of equilibrium (A) with increasing pH results in the decrease of the height of wave $i_{(1+2)}$.

At pH > 7 reduction of the unprotonized form following reaction (E) takes place, followed by a proton-transfer step (F). The radical formed can undergo reduction in steps (C) and (D). Because the potential E_2 is more positive than E_3, only one two-electron wave (i_3) is observed. The decrease in the rate of protonation (F) results in the decrease of wave i_3 at pH > 10. In the buffers used the potential at which the radical anion $ArCOCH_2CH(Ar)SR^-$ undergoes reduction (E_4) was too negative to appear before the reduction of the supporting electrolyte. It is therefore imposssible to draw conclusions about the electron, proton sequence in this step.

The polarographic reduction of β-ketosulphides has proved to be analogous to that of alkyl aryl ketones[12] but rather different from that of β-aminoketones[13]. The separation of waves i_1 and i_2 in acid media (or the appearance of a single one-electron wave at lower pH-values), was observed for alkyl and S-alkyl derivatives but was not observed for N-alkyl derivatives. Moreover the separation of the waves of the protonized and the unprotonized carbonyl found for β-ketosulphides has not been observed for β-aminoketones. The difference can be caused by the presence of another centre, which is more easily protonated than the carbonyl group. For β-aminoketones the reduction would correspond (in the entire pH-range studied)

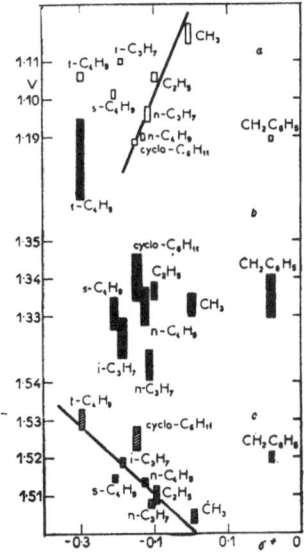

FIG. 4

Dependence of Half-Wave Potentials of 1-Benzoyl-2-phenyl-2-alkylthioethanes (C_6H_5 $.COCH_2CH(C_6H_5)SR$) on the values of Taft Polar Substituent Constants σ_x^*

Half-wave potentials of: *1* Wave i_1 at pH 3·15; *2* wave $i_1 + i_2$ at pH 4·6; *3* wave i_3 at pH 8·6, against s.c.e.

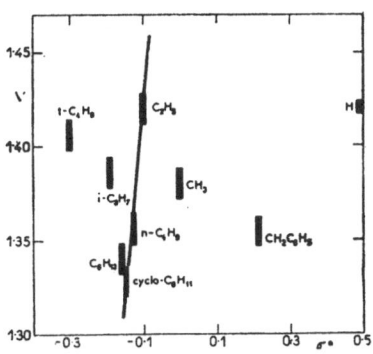

FIG. 5

Dependence of Half-Wave Potentials of N-Alkyl-β-aminopropiophenones ($C_6H_5CO.$ $.CH_2CH_2NHR$) on the values of Taft Polar Substituent Constants σ_x^*

Values against s.c.e.

to the unprotonated carbonyl group and takes place in one two-electron step rather than in two one-electron steps, the latter being characteristic for the reduction of the protonated carbonyl. The shift of half-wave potentials observed for β-aminoketones can then be ascribed to the proton transfer on the amino nitrogen rather than to the preprotonation of the carbonyl group. This view is further substantiated by the fact that formation of a double-positively charged particle, bearing protons on both nitrogen and carbonyl seems rather improbable.

An alternative explanation considered earlier[13], *i.e.* that the amino compounds show a favourable orientation at the electrode surface allowing a practically simultaneous two-electron uptake or two consecutive one-electron processes, cannot be excluded, even when it seems to be less probable. Also it would be necessary to explain, why β-ketosulphides do not show this favourable orientation. The properties of the amino compounds were considered to be affected by the affinity of nitrogen towards mercury. For the analogous sulphur compounds a greater rather than a smaller affinity towards mercury would be expected. Hence the observed facts, concerning β-aminoketones point in favour of the first explanation.

To express the polar effects of substituents, the half-wave potentials of waves i_1 (at pH 3·1$_5$), i_2 (at pH 4·6) and i_3 (at pH 8·6) were plotted (Fig. 4) against the values of the Taft polar substituent constants[14,15] σ^* similarly as has been done for

TABLE II

Half-Wave Potentials of β-Ketosulphides and β-Aminoketones

R	σ^*	$C_6H_5COCH_2CH(C_6H_5)SR$ $-E_{1/2}$, V			$C_6H_5COCH_2CH_2NHR$ $-E_{1/2}$, V pH 10·2
		pH 3·15 i_1	pH 4·6 i_2	pH 8·3 i_3	
CH_3	0	1·118	1·336	1·504	1·38^a
C_2H_5	−0·10	1·106	1·337	1·510	1·42^a
n-C_3H_7	−0·115	1·096	1·318	1·508	—
i-C_3H_7	−0·19	1·110	1·324	1·519	1·385
n-C_4H_9	−0·13	1·090	1·332	1·514	1·355^b
s-C_4H_9	−0·21	1·102	1·332	1·514	—
t-C_4H_9	−0·30	1·106	1·372	1·531	1·405
n-C_6H_{13}	−0·16	—	—	—	1·34^c
cyclo C_6H_{11}	−0·15	1·089	1·341	1·525	1·327^a
H	+0·49	—	—	—	1·42^a
$CH_2C_6H_5$	+0·215	1·089	1·336	1·520	1·355^b
C_6H_5	+0·60	1·103	1·276	1·512	—

$^a\mathrm{d}E_{1/2}/\mathrm{dpH} = 0·08$; $^b\mathrm{d}E_{1/2}/\mathrm{dpH} = 0·07$; $^c\mathrm{d}E_{1/2}/\mathrm{dpH} = 0·02$.

the half-wave potentials of other alkyl derivatives[5]. The correlation was reasonable for wave i_1 and i_3, evaluation of the correlation for wave i_2 was rendered difficult due to the inaccuracy of the measurement of half-wave potentials under these conditions. For all cases deviations for phenyl and benzyl derivatives were observed. Mesomeric contributions can play a role in addition to the polar effect which were considered. The role of adsorption cannot be excluded. Deviations shown for wave i_1 by branched alkyls indicate a steric effect, possibly involving a change in mechanism. Different signs of reaction constant $\varrho_{\pi 1}^{*} = -0.21$ V and $\varrho_{\pi 3}^{*} = +0.09$ V indicate a difference in the reduction mechanisms between the reduction of the protonated and the unprotonated form. A change of the slope of the $E_{1/2}-\sigma^{*}$ plot with pH has been reported[5] for cases in which the effect of alkyl groups attached to various heteroatoms was compared, *e.g.* for 4-alkoxy-1,2-naphthoquinones and 4-alkylamino-1,2-naphthoquinones. The reason, why alkyl substituents on the sulphur atom should affect in opposite directions the attack on protonized carbonyl (A) and on the unprotonized form (E) is not understood.

A similar plot of the approximate values of half-wave potentials of β-aminoketones[16] against polar substituent constants σ^{*} for the two-electron wave shows a wide scattering of data (Fig. 5, Table II). If one restricts the plot to unbranched alkyl groups the sign of the reaction constant $\varrho_{\pi N}^{*} = -1.82$ seems to resemble that of wave i_1 rather than i_3.

LITERATURE

1. Šestáková I.: *Thesis.* Czechoslovak Academy of Sciences, Prague 1967.
2. Lund H.: Acta Chem. Scand. *14*, 1927 (1960).
3. Zuman P., Tang S. Y.: This Journal *28*, 1524 (1963).
4. Saveant J. M.: Compt. Rend. *257*, 448 (1963).
5. Zuman P.: *Substituent Effects in Organic Polarography.* Plenum Press, New York 1967.
6. Holub I., Němec L.: Unpublished results.
7. Thompson R. B.: Ind. Eng. Chem. *43*, 1638 (1951).
8. Kipnis F., Ornfelt J.: J. Am. Chem. Soc. *71*, 3554 (1949).
9. Meguerian G., Clapp L. B.: J. Am. Chem. Soc. *73*, 486 (1951).
10. Nicolet B. H.: J. Am. Chem. Soc. *57*, 1098 (1935).
11. Kalousek M.: This Journal *13*, 105 (1948); Chem. listy *40*, 149 (1946).
12. Zuman P.: This Journal *33*, 2548 (1968).
13. Zuman P., Horák V.: This Journal *26*, 187 (1961).
14. Taft R. W., jr. in the book: *Steric Effects in Organic Chemistry* (M.S. Newman, Ed.). Wiley, New York 1956.
15. Hoefelmeyer A. B., Hancock C. K.: J. Am. Chem. Soc. *77*, 4746 (1955).
16. Šestáková I., Horák V., Zuman P.: This Journal *31*, 3889 (1966).

Translated by the author (P. Z.).

FISSION OF ACTIVATED CARBON–NITROGEN AND CARBON–SULFUR BONDS. VIII.*

ELIMINATION OF β-PIPERIDINOETHYL PHENYL SULFONE METHOIODIDE AND REACTION OF PHENYL VINYL SULFONE WITH HYDROXYL IONS

I.ŠESTÁKOVÁ, P.ZUMAN and V.HORÁK

*J. Heyrovský Institute of Polarography,
Czechoslovak Academy of Sciences
and Institute of Organic Chemistry,
Charles University, Prague*

On the 60th birthday of Academician R. Brdička.

Received May 19th, 1965

Whereas elimination of β-piperidinoethyl phenyl sulfone (*I*) has not been observed, the Hoffmann exhaustive methylation of its methoiodide *II* in contrast proceeds very rapidly. Phenyl vinyl sulfone (*III*) formed in the reaction reacts with hydroxyl ions at pH > 12 ($k_1 = 6\cdot2\,1\,\mathrm{mol}^{-1}\,\mathrm{s}^{-1}$). The behaviour of various carbonyl and sulfonyl compounds is compared.

In connection with the studies of elimination reactions of the β-amino ketone type[1–3] it was of interest to compare the behaviour of β-amino sulfones under similar conditions.

β-Piperidinoethyl phenyl sulfone (*I*) has not shown any polarographic wave in the whole pH-range studied, *i.e.* at pH $2-14$ in buffer solutions containing sodium ions. As phenyl vinyl sulfone, if it were formed, would exhibit a reduction wave, it can be deduced that the elimination reaction does not occur with this compound under the conditions used.

With a β-piperidinoethyl phenyl sulfone methoiodide (*II*) solution, the wave of phenyl vinyl sulfone (*III*) was observed at pH > 7. At pH > 7 and higher the rate of the elimination process was so fast, that the wave-height immediately attained its maximum value. Hence it was impossible to follow the rate of the elimination process.

The study of the elimination process using the polarographic method at pH < 7 was prevented by the fact that the reduction wave of phenyl vinyl sulfone was obscured by the hydrogen evolution. Even when the stock solution of substance *II* was added to buffers of pH $3-5$ and after a short period of time transfered to a solution

---

* Part VII: This Journal *30*, 4316 (1965).

of pH > 7, elimination occured at these higher pH-values so rapidly that no change in limiting current with time could be observed.

Phenyl vinyl sulfone (*III*) formed either by elimination from substance *II* in the solution or prepared synthetically reacts with hydroxyl ions in alkaline solutions. This reaction is described in some detail here.

EXPERIMENTAL

Apparatus

Polarographic curves were recorded using a polarograph Mark LP 55 (produced by Laboratorní přístroje, Prague), using a water-jacketed cell[4] with a calomel reference electrode separated by an agar-bridge and sintered glass (G 3) disc. The capillary used posessed the following characteristics: $t_1 = 4 \cdot 1$ s, $m = 1 \cdot 0$ mg s^{-1} for $h = 40$ cm. pH measurements were carried out using a glass electrode G 200 B in connection with a pH meter PHM 4 c (produced by Radiometer, Copenhagen). For temperature control a Universal Thermostat U 8 according to Wobser (produced by Mechanik Prüfgeräte Medingen) was used.

Substances

β-*Piperidinoethyl phenyl sulfone* (I) (*cf.* ref.[5]) m.p. 78—80°C (benzene) for $C_{13}H_{19}NO_2S$ (253·4) calculated: 61·71% C, 7·57% H, 5·56% N, 12·64% S; found: 61·73% C, 7·52% H, 5·39% N, 12·82% S.

β-*Piperidinoethyl phenyl sulfone methoiodide* (II) m.p. 170—175°C (lit.[5] 175°C),* $C_{14}H_{22}INO_2S$ (395·4) calculated: 32·10% I, found: 32·05% I.

Phenyl vinyl sulfone (III) was prepared from thiophenol *via* β-hydroxyethyl phenyl sulfide[6], oxidized by hydrogen peroxide to β-hydroxyethyl phenyl sulfone (*IV*)[7] which was dehydrated using sulfuric acid[8]. M.p. 64—67°C (tetrachlormethane, water); lit.[6] 68·5°C. For $C_8H_8O_2S$ (168·2) calculated 57·12% C, 4·79% H, 19·02% S; found: 57·05% C, 4·80% H, 19·04% S.

All chemicals used for the preparation of buffers were analytical grade.

Solutions

Ethanolic stock solutions were prepared $4 \cdot 10^{-3}$M from aminosulfones *I* and *II* and $2 \cdot 10^{-2}$M from sulfone *III*. The buffer solutions used for the polarographic investigations of substances *I—III* were Britton-Robinson and acetate buffers. The concentration of sodium carbonate free hydroxide solutions used for kinetic studies was checked using oxalic acid. The ionic strength was kept constant by addition of 2M-NaCl.

Technique of Polarographic Measurements

An aliquot of a stock solution of the substance to be studied was added to the supporting electrolyte in such a proportion that the final depolarizer concentration was $2.5 \cdot 10^{-4}$M. The concentration of ethanol was 1·25% for *III* and 5% for *II*. After deareation the polarographic curve was recorded, *i.e.* some 2 to 3 minutes after mixing with the supporting electrolyte.

---

*   Authors erroneously calculated: 33·40% I; found: 33·49% I.

Technique of Kinetic Measurements

The reaction was followed in the polarographic vessel, after the supporting electrolyte, which was used as the reaction medium, was first deaerated. Methoiodide *II* or sulfone *III* stock solution was then added. The concentration of substances *II* and *III* in the reaction mixture was in most instances $2 \cdot 5 \cdot 10^{-4}$M, that of ethanol $1 \cdot 25\%$ for *III* and 5% for *II*. The temperature was kept constant at 25°C and the ionic strength maintained at $\mu = 1 \cdot 0$. When studying the effect of concentration, the concentration of substance *II* was varied between $2 \cdot 5$ and $7 \cdot 5 \cdot 10^{-4}$M ($5-15\%$ ethanol), of substance *III* between $1 \cdot 10^{-4}$ and $1 \cdot 10^{-3}$M ($0 \cdot 5-5\%$ ethanol).

Potentiometric Titrations

For the determination of the pK_a value 2 ml of $0 \cdot 01$M ethanolic solution of substance *I* was titrated by $0 \cdot 1$M-HCl added from Digipet (Manostat Co., New York) using pH-meter PHM 4c. For the acid dissociation constant the value pK_a $7 \cdot 06$ was found (25°C).

RESULTS

Similarly as with methyl vinyl sulfone[9] the compound *III* undergoes a two-electron polarographic reduction, the half-wave potential $(-1 \cdot 47$ V) being pH-independent between pH 7 and 13. Reduction probably occurs at the C=C bond.

The reduction of the saturated sulfone formed in reduction steps at $-1 \cdot 47$ V occurs at too negative potentials so that it is obscured by the current resulting in the electrolysis of the supporting electrolyte similarly as with other aryl alkyl sulfones[10−14], which are reducible in the $-1 \cdot 9-2 \cdot 0$ V range, no waves were observed with sulphones *I* and *II* in buffer solutions containing sodium ions.

The wave of vinyl sulfone *III* decreased with time at pH > 12 (Fig. 1), the logarithm of wave-height beeing a linear function of time (Fig. 2). The slope of this

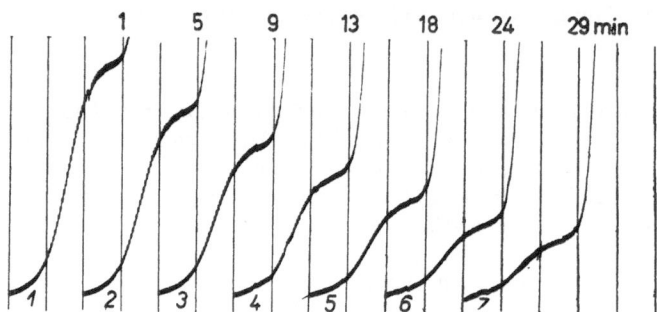

FIG. 1

Decrease of Phenyl Vinyl Sulfone (*III*) Wave in Alkaline Media

$2 \cdot 5 \cdot 10^{-4}$M sulfone *III*, $0 \cdot 19$M-NaOH. Recording of the curves at 25°C started after times given in the polarogram. Curves from $-1 \cdot 2$ V, s.c.e., 200 mV/absc., $h = 40$ cm, full scale sensitivity $2 \cdot 56$ µA.

dependence at a given pH (at which the hydroxyl ion concentration is least a hundred times greater than the concentration of substance *III*) is the effective first order rate constant k' and is practically independent of vinyl sulfone *III* concentration.

The values of the effective rate constants k' are linearly dependent on hydroxyl ion concentration (Fig. 3). The reaction is hence a second order reaction, first order in phenyl vinyl sulfone (*III*) and first order in hydroxyl ions. The rate constant $k = k'/[\text{OH}^-]$ has been found $k = 6{\cdot}2 . 10^{-3} \, \text{l mol}^{-1} \, \text{s}^{-1}$ (25°C) from Fig. 3 for reaction (*A*) probably followed by a rapidly established equilibrium (*B*):

$$C_6H_5SO_2CH{=}CH_2 + OH^{(-)} \xrightarrow{k} C_6H_5SO_2\overset{(-)}{CH}{-}CH_2OH \qquad (A)$$

$$\overset{(-)}{C_6H_5SO_2CHCH_2OH} + H_2O \rightleftharpoons C_6H_5SO_2CH_2CH_2OH + OH^{(-)}. \ (B)$$

The same value of the rate constant k was obtained when a stock solution of metho-

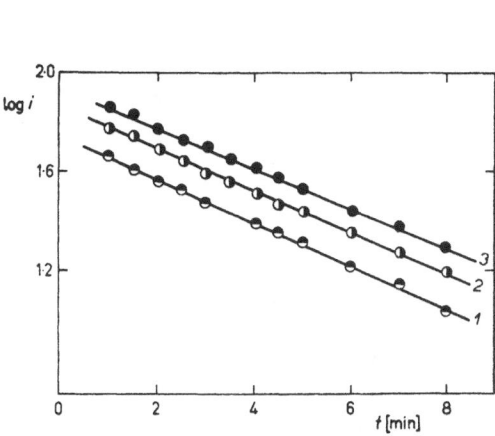

FIG. 2

Time Dependence of the Logarithm of the Limiting Current (*i*) of Phenyl Vinyl Sulfone

Reaction was carried out in $0{\cdot}49_\text{M}$-NaOH at 25°C at phenyl vinyl sulfone concentrations: *1* $4 . 10^{-4}_\text{M}$; *2* $5 . 10^{-4}_\text{M}$; *3* $3 . 10^{-4}_\text{M}$.

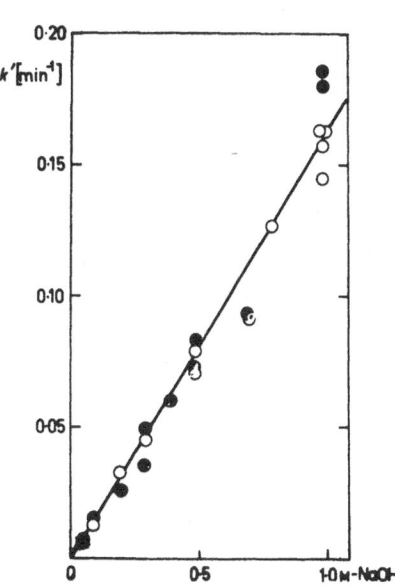

FIG. 3

Dependence of the Formal First Order Rate Constant (*k'*) of the Reaction of Phenyl Vinyl Sulfone on Hydroxyl Ion Concentration

Reaction mixture prepared from: ● $2{\cdot}5 . 10^{-4}_\text{M}$ β-piperidinoethyl phenyl sulfone methoiodide, ○ $2{\cdot}5 . 10^{-4}_\text{M}$ phenyl vinyl sulfone.

iodide *II* was added to hydroxide solutions. This provides an independent proof that phenyl vinyl sulfone is formed in the elimination of methoiodide *II* and also proves that the elimination of methoiodide *II* is a much faster reaction than is the addition (*A*).

Reaction (*A*) ist irreversible. This has been proved by keeping sulfone *III* in sodium hydroxide solutions of concentrations ranging from 10^{-3}M to 10^{-1}M at 50°C in closed bottles for 4 hours. The absence of phenyl vinyl sulfone in all of these solutions proved the irreversibility of reaction (*A*). A control aqueous (neutral) solution of *III* remained unchanged under these conditions.

β-Hydroxyethyl phenyl sulfone (*IV*) formed in reaction (*B*) is not the sole reaction product formed. Formaldehyde has also been detected in 0·1M-NaOH to 1M-NaOH solutions using the colour reaction with 3-methyl-2-benzothiazolon hydrazone[15]. Traces of formaldehyde were detected within 1 minute after the start of the reaction. The formation of formaldehyde has been proved by identification of the corresponding 2,4-dinitrophenyl hydrazone using thin layer chromatography with an authentic sample of formaldehyde 2,4-dinitrophenylhydrazone.

DISCUSSION

It is of interest to compare the studied sulfones *I* − *III* with the corresponding ketones.

Reduction. Only a rough comparison of polarographic behaviour of these substances is possible, as the corresponding reduction processes show different mechanisms. Whereas with ketones of the type $[C_6H_5COCH_2CH_2NC_5H_{10}R^1]^{(+)}$ (for $R^1 = H$ and CH_3) the carbonyl group is reduced[2], no equivalent process can be found for the corresponding sulfones *I* and *II*. The wave at $-1\cdot9$ to $-2\cdot0$ V observed for aryl alkyl sulfones corresponds to a reductive splitting of the C—S bond[11,13,14]. The reduction of phenyl vinyl sulfone is expected to occur at the C=C bond similarly as with the first wave of phenyl vinyl ketone[16]. Whereas the half-wave potentials of the unsaturated keton are pH-dependent (due probably to the protonation of the carbonyl groups at the surface of the electrode), the half-wave potential of the unsaturated sulfone is pH independent. Hence no proton transfer occurs antecedent the electron transfer. This reflects the lower basicity of the sulfonyl group in sulfones when compared with the carbonyl groups in ketones[17,18]. The more negative half-wave potentials of the reduction of the C=C bond in phenyl vinyl sulfone ($-1\cdot47$ V at pH 7−14) than those of the C=C bond in phenyl vinyl ketone ($-0\cdot91$ V in the protonized form at pH 6·3 or at $-1\cdot22$ V in the unprotonized at pH 10·6) is in accordance with the predicted considerably greater conjugation between C=C and COC_6H_5 than between C=C and $SO_2C_6H_5$.

Elimination. The difference in the reactivity of ketones and sulfones can be compared next. Elimination following an E_1 or E_1 mechanism that occured[1,3] with

$C_6H_5COCH_2CH_2NC_5H_{10}$ at pH > 8·3 [*i.e.* at pH > $(pK_{MB} - 1)$, where $pK_{MB} = 9·3$ is the acid dissociation constant of Mannich base] was not observed with sulfone *I* at pH 2 to 14, even when its pK_a value was lower (7·06). Hence the ketone undergoes elimination faster than the sulphone.

Elimination of the methoiodide of the ketone $[C_6H_5COCH_2CH_2N(CH_3)C_5H_{10}]^{(+)}$ occured[1] in a general catalyzed reaction with an E_2 mechanisms at pH > 6. The rate of elimination of the corresponding methoiodide of sulfone *II* at pH > 7 was already too high to be measured; phenyl vinyl sulfone was already present in the reaction mixture in a quantitative yield by the time it was possible to make the first measurement. Owing to that fact that with the applied method it was not possible to measure the phenyl vinyl sulfone formed in this reaction at a lower pH value, no deductions could be made as to the mechanism of this elimination process. Hence it is doubtful whether the comparison of the rates of elimination of the ketone and sulfone methoiodides is strictly possible. It is nevertheless possible to state that at pH 7—9 the rate of elimination of sulfone methoiodide *II* is substantially greater than that of the methoiodide of the corresponding ketone. The consecutive reaction that was observed[19] for $[C_6H_5COCH_2CH_2N(CH_3)C_5H_{10}]^+$ (attributed to the reaction of the phenyl vinyl ketone formed with another molecule of the β-aminoketone) was not observed for *II*. This may be due either to the fact that this consecutive reaction of *II* with *III* is slower than the elimination of *II* (hence no *II* is left available for reaction with *III*), or the C-acid properties of *II* are less pronounced than for the corresponding ketone, or finally phenyl vinyl sulfone *III* is less reactive than phenyl vinyl ketone.

Reactions with hydroxyl ions. Both reactions

$$C_6H_5COCH{=}CH_2 + OH^{(-)} \xrightarrow{k_1} C_6H_5COCH_2CH_2OH \qquad (C)$$

$$C_6H_5SO_2CH{=}CH_2 + OH^{(-)} \xrightarrow{k_2} C_6H_5SO_2CH_2CH_2OH \qquad (D)$$

are virtually irreversible second order reactions, first order in vinyl compound and first order in the hydroxyl ion. The vinyl ketone is more reactive ($k_1 = 4 . 10^{-2}$ l mol^{-1} s^{-1}) than the vinyl sulfone ($k_2 = 6·2 . 10^{-3}$ l mol^{-1}s^{-1}).

Whereas β-hydroxypropiophenone was not isolated but its presence in the reaction mixture was confirmed[17] polarographically, as well as from UV and IR spectra, β-hydroxy ethyl phenyl sulfone *IV* was prepared by an independent method (*cf.* Experimental). This compound *IV* undergoes the retroaldol fission whereas for β-hydroxy prophiophenone the formation of formaldehyde could not have been proved.

In the absence of kinetic studies some conclusions can be drawn from the synthetic work. Addition to α,β-unsaturated sulfone of the type $R^1SO_2CH=CHR^2$ has been observed[20] with water, ethanol in presence of triamylamine, phenol, benzyl-mercaptan and piperidine and for several C-acids[21] in the presence of trimethyl-benzyl ammonium hydroxide. On the other hand with sulfones of the type $R^1SO_2CH=C(CH_3)R^2$ addition has only been obtained[20] with water and ethanol in the presence of sodium, but has not been observed for piperidine, phenol and benzyl-mercaptan. For sulfones of the type $R^1SO_2C(CH_3)=CHR^2$ addition did not occur with any of the studied nucleophils.

From the formation of the less stable *cis*-isomer in the addition of thiophenol to cyclohexenyl-*p*-tolyl sulfone[22] as well as from a comparison[23] of the addition of various alcohols to 2,3-dihydrothiophen-1,1-dioxide with the rate of isomerisation of 3,4-dihydrothiophen-1,1-dioxide it was deduced that the addition follows a synchronous attack of the nucleophilic reagent and the solvent alcohol on the double bond. The kinetics observed in this study does not contradict the proposed mechanism. In order to explain the non-reactivity in the case of $R^1SO_2C(CH_3)=CHR^2$ a more detailed knowledge of the configuration of this type of compound would be necessary.

Comparison. Hence in the reactions of compounds in which the sulfonyl group is adjacent to a methylene group (*e.g.* in Hoffmann degradation of methoiodides or the retroaldolisation of the β-hydroxyderivatives) the polar effect of the $C_6H_5SO_2$ group can be greater than that of the C_6H_5CO group.

In the reactions of compounds, in which the sulfonyl group is adjacent to a double bond, the effect of the $C_6H_5SO_2$ group on the reactivity of the double bond (in the reaction with hydroxyl ions as well as in the polarographic reduction) is smaller than that of the C_6H_5CO group, due to the greater conjugation effect of the latter.

To explain enhanced reactivity of the Mannich base when compared with the analogous sulfonyl compound *I* either a considerable difference in the acid-base properties between the SO_2CH_2 and $COCH_2$ groupings or a participation of the carbonyl oxygen in the transition state formation[17] of the elimination process should be considered.

The authors thank to Drs S. Földeák, B. Matkovics and J. Pórszász, University Szeged, Hungary, for samples. The analyses were carried out in the Analytical Laboratory of the Institute of Chemistry and Biochemistry, Czechoslovak Academy of Sciences, Prague under direction of Dr J. Horáček.

REFERENCES

1. Horák V., Michl J., Zuman P.: Tetrahedron Letters *21*, 744 (1961).
2. Zuman P., Horák V.: This Journal *27*, 187 (1962).
3. Čársky P., Zuman P., Horák V.: This Journal *29*, 3044 (1964).
4. Zuman P., Krupička J.: This Journal *23*, 598 (1958); Chem. listy *51*, 424 (1957).
5. Földeák S., Matkovics B., Pórszász J.: Acta Phys. Chem. Univ. Szeged *6*, 102 (1960).
6. Ford-Moore A. H., Peters R. A., Wakelin R. W.: J. Chem. Soc. *1949*, 1754.

7. Namfara T.: J. Pharm. Soc. Japan *74*, 13 (1954).
8. Kränzlein G., Heyna J., Schumacher W.: German Pat. 842, 198; Chem. Abstr. *47*, 11244 (1953).
9. Johnson C. W., Overberger C. G., Seagers W. J.: J. Am. Chem. Soc. *75*, 1495 (1953).
10. Majranovskij S. G., Neiman M. B.: Dokl. Akad. Nauk SSSR *87*, 805 (1953).
11. Levin E. S., Šestov A. P.: Dokl. Akad. Nauk SSSR *96*, 999 (1954).
12. Horner L., Nickel H.: Chem. Ber. *89*, 1681 (1956).
13. Drushel H. V., Miller J. F.: Anal. Chem. *30*, 1271 (1958).
14. Bowers R. C., Russell H. D.: Anal. Chem. *32*, 405 (1960).
15. Sawicki E., Hauser T. R., Stanley T. W., Elbert W.: Anal. Chem. *33*, 93 (1961).
16. Pasternak R.: Helv. Chim. Acta *31*, 753 (1948).
17. Čárský P., Zuman P., Horák V.: This Journal *30*, 4316 (1965).
18. Arnett E. M. in the book: *Progress in Physical Organic Chemistry*, Vol. 1 (S. G. Cohen, A. Streitwieser, R. W. Taft, Eds), p. 380, 390. Wiley, New York 1963.
19. Michl J.: *Thesis*. Charles University, Prague 1961.
20. Backer H. J., Jong G. J.: Rec. Trav. Chim. *70*, 377 (1951).
21. Truce W. E., Wellisch E.: J. Am. Chem. Soc. *74*, 2881 (1952).
22. Truce W. E., Lewy A. J.: J. Am. Chem. Soc. *83*, 4641 (1961).
23. Procházka M.: *Thesis*. Charles University, Prague 1965.

Translated by the author (P. Z.).

SLOWLY ESTABLISHED ACID–BASE EQUILIBRIUM IN ORGANIC POLAROGRAPHY
TAUTOMERIC CHANGES BETWEEN 3-THIANAPHTHENONE AND 3-HYDROXYTHIANAPHTHENE

N. KUCHARCZYK

Research Institute for Pharmacy and Biochemistry, Prague (Czechoslovakia)

M. ADAMOVSKÝ

Research Institute, Urxovy závody, Ostrava (Czechoslovakia)

V. HORÁK

Department of Organic Chemistry, Charles University, Prague (Czechoslovakia)

P. ZUMAN

J. Heyrovský Institute for Polarography, Czechoslovak Academy of Sciences, Prague (Czechoslovakia)

(Received August 9th, 1965)

Dedicated to Academician A. N. FRUMKIN on his 70th birthday

In the polarographic studies of organic substances, the occurrence of two waves the ratio of which varies with pH, was ascribed, in the early stages of polarography, to the presence of two forms in equilibrium, the equilibrium being either acid–base[1,2] or tautomeric[3]. Nevertheless, it has been shown[4] that in these cases, polarographic limiting currents do not correspond to the equilibrium concentrations (established in the bulk of the solution), but are limited rather by the rate of establishment of the equilibrium which is displaced by electrolysis at the surface of the electrode.

In all the organic systems subsequently reported, where one organic substance showed two waves on the polarographic curves, which changed with pH, the rate of the chemical reaction was high and kinetic currents were observed. Possibly the only exception to this was the waves given by carbonyl compounds in the presence of amine buffers[5].

In the case reported in this paper, 3-thianaphthenone when added to buffer solutions of various pH, shows two polarographic waves which change with pH. With increasing pH, a cathodic wave decreases and an anodic wave increases. The height of both waves is diffusion-controlled even under conditions where the limiting current of the first wave corresponds only to 5% of the total limiting current. This shows that the equilibrium involved, *viz.* between 3-thianaphthenone and the dissociated form of 3-hydroxythianaphthene, has relaxation times that are long when compared with the drop time. Hence, this is a case of a slowly established tautomeric equilibrium.

For 3-thianaphthenone, the isolation of the keto- and enol-forms has not been achieved, although in the case of the related 2-isomer the separation of both forms has been successful[6]. It has been claimed[7,8] on the basis of i.r. spectra that 3-thianaphthenone in the solid state exists mainly in the keto-form. A similar conclusion has been reached for *n*-hexane solutions[9], on the basis of u.v. spectra. The assumption that certain reactions occur preferentially in one form (based on isolation of the products) has often been made[10,11]. Nevertheless, no attempt has been made to treat this type of keto–enol equilibrium quantitatively.

EXPERIMENTAL

Apparatus

The polarographic measurements were carried out using a polarograph Mark LP 55 in connection with an EZ 2 recorder (Laboratorní přístroje, Prague). Full-scale deflection of the recorder (prior to Ayrton shunt settings of the polarograph) was 0.1 μA. The electrolysed solution was placed in a thermostatted Kalousek vessel with a separated calomel electrode (S.C.E.) at $24.8 \pm 0.1°$. The capillary used had the following characteristics (a short-connected circuit, in 1 N KCl): $t_1 = 4.0$ sec, $m = 1.52$ mg sec^{-1} at mercury pressure, $h = 50$ cm. After the recording of the polarographic curve the pH-value was always measured. pH-Measurements were made using an Ionoscope (Křízík, Prague) in connection with a low-resistance glass electrode.

The u.v. spectra were recorded using a Unicam SP 700 spectrophotometer.

Preparative electrolysis was carried out on a manual potentiostat[12]. Reversibility was checked using the commutator method for rectangular voltage polarization[13,14]. Time changes of the instantaneous current (i–t curves) on the first drop were measured on the apparatus built by SMOLER[15].

Materials

3-Thianaphthenone was prepared according to HANSCH AND LINDWALL[16], and then dried under reduced pressure and sublimed at $75°/10$ mm Hg.

3-Methoxythianaphthene was prepared by dissolution of 3-thianaphthenone in ether and the addition of a solution of diazomethane in ether[17]. After 12 h the ether was distilled off, the residue dissolved in tetrachloromethane and chromatographed on alumina (activity II). A yellow oil, b.p. $145–150°/20$ mm Hg, was obtained by distillation of the yellow fraction of the effluent.

All chemicals for the supporting electrolyte preparation were reagent-grade.

Solutions

Solutions of 3-thianaphthenone were prepared in water, and before the recording of the curve, each solution was mixed with the buffer solution, so that the final concentration of the polarographed solution was $5 \cdot 10^{-4}$ M with regard to 3-thianaphthenone.

For the study of 3-methoxythianaphthene, saturated solutions were prepared in Britton-Robinson buffers. Britton-Robinson buffers, 0.05 M borate buffers and dilute sodium hydroxide solutions were used as supporting electrolytes for polarographic studies.

The u.v. spectra were measured in $1 \cdot 10^{-4}$ M solutions in Britton-Robinson buffers.

If traces of oxygen are not carefully removed from the solutions investigated, insoluble red thioindigo is formed and it is, therefore, recommended that all solutions are prepared in the absence of air by using a nitrogen stream for de-aeration.

Electrolytic reduction

     15 mg of 3-thianaphthenone was electrolysed in a nitrogen atmosphere in 100 ml of a buffer solution, pH 7.5, at −1.4 V using a mercury-pool cathode[12]. After at least a 90% reduction had been obtained (in approx. 1 h) the solution was extracted with ether, the extract dried over sodium sulfate and the ether distilled off. 13 mg of the residue was recrystallized from *n*-hexane and yielded 10 mg of a colorless product, mp. 58–59°, that has been proved to be 3-hydroxy-2,3-dihydrothianaphthene[18].

RESULTS

     In buffered solutions of 3-thianaphthenone at pH-values below 7, one single two-electron reduction was observed (Fig. 1). The decrease of this wave at higher pH-

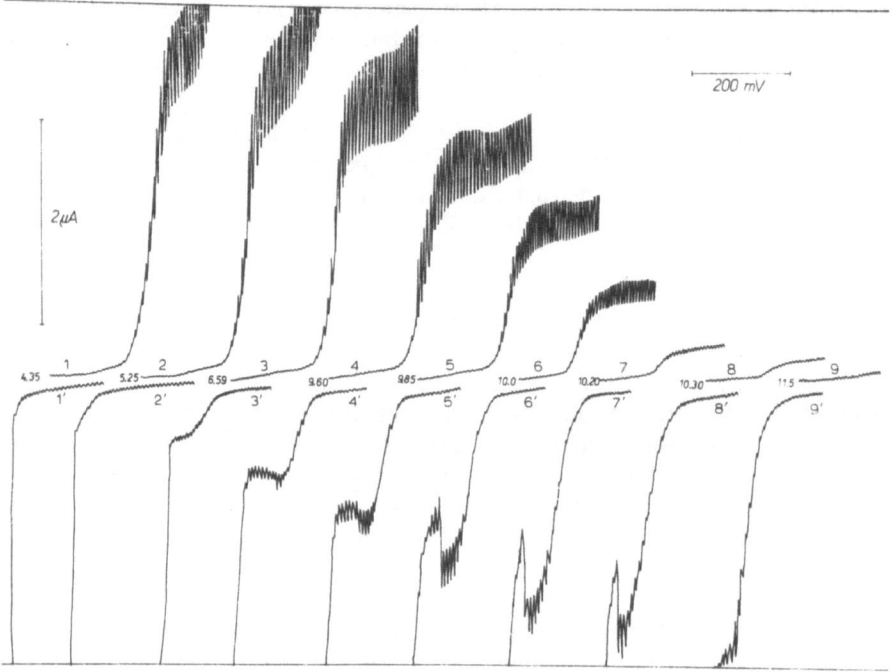

Fig. 1. pH-Dependence of the waves of 3-thianaphthenone ($5:10^{-4}$ M); Britton-Robinson buffers, pH: (1), 4.35; (2), 5.25; (3), 6.6; (4), 9.6; (5), 9.85; (6), 10.0; (7), 10.2; (8), 10.3; (9), 11.5. (1)–(9), reduction waves; (1′)–(9′), oxidation waves. Curves starting at: (1,2), −1.0; (3), −1.1; (4,5), −1.25; (6), −1.3; (7), −1.35; (8), −1.4; (9), −1.45 V. Backward recording of anodic waves started at: (1′–3′), −0.15; (4′), −0.25; (5′), −0.30; (6′)–(9′), −0.25 V. $t_1 = 4.0$ sec.

values corresponds to the dependence predicted for a dissociation curve for an equilibrium

$$AH \rightleftharpoons A^- + H^+.$$

Simultaneously with the decrease of the cathodic wave, an anodic wave at more positive potentials increased. The pH-dependence of the wave-height of the anodic wave also has the form of a dissociation curve, in this case increasing with increasing pH (Fig. 2). At about pH 12, the limiting current of the anodic wave (accompanied by a maximum on the more positive part of the wave) reached the value corresponding to a two-electron oxidation. At pH-values below 9.5, the oxidation wave coincided with a wave corresponding to a mercury salt formation. The anodic wave at pH < 9.5 was no longer linearly proportional to concentration and the form of the i–t curves indicated an adsorption phenomena.

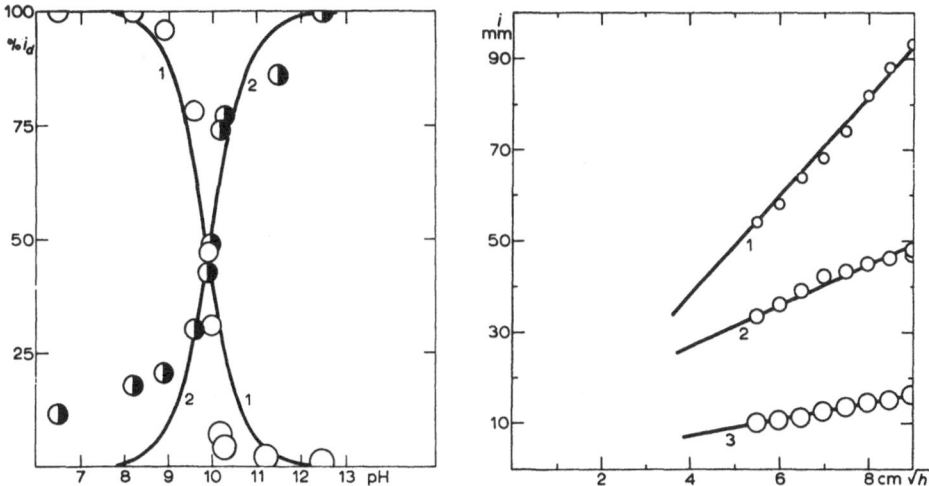

Fig. 2. pH-Dependence of the wave-height (expressed as fraction of the diffusion current—% i_d) of the reduction wave (1) and the anodic wave (2) in the soln. of 3-thianaphthenone.

Fig. 3. Dependence of the wave-height (i) of the cathodic wave of 3-thianaphthenone on the square-root of mercury pressure (h). (1), pH 8.8, $i = i_d$; (2), pH 10.0, $i = 0.3\ i_d$; (3) pH 11.0, $i = 0.05\ i_d$ (the current scale in this case is six times larger than in the other two plots).

For the interpretation of this system, a knowledge of the course of the decrease of the cathodic wave and of the increase of the anodic wave at pH > 9.5 was essential. In the whole pH-region the current was controlled by diffusion. This was proved by the dependence of the wave-height on the square root of the mercury pressure which was linear, not only in the pH-region in which the wave-height corresponded to the maximum value, but also when the limiting current of the cathodic wave was only 5% of the maximum value (Fig. 3). For the anodic wave, the limiting current was proportional to the square root of mercury pressure at pH > 9.5. A further proof of the diffusion-control was given by the i–t curves (Fig. 4), which had the parabolic form $i \sim t^k$ with $k = 0.21$, even for pH-values when the cathodic wave was only 5% of the total current.

An argument in favour of a slowly established equilibrium preceding the electrode process proper, was provided by a comparison of polarographic and spectrophotometric measurements. Absorption spectra show a decrease of the maximum at 235 mμ (corresponding to the keto form) with increasing pH, in the form of a dissociation curve (Fig. 5). Experimental points were in good agreement with the theoretical curve. Similarly, the absorption at 250–260 mμ, corresponding to the anion, increased with increasing pH according to the theoretical curve. The values of the dissociation constants to which the best-fitting dissociation curves correspond, are given in Table 1. The coincidence of the pK-values is within the range of experimental error of both methods and can be considered as a further proof that the ratio of

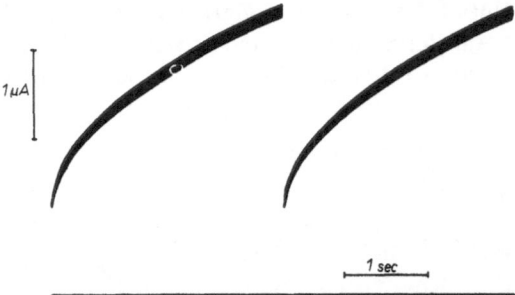

Fig. 4. Change of the instantaneous current with time. $1 \cdot 10^{-3}$ M 3-thianaphthenone; phosphate buffer pH = 11; potential of the limiting current, -1.5 V; zero current recorded; Stylo-galvanometer Mark A 80; swing period, 1/60 sec.

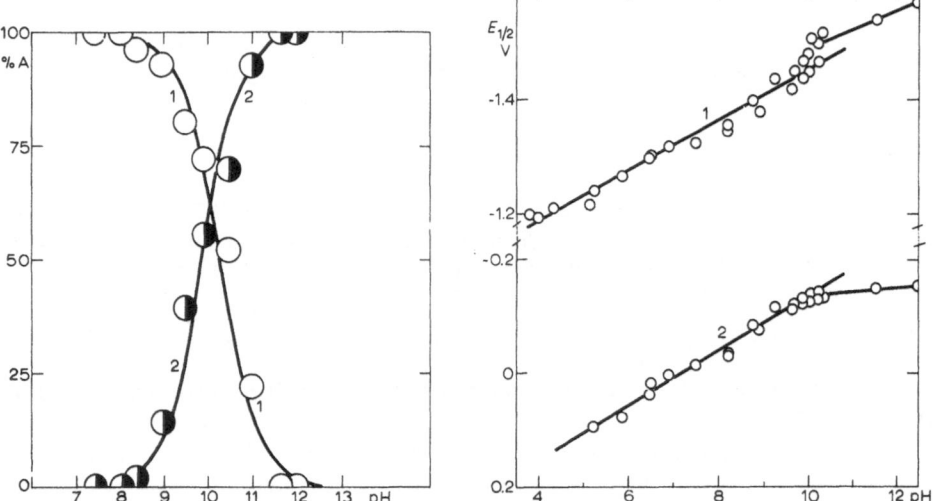

Fig. 5. pH-Dependence of the absorption (% A) of the maximum at (1) 235 mμ; (2), 250–260 mμ of $1 \cdot 10^{-4}$ M 3-thianaphthenone in Britton-Robinson buffers.

Fig. 6. pH-Dependence of the half-wave potentials of the cathodic (1) and anodic (2) wave in $5 \cdot 10^{-4}$ M 3-thianaphthenone, Britton-Robinson buffers. $dE_{1/2}/dpH = 0.055$ V for cathodic and 0.061 V for the anodic wave.

TABLE 1

COMPARISON OF THE pK VALUES DETERMINED

Method*	Quantity measured	Species	pK
Polarography	Cathodic wave	keto	9.85
	Anodic wave	anion	9.97
u.v. spectra	λ_{max}235 mμ	keto	10.25
	λ_{max}250–260 mμ	anion	9.85

* Attempts to measure pK-values potentiometrically failed due to the instability of the anion in the presence of traces of oxygen and to the insolubility of the substance.

polarographic limiting currents actually corresponds to the ratio of concentrations of the two forms in the solution.

Finally, the fact that the pH-dependence of the half-wave potentials shows a discontinuity at pH 10 (Fig. 6) is also in agreement with a slowly established acid–base equilibria. As the spectra were recorded several minutes after dissolution of the substance and no time changes were observed after the first recording*, it can be deduced that the slow establishment corresponds to relaxation periods between one second and several minutes.

Comparison of the wave height with that of benzophenone at pH 8.2, revealed that the limiting current of the cathodic wave at pH < 7 and that of the anodic wave at pH > 12 correspond to a two-electron process. The identification of 3-hydroxy-2,3-dihydrothianaphthene as the reduction product was in agreement with a two-electron process.

The irreversibility of the process was proved by the commutator method of KALOUSEK[13,14]. When, at pH 6, the auxiliary potential was kept at the value of the cathodic limiting current, no anodic wave appeared. When, at pH 11, the auxiliary potential was kept at the potential corresponding to anodic limiting current, a reduction wave at more positive potentials than that of 3-thianaphthenone, appeared. This wave probably corresponds to the reduction of thioindigo.

For 3-methoxythianaphthene, neither cathodic nor anodic waves were observed in the pH-range available.

DISCUSSION

The experimental results are consistent with the following scheme (A)–(E):

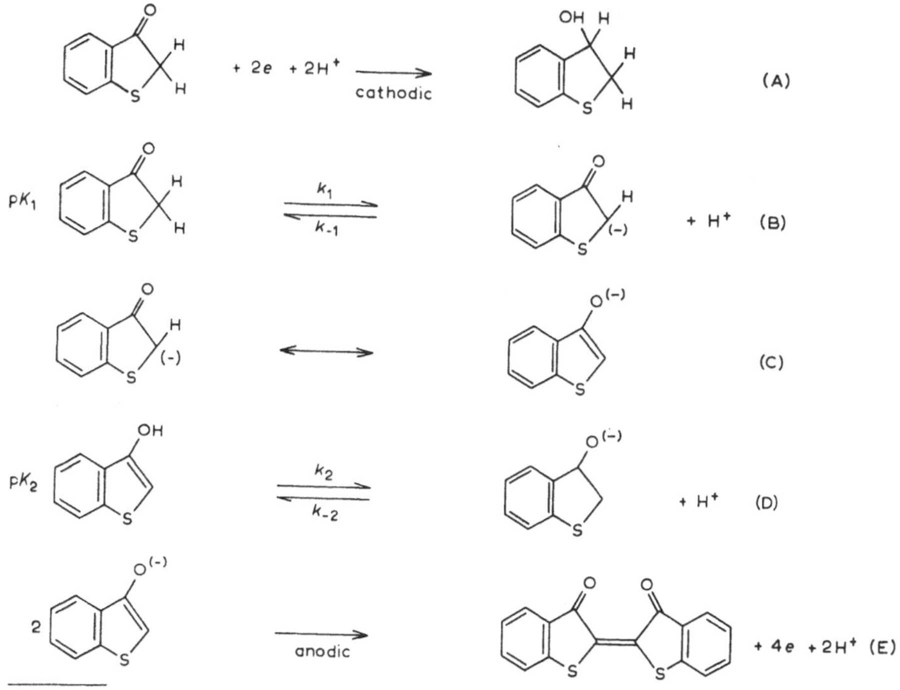

* Similarly, no time changes were observed for the polarographic limiting currents.

The spectrophotometric measurement of the equilibrium concentrations* does not allow any conclusions to be drawn concerning the relative contributions of equilibria (B) and (D) with pK_1 and pK_2. Although it would be logical, in physical organic chemistry, to assume that pK_2 will be smaller than pK_1, no proof based on the equilibrium spectral data solely, is possible.

Polarographic evidence (*i.e.*, the diffusion character of the observed limiting currents) shows that the process of the transformation of the acid form into the basic form is a relatively slow one. This can be understood for a C-H dissociation reaction, but for an O-H dissociation, the rate has, so far, always been found to be high. Moreover, the difference in the spectra of 3-thianaphthenone and 3-methoxythianaphthene in acid media (Fig. 7) as well as the polarographic non-reducibility of the

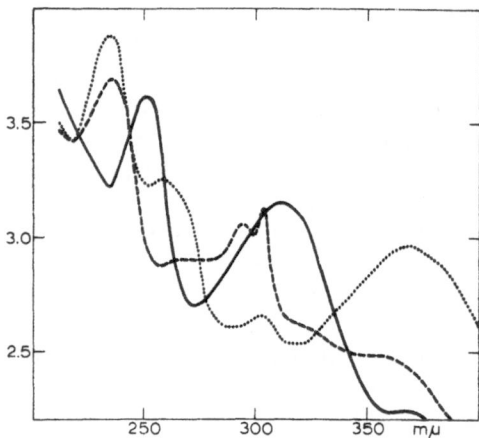

Fig. 7. U.v. absorption spectra. (————), 3-thianaphthenone, pH 12; (. . . . .), 3-thianaphthenone, pH 3.2–7; (- - - - -), 3-methoxythianaphthene, pH 3.2–12. $1 \cdot 10^{-4}$ M 3-thianaphthenone; satd. soln. of 3-methoxythianaphthene; Britton-Robinson buffers.

3-methoxy derivative, indicate that the electroactive form in acid media is the keto-form. It can be assumed that either 3-thianaphthenone predominates in solutions of $pH < 8$ or that the transformation of undissociated 3-hydroxythianaphthene into 3-thianaphthenone is extremely rapid under these conditions. By analogy with other keto–enol equilibria[19], the base-catalysed keto–enol transitions in the scheme (B)–(D) were assumed to take place only *via* the anion. The possibility of a direct transition of type (F) cannot be excluded from the available experimental data, but it is thought to be less probable.

Hence, the sluggish dissociation of a C-acid with the rate constant k_1 is responsible for the unusual polarographic behaviour. With regard to the anion, the present results

---

* Measurements of the rapid time-changes of the u.v. spectra before attainment of equilibrium are being made by Z. GRABOWSKI and his group (Warsaw).

allow no conclusions to be drawn regarding the distribution of electron density in this species. Although the spectra of the anion resemble to a certain degree those of 3-methoxythianaphthene, no conclusions can be drawn from this observation alone.

ACKNOWLEDGEMENT

We thank O. MANOUŠEK and I. SMOLER from the Polarographic Institute, Prague, for their assistance in the controlled-potential electrolysis, commutator method measurements and the recording of the *i–t* curves, and B. KAKÁČ from the Research Institute for Pharmacy and Biochemistry, Prague, for assistance in measurements of u.v. spectra.

SUMMARY

In solutions of 3-thianaphthenone, the cathodic wave of a two-electronreduction decreases and the anodic wave of a two-electron oxidation increases with increasing pH, in accordance with a theoretical acid–base dissociation curve. The limiting currents are diffusion-controlled in the pH-range studied and the polarographically-determined pK-value is in good agreement with the pK-value determined spectrophotometrically. The electrode process is, hence, accompanied by a sluggishly established equilibrium between the reducible keto-form and the oxidizable anion. The slow dissociation of the acid is attributed to the C-acid structure of the keto-form.

REFERENCES

1 A. LANGER, *Ind. Eng. Chem., Anal. Ed.*, 14 (1942) 283.
2 N. H. FURMAN AND C. E. BRICKER, *J. Am. Chem. Soc.*, 64 (1942) 660.
3 J. P. BAUMBERGER AND O. H. MÜLLER, *J. Am. Chem. Soc.*, 61 (1939) 590.
4 R. BRDIČKA, *Collection Czech. Chem. Commun.*, 19, Suppl. 2 (1954) 41.
5 P. ZUMAN, *Collection Czech. Chem. Commun.*, 15 (1950) 839.
6 C. MARSCHALK, *J. Prakt. Chem.*, 88 (1913) 227.
7 K. V. AUWERS AND W. THEISS, *Ber.*, B53 (1920) 2285.
8 F. KRÖHNKE, *Chem. Ber.*, 92 (1959) CXIV.
9 G. M. OKSENGENDLER AND M. A. MOSTOSLAVSKII, *Ukr. Khim. Zh.*, 26 (1960) 69.
10 H. D. HARTOUGH AND S. L. MEISEL, *Compounds with Condensed Thiophene Rings*, Interscience Publishing Co., New York, 1954.
11 A. R. KATRITZKY, *Advances in Heterocyclic Chemistry*, Vols. 1 and 2, Academic Press, New York-London, 1963.
12 O. MANOUŠEK AND P. ZUMAN, *Collection Czech. Chem. Commun.*, 29 (1964) 1718.
13 M. KALOUSEK, *Collection Czech. Chem. Commun.*, 13 (1948) 105.
14 M. RÁLEK AND L. NOVÁK, *Collection Czech. Chem. Commun.*, 21 (1956) 248.
15 I. SMOLER, *J. Electroanal. Chem.*, 6 (1963) 465.
16 C. HANSCH AND H. G. LINDWALL, *J. Org. Chem.*, 10 (1945) 383.
17 R. R. BURTNER AND J. M. BROWN, *J. Am. Chem. Soc.*, 73 (1951) 897.
18 N. KUCHARCZYK AND V. HORÁK, *Chem. & Ind.* London, 1964, 976.
19 R. P. BELL, *Acid-Base Catalysis*, Clarendon Press, Oxford, 1941.

Part V.

Polarographic Reduction and some Reactions of α,β-Unsaturated Carbonyl Compounds

17. P. Zuman, and J. Michl: Role of keto-enol tautomerism in the polarographic reduction of some carbonyl compounds. *Nature* 192, 655 (1961).
18. P. Čársky, P. Zuman, and V. Horák: Fission of activated carbon-nitrogen and carbon-sulphur bonds. VII. Kinetics of ketol formation from α,β-unsaturated ketones in alkaline media. *Collect. Czechoslov. Chem. Commun.* 30, 431 (1965).
19. D. Barnes, and P. Zuman: Polarographic reduction of cinnamaldehyde; comparison with 3-phenylpropionaldehyde and phenylpropargylaldehyde. *J. Electroanal. Chem.* 16, 575 (1968).
20. P. Zuman, J. Tenygl, and M. Brezina: The polarography of steroids. I. Directly reducible ketosteroids. (in German) *Collect. Czechoslov. Chem. Commun.* 19, 46 (1954).

Reduction of unsaturated ketones[17,18,20] and aldehydes[19] in the first two-electron step results in the formation of a saturated ketone or aldehyde. The uptake of the first electron in acidic or alkaline media leads to the formation of a radical which can react with mercury. Radical anions formed can react with protons or with alkali metal cations. The product of the reaction of a radical or radical anion with the cation is reduced in a separate wave, which is governed by the rate of formation of this product, and is suitable for investigation of this reaction.

Carbanion resulting from the first two-electron step must be protonated before further reduction. This protonation is fast with aldehydes,[19] but relatively slow for ketones.[17,18] The addition of hydroxyl ions to α,β-unsaturated ketones[18] and the subsequent retroaldolisation fission were studied. For benzalacetone the retroaldolisation fission is faster than the ketol formation, in phenyl vinyl ketone the ketol formed is stable, and for chalkone both rates are comparable.

Role of Keto-Enol Tautomerism in the Polarographical Reduction of Some Carbonyl Compounds

DURING some polarographical investigations of certain α-aminoketones[1] and on the elimination of Mannich bases[2] experimental data were gathered, which can be explained by the occurrence of keto-enol transformation. Though considerable attention has been paid to the mechanism of polarographical reduction of carbonyl compounds[3], no experimental evidence has been adduced for the participation of the keto-enol tautomerism in the reduction of the carbon-oxygen bond. This follows from the fact that, in most cases examined, the keto form is the predominating form, and in buffered solutions used, establishment of keto-enol tautomeric equilibrium is fast. However, the decrease of the limiting current of some keto acids has recently been interpreted[4] as being influenced by the keto-enol equilibrium in addition to hydration. Experimental evidence for the role of enolization in

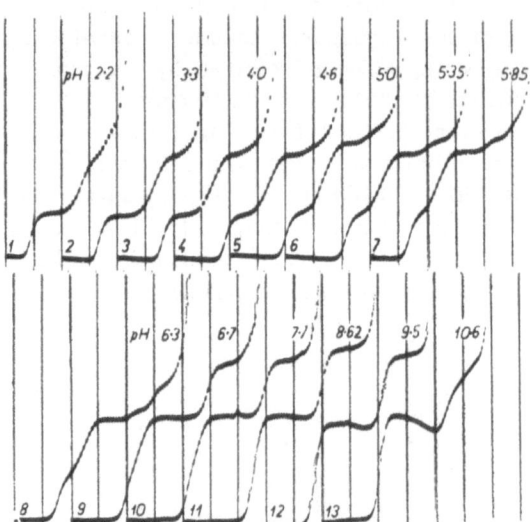

Fig. 1. Dependence on pH of waves fo rphenyl vinyl ketone. Britton–Robinson buffers, $1 \times 10^{-4} M$ phenyl vinyl ketone. Curves starting at: 1–6, -0.4 V.; 7–11, -0.6 V.; 12 and 13, -0.8 V., 200 mV. (absc.), saturated calomel electrode, $h = 56$ cm., $t_1 = 3.0$ sec., $m = 3.3$ mgm./sec., full-scale sensitivity, 2.0 μamp.

the behaviour of substances like acetyl and acetonyl acetone[5] is unconvincing. On the other hand, it has been discovered that the enolate form participates when anodic waves occur, corresponding to the formation of mercury salts, in alkaline solutions of enolizable ketones (Heyrovský, M., private communication). Keto-enol equilibria were involved also in the complicated and, in some ways, questionable scheme derived[6] for the reduction of thenoyl trifluoro-acetone.

Since the conditions for detecting keto-enol tauto-merism in solutions are unfavourable, we have focused attention on cases where the carbonyl compound is formed electrolytically at the surface of the electrode.

As an example of such an electrode process the reduction of phenyl vinyl ketone is first mentioned. In this case (Fig. 1) a two-electron reduction occurs first in two one-electron steps in acid solution, followed by waves at more negative potentials. The half-wave potentials of the latter are practically identical with those for propiophenone. The participation of keto-enol tautomerism is deduced from the dependence of the height of this more negative wave on pH and buffer composition. It is assumed that irreducible enol formed as a reduction product in the first two-electron step is transformed by an acid-base catalysed reaction into the reducible keto form. The rate of formation of the keto form from the enol form governs the limiting current of the more negative wave, which acquires a kinetic character. The assumed irreducibility of the enol form is in agreement with the behaviour of stilbenes. Stilbene itself is reduced at -2.3 V. and the introduction of a hydroxyl group could result ($\rho^* \doteq +0.30$ V., $\sigma'_{OH} = +0.25$) in a shift of maximum by $+0.1$ V.

Hence the mechanism of reduction can be depicted as follows:

$pH < 2$

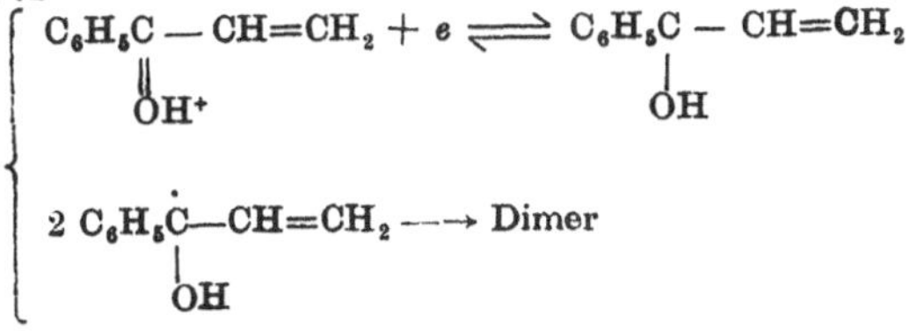

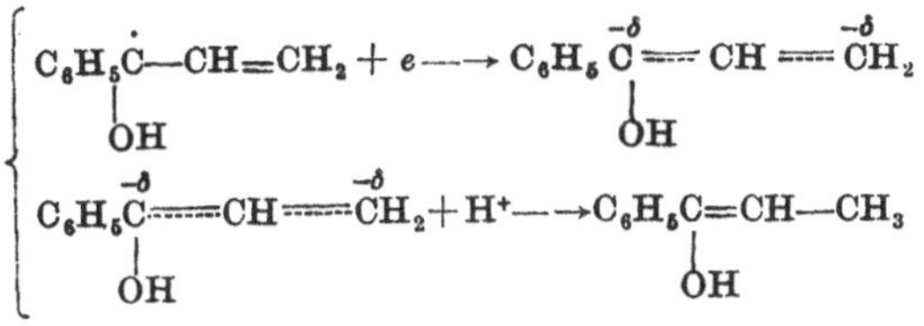

$$C_6H_5C\equiv CH-CH_3 \underset{\text{slow step}}{\overset{\text{acid catalysis}}{\rightleftharpoons}} C_6H_5-C-CH_2\,CH_3$$
$$\underset{OH}{} \qquad\qquad\qquad\qquad \underset{OH^+}{\|}$$

2·1

$$C_6H_5C-CH_2\,CH_3 + e \rightleftharpoons C_6H_5\dot{C}\,CH_2\,CH_3$$
$$\underset{OH^+}{\|} \qquad\qquad\qquad \underset{OH}{|}$$

$$2C_6H_5\dot{C}\,CH_2\,CH_3 \longrightarrow$$
$$\underset{OH}{|}$$

$$\overset{OH}{\underset{}{}} C_6H_5-\overset{|}{\underset{|}{C}}-CH_2\,CH_3$$
$$C_6H_5-\overset{|}{\underset{|}{C}}-CH_2\,CH_3$$
$$\underset{OH}{}$$

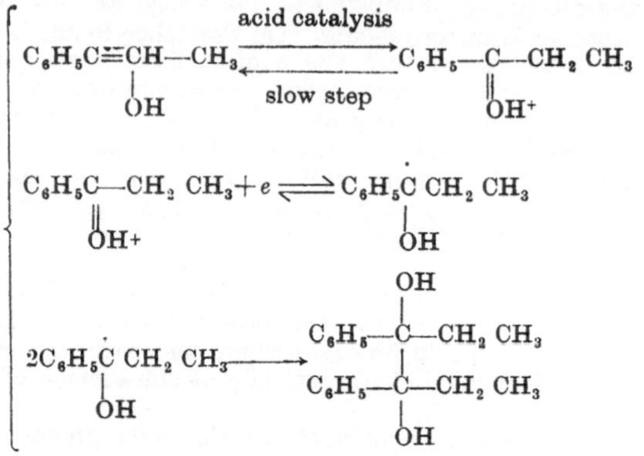

pH 4–6   Waves 1·1 and 1·2 identical as for $pH < 2$.

$$C_6H_5-C=CH-CH_3 \underset{\text{slow step}}{\overset{\text{base catalysis}}{\rightleftharpoons}} C_6H_5\,C\,CH_2\,CH_3$$
$$\underset{OH}{|} \qquad\qquad\qquad\qquad \underset{O}{\|}$$

2·1′

$$C_6H_5C\,CH_2\,CH_3 + H^+ \overset{\text{slow step}}{\longrightarrow} C_6H_5\,C\,CH_2\,CH_3$$
$$\underset{O}{\|} \qquad\qquad\qquad\qquad \underset{OH^+}{\|}$$

$$C_6H_5C\,CH_2\,CH_3 + e \rightleftharpoons C_6H_5\,\dot{C}\,CH_2CH_3$$
$$\underset{OH^+}{\|} \qquad\qquad\qquad \underset{OH}{|}$$

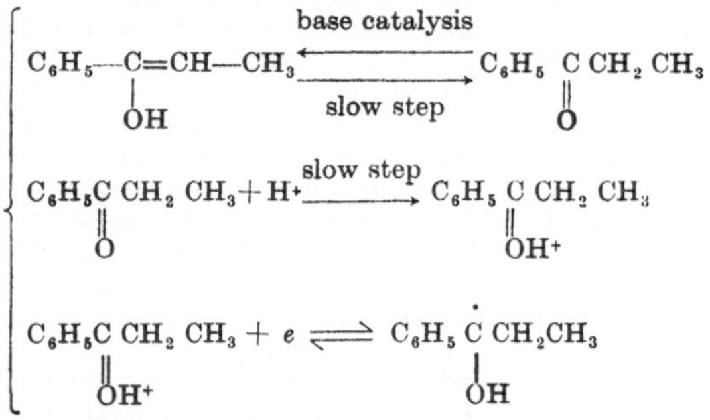

and dimerization as in 2·1

pH 6–9

1·3

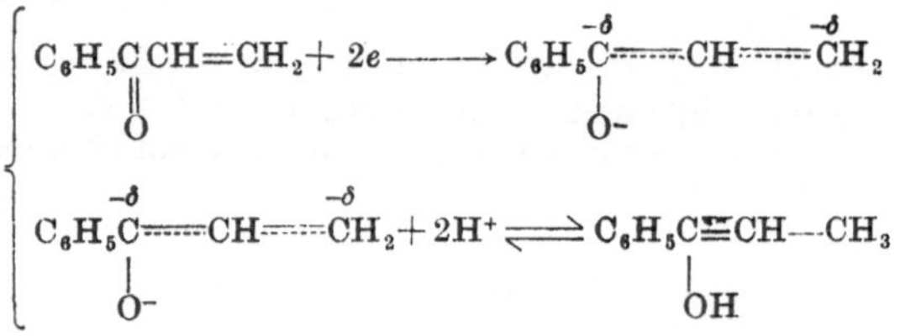

$$C_6H_5CCH=CH_2 + 2e \longrightarrow C_6H_5\overset{-\delta}{C}\cdots CH\cdots\overset{-\delta}{CH_2}$$
$$\underset{O}{\|} \qquad\qquad\qquad\qquad \underset{O^-}{|}$$

$$C_6H_5\overset{-\delta}{C}\cdots CH\cdots\overset{-\delta}{CH_2} + 2H^+ \rightleftharpoons C_6H_5C\equiv CH-CH_3$$
$$\underset{O^-}{|} \qquad\qquad\qquad\qquad \underset{OH}{|}$$

$$C_6H_5\,C=CH-CH_3 \underset{\text{slow step}}{\overset{\text{base catalysis}}{\rightleftharpoons}} C_6H_5\,C\,CH_2\,CH_3$$
$$\underset{OH}{|} \qquad\qquad\qquad\qquad \underset{O}{\|}$$

2·3

$$C_6H_5\,C\,CH_2\,CH_3 + 2e \longrightarrow C_6H_5\,\overset{(-)}{C}\,CH_2\,CH_3$$
$$\underset{O}{\|} \qquad\qquad\qquad\qquad \underset{O^-}{|}$$

$$C_6H_5\,\overset{(-)}{C}\,CH_2\,CH_3 + 2H^+ \longrightarrow C_6H_5\,CH\,CH_2\,CH_3$$
$$\underset{O-}{|} \qquad\qquad\qquad\qquad \underset{OH}{|}$$

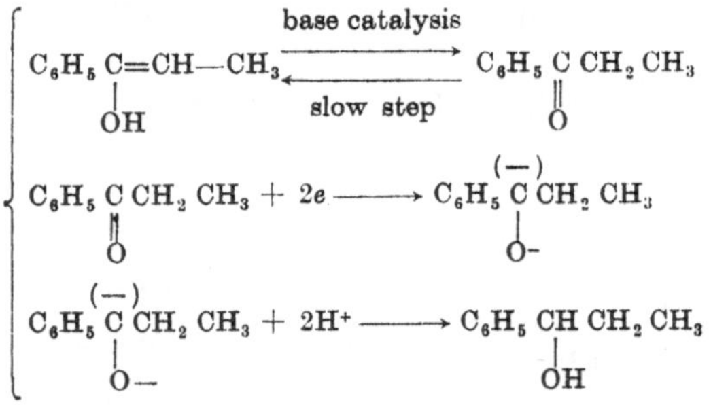

The factors governing the dependence of the limiting current on pH are, in acid solutions, the rate of the acid-catalysed ketone formation, in medium pH-range, either the rate of the base-catalysed ketone formation or of the protonation of the ketone formed, and finally at higher values of pH, base catalysis of the ketone formation.

Essentially the same explanation applies also for the dependence on pH of the limiting currents of acetophenone, when it is formed from $C_6H_5 CO CH_2—X$ (X = halogen, NR_3^+, OR, SR, etc.) observed by Zuman and Horák[1] and Lund[7]. The enolate ion $C_6H_5 CO CH_2(-)$ formed in the first step accepts a proton more quickly on oxygen than on carbon, and the polarographically inactive enol $C_6H_5 C (OH)=CH_2$ is thus the predominating form in the first reduction step.

This enol is transformed either by an acid-catalysed reaction into the protonized form (2·1), when an increase of the one-electron wave with decreasing pH is observed; or by a base-catalysed reaction (2·1′), with subsequent protonation of the ketone formed. The rates of ketone formation and protonation being comparable, the resulting current first increases and then decreases with increase of pH (Fig. 2b). The decreasing part of this dependence, limited by the rate of protonation, is identical with that found for acetophenone. Finally, the unprotonized keto form due to reaction catalysed by a base can be reduced in a two-electron step (2·3) at more negative potentials.

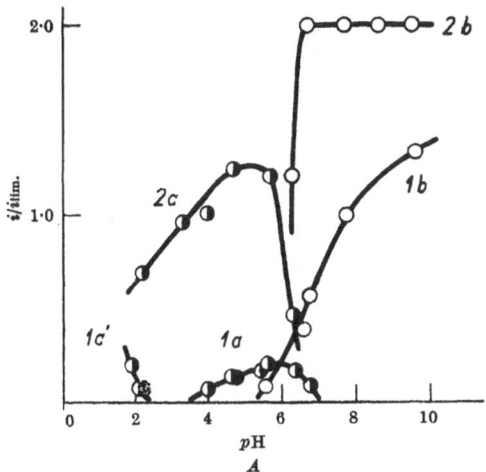

Fig. 2. Dependence on pH of limiting currents
The more negative wave for phenyl vinyl ketone (1a, 1b) compared with waves for propiophenone (2a, 2b). 1a, 2a, one-electron wave; 1b, 2b, two-electron wave

188

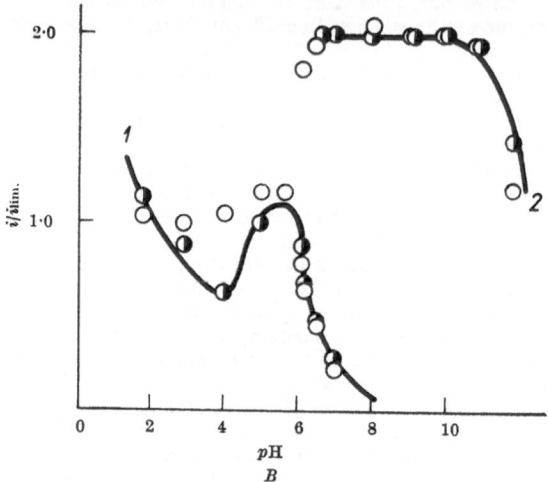

Fig. 2. Dependence on pH of limiting currents
The more negative wave for ω-propiophenone methoiodide
(halved circles) compared with waves for propiophenone (circles).
1, One-electron wave; 2, two-electron wave

In accordance with this supposition the dependence on pH of the more negative wave is little affected by the nature of X in $C_6H_5\,CO\,CH_2$—X, but depends both on substitution in the phenyl ring and on the methylene group[1,7].

The course of the reduction of phenyl vinyl ketone is also consistent with the idea that the reduction of the protonized form leads to inactive $C_6H_5\,CH(OH)CH$ $=CH_2$, while that of the unprotonized form leads to $C_6H_5\,CO\,CH_2\,CH_3$. The increase in height of the wave would be a function of rate of proton transfers only, but this would scarcely explain the increase of the propiophenone wave in acid media.

The effect of keto-enol tautomerism on the polarographic curves of ketones formed at the dropping mercury electrode is being investigated further.

P. ZUMAN

Polarographic Institute,
Czechoslovak Academy of Science.

J. MICHL

Department of Organic Chemistry,
Charles University, Prague.

[1] Zuman, P., and Horák, V., Collect. Czechosloc. Chem. Comm., 26, 176 (1961).
[2] Michl, J., Thesis, Charles University, Prague (1961).
[3] Calzolari, C., and Furlani, C., Ann. Chim. (Roma), 44, 356 (1954). Holleck, L., and Marsen, H., Z. Elektrochem., 57, 301 (1953). Kastening, B., and Holleck, L., ibid., 63, 166 (1959). Elving, P. J., and Leone, J. T., J. Amer. Chem. Soc., 80, 1021 (1958). Nagata, Y., Rev. Polarography, 7, 12 (1957). Mandell, L., Powers, R. M., and Day, jun., R. A., J. Amer. Chem. Soc., 80, 5284 (1958).

[4] Ono, S., Takagi, M., and Wasa, T., *Collect. Czechoslov. Chem. Comm.*, **26**, 141 (1961).
[5] Semerano, G., and Chisini, A., *Gazz. Chim. Ital.*, **66**, 504 (1936). Brdička, R., *Chem. Listy*, **39**, 35 (1945). Vitjajeva, S. I., and Markman, A. L., *Dokl. Akad. Nauk Uzbek, S.S.R.*, No. 8, 33 (1957).
[6] Elving, P. J., and Grodzka, P. G., *Anal. Chem.*, **33**, 2 (1961).
[7] Lund, H., *Acta Chem. Scand.*, **14**, 1927 (1960).

FISSION OF ACTIVATED CARBON–NITROGEN
AND CARBON–SULFUR BONDS. VII.*
KINETICS OF KETOL FORMATION FROM α,β-UNSATURATED
KETONES IN ALKALINE MEDIA

P. Čársky, P. Zuman and V. Horák

J. Heyrovský Institute of Polarography; Czechoslovak Academy of Sciences and Department of Organic Chemistry, Charles University, Prague

Received May 4th, 1965

On the 75 birthday of the Academician J. Heyrovský.

α,β-Unsaturated ketones of the type R—COCH=CHR′ undergo in alkaline media a cleavage described by reactions $(F)-(J)$. According to the nature of groups R and R′ participation of various steps in $(F)-(J)$ can be neglected. Hence with phenyl vinyl ketone an irreversible formation of β-hydroxypropiophenone occurs, whereas with benzalacetone the ketol formed is unstable and benzaldehyde and acetone are produced. For $R = R′ = C_6H_5$, due to a different dependence of $k_1′$, $k_{-1}′$ and $k_2′$ on pH, the reaction scheme varies according to the nature of the starting reactant. When starting with chalcone the cleavage was observed with sodium hydroxide concentrations higher than 0·3M. Under these conditions $k_{-1}′ \ll k_2′$ and hence the ketol VI formed as intermediate splits in a consecutive reaction into benzaldehyde and acetophenone as end products. When starting with ketol VI in 0·003M to 0·1M-NaOH, then $k_1′ \ll k_{-1}′$ and the ketol undergoes cleavage in a competitive reaction partly to chalcone and partly to a mixture of benzaldehyde and acetophenone. At sodium hydroxide concentrations greater than about 0·3M ketol VI yielded practically only benzaldehyde and acetophenone. The reverse aldolisation can be neglected under the conditions used.

In the study of elimination reaction of Mannich bases[1] of the type $C_6H_5COCH_2$. $.CH_2NR_2$ it has been shown that at pH above 10 the phenyl vinyl ketone formed undergoes another reaction besides the addition of amine which takes place even in the absence of amine. In order to avoid this complication in the study of elimination processes so that only the elimination reaction can be studied and not a system of competitive reactions, Mannich bases with pK_a-values well below 10 were selected[1]. For the study of the elimination processes of stronger Mannich bases it was essential that the nature and kinetics of the reactions that phenyl vinyl ketone undergoes be understood. For this reason the cleavage of phenyl vinyl ketone in alkaline solutions has been studied in some detail in this paper.

---

* Part VI: This Journal 29, 2906 (1964).

It has been shown in the preliminary studies that the reaction that phenyl vinyl ketone undergoes is first order in unsaturated ketone and first order in hydroxyl ions. Because the preparation of β-hydroxypropiophenone proved difficult, benzalacetone and chalcone were chosen as further model substances. The latter proved useful not only for the study of the hydration, but of the retroaldolisation fission as well.

Although the hydration of the double bond in α,β-unsaturated ketones is a reaction which is expected to take place generally, relatively little attention has been paid to its kinetics and mechanisms and the interest was devoted mainly to the reactions in acid media. For the behaviour of the system α,β-unsaturated ketone–ketol–(aldehyde + ketone) in alkaline media the following investigations have been reported: Walker and Young[2] studied spectrophotometrically the initial rate of the base-catalysed decomposition of chalcone and the six mononitrochalcones in constant boiling ethanol containing sodium hydroxide. No attempt was made to follow the reaction beyond 20% conversion and hence no experimental evidence for the consecutive and reverse reactions was obtained. Coombs and Evans[3] studied reactions of substituted acetophenones with substituted benzaldehydes using oxime titration in ethanolic media containing ethoxide. They considered dehydration of the ion $C_6H_5CH(O^-)CH_2COC_6H_5$ in one single irreversible step not affecting the kinetics of condensation. Noyce and collaborators[4] started with ketol (VI) and spectrophotometrically measured the rate of both dehydration and retroaldolisation fission. Measurements were carried out in 90% ethanol, but at two hydroxyl ion concentrations only. This limited information prevents us from drawing any deductions on the pH-dependence of the rate constants calculated[4] for the irreversible processes. For aliphatic β-ketols only the fission of the C—C bond was considered (cf.[5–7]).

In addition to these kinetic studies there is a short remark in two papers by Tirouflet[8,9] who studied polarographically the transformation of o-hydroxychalcone to chromanone, that at pH > 9 this reaction is accompanied by a decrease of o-hydroxychalcone concentration which he ascribes to basic hydrolysis yielding aldehyde and ketone. In synthetic studies[10,11] of the condensations of aromatic aldehydes with ketones in alkaline media it has been demonstrated that according to the conditions used either α,β-unsaturated ketone or aldol is the predominating product. A special case of the attack of hydroxyl ions is the formation of hydroxy-p-benzoquinone the kinetics of which has been studied recently[12]. Finally the nucleophilic attack of hydroxyl ions is only a special case of such attacks on C=C bonds that have been recently extensively studied[13–16].

Experimental

Apparatus

The polarograph, electrodes, vessels an pH-meter were the same as were used in Part V (see[1]). UV spectra were recorded on a Uvispek H-700 spectrophotometer (Hilger, Watts, London), using quartz cuvettes with inner diameter 1·00 cm and a quartz prism, IR spectra were recorded on Zeiss UR 10 spectrometer in 0·113 mm NaCl cell.

Substances

Benzaldehyde and acetophenone were pure commercial specimens. Benzaldehyde was freshly distilled. Sodium hydroxide and acetic acid were reagent grade chemicals. Aluminium oxide was neutral with 10% water.

Phenyl vinyl ketone (I) was prepared in aqueous solutions by steam distillation from β-piperidinopropiophenone methoiodide (II)[17] or from bis- β-benzoylethylmethylamine hydrochloride[18]

(*III*). For methoiodide *II* it has already been proved[17] that the amine formed in the elimination from methoiodide *II* is not distilled over with ketone *I*. To prove the absence of methylamine in the distillate of *I* prepared from hydrochloride *III* the resulting solution of ketone *I* was transferred into Britton-Robinson buffer of pH 11·4. At this pH $>$ pK_{amine} methylamine would react with ketone *I* forming Mannich base. Polarographic curves recorded in this solution have shown only two waves[19] of ketone *I*. The wave of Mannich base[20] was absent and appeared only after additon of a small amount of methylamine hydrochloride. Hence the solution of phenyl vinyl ketone (*I*) prepared from hydrochloride *III* is sufficiently pure.

Benzalacetone (IV) was prepared from benzaldehyde and acetone[21]. M.p. 41°C (ref.[21] 40—42°C).

Chalcone (V) was prepared from benzaldehyde and acetophenone[22]. M.p. was in accordance with ref.[22] 55—57°C.

β-*Phenyl-β-hydroxypropiophenone* (VI) was prepared from benzaldehyde and benzoyl acetic acid according to Schöpf[23]. IR spectra were recorded with characteristic absorptions at 3560 cm^{-1} (OH), 1450 cm^{-1} (CH_2), 1580 and 1595 cm^{-1} (C_6H_5) and at 1680 cm^{-1} (C=O).

Product of reaction of phenyl vinyl ketone (I) *with hydroxyl ions*. To an aqueous solution of ketone *I* a solution of sodium hydroxide was added to give a final concentration of 0·07M-NaOH. After standing for 10 minutes the reaction mixture was acidified with acetic acid to pH 5—7. Unreacted ketone *I* was steam distilled off and its absence in the residue was proved polarographically. This solution was used for the study of the polarographic behaviour of the product and for UV-spectra. The acidified solution (prior to steam distillation) was used for measurement of IR-spectra and chromatography. For IR-spectroscopy the aqueous layer was extracted with chloroform which was then dried over magnesium sulphate and in this extract the spectra were measured. For chromatography this solution was further acidified with hydrochloric acid and a saturated solution of 2,4-dinitrophenylhydrazine in 2M-HCl was added. Dinitrophenylhydrazones were separated after 24 hours, dissolved in benzene and chromatographed on a thin layer of aluminium oxide using benzene as solvent.

Product of the retroaldolisation fission. 0·025M chalcone in 0·05M-NaOH containing 50% ethanol was refluxed for 1 hour. The course of the reaction was followed polarographically: Samples were taken and polarographed in a solution corresponding to $2 . 10^{-4}$M chalcone in 0·78M-NaOH containing 20% ethanol. During the reaction the waves of chalcone decreased to some few per cent of original height and the waves of benzaldehyde and acetophenone increased. After cooling the reaction mixture was acidified with hydrochloric acid and an excess of a saturated solution of 2,4-dinitrophenylhydrazine in 2M-HCl added. The 2,4-dinitrophenyl-hydrazones formed were separated and identified chromatographically on a thin layer of aluminium oxide using a system: tetrachloromethane–benzene (17 : 3). The mixture was identified using authentic 2,4-dinitrophenylhydrazones of benzaldehyde (R_F 0·22) and acetophenone (R_F 0·30).

Solutions

The concentration of ketone *I* in the steam distillate was approximately $2 . 10^{-2}$M. This solution was used as stock solution for kinetic studies. Its accurate concentration was determined by comparison of the wave-height of ketone *I* with the wave of β-morpholinopropiophenone hydrobromide[20].

0·01M stock solutions of benzalacetone (*IV*), chalcone (*V*), acetophenone, and benzaldehyde were prepared in ethanol. When kept in refrigerator solutions of ketones *IV* and *V* did not show changes with time. 0·01M stock solution of β-phenyl-β-hydroxypropiophenone (*VI*) in 40% ethanol was more reactive, but even in this case when kept in refrigerator for one week the deterioration was negligible.

Stock solutions of 2M-NaOH were freshly prepared carbonate free.

Polarographic Behaviour of Studied Substances

Phenyl vinyl ketone (*I*) is reduced principally in two two-electron steps[19]. The half-wave potential of the second wave, its shape and pH-dependence is practically identical with the wave of propiophenone. At pH 9·3 the half-wave potentials are $-1·06$ V and $-1·53_5$ V (s.c.e.). In buffered solutions at pH greater than about 10 the height of the first wave decreases with time. The height of the second wave remains practically constant (Fig. 1*). This was interpreted — on the basis of kinetic measurements (p. 4324) and on identification of the reaction product (p. 4326) — as being due to the formation of β-hydroxypropiophenone (VII). In the course of this reaction the concentration of unsaturated ketone *I* decreases which results in the decrease of the more positive wave. β-Hydroxypropiophenone formed in this reaction is reduced in one single two-electron wave (p. 4326), the half-wave potential of which is similar to that of the second wave of phenyl vinyl ketone. As the total consumption of electrons consumed in the reduction of the potential of second wave remains constant, the height of this wave remains unchanged. Hence in the second wave the propiophenone formed at the electrode surface by electroreduction and the β-hydroxy-propiophenone formed in the chemical reaction in the solution are reduced. In order to simplify the interpretation, most kinetic measurements were carried out in solutions of sodium hydroxide in sufficient excess over the unsaturated ketone (in the absence of buffers). Under these conditions the decrease of the more positive wave was again proportional to the decrease of phenyl vinyl ketone concentration, but the height of the second wave increased in the course of the reaction. This is due to the fact that at this high pH value the height of second wave of phenyl vinyl ketone is already substantially smaller than would be, if it corresponded to a two-electron diffusion controlled process (*cf.* Fig. 1*, ref.[19]). The wave-height of β-hydroxypropiophenone on the other hand remains unchanged and corresponds to a two-electron transfer. Hence the lower second wave of phenyl vinyl ketone is replaced in the course of the reaction by the higher β-hydroxypropiophenone wave, with the observed increase of the second wave as the net result. Comparison of curves 6 and 1 (Fig. 2) indicates that the final wave height of the more negative wave (of the reaction product) is equal to the height of the more positive wave of the phenyl vinyl ketone before the beginning of the reaction. The decrease observed on limiting current on curve 1 (Fig. 2) is due to the change in concentration of ketone *I* during recording of the *i–E* curve.

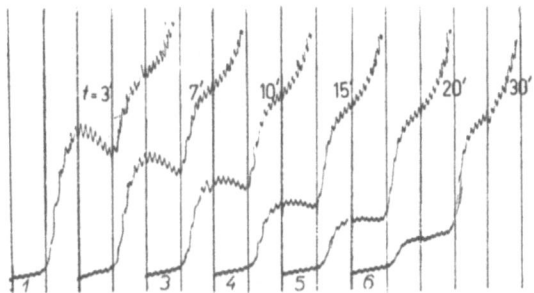

Fig. 2

Reaction of Phenyl Vinyl Ketone (*I*) with Hydroxyl Ions

$2 . 10^{-4}$M ketone *I* in 0·025M-NaOH. Recording started after times given in the polarogram at $-0·8$ V, s.c.e., 200 mV/absc., 400 mV/min, $h = 50$ cm, full scale sensitivity 1·75 μA.

---

*See p. 194

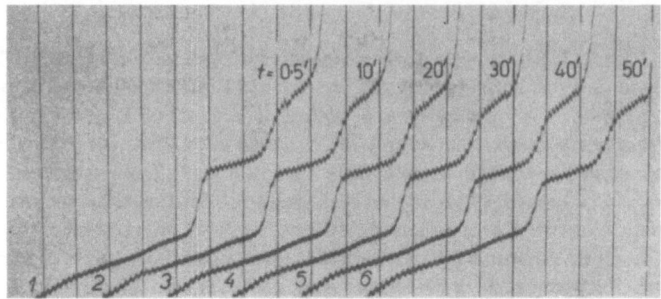

Fig. 1

Reaction of Phenyl Vinyl Ketone (*I*) with Hydroxyl Ions

1 . 10^{-4}M ketone *I*, phosphate buffer pH 11·3. Recording started after times given in the polarogram at 0·0 V, s.c.e., 200 mV/absc., 400 mV/min., *h* = 53 cm, full scale sensitivity 2·45 µA.

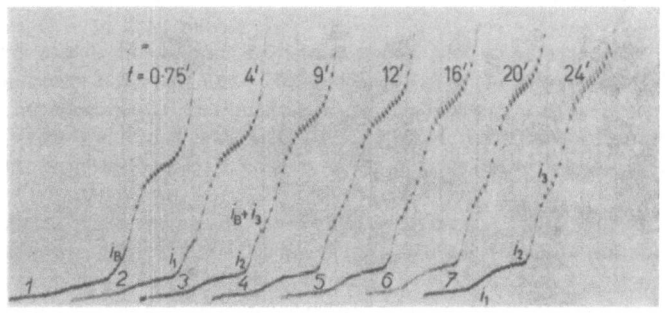

Fig. 8

Changes of Curves in Alkaline Solutions of β-Phenyl-β-hydroxypropiophenone (*VI*)
with Time

1 . 10^{-4}M ketol *VI* in 0·007M-NaOH with 30% ethanol. Recording started after times given in the polarogram at −0·9 V, s.c.e., 200 mV/absc., 400 mV/min, full scale sensitivity 2·45 µA.

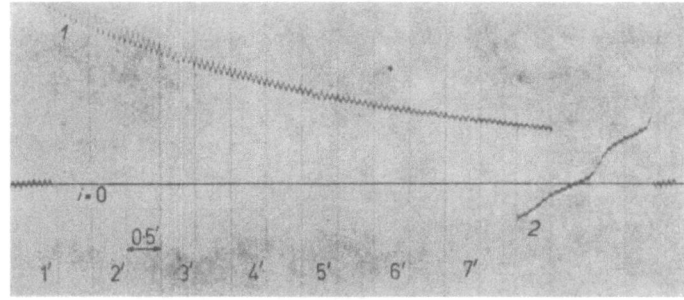

Fig. 9

Reaction of Phenyl Vinyl Ketone (*I*) with Hydroxyl Ions

3 . 10^{-4}M ketone *I* in 0·083M-NaOH, limiting current of *I* recorded at −1·4 V. Recording 1 started one minute after mixing, abscissae recorded each 30 seconds. After 10 minutes complete *i–E* curve 2 was recorded starting at −0·8 V, s.c.e., 200 mV/absc., 400 mV/min, full scale sensitivity 1·05 µA.

The polarographic behaviour of ketone *VII* is analogous to that of unsubstituted propiophenone[20]. In acid solutions only a one-electron wave is observed, the more negative uptake of second electron being obscured by the hydrogen evolution. At pH > 5 the one-electron wave decreases in the form of a steep dissociation curve (Fig. 3a). Another two-electron wave is formed at more negative potentials (Table I). This wave corresponds to a reduction process not involving proton uptake, as it is shown by the practical independence of its half-wave potential*

Table I

Comparison of Half-Wave Potentials ($E_{1/2}$) of β-Hydroxypropiophenone and Propiophenone in Britton-Robinson Buffers

$E_{1/2}$ in volts *vs.* s.c.e.

pH	β-Hydroxypropiophenone		Propiophenone	
	$(E_{1/2})_1$	$(E_{1/2})_2$	$(E_{1/2})_1$	$(E_{1/2})_2$
4·0	−1·23	—	−1·21	—
4·75	−1·27$_5$	—	−1·30	—
5·6	−1·30$_5$	—	−1·34$_5$	—
6·61	−1·32$_5$	−1·45$_5$	−1·37	−1·53
7·04	—	−1·44	—	−1·51
7·61	—	−1·44$_5$	—	−1·51$_5$
8·04	—	−1·45$_5$	—	−1·52$_5$
8·4	—	−1·45	—	−1·52$_5$
9·1	—	−1·47	—	−1·53
10·35	—	−1·52	—	—
11·22	—	−1·51$_5$	—	−1·57

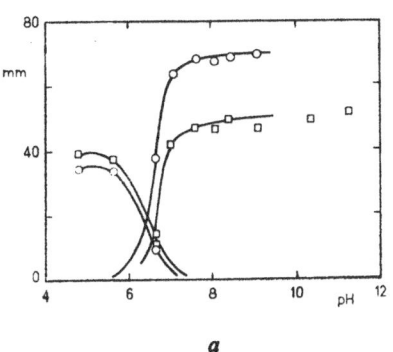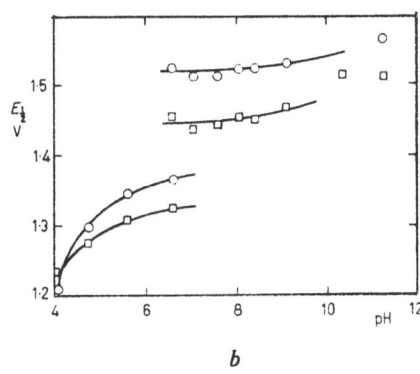

a   b

Fig. 3

Dependence of Wave Heights (*i*) and Half-Wave Potentials ($E_{1/2}$) of β-Hydroxypropiophenone (squares) and Propiophenone (circles) on pH

Wave-heights in Fig. 3a are relative values as exact ketone concentration was unknown. Britton-Robinson buffers, s.c.e.

---

* Small variations observed (Fig. 3b) are probably due to the changes in concentration of cations.

(Fig. 3*b*). The shape of the pH-dependence of the half-wave potential of the one-electron wave in acid media is analogous but not identical with that for propiophenone (Fig. 3*b*).

The polarographic behaviour of benzalacetone at lower pH-values had been found in accordance with published data[24,25]. In alkaline media the two electron wave was split into two waves, the splitting being dependent on the concentration of cations, as observed for other carbonyl compounds[26,27]. In alkaline solutions the first wave decreased, but the height of the more negative wave remained practically constant (Fig. 4). The product of the reaction of benzalacetone with hydroxyl ions ($C_6H_5CH(OH)CH_2COCH_3$) cannot be expected to be active polarographically. The more negative wave has been shown to coalesce with the wave of benzaldehyde in this media.

Chalcone (*V*) is reduced principally in two two-electron steps[28,29]. In solutions of sodium hydro-

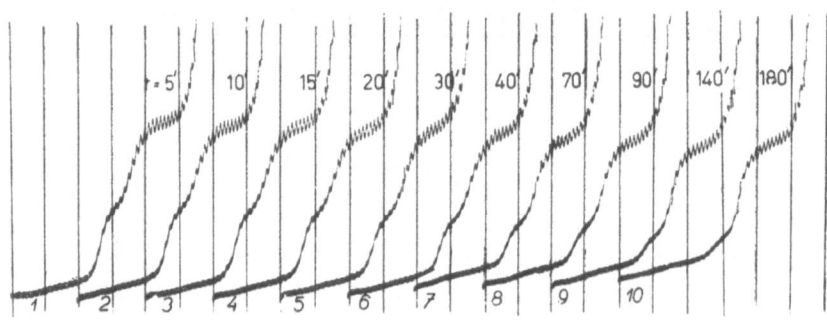

Fig. 4

Reaction of Benzalacetone (*IV*) with Hydroxyl Ions

$2 . 10^{-4}$M ketone *IV* in 1·6M-NaOH with 20% ethanol. Recording started after times given in the polarogram at -0.6 V, s.c.e., 200 mV/absc., 400 mV/min, $h = 35$ cm, full scale sensitivity 1·75 μA.

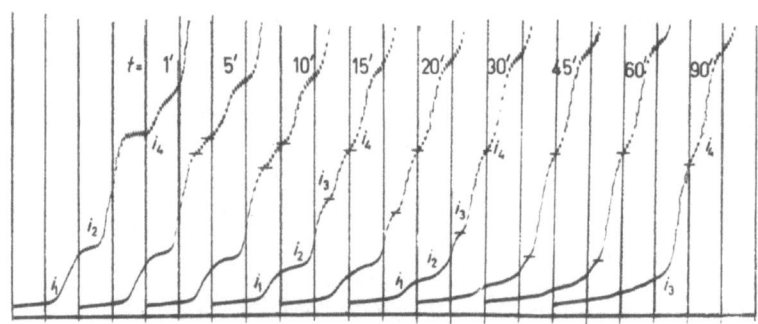

Fig. 5

Reaction of Chalcone (*V*) with Hydroxyl Ions

$2 . 10^{-4}$M ketone *V* in 2M-NaOH with 20% ethanol. Recording started after times given in the polarogram at -0.6 V, s.c.e., 200 mV/absc., 400 mV/min, $h = 77$ cm, full scale sensitivity 3·5 μA.

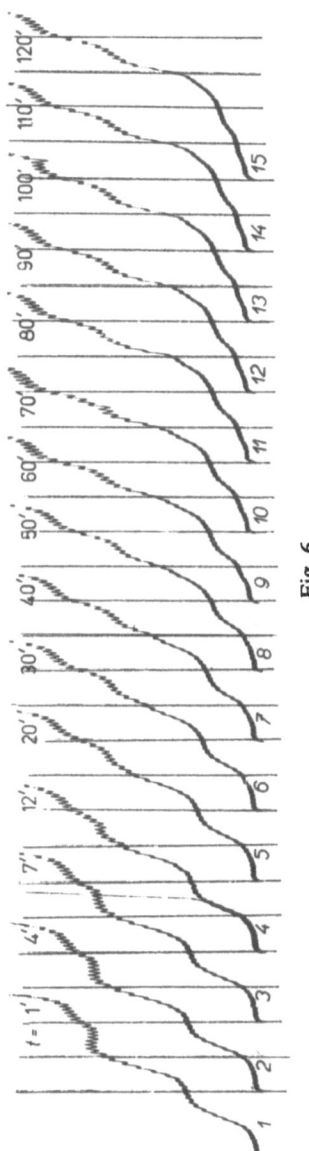

Fig. 6

Reaction of Chalcone (*V*) with Hydroxyl Ions

$2 . 10^{-4}$M ketone *V* in 1M-NaOH with 30% ethanol. Recording started after times given in the polarogram at -0.8 V, s.c.e., 200 mV/absc., 400 mV/min., $h = 53$ cm, full scale sensitivity 1·75 μA.

xide these waves showed changes with time. Even when extrapolated to $t = 0$ this behaviour was still complicated and will be discussed in detail elsewhere[30]. In 0·3M to 2M-NaOH in 20% ethanol chalcone is reduced in two steps (i_1, i_2, Fig. 5) and both of these waves decrease with time and their ratio remains unchanged. Wave i_4 (attributed to ketol *VI*) initially increases; later on wave i_3 (corresponding to benzaldehyde) appears and i_4 (corresponding now to both ketol *VI* and acetophenone) further increases. The ratio of waves $i_3 : i_4$ changes, since at the beginning of the reaction wave i_4 corresponds almost only to ketol and at the end only to acetophenone. When all of the chalcone was consumed, waves i_3 and i_4 were in the ratio of approximately 1 : 1 and corresponded only to benzaldehyde and acetophenone. At higher concentrations of ethanol (30—50%) waves i_2, i_3 and i_4 are less well separated and the time-change is less marked (Fig. 6).

Similarly as with other ketones containing grouping $C_6H_5COCH_2R$, the substance *VI* gives two one-electron waves in acid media, the more negative being overlapped by hydrogen evolution at lower pH-values. The more positive wave is accompanied by an adsorption pre-wave, the height of which limits at $5 . 10^{-4}$M depolarizer. At pH above approximately 5 both one-electron waves merge into one single two-electron wave (i_A). At pH > 6 the height of this wave decreases with increasing pH in a form of a dissociation curve with $pK' = 7.3$, similarly as for acetophenone[31] ($pK' \doteq 6.3$), propiophenone ($pK' \doteq 6.5$) and β-hydroxypropiophenone ($pK' \doteq 6.7$). A more negative two-electron wave (i_B) is formed at the expenses of wave i_A, accompanied by an adsorption pre-wave. The half-wave potentials of the wave i_A are shifted to more negative values ($dE_{1/2}/dpH = 80$ mV/pH) to pH approximately 7. At higher pH values this half-wave potential is pH-independent, similar to the half-wave potential of the more negative two-electron step (i_B) (Fig. 7), the scattering of values in Fig. 7 being result of the differences in buffer composition.

At pH above 13 in 30% ethanol the height of the wave i_B decreases with time and waves of chalcone (i_1 and i_3), benzaldehyde (i_2) and acetophenone (i_3) are formed (Fig. 8). In the most negative wave currents i_B and i_3 coalesce, i_3 being composite of the acetophenone wave and of the second wave of chalcone.

Techniques of Kinetic Measurements

10 ml of reaction mixture was prepared in the polarographic cell from stock solutions to contain $1-2 . 10^{-4}$M ketone, the given concentration of sodium hydroxide, and ethanol (for phenyl vinyl ketone nil, for benzalacetone 20%, for chalcone and ketol VI 20–50%, to increase solubility).

For chalcone and ketol VI polarographic $i-E$ curves were recorded after selected time-intervals (Figs. 5, 6 and 8*). Time changes of waves i_1 and i_2 (of chalcone — description of Fig. 5), i_3 (benzaldehyde) and i_4 (ketol VI and acetophenone) were measured.

For phenyl vinyl ketone and benzalacetone time-changes of the limiting current of the first wave, corresponding to a concentration change of the unsaturated ketone was recorded continuously (Fig. 9*). The condenser current before the addition of ketones I or IV was recorded. To check the appropriate choice of potential an $i-E$ curve was recorded after finishing the record of the current-time curve (Fig. 9*).

Results

Reaction of Phenyl Vinyl Ketone (I)

The time-change of the height of the first wave of ketone I at a given concentration of hydroxyl ions in excess follows first order kinetics. The formal rate constant k' (dependent on hydroxyl ion concentration) was computed using equation: $k't = 2.3 \log h_0 - 2.3 \log h_i$, where h_0 is the limiting current of the first wave of ketone I for $t = 0$, h_i for $t = t_i$. The value of k' was determined graphically from the slope of the

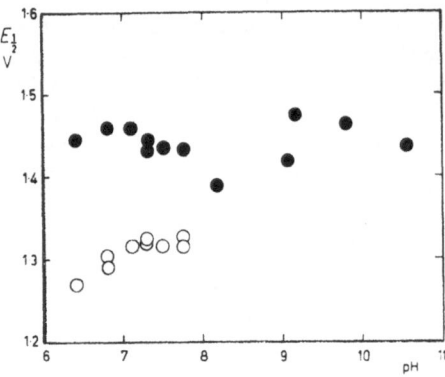

Fig. 7

pH-Dependence of Half-Wave Potentials of β-Phenyl-β-hydroxypropiophenone (VI)

$2 . 10^{-4}$M ketol VI; pH < 8 phosphate, pH > 8 Britton-Robinson buffers, s.c.e., ○ wave i_A, ● wave i_B.

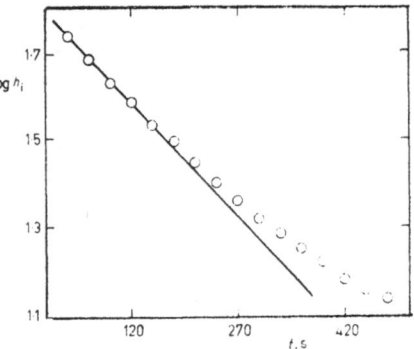

Fig. 10

Reaction of Phenyl Vinyl Ketone (I) with Hydroxyl Ions

Time dependence of logarithm of height of first phenyl vinyl ketone wave. $3 . 10^{-4}$M ketone I in 0·083M-NaOH.

*See p. 194

$\log h_1 - t$ plot (Fig. 10). The linear course was followed to at least 50 per cent conversion, deviations observed at higher conversions probably being due to the reversibility of the reaction. The formal rate constant k' at pH 12·9 is independent of the initial phenyl vinyl ketone concentration (Fig. 11). This, together with the first order kinetics observed at these conditions, makes dimerization as the first and rate determining step improbable. On the other hand a slow dimerization reaction consecutive to the attack of hydroxyl ions cannot be excluded. The formal rate constants k' are a linear function of the hydroxyl ion concentration (Fig. 12). This dependence showing no intercept on the y axis indicates first order reaction in hydroxyl ions and hence the kinetic evidence shows that reaction (A) takes place:

$$C_6H_5CO—CH=CH_2 + OH^- \xrightarrow[(+H^+)]{k_1} C_6H_5COCH_2CH_2OH \qquad (A)$$

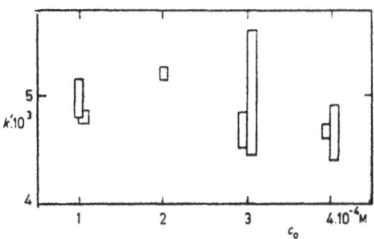

Fig. 11

Reaction of Phenyl Vinyl Ketone (I) with Hydroxyl Ions

Formal rate constant k' as function of the initial concentration (c_0) of ketone I in 0·1M-NaOH.

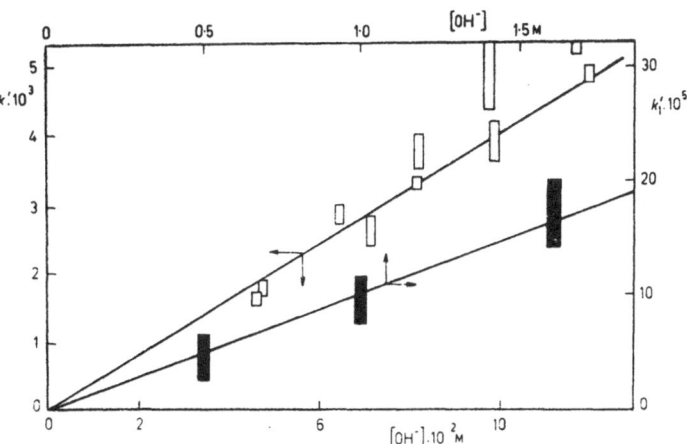

Fig. 12

Reaction of Phenyl Vinyl Ketone (I) and Benzalacetone (IV) with Hydroxyl Ions

Dependence of the formal rate constant k' on hydroxyl ions concentration in sodium hydroxide. $3 . 10^{-4}$M ketone I ($\square$) left and lower axis, $2 . 10^{-4}$M ketone IV 20% ethanol ($\blacksquare$) right and upper axis.

From the slope of the linear plot in Fig. 12 a value of $k_1 = 4 \cdot 10^{-2}$ $1 \text{ mol}^{-1} \text{ s}^{-1}$ was computed.

Attempts to prepare β-hydroxypropiophenone (*VII*) formed in reaction (*A*) by independent routes failed. The product of reaction of *I* in alkaline media was hence

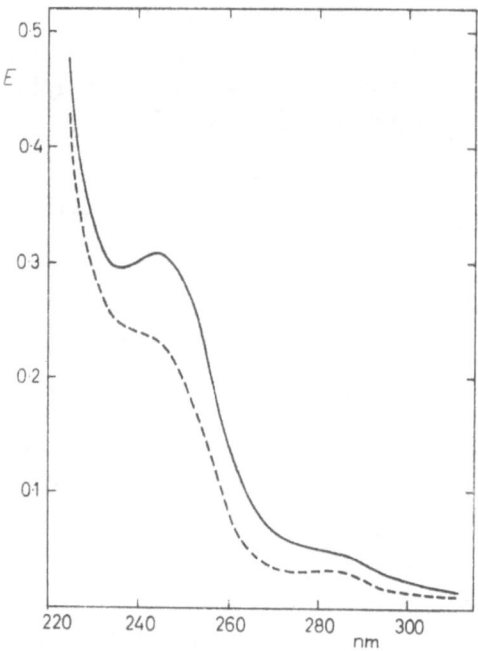

Fig. 13

Electronic Absorption Spectrum of the Product of Reaction of Phenyl Vinyl Ketone with Hydroxyl Ions (full line) and of Propiophenone (dotted line)

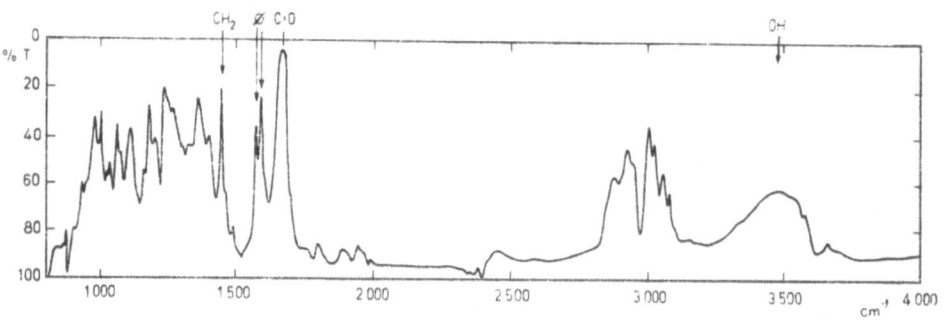

Fig. 14

IR Spectrum of β-Hydroxypropiophenone
Chloroform, 0·113 mm NaCl.

isolated. Its polarographic behaviour was in accordance with that assumed for β-hydroxypropiophenone — i.e. similar to but not identical with that of unsubstituted propiophenone. Such analogy has been proved for both pH-dependence of limiting current (Fig. 3a) and half-wave potentials (Fig. 3b). Similarly absorption spectra in UV-region of the product formed in alkaline media and of propiophenone (Fig. 13) are analogous; this shows that the chromophore part (C_6H_5CO) remained unchanged in the reaction. Chromatographic analysis has proved the absence of formaldehyde and acetophenone in the reaction mixture, showing that ketol *VII* does not undergo retroaldolisation under the conditions used. The intensity of the colouring of the spot for phenyl vinyl ketone decreased during reaction; in the vicinity of the start another spot appeared. This spot was ascribed to ketol *VII*, the change in R_F when compared with propiophenone was ascribed to the interaction between aluminium oxide and the hydroxyl group. But the most conclusive proof (in addition to analogy with chalcone, where the ketol was synthesized – p. 4328) has been given by IR-spectra (Fig. 14). Absorption peaks for hydroxy and methylene groups are present.

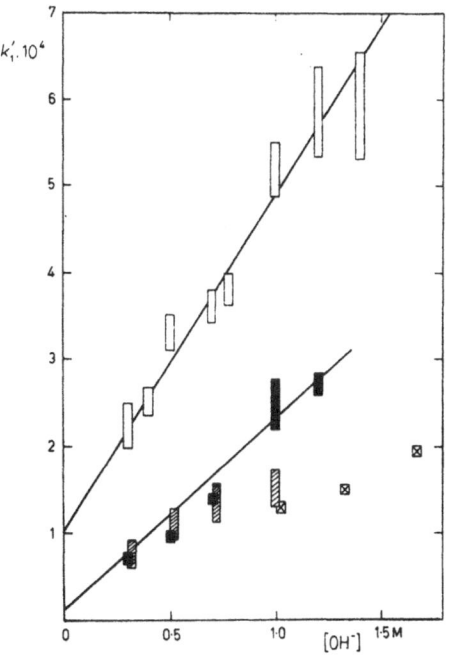

Fig. 15

Reaction of Chalcone with Hydroxyl Ions

Dependence of formal rate constant k_1' on hydroxyl ions concentration, $2 \cdot 10^{-4}$M chalcone.
Ethanol: □ 20%; ■ 30%; ▨ 40%; ⊠ 50%.

Reaction of Benzalacetone (IV)

The decrease in the concentration of ketone *IV* in the reaction with hydroxyl ions using the measurement of the first wave was treated in an analogous way as for ketone *I*. The kinetic behaviour of ketone *IV* resembled that of vinyl ketone *I* in that it was first order in unsaturated ketone *IV* and first order in hydroxyl ions (Fig. 12). From the slope of the linear dependence in Fig. 12 a value for the second order rate constant $k_1 = 1 . 10^{-4}$ 1 mol^{-1} s^{-1} was determined.

Table II

Formal Rate Constants of Reaction of Chalcone in Alkaline Media at Low and High Hydroxide Concentration in 20% Ethanolic Solutions $(k' . 10^4 \text{ in s}^{-1})$

Constant	$5 . 10^{-2}$M-NaOH	1M-NaOH
k'_1	<0·1	3·8
k'_{-1}	2—3	2—3
k'_2	8	30—40

Table III

Second Order Rate Constants (k_1) for Reaction of α,β-Unsaturated Ketones with Hydroxyl Ions (20% ethanol at 25°C, k_1 in 1 mol^{-1} s^{-1})

Ketone		$k_1 . 10^4$
Phenyl Vinyl Ketone	*I*	400[a]
Benzalacetone	*IV*	1
Chalcone	*V*	3·8

[a] Aqueous solution.

Similarly as with phenyl vinyl ketone for the reaction of benzalacetone with hydroxyl ions deviations from simple first order kinetics were observed at higher conversions. Nevertheless, these deviations may also in this case be affected by consecutive reactions. The ketol $C_6H_5CH(OH)CH_2COCH_3$ (*VIII*), the formation of which can be concluded from the first order kinetics and by analogy of the reaction of ketone *IV* with the reactions of ketones *I* and *V*, can be expected to be polarographically inactive. Nevertheless the height of the second wave in Fig. 4 remains unchanged. This demonstrates the formation of a different final product than *VIII*. By coincidence of the half-wave potentials this final product has been shown to be benzaldehyde (the other final product — acetone — being polariographically inactive in the given media). It has been shown that it is possible to separate in the reaction mixture volatile benzaldehyde from nonvolatile ketone *IV* using a stream of nitrogen.

Into a $4 . 10^{-4}$M solution of pure ketone *IV* and benzaldehyde in 0·1M-NaOH (where the reaction is slow compared with that in Fig. 4) nitrogen was purged by a vigorous stream for 20 minutes and the gaseous effluent introduced into washbottles containing 0·1M-NaOH. Polarographic analyses of the washbottles contents have shown no wave in the case of *IV*, but a well developed wave for benzaldehyde.

In order to obtain similar conditions of volatility to the previous experiment, benzalacetone (*IV*) was allowed to react in 0·1M-NaOH. After 15 hours, when 50% conversion of ketone *IV* was achieved, a nitrogen stream was introduced into reaction mixture for 10 minutes and the volatile products collected in 0·1M-NaOH. The benzaldehyde wave was detected in this solution.

Hence in the second wave (Fig. 4) the second reduction step of benzalacetone (IV) coincides with the wave of benzaldehyde and reaction (B) is to be considered*:

$$C_6H_5CH{=}CHCOCH_3 + OH^- \underset{k'_{-1}}{\overset{k'_1}{\rightleftharpoons}} \underset{\underset{OH}{|}}{C_6H_5CHCH_2COCH_3} \underset{k'_{-2}}{\overset{k'_2}{\rightleftharpoons}}$$

$$\underset{k'_{-2}}{\overset{k'_2}{\rightleftharpoons}} C_6H_5CHO + CH_3COCH_3 \tag{B}$$

Because ketol VIII is polarographically inactive and the wave for benzaldehyde is not separated (analyses at other pH-values, at different cation concentrations (cf. p. 4321) or using difference in volatility for separate determination of the change of the benzaldehyde concentration with time were not carried out) no quantitative treatment of reactions with constants k'_{-1}, k'_2 and k'_{-2} can be given. First order irreversible kinetics to at least 50 per cent conversion shows that reaction with constant k'_{-1} does not considerably affect the measured reaction rate. From the fact that the second wave in Fig. 4 remains practically constant it can be deduced that the reaction with constant k'_2 is not only faster than the reaction with constant k'_{-1} ($k'_2 \gg k'_{-1}$), but is also faster than the reaction with constant k'_1 ($k_2 \gg k'_1$).

Reaction of Chalcone (V) and its Ketol VI

The most comprehensive kinetic analysis was carried out for chalcone, as in this case the reactants, intermediates and reaction products could be determined simultaneously. The reaction rates from these separate measures were comparable and the reaction was studied both with the unsaturated ketone V and with ketol VI which had also been prepared synthetically by an independent route.

From a consideration of the kinetic evidence given below the changes of chalcone in alkaline media were described by reaction (C):*

$$C_6H_5COCH{=}CHC_6H_5 + OH^- \underset{k_{-1}}{\overset{k_1}{\rightleftharpoons}} \underset{\underset{OH}{|}}{C_6H_5COCH_2CHC_6H_5} \underset{k_{-2}}{\overset{k_2}{\rightleftharpoons}}$$

$$\underset{k_{-2}}{\overset{k_2}{\rightleftharpoons}} C_6H_5CHO + C_6H_5COCH_3. \tag{C}$$

Since in an equimolar mixture of $2 . 10^{-4}$M benzaldehyde and acetophenone in 1M-NaOH containing 30% ethanol no appreciable formation either of ketol VI or of chalcone has been

---

* Only summarizing formulation, with proton-transfer on RCOCHCH(OH)R′ omitted.
$$\overset{(-)}{}$$

detected, aldolisation reaction with constant k_{-2} can be neglected in all following considerations.

When starting with chalcone changes of polarographic curves with time (Fig. 5) (described in some detail on p. 4322) are only observed when the concentration of hydroxyl ions is greater than about 0·3M. Decrease of both the first (i_1) and second

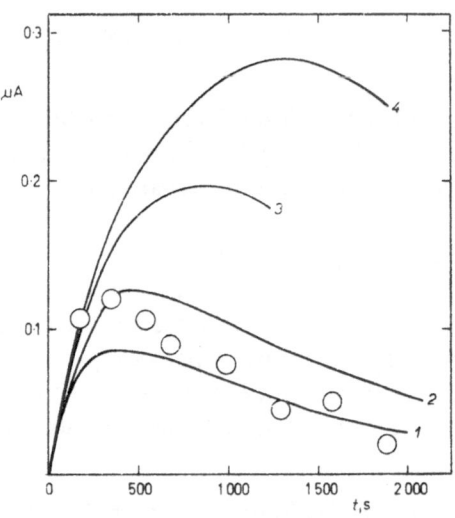

Fig. 16

Time Changes of Limiting Current of Ketol *VI* During Reaction of Chalcone with Hydroxyl Ions

$2 . 10^{-4}$M chalcone in 1·4M-NaOH with 20% ethanol. ○ Experimental values for current (i) corresponding to concentration of *VI* computed from difference $i_4 - i_3$ (*cf*. Fig. 5); *1—4* theoretical curves computed from equation (*1*) using calue $k_1' = 6·5 . 10^{-4} s^{-1}$ and $[Ch]_0$ from average of limiting current i_2 (before reaction) and i_3 or i_4 (after reaction was carried out) (*cf*. Fig. 5) for k_2': *1* $4 . 10^{-3} s^{-1}$; *2* $3 . 10^{-3} s^{-1}$; *3* $2 . 10^{-3} s^{-1}$; *4* $1 . 10^{-3} s^{-1}$.

(i_2) chalcone waves at a given hydroxyl ion concentration followed first order kinetics. Values of formal rate constants k_1' defined in the same way as for phenyl vinyl ketone and obtained from the log $h–t$ plot were practically identical for the first and second waves. The value of the second order rate constant $k_1 = k_1'/[OH^-]$ was determined from the linear $k_1'–[OH^-]$ dependence (Fig. 15) $(k_1 = 3·8 . 10^{-4}$ l mol^{-1} s^{-1} in 20% ethanol).

The concentration of ethanol has been observed to have a considerable effect on the values of rate constant k_1' (Fig. 15), the rate decreasing with increasing ethanol concentration. Moreover, the rate constant k_1' is a linear function of hydroxyl ion concentration, but contrary to ketones *I* and *IV* for 20% ethanol seems to show an intercept on the y axis (Fig. 15). If this dependence is not affected by the uncertainties of measurement of particular rate constants, it would be necessary to interpret it as participation of water as nucleophilic reagent attacking the double bond. At higher ethanol concentration this effect was negligible.

From the nature of the time-dependence of the concentration of reactants, intermediates, and products, it was concluded that when starting with chalcone (Ch),

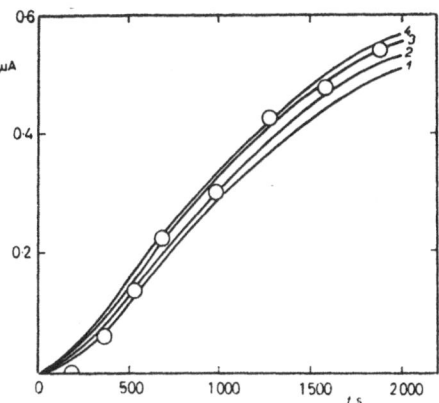

Fig. 17
Time Changes of Limiting Current of Benzaldehyde During Reaction of Chalcone with Hydroxyl Ions

$2 \cdot 10^{-4}$M chalcone in $1 \cdot 4$M-NaOH with 20% ethanol. ○ experimental values for current (i) corresponding to benzaldehyde concentration — i_3 in Fig. 5. $1-4$ theoretical curves computed for k'_1 and $[Ch]_0$ as in Fig. 16 using equation (2) for k'_2: $1\ 3 \cdot 10^{-3}\ s^{-1}$; $2\ 4 \cdot 10^{-3}\ s^{-1}$; $3\ 5 \cdot 10^{-3}\ s^{-1}$; $4\ 6 \cdot 10^{-3}\ s^{-1}$.

reaction with constant k_{-1} can be neglected because $k_{-1} \ll k_2$. Hence equation (C) reduced * to (D):

$$Ch + OH^- \xrightarrow{k'_1} Ke \xrightarrow{k'_2} Bz + Ac. \qquad (D)$$

For time-changes of ketol (Ke) and benzaldehyde (Bz) it was possible to apply rate equations (1) and (2):

$$[Ke] = \frac{[Ch]_0\ k'_1}{k'_2 - k'_1} \left(e^{-k'_1 t} - e^{-k'_2 t} \right) \qquad (1)$$

$$[Bz] = [Ch]_0 \left\{ 1 + \frac{1}{k'_1 - k'_2} \left(k'_2\, e^{-k'_1 t} - k'_1\, e^{-k'_2 t} \right) \right\} \qquad (2)$$

where $[Ch]_0$ is the initial chalcone concentration.

---

*cf. note on p. 203

Changes of [Ke] and especially of [Bz] with time were known but the complexity of the equations prevented application of simple graphical methods. It proved most convenient to determine the best value of the rate constant k_2' by the method of "trial and error". Time-changes of [Ke] and [Bz] were computed and theoretical curves were compared with experimental points (Figs 16 and 17).

The value of the rate constant k_2' determined from equations (1) and (2) increases strongly with increasing hydroxyl ion concentration at pH > 13 (Fig. 18) and the shape of the graphical presentation of this dependence resembles a part of a dissociation curve.

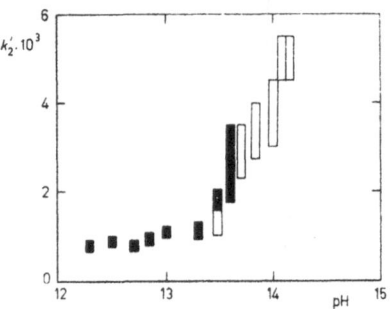

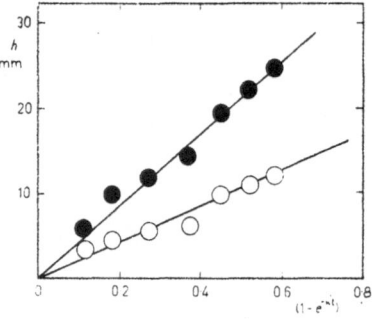

Fig. 18

pH-Dependence of the Formal Rate Constant (k_2') of the Retroaldolisation Fission of Ketol *VI*

Sodium hydroxide solutions, 20% ethanol; ■ experimental values for reaction of $2 . 10^{-4}$M ketol *VI*; □ experimental values for reaction of $2 . 10^{-4}$M chalcone (*V*).

Fig. 19

Reaction of β-Phenyl-β-hydroxypropiophenone (*VI*) in Alkaline Media

$2 . 10^{-4}$M ketol *VI* in 0·005M-NaOH with 30% ethanol. Graphical treatment of equations (5) and (6), k' determined using equation (4). ○ wave-height (h_1) for chalcone; ● wave-height (h_2) for benzaldehyde.

The time-changes of the waves of ketol *VI* were observed at substantially lower hydroxyl ion concentrations than for chalcone (*V*). Measurable changes were observed for hydroxyl ion concentrations greater than approximately $3 . 10^{-3}$M. Inspection of Fig. 15 reveals that under such conditions the value of rate constant k_1' will be very small and participation of reaction with constant k_1' in reaction (*C*) improbable. To prove that it was possible to neglect k_1', $2 . 10^{-4}$M ketol was added to 0·01M-NaOH in 30% per cent ethanol. Polarographic curves in these solutions have shown only waves for chalcone and a mixture of benzaldehyde and acetophenone. The ratio of wave heights confirmed that the formation of benzaldehyde and acetophenone predominated over that of chalcone. Hence even under these conditions $k_2' > k_{-1}'$. The wave-heights and their ratio, when recorded after 30, 48 and 70 hours remained unchanged. If reaction with constant k_1' had taken place even very slowly then only benzaldehyde and acetophenone would be finally found in the reaction mixture. The presence of an unchanged concentration of chalcone shows, that reaction with constant k_1' can be neglected $(k_1' < k_{-1}')$ and scheme (C) reduces* to (E):

$$Ch + OH^- \underset{k_{-1}'}{\rightleftharpoons} Ke \xrightarrow{k_2'} Bz + Ac \qquad (E)$$

To prove this scheme wave-heights of chalcone and benzaldehyde were plotted against $(1 - e^{-k't})$, where $k't = 2 \cdot 3 \log [Ke]_0 - 2 \cdot 3 \log [Ke]$ and k' is given by equation (3). These dependences gave linear plots (Fig. 19) for sodium hydroxide concentrations lower** than about $2 . 10^{-2}$M (in 30 per cent ethanol). This means that ratio of the concentrations of chalcone and benzaldehyde during the course of the reaction remained constant and was equal to the ratio k_{-1}'/k_2'.

To determine the values of the particular constants k_2' and k_{-1}', equations $(3)-(6)$ were used:

$$k' = k_{-1}' + k_2' \qquad (3)$$

$$k't = 2 \cdot 3 \log h_0 - 2 \cdot 3 \log h_i \qquad (4)$$

$$h_1 = \frac{k_{-1}' h_0}{k'} (1 - e^{-k't}) \qquad (5)$$

$$h_2 = \frac{k_2' h_0}{k'} (1 - e^{-k't}) \qquad (6)$$

where h_0 is the extrapolated value of wave height of ketol VI for $t = 0$, h_i its height for $t = t_i$, h_1 of the wave of chalcone at $t = t_i$ and h_2 of the wave of benzaldehyde at $t = t_i$. The determination of values of k_{-1} and k_2 is less reliable and accurate than due to difficulties in the measurement of the wave-heights on the polarograms.

The dependence of ratio of chalcone and benzaldehyde (produced in the reaction) on pH was further studied. Recording of whole $i-E$ curves was necessary to allow measurement of the waves of chalcone and benzaldehyde simultaneously in the course of the reaction and hence the study was restricted to concentrations of sodium hydroxide lower than about $2 . 10^{-2}$M. Under more alkaline conditions the reaction proceeded too fast for the whole $i-E$ curves to be recorded.** Hence to study the dependence of the ratio k_2'/k_{-1}' in a broader pH-range it was sufficient only to measure the wave-heights of the reaction products after the completion of the run. This technique enabled the range of sodium hydroxide concentrations to be extended from $3 . 10^{-3}$M to $3 . 10^{-1}$M. At higher hydroxide concentrations the reaction proceeds almost completely in the direction k_2' but the wave for chalcone was too small for the measurement and determination of the ratio k_2'/k_{-1}'.

---

*cf. note on p. 203
** At higher hydroxide concentrations the reaction was too fast for accurate measurements.

In the pH-range ($< 2.10^{-2}$M-NaOH) where it was possible using the treatment given in Fig. 19 to obtain separate values for k'_{-1} and k'_2, it has been shown that the value of k'_{-1} is practically pH-independent ($2 - 3 . 10^{-4}$ s^{-1}). It has been assumed that this value is pH-independent also at higher pH values. This assumption has been proved by the fact that values of k'_2 calculated from experimental values of the ratio k'_2/k'_{-1}, fitted closely the pH-dependence of k'_2 obtained from chalcone decomposition, *i.e.* by completely independent measurements and computations (Fig. 18). In confirmation of this assumption are also the kinetics of alkaline chalcone decomposition in which the reaction with constant k'_{-1} can be neglected in comparison with the reaction with constant k'_2. If the rate with k'_{-1} had increased with increasing pH equation (D) would not have been observed. Also the formation of only benzaldehyde and acetophenone from ketol *VI* at pH > 13.5 shows that $k'_2 \gg k'_{-1}$. As under this condition k'_2 is of the order 10^{-3} s^{-1}, constant k_{-1} must be of the order 10^{-4} s^{-1} and hence of the same order as determined at pH < 12.3.

Discussion

For reactions of α,β-unsaturated ketones of the type R—CO—CH=CH—R' in alkaline media the following general scheme, based on the experimental evidence provided by the ketones studied, can be considered:

$$\text{RCOCH=CHR'} + \text{OH}^- \underset{k_{-1}}{\overset{k_1}{\rightleftharpoons}} \overset{(-)}{\text{RCOCHCHR'}} \quad (F)$$
$$\text{OH}$$

$$\underset{(IX)}{\overset{(-)}{\text{RCOCHCHR'}}} + \text{H}_2\text{O} \underset{\text{fast}}{\overset{K_1}{\rightleftharpoons}} \text{RCOCH}_2\text{CHR'} + \text{OH}^- \quad (G)$$
$$\text{OH} \qquad\qquad\qquad \text{OH}$$

$$\text{RCOCH}_2\text{CHR'} + \text{OH}^- \underset{\text{fast}}{\overset{K_2}{\rightleftharpoons}} \text{RCOCH}_2\text{CHR'} + \text{H}_2\text{O} \quad (H)$$
$$\text{OH} \qquad\qquad\qquad (X) \quad \text{O}_{(-)}$$

$$\text{RCOCH}_2\text{CHR'} \underset{k_{-2}}{\overset{k_2}{\rightleftharpoons}} \text{RCOCH}_2^{(-)} + \text{R'CHO} \quad (I)$$
$$\text{O}_{(-)}$$

$$\text{RCOCH}_2^{(-)} + \text{H}_2\text{O} \rightleftharpoons \text{RCOCH}_3 + \text{OH}^- . \quad (J)$$

This scheme corresponds to the observed kinetics at a given pH and to the linear dependence of the formal rate constant k'_1 on hydroxyl ion concentration. To explain the effect of pH on the values of the rate constants k'_{-1} and k'_2 rapidly established equilibria (G) and (H) with constant K_1 and K_2 are considered.

The increase of the rate constant k_2' with increasing pH in the form of a part of a dissociation curve with pK_a about 14 or greater (Fig. 18 in which values obtained from two independent measurements with chalcone and ketol are collected) can be explained by the role of equilibrium (H) preceding the rate determining step (I). The virtual independence of values of k_{-1}' on pH in the studied region demonstrates that in this region equilibrium (G) is shifted so that the undissociated ketol predominates. This involves assumption that $pK_2 < pK_1$ which is in agreement with the well-known fact that O-acids are usually stronger than C-acids. Moreover the region of pK_2 value that would fit experimentally observed values for pK_a (Fig. 18) is in agreement with pK-values for dissociation of other alcoholic groups.

It cannot be excluded that instead of steps (G) and (H) transformation of IX to X occurs in a cyclic mechanism, but the observed dependence of k_2' on pH would be difficult to explain.

This scheme is in agreement with that suggested by Walker and Young[2] who confirmed the course of the step (F) and deduced steps analogous to $(G)-(J)$ basing their assumptions on proposals made in literature for similar reaction and also on intuition. This is an example of the rather infrequent happening that a suggested reaction mechanism is later completely confirmed by kinetic evidence. Steps $(H)-(J)$ are in accordance with accepted views[7,32] on alkaline β-ketol cleavage and retroaldolisation fission.* The varied dependence of the formal rate constants on hydroxide concentration explains, why according to the pH-region one or another reaction can be neglected and why the reaction scheme varies according to starting reactants. Hence when starting with chalcone it is necessary to work at $[OH^-] > 0.3$ M in order to increase sufficiently the rate of reaction with constant k_1'. But due to the fact that at the same time the rate of reaction with constant k_2' sharply increases whereas that of reaction with constant k_{-1}' remains practically unchanged, k_{-1}' can be neglected (cf. values for 1M-NaOH, Table II) and system of irreversible successive reactions (C) results.

On the other hand when starting with ketol VI the reaction rates at high hydroxide concentration are too high for practical measurements. But when working at lower hydroxyl ion concentration, (cf. values for $5 . 10^{-2}$M-NaOH, Table II), reaction with constant k_1' becomes negligible and a system of competitive reactions (D) results.

Participation of various steps in the scheme $(F)-(J)$ depends on the nature of R and R'. For the limited number of compounds studied so far it was possible to show that for R = C_6H_5, R' = H (I) k_{-1}', and k_2' can be neglected against k_1. For R = R' = C_6H_5 (V) k_{-2}' was negligible. Finally for R = CH_3, R' = C_6H_5 accumulation of ketol or another intermediate was excluded. Hence k_{-1}' and k_{-2}'

---

* In the scheme for condensation of steroid ketones with benzaldehyde[33] reaction of benzylidene ketone with hydroxyl ions has not been considered and detected, probably due to high alcohol and low hydroxyl ions concentration.

were negligible. The difference between I and V can be explained as an effect of the phenyl group on equilibrium (H), where the pK-value of the unsubstituted compound would be higher than that of the phenyl derivative. In order to elucidate the structural reasons for the predominance of particular reactions a more detailed study of substituent effects is planned.

The structure of the ketone substantially affects its reactivity towards nucleophilic attack of hydroxyl ions (Table III). Here again to separate inductive, resonance and steric contributions to the change in reaction rate a more detailed study is needed.

References

1. Čársky P., Zuman P., Horák V.: This Journal 29, 3044 (1964).
2. Walker E. A., Young J. R.: J. Chem. Soc. 1957, 2045.
3. Coombs E., Evans D.: J. Chem. Soc. 1940, 1295.
4. Noyce D. S., Pryor W. A., Bottini A. H.: J. Am. Chem. Soc. 77, 1402 (1955).
5. Barthel J., Dubois J. E.: Z. Physikal. Chem. N. F. 23, 37 (1960); where older references are quoted.
6. Barthel J., Dubois J. E.: Z. Physikal. Chem., N. F. 32, 296 (1962).
7. Dubois J. E., Viellard H.: Tetrahedron Letters 27, 1809 (1964).
8. Tirouflet J., Corvaisier A.: Bull. Soc. Chim. France 1962, 540.
9. Tirouflet J., Peng Li Cheng: Bull. Soc. Chim. France 1963, 2252.
10. Midorikawa H.: Bull. Chem. Soc. Japan 26, 460 (1953).
11. Midorikawa H.: Bull. Chem. Soc. Japan 27, 131 (1954).
12. Eigen M., Matthies P.: Chem. Ber. 94, 3309 (1961).
13. Patai S., Rappoport Z.: J. Chem. Soc. 1962, 377.
14. Rappoport Z.: J. Chem. Soc. 1962, 396.
15. Rappoport Z., Degani Ch., Patai S.: J. Chem. Soc. 1963, 4513.
16. Patai S., Weinstein S., Rappoport Z.: J. Chem. Soc. 1962, 1741.
17. Michl J.: Thesis. Charles University, Prague 1961.
18. Uchino N.: Bull. Chem. Soc. Japan 32, 1012 (1959).
19. Zuman P., Michl J.: Nature 192, 655 (1961).
20. Zuman P., Horák V.: This Journal 27, 187 (1962).
21. Drake N. L., Allen P.: Org. Syntheses, Second Ed., Coll. Vol. I, p. 77. Wiley, New York 1947.
22. Kohler E. P., Chadwell H. M.: Org. Syntheses, Second Ed., Coll. Vol. I, p. 78. Wiley, New York 1947.
23. Schöpf C., Thierfelder K.: Ann. 518, 149 (1935).
24. Koršunov I. A., Vodzinskij J.: Ž. Fiz. Chim. 27, 1152 (1953); Chem. Abstr. 48, 5674 (1954).
25. Pasternak R.: Helv. Chim. Acta 31, 753 (1948).
26. Ashworth H.: This Journal 13, 229 (1948).
27. Kastening B., Holleck L.: Z. Elektrochem. 63, 166 (1959).
28. Geissman T. A., Friess S. L.: J. Am. Chem. Soc. 71, 3893 (1949).
29. Schraufstätter E.: Experientia 4, 192 (1948).
30. Kejharová A.: Unpublished results.
31. Zuman P., Horák V.: This Journal 26, 176 (1961).
32. Frost A. A., Pearson R. G.: Kinetics and Mechanism, IInd Ed., p. 335. Wiley, New York 1961.
33. Barton D. H. R., McCapra F., May P. J., Thudium F.: J. Chem. Soc. 1960, 1297.

Translated by the author (P. Z.).

POLAROGRAPHIC REDUCTION OF CINNAMALDEHYDE; COMPARISON WITH 3-PHENYLPROPIONALDEHYDE AND PHENYLPROPARGYLALDEHYDE

D. BARNES AND P. ZUMAN*

Department of Chemistry, University of Birmingham (Great Britain)

(Received July 10th, 1967)

For the polarographic reduction of α,β-unsaturated ketones, it was established fairly early[1] that the first two-electron step corresponds to the hydrogenation of the C=C bond in contrast with the rather vague ideas proposed to explain the reduction of α,β-unsaturated aldehydes, especially cinnamaldehyde.

Cinnamaldehyde is reduced in a two-electron process, followed by a further reduction step at more negative potentials which was either not recorded or missed by previous investigators who concentrated their attention on the first two-electron process; this led them to make erroneous deductions. In the early papers[2,3], the waves of the first two-electron process were compared with those of benzaldehyde and because of the formal resemblance observed, the first two-electron process was attributed to the reduction of the carbonyl group.

In later studies[4], the waves of cinnamaldehyde were compared with those of phenylpropargylaldehyde. Because these authors[4] restricted their study to the lower pH-region and to the more positive potential range, they again concluded that the product of the two-electron reduction is an unsaturated alcohol.

Controlled-potential electrolysis at pH 7.75 in the potential range corresponding to the limiting current of the two-electron process is said[5] to yield cinnamic alcohol. Only SATO[6] considered the possibility of reduction in the side chain C=C bond, and this was based on circumstantial evidence.

Polarography has been used in the determination of cinnamaldehyde in a variety of mixtures[7-9], and it is therefore of interest to have a better understanding of the reduction sequence.

In connection with the study of reactions of nucleophilic reagents with unsaturated carbonyl compounds[10-12] (which is to be extended to the reactions of cinnamaldehyde) it was of importance to have a detailed knowledge of the course of the reduction of cinnamaldehyde. This type of information was also needed for an extension of our studies on the reduction of α,β-unsaturated carbonyl compounds[12-14].

* On leave from J. Heyrovský Institute of Polarography, Czechoslovak Academy of Sciences, Prague.

EXPERIMENTAL

Apparatus

The polarographic curves were recorded on a Cambridge pen-recording polarograph. A Kalousek vessel with a separated calomel electrode (SCE) was used: The capillary used had, at the potential of the SCE, an outflow velocity $m = 2.96$ mg sec^{-1} and drop-time $t_1 = 3.1$ sec for a mercury pressure, $h = 60$ cm. The pH of the buffer solutions were measured using a Pye Dynacap pH-meter with a general purpose glass electrode.

Controlled-potential electrolysis using a mercury dropping electrode was carried out as recommended by MANOUŠEK[15], using the potentiostatted three-electrode system of a Beckman Electroscan 30 instrument.

Ultraviolet measurements were made using a Unicam S.P. 800, the electrolysis product being diluted to produce approximately $5 \cdot 10^{-5}$ M solutions.

Gas-liquid chromatography (G.L.C.) was carried out as described previously[16], using a polyethyleneglycol adipate column at 180° and a flow rate of 10 ml sec^{-1} using the Pye 104 chromatograph with a flame ionization detector.

Solutions

All buffer solutions were prepared from AnalaR chemicals.

Organic compounds were redistilled under reduced pressure and then used for preparing stock 0.01 M solutions in spectroscopic ethanol.

Techniques

9.8 ml of the appropriate buffer were mixed with 0.2 ml of the 0.01 M stock solution; the solutions were de-aerated and the polarographic curves recorded and then checked to determine whether the curves changed with time. Most of the experiments were carried out in buffers containing 2% ethanol in the final solutions. When it was necessary to avoid adsorption waves, curves were recorded in buffer solutions containing 25% ethanol.

For controlled-potential electrolysis, 10^{-3} M solutions were prepared in the appropriate buffer so that the final solution contained 5% ethanol. Usually, 2.5 ml of this solution were electrolysed to 60–70% completion during 8–12 h. For the electrolysis in alkaline media, 0.5 ml were electrolysed at 0° during 4 h.

The numbers of electrons transferred were estimated by comparison with the known two-electron wave of benzophenone, and were determined coulometrically from the decrease of limiting current during the electrolysis, using a dropping mercury electrode in a small volume which was stirred by the falling of the drops.

RESULTS AND DISCUSSION

The reduction of cinnamaldehyde in aqueous solutions containing a small amount of ethanol, proceeds principally in two steps (Figs. 1 and 2). The first step, i_I, is a two-electron reduction and is followed by a further reduction step, i_{II}.

In the elucidation of the electrode process, the nature of the more negative wave, i_{II}, is of primary importance. It has been shown that the half-wave potential of wave i_{II}, its shift with pH and the pH-dependence of its height, are identical with those observed for 3-phenylpropionaldehyde. Additional proof was obtained after controlled-potential electrolysis at the limiting current of wave i_I, using a normal dropping mercury electrode as the working electrode[15]. The u.v. spectra and G.L.C. retention times[16] of the product were identical with those of 3-phenylpropionaldehyde. Submicro-titration of the electrolysed solution[16] with bromine monochloride indicated no C=C bond in the product. Finally, a solution of the electrolysis product was studied polarographically at various pH-values in the presence of lithium ions.

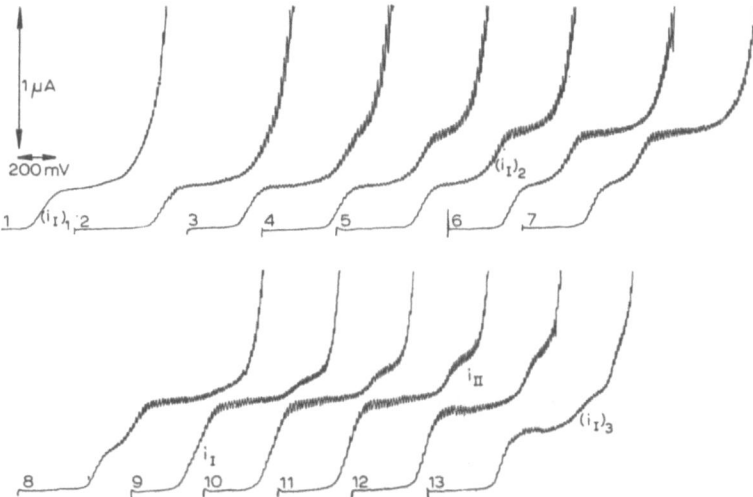

Fig. 1. Dependence of reduction waves of cinnamaldehyde on pH; $2\cdot10^{-4}$ M cinnamaldehyde in Britton-Robinson buffers containing 25% ethanol. pH-values: (1), 2.3; (2), 2.7; (3), 3.55; (4), 4.6; (5), 5.3; (6), 6.5; (7), 7.0; (8), 7.7; (9), 8.85; (10), 9.5; (11), 9.9; (12), 10.4; (13), 11.8. Curves starting at: (1), −0.4; (2), −0.2; (3–5), −0.4; (6–8), −0.6; (9–13), −0.8 V SCE.

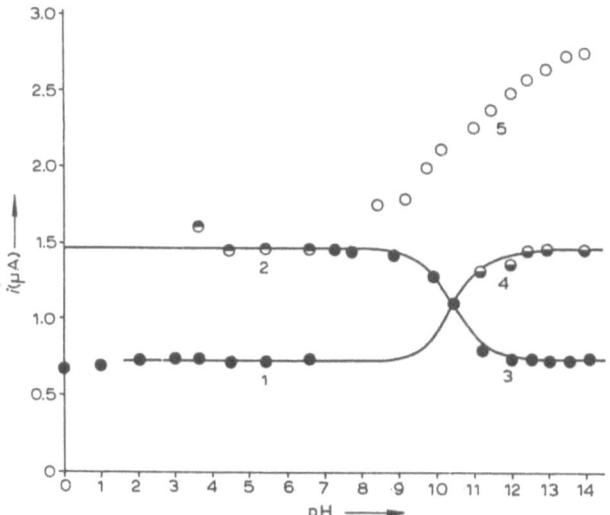

Fig. 2. Dependence of limiting current on pH; $2\cdot10^{-4}$ M cinnamaldehyde in Britton-Robinson buffers (pH 2–12), sulphuric acid (pH 0–2) and lithium hydroxide (pH 12–14), containing 2% ethanol. (1, ●), wave $(i_I)_1$; (2, ◓), wave $(i_I)_2$; (3, ●), wave i_I; (4, ◖), wave $(i_I)_3$; (5, ○), wave i_{II}.

Over a wide pH-range, the waves obtained for the product had half-wave potentials identical with 3-phenylpropionaldehyde and showed a similar dependence of wave height on pH.

The waves of 3-phenylpropionaldehyde increase with increasing pH, until at pH > 12, they reach a limiting value* corresponding to a two-electron process. The exponent, p, of the dependence of the instantaneous current on time $(i \sim t^p)$ increases with decreasing pH. Thus, the wave which at pH 12 is diffusion-controlled, at lower pH-values is partly governed by the rate of a chemical reaction. In the rising part of the current–pH dependence, the current depends on buffer type and concentration, as with formaldehyde[17]. It is assumed that the chemical reaction that precedes the electrode process proper is a general base-catalysed dehydration of the methylene glycol form of 3-phenylpropionaldehyde.

$$RCHO + H_2O \rightleftharpoons RCH \begin{matrix} OH \\ \\ OH \end{matrix}$$

The essential difference, compared with formaldehyde, is the position of the hydration–dehydration equilibria which with 3-phenylpropionaldehyde is shifted towards the non-hydrated form. The sensitivity to bases other than hydroxyl ions is considerably smaller for 3-phenylpropionaldehyde than for formaldehyde.

The half-wave potentials of the 3-phenylpropionaldehyde wave depend on the nature and concentration of the cations present in the solution. For a given concentration, lithium and caesium ions cause the greatest shifts toward more positive potentials, anions having little or no effect. Apparently, two antagonistic effects govern the effect of cations on the half-wave potential**.

Confirmation of the course of the reduction of cinnamaldehyde is given by a comparison of its waves with those of phenylpropargylaldehyde (Fig. 3). Under suitable conditions, phenylpropargylaldehyde is reduced at pH 9.3 in two two-electron steps, followed by a wave of 3-phenylpropionaldehyde. The second wave has a half-wave potential identical with that of cinnamaldehyde. Hence, under these conditions, where phenylpropargylaldehyde is reduced in the unprotonated form, the reduction follows the sequence $-C \equiv C-$, $>C = C<$, CHO according to scheme (1–3)

$$C_6H_5C \equiv CCHO + 2\ e^- + 2\ H^+ \rightarrow C_6H_5CH = CHCHO \tag{1}$$

$$C_6H_5CH = CHCHO + 2\ e^- + 2\ H^+ \rightarrow C_6H_5CH_2CH_2CHO \tag{2}$$

$$C_6H_5CH_2CH_2CHO + 2\ e^- + 2\ H^+ \rightarrow C_6H_5CH_2CH_2CH_2OH \tag{3}$$

At lower pH-values, the reduction of phenylpropargylaldehyde follows a more complex pattern which will be discussed elsewhere.

---

* No decrease in the current was observed up to pH 14, provided that in alkaline solutions the values are obtained either by extrapolation to zero-time or by working at 0° to exclude the effect of aldolisation. As $RCH(OH)_2$ for formaldehyde, and 3-phenylpropionaldehyde are acids of comparable strength, the decrease of current at pH > 13 observed for formaldehyde is probably caused by a phenomenon other than the proposed[17] position of the acid–base equilibria.
** Both the change in the value of the capacity current in d.c. polarography, and the tensammetric peaks in a.c. polarography indicate a strong adsorption at a potential of -0.2 V and desorption at -1.4 V; the desorption therefore occurs at potentials more positive than that of the reduction process[18].

Behaviour in acid media

The first two-electron wave, i_I, of cinnamaldehyde at pH < 8 is split into two one-electron steps, $(i_I)_1$ and $(i_I)_2$. The coalescence of the one-electron waves $(i_I)_1$ and $(i_I)_2$, with increasing pH, is due to the shifting of the half-wave potential of wave $(i_I)_1$ to more negative potentials, while wave $(i_I)_2$ is practically pH-independent (Fig. 4). In agreement with previous deductions[2-4] and the results of controlled-

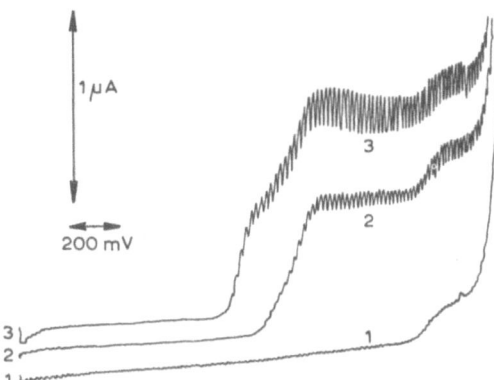

Fig. 3. Comparison of the reduction waves of unsaturated and saturated aldehydes; $2 \cdot 10^{-4}$ M aldehyde in borate buffer pH 9.2. (1), 3-Phenylpropionaldehyde; (2), cinnamaldehyde; (3), phenylpropargylaldehyde. Curves starting at 0.0 V SCE.

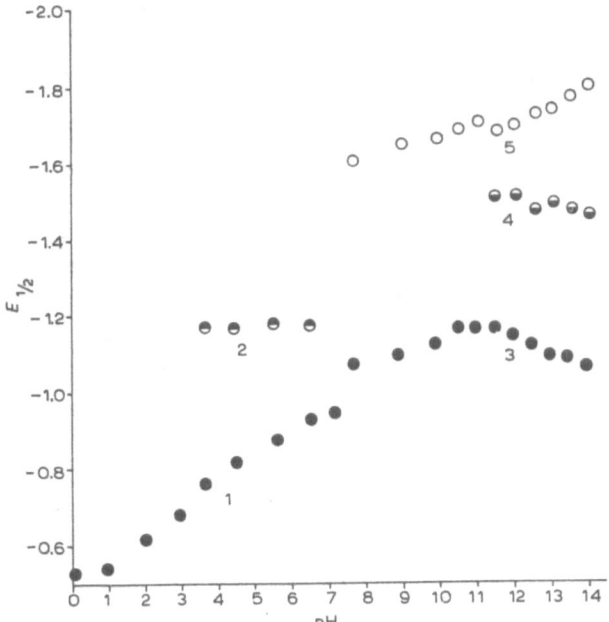

Fig. 4. Effect of pH on half-wave potentials for cinnamaldehyde; $2 \cdot 10^{-4}$ M cinnamaldehyde in Britton-Robinson buffers (pH 2–12), sulphuric acid (pH 0–2) and lithium hydroxide (pH 12–14), containing 2% ethanol. (1, ●), wave $(i_I)_1$; (2, ◑), wave $(i_I)_2$; (3, ●), wave i_I; (4, ◐), wave $(i_I)_3$; (5, ○), wave (i_{II}).

potential electrolysis[3], the predominating product in the first wave was proved to be a dimer.

In acid media, scheme (4)–(10) can therefore be suggested for the polarographic reduction of cinnamaldehyde:

$$C_6H_5CH{=}CHCHO + H^+ \rightleftharpoons \underset{H^{(+)}}{\underline{C_6H_5CH{=}CHCHO}} \tag{4}$$

$$\underset{H^{(+)}}{\underline{C_6H_5CH{=}CHCHO}} + e^- \overset{(i_I)_1}{\rightleftharpoons} \underset{H^{(+)}}{\underline{C_6H_5\overset{\cdot}{CH}{=}CHCHO}} \tag{5}$$

$$2\ \underset{H^{(+)}}{\underline{C_6H_5\overset{\cdot}{CH}{=}CHCHO}} \longrightarrow \text{dimer} \tag{6}$$

$$\underset{H^{(+)}}{\underline{C_6H_5\overset{\cdot}{CH}{=}CHCHO}} + e^- \overset{(i_I)_2}{\longrightarrow} \overset{(-)}{C_6H_5CH_2CHCHO} \tag{7}$$

$$\overset{(-)}{C_6H_5CH_2CHCHO} + H_2O \rightleftharpoons C_6H_5CH_2CH_2CHO + OH^- \tag{8}$$

$$C_6H_5CH_2CH_2CHO + H_2O \rightleftharpoons C_6H_5CH_2CH_2CH(OH)_2 \tag{9}$$

$$C_6H_5CH_2CH_2CHO + 2\ e^- + 2\ H^+ \overset{i_{II}}{\longrightarrow} C_6H_5CH_2CH_2CH_2OH \tag{10}$$

Behaviour in alkaline media

In alkaline media*, wave i_I decreases with increasing pH in the form of a dissociation curve with pK' 10.5, until it reaches a height corresponding to a one-electron transfer. Simultaneously, at more negative potentials, another wave appears, $(i_I)_3$; the sum of these two waves remains sensibly constant. This behaviour formally resembles that of benzaldehyde, but does not correspond to a change in mechanism. Even at pH $>$ 12, when two one-electron waves, i_I and $(i_I)_3$, are observed, the height and half-wave potential of wave i_{II} corresponds to the reduction of 3-phenylpropion-aldehyde. Moreover, controlled-potential electrolysis on the two-electron limiting current of $(i_I)_3$ produced a product similar to 3-phenylpropionaldehyde regarding u.v. spectra, G.L.C. retention time, and polarographic behaviour.

In alkaline media, at pH-values greater than about 9, where the half-wave potentials are practically pH-independent (Fig. 4), the following scheme, (11)–(15), can therefore be proposed:

$$C_6H_5CH{=}CHCHO + e^- \overset{i_I}{\rightleftharpoons} \underset{(-)}{\underline{C_6H_5CH{=}CHCHO}} \tag{11}$$

$$\underset{(-)}{\underline{C_6H_5CH{=}CHCHO}} + H^+ \rightleftharpoons \underset{H^{(+)}}{\underline{C_6H_5\overset{\cdot}{CH}{=}CHCHO}} \tag{12}$$

---

* Values at pH $>$ 12 were extrapolated to zero-time to eliminate the effect of hydration of the double bond[12].

$$C_6H_5\overset{\centerdot}{\overbrace{CH=CHCHO}} + e^- \xrightarrow{\ i_I\ } \overset{(-)}{C_8H_5CH_2CHCHO} \qquad (13)$$
$$\underbrace{}_{H^{(+)}}$$

$$\overset{(-)}{C_6H_5CH_2CHCHO} + H_2O \rightleftharpoons C_6H_5CH_2CH_2CHO + OH^- \qquad (14)$$

$$C_6H_5CH_2CH_2CHO + H_2O \rightleftharpoons C_6H_5CH_2CH_2CH(OH)_2 \qquad (15)$$

$$C_6H_5CH_2CH_2CHO + 2\,e^- + 2\,H^+ \xrightarrow{\ i_{III}\ } C_6H_5CH_2CH_2CH_2OH \qquad (16)$$

$$C_6H_5CH=CHCHO + e^- \xrightarrow{\ (i_I)_3\ } \underset{2^{(-)}}{\underbrace{C_6H_5CHCHCHO}} \qquad (17)$$
$$\underset{(-)}{\underbrace{}}$$

$$\underset{2^{(-)}}{\underbrace{C_6H_5CH-CHCHO}} + 2\,H_2O \rightleftharpoons C_6H_5CH_2CH_2CHO + 2\,OH^- \qquad (18)$$

The half-wave potential of process (13) is either not too different from that of process (11) or is more positive than that of process (11); hence only one two-electron wave (i_I) is observed at pH-values where reaction (12) is fast enough to transfer most

of the radical anion $\underset{(-)}{\underbrace{C_6H_5CH=CHCHO}}$ into the radical $C_6H_5\overset{\centerdot}{\overbrace{CH=CHCHO}}$.
$$\underset{H^{(+)}}{\underbrace{}}$$

With increasing pH, the rate of reaction (12) is decreased and the height of wave i_I decreases. The radical anion that was not transformed into the radical, undergoes reduction in wave $(i_I)_3$ according to eqn. (17), followed by protonation (18). The reduction of 3-phenylpropionaldehyde given by (16) is affected by hydration (15).

Hence, in the polarographic reduction of cinnamaldehyde, the product of the two-electron reduction over the entire pH-range, is the saturated aldehyde, even when the product of the dimerisation following the one-electron process may be a pinacol. In the reduction of the system $C = $ -CO, the carbon–carbon double bond is hydrogenated first, whether the system con rned is an aldehyde or a ketone. The separation of two one-electron waves in acid or alkali cannot be taken as proof of the reduction of a carbonyl group.

ACKNOWLEDGEMENTS

We thank Professor R. BELCHER for his interest and support.

One of us (P.Z.) expresses his thanks to the Scientific Research Council and the other (D.B.) to J. Lyons and Co. Ltd., for the grants which made this study possible.

We would both like to thank Dr. W. BOARDMAN of Hatfield College of Technology for his measurements of the i–t curves of 3-phenylpropionaldehyde, Dr. P. C.

UDEN for his advice on the G.L.C. experiments, Dr. P. L. COE and his colleagues for the use of the G.L.C. apparatus, Esso Research for the loan of the Cambridge Polarograph, and Beckman Instruments for the use of the Electroscan 30.

SUMMARY

Cinnamaldehyde is reduced at the dropping mercury electrode in a two-electron step which leads to the formation of 3-phenylpropionaldehyde. This has been proved by controlled-potential electrolysis and identification of the products using u.v. spectra, gas-liquid chromatography, submicro-methods, and electrochemical techniques. The reduction wave of 3-phenylpropionaldehyde occurs at potentials where the compound is desorbed and the wave height is limited by an antecedent dehydration reaction. In acid media, the reduction of the protonated form of cinnamaldehyde occurs in two one-electron steps. Dimer is formed at the potentials of the first step. At sufficiently high pH-values, the unprotonated form of cinnamaldehyde is reduced in a one-electron step to a radical anion that dimerizes and reduces only at more negative potentials. At pH 9.3, the carbon–carbon triple bond in phenylpropargylaldehyde is first reduced to a double bond; in subsequent steps the C=C bond and then the CHO group are reduced. Even in α,β-unsaturated aldehydes the hydrogenation of the double bond takes place prior to the reduction of the carbonyl group.

NOTE ADDED IN PROOF

After the manuscript of this paper was submitted, the paper by G. CAPOBIANCO, E. VIANELLO AND G. GIACOMETTI (*Gazz. Chim. Ital.*, 97 (1967) 243) on the reduction of phenylpropargylaldehyde came to our attention. We agree with the views on the triple bond reduction, a difference exists in the interpretation of the cinnamaldehyde reduction.

REFERENCES

1 R. PASTERNAK, *Helv. Chim. Acta*, 31 (1948) 753.
2 G. SEMERANO AND A. CHISINI, *Gazz. Chim. Ital.*, 66 (1936) 510.
3 A. KORSHUNOV, Z. B. KUZNETSOVA, L. N. SAZANOVA AND A. S. KIRILLOVA, *Zavodsk. Lab.*, 16 (1950) 144.
4 CH. PRÉVOST, P. SOUCHAY AND J. CHAUVELIER, *Bull. Soc. Chim. France*, 18 (1951) 715.
5 S. ONO, *Nippon Kagaku Zasshi*, 76 (1955) 631.
6 H. SATO, *Eisei Shikenjo Hokoku*, 77 (1959) 51.
7 M. SHIKATA AND K. SHOJI, *Mem. Coll. Agr. Kyoto Univ.*, 4 (1927) 75.
8 B. BITTER, O. HANČ AND F. ŠANTAVÝ, *Chem. Listy*, 43 (1949) 137.
9 G. DUŠINSKÝ, M. TYLLOVÁ AND Z. GRUNTOVÁ, *Česk. Farm.*, 6 (1957) 87.
10 V. HORÁK AND P. ZUMAN, *Collection Czech. Chem. Commun.*, 26 (1961) 173.
11 I. ŠESTÁKOVÁ, V. HORÁK AND P. ZUMAN, *Collection Czech. Chem. Commun.*, 31 (1966) 3889.
12 P. ČÁRSKY, P. ZUMAN AND V. HORÁK, *Collection Czech. Chem. Commun.*, 30 (1965) 4316.
13 P. ZUMAN AND J. MICHL, *Nature*, 192 (1961) 655.
14 A. RYVOLOVÁ-KEJHAROVÁ, unpublished results.
15 O. MANOUŠEK, unpublished results.
16 D. BARNES, R. BELCHER AND P. ZUMAN, *Talanta*, 14 (1967) 1197.
17 R. BRDIČKA AND K. VESELÝ, *Collection Czech. Chem. Commun.*, 12 (1947) 313.
18 P. ZUMAN AND D. BARNES, *Nature*, 215 (1967) 1269.

POLAROGRAPHIE DER STEROIDE I.*

DIREKT REDUZIERBARE KETOSTEROIDE: Δ⁴-CHOLESTENON, METHYLTESTOSTERON, TESTOSTERON, PROGESTERON UND DESOXYCORTICOSTERON

P. Zuman, J. Tenygl und M. Březina

Polarographisches Institut, Tschechoslowakische Akademie der Wissenschaften, Praha

Eingegangen am 5. Januar 1953

Es wurden die Reduktionsstufen der Δ⁴-3-Ketosteroide in gepufferten wässrig-alkoholischen Lösungen verfolgt. In diesen Lösungen wurde polarographisch die Anwesenheit von zwei Formen bewiesen, deren Verhältnis sich mit dem pH ändert. Im alkalischen Bereich, und beim Δ⁴-Cholesten-3-on und Desoxycorticosteron auch im sauren Bereich, wurden Adsorptionsstufen gefunden. Im sauren Gebiet wurde bei kleineren Äthanolkonzentrationen eine katalytische Welle beobachtet.

Es wurden die günstigsten Bedingungen bestimmt, die einerseits die Bestimmung des Gesamtgehaltes an Δ⁴-3-Ketosteroiden, andrerseits die Bestimmung in einigen Gemischen, z. B. Desoxycorticosteron neben Testosteron oder Progesteron neben Methyltestosteron, ermöglichen.

3-Keto-4,5-ungesättigte Steroide unterliegen der polarographischen Reduktion, wie Eisenbrand und Picher[1] beim Progesteron, Testosteron und Desoxycorticosteron festgestellt haben. Diese Autoren arbeiteten ähnlich wie Eckwall, Lundsten und Sjöblom[2], welche dieselben Stoffe in Gegenwart von Oleaten, Myristaten u. ä. bestimmten, in ungepufferten Lithiumchloridlösungen. In den Arbeiten dieser Autoren[1,2] sind an den polarographischen Kurven zwei Stufen sichtbar, die von ihnen nicht weiter erklärt werden. So führen ebenfalls Sartori und Bianchi[3], welche die Abhängigkeit des Halbstufenpotentials vom pH beim Methyltestosteron und Pregnenin-3-on-17-ol verfolgten, nichts Näheres über den Charakter der untersuchten Wellen an.

Endlich polarographierten Wolfe, Hershberg und Fieser[4] Δ⁴-3-Ketosteroide in einer 40% Isopropanol und 0,1m-Tetraäthylammoniumhydroxyd enthaltenden Lösung. Sie fanden eine Verschiebung des Halbstufenpotentials mit der Konzentration und stellten fest, dass die Lösungen aller untersuchten Steroide im alkalischen Medium zumindest 1 Stunde lang beständig sind. Sie stellten fest, dass die in Stellung 20 befindliche Ketogruppe beim Corticosteron und Progesteron das polarographische Verhalten nicht beeinflussen und dass zwischen dem Verhalten des Desoxycorticosterons und seinem Acetat kein Unterschied besteht. Die Stufen des Δ⁴-Cholesten-3-on und Cortisons (das eine isolierte Ketogruppe in Stellung 11 besitzt) sind bedeutend niedriger als die Stufen der übrigen Steroide. Das Halbstufenpotential des Δ¹-Cholesten-3-on war beiläufig um 150 mV positiver als die Potentiale der Δ⁴-3-Ketosteroide. Die Autoren empfehlen jedoch vor der Bestimmung eine Überführung in die Betainylhydrazone vorzunehmen.

$$I: R^1 = CH \cdot (CH_2)_3 \cdot CH \cdot (CH_3)_2, \quad R^2 = H,$$
$$\quad\quad\quad\quad\quad | $$
$$\quad\quad\quad\quad\quad CH_3$$
$$II: R^1 = OH, \quad R^2 = CH_3,$$
$$III: R^1 = OH, \quad R^2 = H,$$
$$IV: R^1 = COCH_3, \quad R^2 = H,$$
$$V: R^1 = CO \cdot CH_2OH, \quad R^2 = H.$$

---

* Diese Arbeit erschien bereits in tschechischer Sprache in Chem. listy *47*, 1152 (1953).

In dieser Arbeit wurde das polarographische Verhalten von Δ^4-Cholesten-3-on (I), Methyltestosteron (II), Testosteron (III), Progesteron (IV) und Desoxycorticosteron (V) in gepufferten Lösungen untersucht.

Experimentelles

Es wurde die übliche polarographische Einrichtung mit dem Gefäss nach Kalousek mit getrennter Kalomelelektrode benützt.

Folgende Präparate wurden verwendet: Progesteron, Desoxycorticosteron-acetat (Schering), Methyltestosteron (Boehring), Testosteron-propionat, Δ^4-Cholesten-3-on, Diese Ketosteroide erhielten wir von der Firma Spofa n. p., Ústí n. Labem und vom Institut für organische Chemie der Tschechoslowakischen Akademie der Wissenschaften.

Aus den kristallinen Stoffen wurden 0,01M Lösungen in 96%igem Äthanol hergestellt.

Die grössten Schwierigkeiten verursachte die geringe Löslichkeit der Steroide. Für Konzentrationen von der Grössenordnung $2,5 \cdot 10^{-4}$M mussten folgende Lösungen verwendet werden: beim Methyltestosteron 10% Äthanol, beim Testosteron-propionat und Progesteron 20% Äthanol, beim Desoxycorticosteron-acetat 30% Äthanol und beim Δ^4-Cholestenon 50% Äthanol. Die alkoholischen Lösungen waren ebenso wie diejenigen in Puffergemischen während einiger Tage beständig. Es wurde kein Unterschied zwischen dem polarographischen Verhalten von Testosteron und Testosteron-propionat beobachtet.

Bei der verwendeten Kapillare waren $m = 3,06$ mg und $t = 2,35$ s bei $h = 50$ cm.

Beim Studium der pH-Abhängigkeit wurden die Puffergemische nach Britton-Robinson verwendet, die derart mit Äthanol verdünnt wurden, dass die Alkoholkonzentration 30—60% betrug.

Das pH der Lösungen (auch der äthanolischen) wurde mit der Glaselektrode und (wässrigen) Kalomelelektrode unter Verwendung des pH-Meters der Firma Cambridge Instruments Ltd. gemessen. Alle pH-Werte müssen also mit Rücksicht auf die Anwesenheit des Äthanols als relativ betrachtet werden.

Übersicht der Ergebnisse

Abhängigkeit der Stufenhöhe vom pH

Methyltestosteron, Testosteron, Progesteron

Wenn die polarographischen Kurven in Britton-Robinson Puffergemischen, die 50% Äthanol enthalten, verzeichnet werden, werden an den polarographischen Kurven zwei Stufen beobachtet, deren Summe konstant verbleibt. Diese Summe wird als „Hauptwelle" (Abb. 1) bezeichnet. Die erste Welle, die bei pH-Werten grösser als 6—7 zu sinken beginnt, wird mit HA, die zweite Welle bei negativeren Potentialen mit A' bezeichnet. (Die Bezeichnung der Wellen ist aus der schema-

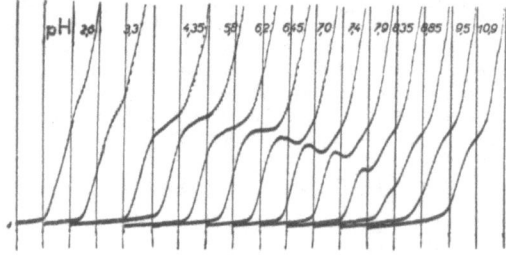

Abb. 1

Progesteron, Abhängigkeit vom pH

5 ml Puffergemisch nach Britton und Robinson, 50% Äthanol + 0,3 ml 0,01M-Progesteron. Kurven von der 6. Drahtwindung (*1—6*) und von der 7. Drahtwindung (weiter) an, Mercurosulfatelektrode, 200 mV/Absz., Empf. 1 : 20, $h = 50$ cm.

tischen Aufzeichnung in Abb. 2 ersichtlich). Der Abfall der ersten Welle HA besitzt die Gestalt der Dissoziationskurve (Abb. 3), die bei den einzelnen untersuchten Steroiden gegeneinander verschoben sind. Die scheinbaren Dissoziationskonstanten (K') wurden aus den Schnittpunkten der Abhängigkeit $\log [\%i_{HA}/(100 — \%i_{HA})]$ vom pH mit dem Nullwert des Logarithmus (Abb. 4) bestimmt.

Beim Testosteron wurden, ähnlich wie beim Desoxycorticosteron, bei niedrigeren pH-Werten Abweichungen vom theoretischen Verlauf beobachtet.

Bei kleineren Äthanolkonzentrationen (20—30%) wurden ausser diesen Stufen noch zwei weitere Wellen beobachtet.

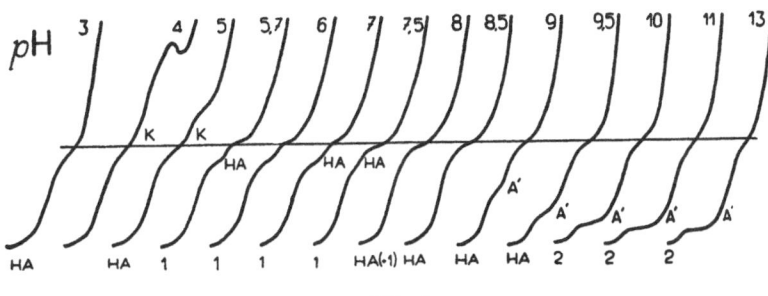

Abb. 2

Schematische Veranschaulichung der Veränderung der Steroidwellen mit dem pH HA, A' — Hauptwelle, K — katalytische, 2 — Adsorptionswelle im alkalischen, 1 — im sauren Medium. Schema — da die Welle 2 bei grösseren Äthanolkonzentrationen als 30% nicht auftritt, die Welle 1 sich jedoch bei niedrigeren Äthanolkonzentrationen als 40% nicht abtrennt.

Im sauren Gebiet, bei einem niedrigeren pH als cca 5, ist an den Kurven (Abb. 5) eine Welle mit einem Maximum bei negativeren Potentialen als die Welle HA sichtbar. Ihre Höhe wächst steil mit sinkendem pH (Abb. 6); wir bezeichnen sie mit K.

Im alkalischen Gebiet ist bei geringeren Äthanolkonzentrationen eine Welle bei positiveren Potentialen als die Welle A' sichtbar. Das Halbstufenpotential dieser Welle, die als Welle 2 bezeichnet wird, stimmt mit dem Halbstufenpotential der Welle HA bei ihrem Abfall überein. Infolgedessen wird bei Steigerung des pH der Abfall der Welle HA nur bis zu einem gewissen Bruchteil des gesamten Grenzstromes beobachtet. Bei weiterer Steigerung des pH verändert sich die Höhe der Welle 2 bereits nicht mehr (Abb. 6).

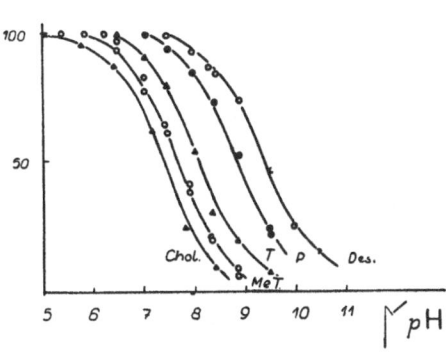

Abb. 3

Dissoziationskurven der Steroide

Höhe der ersten Welle in Prozenten des gesamten Grenzstromes in Abhängigkeit vom pH beim Δ^4-Cholestenon (*Chol*), Methyltestosteron (*Met*), Testosteron (*T*), Progesteron (*P*) und Desoxycorticosteron (*Des*).

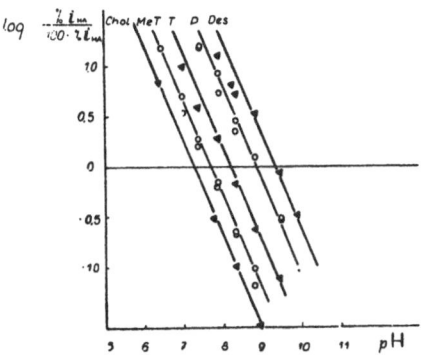

Abb. 4

Abhängigkeit des log $[\%i_{HA}/(100 - \%i_{HA})]$ vom pH

Der Schnittpunkt mit der Nullinie gibt die Dissoziationskonstante K an. Abkürzungen wie in Abb. 3.

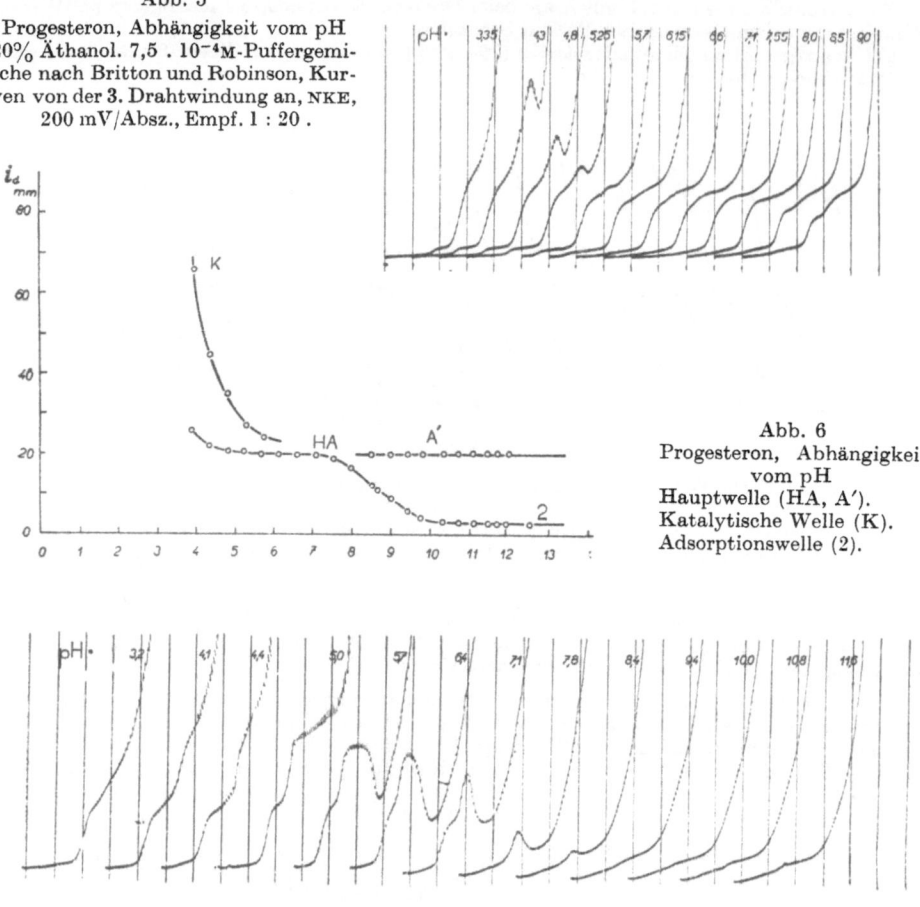

Abb. 5

Progesteron, Abhängigkeit vom pH
20% Äthanol. 7,5 . 10^{-4}M-Puffergemi-
sche nach Britton und Robinson, Kurven von der 3. Drahtwindung an, NKE,
200 mV/Absz., Empf. 1 : 20 .

Abb. 6
Progesteron, Abhängigkeit
vom pH
Hauptwelle (HA, A').
Katalytische Welle (K).
Adsorptionswelle (2).

Abb. 7
Δ^4-Cholestenon, Abhängigkeit vom pH

7 . 10^{-4}M-Δ^4-Cholestenon, Pufferlösungen nach Britton und Robinson. Kurven von der 5. Drahtwindung an, NKE, 200 mV/Absz., Empf. 1 : 5.

Desoxycorticosteron

In Gegenwart von 50% Äthanol wird im sauren Bereich neben den Wellen HA und A' noch
eine weitere Welle beobachtet, die mit 1 bezeichnet wird (Abb. 2 und 10). Ihre Höhe ändert sich
nicht mit dem pH, insofern kein Abfall der Welle HA eintritt. Im sauren Bereich sind die Wellen
schlecht ausgebildet. Bei geringeren Äthanolkonzentrationen fällt die Welle 1 mit der Welle HA
zusammen; an den Kurven sind wiederum die Stufen K und 2 sichtbar.

Δ^4-Cholestenon

In 50%igem Äthanol sinkt die Stufe wiederum mit wachsendem pH (Abb. 7). An der Kurve
ist im sauren Bereich die Welle 1 erkenntlich und hier ist besonders gut zu sehen, dass ihre Höhe
erst dann zu sinken beginnt, wenn sich die Welle HA genügend verkleinert hat. An der Welle
HA sind charakteristische nichtwirbelnde Maxima beobachtbar. Bei einem pH-Wert von cca
5 wird auch die Welle K sichtbar.

Die Welle A′ ist im alkalischen Gebiet nicht erkennbar, am wahrscheinlichsten ist sie durch die Reduktionsstufe der Ionen des Grundelektrolyten verdeckt.

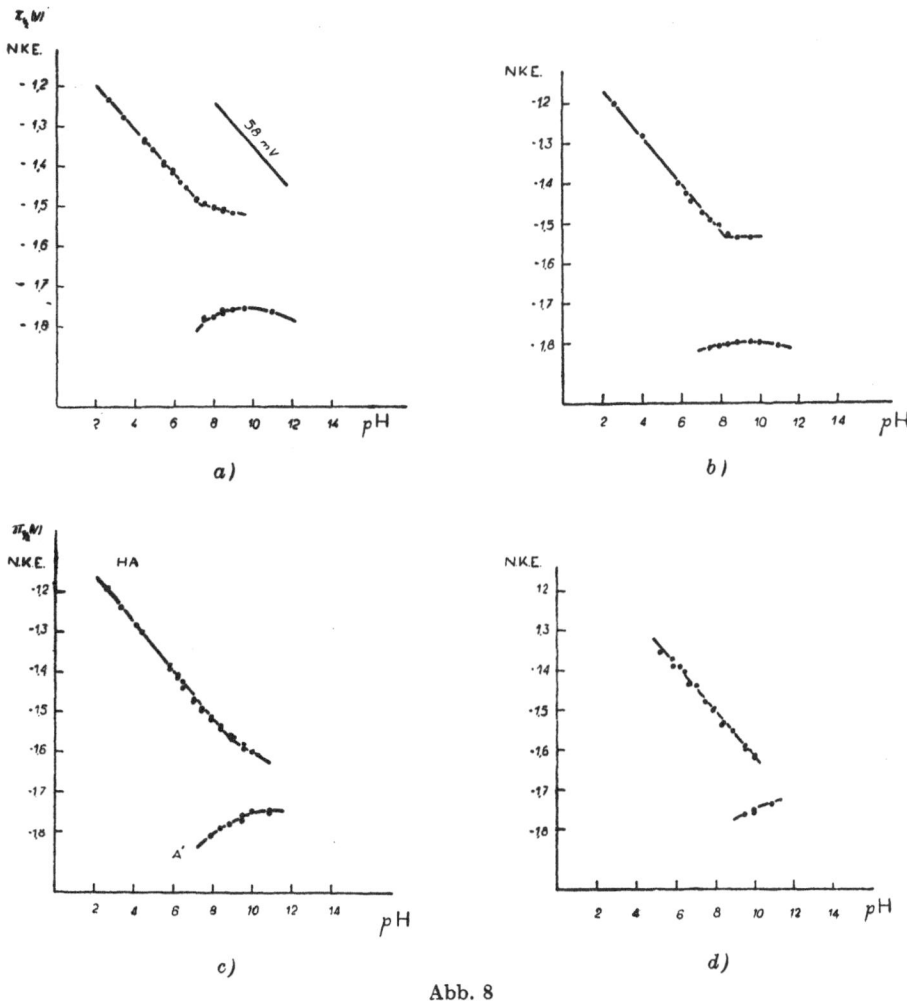

Abb. 8

Abhängigkeit der Halbstufenpotentiale vom pH

a) Methyltestosteron, *b)* Testosteron, *c)* Progesteron, *d)* Desoxycorticosteron.

Abhängigkeit des Halbstufenpotentials vom pH

Die Halbstufenpotentiale der Wellen HA und A′ sind in der Abb. 8*a—c* angeführt. Beim Desoxycorticosteron wurde das Halbstufenpotential der Welle 1 bzw. HA, die ineinander übergehen, gemessen (Abb. 8d). Das Halbstufenpotential der Welle 2 ändert sich nicht mit dem pH. Die Neigung der Abhängigkeit beträgt bei der Welle HA 58 mV/pH bis zum pH 7—9, bei grösseren pH-Werten ist das Halbstufenpotential weniger, eventuell unabhängig vom pH.

Abhängigkeit der Grenzströme von der Konzentration

Methyltestosteron, Testosteron, Progesteron. Die Höhen der Wellen HA und A′ sind im sauren und alkalischen Bereich ebenso wie im Existenzbereich beider Wellen in 50%igem Äthanol der

Konzentration direkt proportional. In 30%igem Äthanol mit 0,1N-KOH ist die Höhe der Welle 2 praktisch von der Konzentration unabhängig, wogegen die Summe beider Wellen linear mit der Konzentration anwächst. Die Höhe der Welle 2 ist praktisch bei allen untersuchten Steroiden gleich (Abb. 9). Die Höhe der Welle K wächst bedeutend stärker als die Höhe der Welle HA.

Desoxycorticosteron. Die Welle 1 wächst zunächst gleichzeitig mit der Welle HA an (Abb. 10, 11), verändert sich jedoch nach Erreichung einer gewissen Grenzkonzentration bei weiterer Konzentrationssteigerung nicht mehr.

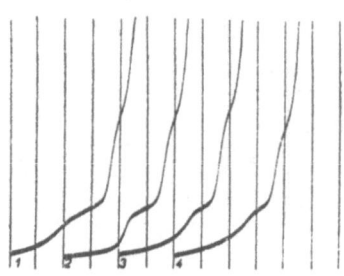

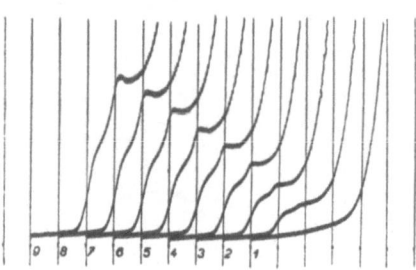

Abb. 9

Vergleich verschiedener Steroide in 0,1N-KOH mit 30% Äthanol
$2 \cdot 10^{-4}$M-Steroid: (*1*) Methyltestosteron, (*2*) Testosteron, (*3*) Progesteron, (*4*) Desoxycorticosteron. Kurven von der 7. Drahtwindung an, Mercurosulfatelektrode, 200 mV/Absz., Empf. 1 : 10, $h = 80$ cm.

Abb. 10

Desoxycorticosteron, Konzentrationsabhängigkeit der Welle 1 und der Hauptwelle (1): 5 ml Pufferlösung nach Britton und Robinson, pH 7,4; (2) bis (9): je 0,1 ml 0,01M--Desoxycorticosteron zugefügt. Kurven von der 7. Drahtwindung an, Mercurosulfatelektrode, Empf. 1: 20, $h = 50$ cm.

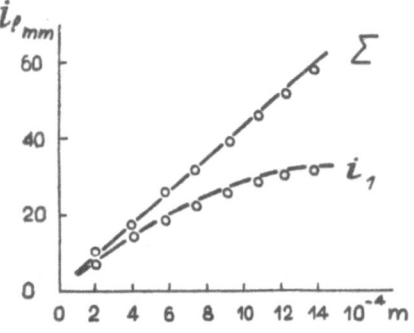

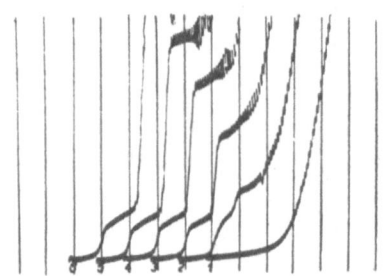

Abb. 11

Desoxycorticosteron, Abhängigkeit der ersten Welle 1 und des gesamten Grenzstromes Σ von der Konzentration

Abb. 12

Cholestenon, Konzentrationsabhängigkeit bei pH 5,0
5 ml Pufferlösung nach Britton und Robinson mit 50% Äthanol; pH 5,50. Je 0,1 ml 0,01M--Cholestenon zugegeben. Kurven von der 5. Drahtwindung an, NKE, 200 mV/Absz., Empf. 1 : 7.

Δ^4-*Cholestenon.* Die Höhe der Welle 1 erreicht beim Cholestenon ebenfalls bald einen Grenzwert (Abb. 12). Beim Cholestenon beträgt die Höhe der Welle 1 ungefähr 40% der Höhe der Welle 1 des Desoxycorticosterons.

Abhängigkeit der Grenzströme von der Behälterhöhe

Methyltestosteron, Testosteron, Progesteron. Die Summe der Stufen HA und A' bzw. der Stufen HA resp. A' in demjenigen pH-Bereich, wo an den Kurven eine einzige Welle sichtbar ist, sind der Wurzel aus h proportional und sind daher durch Diffusion bestimmt. Das Maximum an der ersten Welle ist bei höheren Behälterhöhen deutlicher (Abb. 13). Wenn die Welle HA nur cca 20% des gesamten Grenzstromes beträgt, ist ihre Höhe praktisch von h unabhängig (Abb. 14). Die Höhe der Vorstufe 2 ist im alkalischen Bereich der Behälterhöhe h direkt proportional (Abb. 15). Die Welle K, deren Höhe sich mit der Konzentration des Puffergemisches ändert, wächst mit sinkender Behälterhöhe.

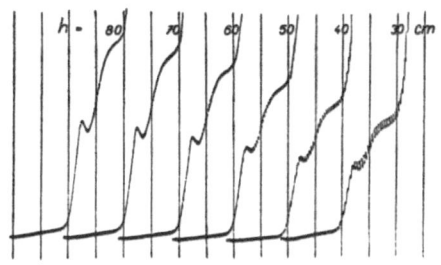

Abb. 13

Progesteron, Abhängigkeit von der Behälterhöhe bei pH 9,35
5 ml Pufferlösung nach Britton und Robinson mit 25% Äthanol, pH 9,35, 2 . 10^{-2}M-Progesteron. Kurven von der 6. Drahtwindung an, NKE, 200 mV/Absz., Empf. 1 : 10.

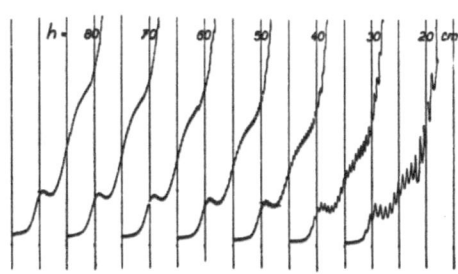

Abb. 14

Methyltestosteron, Abhängigkeit von der Behälterhöhe bei pH 8,5
5 ml Pufferlösung nach Britton und Robinson mit 30% Äthanol, pH 8,5. 1 . 10^{-3}M--Methyltestosteron. Kurven von der 6. Drahtwindung an, NKE, 200 mV/Absz., Empf. 1 : 20.

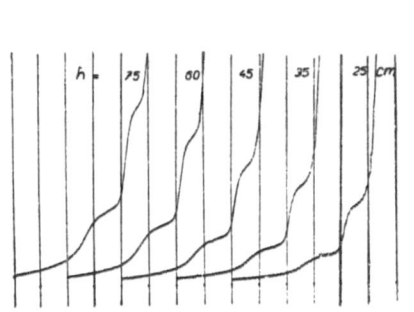

Abb. 15

Methyltestosteron, Abhängigkeit von der Behälterhöhe in 0,3N-LiOH. 2 . 10^{-4}M-Methyltestosteron.
Kurven von der 6. Drahtwindung an, Mercurosulfatelektrode, 200 mV/Absz., Empf. 1 : 10.

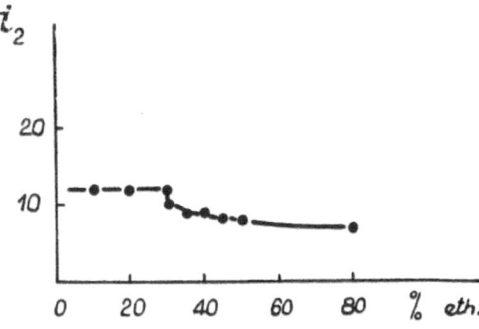

Abb. 16

Abhängigkeit der Stufenhöhe der Adsorptionswelle 2 beim Methyltestosteron von der Konzentration des Äthanols in 0,1N-KOH.

Desoxycorticosteron. In einer 1,4 . 10^{-3}M Lösung von Desoxycorticosteron ändert sich zwar das Verhältnis beider Wellen mit steigender Behälterhöhe zugunsten der Welle 1, die lineare Abhängigkeit von h ist hier jedoch noch nicht für alle Behälterhöhen h erreicht.

Abhängigkeit der Grenzströme von der Äthanolkonzentration

Methyltestosteron. Im alkalischen Medium sinkt die Vorstufe 2 mit steigender Äthanolkonzentration bei Überschreitung von 30% Äthanol und wird undeutlicher (Abb. 16).

Desoxycorticosteron. Die Höhe der Welle 1 sinkt ebenfalls mit steigender Äthanolkonzentration. Die Abspaltung der Stufe 1 wird bei Überschreitung von 42% Äthanol sichtbar.

Elektrokapillarkurven

Methyltestosteron. An den Elektrokapillarkurven wurde in einer 0,1M-NaOH-Lösung mit steigender Konzentration des Methyltestosterons eine Verflachung der Kurve bis zum Potential von — 1,7 V beobachtet, wo die Reduktion einsetzt.

Desoxycorticosteron. In einer Lösung vom pH 7,3 (Britton-Robinson Puffergemisch) wurden nach Zugabe von Desoxycorticosteron eine Verflachung der Elektrokapillarkurve, hierauf jedoch bedeutende Deformationen beobachtet, die für komplizierte Adsorptionsvorgänge sprechen.

Diskussion

Die Hauptwelle

Die Teilung der Hauptwelle in die Wellen HA und A' und der Abfall der Welle HA in Gestalt einer Dissoziationskurve (Abb. 1—3), wobei die Summe beider Wellen, deren Verhältnis von der Konzentration des Steroids unabhängig ist, erhalten bleibt, spricht dafür, dass der polarographischen Reduktion zwei Formen unterliegen. Die eine von ihnen ist im sauren, die andere im alkalischen Medium beständig. Es handelt sich hier wahrscheinlich eher um Dissoziation als um Keto-Enol-Tautomerie. Bei Berücksichtigung der relativen pH-Werte in den äthanolischen Lösungen kann aus den polarographischen Messungen die Dissoziationskonstante K' (Abb. 4, Tab. 1) bestimmt werden, deren Werte von den Konstanten der Kapillare und von der Äthanolkonzentration abhängig sind. Da beim Elektrodenvorgang auch Rekombination eintritt, wie aus der Abhängigkeit von der Behälterhöhe ersichtlich ist (Abb. 14), besitzen diese Dissoziationskonstanten (K') nur scheinbare Werte und sind um einige Grössenordnungen höher als die tatsächlichen Gleichgewichtskonstanten[5,6].

Die Schnittpunkte der linearen Teile der Abhängigkeiten der Halbstufenpotentiale der Welle HA vom pH (Abb. 8a—c) entsprechen ebenfalls den scheinbaren Dissoziationskonstanten (K''). Für die Werte der scheinbaren Konstanten wurden in Gegenwart von 50% Äthanol auf beide Arten die in der Tabelle 1 angeführten Werte gefunden.

Tabelle 1

Polarographische Dissoziationskonstanten

	pK' (aus i_{HA})	pK'' (aus $\pi_{\frac{1}{2}}$—pH)
Δ^4-Cholestenon	7,3	—
Methyltestosteron	7,7	7,3
Testosteron	8,2	8,2
Progesteron	8,9	8,8
Desoxycorticosteron	9,4	—

Die Abhängigkeit des Halbstufenpotentials vom pH spricht ebenfalls für die Erklärung durch Dissoziation, da die Abhängigkeit bei der Welle A' ähnlich wie bei den Säureanionen[7-11] ist.

Auch der Abfall der Kurven in Form eines unechten Maximums (Abb. 1, 7, 13, 14), der ähnlich wie bei der Phenylglyoxylsäure[8] und bei Oximen[9] verläuft, zeugt für eine Dissoziation. Es gelangt hier am ehesten zu einer Geschwindigkeitsveränderung der Rekombination bei Veränderung des Potentials. Das Auftreten dieser unechten Maxima erschwerte die Messung, so dass die Höhen der Wellen im Bereich der Dissoziation nur angenähert bestimmt werden können; so können auch die Abweichungen beim Testosteron und Desoxycorticosteron erklärt werden. Da es nicht möglich war, die Gleichgewichtskonstante der Reaktion potentiometrisch zu bestimmen, wurden die Beziehungen für den Reaktionsstrom[5,8,11] an die Versuchsergebnisse nicht angewandt.

Da wir voraussetzen, dass der Reduktion die mit der Doppelbindung in Stellung 4,5 konjugierte Carbonylgruppe in Stellung 3 unterliegt, kann bis jetzt nicht eindeutig über die Struktur der im sauren und der im alkalischen Bereich beständigen Form entschieden werden. Aus den angeführten Tatsachen geht jedoch hervor, dass sich beide Formen zu einander am ehesten wie Säure und konjugierte Base* verhalten.

Die reduzierbare Form kann in der Lösung jedoch in geringer Gleichgewichtskonzentration vorliegen (unter Voraussetzung eines dem Elektrodenvorgang vorgeschalteten beweglichen Gleichgewichts), da an den potentiometrischen Kurven kein jäher Potentialsprung, sondern bloss eine Deformation der Kurve zu sehen ist.

Die Adsorptionsvorstufe im alkalischen Bereich

Die Welle 2, die bei niedrigeren Äthanolkonzentrationen sichtbar ist und die an den Abfall der Welle HA anschliesst, war im untersuchten Konzentrationsbereich praktisch von der Konzentration unabhängig (bereits von $5 . 10^{-5}$M) und ihre Höhe ist der Behälterhöhe direkt proportional (Abb. 15); diese Welle gehört also der Reduktion des adsorbierten Stoffes an und ihre Höhe ist der Oberfläche der Elektrode proportional[12,13]. Aus der Beziehung[12,13]

$$i_a = 0{,}85 . n . F . z . m^{\frac{2}{3}} . t^{-\frac{1}{3}}$$

wurden aus dem Wert z (Anzahl der an 1 cm² der Elektrodenoberfläche adsorbierten Mole), für die durch ein Molekül des Steroids in 30%igem Äthanol besetzte Fläche A, folgende Werte berechnet (Abb. 9): Methyltestosteron 310 Å²; Testosteron 210 Å²; Progesteron 250 Å² und Desoxycorticosteron 310 Å². Beim Cholestenon ist — ähnlich wie bei einigen Herzglykosiden[14] — im alkalischen Bereich nur diese Adsorptionswelle sichtbar.

In Gegenwart der Ketosteroide wurde ein Abfall des Kapazitätsstromes beobachtet — ähnlich wie bei den Herzglykosiden[14] — was ebenfalls für die Adsorbierbarkeit dieser Stoffe spricht. Dieser Einfluss ist auch an den Elektrokapillarkurven sichtbar.

Auch der Abfall der Vorstufen im alkalischen Bereich mit steigender Äthanolkonzentration spricht für den Adsorptionscharakter dieser Stufen. Kaye und Stonehill[15] beobachteten eine ähnliche Erscheinung bei einigen Derivaten des Acridins.

---

* Die Möglichkeit der Formulierung der Ketosteroide als C-Säuren, auf welche uns Herr Dr E. Knobloch aufmerksam machte, wird später untersucht werden.

Die Adsorptionsvorstufe im sauren Bereich beim Δ^4-Cholestenon

Die Vorstufe 1, die beim Cholestenon im sauren Bereich sichtbar ist (Abb. 7), deren Höhe sich nicht mit dem pH ändert, zeigt ebenfalls eine solche Abhängigkeit von der Steroidkonzentration und der Behälterhöhe, wie sie bei Adsorptionsströmen[12,13] beobachtet wird. Diese Welle entspricht am ehesten der Reduktion der adsorbierten Form HA, denn nachdem die Höhe der Reduktionsstufe HA auf die Höhe der Vorstufe abgefallen ist, beginnt die Vorstufe zu sinken (Abb. 7). Die Konzentrationsgrenze beträgt bei $h = 50$ cm, $m = 3,06$ mg/s, $t = 2,3$ s rund $2,0 \cdot 10^{-4}$M und diese Stufe ist ungefähr zweimal so gross wie die Adsorptionsstufe im alkalischen Bereich. Für die besetzte Fläche wurde der Wert $A = 89$ Å^2 gefunden.

Die Vorstufe im sauren Bereich beim Desoxycorticosteron

Die im sauren Bereich beim Desoxycorticosteron beobachtete Vorstufe 1, deren Höhe sich nicht mit dem pH ändert, fällt ebenfalls mit steigender Äthanolkonzentration.* Die Konzentrationsabhängigkeit, die die Gestalt einer Exponentialfunktion hat und keine scharfe Grenze besitzt, wann beide Wellen gleichzeitig zu wachsen beginnen, entspricht den Abhängigkeiten, die beim 1-Hydroxyphenazin[16] und Chinin[17] beobachtet wurden. Die Abhängigkeit von der Behälterhöhe liegt ebenfalls zwischen $i_d = k \cdot \sqrt{h}$ und $i_a = k \cdot h$. Es handelt sich am ehesten um eine andere Art der Adsorption. Die Grösse des Stromes (das Zweifache der Welle 1 beim Cholestenon) spricht für eine abweichende Orientierung an der Elektrodenoberfläche.

Die katalytische Welle

Bei kleineren pH-Werten als cca 6 wird an den polarographischen Kurven eine Welle (K) sichtbar, die ähnlich wie bei den Dissoziationskurven mit fallendem pH wächst. Aus dieser Abhängigkeit kann, ebenso wie aus der Veränderung der Höhe mit der Konzentration des Puffers, mit der Behälterhöhe und aus der Gestalt der Abhängigkeit von der Konzentration des Ketosteroids, geschlossen werden, dass es sich um die Reduktion von Wasserstoffionen handelt, deren Überspannung katalytisch erniedrigt wurde.

Analytische Anwendbarkeit

Zur Bestimmung der Summe der untersuchten Steroide ist ein pH von 5—6, gegebenenfalls eine Lösung von 0,1N-LiOH am geeignetsten. Wenn auch die Wellen der untersuchten Steroide praktisch gleich gross sind, ist es dennoch günstiger für jedes einzelne Steroid eine gesonderte Eichkurve anzufertigen.

Aus der verschiedenartigen Abhängigkeit vom pH ergibt sich die Möglichkeit einer gleichzeitigen Bestimmung einiger der untersuchten Ketosteroide. So können Methyltestosteron, Testosteron, Progesteron und Desoxycorticosteron bzw. ihre Summe auch in Gegenwart eines bedeutenden Überschusses an Cholestenon im alkalischen Bereich bestimmt werden, wo die Welle der dissoziierten Form des Cholestenons nicht sichtbar ist.

In Anbetracht der verschiedenen Dissoziationskonstanten kann in einigen Fällen ein solches pH gewählt werden, bei welchem das eine Steroid bereits völlig dissoziiert ist, das andere hingegen nur teilweise. Die Höhe der positiveren Welle der undissoziierten Form (HA) gibt dann die Konzentration desjenigen Ketosteroids an, das die höhere Dissoziationskonstante besitzt.

---

* Es handelt sich am ehesten um Verdrängung des adsorbierten Stoffes aus der Elektrodenzwischenphase (Dr J. Koryta — Privatmitteilung).

So kann Desoxycorticosteron neben Testosteron bei einem pH von 9,5 und neben Methyltestosteron bei pH 9,2 (Abb. 17) und Progesteron neben Methyltestosteron bei einem pH von 9,2 bestimmt werden.

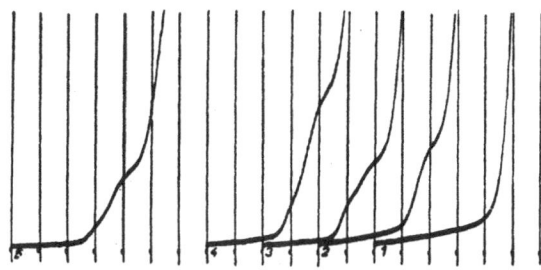

Abb. 17

Bestimmung von Desoxycorticosteron neben Methyltestosteron

5 ml Pufferlösung nach Britton und Robinson, 50% Äthanol, pH 9,5. (*1*) leere Lösung, (*2*) + 0,3 ml 0,01m-Methyltestosteron, (*3*) + 0,3 ml 0,01m-Desoxycorticosteron, (*4*) + 0,3 ml 0,01m-Methyltestosteron + 0,3 ml 0,01m-Desoxycorticosteron, (*5*) dasselbe bei Empf. 1 : 40. Kurven von der 7. Drahtwindung an, Mercurosulfatelektrode, 200 mV/Absz. Empf. 1: 20 [ausser (*5*)], $h = 50$ cm.

Schlussfolgerungen

Es wurde eine Erklärung für das Auftreten von zwei Wellen an den polarographischen Kurven der Δ^4-3-Ketosteroide gegeben, die von den älteren Autoren[1-3] unbeachtet blieben. Die Existenz der dissoziierten Form dieser Steroide in alkalischen äthanolischen Lösungen kann polarographisch bewiesen werden.

Ausser diesen Wellen wurden bei allen Steroiden im alkalischen Gebiet bei niedrigeren Äthanolkonzentrationen Adsorptionsströme beobachtet, ähnlich wie beim Cholestenon und Desoxycorticosteron in 50%igem Alkohol im sauren Bereich. Bei pH < 5 wurden bei kleineren Äthanolkonzentrationen katalytische Wellen beobachtet.

Die einzelnen Hormone unterscheiden sich voneinander durch ihre Dissoziationskonstanten, das Cholestenon überdies noch durch seine Unreduzierbarkeit in alkalischen Puffergemischen. Diese Unterschiede konnten zur gleichzeitigen Bestimmung einiger Gemische der angeführten Steroide ausgenützt werden.

Für zahlreiche Diskussionen danken wir Herrn Dr J. Koryta, für die technische Hilfe den Frl. A. Bittermann und H. Jelinek sowie Herrn M. Kabát. Für die Δ^4-Cholestenon und Testosteron-propionat-Präparate danken wir Herrn Dr V. Černý aus dem Institut für organische Chemie der ČSAV.

Literatur

1. Eisenbrand J., Picher H.: Z. physiol. Chem. *260*, 283 (1949).
2. Ekwall P., Lundsten T., Sjöblom L.: Acta chem. scand. *5*, 1383 (1951).
3. Sartori G., Bianchi E.: Gazz. chim. ital. *74*, 8 (1944).
4. Wolfe J. K., Hershberg E. B., Fieser L. F.: J. Biol. Chem. *136*, 633 (1940).
5. Brdička R.: Collection *12*, 212 (1947).
6. Brdička R., Wiesner K.: Collection, *12*, 138 (1947).
7. Hanuš V., Brdička R.: Chem. listy *44*, 291 (1950).

8. Hanuš V.: *Abhandlungen des I. int. polarogr. Kongresses*, Band I, S. 811, Naturwissenschaftliche Verlagsanstalt, Praha 1951.
9. Souchay P., Ser S.: J. Chim. phys. *49*, C 172 (1952).
10. Elving P. J., Teiltelbaum Ch.: J. Am. Chem. Soc. *71*, 3916 (1949).
11. Koutecký J., Brdička R.: Collection *12*, 337 (1947).
12. Brdička R.: Z. Elektrochemie *48*, 278 (1942).
13. Brdička R.: Collection *12*, 522 (1947).
14. Zuman P., Šantavý F.: Chem. listy *46*, 393 (1952).
15. Kaye R. C., Stonehill H. J.: J. Chem. Soc. 1951, 2638.
16. Müller O. H.: J. Biol. Chem. *145*, 425 (1942); Trans. Electrochem. Soc. *87*, 441 (1945).
17. Zuman P.: Im Buche Březina M., Zuman P.: „*Polarographie in der Medizin, Pharmazie und Biochemie*", S. 194. Verlagsanstalt für Gesundheitswesen, Praha 1952.

Übersetzt von H. Bažantová.

Part VI

Reduction of the Activated C – N, C – S and C – O Bonds

21. V. Horák, and P. Zuman: Fission of activated carbon-nitrogen and carbon-sulphur bonds. Introductory remarks. *Collect. Czechoslov. Chem. Commun.* **26**, 173 (1961).

22. P. Zuman, and V. Horák: Fission of activated carbon-nitrogen and carbon-sulphur bonds. I. Polarographic reduction of the single C – N bond. *Collect. Czechoslov. Chem. Commun.* **26**, 176 (1961).

23. P. Zuman, O. Manoušek, and V. Horák: Fission of activated carbon-nitrogen and carbon-sulphur bonds. VI. Polarographic reduction of the C – N$^{(+)}$ bond in a-aminonitriles. *Collect. Czechoslov. Chem. Commun.* **29**, 2906 (1964).

24. P. Zuman, and S. Tang: Fission of activated carbon-nitrogen and carbon-sulphur bonds. III. Reduction of the C–S$^{(+)}$ bond in methyl butyl phenacyl sulphonium perchlorate. *Collect. Czechoslov. Chem. Commun.* **28**, 829 (1963).

25. S. Tang, and P. Zuman: A new type of maximum on the limiting current of the reduction wave of phenacyl sulphonium ions. *Collect. Czechoslov. Chem. Commun.* **28**, 1524 (1963).

26. O. Manoušek, and P. Zuman: Polarography of pyridoxine and some of its derivatives. *Collect. Czechoslov. Chem. Commun.* **29**, 1432 (1964).

27. O. Manoušek, and P. Zuman: Polarographic reduction of 2-(2-methyl 3-hydroxy-5-hydroxymethyl-4-pyridyl)thiazolidine-4-carboxylic acid. *Collect. Czechoslov. Chem. Commun.* **29**, 17 (1964).

28. O. Manoušek, O. Exner, and P. Zuman: Fission of activated carbon-nitrogen and carbon-sulphur bonds. X. Polarographic reduction of substituted methyl phenyl sulphones. *Collect. Czechoslov. Chem. Commun.* **33**, 3988 (1968).

29. O. Manoušek, O. Exner, and P. Zuman: Fission of activated carbon-nitrogen and carbon-sulphur bonds. XI. Polarographic reduction of substituted benzene sulphonamides. *Collect. Czechoslov. Chem. Commun.* **33**, 4000 (1968).

Reduction of C – N, C – S and C – O single bonds takes place, if the saturated carbon of the bond is attached to groups such as RCO, CN or pyridine.[21-27] The reactivities of the above bonds in pyridoxal, pyridoxamine and pyridoxthiol were compared.[26] In the thiazolidine derivative resulting from the reaction of pyridoxal with cysteine, the C – S bond is more reactive[27] towards reduction than C – N. Whereas a-aminoketones[22] are very weak acids, phenacyl sulphonium ions are acids of medium strength where the equilibrium between the parent compound and ylide is rapidly established.

In the reduction of substituted methyl phenyl sulphones[28] and benzene sulphonamides[29] fission of the C – S bond occurs, before that of the C – N bond. Deductions on the structure of the transition state in these aromatic nucleophilic substitutions can be made from the substituent effects, by means of the Hammett equation.

FISSION OF ACTIVATED CARBON–NITROGEN AND CARBON–SULPHUR BONDS

INTRODUCTORY REMARKS

V. Horák and P. Zuman

Department of Organic Chemistry, Charles University and Polarographic Institute,
Czechoslovak Academy of Science, Prague

Received June 6th, 1960

To Academician J. Heyrovský on his 70th birthday.

The problem of carbon–nitrogen and carbon–sulphur bond fission has a direct bearing on the practically important applications of Mannich bases[1], their quaternary derivatives, and analogous compounds as alkylating reagents.

The fundamental type of Mannich bases[2] are the β-aminoketones[3] in which it is the ketonic carbonyl which acts as the activating group. In principle, the keto group may be replaced by any other electronegative group to give other types, such as phenolic and heterocyclic Mannich bases in which the activating grouping is incorporated in an aromatic system carrying a phenolic hydroxyl or heterocyclic imino group in β-position to the amino group. All these compounds are accessible by direct synthesis from the appropriate amines, formaldehyde, and a methyl ketone, phenol, or heterocyclic compound, as the case may be. Alkylation by such derivatives has been variously interpreted as proceeding by nucleophilic substitution of the nitrogen or by elimination followed by Michael-type addition to the activated olefin formed, or (in special cases) by reversal of the Mannich reaction followed by a second condensation of the same type[4].

However, pronounced alkylating properties are well known to be exhibited by certain simple benzylammonium salts[5] which cannot be prepared by straightforward synthetic reactions involving formaldehyde and an amine. These alkylating properties are doubtlessly due to the activating effect of the phenyl group as such, together with the stability of the benzyl cation; an elimination addition mechanism is, of course, excluded here.

Other compounds related to the benzylammonium salts by replacement of the phenyl group with other electronegative substituents such as ketonic carbonyl or nitrile groups (α-aminoketones, α-aminonitriles) merit consideration in this connection. So far such compounds have been found to display alkylating properties only quite exceptionally[6], though the corresponding fragments should obviously show high cation stability. Unlike the β-aminoketones which are reduced by chemical reducing agents at the carbonyl group[7], the α-aminoketones are reductively cleaved at the C—N bond and therein resemble the benzyl derivatives[8,9].

The closely related compounds of the phenacylpyridinium type in addition to reductive fission of the C—N bond and negligible alkylating properties exhibit a marked tendency to fission of the C—C bond under alkaline conditions[10].

Whereas there can be little doubt concerning the course of the reactions undergone by α-aminoketones and related compounds, the mechanism of alkylation by Mannich bases is still unsettled. In order to extend our understanding of the general rules which govern the fission of compounds containing activated carbon–nitrogen and carbon–sulphur bonds[11-13] we have undertaken an investigation of the following groups of compounds:

1. Substances of the type I—III, where R^1 and R^2 are alkyl or aryl groups and X is $CO.R^3$, $SO_2.R^3$, CN, $CO.NH_2$, $COOR^3$, NO_2, aryl, or a cationic group.

2. Substances of the type IV—VII, where R^1, R^2, and X are the same as in the first group and R^4 is an alkyl or aryl grouping. This class also includes compounds of the type VIII, where again X represents the same substituents as before.

3. Vinyl derivatives of the types IX (quaternary ammonium compounds) or X (tertiary sulphonium compounds), with R^1, R^2, R^4, and X representing the same substituents as before.

4. Some ylides, included for comparison.

The substituents R were chosen so as to give a variety of polar and steric effects and the substituents X for a wide range of polar, steric, and mesomeric (resonance) effects. Since the inner inductive effect decreases rapidly with increasing number of the interpolated methylenegroups we restricted ourselves in most cases to substances with n = 1 and 2; in cyclic compounds of the type II, VI and VII, with m > 3.

$$R^1R^2N\text{—}(CH_2)_n\text{—}X \qquad (CH_2)_m\overset{\frown}{N}\text{—}(CH_2)_n\text{—}X \qquad R^1S\text{—}(CH_2)_n\text{—}X$$

$$I \qquad\qquad\qquad II \qquad\qquad\qquad III$$

$$R^1R^2R^4\overset{(+)}{N}\text{—}(CH_2)_n\text{—}X \qquad R^1R^4\overset{(+)}{S}\text{—}(CH_2)_n\text{—}X$$

$$IV \qquad\qquad\qquad V$$

$$(CH_2)_m\overset{(+)}{N}\text{—}(CH_2)_n\text{—}X \qquad (CH_2)_m\overset{(+)}{S}\text{—}(CH_2)_n\text{—}X$$
$$\overset{|}{R^4}$$
$$VI \qquad\qquad\qquad VII$$

$$\overset{(+)}{N}\text{—}(CH_2)_n\text{—}X \qquad R^1R^2\overset{(+)}{N}\text{—}CH{:}CH\text{—}X \qquad R^1\overset{(+)}{S}\text{—}CH{:}CH\text{—}X$$
$$\overset{|}{R^4} \qquad\qquad\qquad \overset{|}{R^4}$$
$$VIII \qquad\qquad\qquad IX \qquad\qquad\qquad X$$

The first studies of the present series were concerned with the reactivity of activated carbon–nitrogen bonds under the conditions of polarographic electrolysis. In the first paper it will be shown that α-aminoketones and their methoiodides, as well as compounds of the type *VIII* and *IX*, undergo reduction of the C−N bond. The second paper demonstrates that, in agreement with their chemical behaviour, β-amino-ketones undergo reduction at the carbonyl group. The information gained with these model substances is then applied in a study of the polarographic behaviour of tro-pinone, its methiodide, and the corresponding sulphur analogues; the products formed from these compounds under alkaline conditions are identified. Finally, in the fourth paper of the present group the polarographic waves given by Mannich bases and the corresponding α,β-unsaturated ketones are used to follow the kinetics and mechanism of the elimination reaction and certain competitive reactions.

References

1. Brewster J. H., Eliel E. L.: Org. Reactions *7*, 99 (1953).
2. Blicke F. F.: Org. Reactions *1*, 303 (1942).
3. Brewster J. H., Eliel E. L.: see 1, p. 178.
4. Brewster J. H., Eliel E. L.: see 1, p. 126.
5. Brewster J. H., Eliel E. L.: see 1, p. 163.
6. Eliel E. L., Murphy N. J.: J. Am. Chem. Soc. *75*, 3589 (1953).
7. Blicke F. F.: see 2, p. 323.
8. Stoermer R., Dzimski O.: Ber. *28*, 2220 (1895).
9. Van Ark H. H.: Arch. Pharm. *238*, 321 (1900).
10. Kröhnke F.: Ber. *92*, CIV (1959).
11. Parham W. E., Ramp F. L.: J. Am. Chem. Soc. *73*, 1293 (1951).
12. Horák V., Černý M.: This Journal *18*, 379 (1953).
13. Poppelsdorf F., Holt S. J.: J. Chem. Soc. *1954*, 1124.

FISSION OF ACTIVATED CARBON–NITROGEN AND CARBON–SULPHUR BONDS. I.

POLAROGRAPHIC REDUCTION OF THE SINGLE C–N-BOND

P. Zuman and V. Horák

Polarographic Institute, Czechoslovak Academy of Science and Department of Organic Chemistry,
Charles University, Prague

Received June 3rd, 1960

To Academician J. Heyrovský on his 70th birthday.

In methoiodides of α-aminoketones and in aminoaldehydes with the grouping $-COCH_2-NR_3^{(+)}$ or $-COCH=CH-NR_3^{(+)}$ the $C-N^{(+)}$-bond is reduced. The products of the two-electron reduction of α-aminoketones are the ketone $-COCH_3$ and the amine NR_3. When the carbonyl compound is reducible (as it is in the case of acetophenone formed in the reduction of ω-piperidinoacetophenone methoiodide (*I*) and hydrochloride (*II*) or phenacylpyridinium chloride (*VII*) or in the case of acrolein formed in the reduction of β-N-triethylaminoacrolein perchlorate (*VII*)), another two-electron step, corresponding to the reduction of a ketone formed in the first step, appears. In hydrochlorides of α-aminoketones the $C-N^{(+)}$ bond is reduced in the protonized form only, with the grouping $-COCH_2NHR_2^{(+)}$, but not in the free base. A comparison with other reductions of single bonds is given.

In the course of study of the polarographic behaviour of aminoketones some simple compounds containing the grouping $-COCH_2NR^1R^2R^{3(+)}$ were studied. It is the aim of the present paper to prove that polarographic reductive cleavage of the $C-N^{(+)}$-bond occurs in these compounds. For this purpose ω-piperidinoaceto-phenone methoiodide (*I*) has been chosen as a suitable model substance. In this compound, besides the reduction of the activated $C-N^{(+)}$-bond, the reduction of the carbonyl group conjugated with a phenyl group may be studied as well. The role of the phenyl group on the reactivity of the $C-N^{(+)}$-bond was distinguished by comparison with piperidinoacetone methoiodide (*III*). In the cyclopentanone derivative *V* the relative positions of the carbonyl grouping and the nitrogen atom were fixed. To compare the effect of a methyl group in the quaternary ammonium grouping with the effect of a hydrogen atom, the corresponding hydrochlorides of all three types (*II*, *IV* and *VI*) were studied as well.

For the sake of comparison phenacylpyridinium chloride (*VII*) bearing the quaternary nitrogen bond in a pyridine ring as well as unsaturated aminoaldehydes *VIII* and *IX* were included. Anylid *X* with separated integral charges on carbon and nitrogen $^{(-)}C-N^{(+)}$ should show the reducibility of this bond.

Experimental

The polarographic curves were recorded on a Heyrovský Polarograph, Type VII (Nejedlý) and Type V 301 (TOS Služba, Prague), with a galvanometer of sensitivity $2 . 10^{-9}$ A/mm. A Kalousek vessel with a separated saturated calomel electrode (s. c. e) was used. The capillary used had the following constants: outflow rate $m = 2 \cdot 1$ mg/s, and drop-time $t_1 = 3 \cdot 7$ s at the potential of s. c. e. at mercury pressure $h = 65$ cm. All the values of half-wave potentials are expressed *versus* s. c. e. The pH-values of buffer solutions were checked using the Autojonometer Type PHM 21d (Radiometer-København) pH-meter with a glass electrode type G 200 B.

Reagents

The substances used in this study together with some important data are given in Table I.

Table I

Substances Studied

Substance	No.	Substituent	M. p., °C (Kofler)	References
$C_6H_5COCH_2\overset{(+)}{N}$ ◇ $X^{(-)}$ R	*I*	$R = CH_3; X = I$	184°	1
	II	$R = H; \quad X = Cl$	221°^{a}	2
$CH_3COCH_2\overset{(+)}{N}$ ◇ $X^{(-)}$ R	*III*	$R = CH_3; X = I$	127°	3
	IV	$R = H; \quad X = Cl$	b	3
(cyclopentanone-N ring) $X^{(-)}$ R	*V*	$R = CH_3; X = I$	171°	e
	VI	$R = H; \quad X = Cl$	c	e
$C_6H_5COCH_2\overset{(+)}{N}$ ⬡ $Cl^{(-)}$	*VII*		204°^{d}	4
$OHC{-}CH{=}CH{-}\overset{(+)}{N}(C_2H_5)_3 .$ $.ClO_4^{(-)}$	*VIII*			4
$OHC{-}CH{=}CH{-}N(CH_3)_2$	*IX*		$-2°$	5
H_5C_2OOC ＼ $\overset{(-)}{C}{-}\overset{(+)}{N}$ ⬡ H_5C_2OOC ／	*X*			5

^{a} Very hygroscopic substance; ^{b} very hygroscopic substance, prepared as a viscose liquid; ^{c} very hygroscopic half solid mass; ^{d} literature^4 196—198°; ^{e} see Experimental.

2-*Piperidinocyclopentanone*[7]. 13·3 g of 2-chloro-cyclopentanone and 30·0 g of piperidine reacted exothermically in ethylether. Isolated bases (60%) were distilled at 110—120°/10 mm. By paper chromatography (paper Whatman No. 1, n-butanol—acetic acid—water 4 : 1 : 5, descending, detection by Dragendorff reagens, the presence of a small quantity of another base with $R_F = 0·99$ was detected besides the substantially greater spot of *VI* with $R_F = 0·69$; the impurity could not be separated after treble distillation.

The hydrochloride was prepared by reaction with gaseous hydrochloric acid in desiccator (m. p. 140—160°). The methoiodide was prepared in the usual way in analytical purity (m. p. 171° methanol, ethanol, small part of ether).

For $C_{11}H_{20}INO$ (309·2) found, (calculated): 42·76 (42·8)% C; 6·70 (6·74)% H; 40·92 (41·1)% I; 4·32 (4·53)% N.

β-N-*Triethylaminoacrolein perchlorate* (VIII) and β-N-*diethylaminoacrolein* (IX) were kindly presented by Dr Z. Arnold (Chemical Institute, Czechoslovak Academy of Science, Prague), *diethyl-malonesterpyridiniumylid* (X) by Doz. H. Schläfer (Institute of Physical Chemistry, the University, Frankfurt am Main). *Acetophenone* (XI) and *acrolein* (XII) (B. D. H.) were freshly distilled. Solutions of acrolein are to be prepared and submitted to the polarographic electrolysis on the same day as the distillation is performed.

0·01M stock solutions were prepared by dissolving a weighed amount of the substance in water at 50—70°C. The buffer solutions used (Britton-Robinson universal buffer, acetate and phosphate buffers) were prepared from analytical reagent grade chemicals.

Technique of Measurements

9·8 ml of the appropriate buffer solution were deaerated by a stream of pure nitrogen. 0·2 ml of a 0·01M solution of the substance under study was added and the deaeration was continued for a further 30 seconds. The curves were then recorded. Only when explicitly stated, 0·01% to 0·015% gelatine was added to suppress the maxima of the first kind.

The change of the curves with time was followed mainly in alkaline solutions.

Constant Potential Electrolysis

The electrolysis at a controlled potential was performed in 10 ml of a 0·01M solution using the dropping mercury electrode. The potentiostat and the electrolysis vessel, enabling the three electrode system to be used, were designed by Peizker[8]. The appropriate choice of the controlled potential, in the potential region corresponding to the limiting current, was controlled by scanning the polarographic curve. The decrease of the current with time during the electrolysis was followed continually by a recording instrument. At chosen time intervals complete polarographic curves were recorded.

During the whole electrolysis a nitrogen atmosphere was maintained in the electrolysis cell. Volatile ketones and amines formed in the electrolysis were trapped in washing bottles containing a saturated solution of 2·4-dinitrophenylhydrazine in 0·1M-HCl and 0·05M-HCl respectively.

The identification of the products was performed by polarography of the electrolyzed solution, by paper chromatography of the 2,4-dinitrophenylhydrazone formed and by retitration of the unconsummed acid in the washing bottle.

Results

ω-Piperidinoacetophenone Methoiodide (*I*) and Hydrochloride (*II*)

ω-*Piperidinoacetophenone methoiodide* (I) is reduced at the dropping mercury electrode in two two-electron steps (Fig. 1).

1. In the more positive wave the C–N$^{(+)}$-bond is reduced. The wave height corresponds to a two electron transfer and remains unchanged in the whole pH-range studied. A sharp maximum was observed on the limiting current (Fig. 2). For the measurement of the half-wave potentials the presence of 0·015% gelatine was necessary. The dependence of the half-wave potentials on pH is given in Fig. 3. At pH-values lower than about 9·2 the half-wave potential are shifted by 30 mV/pH.

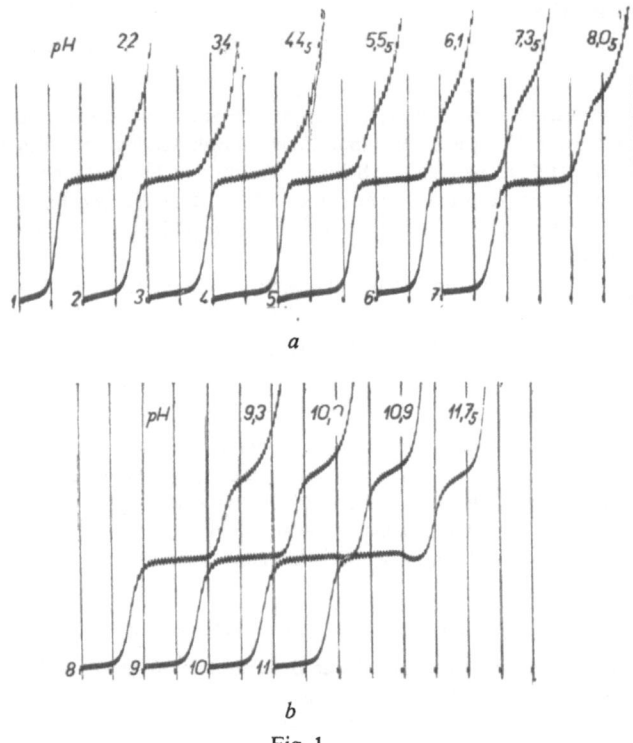

Fig. 1

The pH-Dependence of the Waves for ω-Piperidinoacetophenone (*I*) Methoiodide in Britton-Robinson Buffers

$2 . 10^{-4}$M depolarizer, 0·015% gelatine, pH-values given on the polarogram. Curves *1—5* starting at −0·4 V, *6—11* at −0·6 V, s. c. E, 200 mV/absc., $h = 65$ cm, sens. 1 : 20.

2. The more negative reduction step corresponds to the reduction of acetophenone, formed in the first step by reductive substitution of substance *I.*

The following proofs were given for this assignment:

1. The reduction of the C–N$^{(+)}$-bond in the more positive wave was proved by the identification of the more negative wave as the wave for acetophenone. Furthermore, this assumption was verified by controlled potential electrolysis. By electrolysis at the potential of the limiting current of the first wave, acetophenone and a corresponding amount of an amine were formed and proved among the electrolysis products.

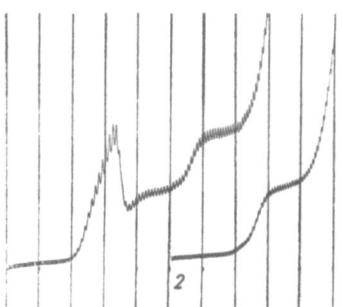

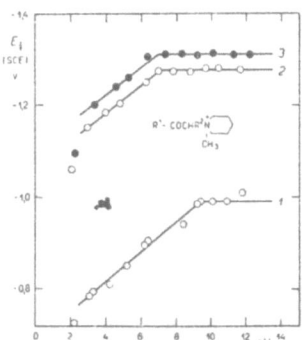

Fig. 2

Comparison of Waves for ω-Piperidinoaceto-phenone Methoiodide (*I*) and Acetophenone
2.10^{-4}M depolarizer, phosphate buffer pH 6·95. *1* ω-piperidinoacetophenone methoiodide; *2* acetophenone. *1* starting at $-0·6$ V; *2* at $-1·2$ V, s. c. e., 200 mV/absc., $h = 65$ cm, sens. 1 : 30.

Fig. 3

The pH-Dependence of the Half-Wave Potentials of α-Aminoketones Methoiodides
1 ω-piperidinoacetophenone methoiodide (*I*), $R^1 = C_6H_5$, $R^2 = H$; *2* piperidinoacetone methoiodide (*III*), $R^1 = CH_3$, $R^2 = H$; *3* 2-piperidinocyclopentanone methoiodide (*V*), R^1, $R^2 = CH_2CH_2CH_2$. Britton-Robinson buffer, 2.10^{-4}M depolarizer, 0·015% gelatine.

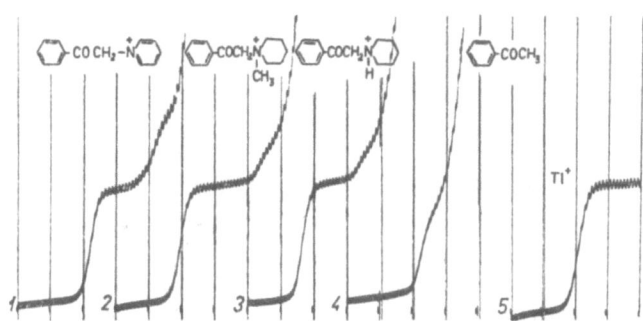

Fig. 4

Comparison of α-Aminoketones
2.10^{-4}M α-aminoketone, acetate buffer pH 4·45, 0·015% gelatine. *1* phenacylpyridinium chloride (*VII*); *2* ω-piperidinoacetophenone methoiodide (*I*); *3* ω-piperidinoacetophenone hydrochloride (*II*); *4* acetophenone; *5* Tl$^+$. Curves *1* and *2* starting at $-0·4$V; *3* and *4* at $-0·6$ V; *5* at 0·0 V, s. c. e., 200 mV per absc., $h = 65$ cm, sens. 1 : 20.

2. The more negative wave for substance *I* is identical in height, in the dependence of the wave-height on pH, in the half-wave potentials and in the form of the dependence of the half-wave potentials on pH, with the wave of acetophenone (Figs. 2, 4).

Both the more negative wave for *I* and the wave for acetophenone show a one-electron wave in phosphate buffers at pH < 3 and a two-electron wave at pH 8 — 10.

Furthermore both waves show a decrease in the wave-height at pH greater than about 10·(Fig. 5).

The waves in acid solutions, corresponding to the reduction of the protonized form, show at pH higher than about 6 a steep decrease in the form of a dissociation curve. The shape of this dissociation curve and its pK'-value are identical for both acetophenone and the more negative wave of methoiodide I (Fig. 5). Moreover the influence of buffer composition and of the presence of gelatine (influencing proton transfer by its aminogroups as with other recombinations[9]) is the same for both waves.

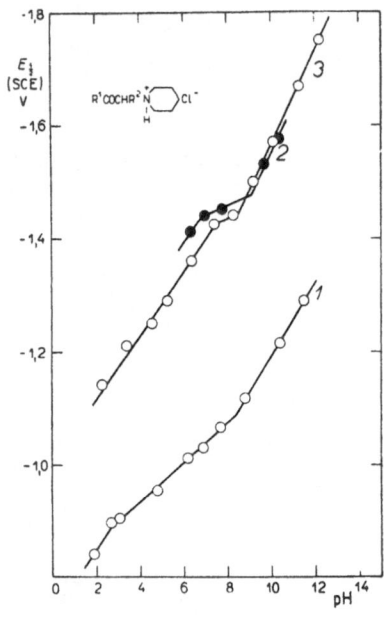

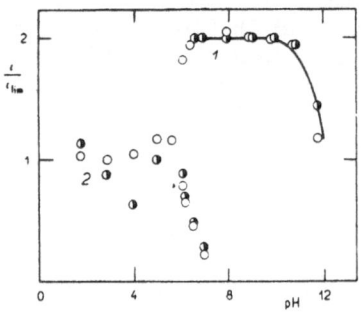

Fig. 5

The pH-Dependence of the Limiting Current of Acetophenone (○) and of the Second Wave for ω-Piperidinoacetophenone Methoiodide (◑) Compared with the Limiting Current, Corresponding to a One-Electron Reduction Britton-Robinson buffers, $t_1 = 2\cdot7$ s, $2\cdot10^{-4}$M depolarizer.

Fig. 6

The pH-Dependence of the Half-Wave Potentials of α-Aminoketones Hydrochlorides 1 ω-Piperidinoacetophenone hydrochloride(II), $R^1 = C_6H_5$, $R^2 = H$; 2 piperidinoacetone hydrochloride (IV), $R^1 = CH_3$, $R^2 = H$; 3 2-piperidinocyclopetnanone hydrochloride (VI), $R^1, R^2 = CH_2CH_2CH_2$. Britton-Robinson buffers, $2\cdot10^{-4}$M depolarizer.

The second wave of substance I *differs* from that of acetophenone in the change of the wave-height in the region between pH 2 and 5. Whereas in this pH region the half-wave potentials of both waves are identical, the limiting current of the one-electron wave of the protonized form of acetophenone remains unchanged and the limiting current of the second wave for substance I shows a decrease at pH 2 followed by an increase at pH about 4 (Fig. 5).

Table II

Half-Wave Potentials of Substances Studied at Selected[a] pH-values in Britton-Robinson Buffers (in volts, vs. s. c. e.)

No.	Compound	Half-Wave Potentials				Gelatine
I	$C_6H_5COCH_2N$ (+) ring I⁻, CH$_3$	pH $\quad$ 3·3 $\quad$ 6·4 $\quad$ 9·4 $\quad$ 10·2 $E_{\frac{1}{2}}$ $\quad$ −0·79$_5$ −0·90$_2$ −0·99 −0·98$_6$				0·015%
II	$C_6H_5COCH_2H$ (+) ring Cl⁻, N	pH $\quad$ 3·1 $\quad$ 6·2 $\quad$ 8·8 $\quad$ 10·4 $E_{\frac{1}{2}}$ $\quad$ −0·90$_4$ −1·01$_6$ −1·11$_7$ −1·21$_4$ $\quad$ −1·10$_0$ −1·20$_3$				0·015% 0
III	CH_3COCH_2N (+) ring I⁻, CH$_3$	pH $\quad$ 4·0 $\quad$ 6·3 $\quad$ 7·0 $\quad$ 10·4 $E_{\frac{1}{2}}$ $\quad$ −1·18$_3$ −1·25$_0$ −1·27$_4$ −1·28				0
IV	CH_3COCH_2N (+) ring Cl⁻, H	pH $\quad$ 6·3 $\quad$ 7·8 $\quad$ 10·4 $E_{\frac{1}{2}}$ $\quad$ −1·41$_2$ −1·45$_3$ −1·57$_5$				0
V	cyclic structure O (+) N I⁻, CH$_3$	pH $\quad$ 6·4 $\quad$ 8·3 $\quad$ 10·1 $E_{\frac{1}{2}}$ $\quad$ −1·30$_6$ −1·31$_1$ −1·31$_6$				0·015%
VI	cyclic structure O (+) N Cl⁻, H	pH $\quad$ 6·4 $\quad$ 8·3 $\quad$ 10·1 $E_{\frac{1}{2}}$ $\quad$ −1·36$_4$ −1·43$_2$ −1·56$_1$				0·015%
VII	$C_6H_5\ COCH_2N$ (+) ring Cl⁻	pH $\quad$ 3·1$_5$ $\quad$ 6·4 $\quad$ 9·4 $E_{\frac{1}{2}}$ $\quad$ −0·87 −0·99$_2$ −1·09 $\quad$ −1·21				0·015%
VIII	OHC—CH=CH—$\overset{(+)}{N}(C_2H_5)_3$ $ClO_4^{(-)}$	pH $\quad$ 2·1 $\quad$ 5·2 $\quad$ 7·2 $\quad$ 9·25 $E_{\frac{1}{2}}$ $\quad$ −0·68$_1$ −0·77$_4$ −0·84$_4$ −0·91$_4$				0
IX	OCH—CH=CH—N(CH$_3$)$_2$	pH $\quad$ 2·1 $\quad$ 5·0 $\quad$ 7·8 $E_{\frac{1}{2}}$ $\quad$ −1·12^b −1·51^b −1·67^b				0
X	H_5C_2OOC (−) (+) $\,$C—N ring H_5C_2OOC	pH $\qquad\qquad$ 9·4 $E_{\frac{1}{2}}$ $\qquad\qquad$ −1·50$_5$ $\qquad\qquad$ −1·50$_7$				0·01% 0

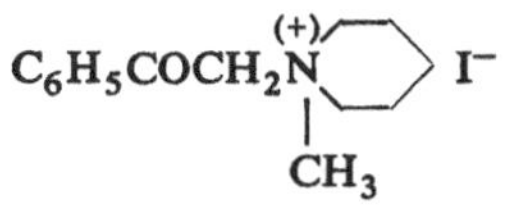

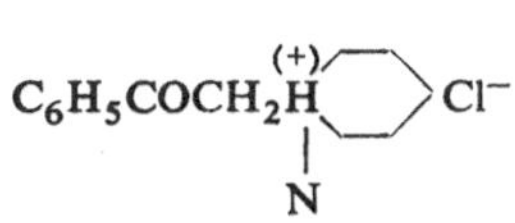

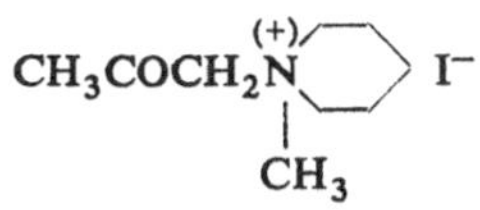

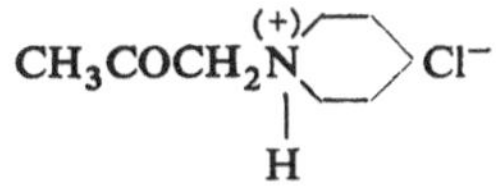

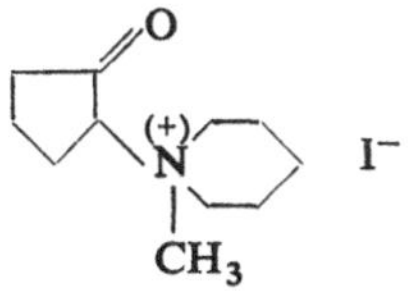

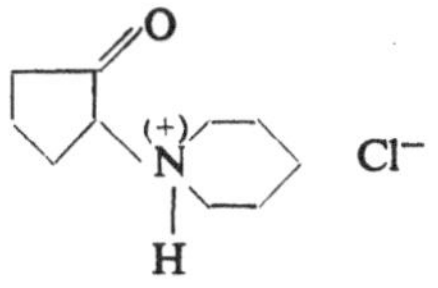

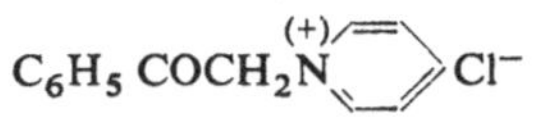

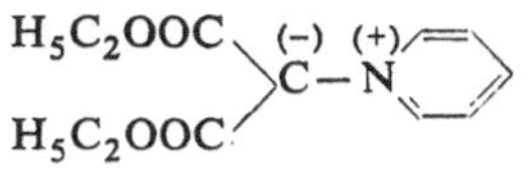

[a] The pH-values were selected according to the most important points on the $E_{\frac{1}{2}}$–pH-plot.
[b] Approximate values, ill-developed waves.

The behaviour of acetophenone undergoing reduction after transport from the bulk of solution is thus not identical with that of acetophenone formed at the surface of the electrode in the reduction of *I*. In addition to the orientation at the electrode surface, the state of hydration or keto-enol tautomeric changes and the different conditions of protonation with the acetophenone supplied by diffusion and with the acetophenone formed may cause this difference. Whatever the reason is, it is dependent on the structure of the parent aminoketone, as demonstrated by different heights of the second wave for different phenacylamines (Fig. 4).

The behaviour of ω-*piperidinoacetophenone hydrochloride* (II) is analogous to the corresponding methoiodide *I* (Fig. 4) at pH lower than about 9·5. Only the first wave is shifted towards more negative potentials (Tab. II) and the dependence of the half-wave potentials on pH-values has a different shape (Fig. 6). At pH greater than 10 decomposition of substance *II* was observed, showing changes on polarographic curves with time. With the methoiodide *I* no changes were observed even in 0·1N-NaOH.

Phenacylpyridinium Chloride (*VII*)

A similar behaviour was shown also for phenacylpyridinium chloride (*VII*), reduced at lower pH-values in two two-electron waves (Fig. 4). The more positive wave corresponds again to the reduction of the C–N$^{(+)}$-bond; in the second wave acetophenone formed is reduced. The dependence of the half-wave potentials on pH-values is even more complicated (Fig. 7). The height of the protonized form changes much less with pH in the region pH 2–5 than with *I* and *II*.

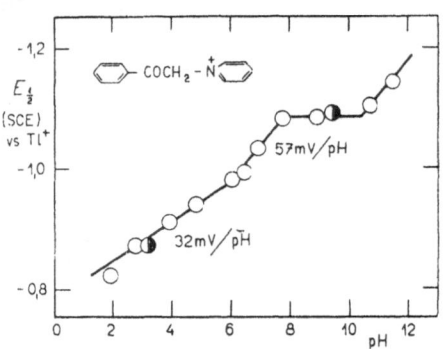

Fig. 7

The pH-Dependence of the Half-Wave Potentials of Phenacylpyridinium Chloride Britton-Robinson buffers, 2 . 10^{-4}M depolarizer, 0·015% gelatine.

At pH 6·2 the wave-height and -shape is practically time-independent. At pH 8 changes with time are observed on polarographic curves.

At lower pH values a sharp turbulent maximum is observed that can be suppressed by addition of 0·01% gelatine. This concentration was used in measurements of half-wave potentials. The half-wave potential of the first wave is practically unaffected by the presence of gelatine. The wave for acetophenone on the other hand is increased (due to the influence on protonation[9]) and shifted to more negative potentials.

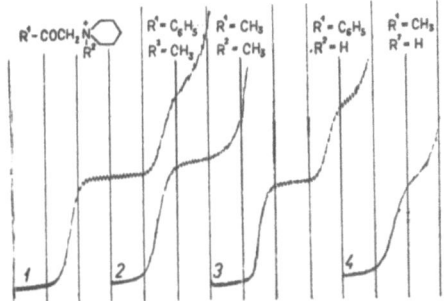

Fig. 8

Comparison of Acetophenone and Acetone Derivatives

$2 \cdot 10^{-4}$M α-aminoketone, Britton-Robinson buffer pH 8·5, 0·015% gelatine.

1 ω-Piperidinoacetophenone methoiodide (I); 2 piperidinoacetone methoiodide (III); 3 ω-piperidino-acetophenone hydrochloride (II); 4 piperidinoacetone hydrochloride (IV). Curve 1 starting at −0·6 V, 2 at −1·0 V, 3 at −0·8 V, 4 at −1·2 V, s. c. e., 200 mV/absc., $h = 65$ cm, sens. 1 : 20.

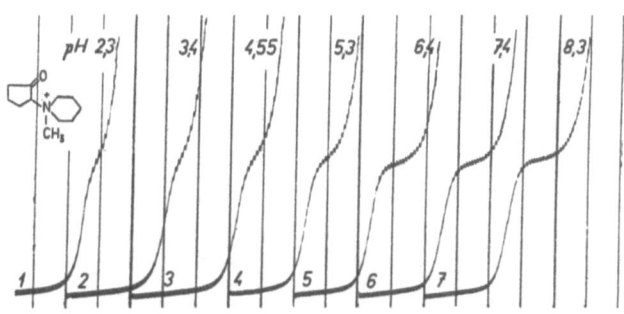

a

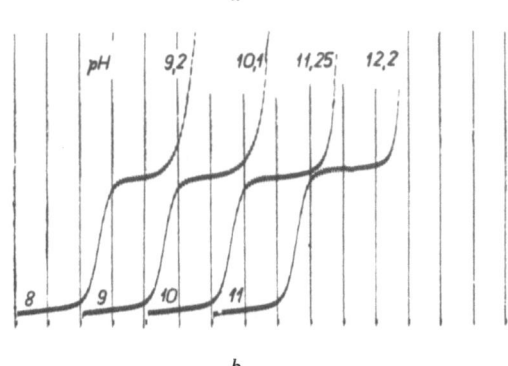

b

Fig. 9

The pH-Dependence of the Waves for 2·Piperidinocyclopentanone Methoiodide (V) in Britton-Robinson Buffers

$2 \cdot 10^{-4}$M depolarizer, 0·015% gelatine, pH-values given on the polarogram. Curves 1−3 starting at −0·6 V, 4−11 at −0·8 V, s. c. e., 200 mV/absc., $h = 65$ cm, sens. 1 : 20.

244

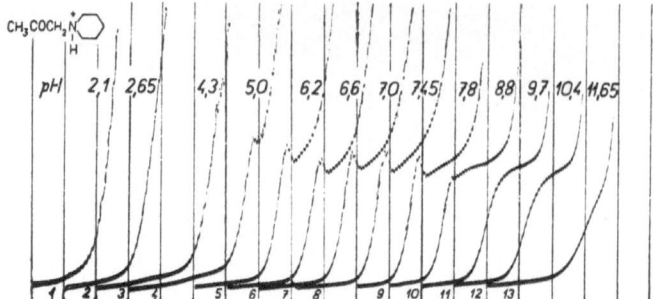

Fig. 10

The pH-Dependence of the Waves for Diperidinoacetone Hydrochloride (*IV*) in Britton-Robinson Buffers

$2 \cdot 10^{-4}$M depolarizer, pH-values given on the polarogram. Curves *1—4* starting at -0.8 V; *5—8* at -1.0 V; *9—13* at -1.2 V, s. c. e., 200 mV/absc., $h = 65$ cm, sens. 1 : 20.

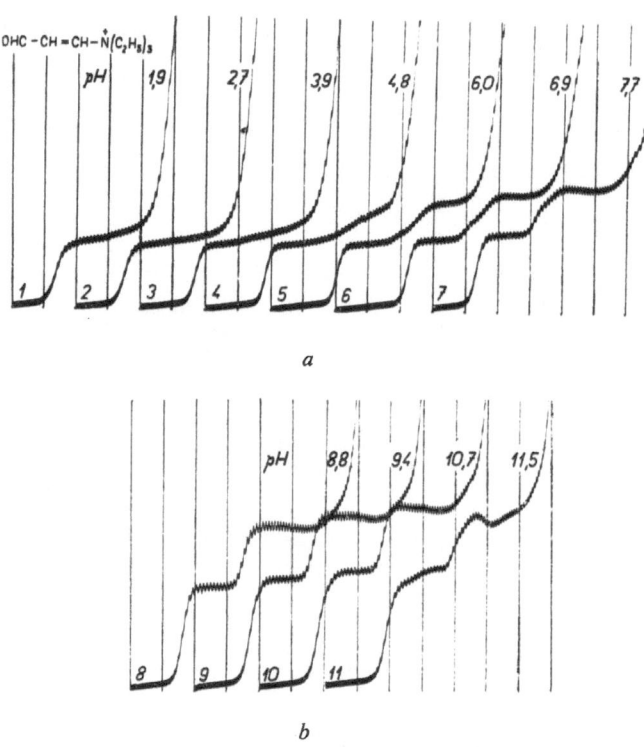

a

b

Fig. 11

The pH-Dependence of the Waves for β-N-Triethylaminoacroleinperchlorate (*VIII*) in Britton-Robinson Buffers

$2 \cdot 10^{-4}$M depolarizer, pH-values given on the polarogram. Curves *1—6* starting at -0.4 V; *7—11* at -0.6 V, s. c. e., 200 mV/absc., $h = 65$ cm, sens. 1 : 20.

Methoiodides and Hydrochlorides of Piperidinoacetone and 2-Piperidinocyclopentanone

Piperidinoacetone methoiodide (*III*) and hydrochloride (*IV*) and 2-piperidinocyclopentanone methoiodide (*V*) and hydrochloride (*VI*) are reduced polarographically in one two-electron wave. The number of electrons transferred was proved by comparison with corresponding acetophenone derivatives (Fig. 8). With methoiodides *III* and *V* the wave-height is pH-independent in the whole pH-range (Fig. 9), with hydrochlorides *IV* and *VI* at pH lower than about 11 (Fig. 10). At higher pH-values a decrease in the form of a dissociation curve is observed. The limiting current of waves (at high pH-values) that are lower than about 20% of the original height, possess a kinetic character.

The half-wave potentials are more negative than that of *I* and *II* (Tab. II) but the shape of the pH-dependence of methoiodides *III* and *V* is analogous to that of the more positive wave of methoiodide *I* (Fig. 3). The pH-dependences of half-wave potentials of hydrochlorides *IV* and *VI* are analogous to that of the more positive wave of hydrochloride *II* (Fig. 6), only the intersections pK' and pK'' have different values.

This analogy between waves for *III*—*VI* and first waves for *I* and *II* proves, that also with *III*—*VI* the $C-N^{(+)}$-bond is reduced. Further proof was given by controlled potential electrolysis of *IV* and detection of acetone formed.

β-N-Triethylaminoacrolein Perchlorate (*VIII*) and β-N-Dimethylaminoacrolein (*IX*)

β-N-*triethylaminoacrolein perchlorate* (VIII) is reduced principally in two steps (Fig. 11): The more positive step in acid solutions corresponds to an uptake of one electron. This was proved by comparison with ω-piperidinoacetophenone methoiodide (*I*). With increasing pH the height of this wave increases in the form of a dissociation curve with a $pK' = 8.7$ (at $t_1 = 3.2$ s). At pH 10 the height of the more positive wave corresponds to a transfer of two electrons. At pH greater than about 11 decomposition accompanying the change of polarographic curves with time was observed. The half-wave potentials (Table II) are shifted to more negative potentials with increasing pH with a slope of about 32 mV/pH.

The second wave is a composite one. At pH-values about 4 a more negative wave increases with increasing pH. At pH about 6 another wave, ill-separated at potentials about 0·15 V more positive, increases with increasing pH-value at the expense of the more negative wave. This wave has been shown to be identical with the wave for acrolein (Fig. 12). The increase of this wave with increasing pH-value again has the form of a dissociation curve, the pK'-value being about 8·5 (at $t_1 = 3.0$ s).

Based on the coincidence of the second wave at a pH greater than about 8 with the wave for acrolein a reduction of the $C-N^{(+)}$ bond in the first wave of *VIII* can, under these conditions, be taken for granted. As the half-wave potentials show no

246

substantial change in the $dE_{1/2}/dpH$ slope it can be deduced that the attack of one electron accompanied by the formation of a radical in acid solution can proceed on the same bond as well, even if the $\pi-p$ and $\pi-\pi$ conjugation makes a decision difficult.

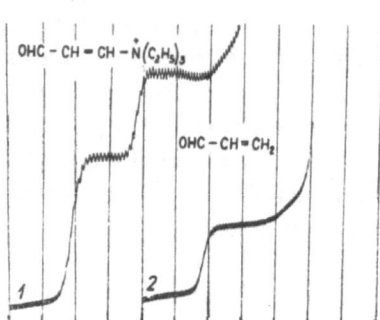

Fig. 12

Comparison of Waves for β-N-Triethylaminoacrolein Perchlorate (*VIII*) and Acrolein

$2 . 10^{-4}$M depolarizer, Britton-Robinson buffer pH 10·7. *1* β-N-triethylaminoacrolein perchlorate (*VIII*), *2* acrolein. *1* starting at -0.6 V, *2* at -1.0 V, s. c. e., 200 mV/absc., $h = 65$ cm, sens. 1 : 15.

β-N-*Dimethylaminoacrolein* (IX) is reduced giving a poorly developed wave at lower pH-values. Well defined two-electron waves were obtained at pH 6−8, where the wave decreases in the form of a very steep dissociation curve with a $pK' = 7.4$. Only the protonized form is thus reduced. Due to the ill-developed limiting current it was not decided which bond is reduced in this electrode process. The half-wave potentials are shifted by some 140 mV/pH in the pH-range 2−5 and by 55 mV/pH at higher pH-values.

Acrolein (XII). Due to the unusual pH-dependence of the second wave of substance *VIII*, the polarographic behaviour of acrolein was revised and deviations found from that described in the literature[10].

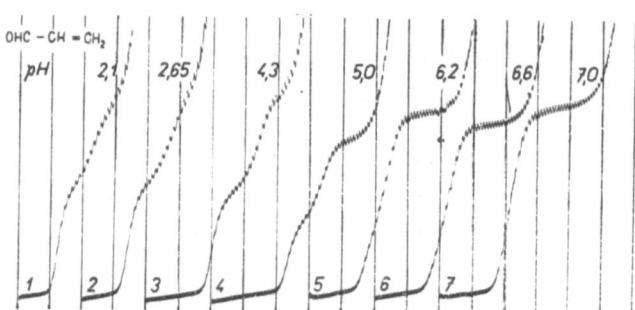

Fig. 13

The pH-Dependence of the Waves for Acrolein in Britton-Robinson Buffers

$2 . 10^{-4}$M depolarizer, pH-values given on the polarogram. Curves registered 2 minutes after mixing of the stock solution with the buffer *1 − 4* starting at -0.6 V, *5 − 7* starting at -0.8 V, s. c. e., 200 mV per absc., $h = 65$ cm, sens. 1 : 15.

Acrolein is reduced at higher pH-values in one two-electron wave (Fig. 13). At lower pH-values this wave divides into two of about the same height. The height of the first wave decreases sharply with increasing pH in the region about pH 5·0. This is shown by the decrease of the ratio $i_1 : i_2$ as well as by the course of the dependence $E_{\frac{1}{2}}$—pH, where a marked change in slope is observed in this region.

Both waves in acid solution and the two-electron wave at pH higher than about 6 correspond most probably to the reduction of the double bond. The reduction of the protonized form in two steps can be presumed in acid solution. In accordance with this interpretation is a small wave at $-1·7$ V, the height of which increases with increasing pH. This wave can be interpreted as the wave for propionaldehyde formed at the electrode and the pH-dependence can be explained as due to dehydration. As acrolein decomposed rapidly in aqueous solutions (with a pH-independent rate of the first order) interference due to decomposition (polymerization) products cannot be excluded.

Diethyl Malonesterpyridiniumylid (X)

Diethyl malonesterpyridiniumylid is reduced at pH greater than about 8 in a single one-electron wave. Both the height of this wave and its half-wave potential ($-1·51$ V) are pH-independent (Fig. 14). A radical is thus formed, probably involving electron uptake by the pyridine ring.

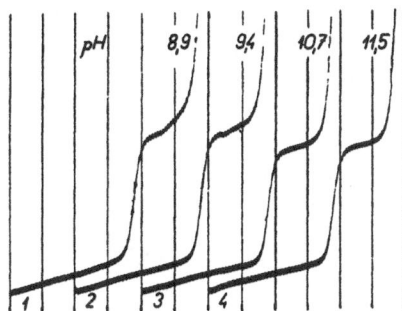

Fig. 14

The pH-Dependence of the Waves for Diethyl Malonesterpyridiniumylid (X) in Britton-Robinson Buffers $2 . 10^{-4}$M depolarizer, pH-values given on the polarogram. Curves starting at $-0·8$ V. s.c.e., 200 mV per absc., $h = 65$ cm, sens. 1 : 10.

At lower pH-values a catalytic wave obscures the reduction current. There is another ill-developed wave at $-0·95$ V, the half-wave potential of which is similarly pH-independent; but its limiting current is a function of pH.

Discussion

Considerable attention has recently been paid to the polarographic reduction of certain activated single C–X bonds, previously considered as unreducible.

Thus in α-hydroxyketones[11,12] the reduction of the C–OH bond was claimed, as for the C–OR bonds in phenoxyacetamidines[13]. A similar cleavage of the C–O bond occurs in the reduction of 2,2',4,4'-tetraphenyl-3-oxetanone[14] and in mandelonitriles and their esters[15]. For other esters[16,17] the course of the reduction has not been proved yet, but a similar cleavage of an activated C–O bond can be supposed.

A likely analogy is the reductive cleavage of the activated C–S bond in sulfones[18,19]. Several compounds of the type Ar COCRR'–X–Y (where the reduction of the C–O or C–S bond is followed in compounds with X = O or S and Y = C_6H_5, H, C_6H_5 CO, CH_3 CO, CN and NO_2) are in the course of investigation[20].

For the C—F bond reduction was observed when activating groups in phenacyl fluoride[21] and 6-trifluoromethyl-7-sulfamyl-3,4-dihydro-1,2,4-benzothiadiazine-1,1-dioxide[22] were present. Both reduction processes are pH-dependent. This is probably due to a proton transfer on the sulfamyl or carbonyl part of the molecule.

The reduction of the single C—N bond has not been proved yet. Such a process was assumed for the second wave of p-trimethylammonium benzaldehyde[23], for the reduction process in ethylenimonium ion[24,25] as well as for benzamidine and related substances[13], where also a multiple carbon—nitrogen bond may be involved. It appears that reduction of the cation of aminomethylpyridine[26] $C_5H_4N.CH_2NH_3^{(+)}$ also involves the cleavage of the C—N(+) bond.

α-Aminoketones. Based on identification of the product reduced further in the second reduction wave on polarographic curves of the acetophenone derivatives *I*, *II* and *VII*, on analogies of the pH-dependence of the half-wave potentials and on the detection of the product of the controlled potential electrolysis, it may be presumed that in α-aminoketones the C–N(+)-bond is reduced at more positive potentials than the C=O group.

During the proper reduction step of the C–N(+)-bond no proton transfer occurs, as in the reductive substitution of the C–X-bonds in alkylhalides. The shifts of half-wave potentials are due to proton-transfers preceding the electron transfer. In methoiodides *I*, *III* and *V* a protonation on nitrogen is excluded and the course of the reduction at pH lower than about 7 can be expressed as follows:

$$R^1\text{—CO—CHR}^2\text{—}\overset{(+)}{N}XYZ + H^+ \rightleftarrows [R^1\text{—CO—CHR}^2\text{—NXYZ}]^{2+} \quad (1)$$
$$\underset{H}{\underbrace{\phantom{R^1\text{—CO—CHR}^2}}}$$

$$[R^1\text{—CO—CHR}^2\text{—NXYZ}]^{2+} + 2\,e \rightarrow R^1\text{—CO—CH}_2R^2 + NXYZ \quad (2)$$
$$\underset{H}{\underbrace{\phantom{R^1\text{—CO—CHR}^2}}}$$

$$R^1\text{—CO—CH}_2R^2 \xrightarrow{\text{Reduction}} \text{Products} \quad (3)$$

The equilibrium (*1*) is established fast. The second reduction process (*3*) occuring in the more negative wave, proceeds in a similar way as the reduction of the carbonyl group in the ketone R^1—CO—CH$_2R^2$ under the same conditions. It involves protonation followed by a one- or two-electron process with the formation of a dimer or alcohol as the final reduction product. With ketones *III* and *V* bearing no phenyl group this process is obscured by the current of the supporting electrolyte.

At pH-values greater than about 9 a pH-independent wave is observed, corresponding to the reactions (*4*), (*5*) and (*3*).

$$R^1\text{—CO—CHR}^2\text{—}\overset{(+)}{N}XYZ + 2\,e \rightarrow [R^1\text{—CO—CHR}^2]^{(-)} + NXYZ \quad (4)$$
$$[R^1\text{—CO—CHR}^2]^{(-)} + H^+ \rightleftarrows R^1\text{—CO—CH}_2R^2 \quad (5)$$

Hence reactions (*1*)–(*5*) can explain the course of the pH-dependence of the half-wave potentials given in Fig. 3.

Hydrochlorides *II*, *IV* and *VI* undergo the reduction only in the form bearing quaternary nitrogen $R^1 COCHR^2—NXYH^{(+)}$. The shifts of the half-wave potentials with pH-values (Fig. 6) show that the protonation of the CO—CHR^2 part of the molecule is also involved.

When comparing half-wave potentials of methoiodides *I*, *III* and *V* at any pH where the $dE_{\frac{1}{2}}/dpH$ is the same for all three compounds studied — Fig. 3) and of hydrochlorides *II*, *IV*, *VI* at pH 8 (where the slope $dE_{\frac{1}{2}}/dpH$ and thus the reduction path are comparable — Fig. 6), the positive shift due to the phenyl group is evident. The effect of other groups R^1 (to study correlations using the modified Taft equation[27]) and of the ring-size on polar and steric contributions is under course of study. But from the direction of the shift caused by the exchange of methyl for phenyl (Fig. 8, Tab. II) preliminary deduction can be drawn that the sign of the reaction constant is positive. Thus a nucleophilic mechanism[27] can be expected to operate in the potential determining step. A nucleophilic attack of electron on carbon of the C—N bond is a plausible explanation. Such a type of S_N-mechanism was suggested among other possibilities for the explanation of the reduction of the carbon-halogen bond[28].

Due to the different type of pH influence (and thus of mechanism of the reduction process) no quantitative comparison of the reduction of methoiodides R^1—CO— —CHR$^2 \div NXYZ^{(+)}$ and hydrochlorides R^1—CO—CHR^2—NXYH$^{(+)}$ can be given. As a very rough approximation only it can be deduced based on the comparison of Fig. 3 and 6 (cf. also Tab. II) that hydrochlorides *II*, *IV* and *VI* are reduced generally at more negative potentials than methoiodides *I*, *III*, and *V*.

The activating effect of the R^1CO group in molecules R^1—CO—C—X (where X = O, S, N or F, cf. p. 188) may be caused both by polar and mesomeric effect. The shift

$$O \overset{\frown}{=} C \overset{\perp}{\underset{\underset{R^1}{|}}{\frown}} C \underset{+\delta \quad -\delta}{\overset{\frown}{X}}$$

is analogous to that

$$C \overset{\frown}{=} C \overset{\perp}{\frown} C \underset{+\delta \quad -\delta}{\overset{\frown}{X}}$$

explaining the positive reduction potential of allylhalogenides[29], To distinguish between the contributions of polar and mesomeric effects of the activating groups a study using the modified treatment by Taft[27] is planned.

Acetophenone, acrolein and malonesterylid (X). The experimental results observed with aceto-phenone (e. g. the influence of buffer composition) and also as with acrolein (i. e. the change of the wave height with pH) have shown that a special detailed study of these compounds would be important.

The behaviour of the ylide (*X*) differs from that of α-aminoketones. The one-electron process presumably followed by dimerization resembles rather the process suggested for the reduction of alkylpyridinium ions[30].

β-N-*Triethylaminoacrolein perchlorate* (VIII). The behaviour of this substance is substantially different from that of α-aminoketones in acid media where a one-electron process occurs.

A possible explanation of the pH-dependence of the wave for *VIII* is a proton transfer preceding the electrode process. The protonized form (after uptake of one-electron) forms a radical that readily dimerizes. The unprotonized form takes up two electrons and splits off the $N(C_2H_5)_3$ group in the first wave. In the second wave at higher pH-values acrolein formed in this way is reduced.

Another possible explanation is that the proton transfer proceeds on the radical formed in the uptake of the first electron. The protonized radical readily dimerizes, the unprotonized is reduced further under splitting off the triethylamine.

Both these explanations give account for the observed one-electron wave height of the positive reduction process, for the absence of the acrolein wave in acid solutions and for the increase of the acrolein wave accompanying the increase of the limiting current of the first wave with increasing pH-values.

The comparison with acrolein enables us to distinguish the product reduced in the more positive part of the second wave. The process in the more negative part in the medium pH-range, which corresponds to the reduction of a product of the one-electron process, has not been identified yet.

The structure of the radical, formed in acid solutions, is hardly distinguishable due to the $\pi-\pi$ and $\pi-p$-conjugation in the system. But the course of the half-wave potential of the wave, showing no substantial change in the pH-range where the wave increases from a one-electron to a two-electron process, seems to prove the attack of the $C-N^{(+)}$-bond by the first electron.

References

1. Van Ark H. H.: Arch. Pharm. *238*, 330 (1900).
2. Rabe P., Schneider W.: Ber. *41*, 872 (1908).
3. Stroemer R., Burkert O.: Ber. *28*, 1250 (1895).
4. Van Ark H. H.: Arch. Pharm. *238*, 321 (1900).
5. Arnold Z., Šorm F.: This Journal *23*, 452 (1958).
6. Hartmann H., Gossel H.: Z. Elektrochem. *61*, 337 (1957).
7. Walther H., Treibs W., Michaelis K.: Chem. Ber. *89*, 60 (1956).
8. Peizker J.: This Journal *26*, 230 (1961).
9. Volková V.: Nature *185*, 743 (1960).
10. Kolthoff I. M., Lingane J. J.: *Polarography*, 2nd Ed., p. 657. Interscience, New York 1952.
11. Kabasakalian P., McGlotten J.: Anal. Chem. *31*, 1091 (1959).
12. Fedoroňko M.: Chem. zvesti *12*, 17 (1958).
13. Kane P. O.: Z. anal. Chem. *173*, 50 (1960).
14. Powers R. M., Day R. A. Jr.: J. Org. Chem. *24*, 722 (1958).
15. Wawzonek S., Frederikson J. D.: J. Electrochem. Soc. *106*, 325 (1959).
16. Imoto E., Otsuji Y.: Bull. Univ. Osaka Prefect. Ser. A *6*, 115 (1958).

17. Bezuglyj V. D.: Aptečnoje delo *2*, 17 (1959).
18. Drushel H. V., Miller J. F.: Anal. Chem. *30*, 1271 (1958).
19. Horner L., Nickel H.: Chem. Ber. *89*, 1681 (1956).
20. Lund H.: Private communication.
21. Elving P. J., Leone J. T.: J. Am. Chem. Soc. *79*, 1546 (1957).
22. Lund H.: Acta Chem. Scand. *13*, 192 (1959).
23. Bartel E., Grabowski Z. R., Kemula W., Turnowska-Rubaszewska W.: Roczniki Chem. *31*, 13, 27 (1957).
24. Lordi N. G., Christian J. E.: J. Am. Pharm. Assoc. *45*, 530 (1956).
25. Matsavinos R., Christian J. E.: Anal. Chem. *30*, 1071 (1958).
26. Volke J., Kubíček R., Šantavý F.: This Journal *25*, 871 (1960).
27. Zuman P.: This Journal *25*, 3225 (1960).
28. Elving P. J.: Kresge-Hooker Sci. Lib., Record Chem. Progr. *14*, 99 (1953).
29. Zuman P.: Chem listy *48*, 94 (1954).
30. Mairanovskiy S. G.: Dokl. Akad. nauk SSSR *110*, 593 (1956).

Translated by the Author (P. Z.).

FISSION OF ACTIVATED CARBON–NITROGEN AND CARBON–SULPHUR BONDS. VI.*

POLAROGRAPHIC REDUCTION OF THE C—N$^{(+)}$ BOND IN α-AMINONITRILES

P. Zuman, O. Manoušek and V. Horák

Jaroslav Heyrovský Institute of Polarography, Czechoslovak Academy of Sciences and Institute of Organic Chemistry, Charles University, Prague

Received August 27th, 1962

C—N$^{(+)}$-Bond is reduced in α-aminonitriles with quaternary amino group. α-Aminonitriles with a secondary amino group show catalytic hydrogen evolution. Unusual dependence on buffer capacity and concentration of the depolarizer was observed.

In a previous paper of this series[1] the activation of the C—N$^{(+)}$-bond caused by a ketogroup in α-position has been demonstrated. In this paper the possibility of the activation of such a bond by a CN group in quaternary α-aminonitriles is shown.

Experimental

The experimental conditions for classical polarography and for the pH-measurement and the technique of measurements were the same as used in the Part I of this series[1]. Polarisation using a periodically changed rectangular voltage was carried out using a commutator designed by Kalousek and Rálek[2]. The hanging mercury drop electrode was of the syringe type designed by Vogel[3].

The substances $I - XVI$ were used in this study. The substances $I - XIII$ were prepared according to the literature (see Table I); V, VII, VIII, IX and XII were prepared by D. J. Voaden, VI and XI by G. Sugerman and M. I. Hunsberger in the Department of Chemistry, Fordham University, New York, XIII by S. Féldéak and B. Matkovics. Substance XIV was prepared by the reaction of N-methyl piperidine and chloro acetamide in usual manner; for $C_8H_{17}ClN_2O$ (192·7) calculated: 49·9% C, 8·9% H, 14·54% N, 18·40% Cl; found: 49·47% C, 8·64% H, 14·31% N, 18·28% Cl. The experiments with the commutator and the electrolysis at the hanging mercury drop were carried out with chlorides of substances I and II, because the anodic wave of iodides interfered in the detection of cyanides.

The controlled potential electrolysis was carried out using a previously[14] described circuit and a separated anode[15]. 0·001M solutions of aminonitrile were electrolyzed in 0·05M borax. To prevent effects of side-reactions the electrolysis was carried out at potentials more positive than that of limiting current, *viz.* at $E_{1/4}$, *i.e.* at $-1\cdot8$ V for I and at $-1\cdot7$ V (s.c.e.) for II. The limiting current decreased with duration of electrolysis (Fig. 1).

---

* Part V.: This Journal *29*, 3044 (1964).

Table I

Substances Used

No	Compound	M. P., °C B. P., °C/mm Hg	References
I	$(C_2H_5)_2\overset{(+)}{N}-CH_2CN$ $J^{(-)}$ CH_3	199	4
II	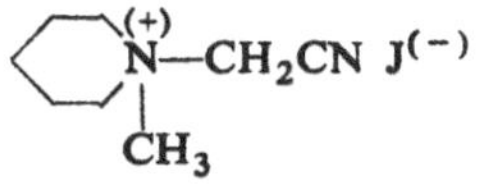 $\overset{(+)}{N}-CH_2CN$ $J^{(-)}$ CH_3	206	4
III	$(C_2H_5)_2N-CH_2CN$	53	4
IV	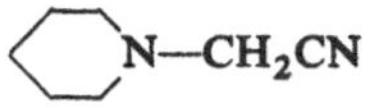 $N-CH_2CN$	83	4
V	$n\text{-}C_4H_9NH-CH_2CN$	65–66	5, 6
VI	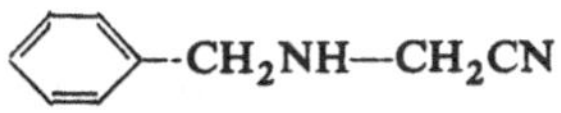 $-CH_2NH-CH_2CN$	^{a}	7
VII	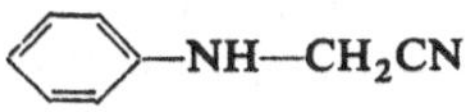 $-NH-CH_2CN$	48	8, 9
VIII	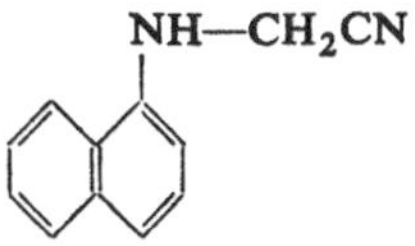 $NH-CH_2CN$	92	8, 9
IX	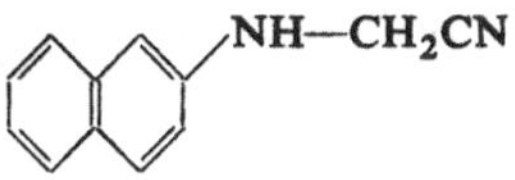 $NH-CH_2CN$	105	8–10
X	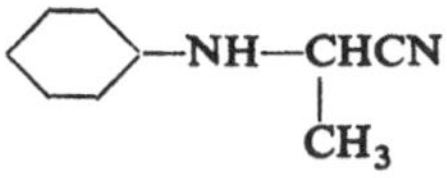 $-NH-CHCN$ CH_3	109–111/20	11
XI	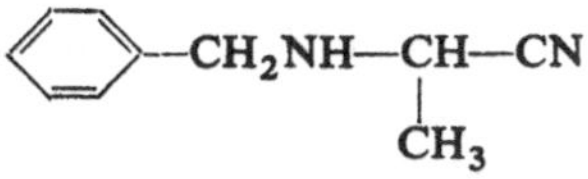 $-CH_2NH-CH-CN$ CH_3		12
XII	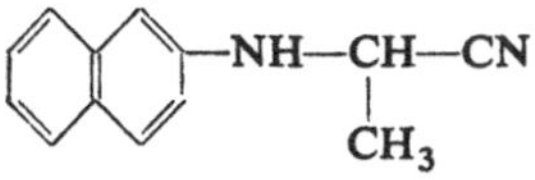 $-NH-CH-CN$ CH_3	138–139	12
XIII	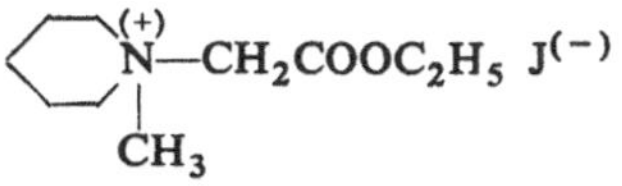 $\overset{(+)}{N}-CH_2COOC_2H_5$ $J^{(-)}$ CH_3	160–160·3	13
XIV	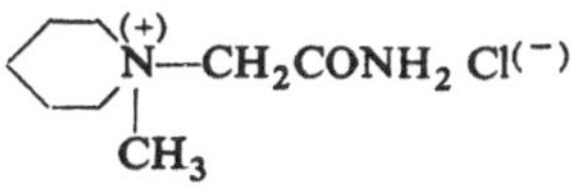 $\overset{(+)}{N}-CH_2CONH_2$ $Cl^{(-)}$ CH_3	213–215	^{b}

^{a} Hydrochloride, m. p. 171°C (decomp.); ^{b} cf. the text.

The products of electrolysis were identified by gas chromatography using acidification, distillation and detection of acetonitrile (in mixture with acetic acid formed by saponification) using gas–liquid chromatograph Fraktovap, Model C, Milano on 80 m capillary column using ethylene glycol adipate as stationary phase at 90°C. Similarly the reaction mixture was analysed directly in the borate buffer pH 9·2 for dimethyl ethyl amine.

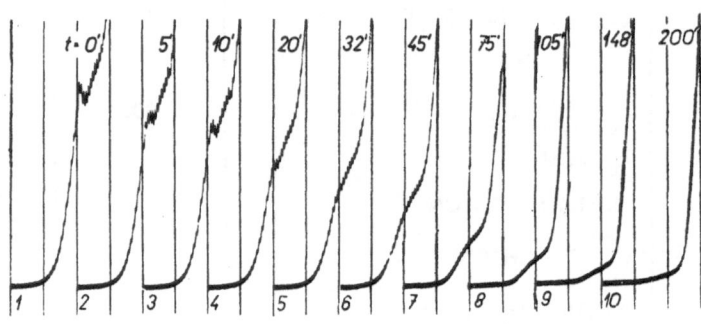

Fig. 1

Controlled Potential Electrolysis of Diethylamino Acetonitrile Methoiodide (*I*)

$1 . 10^{-3}$M depolarizer in 0·05M borax pH 9·2, total volume 110 ml, electrolyzed at 1·8 V for times given on the polarogram. Curves recorded starting at $-1\cdot26$ V, 190 mV/absc., s.c.e., full scale sensitivity 20·1 μA. ·

Results and Discussion

Nitriles with a quaternary aminogroup. Reduction of substances *I* and *II* occurs in one two-electron step (Figs 2 and 3). The height of the wave is limited by diffusion, as was shown by the linear dependence on concentration (Fig. 4) and the square root of mercury column. The reduction could theoretically proceed either on the C—CN grouping with amino group as activating or on the $C-N^{(+)}$-bond with nitrilo group as activating. Using either commutator method or hanging mercury drop method, formation of cyanides could not have been proved. Furthermore presence of CH_2NH_2 grouping as resulting from the electrolytic reduction of nitrilo group was not detected. On the other hand controlled potential electrolysis in connection with gas chromatography enabled identification of trialkylamine and acetonitrile as reduction products. Thus nitrilo group plays the role of an activator and the mechanism of the reduction process is assumed to be:

$$R_3\overset{(+)}{N}-\overset{.}{C}H_2CN + 2\,e \xrightarrow{+H^{(+)}} R_3N + CH_3CN. \tag{A}$$

The wave-height remains in the pH-range $8-10$ practically constant, when extrapolated for time $t = 0$. At pH < 8 the half-wave potential of the nitrile lies in the vicinity of the reduction potential of the components of supporting electrolytes. This causes the apparent increase of the limiting current observed in Fig. 2.

At pH > 10 the height of the reduction waves changes with time. Thus when curves were recorded after a constant period after the stock solution of the depolarizer was added to the buffer or sodium hydroxide, a decrease of the limiting current with increasing pH-value was observed (Figs 2 and 3). This corresponds to a base catalysed hydrolysis of the nitrile.

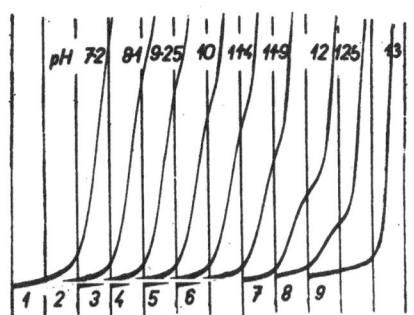

Fig. 2

Dependence of the Wave of Diethylamino Acetonitrile Methoiodide (I) on pH

$2 . 10^{-4}$M depolarizer , 1−6 Britton-Robinson buffers and 7 0·01N-NaOH, 8 0·03N-NaOH; 9 0·1N-NaOH. Approximate pH-values given on the polarogram. Curves recorded 2 minutes after addition of the stock solution to the supporting electrolyte, 1−6 starting at −1·2 V, 7−9 at −1·4 V, 200 mV/absc., s c E., Smoler's capillary, full scale sensitivity 4·2 μA.

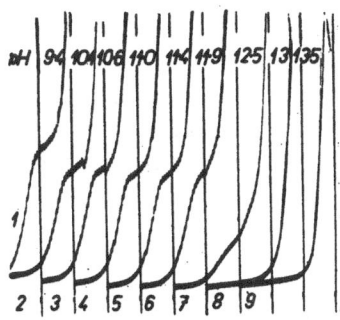

Fig. 3

Dependence of the Wave of Piperidino Acetonitrile Methochloride (II) on pH

$4 . 10^{-4}$M depolarizer, 1−6 Britton-Robinson buffers: 7 0·03N-NaOH; 8 0·1N-NaOH; 9 0·3N-NaOH. Approximate pH-values given on the polarogram. Curves recorded 1 minute after addition of the stock solution to the supporting electrolyte, at pH above 12 a change with time was observed. Curves starting at −1·4 V, 200 mV/Absc., s.c.e., full scale sensitivity 8·4 μA.

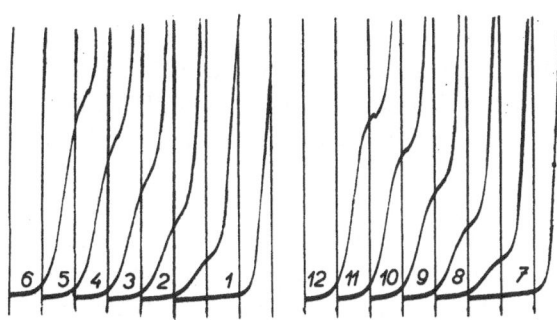

Fig. 4

Dependence of Waves of Substances I and II on Concentration

Britton-Robinson buffer pH 9·8; 2−6 diethylamino acetonitrile methoiodide (I); 8−12 piperidino acetonitrile methoiodide (II) Concentration of nitrile: 1,7 0; 2,8 1 . 10^{-4}M; 3,9 2 . 10^{-4}M; 4,10 3 . 10^{-4}M; 5,11 4 . 10^{-4}M; 6,12 5 . 10^{-4}M. Curves starting at −1·4 V, 200 mV/absc., s.c.e., Smoler's capillary, full scale sensitivity 20 μA.

In agreement with our supposition, CN-group in α-position activates the C—N$^{(+)}$-bond in a similar direction as the RCO-group. In a rough comparison* the activating (inductive and mesomeric) effects of these groups decrease in the following sequence:

$$C_6H_5CO > CH_3CO > CN > COOH, COOC_2H_5, CONH_2.$$

This sequence differs from that caused by total polar effects expressed by σ_m and σ_p constants. Additional conjugation effects of keto groups are probably operating in electroreduction.

Similarly as for comparison of $R^1CH_2OR^2$ and $R^1CH_2NR^3R^4$ in phenacyl compounds[1,16] (where $R^1 = C_6H_5CO$), also for nitriles (where $R^1 = CN$) analogies between the activation of the reducibility of the C—O and C—N bonds can be detected. The reducibility of the C—O bond in α-position relative to the nitrile group in mandelonitriles has been namely described recently[17].

The half-wave potentials, $E_{1/2} = -1.76_3$ V for substances *I* and -1.74_4V for substance *II* (both *versus* S.C.E.) are practically independent in the whole pH-range studied. The independence of the half-wave potentials of substances *I* and *II* of pH shows that the protonation in equation (*A*) occurs after a irreversible potential determining step. This finding is another proof for our conclusion[1] that the shift of half-wave potentials of the C—N bond of α-amino ketones and of the C—X bond of α-halogeno ketones is due to the protonation of the carbonyl group.

Polarography offers a method for the quantitative study of the hydrolysis of compounds of that type. Whereas aqueous stock solutions of substances *I* and *II* are relatively stable, ethanolic solutions exposed to light and air were less stable. During a week the solution was decomposed. On polarographic curve two waves were observed, resembling those of α-amino ketones. The product of this reaction was not further studied.

Nitriles with a secondary amino group. Substances *V — XII* show on polarographic curves currents due to the catalytic hydrogen evolution. If the reduction of the C—N$^{(+)}$-bond occurs, it is overlapped by the catalytic effects. For substances *VII — IX* and *XII* the catalytic currents were observed at pH < 5, due to the effect of aryl groups on the dissociation constants. At these lower pH-values the catalytic hydrogen evolution occurs at such negative potentials that no well developed limiting current was observed.

Substances *V* and *VI* show one wave at pH 5—7, substance *XI* two waves at pH 8 to 10, compound *X* two waves at pH 5·5—9·5. All these waves decreased with increasing pH-value of the buffer in the form of a part a dissociation curve. For those cases, where two catalytic waves were observed, both waves decreases with increasing pH, similarly as observed for the catalytic waves of 2,4-dithiopyrimidine[18]. The two waves at -1.5 V and -1.7 V for substance *X* were studied in some detail. The height of the wave at -1.5 V is practically independent of mercury pressure, the sum of the waves

---

* The half-wave potentials of amino ketones are pH-dependent, those of amino nitriles not, hence only a roughly approximative comparison is possible.

is almost proportional to the square root of mercury column. Both waves are 10—100 times greater than predicted for a diffusion controlled two-electron process. The waves are time independent (Fig. 5, curves 5—7).

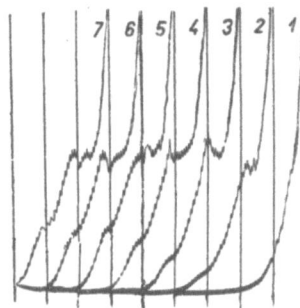

Fig. 5

Dependence of the Catalytic Waves of Substance X on the Concentration of the Nitrile

0·2M acetate buffer pH 5·75, concentration of nitrile X: 1 0; 2 1.10^{-4}M; 3 2.10^{-4}M; 4 3.10^{-4}M; 5—7 4.10^{-4}M. Curve 5 immediately, 6 after 5 minutes; 7 after 15 minutes. Curves starting at −1·2 V, 200 mV/absc., s.c.e., full scale sensitivity 20 μA.

Extraordinary is the effect of buffer capacity. The wave at −1·7 V decreases strongly with decreasing buffer capacity in acetate or phosphate buffers at pH 5 to 7 (Fig. 6). In acetate buffer pH 5·7$_5$ the wave at −1·5 V is only slightly dependent, in phosphate buffer pH 5·9 or 7·4 (Fig. 6) this wave increases with decreasing buffer capacity. The effect of ionic strength and of nature of the cations and anions present, discussed recently for catalytic waves[19], has not been studied in the present case.

With increasing concentration of substance X in 0·2M acetate buffer pH 5·7 the height of the wave at −1·5 V increases first linearly. At concentrations above 4.10^{-4}M systematic deviations from linear of the dependence of the wave height at −1·5 V on concentration occurs, but the sum of the waves at −1·5 V and −1·7 V remains practically constant (Fig. 5). This implies that the height of the wave at −1·7 V

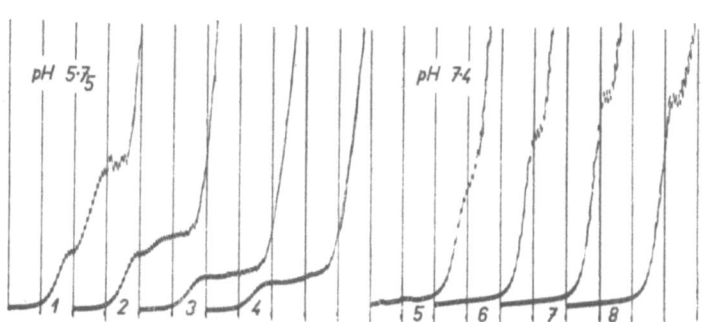

Fig. 6

Dependence of the Catalytic Waves of Substance X on the Buffer Capacity

4.10^{-4}M nitrile X, concentration of the buffer: 1—4 acetate pH 5·75; 5—8 phosphate pH 7·4; 1 0·2M; 2 0·1M; 3 0·05M; 4 0·025M; 5 0·2M; 6 0·1M; 7 0·05M; 8 0·025M. Curves 1—4 starting at −1·2 V. 5—8 at −1·0 V, 200 mV/absc., s.c.e., full scale sensitivity 1—4: 20 μA, 5—8 4·2 μA.

actually decreases with increasing concentration. In a more diluted acetate buffer (say 0·025M, *cf*. Fig. 6, curve *4*) the height of the wave at $-1·7$ V is considerably lower and practically only the wave at $-1·5$ V is observed already in solutions containing $2 \cdot 10^{-4}$M substance *X*. With a further increase of concentration of substance *X* the wave at $-1·5$ V increases only little and at $c > 5 \cdot 10^{-4}$M even a small decrease was observed. This behaviour can be explained, when it is assumed that a product of the electrode reaction occuring at $-1·5$ V reacts with the form of depolarizer which shows a wave at $-1·7$ V.

Nitriles with tertiary aminogroup *III* and *IV* as well as ester *XIII*, and amide *XIV* show neither reduction nor catalytic waves. The protonation of substances *III* and *IV* occurs probably at such low pH-values that the reduction wave (in analogy with aminoketones[1] assumed to be more negative than for corresponding methoiodides *I* and *II*) is overlapped by the current of reduction of supporting electrolyte. The $R^1 = COOC_2H_5$ and $CONH_2$ in $R^1{-}CH_2NR^2R^3R^{4(+)}$ is thus not sufficient to shift the reduction into the available potential range.

References

1. Zuman P., Horák V.: This Journal *26*, 176 (1961).
2. Kalousek M., Rálek M.: This Journal *19*, 1099 (1954); Chem. listy *48*, 808 (1954).
3. Březina M.: *Advances in Polarography, Proc. IInd Internat. Congr. Polarogr., Cambridge 1959* (I. S. Longmuir Ed.) Vol. 3, p. 933. Pergamon Press, Oxford 1960.
4. Luten D. B. Jr.: J. Org. Chem. *3*, 594 (1939).
5. Luskin L. S., Culver M. J., Gantert G. E., Craig W. E. Cook R. S.: J. Am. Chem. Soc. *78*, 4042 (1956).
6. Exner L. J., Luskin L. S., De Benneville P. L.: J. Am. Chem. Soc. *75*, 4841 (1953).
7. Baker W., Ollis W. D., Poole V. D.: J. Chem. Soc. *1949*, 307.
8. Knoevenagel E.: Ber. *37*, 4073 (1904).
9. Bucherer H., Schwalbe A.: Ber. *39*, 2808 (1906).
10. Bucherer H.: DRP 157 840; Chem. Zentr. *1905*, 316.
11. Wagner-Jauregg T.: Helv. Chim. Acta *44*, 1237 (1961).
12. Voaden D. J.: Unpublished results.
13. Féldéak S., Matkovics B.: Acta Univ. Szegedensis, Acta Phys. et Chem. *5*, 43 (1959).
14. Manoušek O., Zuman P.: This Journal *29*, 1718 (1964).
15. Manoušek O.: Chem. listy, in the press.
16. Lund H.: Acta Chem. Scand. *14*, 1927 (1960).
17. Wawzonek S., Fredrickson J. D.: J. Electrochem. Soc. *106*, 325 (1959).
18. Zuman P., Kuik M.: This Journal *24*, 3861 (1959).
19. Ždanov S. I., Zuman P.: This Journal *29*, 960 (1964).

Translated by the author (P. Z.).

FISSION OF ACTIVATED CARBON–NITROGEN AND CARBON–SULPHUR BONDS. III.*

(+)
REDUCTION OF THE C–S BOND IN METHYL BUTYL PHENACYL SULFONIUM PERCHLORATE

P. Zuman and Süe-yuan Tang**

Polarographic Institute, Czechoslovak Academy of Science, Prague

Received February 20th, 1962

In dialkyl phenacyl sulfonium ions the $C-S^{(+)}$ bond is reduced with uptake of two electrons at relatively positive potentials. Acetophenone formed in this process is reduced in another more negative step. A round maximum was observed on the limiting current of the first step, followed by desorption phenomena. Methyl butyl phenacyl sulfonium hydroxide is a weak base (pK 7·7).

As a part of our studies on the polarographic reduction of the C—N and C—S bonds[1-3], the polarographic behaviour of phenacyl sulfonium salts was followed. Among the polarographic behaviour of related compounds, the polarographic curves for phenacyl sulfides[4], trialkyl[5] and triaryl[6] sulfonium salts were reported. In addition to the extension of our knowledge of the course of polarographic reductions of organic compounds, a description of the polarographic behaviour of phenacyl sulfonium salts was aimed to form a base for the study of the reaction of phenacyl bromide with organic sulfides[7].

Experimental

Apparatus. The polarographic curves were recorded using a Polarograph LP 55 (produced by Laboratorní přístroje, Prague) with photographic recording with a galvanometer of sensitivity $2·7 . 10^{-9}$ A/mm. A Kalousek vessel with a separated saturated calomel electrode (s. c. e.) was used. The capillary used in most experiments had the following constants: the outflow rate $m = = 2·5$ mg/s, and the drop-time $t_1 = 2·9$ s at the potential of s.c.e. in saturated potassium chloride at mercury pressure $h = 65$ cm. All the values of half-wave potentials are expressed against the s.c.e. The pH-values of buffer solutions were checked using the Autojonometer Type PHM 21d (Radiometer-København) pH-meter with a glass electrode type G 200 B.

Reagents. As it has been proved that the polarographic behaviour of dialkyl phenacyl sulfonium ions follows principally the same pattern, methyl butyl phenacyl sulfonium perchlorate (m. p. 85°C) has been chosen as a model substance in this work. The effect of alkyls on half-wave potentials is discussed in another part of this series.

---

* Part II.: This Journal *27*, 187 (1962).

** Standing adress: Petroleum Research Institute, Chinese Academy of Science, Talien, China.

Solutions. 0·01M stock solutions were prepared by dissolving a weighed amount of the substance in 96% ethanol. The buffer solutions used (Britton-Robinson and acetate buffers) as well as other supporting electrolytes were prepared from reagent grade chemicals.

Technique of measurements. The technique was the same as in the previous paper[3]. The concentration of ethanol in the final solutions was usually 2%. A decrease of ethanol concentration to 0·2% is practically without influence on polarographic curves.

Results

The reduction of methyl butyl phenacyl sulfonium perchlorate proceeds principally in two steps A and B (Fig. 1): The first two-electron reduction wave A is diffusion controlled. The height of this wave is pH-independent (pH 1 to 11·5) (Fig. 2), the pH-dependence of the half-wave potentials is given in Fig. 3 (the slope above pH 7·7 is 0·076 V/pH), the exact values in Table I. The measurement of the slope of the wave is affected by the presence of maxima; between pH 3·0 and 7·0 the value of $RT/n\alpha F$ was approximately 0·06 to 0·07.

The half-wave potentials of the more negative reduction steps B_1 and B_2 are practically identical with those obtained with acetophenone under the same conditions

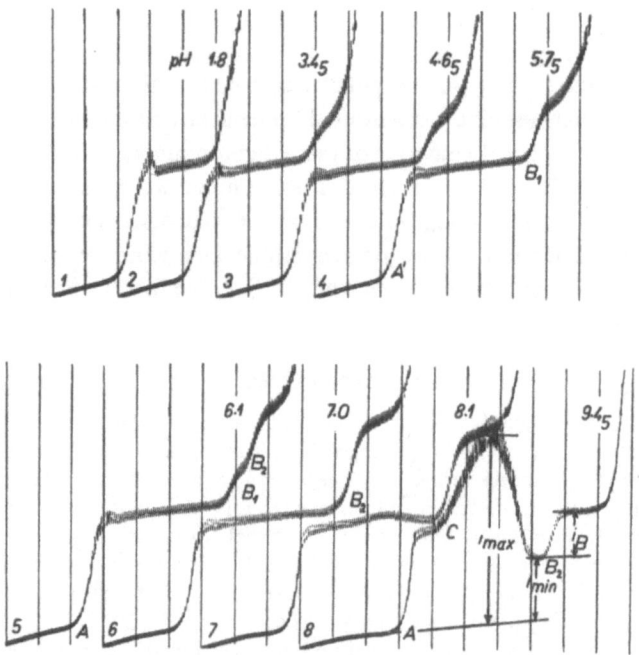

Fig. 1
Dependence of Reduction Waves of Methyl Butyl Phenacyl Sulfonium Perchlorate on pH
$2 . 10^{-4}$M depolarizer in Britton-Robinson buffers containing 0·2% ethanol, pH-value: *1* 1·8; *2* 3·4$_5$; *3* 4·6$_5$; *4* 5·7$_5$; *5* 6·1; *6* 7·0$_5$; *7* 8·1; *8* 9·4$_5$. Curves starting at 0 V, s.c.e., 200 mV/absc., $h = 65$ cm, $t_1 = 2·9$ s, $m = 2·5$ mg/s, sens. 1 : 10, full scale deflection 2·7 μA.

in the whole pH-range studied[3]. In acid solution the reduction proceeds in two one-electron steps, the one-electron wave B_1 and the second which is obscured by reduction of supporting electrolyte. At a pH above 5 the height of the first one-electron step B_1 decreases in the form of a steep dissociation curve (Fig. 2). The form of this decrease as well as the pH-value, at which the wave-height arrives at 50% of the original value, are identical with that for acetophenone[3]. The decrease of the one-electron step B_1 is accompanied with an increase of a two-electron reduction wave B_2 at more negative potentials (Fig. 1, 2). At pH above 8 this wave decreases in the form of a flat dissociation curve, until at pH above 12 it practically vanishes.

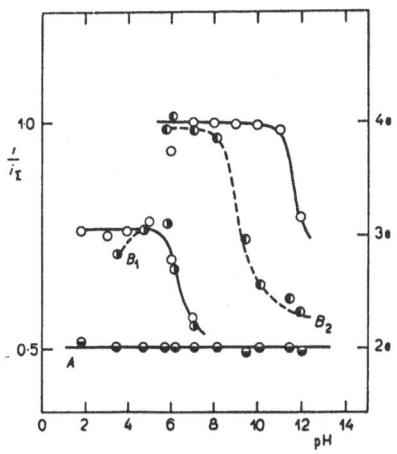

Fig. 2

Dependence of Limiting Currents on pH $2 \cdot 10^{-4}$M depolarizer in Britton-Robinson buffers containing 0·2% ethanol. Acetophenone[3] (○); methyl butyl phenacyl sulfonium perchlorate: ⊖ first two-electron wave A, ◑ one-electron B_1, ◔ two-electron step B_2. Particular limiting current (i) compared with the total four-electron limiting current (i_Σ). $t_1 = 2\cdot9$ s, $m = 2\cdot5$ mg/s, $h = 65$ cm.

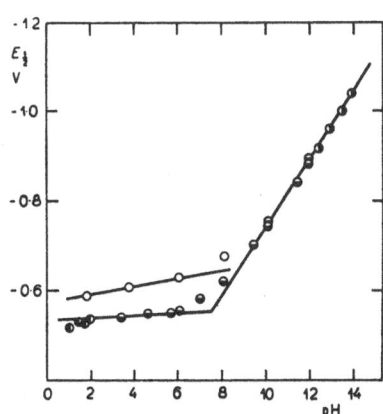

Fig. 3

Effect of pH on Half-Wave Potentials of Methyl Butyl Phenacyl Sulfonium Perchlorate $2 \cdot 10^{-4}$M depolarizer, 0·2% ethanol, Britton-Robinson buffers (⊖), H_2SO_4 (◑) and NaOH (◔), Britton-Robinson buffers with 0·01% gelatin (○).

At a pH above 8 a maximum C appears on the limiting current of the first wave A (Fig. 1, curve 8). The character of maximum C is described separately[8]. The maximum is followed by a decrease of current below the original limiting current. The pH-dependence of the current in minimum (i_{min}) possesses the same form as that for the wave of acetophenone. When the wave-height of the acetophenone wave (i_B) is expressed as the ratio i_B/i_{min} (where i_{min} is the current in the potential region where the decrease of current is observed (Fig. 1)), the ratio i_B/i_{min} is almost pH-independent until pH 11 (Fig. 2). The maximum is supressed by the addition of gelatin (Fig. 4)

Table I

Half-Wave Potentials of Methyl Butyl Phenacyl Sulfonium Perchlorate (*vs.* s.c e., in the Absence of Gelatin)

Solution	$E_{1/2}$, V	Solution	$E_{1/2}$, V
$1·0$N-H_2SO_4	$-0·52$	buffer[a] pH 9·45	$-0·79$
$0·3$N-H_2SO_4	$-0·53$	buffer[a] pH 10·1	$-0·74_5$
$0·1$N-H_2SO_4	$-0·54$	buffer[a] pH 11·4_5	$-0·84$
Buffer[a] pH 1·8	$-0·52_5$	buffer[a] pH 11·9_5	$-0·88$
Buffer[a] pH 3·4_5	$-0·54$	$0·01$N-NaOH	$-0·88$
Buffer[a] pH 4·6_5	$-0·55$	$0·03$N-NaOH	$-0·92$
Buffer[a] pH 5·7_5	$-0·55$	$0·1$N-NaOH	$-0·96_5$
Buffer[a] pH 6·1	$-0·56_5$	$0·3$N-NaOH	$-1·00_5$
Buffer[a] pH 7·0_5	$-0·58$	$1·0$N-NaOH	$-1·04_5$
Buffer[a] pH 8·1	$-0·62_5$		

[a] Britton-Robinson buffer.

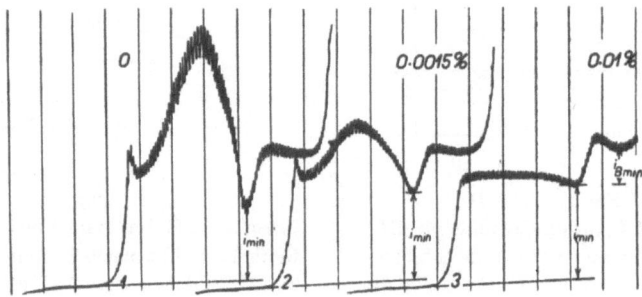

Fig. 4

Effect of Gelatin on the Reduction Curve of Methyl Butyl Phenacyl Sulfonium Perchlorate $6 . 10^{-4}$M depolarizer in $0·05$M borax containing $0·6\%$ ethanol. Gelatin added: *1* 0; *2* 0·0015%; *3* 0·01%. Curves starting at 0·0 V, s.c.e. 200 mV/absc., $h = 65$ cm, $t_1 = 2·9$ s, $m = 2·5$ mg/s, sens. 1 : 30, full scale deflection 8·4 μA.

or of tetrabutylammonium iodide.* The decrease (minimum) on the current–voltage curve A (i_{min}) is also eliminated (Fig. 4).

The limiting current of the acetophenone wave in the presence of gelatin (Fig. 4, curve *3*), tetrabutylammonium ions or in ammonia–ammonium chloride buffers, was

---

* In the presence of tetrabutylammonium iodide or bromide the height of the first reduction wave for the phenacyl sulfonium salt decreased somewhat, perhaps due to a chemical reaction in the solution.

not parallel to the condenser current, but a decrease of current (minimum) $i^{\bullet}_{Bmin}$ was observed. Hence ammonium-type ions cause an elimination of the minimum i_{min} on the limiting current of the sulfonium salt A but at the same time an increase of a minimum i_{Bmin} on the acetophenone wave B_2. A minimum i_{Bmin} on the acetophenone wave B_2 was also observed at higher concentrations of depolarizer ($6 . 10^{-4}$M) in the absence of ammonium ions. The decrease (minimum) i_{min} on the limiting current of the more positive wave A is also diminished with a simultaneous increase of the total limiting current of acetophenone when a sufficiently high concentration of ions of alkali metals is added (Fig. 5). The effect of the ionic strength is primarily an effect of cations. Ammonium ions are more effective than potassium or sodium ions.

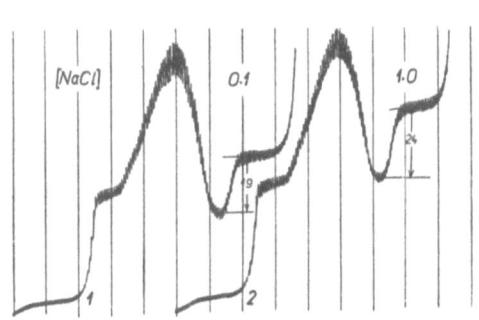

Fig. 5

Effect od Addition of Sodium Chloride on Reduction Curve of Methyl Butyl Phenacyl Sulfonium Perchlorate

$2 . 10^{-4}$M depolarizer in 0·05M borax containing 2% ethanol. 1 0·1M-NaCl; 2 1·0M-NaCl. Curves starting at $-0·2$ V, s.c.e., 200 mV/absc., $h = 65$ cm, $t_1 = 2·9$ s, $m = 2·5$ mg/s, sens. 1 : 10, full scale deflection 2·7 μA.

The course of the pH-dependence of polarographic half-wave potentials pointed out the possibility of acid-base reactions. Methyl butyl phenacyl sulfonium perchlorate was thus titrated by a standard solution of sodium hydroxide. A theoretical titration curve corresponding to the transfer of one proton (Fig. 6) with pK 7·7 was obtained. The reversibility of the acid–base reaction was demonstrated by retitration using a standard solution of hydrochloric acid. The observed difference in pK-values can be influenced by the change in ethanol concentration.

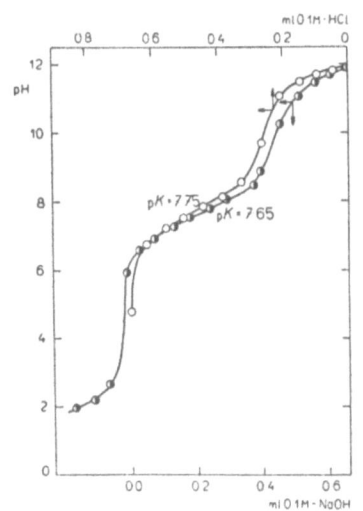

Fig. 6

Titration Curve of Methyl Butyl Phenacyl Sulfonium Perchlorate

10 ml of 0·005M sulfonium salt in 5% ethanol ○ titrated with 0·1M-NaOH, ◑ back-titrated with 0·1N-HCl.

Discussion

Acid–Base Properties

Sulfonium salts are reported[9] to be strong bases. The observed behaviour of dialkyl phenacyl sulfonium salts, showing weakly basic properties, can be caused by the substituent effect of the phenacyl group on acid-base properties of the sulfonium group or by inductive and resonance effects of the sulfonium group on the dissociation of the enol form (the latter including the effect of d-orbital resonance):

$$C_6H_5{-}CO{-}CH_2SR_2^{(+)} + OH^{(-)} \rightleftarrows C_6H_5{-}CO{-}CH_2S(OH)R_2 \qquad (1)$$

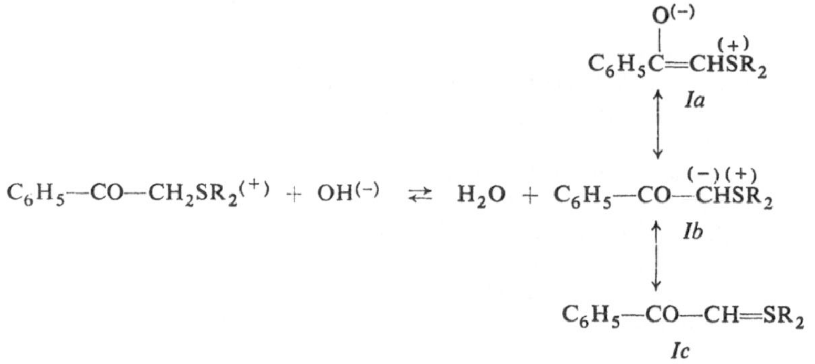

$$\begin{array}{c} O^{(-)} \\ | \\ C_6H_5\overset{(+)}{C}{=}CH\overset{(+)}{S}R_2 \\ \uparrow \quad Ia \\ \downarrow \end{array}$$

$$C_6H_5{-}CO{-}CH_2SR_2^{(+)} + OH^{(-)} \rightleftarrows H_2O + C_6H_5{-}CO{-}\overset{(-)(+)}{CHSR_2} \qquad (2)$$

$$\begin{array}{c} \uparrow \quad Ib \\ \downarrow \\ C_6H_5{-}CO{-}CH{=}SR_2 \\ Ic \end{array}$$

The scheme (2) is analogous with that suggested for dissociation of dimethylsulfoniopyruvic acid bromide[10]. The effects of $COO^{(-)}$ group (pK about 8·0, approximately determined from Fig. 1 in[10]) and of $COOCH_3$ and $COOC_2H_5$ groups respectively (pK about 5·6, cf. Fig. 1 in[10]) are comparable with those of C_6H_5CO (pK 7·7 for methyl butyl derivative). The sequence COOR > > COC_6H_5 > $COO^{(-)}$ is the same as for the polar effects expressed in Hammetts σ_m or σ_p constants[11].

The available data do not allow to distinguish between possible schemes (1) and (2), even when (2) seems more plausible. For the sake of convenience form Ib is used in further explanations.

Weak acid–base properties have recently been demonstrated using titration curves and also kinetic measurement for isothiouronium salts[12], but in this case the possibility of participation of proton-transfer on nitrogen atoms could not be neglected.

Course of the Polarographic Reduction

The course of the polarographic reduction of dialkyl phenacyl sulfonium ions follows closely the course of reduction of methoiodides of phenacyl amines[1-3], phenacyl sulfides[4] and related compounds[4]. In the first two-electron wave A the hydrogenolysis of the $C{-}S^{(+)}$ bond occurs, in the second step B_1 or B_2 reduction of acetophenone (formed in the first step) takes place. This was proved by the identity of half-wave potentials. The difference in the pH-dependence of the wave-heights of acetophenone and of the product of reduction of the phenacyl sulfonium ion (waves B below pH 8 — Fig. 2) indicate that the primary product of the electrode process in the first wave A of phenacyl sulfonium ion is the enolate form of acetophenone.* The height of the second step B_1 is limited[13] by the rate of ketone formation (4), as only the keto form of acetophenone is expected to be reduced in the potential range available. The rate of ketone formation is acid-base catalysed which explains the pH-dependence of the second step B_1.

Our present knowledge can be summarized as follows:

For the range pH 2 to 6:

$$C_6H_5-CO-CH_2SR_2^{(+)} + 2e \rightarrow C_6H_5-\underset{\underset{O}{\|}}{C}=CH_2^{(-)} + SR_2 \qquad (3)$$

$$C_6H_5-\underset{\underset{O}{\|}}{C}=CH_2^{(-)} + H^{(+)} \underset{\overrightarrow{\text{not very fast}}}{\overset{\text{acid-base catalysed}}{\rightleftharpoons}} C_6H_5-\underset{\underset{O}{\|}}{C}-CH_3 \qquad (4)$$

$$C_6H_5-\underset{\underset{O}{\|}}{C}-CH_3 + H^{(+)} \underset{\overrightarrow{\text{not very fast}}}{\rightleftharpoons} C_6H_5-\underset{\underset{OH^{(+)}}{\|}}{C}-CH_3 \qquad (5)$$

$$C_6H_5-\underset{\underset{OH^{(+)}}{\|}}{C}-CH_3 + e \rightleftharpoons C_6H_5-\underset{\underset{OH}{|}}{\overset{\cdot}{C}}-CH_3 \qquad (6)$$

$$2\, C_6H_5-\underset{\underset{OH}{|}}{\overset{\cdot}{C}}-CH_3 \longrightarrow \begin{array}{c} C_6H_5-C\overset{\displaystyle OH}{\underset{\displaystyle CH_3}{<}} \\[4pt] \overset{\displaystyle |}{} \\[-2pt] C_6H_5-C\overset{\displaystyle OH}{\underset{\displaystyle CH_3}{<}} \end{array} \qquad (7)$$

The reduction process in the first step A corresponds to (3). The height of this step is pH-independent. Equation (3) is also in agreement with the observed pH-independence of the half-wave potential of the first step A in this region. Decrease of the limiting current of the second (here one-electron) step B_1 of sulfonium salt at pH below 5 is limited by the rate of reaction (4). The formation of a radical is stabilised only in the protonised form of acetophenone. The decrease of the one-electron step of acetophenone (and of the second step B_1 of the phenacyl sulfonium ion) at pH above 6 is due to the decrease in rate of reaction (5).

For the range pH 6 to 8:

$$C_6H_5-CO-\overset{(-)(+)}{C}HSR_2 + H^{(+)} \rightleftharpoons C_6H_5-CO-CH_2\overset{(+)}{S}R_2 \qquad (8)$$

* The difference at pH above 8 are interpreted as involving desorption phenomena and are discussed separately (p. 266).

followed by reaction (*3*) and by fastly established base catalysed reaction (*4*). Then

$$C_6H_5-CO-CH_3 + 2e + 2H^{(+)} \rightarrow C_6H_5-CH(OH)-CH_3 \qquad (9)$$

Reaction (*8*) explains the pH-dependence of the half-wave potential of the first step *A* above pH 7. Fast base catalysed reaction (*4*) explains the pH-independence of the height of the second two-electron step B_2 betwen pH 6 and 8. Irreversibility of electrode reaction (*9*) with a subsequent uptake of protons explains the pH-independence of half wave potentials of acetophenone and of the second wave for the sulfonium salt B_2 in this pH-region.

The decrease of the current of the second (one-electron) step B_1 below pH 5 is smaller than for ω-piperidinoacetophenone[3]. This can be explained by a greater increase in rate of reaction (*4*) in the presence of sulfides SR_2 formed in reaction (*3*) at the surface of the electrode than in the presence of piperidine formed in the reduction of ω-piperidinoacetophenone[3].

Repulsion Phenomena

The decrease of current at −1·4 V, following the decrease of the round maximum (Fig. 1, curve *8*; Figs 4 and 5) can be attributed to the desorption of a negatively charged particle from a negatively charged double-layer, according to the theory developed by Frumkin[14]. The elimination of this decrease in the presence of tetraalkylammonium ions or at higher concentration of alkali ions can be explained by the changes in ψ_1-potential. Similarly also the effect of gelatin can be due to its positively charged centres.

The formation of the negatively charged particle, repulsed from the negatively charged double layer is a function of pH. This was demonstrated by the increase of the minimum with increasing pH-value. The idea of repulsion from the surface at negative potentials is further substantiated by the change in the height of the wave B_2 of acetophenone formed in reaction (*3*). As the surface concentration of the repelled particle is lowered at −1·4 V, the wave B_2 corresponding to acetophenone reduction according to equation (*9*) is also decreased, as the surface concentration of acetophenone is lowered. The decrease of height of the acetophenone wave B_2 at pH above 8 (Fig. 2) follows closely the decrease of the current i_{min} at −1·4 V (at the minimum, before the rise of the acetophenone wave). The addition of potassium ions or tetraalkylammonium ions caused an increase (Fig. 5) of the acetophenone wave (parallelling the increase of current at −1·4 V).

No conclusive evidence can be given for the nature of the negatively charged particle, repelled from the surface. The repelled particle can be either the enolate or a negatively charged intermediate in reaction (*4*). The former seems to be supported by the increasing repulsion effect with increasing pH-values and by the absence of such an effect with the corresponding α-aminoketones[1−3]. It could be mentioned here that a similar decrease (and a similar effect of cations) was observed with reduction waves of elemental sulphur[15].

The formation of a minimum on the limiting current of the reduction step B_2 in the presence of ammonium ions and of gelatin can be explained either by an inhibiting effect on the rate of reaction (4) or by repulsion of positively charged sulfonium ions from a double layer, where the ψ_1-potential or even the sign of the charge was changed.

We express our thanks to Drs J. Gasparič and M. Večeřa (Research Institute for Organic Syntheses, Pardubice-Rybitví) for the specimens of methyl butyl phenacyl sulfonium perchlorate, Dr O. Exner and Dr J. Volke, from our Institute, for valuable discussion, and to Miss H. Jänchenová for technical assistance.

References

1. Zuman P., Horák V.: *Advances in Polarography, Proc. IInd Internat. Congr. Polarography, Cambridge 1959*, Vol. III, p. 804. Pergamon Press, London 1960.
2. Horák V., Zuman P.: This Journal *26*, 173 (1961).
3. Zuman P., Horák V.: This Journal *26*, 176 (1961).
4. Lund H.: Acta Chem. Scand. *14*, 1927 (1960).
5. Colichman E. L., Love D. L.: J. Org. Chem. *18*, 40 (1953).
6. Matsuo H.: J. Sci. Hiroshima Univ., Ser. *A*, *22*, 281 (1958).
7. Zuman P., Horák V.: Unpublished results.
8. Tang Süe-Yuan, Zuman P.: This Journal, in the press.
9. Rodd E. H. (Ed.): *Chemistry of Carbon Compounds.* Vol. IA, p. 350. Elsevier, Amsterdam 1951.
10. Blau N. F., Stuckwisch C. G.: J. Org. Chem. *22*, 82 (1957).
11. Jaffé H. H.: Chem. Revs *53*, 191 (1953).
12. Zuman P., Fedoroňko M.: Z. physik. Chem. (Leipzig) *209*, 376 (1958) and unpublished results.
13. Zuman P., Michl J.: Nature *192*, 655 (1961).
14. Frumkin A. N.: Trans. Faraday Soc. *55*, 156 (1959), where earlier references are quoted.
15. Stackelberg M. v., Hans W., Hauck G.: Z. Elektrochem. *61*, 473 (1957).

Translated by the Author (P. Z.).

A NEW TYPE OF MAXIMUM ON THE LIMITING CURRENT
OF THE REDUCTION WAVE OF PHENACYL SULFONIUM IONS

SÜE-YUAN TANG* and P. ZUMAN

Polarographic Institute, Czechoslovak Academy of Sciences, Prague

Received February 20th, 1962

A typical round shaped streaming maximum was observed in solutions of methyl butyl phenacyl sulfonium perchlorate at pH 8— 12. Its properties differ from most commonly known catalytic and streaming maxima. Some similarity exists between the observed maximum and the maxima of second kind, the difference being mainly in the effect of capillary characteristics.

When studying the course of the reduction of phenacyl sulfonium ions[1], a characteristic maximum was observed in the middle of the limiting current, corresponding to the reduction of the $C—S^{(+)}$ bond. The form of maximum differed from that observed with most streaming maxima and resembled more that obtained for catalytic evolution of hydrogen in ammoniacal cystine solutions[2,3].

It was thus necessary to study the effect of experimental conditions on the height and shape of this maximum in more detail and to compare the observed behaviour with that of other catalytic and streaming maxima.

Experimental

Apparatus, reagents. The polarographic equipment, reagents and technique of measurement were the same as in the preceding paper[1]. The greater part of the measurements was carried out with solutions of methyl butyl phenacyl sulfonium perchlorate. In some preliminary experiments with dibutyl phenacyl sulfonium ions and methyl ethyl p-bromophenacyl sulfonium ions it has been shown that the behaviour is in so far analogous that the behaviour of methyl butyl phenacyl sulfonium ion can be taken as representative for the whole class of compounds.

Streaming. The streaming of the electrolyte in the neighbourhood of the dropping electrode was observed in the Kalousek cell using a binocular microscope. Small amount of active charcoal was added to the solution studied. Little effect was observed on polarographic *i–E* curves on both limiting and maximum current in the presence of used amounts of charcoal.

Controlled drop-time. In addition to the effect of mercury pressure h the rôle of drop-time t_1 and out-flow velocity m was studied separately using an electrode with controlled drop-time. The drop-time was regulated using a device described by Mojžiš[4]. The drop-time was changed in the region from 0·5 s to 3·9 s at constant $m = 1·55$ mg/s, when $h = 30$ cm and unregulated drop-time was $t_1 = 5·6$ s. The out-flow velocity was changed from $m = 1·38$ mg/s to $m = 3·33$ mg/s at regulated constant $t_1 = 1·43$ s. From the measurements with regulated drop-time the

---

* Present adress: Petroleum Research Institute, Chinese Academy of Sciences, Talien, China.

dependences of the mean current on drop-time ($\bar{\imath}$–t_1 curves) and the dependence of this current on out-flow velocity m were plotted.

i–t Curves. The change of the instantaneous current during the drop-life was measured on first and second drop using a device described by Němec and Smoler[5]. The i–t-curves were recorded on a photographic paper. The i–t curves were recorded at several chosen potentials corresponding to typical points on the polarographic i–E curves.

Hanging drop. The hanging mercury drop current-voltage curves which have been used for determination of the course of electrolysis recently by Kemula[6], were recorded photographically using the syringe type electrode devised by Vogel[7]. The curves, starting at different potentials, were traced from negative to positive values either immediately after a given negative value of potential (usually that of alkali ions reduction) was reached or after a chosen time-period.

Commutator method. The rectangular voltage polarisation in the commutator method devised by Kalousek[8] was applied using the relais commutator[9]. The auxiliary potential was chosen at selected potentials corresponding to characteristic points on the limiting current.

Constant potential electrolysis was carried out using a mercury dropping electrode at the potential of the maximum current (i_M). 10 ml of borate buffer pH 9·3 containing $1 . 10^{-3}$M methyl butyl phenacyl sulfonium perchlorate was electrolysed in a Kalousek vessel for 40—70 hours. The central stopcock connecting the electrolytic compartment and the reference electrode was closed to avoid diffusion. The appropriate voltage to secure the potential of i_M was chosen by recording polarographic i–E curves. Instead of lower stopcock in the Kalousek cell a mercury seal was provided to ensure the out-flow of mercury.

Oscillopolarography. The oscillographic dE/dt–E curves were recorded using Polaroscop, manufactured by Křižík (Prague). Electrocapillary curves were recorded using the drop-time method.

Results

Effect of pH, *Kind of Buffer and Ionic Strength*

When the curves were recorded in buffers containing 0·05M sodium borate and 0·2 per cent ethanol and varying amounts of boric acid (thus at varying pH-values at approximately constant ionic strength) at pH 7·5 to 10·3, a bell-shaped change of the maximum current with pH-value was observed (Fig. 1, curve *1* and Figs 2). The increasing part of this dependence possesses the form of a disscioation curve with inflexion point about pH 8·3. When 0·01M borate was used and concentration of boric acid changed, the maximum was substantially lower (predominantly an effect of the ionic strength, cf. p. 1528). An S-shaped increase of the maximum current was observed again (Fig. 1, curve 2), but was less steep and shifted towards higher pH-values (with inflection point at about pH 9·2). In buffers containing oppositely constant concentration of the boric acid and changed concentration of borate, a steep S-shaped increase of the maximum current was also observed. The pH-dependence was here complicated by the effects of changing ionic strength. This has been proved by comparison with the effect of neutral salt addition (cf. p. 1528). The predominating effect of ionic strength at given pH is also obvious from the fact that the change in concentration (capacity) of the borate buffer at constant ionic strength (regulated by additions of sodium chloride) has little effect on the maximum current.

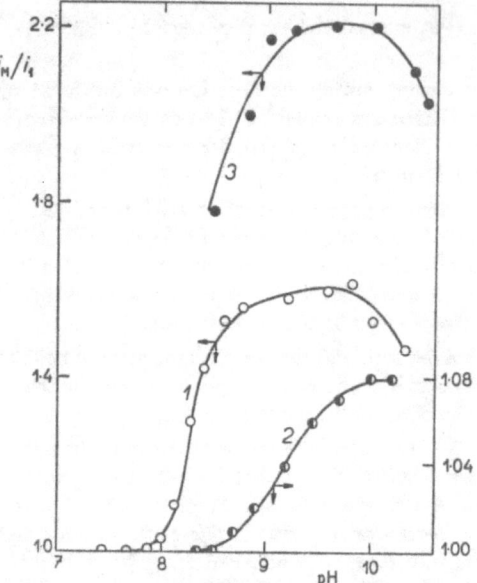

Fig. 1

pH-Dependence of the Total Maximum Current (i_M) of Methyl Butyl Phenacyl Sulfonium Perchlorate Compared with the Diffusion Current of the First Reduction Wave (i_1)

2 . 10^{-4}M Depolarizer, 2 % ethanol, 1, 2 Borate buffer, constant concentration of borate: 1 0·05M; 2 0·01M. Concentration of the boric acid changed. 3 Buffer composed of 0·1M-NH_4Cl and changing amounts of ammonia. Current i_M measured at the potential of the maximum, $m = 2·5$ mg/s, $t_1 = 2·9$ s, $h = 65$ cm.

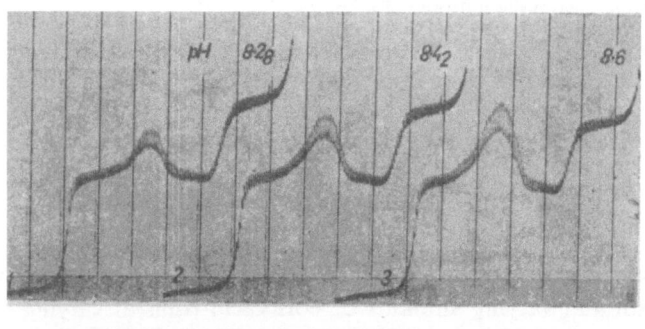

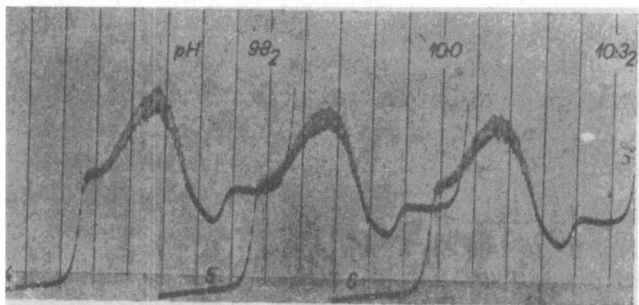

Fig. 2

pH-Dependence of Waves of Methyl Butyl Phenacyl Sulfonium Perchlorate

2 . 10^{-4}M Depolarizer, 2 % ethanol. Borate buffer, constant concentration of borate (0·05M) and changed content of boric acid, pH: 1 8·3; 2 8·4; 3 8·6; 4 9·8; 5 10·0; 6 10·3. Curves starting at −0·2 V, 200 mV/absc., s.c.e., $m = 2·5$ mg/s, $t_1 = 2·9$ s, $h = 65$ cm, full scale deflection 2·7 μA.

Similar behaviour as in 0·05M borate buffers was observed in Britton-Robinson buffers. The maximum was obtained even in diluted solutions of sodium hydroxide. The pH-dependence observed with ammonia buffers containing 0·1M-NH₄Cl and changed amounts of ammonia is given in Fig. 1 (curve 3). The S-shaped increase of the maximum current arrives at the highest value at pH about 9·0 and the decrease of the maximum current was observed above pH 10·0. The highest value in a buffer containing 0·1M-NH₄Cl of pH 9·6 is about twice as high as the corresponding current in 0·05M borate buffer. Comparison of Fig. 1 and Fig. 3 shows that the difference of curves *1* and *3* in Fig. 1 is predominately caused by the change in buffer capacity (ionic strength).

Addition of ethanol to borate buffers changed the height of the maximum in a way which could be predicted from the apparent shift of the pH-values, caused by

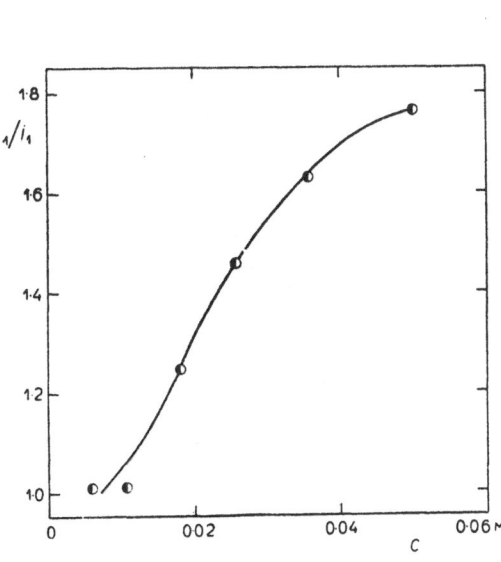

Fig. 3

Effect of Buffer Capacity on Total Maximum Current (i_M) of Methyl Butyl Phenacyl Sulfonium Perchlorate Compared with the Diffusion Current of the First Wave (i_1)

$2 . 10^{-4}$M Depolarizer, 2 % ethanol, borate buffer pH 9·6 diluted with water (ionic strength changed). Current i_M measured at the potential of maximum, $m = 2·5$ mg/s, $t_1 = 2·9$ s, $h = 65$ cm.

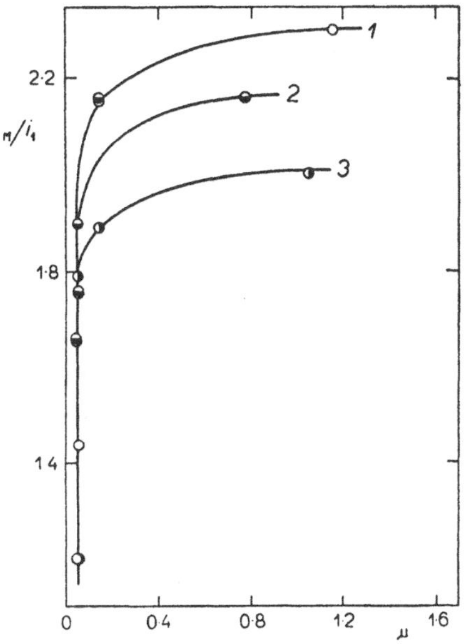

Fig. 4

Effect of Ionic Strength on Total Maximum Current (i_M) of Methyl Butyl Phenacyl Sulfonium Perchlorate Compared with the Diffusion Current of the First Wave (i_1)

$2 . 10^{-4}$M Depolarizer, 2 % ethanol, borate buffer (0·05M borate, changed concentration of boric acid) pH: *1* 8·3; *2* 9·2; *3* 10·3. Ionic strength changed by addition of sodium chloride. Current i_M measured at the potential of maximum, $m = 2·5$ mg/s, $t_1 = 2·9$ s, $h = 65$ cm.

the increase of pH-values with increasing ethanol concentration. Addition of dibutyl sulfide or acetophenone (expected reduction products) was without effect on the maximum current.

With increasing concentration of sodium chloride, added to borate buffers containing 0·05M borate ion at three selected pH-values (Fig. 4), the maximum current shows the tendency to reach a limiting value, dependent on the pH-value of the buffer used. Dilution of borate of pH 9·6 with water caused a decrease of the maximum current given in Fig. 3. The increase of the maximum current with increasing ionic strength (in a region complementary to that shown in Fig. 4) reaches again a limitting value.

In ammonium chloride-ammonia buffer pH 9·3 also an S-shaped dependence of the maximum current on buffer concentration was observed. The limiting value of $i_M/i_1 = 2·8$ was higher than for borate buffers of the same pH-value (limiting value of i_M/i_1 did not exceed 2·3) at constant ionic strength $\mu = 0·2$.

Effect of Concentration of Sulfonium Cation

With increasing concentration of the methyl butyl phenacyl sulfonium cation, the height of the maximum reaches a limiting value (Fig. 5). This limiting value depends on pH in a similar way as the maximum current in $2 . 10^{-4}$M solutions (Fig. 1). At pH above 10 the measurement of the round maximum is rendred difficult at higher concentrations of the sulfonium cation, as the sharp maximum of the first kind, observed at the beginning of the limiting current, increases significantly with increasing concentration of the depolarizer. Thus at higher concentration of sulfonium ions the sharp maximum of the first kind is high (and extended over a broad range of the applied voltage) and coalesces with the round maximum.

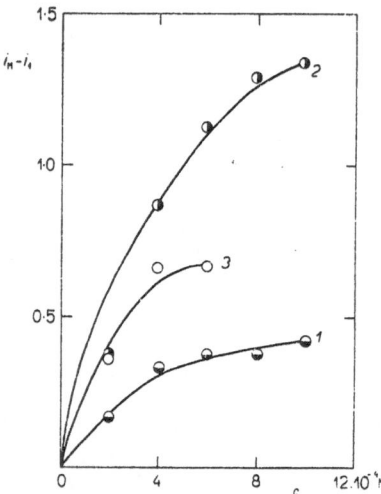

Fig. 5

Effect of Concentration (c) of Methyl Butyl Phenacyl Sulfonium Perchlorate on the Difference Between the Total Height of the Maximum Current (i_M) and the Diffusion Current of the First Wave (i_1)

$2 . 10^{-2}$M Stock solution of depolarizer in 10 % ethanolic solution added to borate buffer (0·05M borate, changed concentration of boric acid) pH: 1 8·2; 2 9·1; 3 10·1. Current i_M measured at the potential of maximum, $m = 3·5$ mg/s, $t_1 = 2·5$ s, $h = 65$ cm.

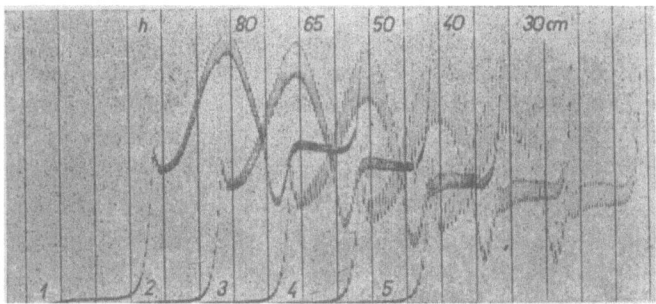

Fig. 6 ·
Effect of Mercury Pressure on the Wave of Methyl Butyl Sulfonium Perchlorate
$6 . 10^{-4}$M Depolarizer, 0·2 % ethanol, 0·05M borax, pH 9·3; $h = 1$ 80, 2 60; 3 50; 4 40;
5 30 cm. Curves starting at 0V, s.c.e., 200 mV/absc., full scale deflection 8·4 μA.

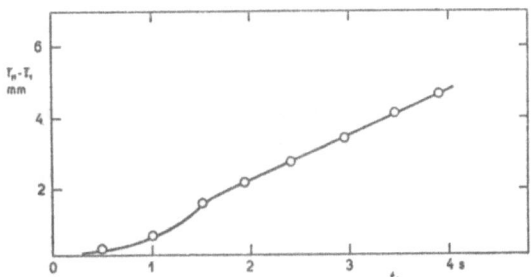

Fig. 7
Dependence of the Difference of the Mean Total Maximum Height (i_M) and the Diffusion Current
of the First Wave (i_1) on Drop-Time (t_1) Using Controlled Dropping Electrode
$2 . 10^{-4}$M Methyl butyl phenacyl sulfonium perchlorate, 2 % ethanol, 0·05 M borax, pH 9·3.
Current i_M measured at the potential of the maximum, regulated drop-time, constant $h = 30$ cm,
$m = 1·55$ mg/s, unregulated drop-time $t_1 = 5·6$ s.

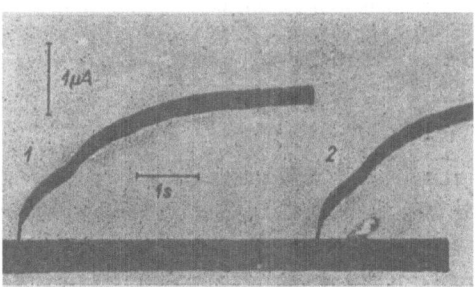

Fig. 8
Dependence of the Instantaneous Current on Time $(i-t$ Curves)
$6 . 10^{-4}$M Methyl butyl phenacyl sulfonium perchlorate, 0·6% ethanol, 0·05 M borax, pH 9·3,
$i-t$ curve recorded at the potential of maximum ($-1·2$ V), $t_1 = 5$ s. 1 First, 2 second drop.

Effect of Capillary Constants and Electrocapillary Curves

When using an ordinary mercury dropping electrode and changing the mercury pressure, the maximum current i_M is very little dependent on the height of the mercury head h (Fig. 6). The dependence of the mean maximum current on regulated drop-time is almost linear with an inflection (Fig. 7). Similarly also the instantaneous current i at the potential of the maximum (where it corresponds to the sum of both diffusion controlled i_1 and the maximum current) shows an inflexion (Fig. 8). The mean maximum current is almost proportional to m.

Electrocapillary curves in buffered solutions of $2 . 10^{-4}$M methyl butyl phenacyl sulfonium ion show an adsorption of the oxidised form at potentials more positive than about -0.7 V, with $1 . 10^{-3}$M this range was extended to about -0.9 V. At potentials more negative than that of the reduction of the $C—S^{(+)}$ bond the electrocapillary curve was practically identical with those for pure supporting electrolyte (Fig. 10).

Streaming of Solution

In the potential region, where the maximum appears, a streaming of the solution in the direction given in Fig. 9 was observed. The change in intensity of the streaming paralleled the change in intensity of the maximum. When a low mercury pressure was used, it was possible to observe that at the point A (Fig. 9) carbon particles remained attached to the surface of the electrode near to the neck. These particles oscillated on the surface of the electrode. At potentials corresponding to the rising

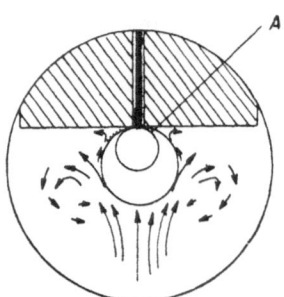

Fig. 9

Streaming of Solution at the Potential of the Maximum $1 . 10^{-3}$M Methyl butyl phenacyl sulfonium perchlorate, 0·05M borax, pH 9·3, the direction and intensity of streaming at $h = 80 - 25$ cm shown by arrows, the oscillations most distinct at point A at $h = 25$ cm.

part of the maximum, it can be observed that during the early stages of the drop-life carbon particles attached to the surface at the top region of the drop are carried towards the neck, where they start to oscillate during the second half of the drop-life.

In the potential region at the beginning of the reduction wave of the $C—S^{(+)}$ bond, where regular maxima of the first kind were observed, the solution streamed in the same direction as with the round maximum.

The streaming is suppressed by addition of gelatin. On the polarographic $i–E$ curve both the sharp maximum of the first kind and the round maximum are competely suppressed[1]. The round maximum at $-1\cdot2$ V cannot be observed when the hanging mercury drop electrode is used, if the voltage is gradually increased (using the usual rate of voltage scanning). When the voltage was suddenly changed from zero to $-1\cdot2$ V a current decreasing with time was found. In microscopic observations, oscillations of carbon particles at the surface of the mercury electrode were observed, the frequency of which decreased with time (parallelled by the decrease in current).

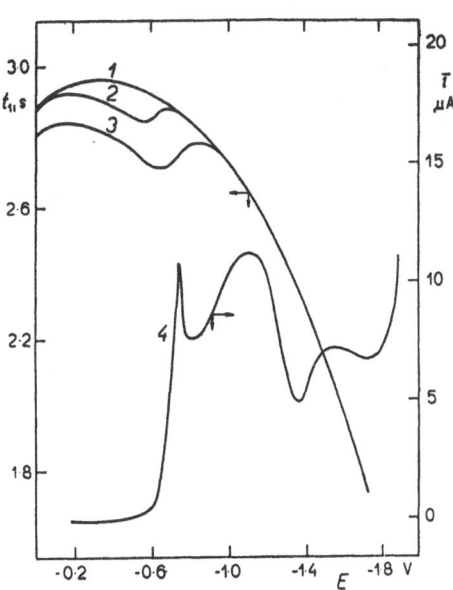

Fig. 10

Electrocapillary Curves

0·05M Borax pH 9·3, concentration of methyl butyl phenacyl sulfonium perchlorate: *1* 0; *2* $2\cdot10^{-4}$M; *3* and *4* $1\cdot10^{-3}$M. Concentration of ethanol: *1* 0; *2* 0·2%; *3* and *4* 1·0%. *1—3* electrocapillary curves, *4* polarographic *i–E* curve. *E* against s.c.e.

Controlled Potential Electrolysis

Because the results obtained with Kalousek's commutator method[8,9] were rather complicated, a prolonged controlled potential electrolysis was carried out using dropping mercury electrode at the potential of the maximum current (i_M).

In the electrolysis the potential was applied to two Kalousek cells with the electrolysed solution (p. 1527) connected parallelly. In one solution the maximum was suppressed by addition of 0·01% gelatin. Thus the electrolysis was carried out once in the presence and once in the absence of the round maximum, the other conditions being identical. 1 ml samples were taken after different time-intervals after occasional stiring with nitrogen. To the sample, where the electrolysis was carried out in the presence of maximum, gelatin was added so that the final concentration equals to that in the other sample. After dilution to 10 ml with borate buffer pH 9·3 both curves were recorded and increase of the acetophenone wave B_2 was measured. This wave is composed of the acetophenone just freshly formed at the surface of the dropping electrode in the wave A and of the acetophenone present in the bulk of the solution as a product of the prolonged electrolysis.

The production of acetophenone in solutions where the maximum occured (Fig. 11, curve *1*) was greater than that observed in the absence of maximum (Fig. 11, curve *2*). When from the height of the wave B_2 the height corresponding to surface formed acetophenone (11 mm on Fig. 11) was subtracted, it is seen that the acetophenone formed in the presence of gelatin is about 70% of the amount formed in solutions where maximum occured. When the amount of electricity was estimated from the dependence i_B/i_A on time, the number of coulombs consumed was again lower (65·6% and 70·5% in two parallel runs) for the solution containing gelatin. These results show that when the current rises in the form of the maximum at $-1·2$ V

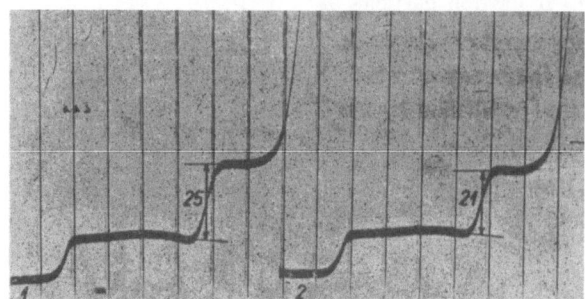

Fig. 11
Controlled Potential Electrolysis
Electrolysis of $1 . 10^{-3}$M methyl butyl phenacyl sulfonium perchlorate was carried out in borax pH 9·3 at $-1·2$ V for 70 hours. *1* Electrolysis without gelatin; *2* with 0·01% gelatin. After electrolysis gelatin added also to *1* and curves recorded from $-0·4$ V, s.c.e., 200 mV/absc., full scale deflection 19·6 μA.

only phenacyl sulfonium ions are reduced in increased amount and acetophenone produced. A further reduction step or a reduction of hydrogen ions can be excluded.

Discussion

The shape of the maximum on the limiting cirrent of methyl butyl phenacyl sulfonium perchlorate and of other sulfonium ions resembles most closely that of catalytic hydrogen waves. Thus the comparison of the behaviour of the maximum with catalytic maxima will be discussed first.

Only the effect of concentration of the sulfonium compound resembled the limiting dependence observed with catalytic waves. Effects of mercury pressure and ionic strength are of the type observed sometimes with catalytic currents, but present no conclusive evidence. On the other hand the dependence on pH, little dependence on buffer capacity, the streaming of solution,* the absence of hydrogen bubbles on the surface of mercury drop** and suppression by gelatin are different from those observed with catalytic currents.

Shape of the maximum, position in the middle of limiting current and effect of

---

* Even when Březina[14] recently observed a weak streaming during a short period after addition of charcoal for catalytic waves of cystine, under comparable conditions when for sulfonium ion streaming was observed, no streaming was observed in the presence of cystine.

** For catalytic hydrogen waves of cystine giving currents of the same order as maximum on sulfonium wave evolution of hydrogen was observed microscopically.

ionic strength excluded the possibility of a maximum of first kind. Furthermore the round maximum at $-1 \cdot 2$ V is observed in addition to the maximum of the first kind at $-0 \cdot 8$ V.

The studied maximum on sulfonium waves seems to be nearest to maxima of second kind. The analogies lie in kind of streaming of the solution, in the round shape, in the effect of gelatin and of ionic strength. The observations of the type of streaming and the absence of the maximum, when hanging mercury drop electrode is used, show that the outflow velocity and the movement of the mercury surface due to this streaming effect are of primary importance. Effects of pH and of concentration of the depolarizer were not studied systematically for maxima of second kind and cannot be thus used as arguments neither pro nor con. On the other hand the most important difference between sulfonium salt maximum and regular maxima of second kind is the effect of capillary constants.

Whereas the sulfonium salt maximum shows dependence $i_M \sim h^n$, where n is considerably smaller than $0 \cdot 5$ (Fig. 6), for maxima of second kind n = $3 \cdot 5$ was found by Vintera (cf.[10]) [or at least $(i_d + i_M) \sim h^{1 \cdot 0}$ was found by Dvořák[11]]. Differences exist in the effect of drop-time and in the i–t curves. Out-flow velocity shows for our maximum a dependence $i_M \sim m^{1 \cdot 0}$, whereas for maximum of second kind[11] the relation $(i_d + i_M) \sim m^{7/6}$ was found.

In addition to catalytic currents maxima in the middle of limiting current were rarely observed. The case of tellurite maximum[12] can be mentioned here, but this is a sharp maximum. Round maxima similar to that obtained with sulfonium ions were claimed to be obtained with some naphthoquinone derivatives even with low buffer concentrations[13].

The limiting character of the observed dependence shows that the current is a function of the area of the mercury electrode. Microscopic observations have indicated the possibility of participation of adsorption. The round shape of the maximum can be explained by two overlapping factors: (i) concurrence adsorption (possibly even of the depolarizer or of surface active impurities) shifting the increase of current into region of more negative potentials and (ii) repulsion of the electroactive form (or a form in equilibrium with this form) from the negatively charged double-layer.

In the mechanism of the process causing the increase of current also two factors at least play a rôle: One of them is the streaming out of mercury and the transport from the top towards the neck of the mercury drop. The oscillations of the attached particles are the result of these factors. The change of the maximum height with pH can be explained by participation of a certain ionic form. The difference between the activity of different ionic forms can be explained e.g. by their different adsorbability. The participation of the conjugate base I (cf.[1]) could explain the desorption phenomena.

We express our thanks to Dr J. Kůta for his help in recording of the i t curves, to Dr M. Březina and Dr K. Bláha for their review of the manuscript, to Drs J. Gasparič and M. Večera for the samples of sulfonium salts and to Miss H. Jänchenová for her careful technical assistance.

278

References

1. Zuman P., Tang Süe-Yuan: This Journal *28*, 829 (1963).
2. Brdička R.: This Journal *5*, 112, 148 (1933).
3. Březina M., Zuman P.: *Polarography in Medicine, Biochemistry and Pharmacy*, p. 585. Interscience, New York 1958.
4. Mojžíš J.: Chem. listy *50*, 318 (1956).
5. Němec L., Smoleŕ I.: Chem. listy *51*, 1958 (1957).
6. Kemula W.: *Advances in Polarography, Proc. IInd Congr. Polarography, Cambridge 1959*, Part I, p. 105. Pergamon Press, Oxford 1960.
7. Vogel J.: A paper read before Czechoslovak-Hungarian Polarographic Conference, Prague, July 1959. See also J. Říha: *Progress in Polarography*, Vol. II, p. 383. Interscience, New York 1962.
8. Kalousek M.: This Journal *13*, 105 (1948.)
9. Rálek M., Novák L.: This Journal *21*, 248 (1956).
10. Zuman P.: Chem. listy *44*, 162 (1950).
11. Dvořák J.: This Journal *19*, 39 (1954); Chem. listy *47*, 969 (1953).
12. Hans W., Stackelberg M. v.: Z. Elektrochem. *54*, 62 (1950).
13. Vintera J.: *Thesis*. Charles University, Prague 1950.
14. Březina M.: Private communication.

Translated by the Author (P. Z.).

POLAROGRAPHY OF PYRIDOXINE AND SOME
OF ITS DERIVATIVES

O. Manoušek* and P. Zuman

*Central Research Institute of Food Industry and Polarographic Institute,
Czechoslovak Academy of Sciences, Prague*

Received November 27th, 1963

Polarographic curves for pyridoxol, 4-methoxymethyl-5-hydroxymethyl-3-hydroxy-2-methylpyridine, pyridoxamine, pyridoxamine-5-phosphate, 4-pyridoxthiol, bis-4-pyridoxyldisulphide, pyridoxal, pyridoxal-5-phosphate, pyridoxaloxime and pyridoxaloxime-5-phosphate in buffered media were described and discussed. Reduction of the C—O, C—N$^{(+)}$ and C—S bonds in pyridoxol, pyridoxamine and pyridoxthiol has been proved. The reactivity of these bonds in polarographic reductions decreases in sequence: C—SH > > C—N$^{(+)}$H$_3$ > C—OCH$_3$, C—OH. The effect of esterification by phosphoric acid in position 5 in pyridoxal, pyridoxamine and pyridoxaloxime and the participation of the phosphoric acid grouping in acid-base reactions was shown. Two types of polarographic dissociation curves of dibasic acids were demonstrated. Most of the acid-base reactions encountered in this group of substances involve not only hydroxonium but also other protondonors.

Inspite of the importance of pyridoxine** in biochemistry in general and particularly in enzymic reactions, the polarography of most of these substances has not been studied systematically so far.

For the polarographic reduction of pyridoxol Lingane and Davis[2] claim the occurrence of two waves at −1·8 and −2·0 V in tetramethylammonium iodide. The nature of the electrode process has not been elucidated, similarly as for pyridoxamine, where a two-electron wave has been found[3]. The pH-dependence of the wave- height for pyridoxal has been attributed[4,5] to the rate of the ring opening of the hemiacetal form.

For analytical applications in biological material[6−9] as well as for applications in kinetic studies of some biochemically important reactions[7−13] a more detailed study of the polarographic behaviour of the particular compounds was of importance.

---

* Present address: Polarographic Institute, Czechoslovak Academy of Sciences, Prague.

** According to a decision[1] of the Commision on the Nomenclature of Biological Chemistry of the International Union of Pure and Applied Chemistry, pyridoxol, pyridoxal and pyridoxamine as well as their phosphoesters should be denoted summarily as pyridoxine (instead of using the term "vitamin B$_6$").

As pyridoxaloxime has found important analytical applications[8] and as the course of oxime formation was studied[12,13], a study of pyridoxaloxime and its phosphate has been included as well.

Experimental

Apparatus

Polarographic curves were recorded using a Heyrovsky type polarograph Mark V 301, produced by Zbrojovka (Brno). The electrolysed solution was placed in a Kalousek vessel with a separated calomel electrode. The capillary used possessed at a zero applied voltage the following characteristics: $t_1 = 2.9$ s, $m = 3.1$ mg/s. The oscillographic $dE/dt = f(E)$ curves were recorded using oscilloscope P 524 produced by Křižík (Prague).

For the determination of the pH values of buffer solutions used and of the pK values of the substances studied the glass electrodes G 200 A and G 200 B in connection with the pH-meter Radiometer PHM 3h and the glass electrodes of the type U and H in connection with the pH-meter Metrohm E 148e were used. pK values were determined from the potentiometric titration curves using 0·02M solutions of the pyridine derivative. Approximate pK values (± 0.1) measured are given in Table I.

Absorption spectra were measured in acetate and phosphate buffers in a 1 cm quartz cell using the spectrophotometer SF-4. As the wave lengths of absorption maxima of the conjugated

Table I

Dissociation Constants of Pyridoxine and Some of its Derivatives

Substance	$pK_1{}^a$	$pK_2{}^b$	$pK_3{}^a$	$pK_4{}^b$	pK_5
I	—	5·00	—	8·96	—
III	—	4·23	—	8·70	—
IV	—	3·54	—	8·21	$10·63^c$
V	$0·90^d$	$3·65^e$	6·20	8·69	—
VI	2·5	3·69	5·76	8·61	$10·92^c$
VII	—	$4·10^e$	—	$8·10^e$	10^f
VIII	2·5	$3·70^e$	$6·10^e$	$8·60^e$	10^f
IX	—	$5·37^g$	—	$9·10^g$	—
X	—	$4·45^e$	—	$8·85^e$	—
XI	—	$4·08^e$	—	$8·80^e$	—
XII	—	$6·15^g$	—	—	—
XIII	—	$5·33^g$	—	—	—

[a] Dissociation constants of the phosphoric acid part; [b] dissociation constants of pyridine nitrogen and phenolic oxygen; [c] dissociation constants of the side-chain nitrogen; [d] values obtained from kinetic measurement[11]; [e] values determined potentiometrically in this paper; [f] dissociation constants of the oxime grouping; [g] values determined spectrophotometrically in this paper; other values according to Williams and Neilands[14].

acid-base couples of the studied substances were not sufficiently separated, the pK values were determined from the change of extinction coefficients at 265 mμ with pH that possessed the form of a dissociation curve.

Chemicals

Pyridoxol hydrochloride (I) (m. p. 206°C, decomp., ref.[15] 204—208°C) was a commercial sample supplied by Farmakon (Olomouc). The free base was prepared by neutralization with sodium hydroxide, evaporation and extraction of the base into methanol. The resublimed specimen had a m. p. of 158°C (ref.[15] 160°C).

4-Methoxymethyl-5-hydroxymethyl-3-hydroxy-2-methyl pyridine hydrochloride (II) (m. p. 180—181°C) was an intermediate in the pyridoxol synthesis by Farmakon (Olomouc).

Pyridoxal hydrochloride (III) (174°C decomp., ref.[15] 173—174°C) and *pyridoxamine dihydrochloride* (IV) (229—230°C decomp., ref.[15] 226—227°C) were supplied by Merck-Darmstadt, *pyridoxal-5-phosphate monohydrate* (V) (m. p. 143°C) by Light (Colnbrook) and Fluka (Buchs).

Pyridoxamine-5-phosphate hydrochloride (VI), (m. p. 224°C decomp., ref.[15] 222—224°C), was prepared from *IV* by esterification by metaphosphoric acid. Pyridoxamine-5-triphosphate formed in this reaction was hydrolysed in acid solution to *VI*.

Pyridoxaloxime (VII) was prepared from *III* and hydroxylamine hydrochloride in acetate buffer pH 4·5 (m. p. 225—226°C, 70% ethanol, ref.[15] 225—230°C). Another route is oxidation of pyridoxol hydrochloride followed by a treatment with hydroxylamine hydrochloride in acetate buffer without isolation of pyridoxal[16]. *Pyridoxaloxime-5-phosphate* (VIII) was prepared by precipitation of the solution of the substance *V* in acetate buffer pH 4·5 by hydroxylamine hydrochloride. After twofold recrystallization from water the m. p. was 229°C decomp. (ref.[15] 229—230°C). The purity of *VII* and *VIII* was proved by nitrogen analysis, electrophoresis and fluorescence spectra.

4-Deoxypyridoxine hydrochloride (IX) was prepared according to Wendt and Bernhart[17] (m. p. 264—265°C) and compared with an authentic sample.

4-Pyridoxthiol hydrochloride (X) was prepared according to *Wendt* and *Bernhart*[17] and also according to Schmidt and Giesselmann[18]. Both specimens had an identical m. p. (176—177°C). *Bis-4-pyridoxyl-disulphide dihydrochloride* (XI) m. p. 222°C, ref.[17], was prepared according to Wendt and Bernhart[17] using the reaction of pyridoxal with cystein and by oxidation of substance X by hydrogen peroxide in the presence of ammonia, ammonium chloride buffer pH 9·0.

4,5-Dihydroxymethyl-2-methylpyridine (XII) m. p. 145—147°C was prepared by Farmakon (Olomouc). *4-Hydroxymethylpyridine* (XIII) m. p. 56°C was prepared by Raschig (Ludwigshafen).

	R¹	R²		R¹	R²
	I, CH₂OH	H		VI, CH₂NH₂	PO(OH)₂
	II, CH₂OCH₃	H		VII, CH=NOH	H
	III, CHO	H		VIII, CH=NOH	PO(OH)₂
	IV, CH₂NH₂	H		IX, CH₃	H
	V, CHO	PO(OH)₂		X, CH₂SH	H

Techniques

Solutions. Aqueous stock solutions $(5 \cdot 10^{-3}\text{M})$ of *I— XIII* were prepared and kept in refriger-ator. Solutions of *V* and *X* were prepared daily fresh. Buffer solutions (0·1 to 0·4M) were prepared from acetic and phosphoric acids, veronal, ammonia and trimethylamine. For orientation pur-poses Britton-Robinson buffers were used. Dilute solutions of perchloric acid and sodium hydro-xide were sometimes used as supporting electrolytes. Ionic strength was maintained by addition of a 5M stock solution of sodium perchlorate or chloride. Chemicals for the preparation of buffer solutions and of other supporting electrolytes were all reagent grade.

Standard procedure. 9 ml of buffer or other supporting electrolyte were deaerated in a Kalousek vessel. 1 ml of the stock solution of the pyridoxine derivative *I— XIII* was added and after final deaeration with nitrogen the curve was recorded.

Hanging mercury drop electrolysis. Formation of reduction products was followed using electro-lysis at hanging mercury drop[19] with the syringe electrode designed by Vogel[20] and preparative electrolysis at constant potential using a mercury pool electrode and graphite reference electrode.

Results and Discussion

PYRIDOXOL DERIVATIVES

Pyridoxol (I). One two-electron wave is observed in ammoniacal buffer solutions of *I* at pH 8·5 to 10 (Fig. 1). At higher pH-values the height of this wave decreases, but due to the overlapping of the current-rise caused by the reduction of the supporting electrolyte the form of this decrease could not be followed in detail. The diffusion character of the limiting current has been proved by the linear dependence of the wave-height on the concentration of *I* and by the proportionality of the wave-height to the square-root of the mercury head.

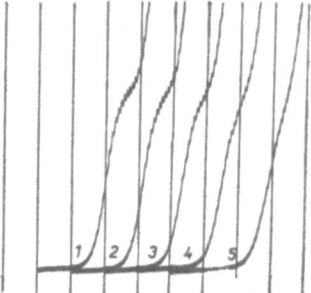

Fig. 1
Dependence of Waves for Pyridoxol on pH
$0\cdot1\text{M-NH}_3$, $0\cdot1\text{M-NH}_4\text{Cl}$; $5 \cdot 10^{-4}\text{M}$ Pyridoxol; pH: *1*
8·7; *2* 9·0; *3* 9·4; *4* 9·8; *5* 10·2. Curves starting at
− 1·85 V, 165 mV/absc., mercurous sulfate electrode,
full scale deflection 5·74 μA.

The reduction wave observed in solutions of *I* is attributed to the reduction of the activated C—OH bond in position 4. This type of process has been proved recently[21] for the C—O activated by a carbonyl[22,23] or nitrile[24] group. The conjugated acid form is most likely the reducible species. The activation of the C—O bond results from the influence of the protonised pyridinium ring. The decrease of the wave-height at higher pH values probably reflects the dissociation of the acid form (considering recombination effects). The half-wave potentials below pH 10·0 are only a little pH-dependent (Table II) and it is anticipated that the electrode process proper (resulting in the rupture of the C—O bond) is pH-independent, as has been proved for other C—O bonds[23].

Table II

Dependence of Half-wave Potential of Pyridoxol (*I*), 4-Methoxymethyl Pyridoxine (*II*) and Pyridoxamine (*IV*) on pH (*vs*. S.C.E.)

pH	*I*	*II*	*IV*	*VI*	
				i_1	i_2
8·7	$-1·70^a$	$-1·68_5{}^a$	$-1·53^b$	$-1·68^b$	—
8·9$_5$	$-1·72^a$	$-1·70^a$	$-1·58^b$	$-1·69^b$	—
9·3	$-1·73^a$	$-1·71^a$	$-1·61^b$	$-1·72^b$	$-1·82^b$
9·8	$-1·75^a$	$-1·75^a$	$-1·67^a$	—	$-1·86^c$
10·1	$-1·79^a$	$-1·77^a$	$-1·69^c$	—	$-1·88^c$

[a] Ammonia–ammonium chloride; [b] veronal buffers; [c] phosphate buffers.

The ammonium ions in the buffer solution not only maintain the pH values but also influence the structure of the double-layer in a similar way as has been observed *e.g.* for phenyl acetyl carbinol[25]. In buffer solutions containing boric acid a decrease of the height of the pyridoxol wave has been observed in the entire pH-range studied. The decrease was a function of concentration of boric acid and was observed even in concentrations of boric acid of the same order as those of pyridoxol. A complex formation is assumed to take place. In veronal buffers pH 8·3−9·5 the diffusion reduction current of pyridoxol is obscurred by a catalytic hydrogen evolution. The total wave-height increases (compared with that in ammonia–ammonium chloride) and at pH > 9·0 two poorly developed waves are separated. The catalytic character of the more negative wave has been proved by a steep increase with decreasing pH-value and with increasing buffer concentration, by a non-linear dependence of the wave-height on concentration as well as by the increase of the wave with lowering of the mercury head.

Also 4,5-dihydroxymethyl-2-methyl-pyridine (*XII*) undergoes a two-electron reduction similarly as pyridoxine, but for 4-hydroxymethyl pyridine (*XIII*) the catalytic wave obscured the wave of the reduction process.

4-Methoxymethyl derivative of pyridoxine (II). In solutions of ammonia–ammonium chloride the 4-methoxymethyl derivative of pyridoxine shows a similar two electron diffusion controlled reduction wave like pyridoxol. The comparison of the half-wave potentials with those for pyridoxol and pyridoxamine is given in Table II. A hydrogenolysis of the OCH_3 group was verified by controlled potential electrolysis; traces of methyl alcohol formed were detected using gas-chromatography.

The presence of boric acid does not influence the wave-height, because in this molecule two adjacent hydroxyl groups enabling the complex formation are not present. Nevertheless in borate buffers containing sodium the limiting current was less well developed than that in ammonia–ammonium chloride solutions, probably due to the cation effect on double-layer structure. In veronal buffers at pH above 8·4 a catalytic wave was observed in addition to the diffusion current (Fig. 2). Whereas the height of the more positive diffusion wave is almost pH-independent between pH 8·5 and 9·1 (and decreases at higher pH values similarly as the waves observed in ammonia–ammonium chloride buffers), the height of the more negative catalytic wave is strongly pH-dependent (Fig. 2). The catalytic wave increases with decreasing mercury pressure.

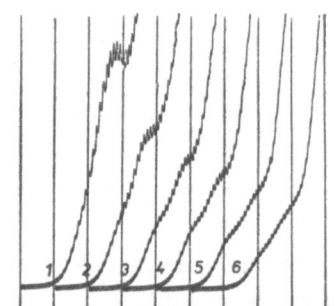

Fig. 2
The Influence of pH on Reduction Waves and Catalytic Waves for 4-Methoxymethyl Pyridoxine Derivative in Veronal Buffers
$5 . 10^{-4}$M 4-Methoxymethyl pyridoxine, pH: *1* 8·3; *2* 8·65; *3* 8·85; *4* 9·0; *5* 9·2; *6* 9·4. Curves starting at −1·72 V; mercurous sulfate electrode, 165 mV/absc. full scale deflection 11·35 μA.

Polarography of dibasic acids. The pyridoxine derivative studied are polybasic acids and due to this fact several waves corresponding to different dissociation steps were observed on polarographic curves. The heights the limiting currents of the particular forms decrease with increasing pH values in the form of dissociation curves.

For dibasic acids H_2A Hanuš and Brdička[26] using the approximate method and later Koutecký[27] using the rigorous treatment derived for the decrease of the more positive wave (i_1) corresponding to the undissociated acid H_2A the relation (*1*), which is applicable, provided that $pK_2 > (pK_1 + 2)$ and $D_{H_2A} = D_{HA^-}$. In equation (*1*)

$$\bar{i}_1/\bar{i}_d = \frac{0·87(t_1 k_1/K_1)^{1/2}\{[H^+]^2/([H^+] + K_2)\}}{1 + 0·87(t_1 k_1/K_1)^{1/2}\{[H^+]^2/([H^+] + K_2)\}} \tag{1}$$

$\bar{i}_d$ is the total diffusion limiting current, t_1 the drop-time, K_1 and K_2 the first and second potentiometric dissociation constants and k_1 the rate constant corresponding to the formation of acid H_2A from anion HA^-. This relation corresponds to a decrease of the limiting current i_1 with increasing pH in the form of a dissociation curve which is twice as steep as the theoretical curve for monobasic acids[28-31], provided that $K_2 > [H^+]$, *i.e.* if $pK_1' > pK_2$. As has been shown by Hanuš and Brdička[26], when $K_2 \ll [H^+]$, *i.e.* if $pK_1' \ll pK_2$, the value of K_2 can be neglected compared with $[H^+]$ and equation (*1*) is reduced to (*2*):

$$i_1/i_d = \frac{0{\cdot}87(t_1 k_1/K_1)^{1/2}\,[H^+]}{1 + 0{\cdot}87(t_1 k_1/K_1)^{1/2}\,[H^+]}.\tag{2}$$

The resulting shape of the pH-dependence of the limiting current (when the difference in diffusion coefficients is neglected) is identical with that for monobasic acid[28-31]. In that pH-range in which $[H^+]$ is comparable with K_2, *i.e.* when $pK_1' \doteq pK_2$ the resulting curve is unsymmetrical and its shape depends on the drop-time (*cf.* Fig. 1 in ref.[20]).

The second wave (i_2) corresponding to the reduction of the species HA^- at more negative potentials than i_1 also decreases with increasing pH values in the form of a dissociation curve. This curve possesses the same shape as that for the monobasic acids, provided* that $D_{H_2A} \doteq D_{HA^-} \doteq D_{A^{2-}}$ and that $pK_2' > (pK_1' + 2)$ and $pK_2' \gg pK_2$. Hence for dibasic acids two dissociation curves (with pK_1' and pK_2') can be observed, either with the same slope** (when $pK_1' \ll pK_2$) or with a greater slope of the first dissociation curve (for $pK_1' > pK_2$).

Pyridoxamine (IV). Pyridoxamine, the reducibility of which was anounced by Volke and co-workers[3], gives in simple buffers of pH 5 to 11 a single two-electron step. The height of this diffusion controlled wave is pH-dependent only to a small degree (Fig. 3). At higher pH values the limiting current decreases with increasing

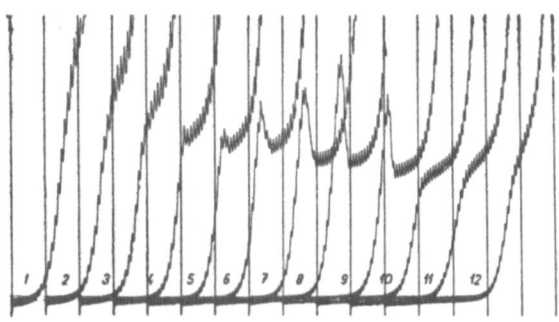

Fig. 3

Waves for Pyridoxamine in Buffer Solutions

$7{\cdot}2 \,.\, 10^{-4}$M Pyridoxamine in buffers of pH: *1* 5·1; *2* 5·7; *3* 6·0$_5$; *4* 6·6; *5* 7·0; *6* 7·3; *7* 7·9; *8* 8·5; *9* 8·9; *10* 9·3$_5$; *11* 10·0; *12* 10·4. Curve *1* and *2* 0·1M acetate buffer; *3—7* 0·1M phosphate buffer; *8—12* 0·2M veronal buffer. Curves starting at $-1{\cdot}49$ V, 166 mV/absc., mercurous sulfate electrode, full scale deflection 8·02 µA.

---

* Values for pK' correspond to the pH values for which the limiting current of i_1 or i_2 respectively [or ($i_1 + i_2$), when $pK_1' > (pK_2' - 2)$] is equal to $i_d/2$.

** Similarly as in a mixture of two monobasic acids with pK_1' and pK_2' respectively.

pH value, but a measurement of this decrease is prevented by the poorly developed form of these waves. Oscillographic $dE/dt = f(E)$ curves show that the reduction is irreversible. The streaming maxima can be suppressed by 0·005% dextrane or gelatine. The shifts of half-wave potentials are given in Fig. 4 (*cf.* Table II).

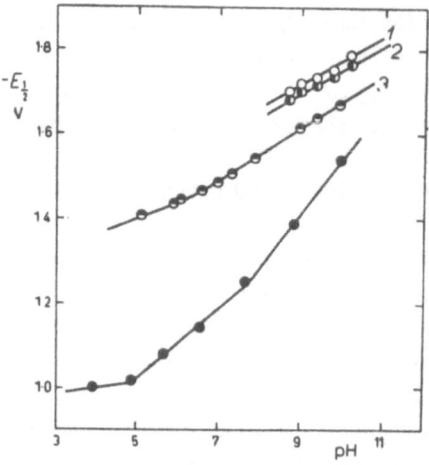

Fig. 4

The pH-Dependence of Half-Wave Potentials
1 Pyridoxol (in 0·1M-NH$_3$, 0·1M-NH$_4$Cl); *2* 4-Methoxy Derivative of Pyridoxine (in 0·1M-NH$_3$, 0·1M-NH$_4$Cl); *3* Pyridoxamine (acetate, phosphate and veronal buffers); *4* 4-Pyridoxthiol (acetate, phosphate and veronal buffers). Half-wave potentials ($E_{1/2}$) against S.C.E.

Fig. 5

Waves for Pyridoxamine-5-Phosphate in Buffer Solutions
$5 . 10^{-4}$M Pyridoxamine-5-phosphate, pH: *1* 7·1$_5$; *2* 7·7; *3* 8·4; *4* 8·7; *5* 9·1; *6* 9·4$_5$. Curves starting at −1·65 V, mercurous sulfate electrode, 165 mV/absc., full scale sensitivity 5·74 μA.

When amine *IV* was reduced at a large scale mercury pool electrode, using a carbon anode in a veronal buffer pH 8·6, ammonia could be traced in the solution electrolysed. The course of the electroreduction has been followed in the electrolysis cell by recording decreasing polarographic curves for pyridoxamine after chosen time-intervals. The relative concentration of ammonia was determined colorimetrically (using Nessler reagent) in samples taken from the electrolysed solution. The increase in ammonia concentration was inversely proportional to the decrease of the polarographic wave for pyridoxamine. The electroreduction can be performed quantitatively.

By analogy with other substances bearing a reducible single carbon-nitrogen bond[32,33] the reductive splitting off the ammonia* results most probably from the protonated form. The reduction of the C—N$^{(+)}$ bond is activated by the presence

---

* Predicted for such cases previously by one of us[32,33].

of the pyridine ring, similarly as is the reduction of the C—O bond in pyridoxol (*cf.* p. 1435).

Pyridoxamine-5-phosphate (VI). In veronal and ammonia–ammonium chloride buffers containing phosphate *VI* a two electron reduction step can be observed. At pH 8 to 10 two waves are separated (Fig. 5). The height of the more positive step decreases with increasing pH value in the form of a dissociation curve, the height of the sum of both waves being practically pH-independent. The macro-scale electrolysis of pyridoxamine-5-phosphate revealed the formation of ammonia, similarly as in the electrolysis of pyridoxamine (*IV*). The half-wave potentials for phosphate *VI* are more negative than those for the parent substance *IV* (Table II).

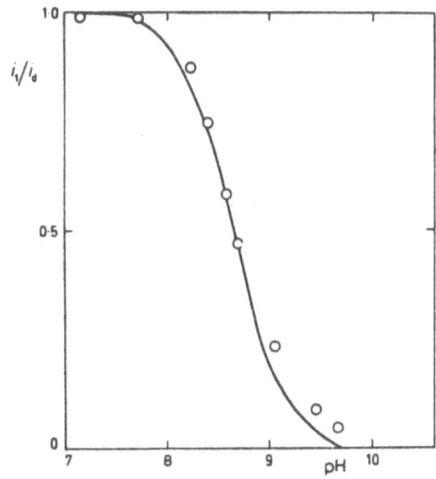

Fig. 6

Polarographic Dissociation Curve of Pyridoxamine-5-phosphate

Theoretical curve computed using equation (*1*) for $pK_1' = 8\cdot7$, $t_1 = 2\cdot1$ s, and $pK_1' > pK_2$; experimental points are given.

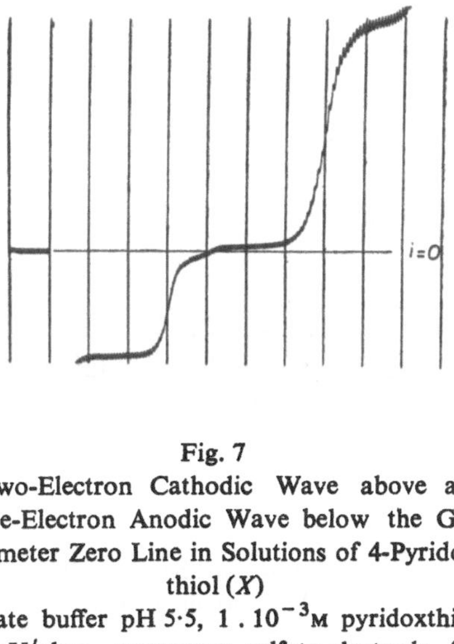

Fig. 7

A Two-Electron Cathodic Wave above and a One-Electron Anodic Wave below the Galvanometer Zero Line in Solutions of 4-Pyridoxthiol (*X*)

Acetate buffer pH 5·5, $1 . 10^{-3}$M pyridoxthiol, 190 mV/absc., mercurous sulfate electrode, full scale deflection 11·40 μA.

The presence of two waves changing with pH reflects the participation of an acid-base reaction in the over-all electrode process. The dependence of i_1/i_d on pH closely follows (Fig. 6) the theoretically computed curve based on equation (*1*), for $pK_1' = 8\cdot7$ and $t_1 = 2\cdot1$ s. As the phosphate *VI* can definitely exist in several ionic forms, the slope of the dependence given in Fig. 6 can be taken as proof (p. 1437) that the pK_2-value involved is substantially smaller than observed pK_1'-value 8·7 (probably the pK_2 is identical with the value $pK_3 = 5\cdot76$ given for substance *VI* in Table I). This, together with the fact that no comparable separation of the two waves was observed with pyridoxamine, allow the deduction that the proton-transfer reaction involves the phosphoric acid part, similar to pyridoxal-5-phosphate (p. 1451). Due to the existence of several forms in acid solutions no attempt is made to assign the structures to the conjugated acid and base involved. For $pK_1 = 3.7$, $pK_2 = 5.76$, $pK_1' = 8.7$ and $t_1 = 2.1$ s (using the symbols used in equation (1)) the value for the formal recombination rate constant $k_1 = 1.2 \cdot 10^{18}\, 1\, mol^{-1}\, s^{-1}$ was computed. This high value allow us to exclude the possibility of a simple homogeneous acid-base reaction[40].

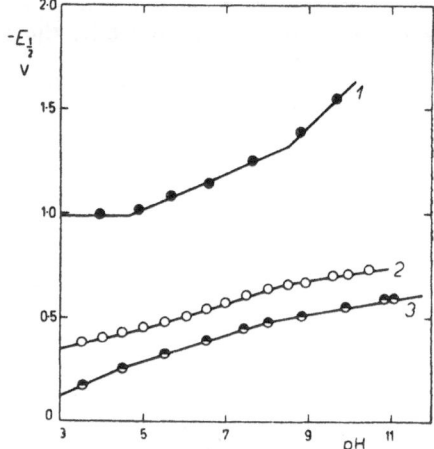

Fig. 8

pH-Dependence of Half-Wave Potentials of Anodic Wave (3) and of the Reduction Wave (1) of 4-Pyridoxthiol, and of the Wave corresponding to Reduction of the Disulphide Bond for Bis-4-pyridoxyl-disulphide (2)

4-*Pyridoxthiol* (X). On polarographic waves of thiol X a two-electron cathodic wave and a one-electron anodic wave is observed (Fig. 7). The half-wave potentials of both cathodic (Fig. 4) and anodic (Fig. 8) waves are shifted towards more negative potentials with increasing pH-value (Table III). The height of the anodic wave is

Table III

Half-wave Potentials of the Cathodic and Anodic Wave of 4-Pyridoxthiol (X) and of the Wave Corresponding to S–S Reduction of Bis-4-pyridoxyl-disulphide (XI)

Britton-Robinson buffers, *vs.* S.C.E.

Substance, Wave	pH							
	3·9	4·9	5·6	6·6	7·6	8·8	9·7	10·7
X, Cathodic	-0.99^a	$-1.01_5{}^a$	-1.08^a	-1.14^a	$-1.26_5{}^a$	-1.39^a	-1.55^b	-1.81^b
X, Anodic	-0.21	-0.29	-0.33	-0.39	-0.46	-0.51	-0.54	-0.58
XI, Disulph-idic Bond	-0.40	-0.45	-0.48	-0.54	-0.61	-0.67	-0.70	-0.73

[a] Wave i_1; [b] wave i_2.

practically pH-independent (Fig. 9), only in alkaline solutions adsorption phenomena at more positive potentials[34] cause a deformation of curves $10-12$ (Fig. 9). The height of the cathodic wave i_1 decreases with increasing pH-value in the form of a dissociation curve (Fig. 10) and a new wave (i_2) is formed at more negative potentials, the half-wave potential of which $(-1·81 \text{ V})$ is practically pH-independent. In the presence of ammonium ions in buffer solutions the decrease of wave i_1 was shifted towards higher pH-values.

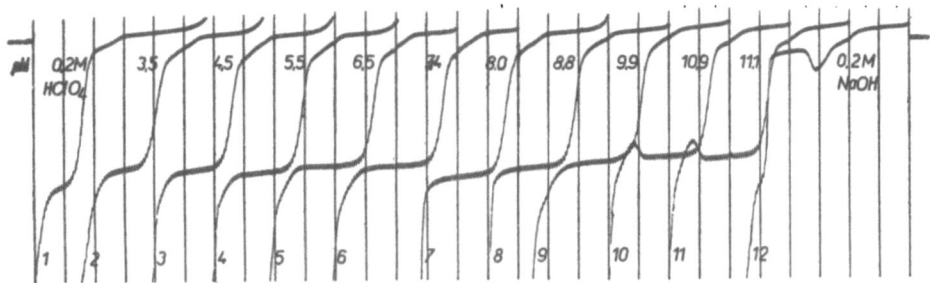

Fig. 9

Dependence of the Anodic Wave for 4-Pyridoxthiol on pH-Values

$1 \cdot 10^{-3}$M Pyridoxthiol in 1 0·2M-HClO$_4$; 2—4 acetate; 5 and 6 and 9—11 phosphate; 7 and 8 veronal buffers and in 12 0·2M-NaOH. pH-given on the polarogram. Curves 1—6 starting at 10th winding of potentiometric wire, curves 7—12 at 11th winding; anod.-cath. polarization, 190 mV/absc., mercurous sulfate electrode, full scale deflection 6·85 μA.

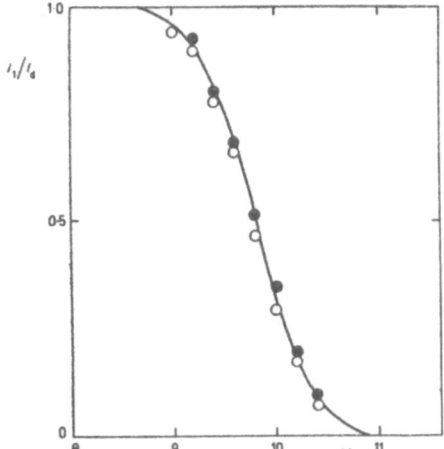

Fig. 10

Polarographic Dissociation Curve for 4-Pyridoxthiol (○) and for the Second Wave of Bis-4-pyridoxyl-disulphide (●)

Theoretical curve computed using equation (I) for $pK_1' = 9·8$ and $t_1 = 2·4$ s for $pK_1' > pK_2$ is given together with experimental points.

The one-electron anodic wave is ascribed to the formation of a compound with mercury[35]. The wave is, similar to other thiols, accompanied by an adsorption pre-wave at more negative potentials.

The two-electron reduction wave corresponds to the hydrogenolysis of the SH group. This has been proved by prolonged electrolysis, after which the hydrogen sulphide could be proved by the anodic wave corresponding to mercury sulphide formation. Similarly the formation of hydrogen sulfide was proved by electrolysis at the hanging mercury drop electrode[19,20]. The more negative* peak A, corresponding to the formation of a mercury salt with sulfide ions increased (Fig. 11) with prolonged electrolysis at potentials corresponding to the limiting current of the cathodic wave as well as after addition of sodium sulfide.

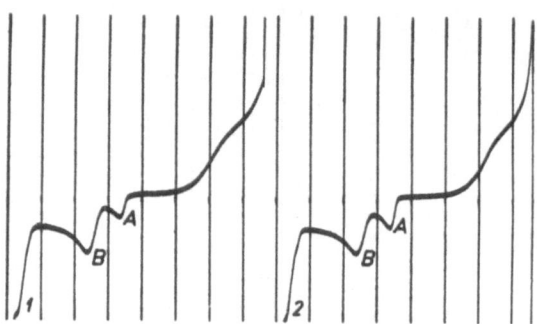

Fig. 11

Effect of Prolonged Polarization on Curves of 4-Pyridoxthiol Using Hanging Mercury Drop Electrode

Acetate buffer, pH 4·7, 2 . 10^{-4}M 4-pyridoxthiol, polarization at −1·5 V for 1 10 s; 2 20 s with successive tracing of the curve starting at −1·5 V towards positive potentials. s.c.e., 200 mV/absc. full scale deflection 3·2 μA.

The decrease of the wave i_1 follows equation (1) for $pK'_1 = 9·81$ and $t_1 = 2·4$ s and thus possesses a steeper form corresponding to the first dissociation curve of a dibasic acid when pK'_1 is substantially greater than the value of pK_2 ($cf.$ p. 1437). Thus K_2 in equation (1) corresponds to K_4 in Table I and for pK_1 involved the value 4·45 can be taken.** Using these values the recombination rate constant $k_1 = 2·4$. 10^{16} l mol^{-1} s^{-1} could be calculated. From the effect of buffer kind and concentration on wave i_1 the participation of proton-donors other than hydroxonium ion[38-51] can be deduced. This observation together with the large value of the calculated rate constants shows that a homogeneous process can be excluded and that the physical meaning of this constant k_1 is dubious.

The small difference in limiting values of the wave heights i_1 and i_2 in acid and alkaline solutions shows that the assumption of comparable values of diffusion

---

* The more positive peak B corresponds to a formation of a mercury salt with the unreacted thiol X.

** Breaks on $E_{\frac{1}{2}}$–pH-plot (Fig. 4) are in accordance with this deduction.

coefficients is fulfilled. The form reduced in wave i_1 is probably *XIV*. The structure of the conjugated base can involve pyridine base, dipolar ion with dissociated phenolic group (*cf.* discussion in[11]) or the thiol group too. The similarity of pK_2 and pK_4 values (Table I) with those for other pyridoxine derivatives seems to exclude the possibility of effect of dissociation of the SH-group, but an accidental coincidence cannot be excluded.

CH$_2$SH

HO CH$_2$OH

H$_3$C N (+)

H   *XIV*

Bis-4-pyridoxyl-disulphide (XI). The reduction of the disulphide *XI* proceeds generally in two steps (Fig. 12). A two electron step i_3 is followed by a four electron step (i_1 and i_2 respectively) at more negative potentials. Both waves are shifted towards more negative potentials with increasing pH value (Table III). The second reduction step of *XI* is identical with the reduction wave of thiol *X*. This has been proved not only by the coincidence of half-wave potentials for both i_1 in acid and i_2 in alkaline solutions, but also by the coincidence of the dependence of the height of the first wave i_1 of the thiol *X* and of the second wave i_1 of the disulphide *XI* on pH (Fig. 10).

Thus it can be concluded that the more positive wave corresponds to the reduction of the S—S bond[52]. Hence in this process two molecules of thiol *X* are formed that are reduced at more negative potentials. Thus the electrode processes can be depicted as follows:

i_3 R—S—S—R + 2 e → 2 RSH

i_1 (or i_2)   2 R—SH + 4 e + 2 H$^+$ → 2 R—H + 2 SH$^-$

CH$_2^-$

HO CH$_2$OH

where R =

H$_3$C N

The irreversibility of the first step i_3 was proved by the difference in the half-wave potentials of the first cathodic wave for disulphide *XI* and of the anodic wave for thiol *X*.

Both reduction waves of the disulphide *XI* are accompanied (mainly in acid solutions) by streaming maxima of the first kind. Moreover, in alkaline solutions pronounced maxima of the second kind (on curve *18*, Fig. 12) appear, even in solutions with ionic strength $\mu = 0.1$. These maxima show a typical large increase with an increase in the height of the mercury head. When the use of surface active substances was undesirable curves were recorded at low mercury pressures.

At a pH above 11 another phenomenon has been observed (Fig. 12c). A trough (minimum) occured on the limiting current. The range, in which this trough occurs depends on the pH value of the buffer used and on the ionic strength. With increasing pH the trough is deeper and occurs in a broader potential range. When we measure

292

the current at a chosen potential, the current decreases in the form of a dissociation curve. The pK'-value depends on the applied potential value, similarly as was observed for periodic acid[53]. With increasing ionic strength the current in the potential

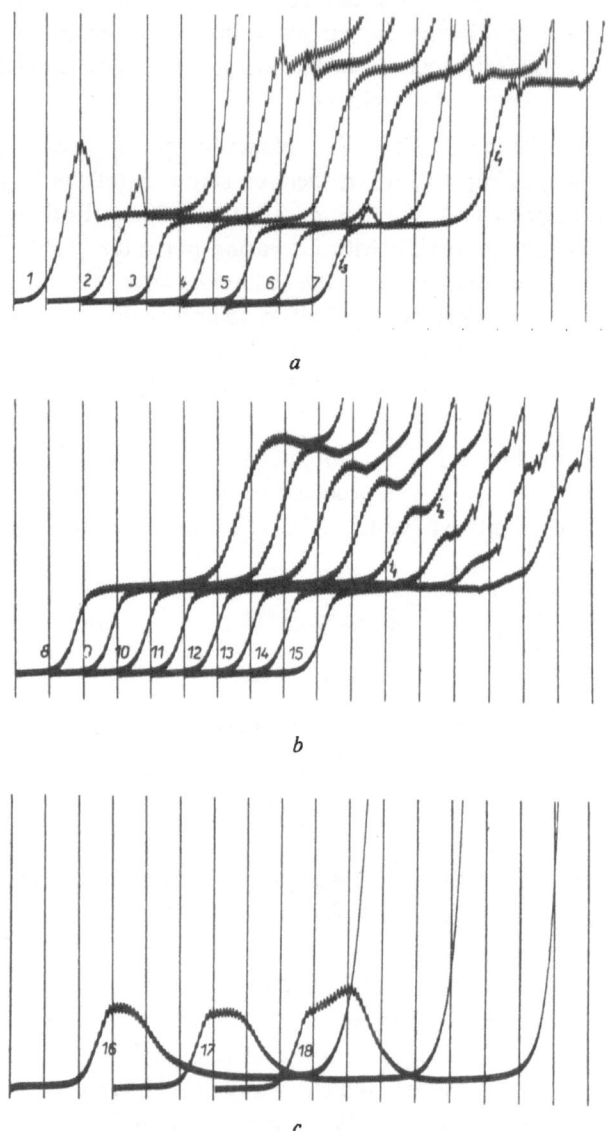

Fig. 12

Dependence of Waves for Bis-4-pyridoxyl Disulfide on pH

$5 \cdot 10^{-4}$M Bis-4-pyridoxyl disulfide a: pH: *1* 0·08M-HClO$_4$; Britton-Robinson Buffers *2* 3·5; *3* 4·5; *4* 5·5; *5* 6·5; *6* 7·35; *7* 8·51; b: *8* 9·0; *9* 9·2; *10* 9·4; *11* 9·6; *12* 9·8; *13* 10·0; *14* 10·2; *15* 10·4; c: *16* 0·02M-NaOH; *17* 0·05M-NaOH; *18* 0·2M-NaOH. Curves *1 — 7* starting at −0·16 V, *8 — 18* at −0·49 V, 165 mV/absc., mercurous sulfate electrode, full scale deflection 11·35 μA.

range, where the decrease is observed, is increased and in 0·2M-CaCl₂ a normal diffusion current is obtained (Fig. 13).

The observed current decrease thus possesses characteristics, described by Frumkin and his co-workers[54-57] for persulfate and several other anions. The decrease of current is ascribed to the repulsion of the anions from the surface on the electrode due to the ψ-potential. In the studied case the anion can possess the structure *XV*.

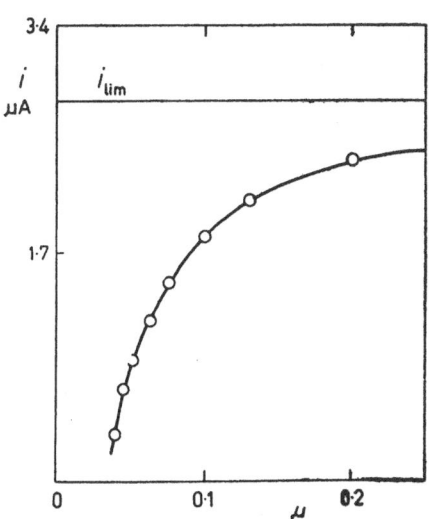

Fig. 13
Influence of Ionic Strength on the Decrease of Limiting Current of Bis-4-pyridoxyl Disulfide in 0·04M-NaOH
$5 . 10^{-4}$M Bis-4-pyridoxyl disulfide, calcium chloride added. Current measured in the region of the decreasing part of the limiting current at −1·3 V.

A non-homogeneous recombination of disulphide *XV* with hydrogen ions and its dependence on potential can explain the observed dependence on pH.

Comparison of Reduction of Substances Bearing the Group —CH₂X *on Pyridine Ring*

The reduction of substances *I, II, IV* and *X* proceeds under hydrogenolysis accompanying the consumption of two electrons. The electrode process proper seems to be pH-independent in all these cases, the pH-dependence of half-wave potentials observed being caused by proton-transfers on parts of the molecule remote from the reactive center. A nucleophilic attack of electrons can be assumed as the potential determining step. In agreement with such a reduction route is also the polarographic inactivity of 4-deoxypyridoxine (*IX*), the expected product of all these reductions.

The hydrogenolysis occurs always in position 4, even when CH₂OH in position 5 is present. The course of polarographic reduction in position 4 is hence in agreement with findings for catalytic hydrogenolysis[58] in which substance *I* yielded 4-deoxypyridoxine (*IX*) but not 4-deoxypyridoxine. In a similar way catalytic reductions of substances *II* and *X* take place[17,58]. Also in other instances pyridine derivatives bearing in position 4 a —C(OH)R$_2$ groups were reduced to corresponding 4-CHR$_2$ groups[59-61], whereas no such reaction was quoted for analogous 3-C(OH)R$_2$ compounds. An increased reactivity of 4-CH$_2$OH is obvious also from the formation of 4-alkoxymethyl derivatives of pyridoxine[58], whereas no formation of 5-alkoxymethyl derivatives was reported. On the other hand in oxidation, even when the oxidation of 4-CH$_2$OH is preferred, an attack of CH$_2$OH group in position 5 cannot be excluded.

When the half-wave potentials of C—X bonds are compared,* the following sequence can be observed:

$$-CH_2OCH_3, \quad -CH_2OH > -CH_2\overset{(+)}{NH_3} \gg -CH_2SH$$

A quantitative correlation would be possible for the first three members, following an analogous reduction mechanism. Due to the different proton-uptake (*cf.* variation in slopes of linear portions in Fig. 4), the comparison with thiol X is only a rough approximation. The observed sequence of reactivity in hydrogenolysis parallels that observed in nucleophilic reactions[62].

PYRIDOXAL DERIVATIVES

Pyridoxal(III). The general course of the pH-dependence of the wave height of pyridoxal in buffers and diluted acid solutions described by Volke[4] has been verified. Since in Britton-Robinson buffers containing borate ions the height of waves is about 20 per cent lower than in buffers of the same pH in the absence of borate ions, acetate, veronal and phosphate buffers were used throughout this study.

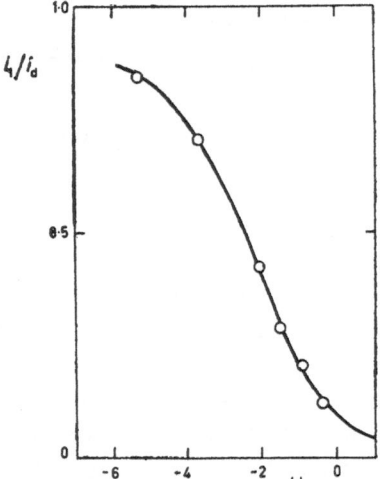

Fig. 14

Dependence of the Wave-Height i_1/i_d for Pyridoxal on Acidity Function H_0 in Perchloric Acid

i_1 is the kinetic current of pyridoxal in acid media.

---

* For pyridoxamine the reduction is attributed to the $CH_2NH_3^{(+)}$-form, by analogy with aminoketones[32,33].

A two-electron wave[4] was observed even in 1M-NaOH. In acid solutions, a single wave increased with increasing concentration of perchloric acid (Fig. 14) in the form of a dissociation curve, arriving at the two-electron value. Only in irradiated solutions another wave at more negative potentials[5] at pH below 2 was observed. The half-wave potentials measured are given in Fig. 15.

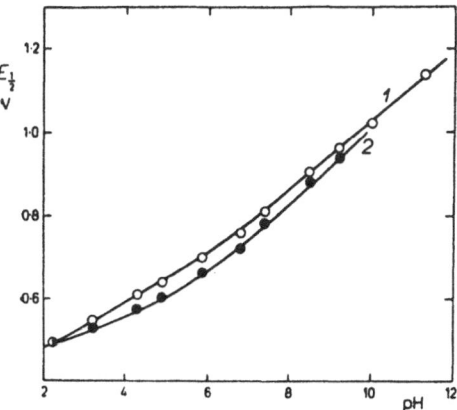

Fig. 15

Dependence of Half-Wave Potentials (*vs.* S.C.E.) for Pyridoxal (*1*) and Pyridoxal-5-phosphate (*2*) on pH

The pH-dependence of the limiting current of pyridoxal was explained by Volke[4] as governed by the rate of the acid-base catalysed ring-opening of the hemiacetal form. As an alternative explanation the rate of the dehydration of the hydrated aldehyde group can be proposed. The spectroscopic evidence[63] quoted for the pre-dominance of the hemiacetal form in aqueous solutions of aldehyde *III* is unequivocal since the observed effects can similarly be explained by the influence of hydration[11].

Some of the polarographic experimental evidence could have been used in the support of the explanation by dehydration: Hence, the pH-dependence of the limiting current of pyridoxal is similar to those found for formyl pyridines[4], the observed difference would be explained by the effects of substituents. In accordance with this assumption is seemingly also the pH-dependence observed for 4-hydroxy-3-formylpyridine.* But the course of the pH-dependence of an acid-base catalysed ring opening and of the dissociation of the pyridinium proton causing the increase of the wave-heights for formyl pyridines[4] can be analogous and this proof is thus not conclusive. Since hemiacetal is supposed to dissociate at pH about 13, a decrease could have been supposed in strongly alkaline media. But the absence of such a decrease can be caused by a recombination reaction.

---

* Waves of 4-hydroxy-3-formylpyridine show a pH-dependence similar to that found for 3-formyl pyridine, only the maximal decrease in acid media is about 30 per cent of the total limiting current (whereas for 3-formylpyridine it is nearer 80 per cent). Thus substitution of an *ortho*-hydroxy group causes either an increase in the rate of dehydration or the equilibrium is shifted more in the favour of the acyclic form.

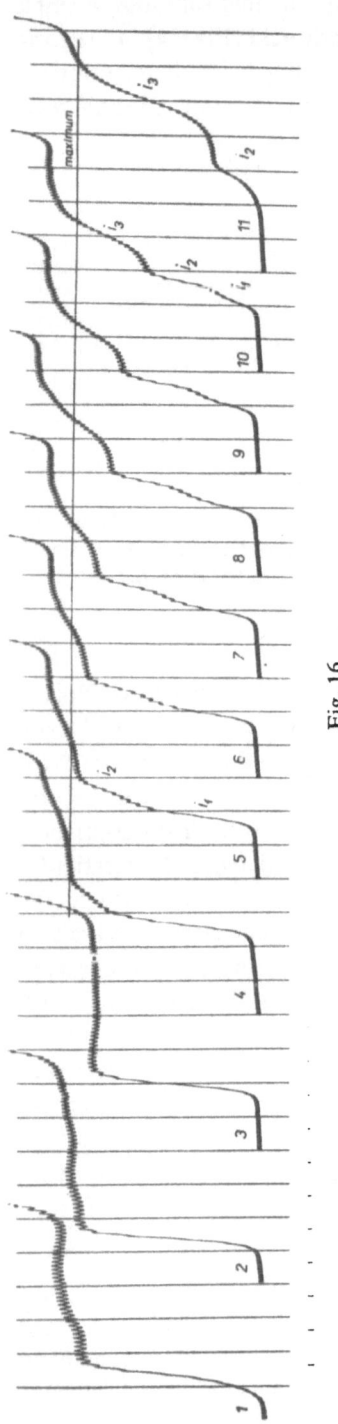

Fig. 16

pH-Dependence of Waves for Pyridoxal-5-phosphate

5 . 10⁻⁴M Pyridoxal-5-phosphate in buffers: 1 acetate pH 3·6; 2 phosphate pH 6·3; 3 veronal pH 8·9; 4—11 phosphate: 4 pH 10·1; 5 pH 10·6; 6 10·8; 7 11·0; 8 11·2; 9 11·4; 10 11·6; 11 11·8; mercurous sulphate electrode, 156 mV/absc., full scale deflection 4·06 μA.

Nevertheless, the most important experimental evidence so far available is the pH-independence of the wave heights for pyridoxal-5-phosphate (*cf.* next section). This behaviour can be interpreted as due to the impossibility of hemiacetal ring formation, on the other hand it would be possible to interpret the independence of the height of the limiting current of pyridoxal-5-phosphate on the pH values also as either due to the acceleration of the dehydration reaction or to a shift of the equilibrium towards the free aldehydic form, caused by vicinal substituents. The former explanation seems more convincing; if it is valid also for pyridoxal, the decrease of current in acid media can be interpreted as due practically only to the ring-opening reaction.

The observed decrease of the wave-height of pyridoxal in the presence of boric acid can be explained as an effect of both complex formation and of the catalytic activity of borate in base catalysed reactions.

Pyridoxal-5-phosphate (V). Polarographic reduction waves for phosphate V were observed in acid solutions (*e.g.* in 2·5M-$HClO_4$), buffer solutions and in solutions of hydroxides up to pH 14 (Fig. 16). Three waves i_1, i_2 and i_3 appear according to the pH. Bellow pH about 10 only a two-electron reduction wave i_1 is observed. The consumption of two-electrons was proved by comparison with the four-electron waves for pyridoxaloxime and by coulometric measurements. The logarithmic analysis for wave i_1 gave the slope $52 - 60$ mV in the pH-range studied. The irreversible character of all waves for phosphate V has been proved by oscillographic $dE/dt = f(E)$ curves.

The height o f wave i_1 below pH 10 is practically pH-independent, but the wave is accompanied by a round maximum of the second kind at more negative potentials. This maximum depends on mercury pressure, on ionic strength and the kind of buffer used. Whereas in phosphate and trimethylamine buffers the wave height remains constant at pH 6 to 8 and 8·5 to 10 and the same as in acid solutions, in buffers containing veronal (Fig. 16, curve 3) the limiting current is some 10−20% lower than in acid solutions. In buffers containing boric acid this decrease is even more pronounced. The diffusion character of wave i_1 was proved by linear concentration dependence, by proportionality to the square root of mercury head and by the temperature coefficient[7] of 1·8% grad^{-1}.

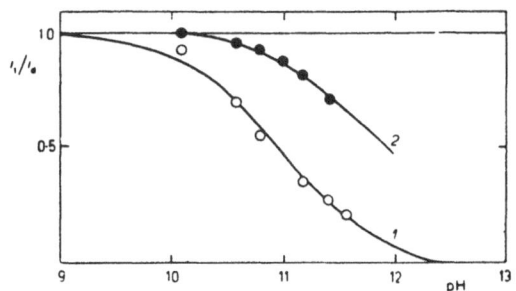

Fig. 17

pH-Dependence of the Limiting Current (as a Fraction of the Diffusion Current i_1/i_d) of the Current i_1 (1) and $(i_1 + i_2)$ (2)

Pyridoxal-5-phosphate in Britton-Robinson buffers (22°C, $t_1 = 2\cdot1$ s).

At pH above 10 wave i_1 decreases with increasing pH at the expense of wave i_2 in the form of a dissociation curve (Fig. 17). At pH above 10·6 the sum of waves $(i_1 + i_2)$ decreases further with increasing pH-value and a new wave at more negative potentials i_3 is formed. The sum of $(i_1 + i_2 + i_3)$ remains practically constant (in Fig. 16 a maximum of the second kind complicates the comparison, the horizontal line gives the height of the total limiting diffusion current). At pH 11·5 both wave i_1 and the current $(i_1 + i_2)$ possess a kinetic character, as has been proved by their independence of the mercury pressure. The height of the limiting current i_1 and of $(i_1 + i_2)$ depends on the kind and concentration of buffer used (at a given pH). The half-wave potentials of wave i_1 are shifted towards more negative potentials with increasing pH-value (Fig. 15). Half-wave potentials for waves i_2 and i_3 are practically pH independent (Table IV).

The polarographic waves for phosphate V can be ascribed to the reduction of the aldehyde group in the whole pH-range studied. For 4-formyl pyridine the reduction occurs in two one-electron steps[4]. The single two-electron wave observed for com- pound V can be caused by the presence of the 3-hydroxy group allowing the contri- butions of the quinone-methine forms[6]. Similarly ortho-hydroxy groups cause

a two-electron reduction of benzaldehydes[65] in acid solutions. — The reduced aldehydic group is either practically unhydrated or the dehydration occurs extremely
rapidly. The decrease of the total limiting current observed with formyl pyridines[4]
and with pyridoxal[4] is absent here. The probable reasons for this were discussed
on p. 1448.

Table IV

Half-Wave Potentials of Pyridoxal-5-phosphate (V)
in Dinatrium and Trinatrium Phosphate Mixtures (vs. s.c.e)

pH	$E_{1/2}$, V		
	i_1	i_2	i_3
8·9	-0.88_5	—	—
10·1	-0.99_5	-1.13_5	—
10·8	-1.03	-1.14	—
11·2	-1.05	-1.14_5	-1.39
11·6	-1.05_5	-1.15	-1.37
11·8	—	-1.16	-1.35

The change in the wave-height with the pH-value shows that at least three differently protonised forms are reducible polarographically. The kinetic character of
these currents shows that the particular protonised forms are formed by recombination. The homogeneity or heterogeneity[36-39] of this reaction was not resolved, but
the effect of buffer composition seems to show that proton donors other than hydroxonium ion can be involved in the recombination reaction.

Since a similar pH-dependence of the wave-heights and occurrence of three waves
i_1, i_2 and i_3 has not been observed with pyridoxal it can be assumed that in the observed acid-base reactions the phosphoric acid part is involved. But the assignment
of the structure to the three dissociation steps involved from the at least five forms
that can exist[11] exceeds our present possibilities. The two dissociation curves with
$pK_1' = 10.9$ and $pK_2' = 11.9$ for $t_1 = 2.1$ s given in Fig. 17 possess the same slope
corresponding to a single-proton transfer in monobasic acids. Hence equation (2)
was used in the computation of curves in Fig. 17. According to the dicsussion given
on p. 1438 this would assume that pK_2 would be greater than pK_1' (i.e. 10·9). As none
of the pK values for substance V was recorded (Table I) in this pH-range, a detailed
discussion of the electroactive forms is impossible.

Pyridoxaloxime (VII). At a pH below 9 oxime *VII* is reduced in one four-electron
diffusion controlled wave (i_1). The number of electrons transferred was proved
by comparison with the wave for pyridoxal-5-phosphate. At higher pH-values, in
addition to the wave i_1, two other waves i_2 and i_3 are observed (Fig. 18). The height
of wave i_1 at a pH below 9 is pH-independent, turbulent maxima at lower pH-values

can be suppressed by addition of 0·005% gelatin or dextran. The pH-dependence of the half-wave potentials of i_1 is given in Fig. 19 and of all three waves in Table V. The half-wave potential for i_1 is shifted about 65 mV/pH, and those for the waves i_2 (−1·25 V) and i_3 (−1·50 V) are practically pH-independent.

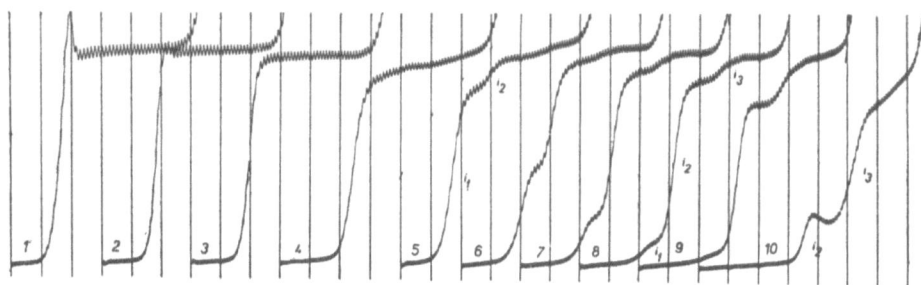

Fig. 18

pH-Dependence of Waves for Pyridoxaloxime

$2·5 . 10^{-4}$M pyridoxaloxime, pH: *1* 3·6; *2* 6·3; *3* 9·0; *4* 9·9; *5* 10·4; *6* 10·8; *7* 11·0; *8* 11·2; *9* 11·4; *10* 11·9. Curve *1* starting at −0·33 V, *2* at −0·5 V, *3−4* at −0·66 V, *5−10* at −0·82 V, mercurous sulphate electrode, 165 mV/absc., full scale sensitivity 4·06 μA.

At a pH above 9 the height of wave i_1 decreases in the form of a dissociation curves at the expense of wave i_2. So does the current $(i_1 + i_2)$ at the expense of wave i_3 (Fig. 18 and 20). Whereas the total limiting current is always diffusion controlled, waves i_1 and i_2 are kinetic currents and are independent of mercury press-

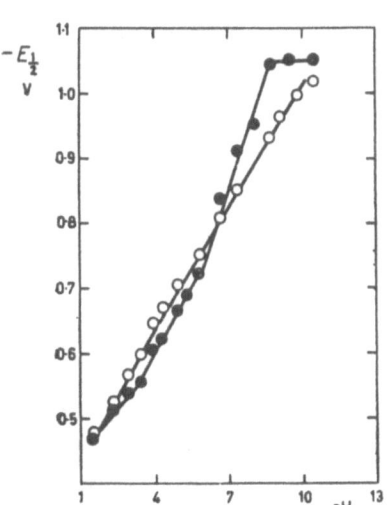

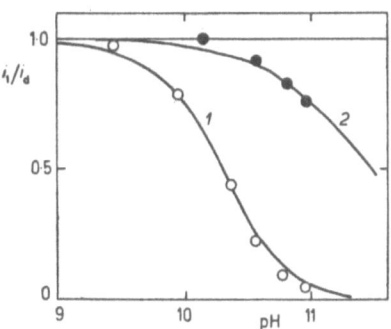

Fig. 19

pH-Dependence of Half-Wave Potentials (*vs.* s.c.e.) for Pyridoxaloxime (●) and Pyridoxaloxime-5-phosphate (○)

Fig. 20

pH-Dependence of Wave-Height of Limiting Currents i_1 (*1*) and $(i_1 + i_2)$ (*2*) for Pyridoxaloxime

Table V

Half-Wave Potentials of Pyridoxaloxime *(VII)* in Dinatrium and Trinatrium Phosphate
Mixtures (vs. S.C.E.)

pH	$E_{1/2}$,V		
	i_1	i_2	i_3
9·9	−0·98	—	—
10·4	−1·02	−1·22	—
10·8	−1·06	−1·23	−1·49
11·0	−1·06	−1·24	−1·50
11·2	−1·06	−1·25	−1·50
11·4	—	−1·26	−1·51

ure when they are lower than about 25% of the total height. In veronal buffers of pH 8 to
9 in solutions of oxime *VII* in addition to the four-electron wave another two-electron
wave at more negative potentials is observed (Fig. 21). The half-wave potentials, and
their pH-dependence as well as the pH-dependence of the limiting current for the
more negative wave of pyridoxaloxime is almost identical with the corresponding
values for pyridoxamine. On oscillographic $dE/dt = f(E)$ curves two irreversible
cathodic incisions were observed. The more negative incision is increased after addition of pyridoxamine.

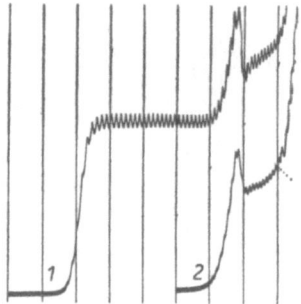

Fig. 21
Comparison of Waves of Pyridoxaloxime (*1*) and Pyridoxamine (*2*)
$5 . 10^{-4}$M Depolarizer in 0·15M veronal buffer pH 8·7. Curve *1*
starting at −0·66 V, *2* at −1·49 V, mercurous sulfate electrode, 165 mV/absc., full scale deflection 8·02 μA.

Similar to pyridine aldoximes[3] the four-electron reduction process is assigned
to the reduction of the C=NOH group to C—NH₂. The coincidence of the more
negative wave with that for the corresponding CH_2NH_2-derivative is for pyridoxaloxime (Fig. 21) even closer than for pyridine aldoximes[3].

The presence of two dissociation curves, the first of which is unsymmetrical,
indicates (p. 1438) a system of dibasic acid for which pK_2 does not differ substantially
from pK_1'. Experimentally found value $pK_1' = 10·3$ suggests that as pK_2 the value
10 (pK_5 in Table I) and for pK_1 the value 8·1 (pK_4 in Table I) can be considered.

Using these data for $t_1 = 2\cdot1$ s the theoretical curves for i_1/i_4 (curve *1*, Fig. 20) using equation (*1*) and for $(i_1 + i_2)/i_4$ with $pK'_2 = 11\cdot5$ and $t_1 = 1\cdot8$ s using a form analogous to equation (*2*) (curve *2*, Fig. 20) were calculated and values $k_1 = 1\cdot8 \cdot 10^{13}$ $1\,mol^{-1}\,s^{-1}$ and $k_2 = 7\cdot3 \cdot 10^{12}\,1\,mol^{-1}\,s^{-1}$ computed. As values for both pK'_1 and pK'_2 are affected by buffer composition and capacity, both the effects of heterogeneity[36-40] and participation of proton donors[38-51] other than hydroxonium are to be considered as an explanation of high values of these formal constants.

It can be deduced that the dissociation step corresponding to pK_5 (Table I) results in a proton transfer on the oxime grouping. In analogy with other oximes[66] it can be assumed that the protonized form of oxime is reduced at more positive potentials than the free base. The anionic form is assumed not to be involved except at extremely high pH-values. Hence the wave i_3 can be attributed to the reduction of form *XVI* and wave i_2 to form *XVII*, whereas the wave i_1 can result in the reduction of a molecule either with undissociated OH group and unprotonised pyridine ring or with dissociated O^- group and protonised pyridine ring.

HC=NOH
$(-)O$ CH_2OH
H_3C N

XIV

H
HC=NOH
$(+)$
$(-)O$ CH_2OH
H_3C N

XVII

Pyridoxaloxime-5-phosphate (VIII). The four-electron reduction step, its pH dependence and the occurrence of three waves i_1, i_2 and i_3 for phosphate *VIII* (Fig. 22) is similar to that for oxime *VII*. The presence of the phosphoric acid part is demonstrated in the pH-dependence of the half-wave potentials of i_1 (Fig. 19, Table VI) where three linear

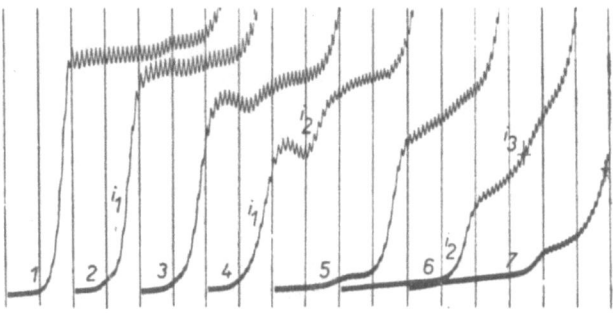

Fig. 22
pH-Dependence of the Waves for Pyridoxaloxime-5-phosphate
$2\cdot5 \cdot 10^{-4}$M Pyridoxaloxime-5-phosphate; pH: *1* 3·75; *2* 8·7; *3* 9·2; *4* 9·8; *5* 10·4; *6* 10·9; *7* 11·45. Curve *1* starting at $-0\cdot32$ V, *2—7* at $-0\cdot80$ V; mercurous sulphate electrode, 160 mV/absc., full scale sensitivity 5·06 μA.

parts with slopes 48 mV/pH (pH 1·5 − 3·5), 67 mV/pH (pH 3·5 − 6) and 98 mV/pH (pH 6 − 10) were observed. Whether $pK'_1 = 10·0$ and $pK'_2 \doteq 10·6$ also involve proton transfer on the phosphoric acid part in addition to those two proton transfer steps observed with oxime *VII*, could not be decided. In the strongly alkaline region the curves for phosphate *VIII* change with time showing chemical reactions (hydrolysis) occurring in the bulk of the solution.

Table VI

Half-Wave Potentials of Pyridoxaloxime-5-phosphate (*VIII*)
in Dinatrium and Trinatrium Phosphate Mixtures (*vs.* s.c.e.)

pH	$E_{1/2}$, V		
	i_1	i_2	i_3
9·2	−1·04	−1·36	−
9·8	−1·05	−1·36	−
10·4	−1·05	−1·36	−
10·9	−	−1·36	−1·70
11·5	−	−1·36	−1·80

The more negative wave, observed in addition to the four-electron step, is in the case of pyridoxaloxime-5-phosphate increased by a catalytic current and thus different from the poorly developed wave observed for phosphate *VI*.

Analytical Applications

The described waves have been applied for solution of the following problems: Pyridoxol in pharmaceutical applications[7], pyridoxol and its 4-methyl ether in industrial synthesis[67], pyridoxamine in the presence of its phosphate[7], pyridoxal in the presence of pyridoxal-5-phosphate from the difference of wave heights in acid media[8], from difference of half-wave potentials in alkaline media[9] and of half-wave potentials of corresponding oximes[8]. These differential determinations were used for the study of the hydrolysis of pyridoxal-5-phosphate[11] and for the determination of the activity of phosphotases of acid and alkaline type[68]. The waves were used for the study of the reaction of pyridoxal and its phosphates with amines[7] and especially with pyridoxamine[7,69] and of the reaction with hydroxylamine[12,13], enabling optimum conditions to be found for the microbiological test[6]. The detailed conditions for particular analytical procedures are given in quoted papers.

Authors express their thanks to Mr J. Weber, Polarographic Institute, Czechoslovak Academy of Sciences, Prague, for the discussion of the part on polarography of dibasic acids.

The authors are indebted to Dr F. W. Bernhart, Wyeth Laboratories, Philadelphia, USA, Prof. Dr. Hotovy, Darmstadt, Dr F. Rasching, Ludwigshafen, Germany, and to Farmakon, Olomouc, for kindly supplying specimens of substances IX, III, IV, XIII, II and XII.

References

1. Commission on the Nomenclature of Biological Chemistry: J. Am. Chem. Soc. *82*, 5575 (1960).
2. Lingane J. J., Davis O. L.: J. Biol. Chem. *137*, 567 (1941).
3. Volke J., Kubíček R., Šantavý F.: This Journal *25*, 871 (1960).
4. Volke J.: Z. physik. Chem. Sonderheft *1958*, 268.
5. Volke J., Valenta P.: This Journal *25*, 1958 (1960).
6. Černá J., Manoušek O.: Folia Microbiol. *5*, 23 (1960).
7. Manoušek O.: *Thesis.* Czechoslovak Academy of Sciences, Prague 1961.
8. Manoušek O., Zuman P.: J. Electroanal. Chem. *1*, 324 (1960).
9. Manoušek O., Zuman P.: This Journal *27*, 486 (1962).
10. Manoušek O., Zuman P.: Biochem. Biophys. Acta *44*, 393 (1960).
11. Zuman P., Manoušek O.: This Journal *26*, 2134 (1961).
12. Manoušek O., Zuman P.: Experientia, in the press.
13. Zuman P., Manoušek O.: This Journal, in the press.
14. Williams V. R., Neilands J. B.: Arch. Biochem. Biophys. *53*, 56 (1960).
15. Vogel H., Knobloch H.: *Chemie und Technik der Vitamine*, Vol. II, *Die wasserlöslichen Vitamine*, Part II, p. 279. F. Enke, Stuttgart 1957.
16. Heyl D.: J. Am. Chem. Soc. *70*, 3434 (1948).
17. Wendt G., Bernhart F. W.: Arch. Biochem. Biophys. *88*, 270 (1960).
18. Schmidt U., Giesselmann G.: Angew. Chem. *72*, 709 (1960).
19. Kemula W.: *Advances in Polarography*, Proc. IInd Intern. Congr. Polarography, Cambridge 1959, Vol. I, p. 105. Pergamon Press, London 1960.
20. Březina M.: *Advances in Polarography*, Proc. IInd Intern. Congr. Polarography, Cambridge 1959, Vol. III, p. 933. Pergamon Press, London 1960.
21. Zuman P., Wawzonek S.: *Trends in Organic Polarography*, in (P. Zuman, I. M. Kolthoff, Eds), *Progress in Polarography*, Vol. I, p. 303. Interscience, New York 1962.
22. Kabasakalian P., McGlotten J.: Anal. Chem. *31*, 1091 (1959).
23. Lund H.: Acta Chem. Scand. *14*, 1927 (1960).
24. Wawzonek S., Frederickson J. D.: J. Electrochem. Soc. *106*, 325 (1959).
25. Fedoroňko M.: Chem. zvesti *12*, 17 (1958).
26. Hanuš V., Brdička R.: Chem. listy *44*, 291 (1950).
29. Koutecký J.: This Journal *19*, 1093 (1954); Chem. listy *48*, 360 (1954).
28. Wiesner K.: Z. Elektrochem. *49*, 164 (1943).
29. Brdička R., Wiesner K.: This Journal *12*, 39 (1947).
30. Wiesner K.: Chem. listy *41*, 6 (1947).
31. Koutecký J.: This Journal *18*, 597 (1953); Chem. listy *47*, 323 (1953).
32. Zuman P., Horák V.: *Advances in Polarography*, Proc. IInd Intern. Congr. Polarography, Cambridge 1959, Vol. III, p. 804. Pergamon Press, Oxford 1960.
33. Zuman P., Horák V.: This Journal *26*, 176 (1961).
34. Zuman P.: This Journal *20*, 876 (1955); Chem. listy *48*, 1025 (1954).
35. Březina M., Zuman P.: *Polarography in Medicine, Biochemistry and Pharmacy*, p. 470. Interscience, New York 1958, where previous references are given.
36. Koryta J.: Z. Elektrochem. *64*, 23 (1960).
37. Grabowski Z. R., Bartel E. T.: Roczniki Chem. *34*, 611 (1960).
38. Strehlow H.: Z. Elektrochem. *64*, 45 (1960).
39. Majranovskij S. G., Liščeta L. I.: This Journal *25*, 3025 (1960).
40. Volková V.: Nature *184*, 743 (1960).
41. Hanuš V.: *Proc. 1st Intern. Polarograph. Congr. Prague 1951*, Part I, p. 881. Published by Přírodovědecké nakladatelství, Prague 1951.

304

42. Wiesner K., Wheatley M., Los J. M.: J. Am. Chem. Soc. *76*, 4858 (1954).
43. Green J. H., Walkley A.: Australian J. Chem. *8*, 51 (1955).
44. Zahradník R., Svátek E., Chvapil M.: This Journal *24*, 347 (1959); Chem. listy *51*, 2232 (1957).
45. Bartel E. T., Grabowski Z. R., Kemula W., Turnowska-Rubaszewska W.: Roczniki Chem. *31*, 13 (1957).
46. Hrdý O.: This Journal *24*, 1180 (1959); Chem. listy *52*, 1058 (1958).
47. Zuman P.: J. Electrochem. Soc. *105*, 758 (1958).
48. Kemula W., Grabowski Z. R., Bartel E. T.: Roczniki Chem. *33*, 1125 (1959).
49. Becker M., Strehlow H.: Z. Elektrochem. *64*, 42 (1960).
50. Wheatley M. S.: Experientia *12*, 339 (1956).
51. Hans W., Henke K. H.: Z. Elektrochem. *57*, 595 (1953).
52. See. 35, p. 517, where further references are given.
53. Zuman P.: Unpublished results.
54. Frumkin A. N., Nikolajeva-Fedorovič N.: *The Electroreduction of Anions* in: P. Zuman, I. M. Kolthoff, Eds: *Progress in Polarography*. Vol. I, p. 223. Interscience, New York 1961; with references to previous papers.
55. Frumkin A.: Nova Acta Leopoldina, Neue Folge *19*, 132 (1957).
56. Frumkin A.: Trans. Faraday Soc. *55*, 156 (1959).
57. Zuman P.: Chem. zvesti *8*, 789 (1954), with references to previous publications.
58. Harris S.: J. Am. Chem. Soc. *62*, 3203 (1940).
59. Clemo G. R., Hoggarth E.: J. Chem. Soc. *1941*, 41.
60. Diels O., Alder K.: Ann. *505*, 103 (1933).
61. Lukeš R., Ernest I.: This Journal *15*, 107 (1950).
62. Hine J.: *Physical Organic Chemistry*, p. 167. McGraw-Hill, New York 1956.
63. Metzler D. E., Snell E. E.: J. Am. Chem. Soc. *77*, 2431 (1955).
64. Heyl D., Luz E., Harris S. A., Folkers K.: J. Am. Chem. Soc. *73*, 3430 (1951).
65. Holleck L., Marsen H.: Z. Elektrochem. *57*, 944 (1953).
66. Souchay P., Ser S.: J. Chim. Phys. *49*, C 172 (1952).
67. Manoušek O., Kočová P.: Mikrochim. Acta *1961*, 754.
68. Manoušek O., Kučerová Z.: Unpublished results.
69. Manoušek O.: Naturwiss. *46*, 323 (1959).

Translated by the author (P. Z.).

POLAROGRAPHIC REDUCTION OF 2-(2-METHYL-3-HYDROXY-5-HYDROXYMETHYL-4-PYRIDYL)THIAZOLIDINE-4-CARBOXYLIC ACID

O. Manoušek and P. Zuman

Polarographic Institute, Czechoslovak Academy of Sciences, Prague

Received November 4th, 1963

In the connection with the systematic polarographic study of pyridoxine derivatives[1] the polarographic behaviour of the condensation product of pyridoxal with cystein, *i.e.* of 2-(2-methyl-3-hydroxy-5-hydroxymethyl-4-pyridyl)thiazolidine-4-carboxylic acid[2] (*I*) was studied. The waves of this substance were interpreted using comparison with waves of pyridoxine derivatives and by the means of controlled potential electrolysis.

Experimental

Polarographic equipment and procedures used in this study were the same as used in previous work[1]. Controlled potential electrolysis was carried out according to Lingane[3] with a manual control, using two polarographs: one as a source of applied potential and the second for control and for recording of polarographic curves in the course of the electrolysis (Fig. 1). Mercury pool electrode used had a surface of $10 \ cm^2$, saturated calomel electrode (*C*) was used for monitoring. First polarographic curve is recorded using polarograph *I* with the dropping mercury electrode *D* and calomel electrode *C* (switch *S* in position *1*). Appropriate potential (usually corresponding to the limiting current) is chosen and the sliding contact of polarograph *I* is set to this potential. Voltage is applied to working electrode *E* and platinum electrode Pt using polarograph *II* as a source. Voltmeter *V* measures voltage-span between *E* and Pt. Potential difference between *E* and *C* should be identical with the potential difference between *D* and *C*. This is achieved by setting the potentiometer of polarograph *II* until the E.M.F. branched off of this polarograph is equal to that branched off of polarograph *I*. This compensation is carried out in position *2* of the switch *S* using galvanometer *G* of polarograph *I* as zero indicator. Current flowing between Pt and *E* is approximately read on milliammeter mA. Using this apparatus and stirring with nitrogen the electrolysis was carried out during 80 minutes, the potential was controlled to ± 20 mV.

2-(2-Methyl-3-hydroxy-5-hydroxymethyl-4-pyridyl)thiazolidine-4-carboxylic acid (I) prepared according to Heyl and co-workers[2] was twice recrystallized from ethanol, m.p. $148-149°C$ (ref.[2] $149-150°C$).

$5 \cdot 10^{-3}$M aqueous stock solution of substance *I* was prepared. Only fresh solutions were used, as hydrolysis occurs in aqueous solutions. Acetate, phosphate and barbital buffers were prepared in concentration 0·2M from reagent grade chemicals. The pH was measured with reproducibility of ± 0.03 pH.

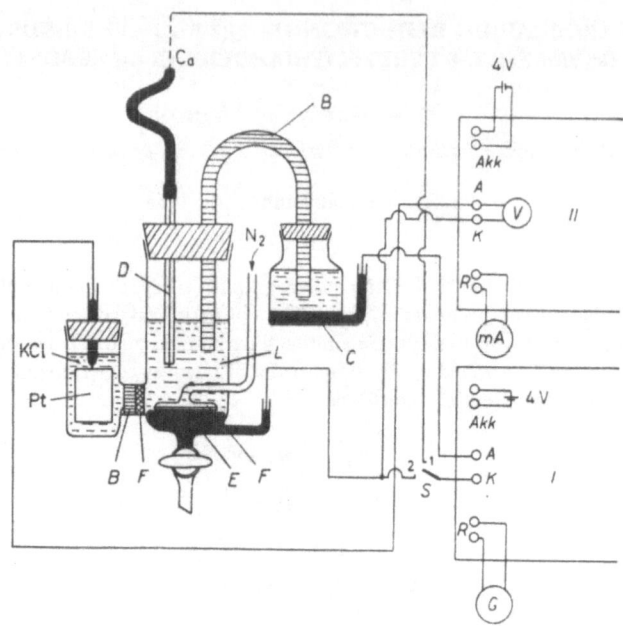

Fig. 1

Scheme of Apparatus for Controlled Potential Electrolysis

I Control polarograph; *II* polarograph used as source of E.M.F.; *A, K, R, Akk* the connections on the particular polarograph; *V* voltmeter; *G* galvanometer; *mA* milliammeter; *B* agar bridge; *C* s.c.e.; *D* mercury dropping electrode with link C_a; *E* mercury pool electrode; *F* sintered glass of density *G 1*; N_2 gas inlet; *KCl* saturated potassium chloride solution; *L* electrolysed solution; *Pt* platinum anode; *S* switch.

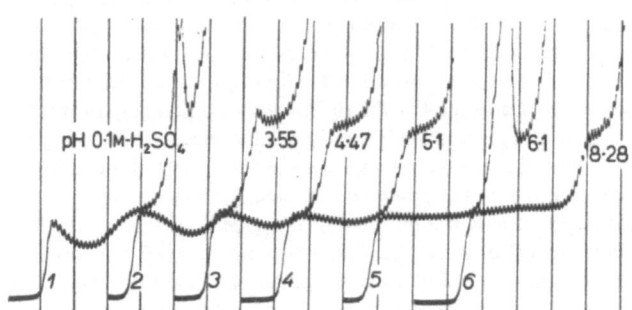

Fig. 2

pH-Dependence of Waves of 2-(2-Methyl-3-hydroxy-5-hydroxymethyl-4-pyridyl)thiazolidine-4-carboxylic Acid

$4.5_5 . 10^{-4}$M depolarizer, curves in: *1* 0.1M-H_2SO_4; *2 — 4* acetate buffer, *5* and *6* phosphate–veronal buffer. pH Values given on polarogram. Curves starting at -0.56 V (*1*), -0.75 V (*2* to *4*), and -0.94 V (*5* and *6*), 188 mV/absc., mercurous sulphate electrode, full scale deflection 7.3 μA.

Results and Discussion

Substance *I* is reduced in the pH-range from pH 1 to 9 in two two-electron steps i_3 and i_4 (Fig. 2, 4). At higher pH-values hydrolysis occurs and wave of pyridoxal[1] appears on polarographic curves increasing with time. In more acidic media the more positive wave shows a decrease (trough) in the potential range of the limiting current. The trough is deeper with increasing concentration of sulphuric acid (Fig. 3). Due to the great change in ionic strength an effect due simply to repulsion from the electrode surface caused by the effect of ψ-potential in the double-layer can be excluded. The half-wave potentials are shifted linearly with the slope 60 mV/pH towards more negative potentials with increasing pH values, selected values being given in Table I. An anodic wave i_1 corresponding to mercury salt formation was observed (Fig. 4, curve *1*).

The course of the reduction has been proved by controlled potential electrolysis in a phosphate buffer of pH 5.8_3 at potential -0.95 V (s.c.e.), *i.e.* at the potential of the limiting current of

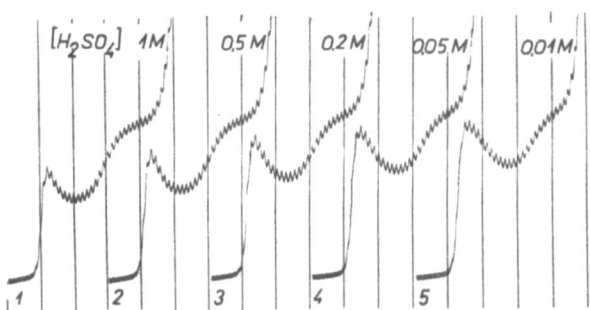

Fig. 3

Waves of Substance *I* in Sulphuric Acid Solutions

$4.5_5 . 10^{-4}$M depolarizer, concentration of sulphuric acid given on the polarogram. Curves starting at -0.65 V, 188 mV/absc., mercurous sulfate electrode, full scale sensitivity 4.9 μA.

Table I

Half-Wave Potentials (in volts, *vs.* s.c.e.) of Substance *I*

pH	Current		
	i_1	i_3	i_4
3.5_5	—	-0.52	-1.09
5.1	$-0.20_5{}^a$	-0.62	-1.23
7.0_5	—	-0.74_5	-1.38_5

^{a} pH 5.8.

wave i_3. During electrolysis wave i_3 decreases whereas the height and half-wave potential of wave i_4 remains practically constant. Anodic wave i_1 also decreases during electrolysis; another more negative wave i_2 at -0.39 V (pH 5·8) increases at the expense of wave i_1 (Fig. 4).

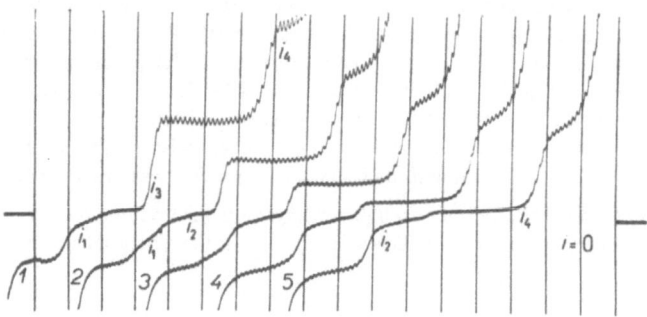

Fig. 4

Curves Recorded During the Controlled Potential Electroreduction of Substance *I*

$5 . 10^{-4}$M depolarizer, phosphate buffer pH 5·8; the reduction was carried out at -0.95 V (*vs.* s.c.e.). Duration of electrolysis: *1* 0; *2* 10; *3* 25; *4* 50 and *5* 80 minutes. Anodic-cathodic polarization, 188 mV/absc., s.c.e., full scale sensitivity 7·3 µA.

The anodic wave i_2 is analogous to that formed in solutions of cysteine that corresponds to a mercury salt formation. As such anodic wave was not observed with S-alkyl cysteine derivatives, it is deduced that in the electrolysis of compound *I* N-pyridoxyl cysteine (*II*) is formed which contains a free SH-group. Hence the first cathodic wave i_3 of acid *I* is attributed to the hydrogenolytic fission of the C—S bond. The second wave i_4 oughts to be attributed to hydrogenolysis of the C—N bond in substance *II* either formed at the surface of the electrode electrolytically from substance *I* or prepared by controlled potential electrolysis. Hence the course of the reduction can be depicted as follows:

$$\begin{array}{ccc}
\begin{array}{c} CH_2\text{---}CH\text{---}COOH \\ | \qquad | \\ S \qquad NH \\ \diagdown \diagup \\ CH \\ | \\ R \end{array}
&
\xrightarrow[i_3]{2e, 2H^+}
&
\begin{array}{c} CH_2\text{---}CH\text{---}COOH \\ | \qquad | \\ SH \quad NH \\ | \\ CH_2 \\ | \\ R \\ II \end{array}
\end{array}$$

$$\xrightarrow[i_4]{2e, 2H^+}
\begin{array}{c} CH_2\text{---}CH\text{---}COOH \\ | \qquad | \\ SH \quad NH_2 \\ \\ + \ CH_3 \\ | \\ R \end{array}$$

where R =

$$\text{HO}\diagdown\!\!\!\bigcirc\!\!\!\diagup\text{CH}_2\text{OH}$$

(pyridoxyl ring: HO and CH₂OH substituents, H₃C and N on ring)

Contrary to some recent statements that in the successive reductions $A \xrightarrow{n_1 e} B \xrightarrow{n_2 e} C$ an alternative route $A \xrightarrow{n_3 e} C'$ (where $n_3 = n_1 + n_2$) cannot be excluded, the constant value of the half-wave potential of wave i_4 in substances *I* and *II* during electrolysis in Fig. 4 strongly suggests a two step process even at negative potentials of wave i_4.

Finally a polarographic curve in the solution of substance *I* was compared with a curve in a solution containing a mixture of 4-pyridoxthiol and pyridoxamine (Fig. 5). In both cases C—S and C—N bonds activated by pyridine ring[1] are splitted, in the former bound on the same carbon atom in one molecule, in the latter in two different molecules. The hydrogenolysis of the C—S

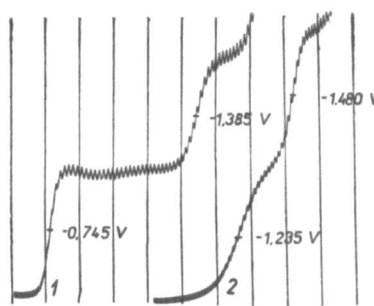

Fig. 5

Comparison of Polarographic Curve of Compound *I* with Curve of a Mixture of 4-Pyridoxthiol and Pyridoxamine

$0.2M$ phosphate buffer pH 7·1; *1* $4.5_5 . 10^{-4}M$ substance *I*; *2* a mixture of $3.6 . 10^{-4}M$ 4-pyridoxthiol and $3.6 . 10^{-4}M$ pyridoxamine. Curves starting at -0.75 V (*1*) and -1.05 V (2), 151 mV/absc., mercurous sulfate electrode, full scale sensitivity 4·9 μA.

(wave i_3) bonds in compound *I* occurs at substantially more positive potentials than in 4-pyridoxthiol. This can be due both to the inductive effect of the nitrogen bound on the same carbon and due to the ring formation. The shift of the wave i_4 corresponding to hydrogenolysis of the C—N bond in N-pyridoxylcysteine (*II*) to more positive potentials when compared with pyridoxamine can be due to the inductive effect of sulphur in β-position.

A preliminary study of 2-(4-pyridyl)thiazolidine-4-carboxylic acid has shown that polarographic reduction of this substance follows practically the same path.

References

1. Manoušek O., Zuman P.: This Journal *29*, 1432 (1964).
2. Heyl D., Stanton A., Harris A., Folkers K.: J. Am. Chem. Soc. *70*, 3429 (1948).
3. Lingane J. J.: Discussions Faraday Soc. *1*, 203 (1947).
4. Frumkin A. N., Nikolajeva-Fedorovič N. in the book: *Progress in Polarography* (P. Zuman, I. M. Kolthoff, Eds), Vol. I, p. 223. Interscience, New York 1962.

Translated by the author (P. Z.).

FISSION OF ACTIVATED CARBON-NITROGEN AND CARBON-SULPHUR BONDS. X.*

POLAROGRAPHIC REDUCTION OF SUBSTITUTED METHYL PHENYL SULPHONES

O.Manoušek, O.Exner and P.Zuman**

J. Heyrovský Institute of Polarography, Czechoslovak Academy of Sciences, Prague 1

Received January 4th, 1968

Reduction of twenty-six *meta*- and *para*-substituted methyl phenyl sulphones at the dropping mercury electrode and at a mercury pool electrode was investigated and by identification of reduction products was decided which of the two functional groupings attached to the benzene ring is reduced. With most derivatives bearing an electron attracting group as a substituent, the methyl sulphonyl group leaves as methane sulphinate anion. A pH-independent, two-electron wave is observed between pH 7 and 14. The reaction represents a kind of aromatic S_N2 substitution and is strongly facilitated by conjugation in the transition state with an acceptor group in the *para*-position. The effect of *para*-substituents is even greater than expressed by the values of σ_{p-x}^{-} substituent constants.

Whereas dialkyl sulphones do not show polarographic reduction waves at potentials more positive than -2.4 V, diaryl and alkyl aryl sulphones undergo[1-6] a polarographic reduction, which for the benzene derivatives without substituents on the phenyl ring takes place from -2.0 V to -2.3 V. The reduction proceeds[2,4-6] according to the Scheme (*A*):

$$ArSO_2R + 2e + H^{(+)} \rightarrow ArSO_2^{(-)} + RH \qquad (A)$$

To prove this reduction scheme, it was shown[2,4-6] that the reduction waves correspond to a two-electron process. The formation of the arenesulphinic acid has been detected[5,6] in the products of controlled potential electrolysis, using UV-spectra[5], but the yield of the sulphinic acid when a mercury pool electrode was used was smaller than theoretical[6]. The number of protons consumed in reaction (*A*) was determined[5] titrimetrically after the controlled potential electrolysis carried out in unbuffered media. The originally proposed[1] reduction scheme indicated the formation of aryl sulphides and anticipated the formation of sulphoxides as intermediates. In addition to the facts mentioned above this four-electron reduction scheme was also ruled out on the evidence[5] that the corresponding sulphoxides are reduced at potentials slightly more negative than

---

* Part IX: This Journal *31*, 3889 (1966).

** Present address: Department of Chemistry, University of Birmingham, Great Britain.

sulphones. The reduction leading to sulphides would therefore anticipate formation of two two-electron waves which has not been observed on the polarographic curves of sulphones.

The half-wave potentials of aryl alkyl sulphones depend on the nature of the alkyl group and the aryl ring and on substituents in this ring. Inductive effects of groups R in $C_6H_5SO_2R$ proved to be decisive for the shifts of half-wave potentials. For half-wave potentials of sulphones for which R = alkyl or aryl the modified Taft equation can be applied[7] using the reaction constant $\varrho^*_{\pi,C-SO_2} = 0.21$ V and $(E_{1/2})_{CH_3}$ for the phenyl methyl sulphone equal to -2.27 V, in a supporting electrolyte of $0.05M$-$N(CH_3)_4Cl$ containing 75% dioxane.

In the present paper the effect of substituents in the phenyl ring on the half-wave potentials was studied. To avoid the use of tetraalkylammonium salts and their specific effects, we had to study phenyl methyl sulphones bearing electronatracting substituents that were able to shift the waves to such positive potentials that it was possible to use aqueous solutions containing sodium or lithium ions. The choice of substituents was further limited by the fact that the substituent must be electroinactive in the given potential range. This excluded the possibility of studying effects of nitro or benzoyl groups.

Reduction of disubstituted benzene derivatives bearing two electron attracting groups which in principle can both undergo reduction, can take place in two possible sequences. The observed sequence does not necessarily follow the sequence observed for monosubstituted derivatives and depends on the mutual interaction of both substituents. Generally, the reduction of an electronattracting group is facilitated in the presence of a second electron attracting group. This mutual interaction and the sequence of the reducibility has been recently studied by us in a series of *meta*- and *para*-substituted benzene derivatives. The investigations reported in this paper were restricted to derivatives in which one reducible electroactive group was the methyl sulphonyl group.

In the studies reported[1−6] the course of the electroreduction was studied for the symmetrical diphenyl sulphone. In the present study it was possible to confirm which of the two C—S bonds in aryl alkyl sulphones undergoes a hydrogenolytic cleavage.

EXPERIMENTAL

Apparatus

The polarographic curves were recorded using a pen-recording polarograph Mark LP 60 (Laboratorní přístroje, Prague). For the study of the polarographic behaviour a Kalousek vessel with a separated saturated calomel electrode (s.c.e.) was used. For the measurement of half-wave potentials using the three-electrode system, an adapted Kalousek vessel was used. The capillary used had a drop-time of $t_1 = 3.6$ s and an out-flow velocity of mercury $m = 2.5$ mg/s at the potential of the s.c.e. in $1M$-KCl. The pH values of the buffer solutions were checked using a pH-meter Mark PHM 4 (Radiometer — København) with a glass electrode type G 200 B. For measurement of half-wave potentials in a potentiometric arrangement, a compensator Mark QTK (Metra, Blansko) was used. The potentiostat used was developed in the J. Heyrovský Institute of Polarography, Czechoslovak Academy of Sciences by J. Holub and L. Němec. It enables the electrolysis to be carried out with a maximum current of 600 mA, against an iR drop of less than 80 V and keeps the voltage in the range $+2$ V to -2 V constant to ± 1 mV. The time-constant of the instrument is 100 μA.

For measurements of the potential of the mercury pool electrode, a universal electronic volt-meter Mark BM 388 (Tesla, Brno) was used. Chromatographic measurements were carried out with Fractovap, Mark C (Carlo Erba, Milano), with an 80 cm capillary column, using ethylene glycol adipate as stationary phase. The working temperature was 155°C.

Substances

Meta- and *para*-substituted derivatives of phenyl methyl sulphone studied are summarized in Table I together with m.p. (corr.). The sulphones *I—XXVI* were prepared partly according to the procedures described in the literature[8−14] or by application of known methods. The synthetic procedures adopted were chosen mostly in such a way as to avoid formation of bivalent sulphur compounds. For this reason some known compounds were prepared using new routes. No precautions were taken to secure the highest yields or to find optimal reaction conditions. Samples for analysis were dried over phosphorus pentoxide at 0·05 mm Hg at 20°C.

3-Methylsulphonylbenzoyl chloride. Boil 3-methylsulphonylbenzoic acid (4·0 g, 0·02 mol) with thionyl chloride and a drop of dimethylformamide. Distill off the excess of thionyl chloride at reduced pressure and recrystalize the crude product (4·37 g, 100%) from a mixture of benzene and cyclohexane; m.p. 84°C. For $C_8H_7ClO_3S$ (218·7) calculated: 43·95% C, 3·23% H; 16·21% Cl; found: 44·20% C, 3·29% H; 16·05% Cl.

3-Methylsulphonylbenzophenone (II). A solution of 3-methylsulphonylbenzoyl chloride (1·09 g, 0·005 mol) in 15 ml of benzene was refluxed with 1·4 g (200%) of anhydrous aluminium chloride for 1 hour. The reaction mixture was decomposed with water and hydrochloric acid, and the benzene layer was separated off. The aqueous layer was extracted with small portions of benzene and the combined benzene extracts were taken to dryness. The crude product thus obtained was boiled with 10% sodium carbonate solution and washed with water; yield 1·08 g (83%), m.p. 90°C (diluted ethanol). For $C_{14}H_{12}O_3S$ (260·3) calculated: 64·59% C, 4·65% H, 12·32% S; found: 64·54% C, 4·70% H, 12·25% S.

4-Methylsulphonylbenzophenone (III). This compound was prepared by exactly the same procedure as in the preceding paragraph starting from 4-methylsulphonylbenzoyl chloride[10]; yield 0·99 g (76%), m.p. 142°C (ethanol). In the literature[15] m.p. 141°C is given for a product obtained in a more complicated way. For $C_{14}H_{12}O_3S$ (260·3) calculated: 64·59% C, 4·65% H, 12·32% S; found: 64·31% C, 4·67% H, 11·94% S.

3-Methylsulphonylbenzoic acid (IV). Warm a solution of 26 g of sodium sulphite (heptahydrate) in 70 ml of water to 80°C. Add successively with vigorous stirring alternatively 3-carboxybenzene-sulpho chloride (11·0 g, 0·5 mol) and sodium hydrogen carbonate (15 g) in small portions during 20 minutes to keep the reaction alkaline to lithmus. Stir for another 10 minutes, and cool the reaction mixture to 0°C. Add hydrochloric acid to precipitate the crude 3-carboxybenzenesulphinic acid and dissolve the wet precipitate in a solution of 10 g of potassium hydrogen carbonate in 30 ml of water. Add 30 ml of ethanol and 5 ml of methyl iodide and boil for three hours. Distil ethanol off at reduced pressure and dilute the residue with water. By acidifying 8·22 g (82%) of the product (m.p. 235°C, ethanol) was precipitated.

We were not able to reproduce the procedure[9], according to which the reduction is carried out in acetone solution and the sequence of the addition of the components is reversed. Under these conditions a mixture of compounds was obtained, the disulphone being probably the principal product.

Methyl 3-methylsulphonylbenzoate (VI). Boil the crude 3-methylsulphonylbenzoyl chloride (1·09 g, 0·005 mol) with 5 ml methanol and separate the product by cooling. Yield 0·86 g (80%), m.p. 101°C (methanol). For $C_9H_{10}O_4S$ (214·2) calculated: 50·46% C, 4·70% H, 14·97% S; found: 50·22% C, 4·84% H, 14·75% S.

Isopropyl 4-methylsulphonylbenzoate (IX). Boil the crude 4-methylsulphonylbenzoyl chloride[10] (1·09 g, 0·005 mol) with 5 ml of absolute 2-propanol; yield 0·96 g (79%), m.p. 82°C (cyclohexane). For $C_{11}H_{14}O_4S$ (242·1) calculated: 54·54% C, 5·82% H, 13·23% S; found: 54·43% C, 5·62% H; 13·25% S.

TABLE I

Characteristics and Half-Wave Potentials of Substituted Aryl Methyl Sulphones ($X-C_6H_4SO_2CH_3$)

No	X	M.p. (corr.)	Ref.	$E_{\frac{1}{2}}{}^a$	n^b	σ_{m_1p}	σ_{p-}
I	H	88	8	—		0	
II	3-COC_6H_5	90	Exp.	$-1·212^e$		0·34	
III	4-COC_6H_5	142	Exp.	$-1·155^e$		0·42	
IV	3-$COOH$	235	Exp.	—		0·35	
V	4-$COOH$	267	9,10	$-1·783$		$\approx 0·0^g$	$\approx 0·3^g$
VI	3-$COOCH_3$	101	Exp.	$-1·726^f$		0·36	
VII	4-$COOCH_3$	119	10	$-1·523$	2·2	0·43	0·76
VIII	4-$COOC_2H_5$	94	11	$-1·520$	2·05	0·43	0·76
IX	4-$COOC_3H_7$-i	82	Exp.	$-1·531$		0·43	
X	3-$CONH_2$	177	Exp.	$-1·792^f$	2·2	0·36	
XI	4-$CONH_2$	224	12	$-1·566^f$	2·15	$0·43^h$	$\approx 0·63$
XII	3-CN	105	9	$-1·786$		0·61	
XIII	4-CN	142	9	$-1·586$	2·05	0·69	0·97
XIV	4-CH_2OH	84	10	—		$-0·01$	
XV	4-CH_2NH_2	271^c	13	—		—	
XVI	4-NH_2	134	14	—		$-0·66$	
XVII	4-$NHCOCH_3$	186	14	—		0·02	
XVIII	3-SO_3H	80^d	Exp.	—		—	
XIX	4-SO_3H	125^d	Exp.	—		—	
XX	3-SO_2NH_2	210	14	$-1·812$		0·53	
XXI	4-SO_2NH_2	244	14	$-1·632$		0·60	0·94
XXII	3-SO_2CH_3	195	Exp.	$-1·777$		0·64	
XXIII	4-SO_2CH_3	261	Exp.	$-1·598$	1·90	0·73	1·05
XXIV	3-SO_2H	108	Exp.	—		—	
XXV	4-SO_2H	180	Exp.	—		—	
XXVI	3,5-Cl_2	114	Exp.	$-1·814$		$2 \times 0·39$	

[a]In volts *vs.* s.c.e., measured in a borate buffer (pH 9·3) with 5% ethanol added; [b]number of electrons transferred; [c]hydrochloride, [d]monohydrate; [e]reduction of the carbonyl group (in 45% ethanol); [f]mixed reductions, [g] for the COO⁻ group; [h] ref.[19]. Exp. Experimental Part.

3-*Methylsulphonylbenzamide* (X). Treat the crude 3-methylsulphonylbenzoyl chloride with aqueous ammonia; yield 85%, m.p. 177°C (ethanol). For $C_8H_9NO_3S$ (199·2) calculated: 48·23% C, 4·55% H, 7·04% N; found: 48·59% C, 4·92% H, 7·04% N.

3-*Methylsulphonylbenzenesulphonic acid* (XVIII). Boil analytical grade 3-methylsulphonyl-benzenesulphochloride[16] (2·54 g, 0·01 mol) with 3 ml of water until dissolved and evaporate at a reduced pressure to dryness. Crystallize the residue (2·52 g, 99% as monohydrate) from a mixture of ethyl acetate and benzene; m.p. 80°C in a sealed capillary. Product is strongly hygroscopic. For $C_7H_8O_5S_2.H_2O$ (254·3) calculated: 33·06% C, 3·96% H, 25·22% S; found: 32·93% C, 4·05% H, 24·79% S.

4-*Methylsulphonylbenzenesulphonic acid* (XIX). Boil analytical grade 4-methylsulphonyl-benzenesulphochloride[14] with water and evaporate to dryness; yield 99% (as monohydrate), m.p. 125°C in a sealed capillary. This substance is even more hygroscopic than the meta isomer. For $C_7H_8O_5S_2.H_2O$ (254·3) calculated: 33·06% C, 3·96% H, 25·22% S; found: 32·95% C, 4·24% H, 25·21% S.

1,3-*Bismethylsulphonylbenzene* (XXII). Boil the solution of 1·10 g (0·005 mol) of the crude 3-methylsulphonylbenzenesulphinic acid (*XXIV*) and 0·50 g of sodium hydrogen carbonate in 5 ml of water and 5 ml of methanol with 1·5 g of methyliodide for 3 hours. Distil ethanol of at reduced pressure, collect the product by suction and wash with a dilute solution of sodium thiosulphate; yield 0·97 g (83%), m.p. 195°C (water). Literature[16] gives m.p. 196°C for a product prepared by oxidation of 1-methylthio-3-methylsulphonylbenzene.

1,4-*Bismethylsulphonylbenzene* (XXIII) was prepared from the crude 4-methyl sulphonyl-benzenesulphinic acid (*XXV*) in the same way as the 1,3-isomer *XXII*; yield 76%, m.p. 261°C (water). Literature[17] gives m.p. 158—260°C for a product prepared by oxidation of 1,4-bismethyl-thiobenzene.

3-*Methylsulphonylbenzenesulphinic acid* (XXIV). Add successively in small portions altern-atively 3-methylsulphonylbenzenesulphochloride[16] (2·55 g, 0·01 mol) and sodium hydrogen carbonate (2 g) during 10 minutes to a solution of 5 g of sodium sulphite (heptahydrate) in 12 ml of water at 80°C, to keep the reaction alkaline to lithmus. Stir for another 10 minutes, cool the reaction mixture, saturate with sodium chloride and precipitate the product by addition of hydro-chloric acid; yield 1·72 g (78%). The substance can be crystallized from water at 0°C by successive evaporation in a desiccator, but the losses are great (m.p. 108°C). For $C_7H_8O_4S_2$ (220·3) calcul-ated: 38·17% C, 3·66% H, 29·11% S; found: 38·06% C, 3·86% H, 29·22% S.

4-*Methylsulphonylbenzenesulphinic acid* (XXV) was prepared from 4-methylsulphonylbenzene-sulphochloride[14] in the same way as the 3-methyl isomer *XXIV*; yield 1·78 g (81%), m.p. 180°C decomp. (water, with considerable losses). For $C_7H_8O_4S_2$ (220·3) calculated: 38·17% C, 3·66% H, 29·11% S; found: 37·96% C, 3·82% H, 28·85% S.

2,6-*Dichloro-4-methylsulphonylaniline*. Add 3 ml of 30% hydrogen peroxide to a solution of 2·08 g (0·01 mol) of 4-methylsulphonylaniline[14] (*XVI*) in 25 ml of hydrochloric acid and 15 ml of water and leave for 24 hours. Terminate the reaction by heating to 50°C, and collect the pre-cipitated product. Yield 2·22 g (92%), m.p. 162°C (benzene–cyclohexane). For $C_7H_7Cl_2NO_2S$ (240·1) calculated: 35·01% C, 2·94% H, 29·53% Cl, 5·83% N; found: 34·86% C, 2·80% H, 29·75% Cl, 5·76% N.

3,5-*Dichlorophenyl methyl sulphone* (XXVI). Add successively with stirring 0·47 g (110%) of solid sodium nitrite to a solution of 1·20 g (0·005 mol) of 2,6-dichloro-4-methylsulphonylani-line in 10 ml of sulphuric acid and 5 ml water at 5°C using external cooling. After 10 minutes add 10 ml of 50% hypophosphorous acid and let the reaction mixture stand at 0°C for 24 hours. Collect by suction 0·90 g (80%) of the crude, coloured product; recrystallisation from cyclohexane gives colourless plates, m.p. 114°C. Literature[18] gives m.p. 116°C for a product prepared by oxidation of the sulphide.

Procedures and Techniques

The procedure used in the three electrode *measurement of half-wave potential*[21,22] has been described elsewhere.[23]

Determination of the number of electrons consumed in the electrode process was carried out in a small volume (0·5 ml) using a dropping mercury electrode under conditions used in classical polarography.

The technique used was a modification of the well known procedure[24,25] in which the small volume of the solution was stirred by falling drops. A drawn out capillary with a narrow orifice and an outside diameter of about 1 mm is placed centrally about 1 mm below the surface of the electrolysed solution. The number of electrons transferred was calculated from the change in concentration of the substance with time, as indicated by the change in the height of polarographic wave. The course of this time-change was always corrected for the change in the concentration of the substance resulting from diffusion towards the reference electrode or a chemical cleavage. For this purpose the change in wave heights with time without an applied external voltage was recorded. For 3—4 hours, during which some 30—40% of the original concentration was electrolysed, the decrease of current without applied voltage was always smaller than 3%. A correction for this parallel decrease in concentration was carried out graphically.

The determinations of numbers of electrons transferred were carried out in a borate buffer solution containing 5% ethanol, and $5 . 10^{-4}$M in the studied substance (prepared from an 0·01M ethanolic stock solution). The calculation was carried out using the expression: $n = k_1 t / [2 \cdot 3 k_2 VF \log (i_0/i_t)]$, where k_1 is the current in ampères corresponding to 1 cm of the wave-height, k_2 is the depolariser molar concentration corresponding to 1 cm of the wave height, t is the time in seconds, V is the volume of the electrolysed solution in litres, F is 96 500 Coulombs, i_0 is the limiting current at the time $t = 0$ (at the beginning of electrolysis), i_t the limiting current in the time $t = t$ (at the end or any chosen moment of electrolysis) (because of the ratio the currents can be measured in microamperes or in centimetres). The reproducibility of the determination of the value of n was better than $\pm 5\%$ for waves with half-wave potentials more positive than about $-1 \cdot 7$ V, and about $\pm 10\%$ for the more negative waves.

Preparative electrolysis for the identification of the electrolysis products was carried out either using the dropping mercury electrode under polarographic conditions, or using a mercury pool electrode with a large surface. The potential was kept constant using a potentiostat.

The electrolysis with a dropping mercury electrode under polarographic conditions was carried out in the vessel[9] used for the determination of the number of electrons. The electrolysis was carried out in 1 ml of a borate buffer (pH 9·3) containing 5% of ethanol and was $1 . 10^{-3}$M in the studied substance. A drawn out capillary was used, with a narrow orifice and a greater out-flow velocity (of about 4—5 mg/s) but with a normal drop-time ($t_1 = 3$ s). To prevent consecutive electrolytic processes from taking place, the potential was kept constant at a value corresponding to the half-wave potential using the polarograph LP 60 as a voltage supply. The electrolysis was carried out until the concentration of the studied substance had decreased to about 30% of the original concentration. The electrolysis time was always about 5 hours. Oxygen was removed from the solution before the beginning of electrolysis using a stream of hydrogen (which contains always less oxygen than the purest nitrogen) and the vessel was carefully sealed. During electrolysis, hydrogen was let neither through nor above the electrolysed solution to prevent losses due to the volatility of the starting material or the electrolysis products.

Mercury pool electrolysis was carried out in 25—100 ml of a borate buffer, pH 9·3, containing 10% ethanol and 1 to 5 . 10⁻³M in the electrolysed sulphone (according to the solubility of the compound). The electrolysis cell was a beaker and the mercury pool electrode at the bottom had a surface of about 20 cm² when unstirred. During electrolysis, nevertheless, the mercury pool electrode was stirred using a magnetic stirrer and its surface somewhat increased. The potential of this electrode was kept constant at a selected value and was controlled against a reference electrode (s.c.e.) using a potentiostat. The electrolytic current flowed between the mercury pool electrode and an auxiliary working anode of platinum mesh. This electrode was immersed in a borate buffer pH 9·3 containing 5—10% ethanol, separated from the electrolysed solution by sintered glass, porosity G 3. The diffusion between anolyte and catholyte solutions during electrolysis was negligible. The 90—95% conversion was achieved during 20—30 minutes electrolysis. The potential difference between the mercury pool electrode and the auxiliary reference electrode was determined from *i–E* curves recorded using the mercury pool electrode. These *i–E* curves were recorded by measuring the current flowing between the mercury pool and working platinum electrodes. These, of course, differed from the shape of the *i–E* curves recorded with the dropping mercury electrode in the classic polarographic arrangement. Because the limiting current of the curves recorded with the mercury pool electrode was indistinct and practically coalesced with the hydrogen evolution, the voltage applied was chosen so as to correspond to about one quarter of the limiting current. In this the possibility of simultaneous hydrogen evolution was considerably reduced. Before beginning the electrolysis oxygen was removed by a stream of hydrogen not only from the catholyte but also from the anolyte. Before the start of the electrolysis the flow of hydrogen was stopped and the electrolytic vessel was carefully sealed (*e.g.* using wax around all contact surfaces between the rubber stopper, the electrolytic cell and the inserts). It is necessary to prevent diffusion of the atmospheric oxygen, which would be reduced at the electrode surface. The course of the electrolysis was controlled, after regular time-intervals, polarographically directly in the electrolysis cell. For this purpose a dropping mercury electrode was immersed into the electrolysed solution and the auxiliary reference electrode (s.c.e.) (otherwise a part of the three electrode potentiostatic system) was used.

When the potential of the mercury pool electrode was controlled independently using a Luggins capillary in various points of the surface of the mercury pool, the difference in the potential value found was smaller than ±5 mV. Measurements of the potential of the mercury pool electrode were carried out using a compensator pH-meter PHM 4 (Radiometer), or a universal electronic voltameter Mark BM 388 (Tesla).

To identify the electrolysis products the electrolysed solutions were subjected to gas–liquid chromatography, titrations and some of the electrolysis products were isolated. Gas–liquid chromatography was used for the analysis of the products of electrolysis carried out under polarographic conditions in 0·5 to 1·0 ml of the solution. A semiquantitative estimation of the yield was carried out by comparison of the areas below the chromatographic peaks obtained with the electrolysed solution with those of standards. The results indicated yields between 50 and 75% with a reasonable reproducibility. 1 μl of the solution after electrolysis was injected into the chromatographic columns. Water, ethanol and the buffer components did not interfere.

The electrolysis product of 4-methylsulphonylbenzoic acid(*V*) was isolated: 50 mg of acid *V* was dissolved in 50 ml of a borate buffer pH 9·3 containing 50% ethanol. After the reduction was terminated, the volume was reduced, and the solution was acidified by five drops of concentrated hydrochloric acid and evaporated to dryness. By ether extraction 25 mg of benzoic acid contaminated by boric acid was isolated. After recrystallization (water) benzoic acid was identified by a mixed melting point (122°C).

Similarly, the formation of benzamide (m.p. 128°C) was proved in the reduction of 4-methyl-sulphonylbenzamide (*XI*).

A potentiometric titration[26] was used to determine the yield of methanesulphinic acid formed in the mercury pool electrolysis. 50 ml of the electrolysed solution (about $5 . 10^{-3}$M in the methane-sulphinic acid in a borate buffer (pH 9·3) containing 10% ethanol) was titrated using a 0·05M-KMnO$_4$ solution. The potential jump at the end-point determination was about 0·2 V; 5—10% of ethanol did not interfere.

RESULTS AND DISCUSSION

Sulphones *V — XIII* and *XX — XXIII* give a two-electron reduction wave at pH 6—13. At pH values less than about 6, the wave is overlapped by the current of the reduction of the supporting electrolyte; in strongly alkaline media hydrolysis and cleavage occurs with some of the compounds.

The wave-height and the half-wave potentials are practically pH-independent over the whole pH-range. For compounds *VIII* and *XIII* which appear at potentials more positive than the others and give well-developed waves, logarithmic analysis was carried out. The slope of the log $i/(i_d-i)$–E plot was found to be 0·044 and 0·045 V respectively; this indicates an irreversible process. The irreversibility was proved using commutator and oscillographic techniques.

The reduction follows the Scheme (*B*) in agreement with results obtained for the unsubstituted compound[2,4-6]:

This has been proved by the isolation of the products, as summarized in Table II.

As far as we are informed this kind of reduction of a sulphone has little analogy in chemical reductions in solutions. Sulphones are generally very stable toward reducing agents[27], and under forced conditions compounds of uncertain constitution were produced. A reaction resembling to a certain degree studied reaction (*B*) is the cleavage of diphenyl sulphone in which benzenesulphinic acid and diphenyl result[27].

The smaller than theoretical yield of the sulphinic acid is in agreement with the observation in the literature[6] and is due probably to a disproportionation of the sulphinic acid. For the bismethylsulphonyl derivatives *XXII* and *XXIII* only one two-electron wave has been observed. Reaction Scheme (*B*) predicts the formation of unsubstituted methyl phenyl sulphone as the reduction product in the first two-electron step. The absence of a second wave is in agreement with the behaviour of methyl phenyl sulphone which is reduced at such negative potentials that its wave cannot appear before the final rise in current in supporting electrolytes containing sodium ions. For sulphonamides *XX* and *XXI* it was necessary to consider two possibilities: either the C—S bond in $C-SO_2CH_3$ or that in $C-SO_2NH_2$ is cleaved. Because it has been proved[14] that in the reduction of sulphonamides ammonia is formed in the solution after electrolysis, the evidence of the absence of ammonia was essential. Among the electrolysis products of both sulphonamides *XX* and *XXI* it was impossible to detect ammonia. It can thus be concluded that even sulphonamides *XX* and *XXI* follow the reduction Scheme (*B*).

The pH-independence of half-wave potentials makes it possible to exclude pre-protonation and elimination—addition mechanisms through a benzyne derivative. If the transfer of the first electron is activation controlled and is the potential determining step, it is nevertheless impossible to distinguish between the sequence (electron, proton, electron) and (electron, electron, proton). Nevertheless, we assume rather that the transition state is the same as with common S_N2 aromatic substitution reactions with the methanesulphinate anion as a leaving group. The nucleophilic reagent can be considered as hydride anion, which is, however, not involved as such but is represented by the two electrons and one proton which are successively accepted by the molecule in a sequence that presently cannot be specified.

TABLE II

Isolation of the Products in the Electrolysis of Substituted Phenyl Methyl Sulphones $(X-C_6H_4SO_2CH_3)$ in Borate Buffer (pH 9·3)

No	X	Electrolysis potential (*vs* S.C.E.)	Type[a]	Product	Analytical method	Yield
V	4-COOH	−1·75	*A*	C_6H_5COOH CH_3SO_2H	Isolation, m.p. KMnO$_4$ Titration	— 70%
VII	4-COOCH$_3$	−1·52	*B*	$C_6H_5COOCH_3$	G.L.C.[b]	—
VIII	4-COOC$_2$H$_5$	−1·52	*C*	CH_3SO_2H	KMnO$_4$ Titration	60%
XI	4-CONH$_2$		*A*	$C_6H_5CONH_2$	Isolation, m.p.	—
XIII	4-CN	−1·55	*B*	C_6H_5CN	G.L.C.	—
XXII	3-SO$_2$CH$_3$	−1·75	*C*	CH_3SO_2H	KMnO$_4$ Titration	qual.
XXIII	4-SO$_2$CH$_3$	−1·56	*C*	CH_3SO_2H	KMnO$_4$ Titration	60%

[a]Types: *A* mercury pool electrode, 20 cm², 50 mg of substance in 100 ml borax with 50% ethanol; *B* dropping mercury electrode, $1 . 10^{-3}$M substance in 1 ml borax with 5% ethanol; *C* mercury pool electrode, 20 cm², $5 . 10^{-3}$M substance in 50 ml borax with 10% ethanol; [b] gas liquid chromatography.

Comparison of the substituent effects on half-wave potentials measured at pH 9·3 (Table I) reveals that even when the value of substituent constant σ_X can be a good guide for prediction of whether a compound will give a reduction wave, it cannot be used for predictions of which of the two electronegative groups present in a molecule will be reduced first. For such a prediction a knowledge of the two substituent constants σ_X involved, the two reaction constants $\varrho_{\pi,R}$ and the two half-wave potentials of the corresponding unsubstituted compounds $(E_{\frac{1}{2}})_R^H$ is in principle necessary. If the parent substance is not reducible in the available potential range, the value $(E_{\frac{1}{2}})_R^H$ must be obtained by extrapolation; this was the case with methyl phenyl sulphones. Because of the uncertainty in the values needed, such predictions are of limited reliability, in particular in the case of compounds such as *XII*, *XIII*, *XX* and *XXI*, where both groups are similar in their electronic effect(σ_X) and reducibility $(E_{\frac{1}{2}})_R^H$. It was thus necessary to determine the reaction site for each individual compound experimentally.

The knowledge of the actual reaction reaction site is also a necessary presumption for attempts to correlate the substituent effects by means of the modified Hammett equation[28]. Firstly, compounds *II* and *III* which are reduced on another reaction centre (the carbonyl group) must be excluded from such a correlation, as it follows

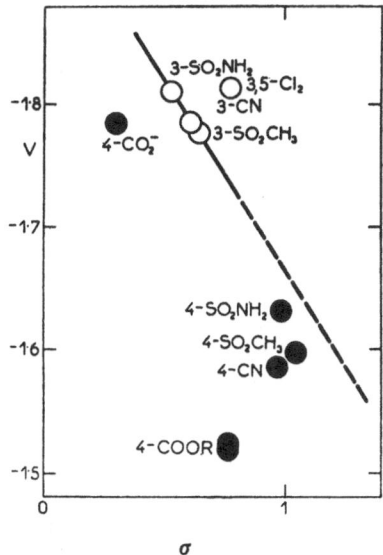

FIG. 1

Dependence of Half-Wave Potentials of Substituted Phenyl Methyl Sulphones in a Borate Buffer (pH 9·3) on Hammett Substituent Constants

Circles: *meta*-derivatives (σ_{m-x}); points: *para*-derivatives (σ_{p-x}^-). The line has a slope of $\varrho_{\pi,SO_2CH_3} = 0\cdot32$ V.

from the characteristic change[23] of waves of these compounds with pH. Furthermore, for quantitative treatment of substituent effects the values of half-wave potentials for *VI*, *X* and *XI* must be excluded, because they undergo a mixed reduction, in which both functional groups can be simultaneously reduced. Such a reduction course is indicated by the reducibility in acidic media which for compounds *X* and *XI* takes place with an uptake of four electrons (corresponding to the reduction of $CONH_2$ to CH_2OH). The reduction at higher pH-values is more complicated and is being studied in more detail; the half-wave potentials of *X* and *XI* were therefore not included into $E_{\frac{1}{2}}-\sigma$ correlation. The wave of compound *VI* occurs at too negative potentials to allow measurement of the number of the transferred electrons. Hence, if the condition[28,29] of identical reaction mechanism for all compared half-wave potentials is to be fulfilled, the selection of experimental data must be restricted. The gross pattern of the substituent effects is shown on Fig. 1. The *meta* derivatives show a moderate effect in the direction expected for an aromatic S_N2 substitution. From three values for substances *XII*, *XX* and *XXII* the approximate value of the reaction constant $(\varrho_{\pi,SO_2CH_3} = 0.32 \text{ V})$ can be estimated. For the deviation shown by the disubstituted derivative *XXVI* no explanation can be advanced. An unusually strong effect of *para*-M substituents cannot be expressed even by using the substituent constants σ_{p-x}^- which express an enhanced M- effect of these acceptor groups when they are conjugated directly with the reaction centre. A large electron excess at the reaction centre is characteristic for the aromatic S_N2 substitution; nevertheless with most homogeneous reactions of that type, the values of constant σ_{p-x}^- are sufficient to account for it. It would be possible to derive a special set of substituent constants expressing the still enhanced conjugation, or to use an equation of the Yukawa-Tsuno type[30] with two variables to express the substituent effects in reaction (*B*). Nevertheless, the limited number of available data does not in our opinion justify such a treatment. We are restricting ourselves to the plot (Fig. 1) which supports the anticipated mechanism and offers the principal information as to which derivatives are reducible.

Elemental analyses were carried out in the Analytical Department, Institute of Organic Chemistry and Biochemistry, Czechoslovak Academy of Sciences, Prague, under the supervision of Dr J. Horáček.

REFERENCES

1. Majranovskij, S. G., Neiman M. B.: Dokl. Akad. Nauk SSSR *87*, 805 (1953).
2. Levin E. S., Šestov A. P.: Dokl. Akad. Nauk SSSR *96*, 999 (1954).
3. Horner L., Nickel H.: Chem. Ber. *89*, 1681 (1956).
4. Drushel H. V., Miller J. F.: Anal. Chem. *30*, 1271 (1958).
5. Bowers R. C., Russell H. D.,: Anal. Chem. *32*, 405 (1960).
6. Levin E. S., Osipova N. A.: Ž. Obšč. Chim. *32*, 2084 (1962).
7. Zuman P.: *Advances in Polarography* (I. Longmuir, Ed.) Vol. 3, p. 814. Pergamon Press, Oxford 1960.

8. Baldwin W. A., Robinson R.: J. Chem. Soc. *1932*, 1445.
9. Oxley P., Partridge M. W., Robson T. D., Short W. F.: J. Chem. Soc. *1946*, 763.
10. Exner O.: This Journal *24*, 3562 (1959).
11. Price C. C., Hydock J. J.: J. Am. Chem. Soc. *74*, 1943 (1952).
12. Buu-Hoï, Lecocq J.: Bull. Soc. Chim. France *1946*, 139.
13. Cymerman J., Koebner A., Short W. F.: J. Chem. Soc. *1948*, 381.
14. Manoušek O., Exner O., Zuman P.: This Journal *33*, 4000 (1968).
15. Balfe M. P., Kenyon J., Searle C. E.: J. Chem. Soc. *1951*, 380.
16. Twist R. F., Smiles S.: J. Chem. Soc. *127*, 1248 (1925).
17. Zincke T., Frohneberg W.: Ber. *42*, 2721 (1909).
18. Dandin S. M., Bhat H. B., Kulkarni G. H., Iyengar H. S. S., Sattur P. B., Nargund K. S.: J. Karnatek Univ. *1*, 64 (1956); Chem. Abstr. *52*, 8071 (1958).
19. Little W. F., Reilley C. N., Johnson J. D., Sanders A. P.: J. Am. Chem. Soc. *86*, 1382 (1964).
21. Gardner H. J.: Nature *167*, 158 (1951).
22. Vlček A. A.: This Journal *19*, 862 (1954), Chem. listy *48*, 189 (1954).
23. Zuman P., Exner O., Nauta W. T.: This Journal *33*, 3213 (1968).
24. Gilbert G. A., Rideal E. K.: Trans. Faraday Soc. *47*, 396 (1951).
25. Weaver R. D., Whitnack G. C. in the book: *Modern Analytical Methods* (G. Charlot, Ed.), p. 51. Elsevier, Amsterdam 1958.
26. Lindberg B.: Acta Chem. Scand. *17*, 383 (1963).
27. Suter C. M.: *The Organic Chemistry of Sulfur. Tetracovalent Sulfur Compounds*, p. 682. Wiley, New York 1944.
28. Zuman P.: This Journal *25*, 3225 (1960).
29. Bunnett J. F., Draper F., Ryason P. R., Noble P., Tonkyn R. G., Zahler R. E.: J. Am. Chem. Soc. *75*, 642 (1953).
30. Yukawa Y., Tsuno Y.: Bull. Chem. Soc. Japan *32*, 971 (1959).

Translated by author (P. Z.).

FISSION OF ACTIVATED CARBON–NITROGEN AND CARBON–SULPHUR BONDS. XI.*

POLAROGRAPHIC REDUCTION OF SUBSTITUTED BENZENESULPHONAMIDES

O.Manoušek, O.Exner and P.Zuman**

*J. Heyrovský Institute of Polarography,
Czechoslovak Academy of Sciences, Prague 1*

Received January 4th, 1968

Substituted benzenesulphonamides bearing strongly electron-attracting substituents can be reduced at the dropping mercury electrode to ammonia, sulphur dioxide and monosubstituted benzene. The products have been identified in a controlled potential electrolysis using a mercury pool electrode. A pH-independent, two-electron wave is observed at pH 7 to 12. Fission of the C—S bond occurs before that of the S—N bond. The reaction represents a kind of aromatic S_N2 substitution with the $-SO_2NH_2$ grouping as a leaving group. The substituent effects on half-wave potentials at pH 9·3 for derivatives $I-XIX$ were examined. The reduction is strongly facilitated by conjugation with acceptor groups in the *para*-position; this effect is even stronger than would be expressed by σ^-_{p-x} constants.

The information available on the polarography of sulphonamides is surprisingly scarce[1-5]. It is restricted to a report of the presence of reduction waves in tetraalkyl-ammonium salts in aqueous[1,2] or dimethylformamide[5] solutions and of waves of 3-trifluoromethyl- and 3-chloro-4,6-disulphonamidoaniline in a borate buffer, pH 8·1. No attempts have been reported to elucidate the course of the electrode process.

In the course of our studies on the polarographic reduction of benzene derivatives bearing two electronattracting groups[6-9] it was necessary to consider the course of the electroreduction of some substituted benzene sulphonamides, to study the effects of substituents on this reduction process and to correlate them with reactivity in homogeneous chemical reactions. For this purpose compounds were chosen bearing such substituents that brought the half-wave potentials into the potential range accessible in aqueous buffer solutions containing sodium or potassium ions. In the present work a group of substituted benzene sulphonamides $I-XIX$ bearing one or more electronegative groups in the *meta*- or *para*- positions was studied

---

*   Part X: This Journal *33*, 3988 (1968).
**   Present address: Department of Chemistry, University of Birmingham, Great Britain.

in order to elucidate the course of the electrode process and to evaluate the polar effects by which these substituents affect the half-wave potential corresponding to the cleavage of the C—S bond.

EXPERIMENTAL

Substances

Meta- and *para*-substituted benzene sulphonamides *I—XV* (Table I) were prepared according to the procedure described in the literature[10-16] or by the application of known methods. No precautions were taken to secure the highest yields, as most attention was paid to the purity of the prepared substances. All substances were purified by crystallization. Melting points are corrected; the samples for analysis were dried over phosphorus pentoxide at 0·05 mm Hg at 20°C. Compounds *XVI—XIX* were kindly provided by the Squibb Institute for Medical Research, New Brunswick.

Methyl 4-sulphamidobenzoate (IV). 4-Sulphamidobenzoic acid[11] (*III*) was converted to the acid chloride by boiling with thionyl chloride[12] in the presence of several drops of dimethylformamide. 4-Sulphamidobenzoylchloride m.p. 141°C (chloroform, literature[12] 139°C) was boiled with methanol, the solvent distilled off and the residue was crystallized from water, m.p. 177°C (not sharp; literature[17] gives 175°—180°C for a product prepared in another way). For $C_8H_9NO_4S$ (215·2), calculated, 44·65% C, 4·21% H and 6·51% N; found, 44·33% C, 4·14% H and 6·70% N.

1,3,5-Benzenetrisulphonamide (XII) was prepared by treating 1,3,5-benzene trisulphochloride with ammonia. The starting chloride (m.p. 184°C) was prepared from sodium 1,2-benzenedisulphonate according to the literature[18], except that the conversion of the sodium salt to the chloride was carried out using phosphorus pentachloride at normal pressure.

3-Methylsulphonylbenzenesulphonamide (XIII) was prepared from 3-methylsulphonylbenzenesulphochloride[19] by heating with an excess of aqueous ammonia, neutralization with hydrochloric acid and evaporation. Yield 88%, m.p. 210°C (water). For $C_7H_9NO_4S_2$ (235·3) calculated: 35·73% C, 3·86% H, 5·96% N; found 35·80% C, 3·73% H, 6·08% N.

4-Methylsulphonylaniline was prepared by hydrolysis of its N-acetylderivative with boiling 10% hydrochloric acid[20]; after evaporation of the solution, the amine was isolated as the hydrochloride, yield 95%. The base has m.p. 134°C (diluted ethanol). The starting 4-methylsulphonylacetanilide m.p. 186°C (water) was prepared according to the literature[21].

4-Methylsulphonylbenzenesulphochloride. 4-Methylsulphonylaniline hydrochloride (4·17 g; 0·02 mol) was dissolved in 8 ml of hydrochloric acid and the ice-cold solution was diazotized in the usual way with a solution of 1·5 g (109%) of sodium nitrite in 4 ml of water. The diazotized solution was slowly added to a cooled solution of about 7 g of sulphur dioxide in 20 ml of glacial acetic acid to which 1 g of cupric chloride (dihydrate) dissolved in the minimal amount of water was added earlier. The reaction was carried out within 5 minutes and the temperature was kept below 30°C. On completion of the reaction, the excess of sulphur dioxide was distilled off, and the reaction mixture was poured into 70 ml of ice-cold water. Yield 4·12 g (81%) of the crude product, m.p. 163°C (benzene). For $C_7H_7ClO_4S_2$ (254·7), calculated 33·01% C, 2·77% H, 13·92% Cl; found 32·97% C, 3·08% H, 13·90% Cl.

4-Methylsulphonylbenzenesulphonamide (XIV) was prepared from 4-methylsulphonylbenzenesulphochloride and aqueous ammonia in a similar way to the *meta* isomer *XIII*. Yield 89%, m.p.

244°C (water). Literature[22] gives the same melting point for a product obtained by sulphide oxidation.

Apparatus, Procedures and Techniques

The polarograph, vessels, capillary, potentionstat, voltmeter and pH-meter used were the same as in Part X (ref.[9]). The procedures used for polarographic examination, for coulometric determination of the number of electrons transferred and in the controlled potential electrolysis for identification of electrolysis products and determination of their yield were also the same[9]. The electrolyses were carried out in a borate buffer (pH 9·3) under polarographic conditions in a volume of 0·5 ml using a dropping mercury electrode and under potentiostatic conditions in a volume of 25—100 ml using a mercury pool electrode.

The final reduction products, ammonium ions and hydrogen sulphite, were detected and sometimes determined. To determine the hydrogen sulphite ions the solution electrolysed using a mercury pool electrode was adjusted to pH 3 to 4; the reduction wave of sulphurous acid was measured in this solution. The concentration of hydrogen sulphite was determined using the standard addition method. Quantitative analysis of the electrolysed solution was carried out for 4-cyanobenzenesulphonamide (VIII), where the yield of the sulphur dioxide corresponds to 92% and for 4-methylsulphonylbenzamide (VI), where the yield of sulphur dioxide was 95%. For all other compounds given in Table I, the presence of sulphur dioxide in the electrolysed solution was shown only qualitatively by the presence of the polarographic wave. The presence of ammonia in the electrolysed solution was proved by Nessler's reagent. The electrolysed solution was purged by a hydrogen stream and the released ammonia was trapped in a wash-bottle containing the Nessler reagent.

RESULTS AND DISCUSSION

For compounds III, IV, VII, VIII, X, XI and XVI—XIX one two-electron wave was observed over the pH range 7 to 12. At lower pH values the wave is superimposed by the reduction of the supporting electrolyte. For compound VIII hydrolysis of the cyano group occurred at pH > 11. Both the wave-height and the half-wave potential were practically pH-independent over the entire pH-range studied. In the solution after controlled potential electrolysis, sulphur dioxide and ammonia were found. We suppose, therefore, that the electrolysis follows the course:

$$ArSO_2NH_2 + 2e + H^+ \rightarrow ArH + SO_2NH_2^{(-)} \qquad (A)$$

$$SO_2NH_2^{(-)} + H_2O \rightarrow HSO_3^{(-)} + NH_3. \qquad (B)$$

An analogous reduction course for a sulphonamide has not been reported either for chemical or for electrochemical reduction. Generally, sulphonamides are very resistant to reduction; only under drastic conditions is the S—N bond broken and ammonia or amine evolved. The formation of various sulphur compounds of lower oxidation state has been reported[24]. In a few instances, the cleavage of the C—S bond was also observed, but in all these examples this reaction took place consecutive with the fission of the C—N bond[24]. The reduction with sodium in amyl

TABLE I

Characteristics, Half-Wave Potentials and Electrolysis Products of Substituted Benzenesulphonamides of the Type $X-C_6H_4SO_2NH_2$. Measurements of half-wave potentials (against s.c.e.) were carried out in a borate buffer ph 9·3; in the same buffer, controlled potential electrolysis was done using a mercury pool electrode with a surface of about 20 cm².

No	X	M.p. °C	Ref.	$E_{\frac{1}{2}}^a$	σ	n^b	Mercury pool electrolysis	
							Potential V	Product
I	H	153	—	→	0	—	—	—
II	3-COOH	241	10	→	0·35	—	—	—
III	4-COOH	281	11	-1·827	$0·42^c$	1·88	-1·84$_0$	SO_2, NH_3
IV	4-COOCH$_3$	177	—	-1·589	0·43	2·1	-1·62$_0$	SO_2, NH_3
V	3-CONH$_2$	177	10	-1·847	0·35	2·2	-1·85$_0$	Mixture^e
VI	4-CONH$_2$	243	12	-1·608	$0·43^d$	2·15	-1·64$_0$	Mixture^e
VII	3-CN	153	10	-1·839	0·61	1·8	-1·85$_0$	SO_2, NH_3
VIII	4-CN	169	11	-1·678	0·69	1·95	-1·72$_0$	SO_2^f, NH_3
IX	4-NH$_2$	163	—	→	-0·66	—	—	—
X	3-SO$_2$NH$_2$	231	13	-1·860	0·53	2·05	-1·86$_0$	SO_2, NH_3
XI	4-SO$_2$NH$_2$	289	14	-1·713	0·60	2·10	-1·75$_0$	SO_2, NH_3
XII	3,5-(SO$_2$NH$_2$)$_2$	~320	15	$-1·702^g$	$2 \times 0·53$	2	-1·74$_0$	SO_2, NH_3, X
XIII	3-SO$_2$CH$_3$	210	—	$-1·812^h$	0·64	—	-1·75	$CH_3SO_2^-$
XIV	4-SO$_2$CH$_3$	244	—	$-1·632^h$	0·73	—	-1·56	$CH_3SO_2^-$
XV	4-Cl	143	16	—	0·22	—	—	—
XVI		—	—	-1·640	—	—	-1·66$_0$	SO_2
XVII		—	—	-1·621	—	2·15	-1·64$_0$	SO_2
XVIII		—	—	-1·715	—	—	-1·73$_0$	SO_2
XIX		—	—	-1·716	—	—	-1·74$_0$	SO_2

^{a} Reproducible to ±2−3 mV; ^{b} number of electrons transferred determined coulometrically; $^c \sigma_{COO^-}$; ^{d} according to ref.[23]; ^{e} 95% yield of SO_2, relative heights of waves in acid and alkali 4 : 3; ^{f} 92% of SO_2; ^{g} this wave is followed by another two-electron wave at $-1·860$ V corresponding to the reduction of compound X; ^{h} corresponds to the reduction of the CH_3SO_2 group.

alcohol proceeds in a similar way[25]. The products of this reaction were the hydro-carbon, sulphur dioxide and hydrogen sulphide; sulphinic acid was proposed as an intermediate[25].

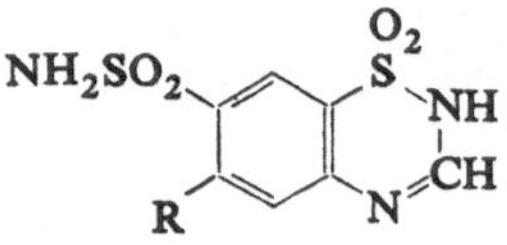

XVI, R = Cl
XVII, R = CF$_3$

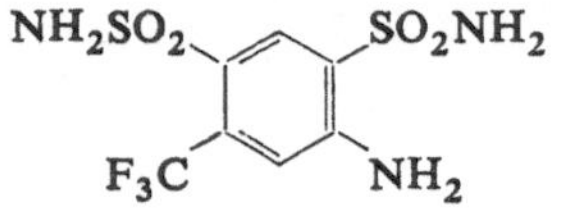

XVIII

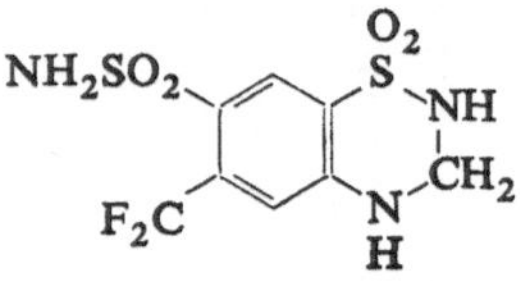

XIX

In the mechanism (A), (B) suggested for the electrolytic reduction, the breaking of the C—S bond is considered to be the first step. The arguments for this reaction course as opposed to the primary splitting of the N—S bond are as follows: The polarographic behaviour is similar in all details to the behaviour of substituted phenyl methyl sulphones[9] which split off reductively $CH_3SO_2^{(-)}$. The existence of sulphurous acid monoamide HSO_2NH_2 as a species of some stability has been proved[26]. In the present series it was impossible to observe any time-changes in the yields of ammonia or the HSO_3^- ion, but chemical reactions of that type have been proved in the electrolysis of saccharin[27]. This indicates that at least an N-acylderiv-ative of the compound HSO_2NH_2 is relatively stable in alkaline solution. Further-more, when the reduction of the S—N bond occurs first, the resulting sulphinic acid would be reducible in the available potential range. However benzene sulphinic acids with electron attracting substituents are polarographically inactive[9].

The polarographic reduction of benzenesulphonamides is hence another example of S_N2 substitution on the benzene ring. The SO_2NH_2 grouping has the role of the leaving group. The independence of the half-wave potential with respect to pH indicates that proton transfer does not occur before the potential determining step in which the electron is transferred. Hence it is possible to exclude protonation or proton-extraction (which would correspond to an elimination-addition mechanism, with benzyne as intermediate) as possible mechanisms. The acceptance of the first electron is the first step in the reduction process, but it is impossible to prove, with the experimental evidence available, if this is followed by another electron-transfer, giving a sequence (e, e, H$^+$) or if the proton-transfer occurs first, with a resulting sequence (e, H$^+$, e). Logarithmic analysis for substances IV and VIII gave a linear plot with a slope of 0·046 V and 0·045 V. This indicates that the electrode process

$$ArSO_2NH_2 + e \rightarrow \text{Radical anion}$$

is irreversible.

For substance *XII*, reduction occurs in two 2-electron steps. The reduction potential of the more negative wave is exactly the same as that of substance *X*. Hence it is possible to conclude that the reduction takes place in two consecutive steps (*C*) and (*D*):

$$\text{(benzene ring with SO}_2\text{NH}_2\text{, NH}_2\text{SO}_2\text{, SO}_2\text{NH}_2\text{)} + 2e + H^+ \xrightarrow{E_1} \text{(benzene ring with SO}_2\text{NH}_2\text{, SO}_2\text{NH}_2\text{)} + SO_2NH_2^{(-)} \quad (C)$$

$$\text{(benzene ring with SO}_2\text{NH}_2\text{, SO}_2\text{NH}_2\text{)} + 2e + H^+ \xrightarrow{E_2} \text{(benzene ring with SO}_2\text{NH}_2\text{)} + SO_2NH_2^{(-)} \quad (D)$$

where potential E_1 is more positive than E_2.

Most complex behaviour has been observed for amides *V* and *VI*. At pH < 6 the compound is reduced in a wave that is higher than the wave at pH > 7. In the range between pH 6 and 7 two waves appear. In acidic media the product of electrolysis on a mercury pool electrode gave a positive reaction with ninhydrin, indicating the presence of the grouping CH_2NH_2. Hence the reduction of the amide group occurs in the protonated form[28]. At pH 6·8 the product of controlled potencial electrolysis at a mercury pool electrode gave positive reactions with ninhydrin, indicating formation of CH_2NH_2 and a polarographic wave for sulphur dioxide, indicating formation of $SO_2NH_2^{(-)}$. Hence it can be concluded that reductions of the sulphamido and carboxamido groups occur in a competitive reaction. Coulometric measurements with a dropping electrode at pH 9·3 indicated a two-electron process for the former and a four-electron process for the formation of the aminomethyl group in acidic media. The ratio of wave-heights in acidic and alkaline media was nevertheless 3 : 2 rather than 4 : 2. The problem whether and under which conditions the competitive reduction takes place at the dropping mercury electrode was subjected to a further, more detailed investigation.

For evaluation of substituent effects on the half-wave potentials, all members of the series must undergo reduction by the same mechanism[29]. Therefore compounds *V*, *VI*, *XIII* and *XIV* which show differences in mechanism must be excluded from structural correlations. For the remaining seven substances the dependence of half-wave potentials on the structural parameter (Fig. 1) is similar to that observed with substituted phenyl methyl sulphones[9].

Through the values of three *meta*-derivatives, a straight line can be plotted with a slope $\varrho_{\pi, SO_2NH_2} = 0·30$ V as compared $-0·32$ V observed for phenyl methyl sulphones[9]. The substituent effect hence operates with *meta* substituents in the expected direction and with a plausible intensity. The effect of *para* substituents is extraordinarily strong and not even constants σ^-_{p-x} are sufficient to express it. Such behaviour indicates very strong conjugation of acceptor substituents with the reaction centre in the transition state. This is compatible with the properties of aromatic S_N2 substitution.

The similarity of the substituent effects in the sulphonamide series with those observed in the reduction of phenyl methyl sulphones is demonstrated when the half-wave potentials of both reaction series are plotted against each other (Fig. 2). Not only the *meta* but also the *para* derivatives give points that are situated near a straight line with a slope of almost unity. A given sulphonamide bearing an electron attracting group is hence reduced at 60 mV more negative than the corresponding methyl sulphone. Derivation of a special set of σ^-_{p-x} constants expressing the extra strong conjugation and appropriate for S_N2 aromatic substitution of the given type or the use of a Yukawa-Tsuno equation[30] with two variables is postponed until further reaction series showing the same behaviour are found.

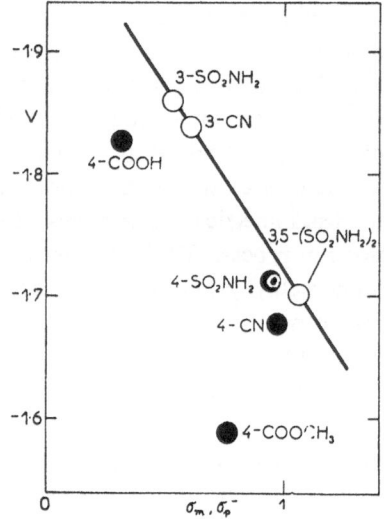

FIG. 1

Dependence of Half-Wave Potentials of Substituted Benzenesulphonamides in a Borate Buffer pH 9·3 on Hammett Substituent Constants.

○ *meta* (σ_{m-x}); ● *para* derivatives (σ^-_{p-x}). The line shows a slope of $\varrho_{\pi, SO_2NH_2} = 0·30$ V.

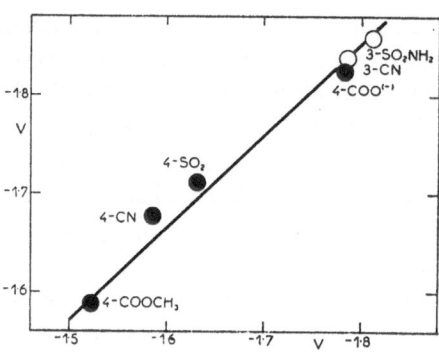

FIG. 2

Dependence of Half-Wave Potentials of Substituted Benzenesulphonamides $(E_{1/2N})$ on Half-Wave Potentials of Similarly Substituted Phenyl Methyl Sulphones $(E_{1/2C})$

○ *meta*; ● *para* derivatives.

Elemental analyses were carried out in the Analytical Department, Institute of Organic Chemistry and Biochemistry, Czechoslovak Academy of Sciences, Prague, under supervision of Dr J. Horáček.

REFERENCES

1. Koršunov I. A., Kirillova A. S., Ščennikova M. K., Sazanova L. N.: Ž. Obšč. Chim. *21*, 565 (1951).
2. Horner L., Nickel H.: Chem. Ber. *89*, 1681 (1956).
3. Urabe N., Yasukochi K.: J. Electrochem. Soc. Japan *25*, 17 (1957).
4. Musil F.: Pracovní lékařství *14*, 426 (1962).
5. Cohen A. I., Keller B. T., Yale H. L.: Anal. Chem., *34*, 216 (1962).
6. Manoušek O., Zuman P.: Chem. Commun. *1965*, 158.
7. Zuman P.: Z. Anal. Chem. *224*, 374 (1967).
8. Kargin Yu., Manoušek O., Zuman P.: J. Electroanal. Chem. *12*, 443 (1966).
9. Manoušek O., Exner O.: Zuman P.: This Journal *33*, 4000 (1968).
10. Delaby R., Harispe J. V., Paris J.: Bull. Soc. Chim. France [5] *12*, 954 (1945).
11. Delaby R., Harispe J. V.: Bull. Soc. Chim. France [5], *10*, 580 (1943).
12. Rodionov V. M., Javorskaja E. V.: Ž. Obšč. Chim. *18*, 110 (1948).
13. Knoevenagel E., Lebach H.: Ber. *37*, 4094 (1904).
14. Kirsanov A. V., Kirsanova N. A.: Ž. Obšč. Chim. *29*, 1802 (1959).
15. Behrend R., Mertelsmann M.: Ann. *378*, 352 (1911).
16. Otto R., Brummer L.: Ann. *143*, 100 (1867).
17. Isozaki M.: Repts. Govt. Chem. Ind. Research Inst. Tokyo *45*, 295 (1950); Chem. Abstr. *46*, 4269 (1952).
18. Sutter C. M., Harrington G. A.: J. Am. Chem. Soc. *59*, 2575 (1937).
19. Twist R. F., Smiles S.: J. Chem. Soc. *127*, 1248 (1925).
20. Jagupolskij L. M., Marenec M. S.: Ž. Obšč. Chim. *27*, 1395 (1957).
21. Child R., Smiles S.: J. Chem. Soc. *1926*, 2696.
22. Blank B., Farina F. A., Kerwin J. F., Saunders H.: J. Org. Chem. *26*, 1551 (1961).
23. Little W. F., Reilley C. N., Johnson J. D., Sanders A. P.: J. Am. Chem. Soc. *86*, 1328 (1964).
24. Searles S., Nukina S.: Chem. Rev. *59*, 1077 (1959).
25. Klamann D., Hofbauer G.: Chem. Ber. *86*, 1246 (1953).
26. Goehring M., Kaloumenos H. W.: Z. Anorg. Chem. *263*, 137 (1950).
27. Manoušek O.: Unpublished results.
28. Lund H.: Acta Chem. Scand. *17*, 2325 (1963).
29. Zuman P.: This Journal *25*, 3225 (1960).
30. Yukawa Y., Tsuno Y.: Bull. Chem. Soc. Japan *32*, 971 (1959).

Translated by the author (P. Z.).

Part VII

Reductions of Azomethine (C=N−) Bond and their Application

30. P. Zuman: Polarographic behaviour of acetone. *Nature* **175**, 585 (1950).
31. P. Zuman: The reaction of carbonyl compounds with primary amines. (in German) *Collect. Czechoslov. Chem. Commun.* **15**, 839 (1950).
32. B. Fleet, and P. Zuman: Quantitative treatment of substituent effects in polarography. V. Polarographic reduction of some semicarbazones. *Collect. Czechoslov. Chem. Commun.* **32**, 2066 (1967).
33. P. Zuman, and O. Exner: Oxime derivatives. VIII. Polarographic reduction of O − and N − substituted oximes. *Collect. Czechoslov. Chem. Commun.* **30**, 1832 (1965).
34. O. Manoušek, and P. Zuman: Polarographic study of the reaction of pyridoxal with hydroxylamine. *Experientia* **20**, 301 (1964).

Aldimines, ketimines and Schiff bases are reduced in the protonated form.[30,31] From the change in wave-height with amine concentration, the stoichiometry and equilibrium constant of the condensation reaction can be determined. The protonated form of semicarbazones is reduced in a four-electron step to form a primary amine and urea.[32] The analogous behaviour of semicarbazones of a wide range of aldehydes and ketones makes it possible to use the polarographic waves for functional group analysis.

In nitrones, the reduction of the N − O bond precedes that of the C=N − bond.[33] In the protonated form the reduction of nitrones closely resembles that of the corresponding O-alkyl oximes. In alkaline media, only nitrones undergo reduction, which enables these two classes to be differentiated. The three forms of pyridoxal also differ in their reactivity towards oximation.[34] In the reaction with hydroxylamine, the fully protonated form ($k_{PH_2^+}$ =500 1 Mole^{-1} min^{-1}) is the most reactive, followed by the anionic form (k_{p^-} = 13.1 1 Mole^{-1} min^{-1}) and then the zwitterionic ($k_{PH_2^+}$ = 3.5 1 Mole^{-1} min^{-1}). No general catalysis was observed, in contrast to the other reported C=N bond-forming reactions.

Polarographic Behaviour of Acetone

IN supporting electrolytes generally used in polaro-graphy, for example, in solutions of potassium, lithium or ammonium salts or lithium hydroxide or in buffer solutions, no polarographic reduction wave of acetone can be observed. Only if electrolytes of tetra-alkylammonium are used can a reduction wave at a half-wave potential of about − 2·1 V. (from

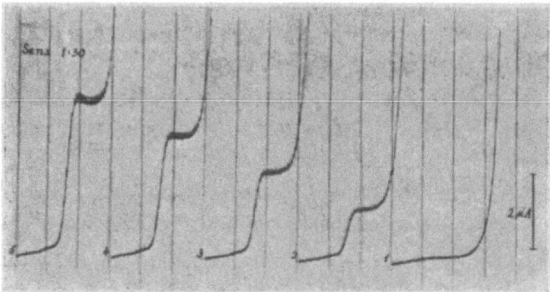

Dependence of the height of wave on concentration of acetone in 1·5 N ammonia, 1·5 N ammonium chloride ; concentration of acetone : (1) 0 ; (2) 0·018 M ; (3) 0·035 M ; (4) 0·051 M. Abscissæ, 0·2 V. ; curves start from − 1·2 V. (S.C.E.) ; rate of flow, m = 2·13 mgm./sec. ; dropping time, t = 3·2 sec.

saturated calomel electrode) be obtained[1,2]. After conversion of acetone to its hydrazone or phenyl-hydrazone, two reduction steps are obtained[3], but none is suitable for analytical purposes ; moreover, the height of these waves is not a linear function of concentration. Thus acetone is polarographically determined in an indirect way in acid sulphite solu-tion, with which it reacts as other ketones and lowers the acid sulphite wave[4].

When studying the reactions of amines with carbonyl compounds, for example, in concentrated solutions of ammonia – ammonium sulphate buffer of pH 9·3, I have found a well-developed wave due to acetone with a half-wave potential of about − 1·6 V. The height of the wave is a linear function of the concentration of acetone (see diagram), and such solutions are therefore well suited for analytical purposes. Since acetone is volatile, the polarographic curves must be recorded after passing an indifferent gas through the solution for a constant time. At greater concentrations of acetone it is possible to record curves even in the presence of air. Another NH_2-compound which is less volatile than ammonia is being sought.

It has been ascertained that the acetone substance reduced at − 1·6 V. is a compound formed by one molecule of acetone and one molecule of ammonia. It is supposed that the polarographic wave is due to the reduction of the imino group. The compound formed is in equilibrium with free acetone, and the height of the wave is influenced by the concentration of free ammonia. Since its dissociation into ammonia and acetone is large and because the concentration of the buffer solution used was limited by the solubilities of ammonium salts, it was impossible to reach the full diffusion current corresponding to acetone according to the Ilkovič equation. In 10 N ammonia saturated with ammonium chloride, the height of the wave was about 20 per cent of the wave due to the same molar concentration of pyruvic acid. This means that acetone may be polarographically estimated in concentrations of 0·1–0·00002 M. The equilibrium state is reached in 15 sec. after the beginning of the reaction. The electrode process is limited only by diffusion, and therefore the limiting current corresponds to the true equilibrium concentration in the bulk of the solution.

Many carbonyl compounds were found to react in a similar manner, the reaction product being reduced at different potentials. The reaction with the carbonyl group was found to take place only with primary amines (or amino-acids), that is, with the amino-group, and has been followed in the case of pyruvic acid, acetaldehyde, benzaldehyde, benzil, diacetyl, phenylglyoxylic acid, coumarindion and dehydroascorbic (or diketogulonic) acid. The study of these reactions and of their equilibrium and rate constants is being continued.

P. ZUMAN

Physico-Chemical Institute,
　　Charles University,
　　　　Prague.
　　Nov. 28.

[1] Neiman, M. B., and Markhina, Z. V., *Zavod. Lab.*, **13**, 1174 (1947).
[2] Stackelberg, M. v., and Stracke, W., *Z. f. Elektrochem.*, **53**, 118 (1949).
[3] Lupton, J. M., and Lynch, C. C., *J. Amer. Chem. Soc.*, **66**, 697 (1944).
[4] Heyrovský, J., in "Polarographie", 368 (Springer, Vienna, 1941).

DIE REAKTION DER KARBONYLVERBINDUNGEN MIT PRIMÄREN AMINEN

von P. ZUMAN.[*])

Bei den Reaktionen der Karbonylverbindungen mit primären Aminen entstehen je nach dem Wesen der reagierenden Karbonylverbindung und nach dem des Amins Aldimine (R—CH=NH), Ketimine (RC—(=NH)—R'), Schiffsche Basen oder Azomethine (R—CH=NR') oder die gut bekannten Verbindungen wie Oxime ($>$C=N—OH), Hydrazone ($>$C=N.NH.R) und Semikarbazone ($>$C=N.NR.CO.NH₂). Öfters reagiert auch noch die so entstandene Verbindung mit einem weiteren Molekül des Amins. Hier werden aber nur diejenigen Kondensationsprodukte verfolgt, welche durch die Reaktion einer Karbonylverbindung mit einem Molekül eines primären Amins entstehen. In allen diesen Fällen entsteht die Doppelbindung $>$C=N—, deren Eigenschaften, insbesondere die Analogie der Gruppe $>$C=O in chemischer und chemisch-physikalischer Hinsicht an einer anderen Stelle behandelt werden sollen.

Polarographisch wurde die Reduzierbarkeit der C=N-Gruppe zusammenfassend noch nicht verfolgt, es wurde nur das Verhalten einiger Verbindungen — vor allem der Oxime und Hydrazone — näher untersucht.

Bei der Ausarbeitung der Methodik einer polarometrischen Titration des Nickels haben I. M. Kolthoff und A. Langer[1]) die Reduzierbarkeit des Dimethylglyoxims in ammoniakalischer Lösung (pH 9,8) in einer Welle bei —1,7 V konstatiert. Bei einer ähnlichen Gelegenheit hat A. Langer[2]) festgestellt, daß Benzoinoxim an der tropfenden Quecksilberelektrode in zwei Stufen, deren Summe konstant blieb und deren Quotient mit einer pH-Änderung sich in der Form einer Dissoziationskurve ändert, reduziert wird. Nach dem analogischen Verhalten des Acetaldoxims und Acetofenoxims schloß er auf die Reduktion der Oximgruppe und hat die Welle der Reduktion von =C=NOH und =C=NO' zugeschrieben.

Weil damals der Einfluß der Ionenrekombination in der Elektrodennähe auf den pK-Wert noch nicht bekannt war[3]), legte A. Langer das gefundene K' = 10⁻⁶ der sauren Dissoziationskonstante gleich. Diese Konstante nähert sich bei Oximen aber der basischen und beide sind ungefähr pK = 9 und deshalb ist die Formulation der Base des Oxims und ihrer konjugierten Säure irrtümlich.

---

[*]) Auszug aus der Dissertation, Karlsuniversität Prag (1950).

Sollte das gefundene pK′ einer saueren Dissoziation entsprechen, müßte diese etwa pK 3—4, höchstens 6 (bei sehr langsamer Rekombination) sein. Wahrscheinlicher scheint die Erklärung, daß der positiveren Welle die Reduktion $(>C=NOH_2)^+$, der negativeren die Reduktion der $=C=NOH$ Gruppe entspricht, ähnlich der Formulation der Hydroxylaminreduktion, verfolgt von F. Petrů.[4] Das gefundene pK′ entspricht dann der basischen Dissoziationskonstante.

Auch die Reduktion der $>C=NOH$ Gruppe in 5-Oxim-methyl-4,5-dioxo-2 pentenoat wurde beschrieben.[5] In dessen Lösungen wurde bei einem pH-Wert größer als 5 eine Welle bei —1,5 V gefunden.

Die Hydrazone einiger aliphatischen Ketonen und Aldehyden verfolgten J. M. Lupton und C. C. Lynch,[6] welche in sauren Lösungen von Hydrazinen zwei Wellen fanden, die bei Aldehyden ein Halbstufenpotential von —0,9 und —1,3 V hatten, während sie bei Ketonen —1,1 und —1,3 V aufwiesen. Von den Abweichungen von einer linearen Abhängigkeit zwischen der Konzentration der Ketone und der Wellenhöhe bemerkten die Autoren, daß es sich um Gleichgewichtszustände handelt (deren Einstellung etwa 30 Minuten dauert), verfolgten sie aber nicht, weil „die chemische Natur der reduzierten Stoffe nicht klar ist".

Quantitative Verhältnisse bei der Bildung von Hydrazonen und Oximen haben G. Sartori und A. Gaudiano[7] verfolgt. Diese Autoren fanden die Reduzierbarkeit des Hydrazons und des Oxims der Mesoxalsäure, die selbst in gewöhnlichen Pufferlösungen nicht reduziert ist, und stellten eine Steigerung der Reaktionsgeschwindigkeit mit steigendem pH sicher. Für die Gleichgewichtskonstante der Reaktion

$$COOH.CO.COOH + NH_2OH \rightleftarrows COOH.C(=NOH).COOH + H_2O$$

haben sie den Wert $K = 3,82 . 10^{-4}$ gefunden.

Relativ die größte Aufmerksamkeit wurde den Betainylhydrazonen, die durch Kondensation von Girard-T-Reagens mit Ketosubstanzen entstehen, gewidmet. Diese Kondensationsprodukte wurden von J. K. Wolfe, E. B. Hershberg und L. F. Fieser,[8] die die Bestimmung einiger Ketosteroiden verfolgten, in die Polarographie eingeführt. Während die gesättigten Ketohormone nicht reduzierbar sind, wurde mittels der Substitution des Karbonyls durch die $C=N$-Gruppe (die durch Kondensation mit der Aminogruppe des Reagens entsteht) die polarographische Bestimmung ermöglicht. Doch auch hier wurden sich einstellende Gleichgewichtszustände, die aus Abweichungen der Wellenhöhe bei größeren Hormonkonzentrationen und aus verschiedenen Stufenhöhen der einzelnen Hormone bei derselben Konzentration des Girardschen Reagens folgen, nicht untersucht. Später wurde eine ganze Reihe von rein analytischen Methoden zur Hormonbestimmung nach Überführung auf Girardsche Kondensationsprodukte ausgearbeitet. Einige andere Hydrazone der Hormone — vor allem Verbindungen, die durch Kondensation der Ketosteroide mit Acidhydraziden entstehen — wurden von J. Barnett und C. J. Morris[9] untersucht.

Eine umfangreiche Studie widmeten V. Prelog und O. Häfliger [10] den Kondensaten des Girardschen Reagens mit Cyklanonen. Sie bestimmten die Beziehungen zwischen der Wellenhöhe und der Quecksilbersäule bei

verschiedenem pH. Weiter verfolgten sie die Abhängigkeit des Halbstufenpotentials vom pH-Wert und von der Größe des hydroaromatischen Ringes.
Sie wiesen auf die Hydrolyse dieser Körper und ihre Abhängigkeit vom pH-Wert
hin; durch verschiedene Hydrolysengeschwindigkeit für Cyklopentanone und
Cyklohexanone konnten sie das verschiedene Verhalten der Vertreter dieser beiden
Gruppen — das schon früher konstatiert wurde[8] — erklären. Zum erstenmal betonten sie die Analogie des polarographischen Verhaltens der C = N-Gruppe mit
der Karbonylgruppe und schlugen ein Reduktionsschema vor. Aber auch diese
Autoren, wenn sie auch aus dem Einfluße des Überschusses des Girardschen Reagens auf die Wellenhöhe erkennen mußten, daß es sich um Gleichgewichtszustände handelt, verfolgten diese Zustände quantitativ nicht.

Die sehr komplizierten Systeme, die bei den Reaktionen des Formaldehyds
mit Aminen entstehen, können auf verschiedene Weise interpretiert werden.
L. Matoušek[11] fand bei einem Überschuß der Amine in Formaldehydlösungen
zwei Wellen, von denen er die erste der Reduktion des Methylolderivats des
Amins, welcher einer katalysierten Dehydratation unterliegt, die zweite der
Reduktion des Formaldehyds, zuschrieb. Das bei der Reaktion des Formaldehyds mit Harnstoff[12] entstehende Methylolderivat wird jedoch polarographisch
nicht reduziert, wie auch durch die Elektrolyse einer krystalisierten Substanz
bestätigt wurde. Die Entstehung einer zweiten Stufe erwähnen G. A. Crowe
und C. C. Lynch[13] nicht.

Endlich sollen noch zwei Arbeiten aus letzter Zeit, in denen die polarographische Reduzierbarkeit der C = N-Bindung verfolgt wurde, erwähnt werden.
V. Sapara[13] hat gefunden, daß Arekaidin-benzalhydrazid

$$CH_3-N\diagdown\begin{matrix}CH_2-CH\\CH_2-CH_2\end{matrix}\diagup C-CO.NH.N=CH-\diagup\diagdown$$

in 0,1 n NH$_4$Cl bei —1,2 V reduzierbar ist, obwohl die Verbindung vor Kondensation mit Benzaldehyd in dieser Lösung polarographisch inaktiv war. Benzaldehyd allein wird unter diesen Bedingungen erst bei negativeren Potentialen
reduziert. Wie beobachtet, unterliegt diese Substanz in saurer Lösung der
Hydrolyse.

E. Knobloch[14] hat in einer Studie der physikalisch-chemischen Eigenschaften des Vitamins K$_5$ (2-Methyl-1,4-aminonaftol) festgestellt, daß dieser
Stoff mit dem entsprechenden Chinonimin ein reversibles Redoxsystem, dessen
Halbstufenpotential um 100 mV positiver als das Potential des entsprechenden
2-Methyl-1,4-naphtochinons liegt, bildet. Ein ganz analoges Verhalten
wurde auch bei p-Aminophenol gefunden, auf dessen Oxydation D. Lester
und L. A. Greenberg[15] hinwiesen. Ähnlich wie bei den Cyklanonen und dem
erwähnten Hydrazid wurde bei Chinonimin eine Hydrolyse festgestellt.

$$\text{(Reaktionsschema Naphtochinonimin)} + H_2O \rightleftharpoons \text{(Naphtochinon)} + NH_3$$

Das Hydrolysenprodukt wurde polarographisch und nach Isolation mit Hilfe der UV-Absorptionsspektren identifiziert. Für die Hydrolyse des 4-Imino-2-methyl-1-naphtochinons wurde bei pH 6,8 die Konstante der monomolekularen Reaktion $k = 2,08 . 10^3$ sec^{-1} bestimmt.

Aus den angeführten Angaben der Literatur geht hervor, daß die Reduktion einer Verbindung mit der C=N-Gruppe gewöhnlich bei positiveren Potentialen als die Reduktion der entsprechenden Karbonylverbindung verläuft. In dieser Arbeit wurde das polarographische Verhalten einiger Karbonylverbindungen in konzentrierten gepufferten Lösungen der primären Amine bei einem pH in der Nähe von pK_b (wo K_b die basische Dissoziationskonstante des Amins ist) verfolgt.

Experimenteller Teil.

Alle Messungen wurden in dem Elektrolysengefäß nach M. Kalousek mit getrennter Kalomelelektrode, mit einer Kapilare, deren Konstanten bei verschieden Queck silberbehälterhöhen h) in einer Lösung von 1 n NH_3 1 n NH_4Cl bei —1,6 V ge messen wurden (siehe folgende Tabelle,

h (cm)	t (sec)	m (mg/sec)
86	1,15	4,50
66	1,73	3,00
46	2,4	2,12
26	3,5	1,45

in Abwesenheit der Luft durchgeführt. Die verwendeten Pufferlösungen sind in einzelnen Fällen angegeben. Bei Verfolgung der Abhängigkeit von der Konzentration des Ammoniaks wurde nach jeder Kurve die Konzentration des Ammoniaks titrimetrisch bestimmt. Die pH-Werte wurden mit Hilfe einer Glaselektrode gemessen.

Übersicht der Ergebnisse.

Die Reaktion der Brenztraubensäure mit Ammoniak.

An der polarographischen Kurve der Brenztraubensäure (im weiteren BH) in konzentrierten Pufferlösungen mit Ammoniak kann man zwei Wellen beobachten (Abb. 1), während wir in gepufferten Lösungen des gleichen pH in der Abwesenheit von Ammoniak nur die Welle des Anions B', die der zweiten Welle in den Lösungen mit Ammoniak entspricht, bemerken. Die Reduktionswelle, die bei Anwesenheit des Ammoniaks bei positiveren Potentialen als die der Reduktion der CO-Gruppeauftritt, entspricht der Depolarisation durch die $C=N$-Gruppe, die durch die Reaktion

$$CH_3.CO.COO' + NH_3 \rightleftharpoons CH_3.\underset{\underset{NH}{\|}}{C}.COO' + H_2O$$

$$\text{(1)}$$

oder $\qquad B' + NH_3 \rightleftharpoons Ko + H_2O$

entsteht. Aus der Abhängigkeit der Wellenhöhe von der Quecksilbersäule ergibt sich, daß die Gleichgewichtszustandseinstellung langsam im Vergleich zu der Geschwindigkeit des elektrolytischen Prozesses

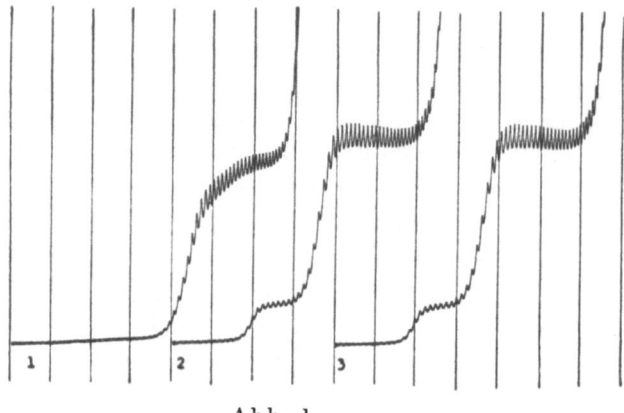

Abb. 1.

Die Welle der Brenztraubensäure (1.10^{-3} m) in einer Boratpufferlösung pH 9,4 (*1*), in ammoniakalischer Pufferlösung pH 9,3 (*2*) und in der Mischung dieser beiden Lösungen (*3*) derselben Konzentration von Ammoniak wie bei der Kurve (*2*). Die Kurven beginnen von —0,6 V, Abszissenabstand 200 mV, h = 32 cm, S = 1:10.

ist. Die Halbzeit der unter Eiskühlung vorgenommen Reaktion, die durch Beobachtung der Höhenänderung der ersten Welle mit der Zeit auf etwa 15 Sekunden geschätzt wurde, ist für die polarographische Bewertung der Reaktionsgeschwindigkeit zu kurz.

Bei Erhöhung der Ammoniakkonzentration (bei konstantem pH) steigt die Höhe der ersten Welle (Abb. 2). Die Abhängigkeit des Logarithmus der Ammoniakkonzentration von der Höhe der ersten Stufe (in % von der gesamten Stufenhöhe) hat die Form einer Dissoziationskurve (Abb. 4) und wird weiter als Dissoziationskurve bezeichnet.

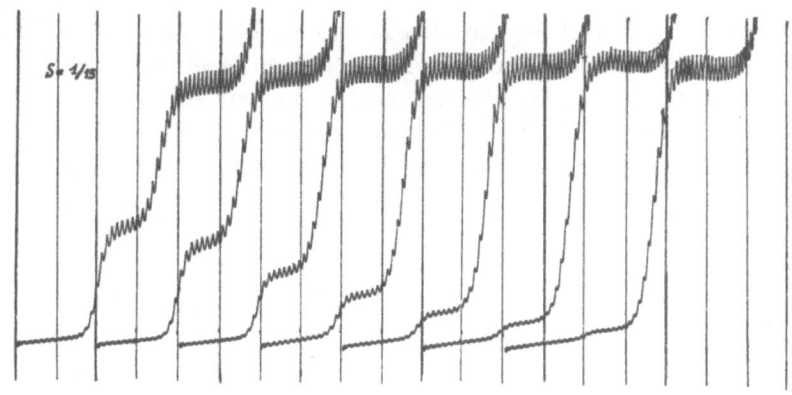

Abb. 2.

Die Reaktion der Brenztraubensäure mit Ammoniak. Die Abhängigkeit der Höhe der ersten Welle von der Ammoniakkonzentration bei $pH = pK_b$. Die Konzentration des Natriumsalzes der Brenztraubensäure war bei allen Kurven $4,7.10^{-4}$ m, die Konzentration des Ammoniaks wurde (von links nach rechts) von 1,94 zu 0,12 n geändert. Die Kurven wurden von —0,6 V an registriert, Abszissenabstand 200 mV, $h = 32$ cm, $S = 1:15$.

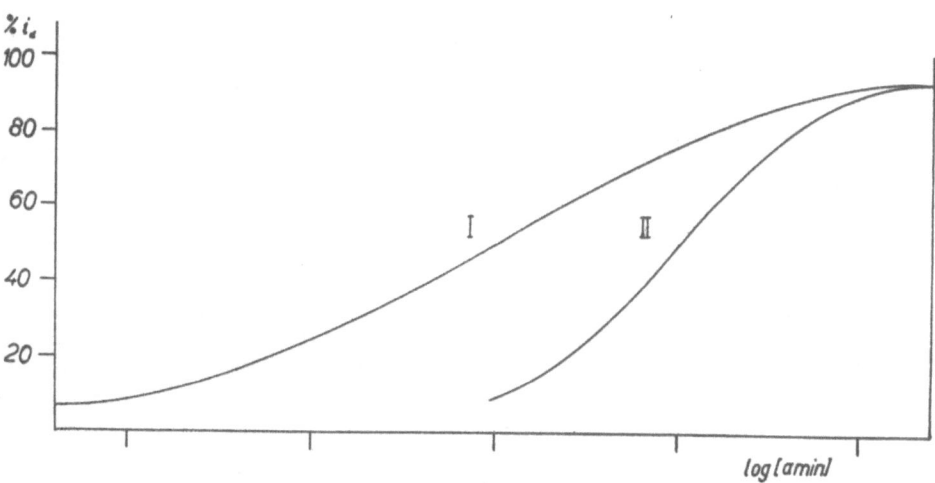

(In demselben Maßstab wie die anderen Diagramme.)

Abb. 3.

Der theoretische Verlauf der Abhängigkeit der Höhe der Welle des Iminoproduktes von der Ammoniakkonzentration (die theoretische Dissoziationskurve). Kurve I: Reaktion von einem Molekül des Amins mit einem Molekül der Karbonylverbindung. Kurve II: Reaktion von zwei Molekülen des Amins mit einem Molekül der Karbonylverbindung.

Wenn wir ihre Steilheit mit den theoretischen Kurven für die Reaktion eines und zweier Moleküle des Amins mit einem Molekül der Karbonylverbindung (Abb. 3) vergleichen, können wir die Übereinstimmung mit der Kurve der Reaktion eines Moleküls des Ammoniaks (Abb. 4) konstatieren, was mit der Stufenhöheabhängigkeit von der BH-Kon-

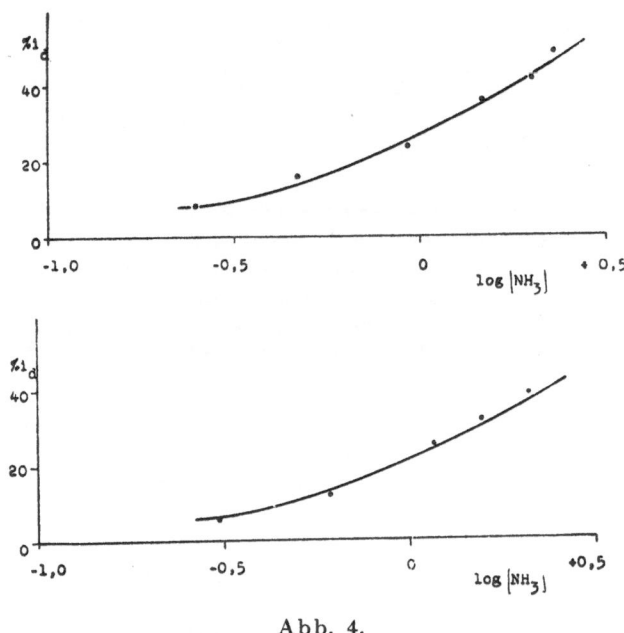

Abb. 4.

Reaktion der Brenztraubensäure mit Ammoniak.

Abhängigkeit der Höhe der ersten Welle (in Prozenten des gesamten Diffusionsstromes) von der Konzentration des Ammoniaks. Gezeichnet ist der theoretische Verlauf und die experimentellen Punkte. Im ersten Falle $(4,7.10^{-5}$ m $CH_3COCOONa)$ entspricht die gezeichnete theoretische Dissoziationskurve der Gleichgewichtskonstante $K = 0,30$, im zweiten Falle $(3.10^{-4}$ m $CH_3COCOONa)$ ist $K = 0,28$.

zentration in völligem Einklang steht. Die erreichbaren Teile der Dissoziationskurven, deren Verlauf durch Löslichkeit der Salze begrenzt war, wurden für BH-Konzentrationen von 5.10^{-5} bis 3.10^{-3} und mit Ammoniakkonzentrationen von 2,4 bis 0,1 n gewonnen. Als zweiter Bestandteil der Pufferlösungen wurden Sulfat und Chlorid des Ammoniums, derselben analytischen Konzentration wie Ammoniak, gebraucht. Bei $pH = pK_b$ wurden für die Gleichgewichtskonstante (K) Werte, von denen einige in dem Diagram auf Abb. 4 und in der Tabelle I zusammengefaßt sind, gefunden.

$$K = \frac{[Ko]}{[NH_3].[B']} = \frac{[Ko]}{[NH_3].(S_B - [Ko])} \tag{2}$$

wo S_B die analytische Konzentration der Brenztraubensäure (BH) bedeutet und für $[NH_3]$ die analytische Konzentration des Ammoniaks, gleich der Konzentration des Ammoniumsalzes, praktisch durch Bildung von Ko nicht verändert, gesetzt wurde.*)

Tab. I.

Gleichgewichtskonstanten der Reaktion von HB mit Amoniak

Anion	[B']	[NH$_3$]	% i$_{d_1}$	[Ko]	K	∅ K	K$_{graph.}$
Sulfat	$1,53.10^{-3}$	2,03	41	$0,63.10^{-3}$	0,35		
		1,40	31	0,48	0,32		
		0,99	22	0,34	0,28		
		0,475	12	0,18	0,28		
		0,24	5	0,08	0,23	0,295	0,29
Chlorid	3.10^{-4}	1,34	26	$0,79.10^{-4}$	0,265		
		1,15	23	0,70	0,26		
		0,89	17	0,51	0,23		
		0,62	11	0,34	0,21		
		0,31	6	0,18	0,20	0,233	0,26
Sulfat	3.10^{-4}	2,44	46	$1,38.10^{-4}$	0,35		
		2,12	38	1,15	0,295		
		1,59	32	0,95	0,29		
		1,18	26	0,77	0,29		
		0,62	12	0,36	0,25		
		0,31	6	0,17	0,20	0,28	0,28

Bemerkungen:

ln allen Fällen ist $pH = pK_b$.

% i$_{d_1}$ Der Beitrag der ersten Welle zu dem gesamten Diffusionsstrom in Prozenten.

∅ K Der Durchschnittswert der Gleichgewichtskonstante.

K$_{graph}$ Graphisch gefundene Konstante.

---

*) Ko entspricht Allgemein dem Kondensationsprodukt der Karbonylverbindung mit dem Amin.

Die gefundenen Werte stimmen in den Fehlergrenzen mit dem erreichbaren Teile der Dissoziationskurve in großem Konzentrationsbereich gut überein. — In stark konzentrierten Lösungen wurden Abweichungen, die durch die Maxima zweiter Art[16]) und durch eine erhebliche Aktivitätsänderung verursacht wurden, beobachtet. In den Chlorid- und Sulfatlösungen differieren die gefundenen Werte der Konstanten einigermaßen.

Als die Abhängigkeit der Höhe der ersten Welle von pH und von Ammoniakkonzentration bei $pH \neq pK_b$ untersucht wurde, mußte bei der Ermittlung der Gleichgewichtskonstanten die Dissoziation des Ammoniaks in Erwägung gezogen werden. Bei Bezeichnung der analytischen HB-Konzentration als S_B und derjenigen des Ammoniaks als S_N und bei Vernachlässigung der [Ko] gegen S_N (ein Unterschied von 3 Größenordnungen) gilt für die Gleichgewichtskonstante:

$$K = \frac{[Ko]\left(1 + \dfrac{[H^+]}{K_b}\right)}{S_N\,(S_B - [Ko])} \tag{3}$$

Es wurden die Wellen in Lösungen von 0,5 n und 1 n NH_3 von pH 8,35 bis 10,05 (mit wechselnder Konzentration des Ammoniumsalzes) verfolgt und es wurde — in Übereinstimmung mit der eingeführten Gleichung — gefunden, daß die Höhe der ersten Welle nur in dem Maße vom pH-Wert beeinflußt wird, wie sie sich mit der pH-Änderung die Konzentration des freien NH_3 ändert (Abb. 5). Der Wert der berechneten Konstanten hängt also vom pH nicht ab, wie aus dem Beispiel in der Tabelle II ersichtlich ist.

Tab. II.

Gleichgewichtskonstante der Reaktion der Brenztraubensäure mit Ammoniak (bei verschiedenem pH gewonnen).

$[NH_3] = 1$ n $6,75 . 10^{-4}$ m $CH_3COCOONa$

pH	S_N	% i_{d_1}	[Ko]	K	Ø K
8,75	6,0	29,3	$1,98 . 10^{-4}$	0,29	
8,84	4,2	25,8	1,73	0,295	
9,04	3,0	23,4	1,58	0,27	
9,10	2,0	20,2	1,36	0,31	
9,29	1,5	19,0	1,27	0,295	
9,70	1,25	17,7	1,19	0,235	0,297

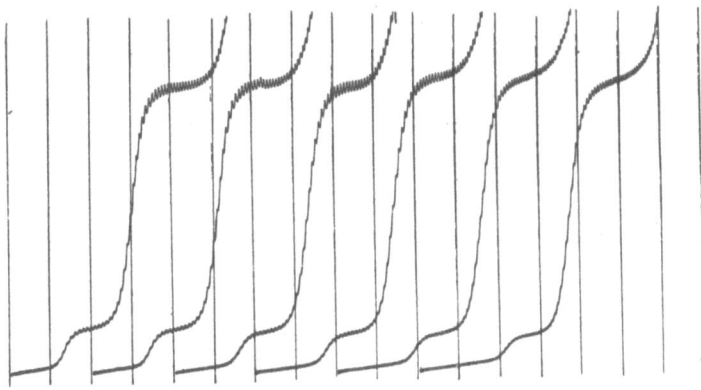

Abb. 5.

Reaktion der Brenztraubensäure mit Ammoniak.

Die Abhängigkeit der Höhe der ersten Welle von pH bei konstanter Konzentration des zugegebenen Ammoniaks (0,5 n). Die Konzentration des NH_4Cl wurde von 3 n bis 0,08 n geändert, wobei die Ionenstärke durch Zugabe von NaCl konstant gehalten wurde. Die Kurven entsprechen (von links nach rechts) dem pH von: 8,35, 8,55, 8,94, 9,22, 9,53, 10,05. In den Lösungen von 4.10^{-4} m $CH_3COCOONa$ sind die Kurven ab 0,8 V registriert, bei h = 42 cm, Abszissenabstand 200 mV und S = 1:15. Durch vielfache Veränderung der analytischen Konzentration des Ammoniaks wird die aktuelle Konzentration von NH_3 etwas vermehrt, wie wir aus diesem Polarogramm in Übereinstimmung mit Formel (3) sehen.

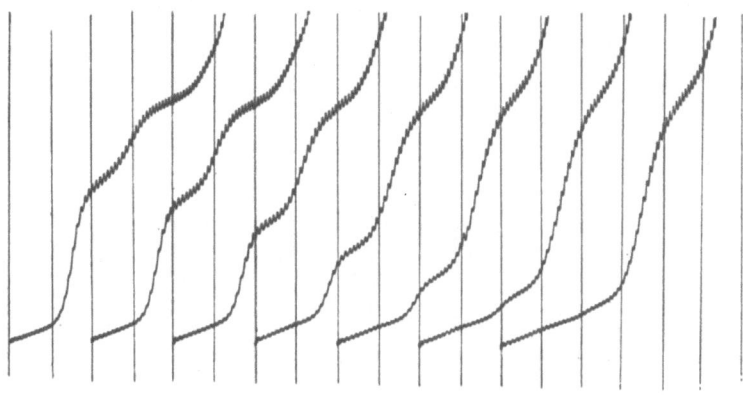

Abb. 6.

Reaktion der Brenztraubensäure mit Glycin.

Abhängigkeit der Höhe der ersten Welle von der Glycinkonzentration. Die Konzentration wurde (von links nach rechts) von 1,94 zu 0,117 herabgesetzt, 9.10^{-4} m $CH_3COCOONa$, Kurven wurden bei 0,8, V begonnen, Abszissenabstand 200 mV, h = 42 cm, S = 1:7.

Die Reaktion der Brenztraubensäure mit Glycin, Alanin und Colamin.

Analog zu den Reaktionen mit Ammoniak ist es möglich die Reaktionen der Brenztraubensäure (BH) mit genügend löslichen Aminen polarographisch zu verfolgen. An der Reaktion nimmt wieder das Anion B′ und die Base des Amins, resp. das Anion der Aminosäure mit undissoziierter Aminogruppe, teil. Die Berechnung der Gleichgewichtskonstante kann nach der Gleichung (3) durchgeführt werden. Beispiele so berechneter Werte sind in der Tabelle III. angeführt.

Die Erhöhung der ersten Welle mit steigender Konzentration des Amins ist auf der Abbildung 6 zu sehen. Graphisch ist die Abhängig-

Tab. III.

Gleichgewichtskonstanten der Reaktionen der Brenztraubensäure mit Glycin, Alanin und Colamin.

$pH = pK_b$ Konstante Ionenstärke

Amin	[B′]	$S_N/2$	K	⌀K
Glycin	$5,6 . 10^{-4}$	0,93	2,50	
		0,59	2,43	
		0,37	2,66	
		0,23	2,66	
		0,14	2,50	
		0,095	2,41	
		0,055	2,47	
		0,037	2,13	2,47
Alanin*)	$9 . 10^{-4}$	0,91	0,92	
		0,54	1,04	
		0,35	1,02	
		0,22	0,99	
		0,138	0,90	0,974
Colamin	$7,4 . 10^{-1}$	0,59	2,76	
		0,37	2,98	
		0,235	2,99	
		0,145	3,10	
		0,087	3,42	
		0,058	3,34	
		0,03	3,17	3,108

*) Das Präparat wurde freundlicherweise von Dr. J. Koštíř überreicht.

keit der Wellenhöhe vom Logarithmus des Amins in Abb. 7 dargestellt. Die Steilheit der Kurve (hier konnte experimentell ein größerer
Teil der Kurve ermittelt werden) entspricht wieder in allen Fällen
der Reaktion von einem Molekül des Amins mit einem Anion B'
(siehe Abb. 3). Die Übereinstimmung der experimentellen Punkte
mit dem theoretischen Verlauf ist am besten bei konstanter Ionenstärke. Bei einem pH kleiner als 9,5 blieb die Gleichgewichtskonstante
vom pH unabhängig und die Änderung der Wellenhöhe war nur der

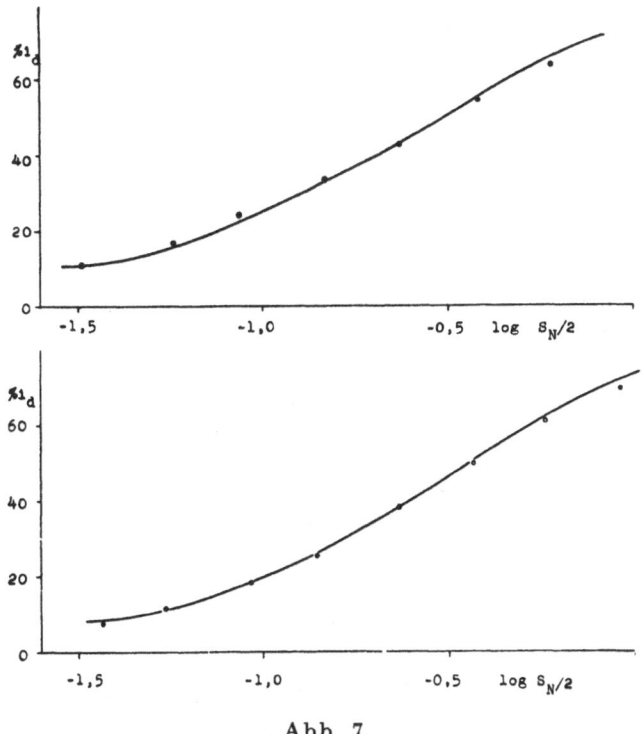

Abb. 7.

Reaktion der Brenztraubensäure mit Glycin und Colamin.

Graphische Darstellung der Höhe der ersten Welle (in Prozenten des gesamten
Diffusionsstromes) in der Abhängigkeit von der Aminokonzentration. Angegeben
sind die theoretischen Kurven (vergleiche Abb. 3) und die experimentellen
Punkte. S_N ist analytische Konzentration des Colamins (oben) und des Glycins
(unten.)

Die höhere Kurve ist für die Reaktion mit Colamin bei konstanter Ionenstärke
in der Lösung von $7,35 \cdot 10^{-4}$ m $CH_3COCOONa$ gezeichnet und entspricht der
Gleichgewichtskonstante $K = 3,20$.

Die zweite Kurve gehört der Reaktion der Brenztraubensäure mit Glycin
bei $pH = pK_b$ mit $5,6 \cdot 10^{-4}$ m $CH_3COCOONa$ an und entspricht einem $K = 2,68$.

Konzentrationsänderung der freien Base gleich. Bei pH 10,2 und größer wurde bei der Reaktion mit Glycin eine Abnahme des Wertes der Konstante beobachtet und durch die Abhängigkeit der Wellenhöhe von der Quecksilbersäule wurde festgestellt, daß diese Abweichung dadurch verursacht wird, daß der elektrolytische Vorgang nicht nur durch bloße Diffusion limitiert ist.

Keine andere der zugänglichen Aminosäuren mit primärer Aminogruppe (Leucin, Arginin, Cystein und Cystin — die weniger löslichen wurden nicht geprüft) war genügend löslich, um die verfolgte Reaktion zu ermöglichen, mit Ausnahme des Histidins, dessen Reaktion sowie diejenige des Histamins, an einer anderen Stelle behandelt wird.[17]

Die Reaktion wurde in der Anwesenheit der sekundären und tertiären Amine, wie Di- und Trimethylamin, Tetramethylammonium Chlorid und Hydroxyd, Di- und Triethanolamin und Diphenylamin, nicht beobachtet. Bei manchen Aminen — wie bei Anilin, p-Aminophenol, o-Phenylendiamin und Benzidin — genügt ihre Löslichkeit in alkalischer Lösung nicht zur Erreicung des notwendigen Überschusses.

Die Oxydation der Ascorbinsäure in Lösungen von Ammoniak.

Bei der Oxydation von verdünnten Lösungen der Ascorbinsäure durch Sauerstoff in Pufferlösungen, die eine höhere Konzentration von Ammoniak enthalten, wurde das Abnehmen der anodischen Oxydationsstufe der Ascorbinsäure und die Entstehung einer Reduktionswelle bei —1,5 V beobachtet (Abb. 8). Außer dieser Welle wurde noch eine andere Welle in dem Potentialgebiet, wo die Peroxydreduktion eintritt, beobachtet. Die Höhe dieser Welle ist in keinem stöchiometrischen Verhältnis zu der Ascorbinsäurekonzentration, ändert sich mit dem pH, der Ionenstärke und der Zusammensetzung der Lösung. In Pufferlösungen, die kein Ammoniak enthalten, können wir nach Oxydation der Ascorbinsäure nur diese Welle, die vermutlich dem, bei der Reduktion von Sauerstoff entstandenen, Hydroperoxyd entspricht, finden. Ein analoges Verhalten wurde auch bei dem Oxydationsprodukt eines anderen Endiols — des Kumarindiols*) beobachtet. Wenn als Oxydationsmittel p-Chinon unter denselben Bedingungen benützt wurde, verläuft die Oxydation nicht n u r in einer Rictung, sondern es entstanden verschiedene Produkte.

---

*) Das Präparat wurde von prof. F. Arndt (Istanbul) hergestellt.

Aus der linearen Beziehung zwischen der Wellenhöhe und der Konzentration der oxydierten Ascorbinsäure ist es klar, daß es sich um die Reduktion einer Verbindung, deren ein Mol aus einem Mol der Ascorbinsäure entsteht, handelt. Die Abhängigkeit der Stufenhöhe von dem Logarithmus der Ammoniakkonzentration (Abb. 9 und 10) bei $pH = pK_b$ hat die Form einer Dissoziationskurve, die mit der theoretischen, deren Steilheit der Reaktion von zwei Molekülen des

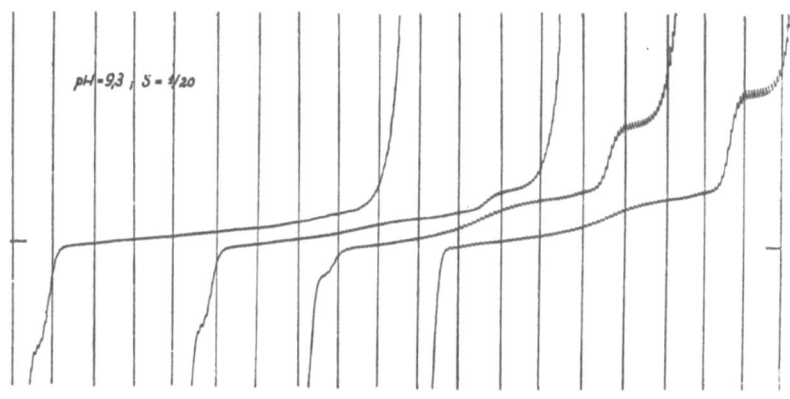

Abb. 8.

Oxydation der Ascorbinsäure in Gegenwart von Ammoniak. 3.10^{-4} m Ascorbinsäure wurde durch Einleiten von Sauerstoff in 1,5 n NH_3 1,5 n NH_4Cl oxydiert. Der Sauerstoff wurde (von links) 0 7 15 und 60 Sekunden in einem gelinden Strom eingeleitet. Die Kurven beginnen bei 0,2 V (Merkurosulfatanode), der Abszissenabstand ist 200 mV, h = 42 cm S = 1:20.

Ammoniaks mit einem Molekül des Oxydationsproduktes der Ascorbinsäure entspricht, übereinstimmt. Bezeichnen wir das Oxydationsprodukt der Ascorbinsäure OxA (wahrscheinlich die 2,3-Diketogulonsäure), dann folgt aus dem Reaktionsschema

$$OxA + 2NH_3 \rightleftarrows Ko$$

für die Gleichgewichtskonstante K — wenn wir die analytische Konzentration des Ammoniaks S_N und diejenige der Ascorbinsäure S_A bezeichnen — der Ausdruck:

$$K = \frac{[Ko] \left(1 + \frac{[H^+]}{K_b}\right)^2}{(S_A - [Ko]) \cdot S_N^2} \tag{4}$$

wenn wir wieder (Ko) gegen S_N vernachlässigen.

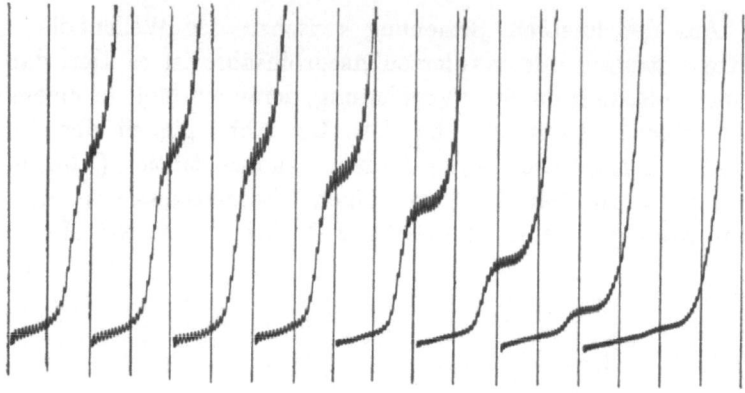

Abb. 9.

Oxydation der Ascorbinsäure in Gegenwart von Ammoniak. Die Lösungen von
$3 . 10^{-4}$ m Ascorbinsäure, in denen die Ammoniakkonzentration bei pH = pK$_b$
2,3 bis 0,03 m (von links) war, wurden durch Sauerstoff oxydiert. Die Ionenstärke
wurde konstant gehalten. Die Kurven beginnen bei 1,4 V, Abszissenabstand
200 mV, h = 37 cm S = 1 : 15.

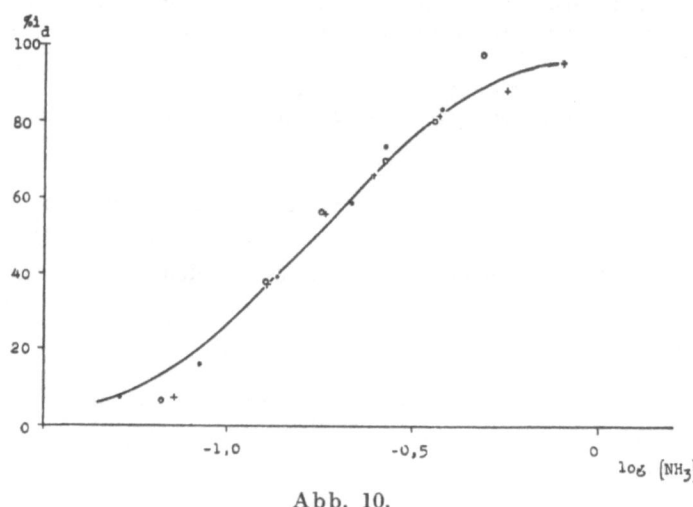

Abb. 10.

Oxydation der Ascorbinsäure in Gegenwart von Ammoniak. Abhängigkeit der
Wellenhöhe von der Ammoniakkonzentration. Angegeben ist die theoretische
Kurve für die Reaktion eines Moleküls der Karbonylverbindung mit zwei
Molekülen des Ammoniaks und die experimentellen Punkte.

Die Bestimmung der Gleichgewichtskonstante wurde bei einem pH = pK_b für 3.10^{-3} und 3.10^{-4} m. Ascorbinsäure in Pufferlösungen mit 1,0 bis 0,03 n NH_3, durchgeführt. Einige Beispiele sind in der Tabelle IV. angegeben. In Anwesenheit von Neutralsalzen wurde eine Deformation der Dissoziationskurven beobachtet.

Die Abhängigkeit von der Ammoniakkonzentration wurde auch bei anderen pH-Werten verfolgt und bei der Berechnung wurde dann die Dissoziation des Ammoniaks beachtet (Abb. 11).

Tab. IV.

Gleichgewichtskonstanten der Reaktion des Ammoniaks mit dem Oxydationsprodukt der Ascorbinsäure.

Konz. der Ascorbinsäure	$[NH_3]$	$[Ko]$	K	∅ K	pH = = pK_b
3.10^{-3}	0,47	$2,38.10^{-3}$	20,8		
	0,31	1,94	21,0		
	0,23	1,61	23,8		
	0,16	1,09	23,4	22,2	
3.10^{-4}	0,47	$2,46.10^{-4}$	25,4		
	0,33	2,19	28,1		
	0,27	1,74	20,6		
	0,17	1,16	23,0		
	0,106	0,49	17,7	23,0	

Die Reaktion der primären Aminen mit Aceton [16a]) und Cyklo-hexanon.*

In Anwesenheit von Ammoniak und primären Aminen wurden in den Lösungen von Aceton und Cyklohexanon Reduktionswellen bei —1,5 V beobachtet, obzwar in einer Boratpufferlösung von demselben pH wie auch in 0,1 n LiOH keine Reduktionswellen auf den polarographischen Kurven bemerkt wurden. Die Höhe der Stufe in den Ammoniaklösungen ist linear von der Konzentration des Acetons bei einem bestimmten Überschuß des Amins und konstanten pH abhängig (Abb. 12) und ist nur ein Teil des gesamten Diffusionsstromes einer Zweielektronenreduktion. Die Wellenhöhe ist praktisch dieselbe in

---

*) Die redestillierten Präparate wurden freundlichst von Dr. M. Vondráček überreicht.

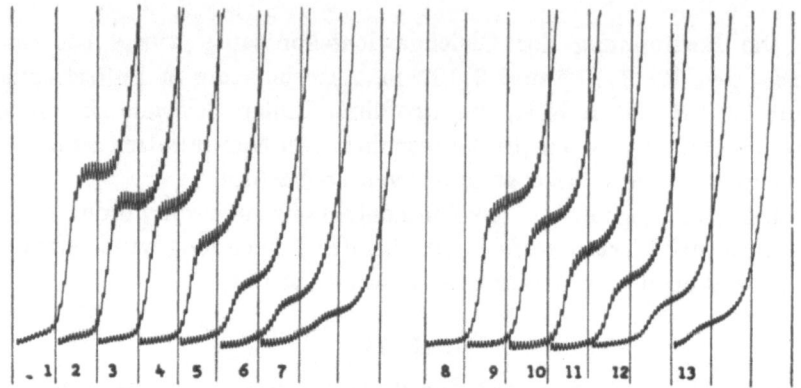

Abb. 11.

Oxydation der Ascorbinsäure in Gegenwart von Ammoniak. Abhängigkeit von der Ammoniakkonzentration bei pH 9,28 und 9,88. Kurve 1—7: 1n NH_3 1 n $(NH_4)_2SO_4$ pH 9,28 (1) wurde so verdünnt, daß die Ammoniakkonzentration allmählich bis auf 0,068n (7) sinkt.

Kurve 8—13: 1 n NH_3 0,25 n $(NH_4)_2SO_4$ pH 9,88 (8) wurde bis auf 0,095 n NH_3 verdünnt.

Die Konzentration der Ascorbinsäure überall 4.10^{-4} m, Oxydation durch Sauerstoff. Anfang der Kurven bei 1,2 V, Abszissenabstand 200 mV, h = 42 cm, S = 1:20.

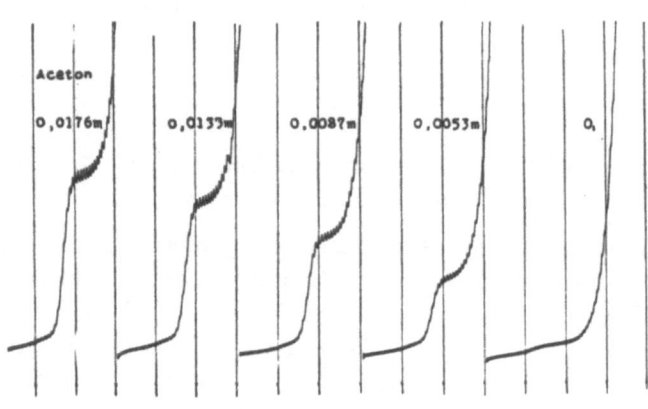

Abb. 12.

Die Reaktion des Acetons mit Ammoniak.

Abhängigkeit von der Acetonkonzentration in 2,5 n NH_3, 2,5 n $(NH_4)_2SO_4$ Konzentration des Acetons:

0,0176 m   0,0133 m   0,0087 m   0,0053 m   0

Die Kurven beginnen bei 1,2 V, Abszissenabstand 200 mV, h = 42 cm, S = 1:20.

Mustern von Aceton verschiedener Herkunft, was die Möglichkeit, daß es sich um eine Verunreinigung handelt, ausschließt und ihre Höhe nimmt dem mit Durchströmen von Stickstoff allmählich ab, was mit der Flüchtigkeit des Acetons übereinstimmt.

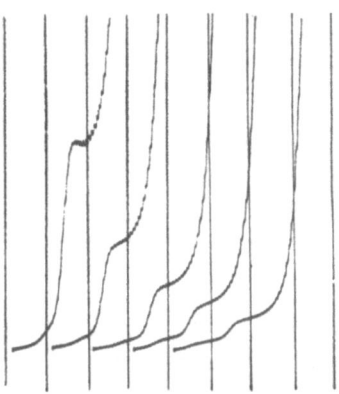

Abb. 13.

Abhängigkeit von der Ammoniakkonzentration. Die Konzentration des Ammoniaks wurde bei einem $pH = pK_b$ von 2,08 bis 0,34n geändert (von links nach rechts). Konzentration von Cyklohexanon $4 . 10^{-3}$ m. Die Kurven beginnen bei 1,2V, Abszissenabstand 200 mV, h = 42 cm, S = 1:20.

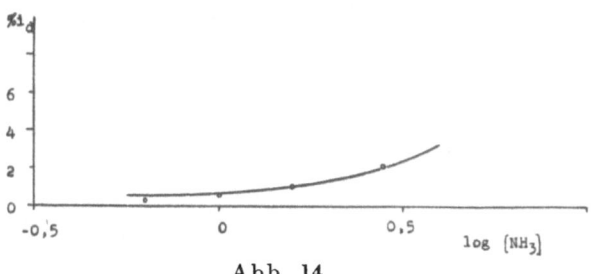

Abb. 14.

Reaktion des Acetons mit Ammoniak.

Abhängigkeit der Wellenhöhe von der Ammoniakkonzentration. Gezeichnet ist die theoretische Kurve und die experimentellen Punkte.

Aus der linearen Beziehung zwischen der Wellenhöhe und der Konzentration folgt, daß auf ein Mol der reduzierten Substanz ein Mol des Acetons kommt und aus dem Vergleich der Abhängigkeit der Wellenhöhe von der Aminkonzentration, die auf Abb. 13 und 14 sichtbar

Tab. V.

Gleichgewichtskonstanten der Reaktion des Acetons mit Ammoniak.

3,7.10^{-2} m Aceton   pH = pK_b

[NH$_3$]	[Ko]	K'	ø K'
2,50	0,077.10^{-2}	0,0084	
1,58	0,039	0,0068	
0,99	0,022	0,0060	
0,63	0,012	0,0049	
0,37	0,007	0,0052	0,0064

Tab. VI.

Gleichgewichtskonstanten der Reaktion des Cyklohexanons mit
Ammoniak.

3,8.10^{-3} m Cyklohexanon   pH = pK_b

% i_d	[NH$_3$]	[Ko]	K'	ø K'
4,5	2,08	0,12.10^{-3}	(0,0157)	
2,25	1,45	0,059	0,0107	
1,22	0,92	0,032	0,0093	
0,75	0,56	0,020	0,0095	
0,42	0,34	0,011	0,0085	0,0095

ist, wurden mittels der Gleichung (3) (nach Modifizierung der Symbole)
die Werte der Gleichgewichtskonstanten berechnet und in Tab. V
und VI zusammengefaßt.

Wenn wir die analytische Konzentration des Glycins konstant halten,
ändert sich die Wellenhöhe mit der pH-Änderung nur in dem
Maße, wie sich die Konzentration der Glycinanionen mit undissoziierter
Aminogruppe ändert (Abb. 15). Die Werte der Konstante, wie sie
bei verschiedenen pH gefunden wurden, sind in Tab. VII und VIII
angegeben.

Die Bestimmung von Aceton (und ähnlich von Cyklohexanon)
wird durch Zugabe eines kleinen Volumen der Probe zu 10 ml. der
Pufferlösung, aus welcher vor der Zugabe der Sauerstoff entfernt wurde.
durchgeführt. Den Rest des Sauerstoffs beseitigen wir durch eine
20 sec dauernde Durchströmung mit Stickstoff. Zu der Bestimmung

Tab. VII.

Gleichgewichtskonstanten der Reaktion des Acetons mit Ammoniak bei verschiedenem pH.

[NH₃] = 1,0 n   3,3 . 10⁻² m Aceton

pH	% i_d	S_N	K'	⌀ K'
8.75	1.47	6,0	0,0094	
8,84	1,03	4,2	0,0081	
9,09	0,74	3,0	0,0056	
9,28	0,56	2,0	0,0050	
9,44	0,455	1,5	0,00455	
9,88	0,43	1,25	0,0040	0,0061

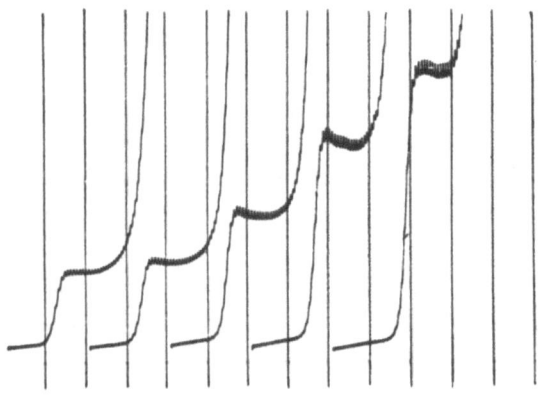

Abb. 15.

Reaktion des Acetons mit Glycin.

Abhängigkeit der Wellenhöhe vom pH bei konstanter analytischen Glycinkonzentration. Das pH wurde von 8,47 bis 10,25 (von links) geändert, die Konzentration des Glycins war stets 2 m, die Konzentration des Acetons 2,6.10⁻² m. Anfang bei 1,2 V, Abszissenabstand 200 mV, h = 42 cm, S = 1:30.

benützen wir die Methode der konstanten Zugabe. Als Pufferlösung wurde 2,5n NH₃ 2,5 n (NH₄)₂SO₄ oder ein 2 m Glycinpuffer mit pH 10 gebraucht. Unter diesen Bedingungen war die Bestimmung etwa 20mal weniger empfindlich als die gewöhnlichen polarographischen Methoden und es war möglich bis 10⁻³ m Lösungen von Aceton zu bestimmen. In mit Ammoniumchlorid gesättigtem 12n NH₃ gelang die Bestimmung von Konzentrationen der Größenordnung 10⁻⁴ m.

Tab. VIII.

Gleichgewichtskonstanten der Reaktion des Acetons mit Glycin
bei verschiedenem pH.

$S_N = 2,0$ m 　　　　　　　　0,017 m Aceton 　　　　　　　　$K_B = 4.10^{-4}$

pH	% i_d	[Ko]	K'	⌀ K'
8,74	2,57	0.044	0,073	
	2,64	0,045	0,076	
8,90	3,00	0,051	0,065	
	3,16	0,054	0,068	
9,25	4,55	0,077	0,057	
	4,57	0,078	0,058	
9,52	7,10	0,121	0,089	
	7,15	0,122	0,089	
10,25	10,0	0,17	0,063	0,071

Da das Halbstufenpotential des Cyklohexanons nur um etwa
20 mV negativer (—1,54) als das Halbstufenpotential des Acetons
(—1,52 V) liegt, ist die gleichzeitige Bestimmung dieser Ketone, auch
mit der Derivationsschaltung nach J. Vogel und J. Říha[18]), unmöglich.

Andere Karbonylverbindungen.

Auch in den Lösungen von Acetaldehyd kann man an polarographischen Kurven in Anwesenheit von Ammoniak zwei Stufen beobachten, von denen die negativere der Welle im Boratpuffer von
demselben pH entspricht. (Abb. 16). Aus der Abhängigkeit der Stufenhöhe von der Quecksilbersäule geht hervor, daß die erste Welle der
Gleichung von Ilkovič entspricht, während die zweite Welle schwer
meßbar ist. Beide Wellen vermindern sich mit der Zeit und zwar in
konzentrierteren Lösungen (etwa 10^{-2} m) viel schneller als in verdünnten. Bei der Konzentration von Acetaldehyd von der Ordnung 10^{-4} m
sind die Wellen im Laufe von einer halben Stunde konstant. Dieses
Verhalten deutet darauf hin, daß in der Lösung zwei reduzierbare
Formen, die sich in einem langsam einstellenden Gleichgewichte befinden
bestehen. Eine dieser Formen unterliegt entweder der Polymerisation
oder der Polykondensation, durch die ein polarographisch inaktives
Produkt vom Typus des Paraldehyds oder des Aldehydammoniaks
entsteht.

Wenn wir die Wellen bei niedrigen Konzentrationen verfolgen, können wir bemerken, daß die Höhe der beiden Wellen linear mit der Konzentration des Acetaldehyds steigt. Aus der Abhängigkeit der Höhe der ersten Welle von der Ammoniakkonzentration folgt, daß an dem, den Elektrodenprozeß bestimmenden, Gleichgewicht ein Molekül des Acetaldehyds und ein Molekül des Ammoniaks teilnimmt.

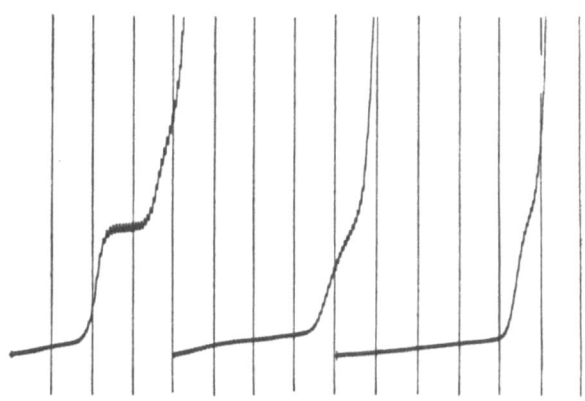

Abb. 16.

Reaktion des Acetaldehyds mit Ammoniak. Cca 0.01% Acetaldehyd in (von links) 1,5 n NH_3 1,5 n NH_4Cl, in Boratpufferlösung von pH 9,4 und in 0,1 n LiOH. Die Kurven beginnen bei 1,0 V, Abszissenabstand 200 mV, h = 42 cm S = 1:20.

Als der Quotient aus den Höhen der ersten Welle und der Summe beide Wellen der prozentuellen Konzentration der Iminoform gleichgesetzt wurde, war es möglich den approximativen Wert der Gleichgewichtskonstante K' zu berechnen. Die Werte, welche hauptsächlich durch Fehler bei der Vermessung der zweiten Stufe belastet sind, sind in der Tabelle IX. angeführt.

Tab. IX.

Gleichgewichtskonstanten der Reaktion des Acetaldehyds mit Ammoniak.

0,01% Acetaldehyd   pH = pK_b

[NH_3]	% i_d	[Ko]	K'	⌀ K'
1,58	50	0,50	0,63	
1,00	42	0,42	0,72	
0,63	33	0,33	0,72	
0,37	13	0,13	0,40	
0,25	9	0,09	0,40	0,54

In ammoniakalischen Lösungen von Benzaldehyd bemerken wir die Welle der Iminoform bei um 210 mV positiveren Potentialen als die Welle der Reduktion des aldehydischen Karbonyls (Abb. 17). Die Höhen der beiden Stufen sind durch Diffusion begrenzt. Das Gleichgewicht in einer Lösung von 1,5n NH_3 1,5 n NH_4Cl, die etwa $1 . 10^{-2}$ m Benzaldehyd enthält, stellt sich langsamer als in allen bis jetzt erwähnten Fällen ein. Hier erfolgt aber noch eine konsekutive Reaktion bei der sich die Wellenhöhe vermindert und ein polarographisch inaktives

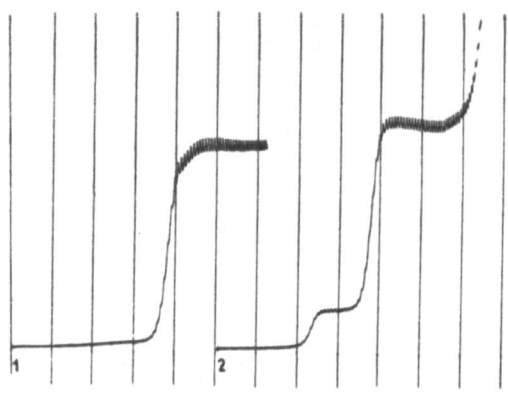

Abb. 17.
Reaktion des Benzaldehyds mit Ammoniak.
Kurve 1: $5 . 10^{-3}$ m Benzaldehyd in Boratpufferlösung pH 9,4
Kurve 2: $5 . 10^{-3}$ m Benzaldehyd in 1,5n NH_3 1,5n NH_4Cl
Die Kurven beginnen bei 0,6 V, Abszissenabstand 200 mV, h = 42 cm, S = 1:70.

Produkt entsteht (Abb. 18). Es handelt sich wahrscheinlich um eine Polykondensation, die bei niedrigen Konzentrationen des Benzaldehyds wesentlich langsamer verläuft. Wenn die Registration der Kurven 2—4 Minuten nach der Zugabe des Benzaldehyds zur gepufferten Aminlösung geschieht, können wir die Abhängigkeit von der Benzaldehyd- und Ammoniakkonzentration vergleichen. Die Höhen der beiden Wellen steigen mit der Benzaldehydkonzentration linear an und ihr Quotient bleibt unverändert, was für die Reaktion von einem Molekül des Benzaldehyds zeugt. Dieses reagiert mit einem Molekül des Ammoniaks, wie aus der Ammoniakkonzentrationsabhängigkeit hervorgeht. Die Gleichgewichtskonstanten dieser Reaktion, nach der modifizierten Gleichung (3) berechnet, sind in der Tabelle X. angegeben.

Gleichgewichtskonstante der Reaktion des Benzaldehyds mit Ammoniak.

Etwa 1.10^{-3} m Benzaldehyd $pH = pK_b$

[NH$_3$]	[K$_0$]	K′	Ø K′
2,50	0,31 . 10^{-3}	0,185	
1,58	0,21	0,17	
	0,22	0,18	
1,00	0,13	0,155	
	0,12	0,135	
0,63	0,07	0,12	
	0,067	0,115	
0,37	0,04	0,115	
	0,044	0,125	0,145

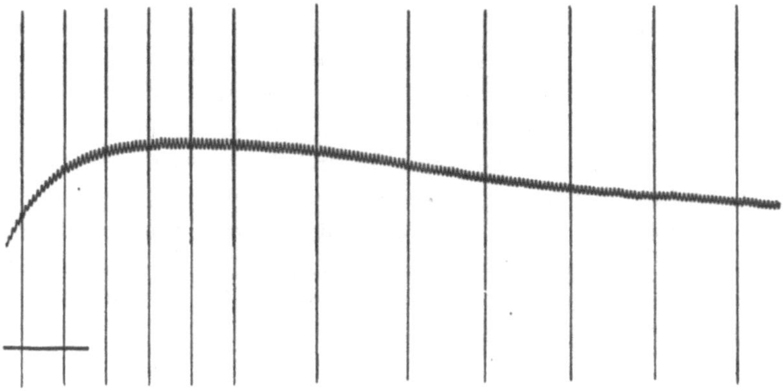

Abb. 18.

Reaktion des Benzaldehyds mit Ammoniak.

Zeitverlauf der Höhe der ersten Welle. 5.10^{-3} m Benzaldehyd in 1,5 n NH$_3$, 1,5 n NH$_4$Cl.

Die Höhe der ersten Welle wurde bei 1,2 V registriert, Abszissenabstand 30 (später 60) Sekunden, h = 42 cm, S = 1:30.

In Gegenwart von Ammoniak können in den Lösungen von Benzil einerseits Polykondensationen, bei denen Stoffe wie Benzilam, Imabenzil und Benzilimid, in welchen zwei bis drei Benzilmoleküle mit einem Molekül des Ammoniaks kondensieren, entstehen, andererseits hydrolytische Spaltungen vorkommen, sodaß die Verhältnisse in der Lösung sehr kompliziert werden. In 1,25 n NH_3 1,25n NH_4Cl mit 50% Ethanol war es unmöglich, die Reaktion zwischen Benzil und Ammoniak zu verfolgen — nur eine allmähliche Zersetzung (wahrscheinlich Hydrolyse des Benzils) wurde in dieser Lösung beobachtet. Wenn die Konzentration des Ethanols auf 20% herabgesetzt wurde, dann verliefen in Gegenwart von 2n NH_3 2 n $(NH_4)_2SO_4$ zwei Reaktionen: Einerseits nahm die gesamte Benzilwelle ab (wahrscheinlich durch Polykondensation oder Hydrolyse) und gleichzeitig ist bei den um 100 mV positiveren Potentialen eine Welle des Iminoderivates entstanden. In 1,2 n NH_3 1,2 n NH_4Cl verlaufen die beiden Reaktionen analog, nur ist die Zersetzung langsamer und die Höhe der ersten Welle kleiner.

In den Lösungen von Diacetyl mit höherer Konzentration von Ammoniak können wir an den polarographischen Kurven drei Wellen beobachten. Die erste von ihnen wächst mit steigender Ammoniakkonzentration auf Kosten der zweiten, die um 110 mV negativer ist. Diese zweite Welle findet man auch in anderen Pufferlösungen von gleichem pH und sie wurde zu analytischen Zwecken gebraucht. Diese Welle ist aber nur ein Teil des gesamten Diffusionsstromes einer Zweielektronenreduktion. Es wurde noch eine dritte Welle bei—1,6 V beobachtet, die wahrscheinlich einer Verunreinigung entspricht. Die ersten zwei Wellen nehmen unter diesen Bedingungen schnell mit der Zeit ab, wahrscheinlich unter Entstehung von Polykondensaten. Wenn die Abhängigkeit von der Ammoniakkonzentration nach einem bestimmten Zeitintervall registriert wurde (½ Min. nach Vermischen), war es möglich, approximativ den Wert der Gleichgewichtskonstante K' nach der Gleichung (3) zu berechnen.

Tab. XI.

Die Gleichgewichtskonstante der Reaktion des Diacetyls mit Ammoniak.

Etwa 0,01% Diacetyl   pH $=pK_b$

NH_3	% i_d	K'	ø K'
2,50	8,0	0,035	
1,57	6,8	0,046	
1,00	5,05	0,053	
0,63	4,1	0,068	
0,375	1,84	0,050	0,050

Das Halbstufenpotential der Reduktion der Iminoform ist in ammoniakalischen Lösungen von Phenylglyoxylsäure um 220 mV positiver als das Potential der Reduktion des Anions. Die Wellenhöhe der Iminoform erreicht ihren Gleichgewichtswert erst im Laufe von einigen Stunden und das Verhältnis der Stufen entspricht der $K' = 0,1$.

Bei Benzophenon wurde die Reaktion in der Lösung von 50% Ethanol nicht einmal nach längerer Zeit beobachtet.

Diskussion.

Wenn auch die polarographische Methode zur Verfolgung der Gleichgewichtszustände in Lösungen sehr geeignet ist, weil es bei ihr gewöhnlich nicht zu einem Eingriff in die Lösung kommt, wurde sie bis jetzt (vor allem bei organischen Systemen) nur selten angewendet.[19]

Bei den Gleichgewichtszuständen, bei denen mindestens eine der beteiligten Komponenten in dem erreichbaren Potentialbereich polarographisch aktiv ist, kann man — wie es in letzter Zeit J. Koryta und I. Kössler[20] betonten — die Gleichgewichtszustände in zweierlei Weise verfolgen: Bei den Systemen, wo der Elektrodenprozess reversibel und die Einstellung des Gleichgewichtes vollkommen mobil ist, kann man die Gleichgewichtskonstante aus der Verschiebung der Halbstufenpotentiale der Welle einer Komponente, bei Änderungen der Konzentration der zweiten Komponente, bestimmen. Die zweite Möglichkeit kommt beim System, wo die Einstellungsgeschwindigkeit des vorgelagerten Gleichgewichtes langsam gegen die Geschwindigkeit des Elektrodenvorganges auf der Quecksilberelektrode ist, in Betracht. Dann können unter geeigneten Bedingungen zwei Wellen auf den polarographischen Kurven beobachtet werden, die einer Komponente und dem Reaktionsprodukt entsprechen. Die Grenzströme beider Wellen sind nur durch Diffusion begrenzt. Das durch Elektrolyse gestörte Gleichgewicht kann während der Lebensdauer eines Tropfens nicht erneuert werden und die Stufenhöhen entsprechen den Konzentrationen der Komponenten in der Lösung.

An dem Beispiel der Brenztraubensäure, der Oxydationsprodukte von Ascorbinsäure und Cumarindiol, des Acetons und Cyklohexanons, des Acetaldehyds und Benzaldehyds, des Benzils, Diacetyls und der Phenylglyoxylsäure wurde gezeigt, daß sich in konzentrierten gepufferten Lösungen von Ammoniak und von primären Aminen bei einem pH nahe pK_b zwischen der Karbonylverbindung und dem Reaktionsprodukt mit dem Amin ein Gleichgewicht, dessen Einstellung im Vergleich mit der Geschwindigkeit des elektrolytischen Vorganges unter den gegebenen Experimentalbedingungen langsam ist (mit einer Halbzeit von der Ordnung 10^1 sec), einstellt.

Auf der polarographischen Kurve bemerken wir zwei Stufen, die beide den Charakter eines Diffusionsstromes haben, deren Summe konstant bleibt und der analytischen Konzentration der Karbonylverbindung proportional ist. Die Welle bei negativeren Potentialen gehört der Reduktion dieser Form der Karbonylverbindung, die dem pH der Lösung entspricht, an und in manchen Fällen (Aceton, Cyklohexanon, der Oxydationsprodukt der Ascorbinsäure) durch die Reduktion der Ionen des Grundelektrolyts verdeckt ist. Die Welle bei positiveren Potentialen, deren Höhe mit steigender Konzentration des Amins auf Kosten der zweiten Welle wächst, ist durch Reduktion des Reaktionsproduktes der Karbonylverbindung und des Amines verursacht.

Da an der Reaktion nur die Amine, welche am Stickstoffatom zwei substituierbare Wasserstoffatome und ein freies Elektronenpaar besitzen (und deshalb nukleophil sind), teilnehmen und da diese Reaktion für die elektrophile Karbonylgruppe charakteristisch ist, geht hervor, daß das Stickstoffatom eine Doppelbindung formen und mit dem Karbonyl reagieren muß. Die Notwendigkeit einer, der Kondensation vorgelagerten Enolisation wurde ausgeschloßen, als die Reaktion auch in den Lösungen der Phenylglyoxylsäure und des Benzils, wo die Existenz einer Enolform ausgeschlossen ist, beobachtet wurde.

Die Zahl der Moleküle der Ketoverbindungen, die in Reaktion mit dem Amin treten, kann man aus der Abhängigkeit von der Konzentration der Karbonylverbindung und aus dem Verhältnis der beiden Wellen bestimmen. Die Anzahl der teilnehmenden Moleküle des Amins erkennen wir aus der Steilheit der Kurve, welche die Dissoziation im Gleichgewichtszustand angibt. Diese erhalten wir aus der Abhängigkeit der Höhe der ersten Welle von der Konzentration des Amins.

Aus der Analogie zwischen der Karbonylgruppe und der Gruppe $C=N-$[21]) und vor allem aus der leichten Hydrogenation der Verbindungen, welche die $C=N-$ Gruppe tragen[22]) und aus den polarographischen Angaben[1-15]) wurde festgestellt, daß die Reduktion an der $C=N-$ Gruppe verläuft. Eine positivere Reduktion dieser Gruppe kann der Elektronegativität des Stickstoffatoms, die kleiner als bei dem Sauerstoffatom ist, zugeschrieben werden. Dies ist im Einklang mit der Beobachtung, daß die $= NR$ Gruppe einen kleineren induktiven und elektromeren Effekt als $= O$ hat (O. Ingold[23]).

Man kann also für das Gleichgewicht, das dem Elektrodenvorgang vorgelagert ist, folgendes Schema vorschlagen:

$$\begin{array}{c} R \\ R \end{array}\!\!\!\!CO + H_2N.R \;\rightleftharpoons\; \begin{array}{c} R \\ R \end{array}\!\!\!\!C\begin{array}{c} OH \\ NRH \end{array} \;\rightleftharpoons\; \begin{array}{c} R \\ R \end{array}\!\!\!\!C=NR + H_2O \qquad (5)$$

$$\qquad\qquad\qquad\qquad\quad I \qquad\qquad\qquad II$$

Aus den Gleichgewichtszuständen kann man nicht entscheiden, ob der langsamste Vorgang die Addition (I) oder die Dehydratation (II) ist, es scheint aber wahrscheinlich, daß im Gleichgewichte mit den Ausgangsstoffen der Addukt, hier als $\begin{array}{c} R \\ R \end{array}\!\!\!\!C\begin{array}{c} OH \\ NRH \end{array}$ formuliert, ist, weil die, zur Zerbrechung der einmal gebildeten festen kovalenten Bindung notwendige Energie wahrscheinlich die Mobilität des Gleichgewichtes (II) hindern könnte. Auch nach der Ionenformulation, die bei der Diskussion der Mechanismen solcher Reaktionen von E. R. Alexander[24]) vorgeschlagen wurde, ist die Mobilität des Dehydratationsgleichgewichtes unwahrscheinlich.

Weil im Laufe dieser Arbeit keine anderen Pufferlösungen als diejenigen, deren basische Komponente das reagierende Amin war, benutzt wurden, so daß nur der pH-Bereich von 8,3 bis 11 zugänglich war, ist es unmöglich gewesen, das Verhalten der Welle des Iminoproduktes in breiterem pH-Bereich zu verfolgen. Es kann also nicht mit Sicherheit entschieden werden, ob es sich um Reduktion der Base $R_2C = N.R$ oder der ihr konjugierten Säure $R_2C = \overset{+}{N}.RH$ handelt. Nach der Analogie mit dem Verhalten der Betainylhydrazone[10]) und des Hydroxylamins[4]) — wo der Reduktion nur die konjugierte Säure unterliegt, während die Base in dem erreichbaren Potentialbereich unreduzierbar ist — kann man folgern, daß auch hier die Säure die reduzierte Form ist. Dem elektrolytischen Vorgang wird also ein zweites Gleichgewicht vorgelagert, das sich in der Elektrodennähe äußert:

$$\begin{array}{c} R \\ R \end{array}\!\!\!\!C=NR + H^+ \;\rightleftharpoons\; \begin{array}{c} R \\ R \end{array}\!\!\!\!C=\overset{+}{N}RH$$

$$\qquad\qquad\qquad\qquad III$$

(oder Mesomer $\begin{array}{c} R \\ R \end{array}\!\!\!\!\overset{+}{C}\!\!-\!\!\overset{-}{N}R$)   (oder Mesomer $\begin{array}{c} R \\ R \end{array}\!\!\!\!\overset{+}{C}\!\!-\!\!NRH$)

So lange die Rekombinationsgeschwindigkeit der Ionen dieser Säure genügend groß ist, beeinflußt sie den Charakter der ersten Welle nicht. Sobald aber diese Geschwindigkeit mit der des elektrolytischen Vorganges unter den gebrauchten Bedingungen vergleichbar zu sein beginnt, nimmt die erste Welle den Charakter eines kinetischen Stromes an. Ein solches Verhalten wurde tatsächlich in den Lösungen von Glycin bei einem pH größer als 10,2 beobachtet. Hier zeigte die Abhängigkeit der Wellenhöhe von der Quecksilbersäule, daß die Wellenhöhe nicht nur durch bloße Diffusion begrenzt wird. Ein ähnliches Verhalten konstatierten V. Prelog und O. Häfliger[10]) bei den Betainylhydrazonen von Cyklanonen. Weil Azomethine schwächere Basen als die korrespondierenden Amine sind, ist es möglich (analog zu der Rekombination bei Säuren[3]) die Verschiebung des pK_b - Wertes zu höheren Werten durch die angegebene vorgerlagerte Rekombination zu erklären. Es kann aber auch die Möglichkeit, daß die Geschwindigkeit der Einstellung der Gleichgewichte (I) und (II) basisch katalysiert sei, nicht ausgeschlossen werden.

V. Prelog und O. Häfliger[10]) haben bei einem pH kleiner als 6,4 eine Zerteilung der Betainylhydrazonwelle in zwei Wellen gefunden und auf Grund der Analogie mit der Reduktion der Karbonylgruppe schlugen sie ein detailiertes Reduktionsschema vor. Da unter den gegebenen Experimentalbedingungen dieser pH-Bereich nicht erreichbar war und da es unmöglich war, den Einfluß der Karbonylgruppe auf die C=N— Gruppe in Betainylhydrazonen auszuschließen, wurde für den verfolgten pH-Bereich, wo eine irreversible Zweielektronenreduktion vorliegt, dieses Schema des elektrolytischen Vorganges vorgeschlagen:

$$\begin{array}{l} R \\ R \end{array} \!\! C=NR + H^+ \rightleftarrows \begin{array}{l} R \\ R \end{array} \!\! C=\overset{+}{N}RH$$

$$\begin{array}{l} R \\ R \end{array} \!\! C=\overset{+}{N}RH + 2e + 2H^+ \longrightarrow \begin{array}{l} R \\ R \end{array} \!\! CH—NRH_2$$

(oder ganz analog mit Hilfe der früher angeführten polaren Formen).

Dieses Schema (mit dem Verbrauch von 3 Protonen) entspricht am besten der gefundenen Abhängigkeit des Halbstufenpotentials vom pH, für die bei konstanter Ionenstärke im pH-Bereich 8,5—10 der Wert 90—100 mV/pH gefunden wurde.

In der Tabelle XII sind die Durchschnittswerte der Gleichgewichtskonstanten der Reaktionen der Amine mit den Karbonylverbindungen und die Differenz der Halbstufenpotentiale der Welle der Reduktion der Karbonylverbindung und der Welle ihres Iminoderivates angegeben.

Tab. XII.

Durchschnittswerte der Gleichgewichtskonstanten und Verschiebung der Halbstufenpotentiale.

Karbonylverbindung	Amin	K	$K_K \cdot .10^{-5}$	$K_S \cdot .10^{-5}$	$\Delta\pi\frac{1}{2}$ mV	$(\pi\frac{1}{2})_1$ V
Brenztraubensäure	Ammoniak	0,25	1,96	—	300	—1,0
Phenylglyoxylsäure	,,	0,1*	—	—	220	—0,84
Acetaldehyd	,,	0,54*	0,34	4,0	300	—1,4
Benzaldehyd	,,	0,15	3,3	0,1	210	—1,08
Aceton	,,	0,0063*	0,0031	0,0025	500	—1,52
Cyklohexanon	,,	0,0095*	0,0047	—	500	—1,54
Diacetyl	,,	0,050	—	—	110	—0,62
Benzil	,,	—	—	—	100	—0,54
Oxydationsprodukt der Ascorbinsäure	,,	22	—	—	400	—1,5 2NH₃
Aceton	Glycin	0,07*			550	—1,46
Brenztraubensäure	,,	2,47			250	—1,05
,,	Alanin	0,93			240	—1,06
,,	Colamin	3,11			300	—1,0
,,	Histidin	1,99			310	—0,99 Ref. 17
,,	Histamin	6,65			310	—0,99

$\Delta\pi\frac{1}{2}$ ist die Halbstufenpotentialdifferenz der ersten und der zweiten Welle

$(\pi\frac{1}{2})_1$ ist das Halbstufenpotential der ersten Welle bei pH = pK$_b$

K ist die Gleichgewichtskonstante der Reaktion der Karbonylverbindung mit dem angeführten Amin

K_K[25] ist die Gleichgewichtskonstante der Reaktion der Karbonylverbindung mit Semikarbazid

K_S[26] ist die Gleichgewichtskonstante der Reaktion der Karbonylverbindung mit Bisulfit

Durch * bezeichnete Werte kann man nur für approximativ halten, weil zu ihrer Bestimmung weniger Messungen benützt wurden oder weil die Werte von dem steilen Teil der Dissoziationskurve unzugänglich waren.

Die Konstanten der Reaktion mit Ammoniak wurden unter der Voraussetzung berechnet, daß die Base des Ammoniaks nur in der Form NH₃, die wegen des freien Elektronenpaares des Stickstoffatoms reaktiv ist, anwesend ist, eventuell daß die anderen Formen mit der reagierenden in mobilem Gleichgewichte stehen. In der Tabelle beobachten wir, daß man die reagierenden Verbindungen in drei Gruppen einteilen kann: Ketone, Aldehyde und Ketosäuren. Das Verhalten der letzteren nähert sich mehr den reaktiveren Aldehyden. Das steht im Einklang mit der Erfahrung, daß die Reaktivität des Karbonyls mit abnehmenden positiven Charakter des elektrophilen Kohlenstoffatoms sinkt (E. R. Alexander[27], E. A. Remick[28]). Eine ähnliche Verteilung finden wir bei den Gleichgewichtskonstanten der analogen Reaktion mit Semikarbazid[25]) und bei der Addition vom Bisulfit[26]). Eine Ausnahmsstellung unter den verfolgten Verbindungen hat das Oxydationsprodukt der Ascorbinsäure, wo an der Reaktion zwei Moleküle des Ammoniaks teilnehmen. Bei den Paaren: Brenztraubensäure-Phenylglyoxylsäure und Acetaldehyd-Benzaldehyd können wir den Einfluß der Substitution von Methyl durch Phenyl, wodurch die Gleichgewichtskonstante auf etwa 30% des ursprünglichen Wertes herabgesetzt wurde, beobachten. Auch die Differenz der Halbstufenpotentiale vermindert sich nach der Substitution von Phenyl im gleichen Maße. Ersetzen wir Ammoniak durch Glycin, vergrößert sich die Konstante etwa zehnmal, wie wir es ebenso bei der Brenztraubensäure wie beim Aceton beobachten können. Zu einer ausführlicheren Diskussion des Einflusses der Substituenten stehen zur Zeit noch wenige Werte für konstitutiv ähnliche Stoffe zur Verfügung.

Wenn wir die Verschiebungen der Halbstufenpotentiale miteinander vergleichen, beobachten wir, daß die größten Verschiebungen bei den Systemen mit den negativsten Reduktionspotentialen vorkommen. Für die Größe der Konstanten der Reaktion mit Ammoniak gewinnen wir eine Reihe:

Aceton $<$ Cyklohexanon $<$ Diacetyl $<$ Phenylglyoxylsäure $<$ Benzaldehyd $<$ Brenztraubensäure $<$ Acetaldehyd $<$ Oxydationsprodukt der Ascorbinsäure

und für die Verschiebung des Halbstufenpotentials eine Reihe:

Benzil $<$ Diacetyl $<$ Benzaldehyd $<$ Phenylglyoxylsäure $<$ Brenztraubensäure $<$ Acetaldehyd $<$ Oxydationsprodukt der Ascorbinsäure, Aceton

Für die Gleichgewichtskonstanten der Reaktion der Brenztraubensäure mit verschiedenen Aminen erhalten wir die Folge:

Ammoniak $<$ Alanin $<$ Histidin $<$ Glycin $<$ Colamin $<$ Histamin

und für die Verschiebung des Halbstufenpotentials bei diesen Reaktionen gilt:

Alanin $<$ Glycin $<$ Ammoniak, Colamin $<$ Histidin, Histamin

In zwei Fällen war es unmöglich, polarographisch die Reaktion der Karbonylverbindungen mit Ammoniak zu verfolgen: Bei Acetessigester und bei Benzophenon. Bei der Reaktion von Acetessigester mit Aminen entsteht lediglich nur die Enaminform.[29]) Man kann also deduzieren, daß entweder das tautomere Gleichgewicht in diesem Falle unmobil ist und daß Enamin der polarographischen Reduktion nicht unterliegt oder daß in alkalischer Lösung praktisch der ganze Acetessig in Enolform, die nicht mit dem Amin reagiert, anwesend ist. Bei Benzophenon, verursachen wahrscheinlich die sterischen Effekte der beiden Phenyle, daß unter gegebenen Bedingungen die Reaktion nicht verläuft. Das ist in Übereinstimmung mit den Angaben der Literatur,[30]) nach welchen die Kondensation in diesem Falle nur bei erhöhter Temperatur und Druck (bis 400 Atm) und in Gegenwart von Katalysatoren, gelingt. Benzophenon kann auch durch die Reaktion mit Bisulfit nicht nachgewiesen werden.[31]) Die Reaktion des Benzophenons mit Ammoniak mußte in einer ethanolischen Lösung verfolgt werden und es ist möglich, daß auch dieses Lösungsmittel die Reaktion hemmt, ähnlich wie bei Benzil beobachtet wurde. Der hemmende Einfluß der Nachbarschaft der Phenylgruppen auf die Reaktionsgeschwindigkeit wurde auch bei der Phenylglyoxylsäure und in geringerem Maße bei Benzaldehyd beobachtet. Die kleinere Reaktivität dieser Stoffe als die des Acetons und Cyklohexanons und die vielmals größere Gleichgewichtskonstanten zeugen dafür, daß hier—ähnlich wie bei der Bildung der Semikarbazide[25]) — keine konsequente Parallelität zwischen der Gleichgewichtskonstante und der Bildungsgeschwindigkeit besteht.

Da die beschriebene Reaktion für primäre Amine und Karbonylverbindungen spezifisch ist, kann sie zur Identifikation beider dieser Gruppen dienen. Die Nachbarschaft der Phenyle kann freilich die Reaktivität des Karbonyls vermindern. Weil die Wellen der Iminoderivate bei wesentlich positiveren Potentialen als die Wellen der korrespondierenden Karbonylverbindungen beobachtet wurden, kann diese Reaktion zur Bestimmung der Karbonylverbindungen dort verwendet werden, wo die Reduktion des Karbonyls bei so negativen Potentialen verläuft, daß die polarographische Bestimmung schwierig ist. So wurden in konzentrierten Pufferlösungen von Ammoniak und primären Aminen das Oxydationsprodukt der Ascorbinsäure, Acetons und Cyklohexanons bestimmt und an der Bestimmung anderer Stoffe wird weiter gearbeitet.

Zusammenfaßung.

Es wurde polarographisch eine Reaktion zwischen den Karbonylverbindungen und Ammoniak oder primären Aminen verfolgt. Es
wurde gefunden, daß an dieser Reaktion nur die basische Form des
Amins teilnimmt. Die Geschwindigkeit der Gleichgewichtseinstellung
in alkalischer Lösung ist gering im Vergleich mit der Elektroden-
prozessgeschwindigkeit. Die durch diese Reaktion entstandene Imino-
verbindung wird bei positiveren Potentialen als die korrespondierende
Karbonylverbindung reduziert. Die meisten der hier studierten Reaktionen wurden noch nicht in der Literatur erwähnt.

Die Gleichsgewichtskonstanten der Reaktion zwischen einer Karbonylverbindung (Brenztraubensäure, Phenylglyoxylsäure, Oxydationsprodukt der Ascorbinsäure und des Kumarindiols, Aceton, Cyklohexanon, Acetaldehyd, Benzaldehyd und Diacetyl) und dem Amin
(Ammoniak, Glycin, Alanin, Colamin, Histamin und Histidin) wurden
bestimmt (siehe Tab. XII.). Die polarographischen Kriterien für die
Bestimmung der Zahl der Molekülen, die an der Reaktion teilnehmen,
wurden angeführt.

In den Lösungen von Ammoniak oder der primären Amine ist
die polarographische Bestimmung von Aceton, Cyklohexanon und
von dem Oxydationsprodukt der Ascorbinsäure möglich. Die beschriebene Reaktion ist für primäre Amine und für die meisten Karbonylverbindungen charakteristisch und kann für ihren Nachweis dienen.

Summary.

The reaction taking place between carbonyl compounds and
ammonia or primary amines was followed polarographically. It was
found, that in this reaction only the base-form of amine takes part.
The rate of this reaction reaching equilibrium in alkaline solution is
slow compared with the rate of the electrode process. The imino-com-
pound formed by this reaction is reduced at more positive potentials
than the original carbonyl compound. Most of the reactions here studied have not yet been mentioned in the literature.

The equilibrium constants of the reaction between a carbonyl
compound (pyruvic acid, phenylglyoxylic acid, oxidation product of
ascorbic acid, acetone, cyclohexanone, acetaldehyde, benzaldehyde
and diacetyl) and amines (ammonia, glycine, alanine, colamine, hist-
amine, and histidine) were determined (Tab. XII.). Polarographic

criteria for the estimation of the number of molecules taking part in the reaction, are given.

In the presence of ammonia or of primary amines the polarographic determination of acetone, cyclohexanone and of the oxidation products of ascorbic acid is possible. The reaction described is characteristic for primary amines and for most of the carbonyl compounds and can be used for their detection.

An dieser Stelle möchte ich meinen besten Dank Herrn Prof. Dr. J. Heyrovský, Herrn Prof. Dr. R. Brdička und Herrn Dr. V. Hanuš für ihr freundliches Interesse ausdrücken.

Eingegangen am 22. September 1950.

Aus dem physikalisch-chemischen Institut der Karls-Universität, Prag.

Schrifttum.

[1]) I. M. Kolthoff und A. Langer, J. Am. Chem. Soc. *62*, 211 (1940).

[2]) A. Langer, Ind. Eng. Chem. Anal. Ed. *14*, 283 (1942).

[3]) R. Brdička, Collection *12*, 212 (1947); Chem. Listy *40*, 232 (1946). R. Brdička und K. Wiesner, Collection *12*, 138 (1947).

[4]) F. Petrů, Collection *12*, 620 (1947).

[5]) E. D. Hartnell und C. E. Bricker, J. Am. Chem. Soc. *70*, 3385 (1948).

[6]) J. M. Lupton und C. C. Lynch, J. Am. Chem. Soc. *66*, 697 (1944).

[7]) G. Sartori und A. Gaudiano, Gaz. *78* 77 (1948), C. A. *42*, 6676 (1948).

[8]) J. K. Wolfe, E. B. Hershberg und L. F. Fieser J. biol. Chem. *136*, 653 (1940).

[9]) J. Barnett und C. J. Morris, Biochem. J. *40*, 450 (1946).

[10]) V. Prelog und O. Häfliger, Helv. Chim. Acta *32*, 2088 (1949).

[11]) L. Matoušek, Dissertation, Karlsuniversität Prag 1949.

[12]) G. A. Crowe und C. C. Lynch, J. Am. Soc. *70*, 3795 (1948).

[13]) V. Sapara, Chem. Listy *43*, 225 (1949).

[14]) E. Knobloch, Collection *14*, 508 (1949); Chem. Listy *44*, 145 (1950).

[15]) D. Lester und L. A. Greenberg, J. Am. Chem. Soc. *66*, 496 (1944).

[16]) P. Zuman, Chem. Listy *44*, 162 (1950).

[16a]) P. Zuman, Nature *165*, 485 (1950).

[17]) P. Zuman, Chem. Listy, *45*, 40 (1951).

[18]) J. Vogel and J. Říha, Tschsl. Patentanmeldung No. P-112-49 von 14. I. 1949.

[19]) R. Brdička, Chem. Listy *39*, 35 (1945); R. Pasternak, Helv. Chim. Acta *30*, 1984 (1947); G. T. Borcherdt und H. Adkins, J. Am. Chem. Soc. *60*, 3 (1938); A. K. Vlček, E. Špalek und L. Krátký Techn. Hlídka Kožel. *24*, 65 (1949); Collection *15*, 340 (1950). Siehe auch Ref. 7, 17, 20.

368

[20]) J. Koryta und I. Kössler, Chem. Listy *44*, 128 (1950); Collection *15*, 241 (1950).

[21]) H. Gilman, Organic Chemistry, Wiley, New York 1948, Vol. I., 658.

[22]) R. Adams, Organic reactions, Wiley, New York 1948, Vol. IV., 174—255.

[23]) O. Ingold, Chem. rev. *15*, 225 (1934).

[24]) E. R. Alexander, Principles of ionic organic reactions, Wiley, New York 1950, 163—165.

[25]) J. B. Conant und W. K. Bartlett, J. Am. Chem. Soc. *54*, 2881 (1932); siehe L. P. Hammett, Physical organic chemistry, McGraw-Hill, New York 1940, 211, 331—336.

[26]) I. M. Kolthoff und V. A. Stenger, Volumetric analysis, Vol. I Interscience New York 1942, 214—222.

[27]) Ref. 24, s. 155.

[28]) E. A. Remick, Electronic interpretation of organic chemistry, Wiley, New York 1947, 124.

[29]) K. v. Auwers und W. Susemihl, Ber. *63*, 1072 (1930); *64*, 2748 (1931); K. v. Auwers und H. Wunderling, Ber. *65*, 70 (1932).

[30]) O. Dimroth und R. Zoepritz, Ber. *35*, 984 (1902); G. Reddelien, Ber. *43*, 2476 (1910), Ann. *388*, 165 (1912); Ber. *42*, 4759 (1909), *46* 2712, 2718; (1913); *47*, 1364 (1914), *53* 334 (1920).

[31]) H. Schwoegler, H. und H. Adkins, J. Am. Chem. Soc. *61*, 3499 (1939), F. Feigl, Quantitative analysis by spot tests. Elsevier, New York 1947, 333—334.

QUANTITATIVE TREATMENT
OF SUBSTITUENT EFFECTS IN POLAROGRAPHY. V.*
POLAROGRAPHIC REDUCTION
OF SOME SEMICARBAZONES

B.Fleet** and P.Zuman

*J. Heyrovský Institute of Polarography,
Czechoslovak Academy of Sciences, Prague*

Received August 22nd, 1966

Polarographic reduction of semicarbazones of aliphatic and aromatic aldehydes and ketones in the protonized form follows predominantly a four-electron course to form a primary amine and urea. Other acids in addition to the hydroxonium ion can participate in the protonation preceding the electrode process proper. Half-wave potentials of benzaldehyde semicarbazones are relatively little sensitive to the substituent effects ($\varrho_\pi = 0.10$ V), and remain practically unchanged by m- and p-CH_3, C_2H_5, i-C_3H_7 and t-C_3H_7 and m-CF_3. Substitution by i-C_3H_7 and t-C_4H_9 in *ortho*-position results in a steric hindrance of coplanarity.

In the course of our study of substituent effects in polarographic behaviour we aimed to study the polar and steric effects of alkyl groups bound in various positions to the phenyl ring in substituted benzaldehyde semicarbazones. In order to allow a reasonable comparison of the measured potentials to be made, it is essential to understand, at least in principle, the course of the electrode process. As in this particular case, the accounts given by other workers are in many aspects contradictory, it is essential first of all to describe the various viewpoints in some detail.***

Semicarbazones are reduced in acid media in a polarographic wave, the height of which over a certain pH range is pH-independent. In some cases the wave is split into two. The total height of this wave gradually decreases with increasing pH. In a few cases the appearance of another wave at more negative potentials was reported, this wave increasing in height at the expense of the more positive wave. This wave can be observed with some aromatic derivatives and appears in the available potential range more often with thiosemicarbazones than with semicarbazones.

*   Part IV: This Journal 27, 2035 (1962).
**   Present address: Department of Chemistry, Imperial College of Science and Technology, London S. W. 7, England.
***   Because of their analogous behaviour thiosemicarbazones will also be included in the discussion.

This more negative wave, the height and half-wave potential of which is pH-independent is unanimously[1-8] attributed to the two-electron reduction of the unprotonised (thio)semicarbazone in which the (thio)semicarbazide derivative $R^1R^2CH_2NHNHCO(S)NH_2$ is formed. The latter compound does not undergo polarographic reduction in the available potential range[8].

The viewpoints concerning the reduction in acid media are, however, rather confusing. In addition to the assumption of a four-electron process[1-3,5,7-11], a two-electron reduction was suggested[3-6,9,12-16], and other instances have been reported[2,6-8,10,11] where the number of electrons transferred was given between 2 and 4. With few exceptions, due to special structural features, most of the evidence indicates a four-electron process following the scheme

$$[{>}C{=}NNHCO(S)NH_2]H^{(+)} + 4\,e + 4\,H^{(+)} \rightarrow \ {-}CH_2NH_3^{(+)} + NH_2CO(S)NH_2\,.$$

This was based on the comparison of the wave-heights of (thio)semicarbazone waves with the waves of equimolar concentrations of aldehydes[9], oximes[2,3,11], phenylhydrazones[2], other (thio)semicarbazones[5,7,10] and benzophenone[9] and also on the identification of the products of the type ${>}CH_2NH_2$ formed in the controlled potential electrolysis, e.g. for benzaldehyde and benzophenone semicarbazones in hydrochloric acid media[2], p-acetamidobenzaldehyde thiosemicarbazone and S-methyl-p-acetamidobenzaldehyde thiosemicarbazone[7] at pH 6·5, products of the type ${-}CH_2NH_3^{(+)}$ were identified. The results of coulometric measurements in hydrochloric acid media[2] and in buffers[8] pH 4·3 are also reported to correspond to a four-electron process.

A two-electron reduction in the first step was proved[3,5,8] for pyridinealdehyde thiosemicarbazones and the product, substituted thiosemicarbazide, isolated[9]. A two-electron reduction has also been suggested[9] without experimental evidence for vanillin semicarbazone, this conclusion being based on comparison of wave-heights at pH 2—4 with benzophenone semicarbazone, even though this semicarbazone was not isolated and the curve of benzophenone in a solution containing an excess of semicarbazide measured. The height of the observed wave was approximately half of the heights of other semicarbazone waves and it cannot be excluded that the equilibrium was not shifted completely towards the semicarbazone as in this paper (p. 2064) equal heights for benzophenone, benzaldehyde and other semicarbazones were reported, in acid media.

A consumption of approximately two electrons per semicarbazone molecule was also deduced from microcoulometric measurements[4,12-14]. These measurements were carried out at a pH of 6—7; in this pH region, however, the limiting current of the protonised form of the semicarbazone is for most derivatives only a small fraction of the maximum limiting current which is observed at lower pH values. Under these conditions the current is controlled by the rate of protonation of the semicarbazone and the participation of a different mechanism from that taking place in acid solutions cannot be disregarded. The authors[12] reported a two-electron reduction in acid media, but later[17,18] have been able to show that (in agreement with the findings of Lund[2]) at pH lower than about 5·5 semicarbazones, thiosemicarbazones and phenylhydrazones are reduced with consumption of four electrons.

The fact that the wave of the protonised form is twice as high as the wave of the unprotonised[1-3,5,7,8] is a strong argument in favour of the reduction of the protonised semicarbazone being a four-electron process rather than a two-electron[6]. Similarly the reduction curves of m-nitrobenzaldehyde semicarbazone quoted[13] as a proof of the two-electron theory, indicate a four-electron process. In acid media the reduction of m-nitrobenzaldehyde semicarbazone proceeds in three steps in the ratio $2:1:2$; for the first step, coulometrically, $n = 4$ and for the second $n = 2$ was found. As the third wave is of equal height to the first it also corresponds to a four-electron step. The first, pH-independent step, corresponds to $-NO_2 + 4e + 4H^+ \to$ $\to -NHOH + H_2O$ and the second wave, which decreases at pH 4 and possesses a shape resembling other phenylhydroxylamine waves corresponds to $-NHOH + + 2e + 2H^+ \to -NH_2 + H_2O$. Hence the third wave corresponds to a four-electron reduction of the semicarbazone group in m-aminobenzaldehyde semicarbazone.

The systems, for which a transfer of less than four but more than two electrons has been reported, can be classified into three types. Firstly, there are the systems where the maximum value of the limiting current has not been reached, $i.e.$ the measurement has not been carried out at sufficiently low values of pH. This appears to be the case with hexose semicarbazones[11]. In the second group are the systems for which the formation of an increasing proportion of semicarbazide derivatives of the type $R^1R^2CHNHNHCONH_2$ was detected when the electrolysis was carried out at the potential of the limiting current of the wave of the protonised form although at higher pH values. Thus for benzophenone semicarbazone in hydrochloric acid medium the product of electrolysis was benzhydrylamine alone[2] but at pH 4 some 30% of 1-benzhydrylsemicarbazide was also found. Similarly, in the electrolysis of p-acetamidobenzaldehyde at pH 6·5 (wave-height corresponds to 4 electrons) 30% of benzylthiosemicarbazide was found, whereas at pH 8·3 (3·3 electrons) 40% and at pH 11·5 (2·9 electrons) 75% of this compound was detected. Secondly, an effect of surface active substances can be anticipated. Kitaev used 0·045% agar, but we have proved that the presence of agar under these conditions, even although it affects the wave-shape, decreases the limiting current only by some 5%.

The third group covers the reports of compounds which give a smaller wave-height. Thus for p-aminoacetophenone[2,6] and p-dimethylaminobenzaldehyde[6] semicarbazones the limiting currents were lower by about $25-30\%$ than the waves for derivatives without amino groups. The wave of methyl isobutyl ketone semicarbazone[10] was found to be 40% lower than the waves of other semicarbazones; in this case it is not certain whether the maximum value of the limiting current was achieved. The waves for semicarbazones are lower by $10-20\%$ than the waves of corresponding thiosemicarbazones[7], oximes and phenylhydrazones[2]. There is also a general statement[10] that the wave-height of aldehyde and ketone semicarbazones vary between $n = 3$ and $n = 4$.

The prime necessity of the present study, therefore, was to check, if the discrepancies in the reports are not caused by specific structural effects and also to find out whether or not semicarbazones of representatively selected types behave in a similar manner. The next requirement was to measure the number of electrons consumed per molecule during electrolysis for several typical compounds under various conditions in order to prove the validity of the suggested reduction scheme. The reduction scheme should take into account the observed deviations from the general picture described in the preceding paragraphs. Only when this picture is completed it is possible to make any study of the structural effects.

EXPERIMENTAL

Preparation of Semicarbazones

Substances *I—V*, *VII* and *VIII* (Tables I and II) and *XXI—XXVII* were prepared as follows[19]: 2·5 g semicarbazide hydrochloride and 4·0 g sodium acetate were mixed together in an extraction tube. The liberated semicarbazide was extracted with 5 ml methanol. The extract was separated using a filter tube with a cotton wool plug and placed in an ice bath together with a further 5 ml of methanol. 30 mmol of the carbonyl compound was then added (solid compounds after solution in the minimum volume of methanol) and the semicarbazone formed filtered and then recrystallised two or three times from methanol or methanol–10% water according to the solubility, and finally dried under reduced pressure.

Benzophenone semicarbazone was prepared by refluxing a pyridine solution of 5·5 g benzophenone and 4·5 g of semicarbazide hydrochloride (total volume 20 ml) for 1 hour and then pouring the mixture into ice cold water. The product was recrystallised three times from ethanol.

Semicarbazones *IX—XX* (Table II) were prepared by Klouwen and Boelens[20]. Purity was checked by comparison of melting points (Tables I and II) and in several cases by thin-layer chromatography using an ethanol–tetrachloromethane 1 : 4 solvent system and detection by spraying with dilute potassium permanganate solution or alternatively by iodine adsorption.

TABLE I

Properties of the Selected Aldehydes and Ketone Semicarbazones of the Type $R^1R^2C=$ $=NNHC\,ONH_2$ in $2 \cdot 10^{-4}$M Solutions

Com-pound	R^1	R^2	M.p., °C detn.	M.p., °C ref.[21]	Limiting currents[a], μA 0·05M-H_2SO_4	Limiting currents[a], μA pH 4·8	Limiting currents[a], μA pH 7·3
I	CH_3	H	167·5	169	1·15[b]	1·15	0·10
II	n-C_6H_{13}	H	107—109	109	0·92[c]	0·81	0·10
III	$C_5H_4N^d$	H	219	—	1·15[b,e]	1·15	0·58
IV	CH_3	CH_3	186·5	190	0·4[f]	0·92	0·08
V	C_6H_5	CH_3	198	203	0·92[c]	1·20	0·12
VI	C_6H_5	C_6H_5	166·5	167	1·22	0·59	0·10
VII	$(CH_2)_5{}^g$		167	167[h,i]	0·20[f]	1·10	0·46

[a] No waves observed in borax, pH 9·25 and 0·1M-NaOH with the exception of the pyridine derivative *III*, for which, in both of these solutions, a wave ($i = 0·57$ μA) was observed; [b] wave split into two waves of approximately the same height; [c] hydrolysis; [d] pyridyl; [e] 0·05M-H_2SO_4; [f] rapid hydrolysis; [g] cyclohexanone semicarbazone; [h] ref.[22]; [i] two waves of unequal heights.

TABLE II

Properties of the Substituted Benzaldehyde Semicarbazones of the Type $X-C_6H_4CH=NNH.CONH_2$ in $2 \cdot 10^{-4}M$ Solutions

Compound	X	M.p., °C		Limiting currents[a], μA		
		detn.	ref.[21]	0·05M-H_2SO_4	pH 4·8	pH 7·3
VIII	H	221·5	224	1·31	1·13[d]	0·06
IX	2-CH_3	—	212[b]	1·31	—	—
X	3-CH_3	—	218[b]	1·35	—	—
XI	4-CH_3	—	216·5[b]	1·35	—	—
XII	2-C_2H_5	—	176[b]	1·35	—	—
XIII	3-C_2H_5	—	211[b]	1·35	—	—
XIV	4-C_2H_5	—	206·5[b]	1·35	—	—
XV	2-i-C_3H_7	—	166[b]	1·31	—	—
XVI	3-i-C_3H_7	—.	181[b]	1·35	—	—
XVII	4-i-C_3H_7	—	208[b]	1·31	—	—
XVIII	2-t-C_4H_9	—	187[b]	1·31	—	—
XIX	3-t-C_4H_9	—	132[b]	1·28	—	—
XX	4-t-C_4H_9	—	212·5[b]	1·31	—	—
XXI	3-Cl	226	228	1·28	1·07[d]	0·10
XXII	4-Cl	223	230	1·24	1·17	0·20
XXIII	3,4-Cl_2	226	—	1·28	1·13[d]	0
XXIV	4-OCH_3	209	210	1·28	1·17[d]	0·21[d]
XXV	3,4-$(OCH_3)_2$	182	177	1·31	1·28[d]	0·92[d]
XXVI	3-OH	184	198	1·17	1·09	0·10
XXVII	4-CN	250[c]	—	1·28[e]	0·77[d]	0·71
XXVIII	3-CF_3	212	—	1·35	1·26[d]	0

[a] No waves observed in borax pH 9·25 and 0·1M-NaOH, with the exception of 4-CN derivative, for which in both solutions a wave ($i = 0.65$ and 0.71 μA, respectively) was observed; [b] authentic substances described in ref.[20]: [c] a dip on the limiting current; [d] another wave at more negative potentials observed.

Solutions

0·01M stock solutions of semicarbazones were freshly prepared in 96% ethanol, with the exception of p-cyanobenzaldehyde semicarbazone where the low solubility only allowed a $2 \cdot 10^{-3}M$ solution to be prepared. Buffer solutions and all other solutions were prepared from Reagent Grade chemicals.

Apparatus

The preliminary experiments were carried out using a photographic recording polarograph LP 55 (Laboratorní přístroje, Prague) with a galvanometer of sensitivity $2 \cdot 4 \cdot 10^{-9}$ A/mm.

Measurements of substituted benzaldehyde semicarbazones were carried out using a pen recording polarograph LP 60 (Laboratorní přístroje, Prague). A Kalousek vessel with a separated
saturated calomel electrode (s.c.e.) was used for polarographic electrolysis. Capillary constants
measured at the potential of the s.c.e. were $t_1 = 3\cdot3$ s, $m = 2\cdot1$ mg s^{-1} at $h = 85$ cm. For the
measurement of half-wave potentials and for the logarithmic analysis a three electrode system[23,24]
was used. The potential of the dropping electrode was measured potentiometrically against
a reference s.c.e. in an auxiliary circuit using a potentiometer Compensator Mark QTK (Metra,
Blansko). A modified Kalousek vessel was used, with a stem junction type calomel electrode
immersed into the studied solution.

Controlled potential electrolysis for microcoulometric measurements and identification of the
electrolysis product was carried out using a transistorized potentiostat[25] and a modified H cell
for a volume of $0\cdot5-2\cdot0$ ml using a dropping mercury electrode.

The relative pH values of buffer solutions containing 50% ethanol were measured using a pH
meter PHM4 (Radiometer, Copenhagen) and a glass electrode type G 200B.

Techniques

5 ml of aqueous buffer solution, 4·8 ml of ethanol and 0·2 ml of the stock solution of semicarbazone were mixed with a stream of nitrogen in a 25 ml beaker for 10 seconds. The solution was
then transferred to the Kalousek vessel or modified Kalousek vessel and after deaeration for
a further 3 minutes the polarographic wave or the half-wave potential measured. In the preliminary experiments the recording was repeated 10 minutes after the start of the first record in order
to detect time changes.

In the controlled potential electrolysis 0·05 ml of the stock solution of semicarbazone is added
to 0·5 ml of the 50% ethanolic buffer solution in the electrolysis cell. After the removal of dissolved
oxygen by passing a stream of pure hydrogen through the solution for 10 minutes, the electrolysis
is carried out at the applied voltage of 1·0 to 1·05 V (*i.e.* at the potential at which the wave
reaches its limiting value) over a period of at least three hours, with *i*–*E* curves being recorded
at intervals of one hour.

The homogeneity of the solution is maintained by the stirring effect of the falling mercury
drops.* The number of electrons, n, was calculated using the relationship (*1*)

$$n = \frac{'k_1 t}{2\cdot3 k_2 VF \log (i_{lim})_0/(i_{lim})_t}. \tag{1}$$

where k_1 is the current in amperes corresponding to a wave-height of 1 cm, k_2 is the concentration of mol.l^{-1} of a solution corresponding to a limiting diffusion current of 1 cm, t is the time
elapsed from the beginning of the electrolysis in seconds, V the volume of the solution in litres,
F 96487 coulombs and $(i_{lim})_0/(i_{lim})_t$ the ratio of the limiting currents at the beginning of the
electrolysis and after time t. The evaluation was carried out graphically by plotting log $(i_{lim})_t$
against t and determining the slope of the linear dependence.

---

* In order to check the validity of this assumption it was proved that the limiting current
measured at any time during the electrolysis shows no change after mixing the solution for a few
seconds with a stream of inert gas.

RESULTS

GENERAL POLAROGRAPHIC BEHAVIOUR OF SEMICARBAZONES

Due to the fact that the polarography of the various types of semicarbazones described in the literature was not carried out under conditions that would allow a direct comparison to be made, semicarbazones *I — VIII* were selected as representing a wide variation in structure. With these compounds preliminary measurements were carried out in $0.05\text{M-H}_2\text{SO}_4$, acetate buffer pH 4·7, phosphate buffer pH 6·8, borax pH 9·2 and 0.1M-NaOH, all in the presence of 50% ethanol.

With all of the compounds studied polarographic reduction waves were observed in acid media, the total height of which was approximately the same for all of the substances compared (Table I). In the case of acetaldehyde, heptaldehyde, 4-pyridine carboxaldehyde, and less markedly for cyclohexanone derivatives, this wave was split into two waves. With the derivatives of aliphatic compounds and also acetophenone a marked decrease in the wave height in sulphuric acid solutions was observed over first ten minutes. For the other substances and for all of the compounds in the other media the wave-height remained practically unchanged (Table II).

All the wave heights decreased with increasing pH values, the region, in which this decrease commenced was observed at lowest pH values (pH > 3) for benzophenone semicarbazone and highest (pH > 6) for the cyclohexanone and acetaldehyde derivatives.

POLAROGRAPHIC BEHAVIOUR OF BENZALDEHYDE AND BENZOPHENONE
SEMICARBAZONES

The polarographic behaviour of benzaldehyde and benzophenone semicarbazones was studied in some detail. Benzaldehyde semicarbazone (*VIII*) was selected as a parent substance of the series for which substituent effects were to be studied while benzophenone semicarbazone (*VI*) was chosen in view of the conflicting reports in the literature (p. 2059) concerning its behaviour.

The measurements were carried out in solutions containing 50% ethanol and 1, 2 and $5 \cdot 10^{-4}\text{M}$ semicarbazone in $0.75 - 0.0025\text{M-H}_2\text{SO}_4$, acetate buffers pH 4·4 to 6·4 and phosphate buffers pH 6·5 to 7·9.

In sulphuric acid media one well developed wave was observed. The wave-height for benzaldehyde semicarbazone (Fig. 1) was practically pH independent, that for the benzophenone derivative *VI* showing a slight decrease at pH > 1·3. Streaming maxima were observed in very strongly acidic media.

The decrease of the wave-height in acetate and phosphate buffers followed essentially the form of a dissociation curve. The decrease in the current which occurs at negative potentials, appears at more positive potentials only at higher pH values. Hence a dip[8] results on the limiting current. The potential region in which this dip

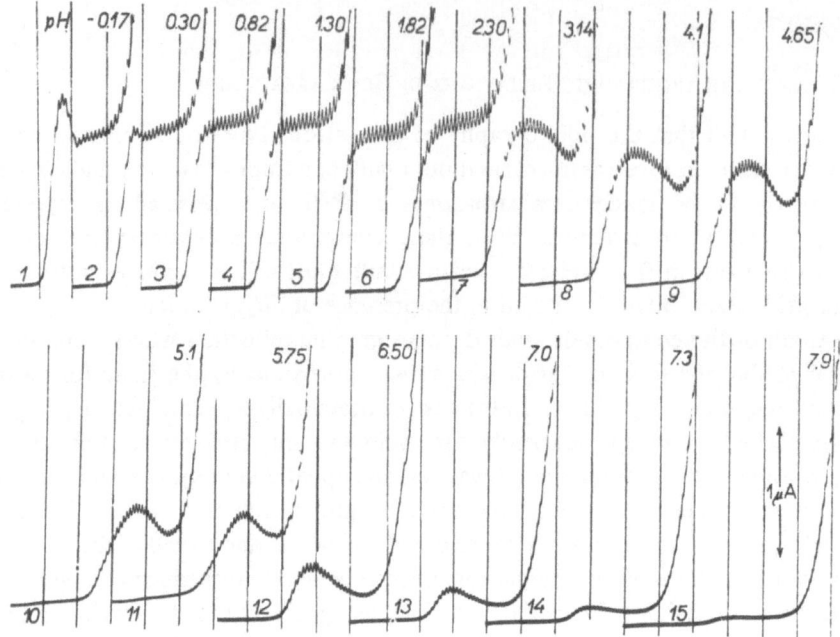

FIG. 1

pH Dependence of Reduction Waves of Benzaldehyde Semicarbazone

$5 \cdot 10^{-4}$M compound *VIII*; *1—6* sulphuric acid; *7—11* 0·1M acetate buffer; *12—15* 0·1M phosphate buffers. pH values are given on the polarogram. All curves start at $-0\cdot4$ V v. s.c.e. 200 mV/absc., $t = 3\cdot3$; $m = 2\cdot1$ mg/s, full scale sensitivity 2·3 μA.

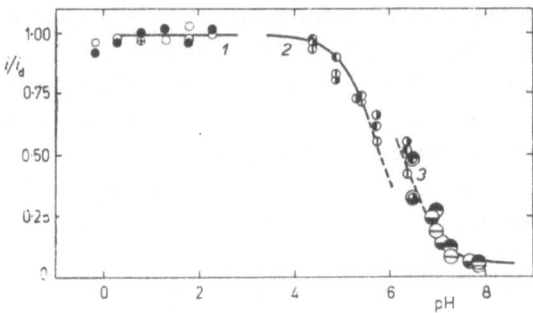

FIG. 2

pH Dependence of Limiting Currents Expressed as a Fraction of the Diffusion Current i/i_d for Benzaldehyde Semicarbazone

Semicarbazone *VIII*: ●, ◐, ◓, $5 \cdot 10^{-4}$M; ○, ◑, ◒, $2 \cdot 10^{-4}$M; ⊕, ◑, ◓, $1 \cdot 10^{-4}$M solutions: *1* sulphuric acid; *2* buffers containing 0·1M sodium acetate; *3* buffers containing 0·1M-Na_2HPO_4.

was observed depends on the structure of the semicarbazone, the part of the curve unaffected by such a decrease being greater for benzophenone than for benzaldehyde semicarbazone.

The discontinuous course of the pH dependence of the limiting current (Fig. 2) where the i/i_d–pH curve is shifted in the region of transition from acetate to phosphate buffers, as well as the observation[13,14] that the limiting current when $i < i_d$ is dependent upon the buffer concentration indicated that proton donors other than hydroxonium ions participate in the protonation. For such a type of chemical reaction which precedes the electrode process proper the experimental results fit the theoretical shape of the i–pH dependence best when simple buffers are used. Acetate buffers were selected as they covered the greatest region for which $i < i_d$ for both semicarbazones *VI* and *VIII*. Moreover, because it has recently been shown[26] that an ideal shape of the i–pH dependence can be obtained by keeping the concentration of the basic buffer component constant and changing the concentration of the acid, the composition of the buffers used in the present study was arranged so that the

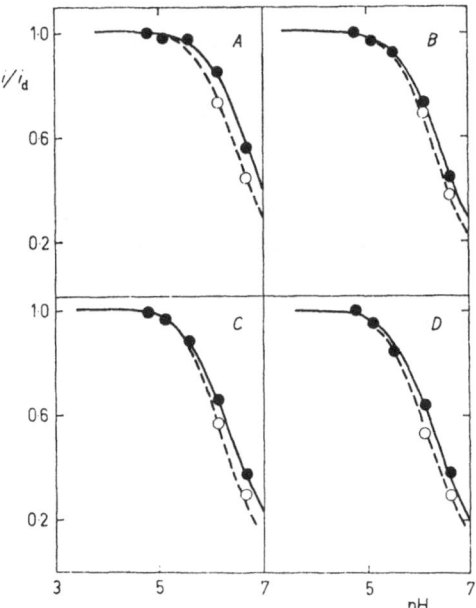

Fig. 3

pH Dependence of Benzaldehyde (A, B) and Benzophenone (C, D) Semicarbazone Waves in Simple Acetate Buffers

A, C 1·0M sodium acetate; B, D 0·2M sodium acetate; varying acetic acid concentration; ● and a full line current measured at maximum value; ○ and a dotted line current measured at minimum value.

TABLE III

Comparison of pH Values at which $i = i_d/2$ for Benzaldehyde and Benzophenone Semicarbazones (25°C $t_1 = 3\cdot3$ s)

Current	i_{max}		i_{min}	
Acetate, mol/l	1	0·2	1	0·2
Benzaldehyde semicarbazone	6·85	6·55	6·60	6·45
Benzophenone semicarbazone	6·45	6·20	6·35	6·20

acetate concentration was kept constant (1M or 0·2M) and the pH changed by adjusting the acetic acid concentration. The ionic strength in such solutions was considered to be effectively constant. Under these conditions a good correlation was obtained between the experimental points* and the theoretical curve derived by Koutecký[27] for a monobasic acid (Fig. 3). The shape of the pH-dependence was similar for the current measured at the maximum and at the minimum of the polarographic wave, the curve for current measured at the minimum being shifted slightly towards lower pH values, corresponding to a slower rate of the protonation (Fig. 3, Table III). This is in agreement with observations made for the reduction of periodic acid[28] and nitrones[29]. Also in agreement with the theory[23] is the observation that the pK″ values (pK″ is the pH at which $i = i_d/2$ for a given buffer concentration) decreased with increasing buffer concentration (Table III).

The shape of the i/i_d–pH dependence in acetate buffers prepared from constant concentrations of acetic acid and varying concentrations of sodium acetate with adjusted ionic strength were also in agreement with the prediction of a theory[26] for a reaction in which other proton donors participate in addition to the hydroxonium ion, being more drawn-out than the theoretical curve for a simple recombination of monobasic acid[27].

The changes of the limiting currents with pH in phosphate buffers also followed the shape of a dissociation curve, but shifted to higher pH values than that obtained for acetate buffers. The specific rate constant $k_{H_2PO_4^-}$ is therefore greater than k_{CH_3COOH} in agreement with the Brønsted relation. At a given pH where $i < i_d$ then the height of the wave is dependent on the nature of the buffer.

---

* The best fit of the experimental data for the i/i_d–pH dependence was obtained when the value of i in acetate buffer pH 4·4 was made equal to i_d. The value of the diffusion current measured in sulphuric acid solutions was approximately 10% higher than that at pH 4·4 and values of i/i_d calculated using the higher value showed some deviations. This difference in i_d values can be attributed to differences in viscosity and solvation.

Various factors influencing the dip on the limiting current of the current–voltage curves were next considered. The relative depth of the dip is greater at higher depolariser concentrations, at lower buffer concentrations and at higher concentrations of added neutral salts. The ratio of i_{min}/i_{max} also decreases slightly with increasing mercury pressure. The effect of neutral salt is practically independent of the nature of the anion and only very slightly dependent on the nature of the cation. The depth of the dip increased for the benzaldehyde derivative *VIII* in the sequence $Li^+ < Cs^+ < Ca^{2+} < NH_4^+ < K^+$ and for the benzophenone derivative *VI* in the sequence $Li^+ < Ca^{2+} < NH_4^+ < K^+ < Cs^+$. At higher concentration of ethanol an increase in the depth of the dip was observed, but this effect was complicated owing to the change in pH.

Hence it can be concluded that in acetate buffers the best developed waves can be obtained with low concentration of semicarbazone, high buffer concentration and in the presence of lithium ions. Even under these conditions it was impossible to obtain regular curves over a range of 2 pH units for all the substances studied in order to allow a comparison of the $dE_{1/2}/dpH$ shifts to be made. Owing to the fact that it was also experimentally impossible to reach a pH region in which the half-wave potentials would be pH independent, it was decided to use sulphuric acid media for the study of structural correlations. Under these conditions all substances studied gave a well developed wave (Fig. 4) corresponding in all cases, with the exception of those mentioned above, to the transfer of the same number of electrons, and possessing a comparable value of $dE_{1/2}/dpH$. When aliphatic compounds were

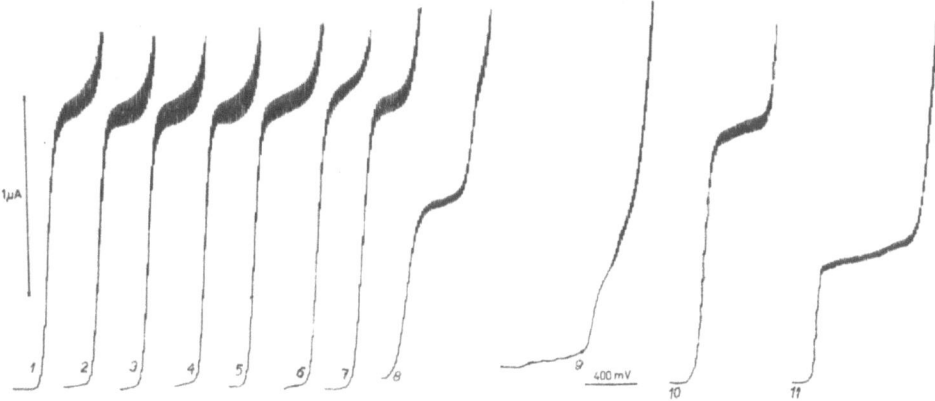

FIG. 4

Comparison of Wave-Heights of Semicarbazones and Model Compounds

$2 . 10^{-4}$M depolariser, *1—8* in 0·05M-H_2SO_4: *1 VIII: 2 XI; 3 XVII; 4 XX; 5 XXIII; 6 XXVIII; 7 VI; 8 III; 9—11* in 0·1M acetate buffer pH 4·81: *9* benzaldehyde; *10* pyridoxal oxime; *11* pyridoxal-5 phosphate. Curves *1—10* start at 0·5 V and *11* at 0·4 V (s.c.e); $t = 3·3$ s; $m = 2·1$ mg/s.

omitted, hydrolysis did not cause any problems, either with the recording of the curve or the measurement of the half-wave potential.

POLAROGRAPHIC BEHAVIOUR OF SUBSTITUTED BENZALDEHYDE SEMICARBAZONES $IX - XXVIII$

With only one exception the polarographic behaviour of compounds $IX - XXVIII$ completely resembled that of benzaldehyde semicarbazone (VIII) which has already been described in the preceding section. The exception was p-cyanobenzaldehyde semicarbazone (XXVII) but the behaviour of this compound will be dealt with separately[30]. Values of half-wave potentials together with n and $dE_{1/2}/dpH$ values are given in Table IV. Whereas the $dE_{1/2}/dpH$ values were reasonably constant

TABLE IV

Half-wave Potentials ($E_{1/2}$ v.s.c.e) of Substituted Benzaldehydes Semicarbazones of the Type $X-C_6H_4CH{=}NNHCONH_2$ in Sulphuric Acid Media Containing 50% Ethanol

Compound	X	0.5M-H_2SO_4 $E_{1/2}$	0.05M-H_2SO_4 $E_{1/2}$	0.005M-H_2SO_4 $E_{1/2}$	αn	$\frac{dE_{1/2}}{dpH}$
VIII	H	—	-0.729^a	-0.791	-0.832	0.81	61
IX	2-CH_3	—	-0.705	-0.765	$--0.826$	0.81	60
X	3-CH_3	-0.069	-0.711	-0.773	-0.834	0.83	61
XI	4-CH_3	-0.17	-0.733^b	-0.792	-0.855	0.80	62
XII	2-C_2H_5	—	-0.7000	-0.762	-0.826	0.83	63
XIII	3-C_2H_5	-0.043	$-0.712^{b,c}$	-0.764	-0.825	0.87	61
XIV	4-C_2H_5	-0.151	-0.719^d	-0.777	-0.838	0.83	61
XV	2-i-C_3H_7	—	-0.712	-0.775	-0.837	0.85	62
XVI	3-i-C_3H_7	—	-0.717	-0.778	-0.840	0.81	62
XVII	4-i-C_3H_7	-0.126	-0.726^e	-0.784	-0.845	0.79	61
XVIII	2-t-C_4H_9	—	-0.741	-0.801	-0.871	0.87	65
XIX	3-t-C_4H_9	-0.120	-0.716^d	-0.776	-0.835	0.87	59
XX	4-t-C_4H_9	-0.197	-0.727^e	-0.784	-0.843	0.85	59
XXII	3-Cl	0.373	-0.689	$--0.755$	-0.820	0.87	65
XXII	4-Cl	0.227	-0.700	-0.764	-0.829	0.85	65
XXII	3,4-Cl_2	0.60	-0.668	-0.730	-0.791	0.83	61
XIV	4-OCH_3	-0.268	—	—	$-0.876^{e,f}$	0.82	—
XXV	3,4-$(OCH_3)_2$	-0.153	—	—	$-0.851^{e,f}$	0.83	—
XXVI	3-OH	-0.002	-0.751	-0.810	-0.870	0.85	60
XXVII	4-CN	0.628	-0.657	-0.720	-0.785	0.82	64
XXVIII	3-CF_3	0.53	-0.711	-0.777	-0.844	0.90	66

^{a} αn 1.28; ^{b} tendency towards maxima formation; ^{c} αn 1.24; ^{d} slight maxima observed; ^{e} large maxima observed; ^{f} maxima observed also in 0.05M-H_2SO_4.

for the group of substances studied, an values showed slightly more scattering. In some cases streaming maxima prevented the measurement of half-wave potentials. These were particularly marked for the alkoxy derivatives in 0·05 and 0·5M-H_2SO_4 and to a lesser extent for the p-alkyl derivatives in 0·5M-H_2SO_4.

DETERMINATION OF THE NUMBER OF ELECTRONS TRANSFERRED IN THE REDUCTION OF THE PROTONISED FORM OF SEMICARBAZONES

From Tables I and II it can be seen that the total limiting current at the maximum value in acid media corresponds to the transfer of the same number of electrons for all the semicarbazones studied. Comparison of the waves of selected semicarbazones in acid media (Table V) with the waves of model compounds for which the number of consumed electrons was known shows that the semicarbazone wave is approximately twice as high as the wave of two-electron processes.

The results of microcoulometric measurements are given in Table VI. Results varied between $n = 3·55$ and 3·9. The presence of substituted benzylamine in the electrolysis product was detected by the ninhydrin reaction and by condensation with pyridoxal. The presence of urea was shown by the biuret reaction. The absence of semicarbazide and of $C_6H_5CH_2NHNHCONH_2$ was shown by the absence of an anodic wave in 0·1M-NaOH. Asahi[7] has shown that $C_6H_5CH_2NHN$.$HCSNH_2$ gives in alkaline solution a four-electron anodic wave.

TABLE V

Comparison of Wave-Heights
2 . 10^{-4}M Depolariser in 50% ethanol.

Compound	Medium	i, µA	n
VIII	0·05M-H_2SO_4	1·31	—
VI	0·05M-H_2SO_4	1·34	—
XI	0·05M-H_2SO_4	1·30	—
XVII	0·05M-H_2SO_4	1·28	—
XXIII	0·05M-H_2SO_4	1·31	—
XXVIII	0·05M-H_2SO_4	1·34	—
Benzaldehyde	acetate pH 4·8	0·65[a]	2
4,4'-Dimethylbenzophenone	acetate pH 4·8	0·64	2
β-Morpholino propiophenone	acetate pH 4·8	0·67	2
Pyridoxal-5-phosphate	acetate pH 4·8	0·55	2
Pyridoxal oxime	acetate pH 4·8	1·21[b]	4
4,4'-Dimethylbenzophenone	0·05M borax pH 9·1	0·71	2

[a] Total wave-height; [b] substance slightly soluble.

TABLE VI

Microcoulometrically Determined Number of Electrons Consumed in the Reduction of the Protonised Form of Semicarbazone

Compound	Medium	n		
XIV	acetate buffer pH 3·9	3·82,	3·88,	3·60
	0·05M-H_2SO_4	3·55		
	0·05M-H_2SO_4, 10^{-1}M semicarbazide hydrochloride	3·40		
XXV	acetate buffer pH 3·9	3·75,	3·7	
XXVIII	0·05M-H_2SO_4	3·62		
	0·05M-H_2SO_4, 10^{-2}M semicarbazide hydrochloride	3·46		
	0·05M-H_2SO_4, 10^{-1}M semicarbazide hydrochloride	3·29		
VIII	0·05M-H_2SO_4, 10^{-1}M semicarbazide hydrochloride	3·82		

In the controlled potential electrolysis where a stream of inert gas is passed above the surface of the solution during the electrolysis, the volatility of ethanol could introduce an error. Instead of using an inert gas stream saturated with ethanol it was decided to use a completely sealed system, *i.e.* after the initial removal of dissolved oxygen the system is closed. The major problem encountered in the controlled potential electrolysis was the hydrolysis of the semicarbazones. This problem can be overcome either by the addition of an excess of semicarbazide or by carrying out the electrolysis at the highest possible pH value at which the wave-height is still at its maximum value. A less satisfactory alternative is to correct the decrease of the current during electrolysis for the decrease in concentration of semicarbazone, due to hydrolysis by carrying out a simultaneous blank experiment. The most suitable model proved to be 3-trifluoromethylbenzaldehyde semicarbazone (*XXVIII*) which only undergoes hydrolysis very slowly. In the selection of the potential for carrying out the electrolysis is should be ensured that the potential chosen is not too negative for the wave of the hydrolysis product, *i.e.* benzaldehyde or substituted benzaldehyde to be reduced as well.

REMARKS ON THE POLAROGRAPHIC BEHAVIOUR OF PYRIDINE-4-CARBOXALDEHYDE SEMICARBAZONE

This compound is reduced in acid media in two steps, the total height of which corresponds approximately to a four-electron reduction (Fig. 5). At pH > 6 the height of the more negative wave decreases with increasing pH. The height of the first wave remains virtually constant over the whole pH range studied (pH < 9·5). Only one wave was previously reported[3,5,8] but the nature of the supporting electrolyte was not mentioned so that it cannot be excluded that the absence of the second wave may be due to a difference in composition of the supporting electrolyte, e.g. due to an effect of surface or catalytically active substances.

DISCUSSION

REDUCTION COURSE

Reduction of the protonised form. Three possible products of the electroreduction of the protonised form of semicarbazones of the type $R^1R^2C = NNHCONH_2$

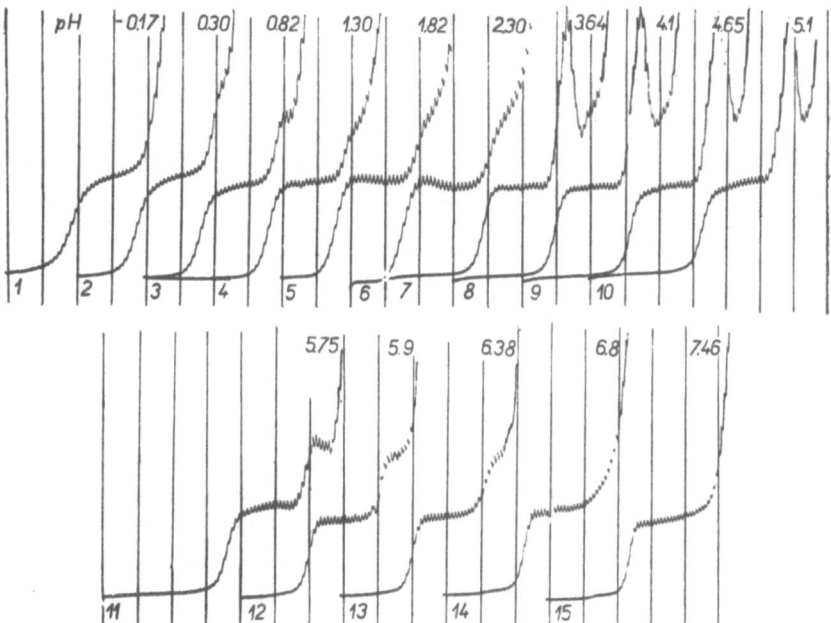

FIG. 5

pH Dependence of Reduction Waves of 4-Pyridine Carboxaldehyde Semicarbazone

2 . 10^{-4}M depolariser; *1—7* sulphuric acid; *8—12* 0·1M acetate buffers; *13—15* 0·1M phosphate buffers. Curves start at −0·45 (s.c.e.); 200 mV/absc.; t_1 = 3·3 s; m = 2·1 mg/s; full scale sensitivity 2·3 μA.

can be suggested,* *viz.* $R^1R^2CHNHNHCONH_2$, $R^1R^2CH_3$ and $R^1R^2CH_2NH_2$. Because the number of electrons in sufficiently acidic solutions approaches four and in any case is greater than two and also because the semicarbazide derivative, $R^1R^2CHNHNHCONH_2$ has not been detected in the products of electrolysis in strongly acidic media[2] and was found[8] to be electroinactive in the potential range studied, this product will not be formed in the main course of the electrolysis.

There are two arguments against the formation of $R^1R^2CH_3$ as the main product of electroreduction. Firstly there is the absence of $R^1R^2CH_3$ and semicarbazide, apart from that formed by hydrolysis, in the electrolysis products in acid media. The second argument is that formation of $R^1R^2CH_3$ would be possible only when a reductive attack on the azomethine bond would take place. This would result in the formation of a semicarbazide derivative $R^1R^2CHNHNHCONH_2$ with a subsequent reductive splitting off of semicarbazide. Because benzylthiosemicarbazide[8] and its *p*-acetamido derivative[7] were found to be polarographically inactive, it is highly improbable that $R^1R^2CHNHNHCONH_2$ would appear during the electroreductions as an intermediate.

The products isolated in strongly acid media[2,7] and the fact that the measured number of electrons approaches four (Table VI) would appear to indicate the formation of $R^1R^2CHNHNHCONH_2$ as the intermediate, or alternatively reductive splitting of the N—N bond. The first possibility is precluded by the deduction made in the preceding paragraph.

Hence the predominating course of the reduction of the protonised form of semicarbazones can be formulated as follows:

$$R^1R^2C\!=\!\underbrace{NNHCONH_2}_{H^+} + 2\,e + 2\,H^+ \;\xrightarrow[E_1]{}\; R^1R^2C\!=\!NH^+ + H_2NCONH_2 \qquad (A)$$

$$R^1R^2C\!=\!NH + 2\,e + 2\,H^+ \;\xrightarrow[E_2]{}\; R^1R^2CHNH_2\,. \qquad\qquad (B)$$

When $R^1 = C_6H_5$ potentials E_1 and E_2 are so close together that only one four-electron step is observed. Certain special cases will be discussed separately.

From oscillographic $dE/dt - E$ curves the absence of an artefact due to imine was explained[6] as ruling out the possibility of imine formation. Apart from the existence of an incompletely understood anodic incut[14] this argument is rather vague, if the rate of the electrolytic or other cleavage of imine is high.

---

* Only one of the possible semicarbazone tautomers is depicted here. It will later be deduced that in all semicarbazones one and the same tautomer is reducible, and as in some derivatives the formation of another tautomer is improbable, the structure with the azomethine group is preferred here.

TABLE VII

Polarographic *ortho*-Shifts of *ortho*- and *para*-Substituted Alkyl Benzaldehyde Semicarbazones

$$\Delta\sigma = (E_{1/2})_{o-x} - (E_{1/2})_{p-x}$$

Alkyl	$\Delta\sigma$, V	
	$0.05\text{M-}H_2SO_4$	$0.005\text{M-}H_2SO_4$
CH_3	0.027	0.029
C_2H_5	0.015	0.012
$i\text{-}C_3H_7$	0.009	0.008
$t\text{-}C_4H_9$	-0.017	-0.028

The effect of the urea part of the molecule on the electroactive bond is rather small, as can be seen from the analogous polarographic behaviour of *p*-acetamidobenzaldehyde thiosemicarbazone and its S-methyl derivative[7]. A four-electron reduction and formation of *p*-acetamidobenzylamine has been proved in the latter case.

Although the scheme following reactions *A* and *B* is considered to be the main course of the reaction there is no experimental evidence (Table VI) that it is necessarily the only possible course.

Mixtures of products formed in the controlled potential electrolysis[2,7] at the mercury pool electrode can be a result either of two competitive reactions of the protonised form or alternatively from simultaneous reduction of the protonised and unprotonised form.* As there was an increasing proportion[7] of the product of a two-electron reduction with increasing pH, the latter explanation seems more probable. The difference in the rate of recombination by which the protonised form is produced at the surface of the dropping mercury electrode and the mercury pool electrode is the reason for the difference in the ratio of i/i_d at the dropping mercury electrode and the fraction of semicarbazide derivative isolated. This can also explain the difference in the number of electrons determined by microcoulometry with a dropping electrode and the composition of the electrolysis products with a pool electrode.

Nevertheless the values of $n = 3.5_5$ to 3.9 obtained with the dropping mercury electrode** as well as the smaller limiting current observed for some derivatives, notably amino[2,6], indicate*** a more complicated course. Two possibilities can be

---

*   The other possibility, simultaneous reduction of bi- and mono- protonised forms which would result in a two- and four-electron process respectively, seems less probable as the polarographic curves showed only one decrease in the current in the form of a dissociation curve, the shape of which corresponded to a monobasic and not a dibasic[31] acid.

**   Also for phenylhydrazones values $n = 3.4$ to 3.8 were determined coulometrically[8].

***   It would be important to prove, by isolation of the reduction products, that the reduction mechanism of amino derivatives does not differ from that for other semicarbazones.

considered, firstly that competitive electrochemical reactions take place and secondly that the intermediate in the reduction undergoes a chemical reaction. The intermediate $R^1R^2C=NH^+$ could possibly undergo hydration, hydrolysis or dissociation. In the hydration process the electroactive form $R^1R^2C=NH^+$ is transformed into the electroinactive $R^1R^2C(OH)NH_2^+$. Hydrolysis in a consecutive reaction to hydration forms R^1R^2CHO and NH_4^+ but this reaction is very little probable as the aldehyde wave would appear on the diminished limiting current of the semicarbazone. Finally, due to the fact that only the protonised form of the imine undergoes reduction, the decrease in the recombination rate would result in an overall decrease in the consumption of electrons. But the effect of a recombination process is usually observed[32] with ketimines and aldimines only at pH > 10. Hence hydration would appear to be the most likely even though the results do not allow the possibility of competitive reactions to be excluded.

Antecedent reaction. The formation of the protonised form of semicarbazones occurs not only by reaction with hydroxonium ions (C) but also with other proton donors HA (D):

$$R^1R^2C=NNHCONH_2 + H_3O^+ \xrightarrow{k_1} \underset{H^+}{R^1R^2C=NNHCONH_2} + H_2O \qquad (C)$$

$$R^1R^2C=NNHCONH_2 + HA \xrightarrow{k_{HA}} \underset{H^+}{R^1R^2C=NNHCONH_2} + A^-. \qquad (D)$$

Even when a correction is made for reactions with constant k_{HA} the values of the calculated[6] rate constants k_1 are too high to correspond to a volume reaction. This observation, together with the dip on the limiting current, indicates that the protonation occurs, at least in part, as a surface reaction[30,33-37]. The dip is due to a decrease in the rate of the protonation reaction that can be caused either by a decrease in the rate constants k_1 and k_{HA} or in the value of the dissociation constant of the semicarbazone or buffer in the electrical field of the electrode or, alternatively, it can be due to the desorption of the semicarbazone or buffer or both. The effect of ionic strength[37] and nature of cation together with the change* in i–t curves[6] indicate that more than one of these effects is operative[6,37]. The rather unusual observation that the current in the region of the dip increases in the presence of added surface active substances such as gelatin[5], agar[4,6,13], imidazole[5], methyl red[5,14] and methylene blue[5] cannot be explained[6] by the catalytic effect of the sur-

---

*   The increase in the value of the exponent k in the relation $i \sim t^k$ after addition of a surface active substance in the potential region of the dip[6] where the current increases after addition of surfactants, is probably due to the suppression of a more negative current with a diffusion character.

factants – this would hold true for gelatin[33,35] and perhaps imidazole, but not for the other surfactants. On the other hand, addition of tetrabutylammonium ion eliminates the wave[5] (whereas tetramethylammonium ion does not affect the shape). The adsorption of a surfactant of the first group apparently prevents the desorption of the species that affects the current decrease. Structural effects, such as the presence of the dip with benzene, thiophene, pyrrole and ferrocene derivatives and its absence with pyridine and thiazole derivatives[3,5] also indicates the role of adsorbability; for the more strongly adsorbable pyridine no dip is observed.

SPECIAL CASES

The different pK'' values for the first and second waves of *aliphatic aldehyde semicarbazones*[15] would seem to indicate that at least in some cases the two separated waves do not correspond to successive electron uptake but to a reduction of two different forms in equilibrium. In this case the possible occurrence of mono- and bi-protonised forms cannot be excluded. The identification of ethylene[15] among the reduction products of acetaldehyde semicarbazone at pH 5·8 indicates that a more complex mechanism is operating in the reduction of these compounds in acid media.

For *aldose semicarbazones*[11] the decrease in the limiting current with pH can probably be attributed to an acid–base reaction, but comparison with an α-hydroxyaldehyde semicarbazone would prove this. The pK'' values are considerably lower than for other carbonyl compounds. The transformation of the cyclic into the reducible acyclic form is probably responsible for the change of the limiting current with time after dissolution. This reaction would be too slow to affect the pH-dependence of the limiting currents after establishment of equilibria in the bulk of the solution.

With *pyridine-4-carboxaldehyde semicarbazone* it can be assumed that the imine intermediate is stabilised and its reduction occurs at substantially more negative potentials than for semicarbazone. The height of the second wave decreases with increasing pH due either to the insufficient rate of protonation of the imine or to the lowering of the rate of the acid catalysis of another transformation of the imine. The semicarbazide formation and reduction cannot be excluded.

Semicarbazones of α,β-unsaturated ketones[2] are reduced in acid media in a four-electron step decreasing first at pH 3 to a two-electron step. At pH > 6 this two-electron step further decreases and another more negative two-electron wave appears. Although it would be possible to explain these observations in terms of the reactions (A)–(D) and conversion of the imine intermediate, in view of the possible electroactivity of the C=C bond it is preferable to have further experimental evidence before suggesting a reaction scheme.

388

Structural Effects

As in many other cases[38] an essentially linear correlation between the half-wave potentials in sulphuric acid media and the Hammett substituent constants σ has been found (Fig. 6). The values for *meta-* and *para-* alkyl and *meta*-trifluoromethyl derivatives deviated significantly from the linear plot obtained for the other substituents. In none of the hundred or so reaction series[38] for which the validity of the modified Hammett equation has been proved for polarographic data, has a substantial variation of the alkyl groups as substituents been reported. In order to decide whether this observed effect is characteristic for polarographic electroreduction or if it is an effect typical of alkyl groups, it would be necessary to have some comparable data, *e.g.* kinetic, available. No obvious correlation between αn and σ values[38] has been found, nor between αn values and deviations from the $E_{\frac{1}{2}}$–σ plot. The observation that alkyl groups exert hardly any effect on half-wave potentials can be interpreted in terms of a factor which is compensating for the polar effects of these groups. It must be borne in mind that the assumption of similar adsorbability for the whole reaction series, which forms the basis of the application of the Hammett equation (derived for homogeneous reactions) to heterogeneous electrode processes may not be true for the larger alkyl groups.

The observed value of reaction constant $\varrho = 0.10$ V, based on the measurement of nine half-wave potentials in sulphuric acid media of substances selected to cover

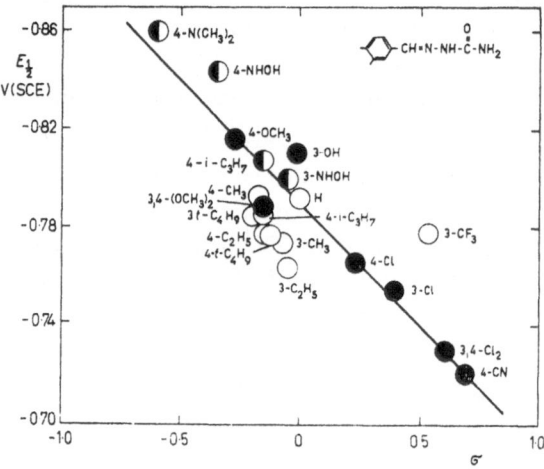

Fig. 6

Dependence of Half-Wave Potentials of Substituted Benzaldehyde Semicarbazones on the Hammett Substituent Constant

• Values used for determination of the reaction constant $\varrho_{\pi,C=NNHCONH_2} = 0.10$ V; ◐ values given by Kitajev and Budnikov; ○ alkyl and CF_3 substituted derivatives.

a wide range of σ values, is in good agreement with the reported value $\varrho = 0.13$ V obtained[4,12] for four half-wave potentials in a pH 6 phosphate buffer. Observations[4] of correlations between pK'' and σ has so far not been extended to alkyl derivatives.

The steric effects of *ortho* substituents[39] expressed by the increasingly negative value of $\Delta\sigma = (E_{1/2})_{o\text{-}X} - (E_{1/2})_{p\text{-}X}$ increase, as expected with increasing size of the alkyl group (Table VII). — When the half-wave potentials measured by Kitaev and co-workers[4,12,14,15] in a phosphate buffer containing 20% methanol were correlated with the sum of the Taft polar substituent constants[38] using the relation $\Delta E_{1/2} = \varrho^* \sum \sigma^*$ a good correlation was observed (Fig. 7) over a wide range of potentials and substituent types. The necessity for an analogous reaction mechanism to operate usually causes the applicability of linear free energy relationships to be limited to rather restricted groups of structurally related compounds. Thus it is possible to compare aliphatic nitro compounds, iodobenzene derivatives, *etc.*, but not aliphatic aldehydes with benzaldehydes or aliphatic ketones with aryl ketones. With semi-carbazones, however, it seems that an analogous reduction mechanism holds for a wide range of compounds, including benzophenone, benzaldehyde, acetophenone, aliphatic aldehydes and aliphatic ketones. Using the value of $\varrho^* = 0.25$ V for the reaction constant approximate values of $\sigma^* = 0.18$ for cyclohexylidene and $\sigma^* = 0.22$ for cyclopentylidene were found. The observed correlation (Fig. 7) indicates that the ideas about the reduction of semicarbazones in various tautomeric forms[14,15,40,41] are doubtful, in agreement with the conclusions that have been reached recently from the infra red and Raman spectra of these groups of compounds[42,43]. If various tautomeric forms do exist in solution, then their transformation into the most easily

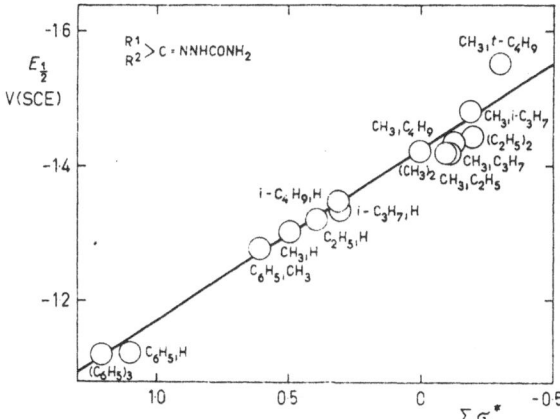

FIG. 7

Dependence of Half-Wave Potentials of Semicarbazones Measured by Kitajev and Cowor-kers[4,12,14,15] at pH 6, on the Sum of the Taft Polar Substituent Constant

reducible form is too fast to be detected on polarographic curves. The effects of ageing of solutions and irradiation[40,41,44] would be more usefully studied from the view point of the reversibility of the chemical process involved.

CONCLUSIONS

The reduction of the semicarbazones of saturated and aromatic aldehydes in the protonised form occurs predominantly by a four-electron course to form primary amine and urea. Other acids in addition to hydroxonium ion can assist in the protonation reaction.

Benzaldehyde semicarbazones are little susceptible to the effects of substituents in the *meta* and *para* positions, the half-wave potential remaining almost unchanged by substitution of *m*- and *p*- methyl, ethyl, isopropyl and tert-butyl as well as *m*-trifluoromethyl.

The essentially analogous mechanism of the electrode process, which is demonstrated by Fig. 7, proves that the analytical methods for the determination of carbonyl compounds in the presence of semicarbazide[8,9,45,46] can be applied for functional group analysis for a wide range of saturated and aromatic aldehydes and ketones[47].

The authors wish to thank to Drs M. H. Klouwen and S. H. Boelens, Research Dept., N. V. Chemische Fabriek "Naarden", Holland for their generous gift of samples of alkylsubstituted benzaldehyde semicarbazones, to Drs O. Manoušek and O. Exner, J. Heyrovský Institute of Polarography, for their assistance in coulometric measurements and discussions of structural effects and to Miss H. Jänchenová for technical assistance. The stay in the J. Heyrovský Institute of Polarography of one of us (B. F.) was enabled by the grant sponsored by the Science Research Council.

REFERENCES

1. Dušinský G.: Pharmazie 8, 897 (1953).
2. Lund H.: Acta Chem. Scand. 13, 249 (1959).
3. Tirouflet J., Person M.: Ric. Sci. 30; Contrib. Teor. Sper. Polarografia 5, 269 (1960).
4. Kitajev Yu. P., Budnikov G. K., Arbuzov A. E.: Izv. Akad. Nauk. SSSR, Otd. Chim. Nauk 1961, 1772.
5. Laviron E., Person M., Fournari P.: Compt. Rend. 255, 2440 (1962).
6. Kitajev Yu. P., Budnikov G. K.: Ž. Obšč. Chim. 33, 1396 (1963).
7. Asahi Y.: Chem. & Pharm. Bull. (Tokyo) 11, 930 (1963).
8. Person M.: Thesis. University of Dijon, 1964.
9. Souchay P., Graizon M.: Chim. Anal. (Paris) 36, 85 (1954).
10. Coulson D. M.: Anal. Chim. Acta 19, 284 (1958).
11. Haas J. W., Storey J. D., Lynch C. C.: Anal. Chem. 34, 145 (1962).
12. Kitajev Yu. P., Budnikov G. K., Trojepolskaja T. V., Arbuzov A. E.: Dokl. Akad. Nauk SSSR 137, 862 (1961).
13. Kitajev Yu. P.: Budnikov G. K.: Arbuzov A. E.: Izv. Akad. Nauk SSSR, Otd. Chim. Nauk 1961, 824.
14. Kitajev Yu. P., Budnikov G. K., Skrebkova I. M.: Izv. Akad. Nauk SSSR, Otd. Chim. Nauk 1962, 244.

15. Kitajev Yu. P., Budnikov G. K.: Izv. Akad. Nauk. SSSR, Ser. Chim. *1964*, 978.
16. Woggon H.: Ernährungsforsch. *6*, 331 (1961).
17. Kitajev Yu. P., Trojepolskaja T. V.; Abstracts of Fifth Conference on Electrochemistry of Organic Compounds, Moscow 1964.
18. Kitajev Yu. P., Budnikov G. K., Trojepolskaya T. V., Skretokova I. M.: Abstracts of Conference on Polarography, Kiev 1965.
19. Horák V.: Personal communication.
20. Klouwen M. H., Boelens S. H.: Rec. Trav. Chim. *79*, 1022 (1960).
21. Frankel M., Patai S.: *Tables for Identification of Organic Compounds*. Chem. Rubber Publ. Co., Cleveland 1960.
22. Sherrill M. L.: J. Am. Chem. Soc. *52*, 1991 (1930).
23. Gardner H. J.: Nature *167*, 158 (1951).
24. Vlček A. A.: This Journal *19*, 862 (1954); Chem. listy *48*, 189 (1954).
25. Manoušek O., Peizker J.: Unpublished results.
26. Turczanyi B., Zuman P.: Unpublished results.
27. Koutecký J.: This Journal *18*, 597 (1953); Chem. listy *47*, 323 (1953).
28. Zuman P.: Unpublished results.
29. Zuman P., Exner O.: This Journal *30*, 1832 (1965).
30. Manoušek O., Zuman P.: This Journal, in the press.
31. Hanuš V., Brdička R.: Chem. listy *44*, 291 (1950); Chimija *1*, 28 (1951).
32. Zuman P.: This Journal *15*, 839, (1950).
33. Volková V.: Nature *185*, 743 (1960).
34. Mairanovskij S. G.: Dokl. Akad. Nauk. *132*, 1352 (1960).
35. Volková V.: *Advanced in Polarography* (I. Langmuir, Ed.), p. 840. Pergamon, Oxford *1960*.
36. Mairanovskij S. G., Liščeta L. I.: This Journal *25*, 3025 (1960).
37. Mairanovskij S. G.: Dokl. Akad. Nauk SSSR *142*, 1120, (1962).
38. Zuman P.: *Substituent Effects in Organic Polarography*. Plenum Press, New York 1967.
39. Zuman P.: This Journal *27*, 648 (1962).
40. Kitajev Yu. P., Budnikov G. K.: Dokl. Akad. Nauk SSSR *127*, 818 (1959).
41. Kitajev Yu. P., Budnikov G. K., Arbuzov A. E.: Izv. Akad. Nauk. SSSR, Otd. Chim. Nauk *1961*, 1222.
42. Shagidullin I. P. P., Rajevskij O. A., Kitajev Yu. P.: Izv. Akad. Nauk, SSSR, Ser. Chem. *1964*, 960; *1966*, 218.
43. Rajevskij O. A., Šagidullin I. P. P., Kitajev Yu. P.: Dokl. Akad. Nauk SSSR *159*, 900 (1964).
44. Kitajev Yu. P., Budnikov G. K., Arbuzov A. E.: Dokl. Akad. Nauk SSSR *127*, 1041 (1959).
45. Chorosin A. V.: Zavodsk. Lab. *28*, 420 (1962).
46. Manoušek O., Kučerová Z.: Věda a výzkum v průmyslu potravinářském *16*, 59 (1965).
47. Fleet B.: Anal. Chim. Acta *36*, 304 (1966).

Translated by the author (P. Z.).

OXIME DERIVATIVES. VIII.*

POLAROGRAPHIC REDUCTION OF O- AND N-SUBSTITUTED OXIMES

P. ZUMAN and O. EXNER

J. Heyrovský Institute of Polarography, Czechoslovak Academy of Sciences, Prague

Received June 25th, 1964

Reduction of nine N-substituted oximes (nitrones) *I—IX* and of seven
O-substituted oximes *XI—XVII* at the dropping mercury electrode was
investigated. Aromatic N-substituted derivatives are reduced in acid as well
as in alkaline media in one four-electron wave to the secondary amine; the
reduction proceeds through the corresponding Schiff base. The behaviour of
aromatic O-alkyl derivatives in acid media is similar to that of corresponding
N-alkyl derivatives whereas in alkaline solutions O-alkyl derivatives do not
show a reduction wave. Aliphatic N-derivatives are reduced in acid media
only and aliphatic O-derivatives are principally polarographically inactive.
In all instances, both isomeric derivatives can be safely distinguished by
polarography.

Isomeric O- and N-alkyl oximes differ in reactivity in several chemical reactions,
especially in their reactivity towards reducing agents. This difference in reactivity
can be used for structure elucidation, as has been shown in Part II of this series[1]
where lithium aluminium hydride has been recommended as suitable reagent.

With this reagent N-alkyl oximes are reduced smoothly to N,N-dialkylhydroxylamines that
show a considerable resistance towards a further attack of hydride, whereas O-alkyl oximes are
attacked only under more drastic conditions where the reduction occurs under fission of the
molecule resulting in alcohol and primary amine formation. Other reducing agents usually attack
only the more reactive N-derivatives. The course of the reaction may be different: Hence, besides
N,N-dialkylhydroxylamines[2,3] also Schiff bases[4,5] and secondary amines[5-7] have been isolated.

The aim of the present study was to find out whether differences in the behaviour
of both isomers persist in the reduction at the dropping mercury electrode. This
being proved, polarography could serve as a reliable method for distinguishing of
both isomers, more rapid than preparative reactions and probably of broader
applicability than *e.g.* ultra violet spectroscopy. The possibility of a rapid distinguishing between the two structural isomers is of importance in the case of alkylated
oximes, as both isomers are frequently formed simultaneously. In addition to the

---

* Part VII.: This Journal *28*, 3150 (1963).

series of N-substituted $(I-IX)$ and O-substituted $(XI-XVII)$ oximes, possible intermediates of electrode processes X and $XVIII-XXI$ were also investigated in order to distinguish between the two possible reduction paths.

Table I

Substances Studied

No	Substance	Substituent	M.P, °C	Reference
I	$C_6H_5CH{=}N{-}R$ $\quad\quad\quad\quad\vert$ $\quad\quad\quad\quad O$	$R = CH_3$	84	8
II		$R = C(CH_3)_3$	75	2
III		$R = $ cyclo-C_6H_{11}	83	3
IV		$R = CH_2C_6H_5$	82	4
V		$R = CH(C_6H_5)_2$	159	1
VI		$R = C_6H_5$	112	9
VII	p-$CH_3O.C_6H_4CH{=}N{-}C_6H_5$ $\quad\quad\quad\quad\quad\quad\quad\quad\vert$ $\quad\quad\quad\quad\quad\quad\quad\quad O$	—	118	9
VIII	$(C_6H_5)_2C{=}N{-}CH_2C_6H_5$ $\quad\quad\quad\quad\quad\vert$ $\quad\quad\quad\quad\quad O$	—	119	4
IX	$CH_3CH{=}N\text{-cyclo-}C_6H_{11}$ $\quad\quad\quad\quad\vert$ $\quad\quad\quad\quad O$	—	86	10
X	$C_6H_5{-}CH{=}N{-}C_6H_5$	—	53	
XI	$C_6H_5{-}C{=}NOCH_3$ $\quad\quad\quad\quad\vert$ $\quad\quad\quad\quad R$	$R = H$	$1{\cdot}5477^a$	11
XII		$R = CH_3$	$1{\cdot}5415^a$	11
XIII		$R = C_6H_5$	62	7
XIV	$(CH_3)_2C{=}NOCH_3$	—	$1{\cdot}4001^a$	11
XV	$(CH_3)_2C{=}NOCH_2C_6H_5$	—	$1{\cdot}4999^a$	1
XVI	⬠$={}NOCH_3$	—	$1{\cdot}4560^a$	11
XVII	⬡$={}NOC_2H_5$	—	$1{\cdot}4630^a$	11
XVIII	C_6H_5NHOH	—	81	
XIX	$C_6H_5.N(OH).CH_2C_6H_5$	—	84	b
XX	$(C_6H_5CH_2)_2NOH$	—	123	1
XXI	$(C_6H_5)_2CH.N(OH)CH_2C_6H_5$	—	107	1

^{a} n_D^{20}; ^{b} see Experimental.

Experimental

Apparatus. The polarographic curves were recorded on a Heyrovský polarograph, Type VII (Nejedlý) with a galvanometer of sensitivity $2 \cdot 10^{-9}$ A/mm. A Kalousek vessel with a separated saturated calomel electrode (s.c.e.) was used. The capillary used had the following constants: outflow velocity $m = 2 \cdot 1$ mg/s, and drop-time $t_1 = 3 \cdot 7$ s at the potential of s.c.e. at mercury pressure $h = 65$ cm. All the values of half-wave potentials were measured using thallous ions as internal standard ($E_{1/2} = -0 \cdot 44_5$ V *versus* s.c.e.) and expressed *versus* s.c.e. The pH-values of buffer solutions used were checked using the Autojonometer Type PHM 21d and the pH-meter Type PHM 4 (both Radiometer — København) with a glass electrode type G 200 B.

Substances. Substances used in this study are given in Table I. N-*Phenyl*-N-*benzylhydroxyl-amine* (XIX) was prepared by reduction of N-phenylbenzaldoxime (*VI*; 1·97 g) with lithium aluminium hydride (0·25 g) in a mixture of anhydrous ether (20 ml) and benzene (20 ml). After boiling for one hour the reaction mixture was decomposed with sodium potassium tartarate solution, yield 1·85 g (93%) m.p. 84°C (dilute ethanol). Literature[12] gives m.p. 86°C.

Solutions. 0·01M ethanolic stock solutions were prepared by dissolving a weighed amount of the substance in 96% ethanol at 25°C. The buffer solutions used and other supporting electrolytes were prepared from reagent grade chemicals.

Techniques. 9·8 ml of the appropriate buffer solution were deaerated by a stream of pure nitrogen. 0·2 ml of a 0·01M solution of the substance under study was added and the deaeration continued for a further 30 seconds. Polarographic curves were then recorded.

Results and Discussion

NITRONES

N-*Substituted Benzaldoximes*

N-*Alkyl benzaldoximes.* The substances $I - V$ are principally reduced at the dropping mercury electrode in one four-electron step in acid media. With increasing pH the current (i_1) decreases in the form of a dissociation curve (Fig. 1). Decrease of current with increasing pH does not occur in the whole potential range of the limiting current simultaneously: It varies according to potential. At more negative potentials the decrease occurs already at lower pH-values, at more positive only at higher pH-values. Such a behaviour results in a formation of a trough (minimum, depression) (Fig. 1) on the limiting current; at high pH-values the first wave can possess the form of a peak (Fig. 2). At lower pH-values an adsorption pre-wave is observed at potentials by about 0·2 V more positive than the main wave (i_1). At higher pH-values a more negative four-electron wave (i_2) is formed (Figs 1, 2).

The trough is independent of addition of various neutral salts and is independent both of the type of anion and of the alkali metal involved. Presence of borate was without influence. With tetramethylammonium chloride (not shown in Fig. 3) the height of the first wave remained unchanged. Due to the shift of wave i_2 towards more positive potentials the decrease of current was somewhat smaller. Tetrabutyl-ammonium chloride in 0·025M solution both shifted the wave i_2 towards more positive potentials and suppressed wave i_1 (Fig. 3, curves 8 and 9) probably due to its surface activity. Hence the observed decreases on the limiting current are not of the type

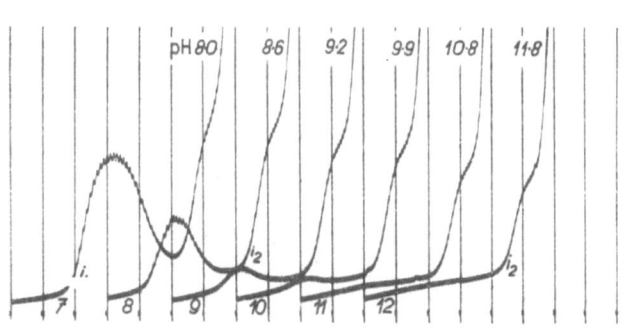

Fig. 1
pH-Dependence of Reduc-
tion Waves of
$C_6H_5CH=N(O)C(CH_3)_3$
(*II*)
$1 \cdot 10^{-4}$M nitrone *II*,
Britton-Robinson buffers,
pH given on the polaro-
gram, 1% ethanol. Curves
starting at: *1—3* —0·2 V;
4—5 —0·4 V; *6—7* —0·6 V;
8—12 —0·8 V, S.C.E.,
200 mV/abcs., $h = 80$ cm,
full scale sensitivity 6·6 μA.

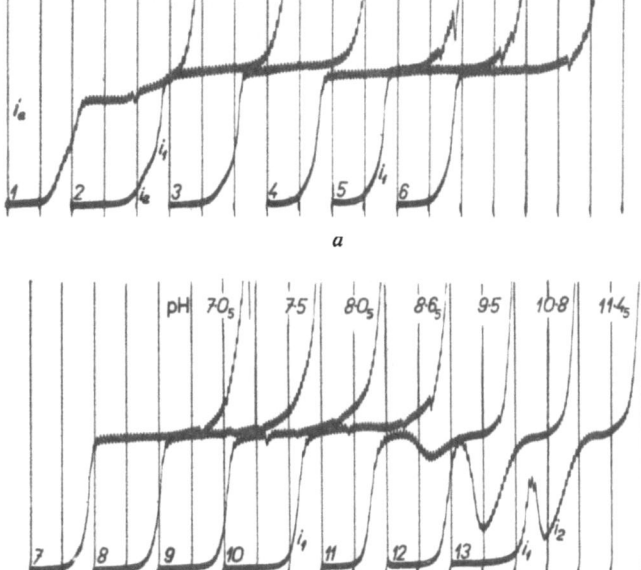

Fig. 2
pH-Dependence of Reduc-
tion Waves of
$C_6H_5CH=N(O)CH_2C_5H_6$
(*IV*)
$2 \cdot 10^{-4}$M substance *IV*,
Britton-Robinson buffers,
pH given on the polaro-
gram, 2% ethanol. Curves
starting at: *1—2* —0·2 V;
3 —0·4 V; *4—10* —0·6 V;
11 —*13* —0·8 V; S.C.E.,
200 mV/absc., $h = 80$ cm,
full scale sensitivity 8·8 μA.

described by Frumkin for reduction of anions[13], but rather of that described for chromates[14] and periodates[15].

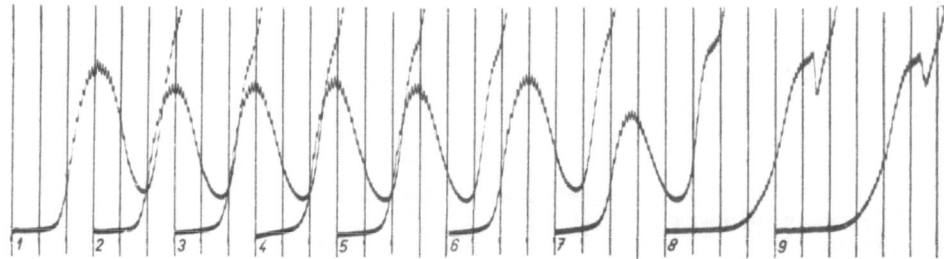

Fig. 3

Effects of Cations on the Trough on the Nitrone Wave

$2 . 10^{-4}$M nitrone *II*, Britton-Robinson buffer pH 7·9, 2% ethanol, added: *1* 0; *2* 0·05M-Na_2SO_4; *3* 0·1M-NaBr; *4* 0·1M-NaI; *5* 0·1M-NaCl; *6* 0; *7* 0·2M-KCl; *8—9* 0·005M-$(C_4H_9)_4$NCl. Curves starting at —0·6 V, s.c.e., 200 mV/absc., *h* = 45 cm, full scale sensitivity 8·8 μA.

The similarity is shown also by the type of pH-dependences. When currents measured at given potential at various pH-values are compared, they fit the theoretical curve derived by Koutecký[16] for recombination of monobasic acids. Curves obtained at various potentials are shifted with respect to each other (Fig. 4). Such dependences were observed for substances *I—IV* in potential range from beginning of the limiting current to the region where second wave i_2 is observed at higher pH-values. At more negative potentials observed i/i_d — pH curves were distorted due to the fact that $(i_1 + i_2)$ was measured instead of i_1.

Since pK_a-values that would allow us to compute[16] values (at least formal) of the rate constants of recombination reaction (k_r) are not available, values pK′, cor-

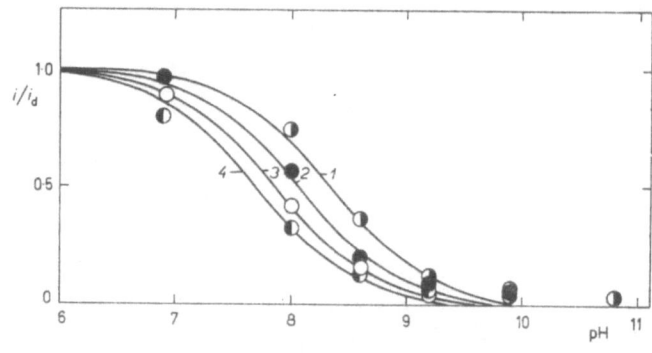

Fig. 4

pH-Dependence of the Ratio i/i_d for Nitrone *II*

Current in solutions of substance *II* in Britton-Robinson buffers measured at various potentials: —1·3 V ◗; —1·4 V ●; —1·45 V ○; —1·5 V ◑. Theoretical curves plotted for *1* pK′ 8·34; *2* pK′ 8·06; *3* pK′ 7·86; *4* pK′ 7·68.

responding to pH for which $i_1 = i_d/2$ were determined. pK'-values, obtained for a series of compounds under identical experimental conditions, are a linear function of changes in the value of $(\log k_r + pK_a)$. pK'-values decreased linearly for all substances studied with increasingly negative potentials (Fig. 5).

This decrease or else decrease of limiting current with increased applied voltage in the particular pH-range can be, in our opinion, explained (i) either by effect of potential on rate and/or equilibrium constants in the field of the electrode (ii) or by decrease in the concentration of one of the particles taking part in the recombination reaction at the electrode surface with increase of negative potential. The reaction antecedent to the electrode process proper is not a "volume reaction" but rather a "surface reaction"[17].

Substance V was too insoluble to be studied in solutions containing 4% ethanol in which the measurements with substances $I-IV$ have been carried out. In 80% ethanol nitrone V showed qualitatively a behaviour similar to that of substances $I-IV$.

In acid solutions smaller limiting currents were observed (Figs 1 and 2). Under such conditions waves change with time (Fig. 6) and a more negative wave is formed, identified as the wave for benzaldehyde.

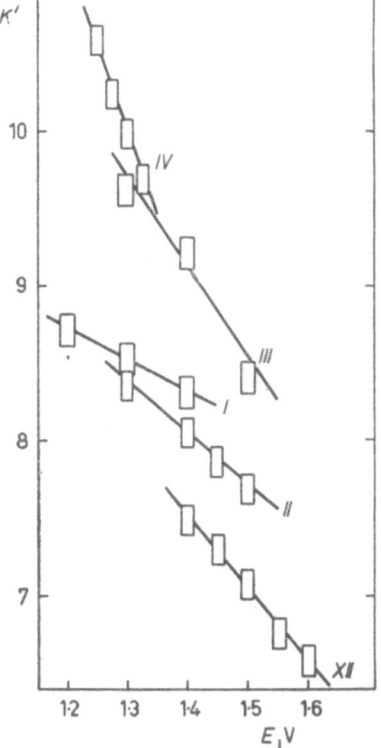

Fig. 5

Dependence of the Rate of Surface Electrode Reaction on Applied Potential

Values of a pH, for which $i = i_d/2$ (i.e. pK') for Current Measured at Various Potentials. Waves of substances I, II, III, IV and XII recorded in Britton-Robinson buffers.

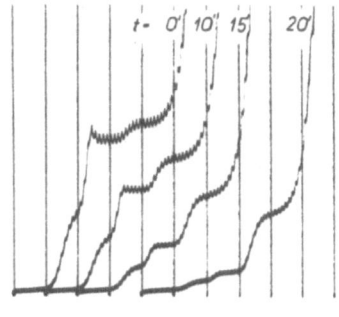

Fig. 6

Acid Hydrolysis of $C_6H_5CH{=}N(O)CH_3$ (I)

$2 . 10^{-4}$M substance I in Britton-Robinson buffer pH 2·0, 2% ethanol. Recording of the curves started at times given on the polarogram at -0.2 V, S.C.E., 200 mV/absc., $h = 80$ cm, full scale sensitivity 6·6 µA.

Table II

pH-Dependence of Half-Wave Potentials of N-Alkylsubstituted Benzaldoximes $C_6H_5CH{=}N{-}R$
Britton-Robinson buffers, 4% ethanol, s.c.e.
$\downarrow$
O

pH	$E_{1/2}$, V			
	$R = CH_3$	$C(CH_3)_3$	cyclo C_6H_{11}	$CH_2C_6H_5$
3·0	−0·74	−0·76$_5$	−0·69$_5$	—
3·9	−0·780	−0·849	−0·810	−0·820
4·6	—	—	—	−0·87
5·1	−0·84	−0·90	−0·86	—
5·6	—	−0·94	—	−0·95
6·0	−0·92	−0·96$_5$	−0·93	—
6·5	−0·95	−1·01	—	—
6·9	−0·983	−1·044	−1·026	−1·001
7·5	—	−1·08	−1·03	−1·05
8·0	−1·03	−1·10	−1·07	−1·06
8·6	−1·06; −1·57	−1·14; −1·749	−1·13; −1·73	−1·11
9·2	— —	−1·16; −1·75	−1·18; −1·73	−1·16
9·9	−1·10; −1·57	— ; −1·76	−1·20; −1·730	−1·19
10·8	— —	— ; −1·75	−1·22$_5$; −1·73	−1·25; −1·59
11·4	— —	— —	— —	−1·29$_5$; —
11·8	— ; −1·57	— ; −1·75	— ; −1·72$_2$	— ; −1·60

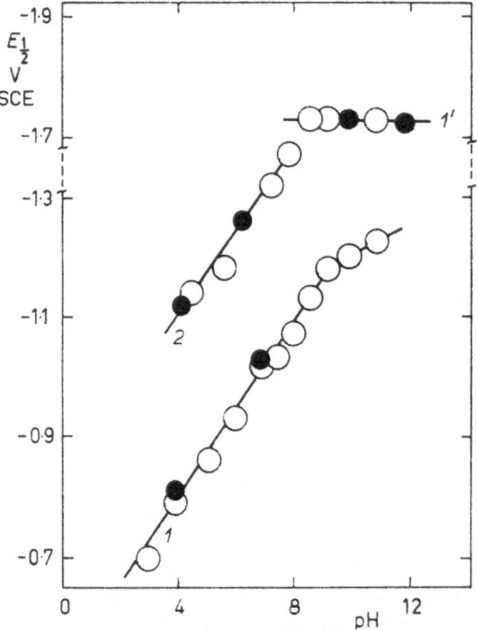

Fig. 7

The Effect of Phenyl Substitution on Half-Wave Potentials of Nitrones

pH-Dependence in Britton-Robinson (○) and simple (●) buffers of *1* phenyl derivative *III* and *2* methyl derivative *IX*. In *1'* half-wave potentials of second wave of *III* are plotted.

Hydrolytic cleavage of the nitrone into benzaldehyde and alkylhydroxylamine hence occurs.

Shifts of half-wave potentials and other important dara are summarized in Tables II and III. Typical $E_{\frac{1}{2}}$ — pH plot is given in Fig. 7. Half-wave potentials of wave i_2 are practically pH-independent.

N-*Phenyl-benzaldoxime* (VI). The four-electron reduction of nitrones $I - V$ can follow principally two courses, both resulting in formation of secondary amine as final product. Either the N—O bond (*1*) or the C=N bond (*2*) are attacked first:

$$C_6H_5CH=NR^{(+)} \xrightarrow[+H^+]{2e} C_6H_5CH=NHR^{(+)} \xrightarrow[+2H^+]{2e} C_6H_5CH_2NH_2R^{(+)} \quad (1)$$
$$\overset{|}{OH} \qquad\qquad\qquad +OH^-$$

$$C_6H_5CH=NR^{(+)} \xrightarrow[2H^+]{2e} C_6H_5CH_2NHR^{(+)} \xrightarrow[+H^+]{2e} C_6H_5CH_2NH_2R^{(+)} \quad (2)$$
$$\overset{|}{OH} \qquad\qquad \overset{|}{OH}$$

Polarography of compound *VI* allows us to exclude possibility (*2*) and by analogy it can be deduced that route (*1*) is more probable even for $I - V$.

Table III

Data on N-Alkylsubstituted Benzaldoximes $C_6H_5CH{=}N{-}R$

$$\overset{|}{O}$$

R	$dE_{1/2}/d\,pH$ mV/pH	pH Range[a]	$\dfrac{RT}{n\alpha F}$[b]	pH[c]
CH$_3$	61·5	3 — 8·4	0·072	3·9
			0·068	4·3
	26	8·4—10·4	0·072	6·9
			0·067	7·3
C(CH$_3$)$_3$	71	3 — 8·1	0·078	3·9
	34	8·1— 9·3	0·074[d]	6·9
cyclo-C$_6$H$_{11}$	73	4 — 9·2	0·063[d]	3·9
	26	9·2—10·8	0·059[d]	6·9
CH$_2$C$_6$H$_5$	72	4 —11·4	—[e]	3·9
			—[e]	6·9
	—	—	0·061	9·9

[a] pH Range in which the value $dE_{1/2}/dpH$ is valid; [b] slope of the logarithmic analysis; [c] individual pH value at which the logarithmic analysis was carried out. [d] Lower part of the curve to $i = 0\cdot6i_d$, at higher part the shape of the curve is affected by a maximum; [e] non-linear logarithmic analysis due to adsorption phenomena.

On polarographic curves of substance *VI* principally two waves are observed at lower pH-values (Fig. 8): A three electron step (i_A) followed by a one electron step (i_B). The height of both these waves remains practically unchanged over the whole pH-range studied. The more positive wave is accompanied by an adsorption pre-wave at potentials by about 150 mV more positive than the main wave. The height of this pre-wave does not change with increasing concentration at concentrations of depolarizer higher than about $1.5 . 10^{-4}$M. The half-wave potential (measured at $5 . 10^{-4}$M to reduce the effect of the adsorption wave) of the more positive wave shifts towards negative potentials with increasing pH-value, whereas the half-wave potential of the negative one-electron step is practically pH-independent. (Fig. 9) Hence in buffers of pH > 11 in the presence of 0·005% gelatine, added to suppress the acute maximum, both waves coalesce and only one single four-electron wave is obtained.

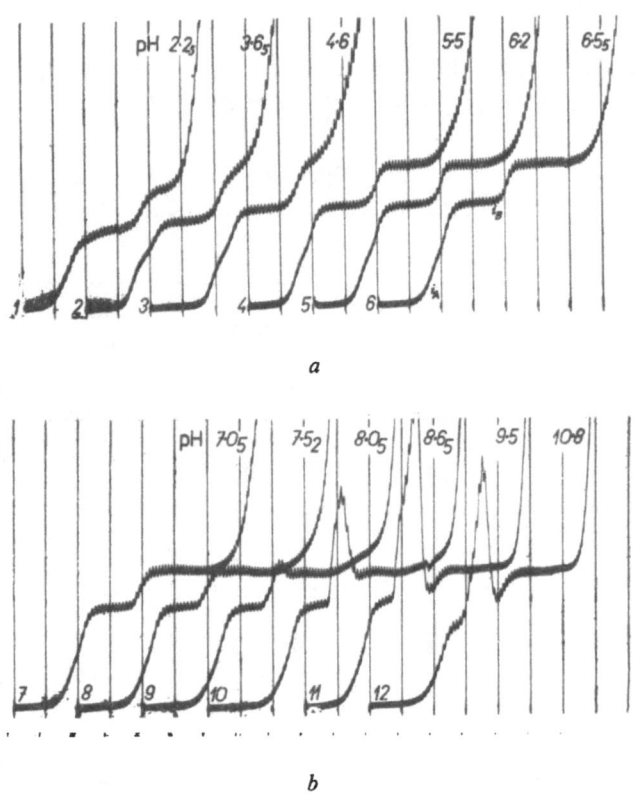

Fig. 8
pH-Dependence of Reduction Waves of $C_6H_5CH{=}N(O)C_6H_5$ (*VI*)
$2 . 10^{-4}$M nitrone *VI*, Britton-Robinson buffers, pH given on the polarogram, 2% ethanol, Curves starting at: *1—3* —0·2 V; *4—10* —0·4 V, *11, 12* —0·6 V, s.c.e., 200 mV/absc., h = 80 cm, full scale sensitivity 8·8 μA.

The two-step reduction is thought to be explained by the sequence of following reactions:

$$C_6H_5CH(O)=NC_6H_5 + H^{(+)} \rightleftharpoons C_6H_5CH(OH)=\overset{(+)}{N}C_6H_5 \qquad (3)$$

$$C_6H_5CH(OH)=\overset{(+)}{N}C_6H_5 + 2e \longrightarrow C_6H_5CH=NC_6H_5 + OH^- \qquad (4)$$

$$C_6H_5CH=NC_6H_5 + H^+ \rightleftharpoons C_6H_5CH=\overset{(+)}{N}HC_6H_5 \qquad (5)$$

$$C_6H_5CH=\overset{(+)}{N}HC_6H_5 + e \longrightarrow C_6H_5\overset{\cdot}{C}H - NHC_6H_5 \qquad (6)$$

$$2\,C_6H_5\overset{\cdot}{C}H - NHC_6H_5 \xrightarrow{k_{dim}} \begin{array}{c} C_6H_5CHNHC_6H_5 \\ | \\ C_6H_5CHNHC_6H_5 \end{array} \qquad (7)$$

$$C_6H_5\overset{\cdot}{C}H - NHC_6H_5 + e \longrightarrow C_6H_5\overset{(-)}{C}HNHC_6H_5 \qquad (8)$$

$$C_6H_5\overset{(-)}{C}HNHC_6H_5 + H^+ \rightleftharpoons C_6H_5CH_2NHC_6H_5 \; . \qquad (9)$$

Reaction (7) can take part in the reaction sequence as a side-reaction. The difference from alkylderivatives $I - V$ can be caused by the stabilisation of the radical formed due to the extension of the conjugated system. The scheme $(3)-(9)$ has been proved as follows:

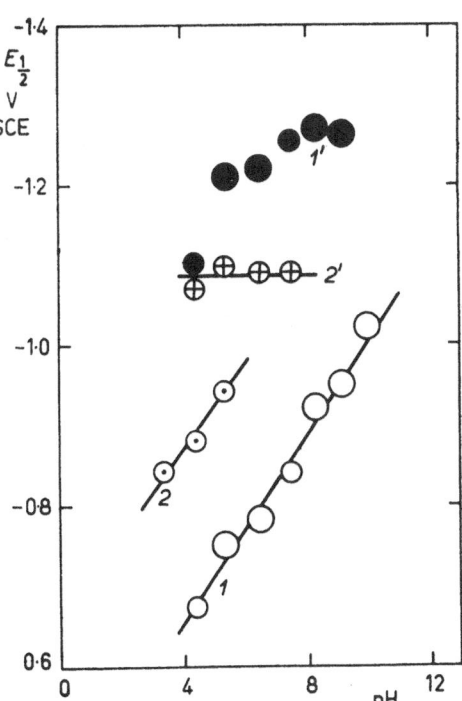

Fig. 9

Comparison of Half-Wave Potentials of $C_6H_5CH{=}N(O)C_6H_5$ (*VI*) and $C_6H_5CH_2N(OH)C_6H_5$ (*XIX*) pH-Dependence in Britton-Robinson and simple buffers of 1, 1' nitrone *VI* (○, ●) and 2, 2' phenyl hydroxylamine derivative *XIX* (⊙, ⊕). 1', 2' refer to second waves.

1. The waves of *VI* have been compared with those for N-phenyl-N-benzylhydro-xylamine (*XIX*). The reduction of *XIX* occurs in two ill-separated waves the height of which decreases with increasing pH. The shape of these drawn-out waves as well as their half-wave potentials differ from those of nitrone *VI*, the first wave of *XIX* being more positive then the first wave of *VI*.

2. The waves of *VI* were compared with those for benzylidene aniline (*X*). Polarography of this compound was studied by Holleck[18] the results of whom we could confirm. At pH > 7 aldimine *X* is reduced in two one-electron waves that coalesce at pH about 11. The half-wave potential of the more positive wave of aldimine *X* nearly coincides with five-sixth potential of the more positive wave of nitrone *VI* (Fig. 10). The half-wave potential of the more negative wave of aldimine *X* coincides with half-wave potential of the more negative one-electron wave of *VI*. Moreover, the pH, at which both waves of *X* coalesce is identical with the pH-value, above which one single wave is observed for *VI*.

At pH between 5 and 7 a direct comparison of half-wave potentials of aldimine *X* and nitrone was made difficult by the fact that in this region hydrolysis of anil already takes place[19]. Moreover the measurement of the half-wave potential of the more negative wave of *X* is made difficult by the fact that benzaldehyde formed in hydrolysis gives a wave that differs little in half-wave potential from those of the more negative wave of *X*.

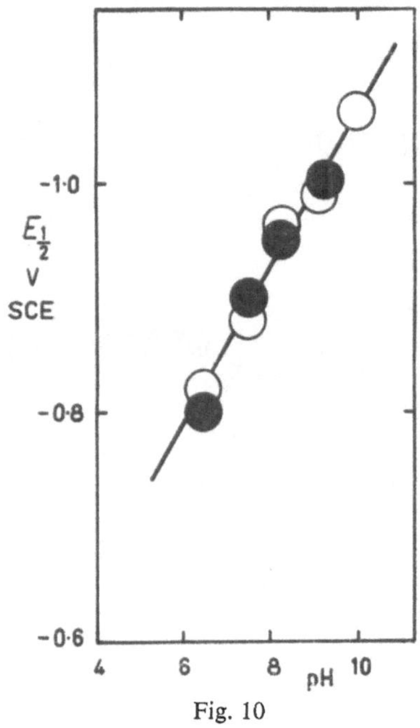

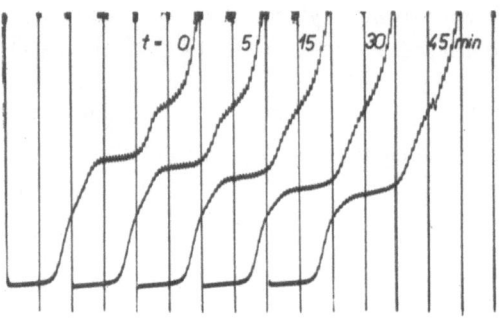

Fig. 10

Comparison of Index Potentials of $C_6H_5CH{=}N(O)C_6H_5$ (*VI*) and of $C_6H_5CH{=}NC_6H_5$ (*X*)

pH-Dependence in Britton-Robinson buffers of $E_{5/6}$ of the first wave of nitrone *VI* (○) and of $E_{1/2}$ of the first wave of aldimine *X* (●).

Fig. 11

Hydrolysis of Nitrone $C_6H_5CH{=}N(O)C_6H_5$ (*VI*) in Acid Media

$2 \cdot 10^{-4}$M nitrone *VI* in acetate buffer pH 4·3. Curves recorded after intervals given on the polarogram, starting at —0·4 V, s.c.e., 200 mV/absc., 400 mV/ /min., full scale sensitivity 4·8 µA.

Coincidence similar as at pH > 7 was found when half-wave potentials were measured for both *VI* and *X* at 0°C.

3. A more indirect additional evidence is given by the changes in the waves of nitrone *VI* with time in acid media. Here again the half-wave potential of the more negative wave for *VI* coincided with that of benzaldehyde.

At lower pH waves of nitrone *VI* decrease with time in the way that the height of first wave decreases by one third and that of the second increases twofold (Fig. 11). This is explained by hydrolysis according to (*10*).

$$C_6H_5CH=NC_6H_5 \quad \xrightarrow{\ H_2O\ } \quad C_6H_5CHO + HNC_6H_5 \qquad (10)$$
$$\underset{O}{|} \qquad\qquad\qquad\qquad\qquad\qquad \underset{OH}{|}$$

Wave of phenylhydroxylamine coincides with first wave of nitrone *VI*, wave of benzaldehyde with second wave of *VI*. The reaction is practically shifted towards products and hence on the last curve (Fig. 11) two two-electron waves can be observed.*
Because in the course of the chemical reaction a substance consuming four electrons (3 + 1) is transformed into two substances each of them consuming two electrons (2 + 2) the total wave-height remains unchanged (Fig. 11).

When benzaldehyde was added to hydrolysis product an increase of the more negative wave was observed accompanied by a smaller decrease in the more positive wave. The latter results from the decrease of phenylhydroxylamine concentration due to reverse condensation reaction.

Half-wave potentials and some other data are summarized in Table IV.

Table IV

Half-Wave Potentials (in Volts) of Substances *VI* and *XIX*
(Britton-Robinson Buffers, 4% ethanol, $5 . 10^{-4}$M depolarizer, 0·005% gelatine, s.c.e.)

pH	VI		XIX	
	$(E_{1/2})_1$	$(E_{1/2})_2$	$(E_{1/2})_1$	$(E_{1/2})_2$
3·35	—	—	—0·823	—
4·4	—0·672[a]	—1·110	—0·88	—1·07
5·4	—0·75	—1·21	—0·94	—1·10
6·5	—0·78	—1·22	—	—1·09
7·5	—0·841[b]	—1·256	—	—1·09
8·3	—0·92	—1·27	—	—
9·15	—0·95	—1·26		
10·0	—1·02	—		
12·2	—1·22			

[a] $RT/n \alpha F = 0.075$, when determined from upper part of the curve, unaffected by adsorption;
[b] $RT/n \alpha F = 0.082$ from the middle part of the curve.

* Benzaldehyde is reduced in one 2 electron step even at these low pH-values due to the presence of a cation of the ammonium type (protonized phenylhydroxylamine).

Structural effects. To compare the effect of substituents R in $C_6H_5C=N(O)R$ in protonized form values of half-wave potentials in simple acetate buffers pH 4·7 containing 4% ethanol were measured. For comparison of structural effects on the

Table V

Effect of Substituent R on Half-Wave Potentials (s.c.e.) of Substances of the Type
$C_6H_5CH=N-R$
$|$
O

| R | $E_{1/2}$, V | | | σ^* |
	acetate pH 4·7 4% ethanol	0·1M-NaOH 4% ethanol	0·1M-NaOH 40% ethanol	
CH_3	0·820	1·596	1·595	0
$C(CH_3)_3$	0·874	1·734	1·740	−0·31
cyclo C_6H_{11}	0·860	1·719	1·727	−0·15
$CH_2C_6H_5$	0·860	$—^a$	1·548	+0·21₅
C_6H_5	0·689^c	(1·322)^{a,b,c}	1·377^c	+0·60

^{a} The measurement influenced by the adsorption pre-wave; ^{b} 0·01% gelatin; ^{c} Half-wave potential of the first, three-electron step was measured.

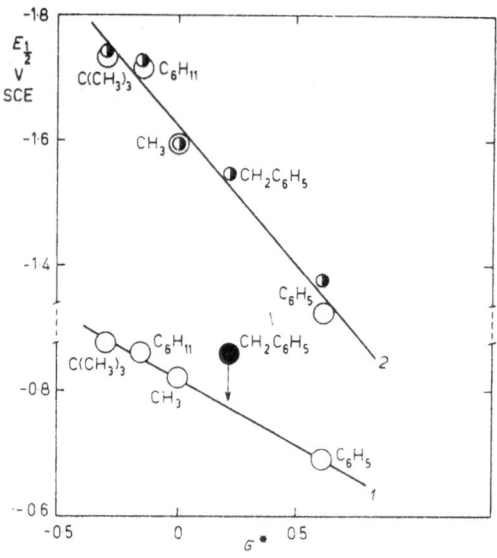

Fig. 12

Substituent Effects on Half-Wave Potentials of N-Substituted Benzoldoximes
1 Acetate buffer pH 4·7; *2* 0·1M-NaOH; σ^* Taft's polar substituent constants; circles 4% ethanol, squares 40% ethanol, full circle deviating.

reduction of the unprotonized form 0·1M-NaOH was chosen. Because in aqueous solutions containing 4% of ethanol adsorption phenomena made the measurement of half-wave potential of benzyl derivative *IV* impossible and of phenyl derivative *VI* difficult, measurements were carried out in 40% ethanol–water mixture. The increase of ethanol concentration affected little the values of half-wave potentials (Table V, Fig. 12).

When plotted against Taft polar substituent constants[20], half-wave potentials of both protonized and unprotonized forms showed a linear dependence (Fig. 12). The deviation for benzyl derivative *IV* in acetate buffer is probably caused by adsorption phenomena. Similar deviation were observed for analogous plots of half-wave potentials in Britton-Robinson buffers pH 3·9 and 6·9. For reaction constant, characterising the slope of $E_{\frac{1}{2}} - \sigma^*$ plot[20,21], values $\varrho_\pi^* = 0.44$ V for 0·1M-NaOH and $\varrho_\pi^* = 0.23$ V for acetate buffer pH $= 4.7$ (in good agreement with $\varrho_\pi^* = 0.22$ V found in Britton-Robinson buffer pH $3-9$ and $\varrho_\pi^* = 0.25$ V at pH 6·9) were determined.

The validity of the linear free energy relationship expressed in $E_{\frac{1}{2}} - \sigma^*$ plot proves that substituent R exerts mainly an inductive effect and that even for the cyclohexyl *III* and tert. butyl *II* derivatives the geometry of the transition state of the electrode process is not substantially affected.

Other N-Substituted Oximes

Polarographic behaviour of *anisaldoxime derivative* (VII) closely resembled that of phenyl derivative *VI*, in the reduction in two (3 + 1 electron) steps and in hydrolysis in acid media. The adsorption phenomena both in acid and in alkaline media were more pronounced with *VII*; in sodium hydroxide more concentrated

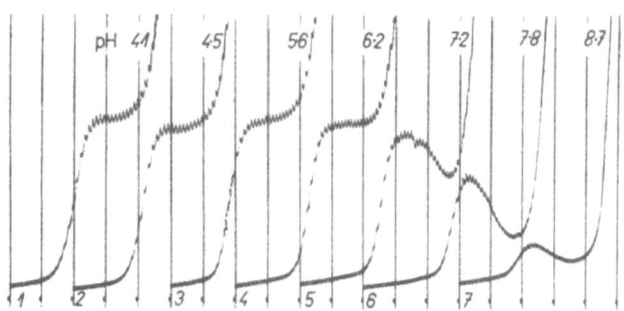

Fig. 13

Reduction Waves of N-Cyclohexyl Acetaldoxime *(IX)*

$1 . 10^{-4}$M nitrone *IX*, Britton-Robinson buffers, pH given on the polarogram, 1% ethanol. Curves 1 and 2 starting at —0·6 V, 3—6 at —0·8 V, 7 at 1·0 V, s.c.e. 200 mV/absc., full scale sensitivity 6·6 μA.

than about 0·3M a cleavage of nitrone *VII* occured. The half-wave potential in 0·1M NaOH with 40% ethanol $(-1·45_8$ V) has shown a theoretically expected shift of $-0·08_1$ V in comparison with unsubstituted phenyl derivative.

Reduction waves of *benzophenone derivative* VIII closely resembled those of the corresponding benzaldehyde derivative *IV*. Again with the less soluble compound *VIII* adsorption phenomena played a more important role than with *IV*.

N-*Cyclohexyl acetaldoxime* (IX). Polarographic behaviour of this compound resembled that of the corresponding benzaldoxime derivative *III* at lower pH only. A four electron wave gradually decreasing with increasing pH and a formation of a trough on the limiting current (Fig. 13) was observed. The hydrolysis in acid solution was marked here even at pH 3·5. However, at higher pH the wave of the unprotonized form did not appear before the rise of current due to electrolysis of the supporting electrolyte. Half-wave potentials (Table VI) of acetaldoxime derivative *IX* are by about $\Delta E_{\frac{1}{2}}$ 0·29 V more negative than those of benzaldoxime derivative *III* (Fig. 7), in qualitative agreement with the theory of substituent effects[21].

Table VI

Half-Wave Potentials of Acetaldoxime Derivative *IX*
Britton-Robinson buffers, $1 . 10^{-4}$M depolarizer.

pH	4·1	4·5	5·6	6·2	7·2	7·8
$E_{1/2}$, V	1·019	1·04	1·18	1·260	1·32	1·37

O-SUBSTITUTED OXIMES

O-*Methyl Derivatives of Aromatic Oximes*

Behaviour and comparison with N-*alkyl derivatives.* At lower pH-values polarographic curves of O-methyl derivatives of aromatic oximes $XI - XIII$ (Fig. 14) closely resemble those of N-alkyl benzaldoximes $I - IV$ (Fig. 1). Not only the shape of the four-electron waves but their decrease with increasing pH with a dip on the limiting current was similar. Current measured at various potentials again fitted theoretical dissociation curves[16] and the pK' value was a linear function of potential (Fig. 5). Inspection of Fig. 5 reveals that pK'-values for *XII* is shifted towards lower pK'-values and hence for O-alkyl derivatives the decrease of current occurs at lower pH-values than for N-alkyl derivatives. This is clearly demonstrated for current measured at $-1·2$ V for isomeric benzaldoximes, where for *I* values p$K' = 8·7$ whereas for *XI* value p$K' = 7·3$ was found. The half-wave potentials (Table VII) of these two compounds were identical and fitted the same linear $E_{\frac{1}{2}}-$ pH dependence (Fig. 15, curve 2).

The difference between O-alkyl and N-alkyl derivatives is shown in alkaline media in which O-alkyl derivatives $XI - XIII$ show no wave of the unprotonized form. Furthermore O-alkyl derivatives are much less sensitive to acid hydrolysis as it is

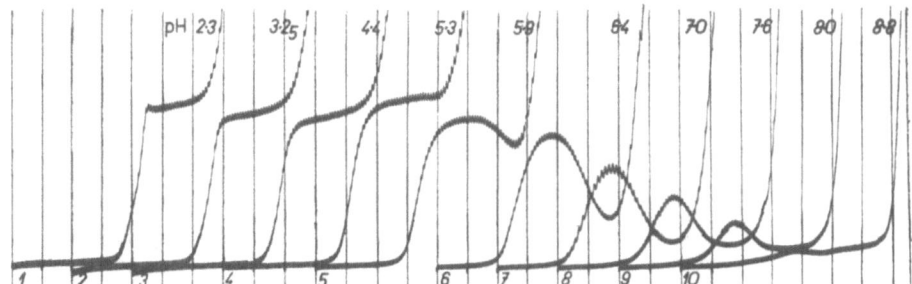

Fig. 14

Reduction Waves of $C_6H_5(CH_3)C{=}NOCH_3$ (*XII*)

$4 \cdot 10^{-4}$M compound *XII*, Britton-Robinson buffers, pH given on the polarogram, 4% ethanol. Curves *1—3* starting at 0·0 V, *4* —0·2 V, *5* —0·4 V; *6—10* —0·6 V, S.C.E. 200 mV/absc., full scale sensitivity 6·6 μA.

Table VII

pH-Dependence of Half-Wave Potentials of O-Methyl Aromatic Oximes $R(C_6H_5)C{=}N{-}OCH_3$, Britton-Robinson buffer, 4% ethanol, S.C.E.

pH	$E_{1/2}$, V		
	$R = H^a$	$R = CH_3{}^a$	$R = C_6H_5{}^b$
2·3	—0·69	—0·800	—0·606
3·2₅	—0·73	—0·86	—0·677
4·4	—0·801^c	—0·912^c	—0·75
5·3	—0·88	—0·982	—0·808^e
5·9	—	(—1·01)^f	—0·858
6·4	—0·984^d	(—1·08)^f	(—0·93)^f

^{a} $dE_{1/2}/dpH = 0.057$ for the whole pH-range; ^{b} $dE_{1/2}/dpH = 0.059$ for the whole pH-range; ^{c} $RT/\alpha\,nF = 0.061$; ^{d} $RT/\alpha\,nF = 0.121$; ^{e} $RT/\alpha\,nF = 0.054$; ^{f} uncertain values due to the dip on the limiting current.

well-known from preparative work[7]. O-Alkyl derivatives are easily volatile from aqueous solution with the purging gas.

The similarity of the behaviour of the protonized forms *Ia* and *XIa* is understood as due to the close resemblance of the electroactive forms:

$$C_6H_5 - CH{=}\overset{(+)}{N} - CH_3$$
$$\vert$$
$$OH$$

Ia

$$C_6H_5 - CH{=}\overset{(+)}{N} - H$$
$$\vert$$
$$OCH_3$$

XIa

On the other hand the electroactivity of *Ib* and inactivity of *XIb* shows that *Ib* is more reactive towards an nucleophilic attack on the N—O bond by electron than *XIb*:

$$\overset{\delta+}{C_6H_5CH}=\overset{\delta-}{N} - CH_3$$
$$\overset{|}{O^{\delta-}}$$

Ib

$$\overset{\delta+}{C_6H_5CH}=\overset{\delta-}{N} - OCH_3$$

XIb

Structural effects. Because the $E_{\frac{1}{2}}$ – pH plots for all three compounds $XI - XIII$ studied were practically parallel, the distance between these linear plots expresses the best average value for the shifts of half-wave potentials $(\Delta E_{\frac{1}{2}})$ due to a substituent R in $R(C_6H_5)C = N{-}OCH_3$ relative to $R = CH_3$. Values $(\Delta E_{\frac{1}{2}})_H = 0.116$ V and $(\Delta E_{\frac{1}{2}})_{C_6H_5} = 0.170$ V were found. These values plotted against polar substituent constants[20] (σ^*) fitted a linear dependence (Fig. 16).

The value of reaction constant $(\varrho_\pi^* = 0.27$ V) expressing the susceptibility towards substituent effects can be compared with those for effects of substituents on nitrogen in N-substituted benzaldoxime derivatives.

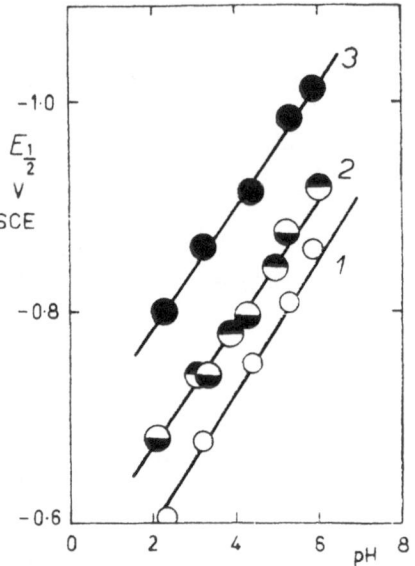

Fig. 15

Comparison of Half-Wave Potentials of Oximes of the Type $C_6H_5C(R)$=NOCH$_3$ $(XI{-}XIII)$ and Corresponding Nitrone *I* pH-dependence in Britton-Robinson buffers of half-wave potentials of oximes for: *1* (○) $R = C_6H_5$; *2* (◐) $R = H$; *3* (●) $R = CH_3$; and *2* (◑) of nitrone C_6H_5CH=N(O)CH$_3$. The slope for *1* is 60 mV/pH, *2* and *3* 56 mV/pH.

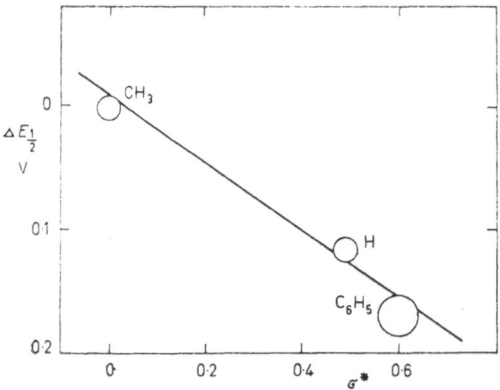

Fig. 16

Substituent Effects on Half-Wave Potentials of Oximes of the Type $C_6H_5C(R)$=NOCH$_3$ $(XI{-}XIII)$

Differences between the half-wave potentials $(\Delta E_{1/2})$ derived from Fig. 15 are plotted against Taft's polar substituent constants σ^* for reaction constant $\varrho_{\pi,R}^* = 0.27$ V.

O-*Alkyl Derivatives of Aliphatic Oximes*

Whereas no developed waves have been obtained for substance *XIV*, some negative waves resembling those for hydroxylamine derivatives [22] were observed for compounds *XV—XVII*. The height of these waves decreased with increasing pH-value. For benzyl derivative *XVI* following values of half-wave potentials were determined: pH 2·3 $E_{\frac{1}{2}}$ = −0·86 V (s.c.e.), pH 4·4 $E_{\frac{1}{2}}$ = −1·25 V (s.c.e.).

HYDROXYLAMINE DERIVATIVES

Phenylhydroxylamines

Phenylhydroxylamine (XVIII) is reduced in one single two electron wave that is steep and well-defined at pH < 3·5 (Fig. 17). At higher pH-values a more negative wave is separated and the height of the first wave decreases with increasing pH in the

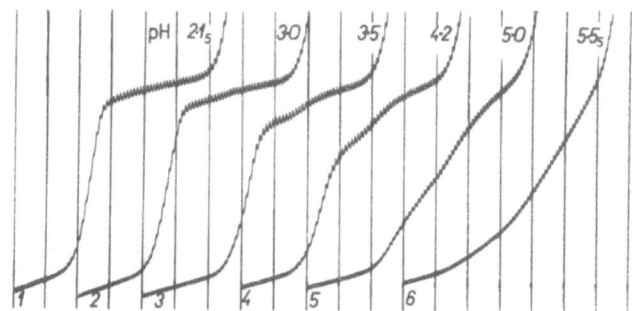

Fig. 17

pH-Dependence of Reduction Waves of Phenylhydroxylamine (*XVIII*)

$2 \cdot 10^{-4}$M compound *XVIII*, Britton-Robinson buffers, pH given on the polarogram, 2% ethanol. Curves *1—3* starting at 0·0 V, *4—6* at —0·2 V, s.c.e., 200 mV/absc., full scale sensitivity 6·6 μA.

Table VIII

pH-Dependence of Half-Wave Potentials (in Volts) of Phenylhydroxylamines C_6H_5NROH
Britton-Robinson buffers, 4% ethanol s.c.e.

R = H		R = CH₂C₆H₅		
pH	$E_{1/2}$	pH	$(E_{1/2})_1$	$(E_{1/2})_2$
2·1₅	−0·451	3·3₅	−0·823	
3·0	−0·519	4·4	−0·88	−1·07
3·5	−0·60	5·4	−0·94	−1·10
4·2	−0·644	6·5	—	−1·09
5·0	−0·72	7·5	—	−1·09

410

form of a theoretical dissociations curve[16] with $pK' \doteq 4.5$. At still higher pH-values the waves became more drawn-out and indistinct. It is assumed that in the more positive wave the protonised form of *XVIII* is reduced. Unusually big shift of half-wave potential of the more positive wave $dE_{\frac{1}{2}}/dpH = 0.096$ V/pH was observed in the pH-range between 2 and 5 (Fig. 18).

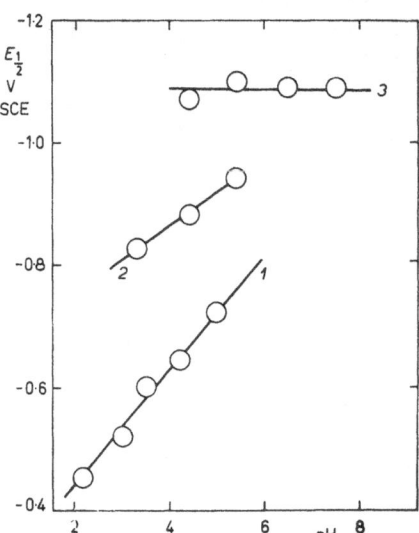

Fig. 18

Comparison of Half-Wave Potentials of Phenyl-hydroxylamines *XVIII* and *XIX*
pH-Dependence in Britton-Robinson buffers of half-wave potentials of *1* phenylhydroxylamine (*XVIII*); *2* first wave of benzylphenylhydroxylamine (*XIX*); *3* second wave of *XIX*.

Benzylphenylhydroxylamine (XIX) shows a similar behaviour, only the first wave is more negative and hence less developed than for *XVIII* and the second wave is better developed so that its half-wave potentials can be measured (Table VIII). pK'-value is about 4.6. The shift of the more positive wave with pH is here normal $(dE_{\frac{1}{2}}/dpH = 0.057$ V/pH) in the pH-range between 3 and 5.5. Due to the difference in the effect of pH on half-wave potentials of *XIX* and *XX* no quantitative expression of the shift of half-wave potentials is possible[21].

At pH > 6 oxidation to nitrone occurs which is at pH > 7 so fast that even in beforehand deaerated solution only curve of nitrone *IV* was recorded. This oxidation reaction is well-known from preparative work[4,5,23], nevertheless, the identity of the product has been proved by comparison of reduction waves at several pH-values with the waves of *IV*.

Benzylhydroxylamines

Benzylhydroxylamine derivatives XX and XXI that bear no phenyl ring directly attached to the nitrogen atom show poorly developed waves at negative potentials resembling those of alkylhydroxylamines[22]. Even for these substances separation of two waves was indicated but the waves were too indistinct to allow measurement of their ratio. Especially *XX* is very sensitive to oxidation into nitrone *V* as stated previously in preparative experiments[5].

Conclusions

In both N- and O-substituted oximes the reduction resembles that of oximes[24,25] in the four-electron transfer of the protonized form producing amines and in the marked dip on the limiting current in the region where the protonation is not fast enough and the transport towards the electrode surface is not governed solely by diffusion. In all these substances reduction of the protonized form occurs at considerably more positive potentials than that of the conjugated base.

It has been shown that an interchange of substituents as in *Ia* and *XIa* does not practically affect the polarographic behaviour, which is almost identical. On the other hand difference in behaviour in alkaline media of the unprotonated forms (*e.g. Ib* and *XIb*) makes polarography a useful tool for distinguishing of these two isomers.

From the study of polarographic reduction of various oximes, hydrazones etc. Lund[25] drew the conclusion that compounds of the general formula $R^1R^2C=N-YR^3$ are reduced by two electrons if Y is hydrogen or carbon atom, but by four electrons if Y is oxygen or nitrogen. Our findings confirm this conclusion in the case of O-alkyl oximes; N-alkyl oximes which contain both a carbon radical and an oxygen atom should be placed into the second category, too.

Finally, a comparison of polarographic and chemical reduction of nitrones and related compounds should be given. As final products of the reduction of nitrones by various agents secondary amines have been obtained; that cannot help in distinquishing between reaction paths (*1*) and (*2*). Nevertheless, isolation of Schiff bases using zinc and ammonium chloride[5] or acetic acid[26] indicates reaction (*1*), whereas formation of N,N-dialkylhydroxylamines using lithium aluminium hydride[1,2] or catalytic hydrogenation [3,27] indicate as possible route (*2*). These and our results are thus consistent with the usual generalization (*cf. e.g.*[28]) that electroreductions at a high overvoltage cathode resemble chemical reductions with a common metal, whereas electroreductions at a low overvoltage cathode are analogous to catalytic hydrogenation or reduction with complex hydrides. The only reaction that does not fit into this scheme is the hydrogenation of benzophenone N-benzyloxime (*VIII*) to the corresponding Schiff base[4].

References

1. Exner O.: This Journal *20*, 202 (1955); Chem. listy *48*, 1543 (1954).
2. Emmons W. D.: J. Am. Chem. Soc. *79*, 5739 (1957).
3. Horner L., Jürgens E.: Chem. Ber. *90*, 2184 (1957).
4. Cope A. C., Haven A. C.: J. Am. Chem. Soc. *72*, 4896 (1950).
5. Alessandri L.: Gazz. Chim. Ital. *51*, I., 75 (1921).
6. Martynoff M.: Ann. Chim. (Paris) [11] *7*, 424 (1937).
7. Semper L., Lichtenstadt L.: Ber. *51*, 928 (1918).
8. Brady O. L., Dunn F. P., Goldstein R. F.: J. Chem. Soc. *1926*, 2386.
9. Beckmann E.: Ann. *367*, 271 (1909).
10. Krimm H.: Chem. Ber. *91*, 1057 (1958).
11. Reiser A., Jehlička V., Dvořák K.: This Journal *16*, 13 (1951).
12. Vavon G., Crajcinovic: Compt. Rend. *187*, 420 (1928).
13. Frumkin A. N.: Trans. Faraday Soc. *55*, 156 (1959); where further references are given.

412

14. Tondeur J. J., Dombret A., **Gierst L.**: **J.** Electroanal. Chem. *3*, 225 (1962).
15. Zuman P.: Unpublished results.
16. Koutecký J.: This Journal *18*, 597 (1953); Chem. listy *47*, 323 (1953).
17. Majranovskij S. G., Lisčeta L. I.: This Journal *25*, 3025 (1960).
18. Holleck L., Kastening B.: Z. Elektrochem. *60*, 127 (1956).
19. Kastening B., Holleck L., Melkonian G. A.: Z. Elektrochem. *60*, 130 (1956).
20. Taft R. W., jr.: *Separation of Polar, Steric, and Resonance Effects in Reactivity* in the book: *Steric Effects in Organic Chemistry* (M. S. Newman, Ed.). Wiley, New York 1956.
21. Zuman P.: This Journal *25*, 3025 (1960).
22. Vodrážka Z.: Chem. listy *46*, 208 (1952).
23. Jones L. W., Sneed M. C.: J. Am. Chem. Soc. *39*, 674 (1917).
24. Souchay P., Ser S.: J. Chim. Phys. *49*, C 172 (1952).
25. Lund H.: Acta Chem. Scand. *13*, 249 (1959).
26. Earl J. C.: Rec. Trav. Chim. *75*, 346 (1956).
27. Cusmano G.: Gazz. Chim. Ital. *51*, II., 306 (1921).
28. Brewster J. H.: J. Am. Chem. Soc. *76*, 6361 (1954).

Translated by the author (P. Z.).

Polarographic Study of the Reaction of Pyridoxal with Hydroxylamine

In the course of the study of biochemically important reactions, the condensation of pyridoxal with hydroxylamine was chosen as a model reaction. The difference in the heights of four-electron diffusion controlled polarographic wave of pyridoxaloxime and two electron kinetically controlled polarographic waves of pyridoxal enabled us to determine the increase of oxime concentration with time.

At pH >3 the equilibrium between pyridoxal and its oxime is shifted towards oxime so that the reaction occurs practically quantitatively already in stoichiometrical ratio. The reaction can thus be treated as irreversible. The reaction rate corresponds to a second order reaction, first order in pyridoxal and first in hydroxylamine. The rate constant is a function of pH (Figure). It is assumed that all three forms of pyridoxal, the protonized (PH_2^+), the neutral molecule or zwitter-ion (PH) and anionic form (P^-) undergo the reaction with constants $k_{PH_2^+}$, k_{PH} and k_{P^-}, and that only unprotonized form of hydroxylamine takes part in the reaction. Hence the measured rate constant k' follows equation (1):

$$k' = \frac{k_{PH_2^+} S_N}{\left(1 + \dfrac{[H^+]}{K_N}\right)\left(1 + \dfrac{K_{PH_2^+}}{[H^+]}\right)} +$$

$$\frac{k_{PH} S_N}{\left(1 + \dfrac{[H^+]}{K_N}\right)\left(1 + \dfrac{[H^+]}{K_{PH_2^+}} + \dfrac{K_{PH}}{[H^+]}\right)} +$$

$$\frac{k_{P^-} S_N}{\left(1 + \dfrac{[H^+]}{K_N}\right)\left(1 + \dfrac{[H^+]}{K_{PH}}\right)}$$

For $pK_{PH_2^+} = 4.23$, $pK_{PH} = 8.7$, $pK_N = 6.0$ and analytical concentration of hydroxylamine (S_N) the curve plotted in the Figure shows a good agreement with experimentally determined values for $k_{PH_2^+} = 500$ l mol⁻¹ min⁻¹, $k_{PH} = 3.5$ l mol⁻¹ min⁻¹ and $k_{P^-} = 13.1$ l mol⁻¹ min⁻¹. Hence all three forms of pyridoxal are reactive but show a marked difference in reactivity. Formation of a hydrated intermediate (which would probably be polarographically inactive) deduced[1] for reaction of hydroxylamine with benzaldehyde, fural etc., was not detected in the reaction with pyridoxal, as no deviations from second order kinetics were observed. General acid-base catalysis, demonstrated for reaction of pyridoxal with semicarbazide[2], practically does not affect the con-

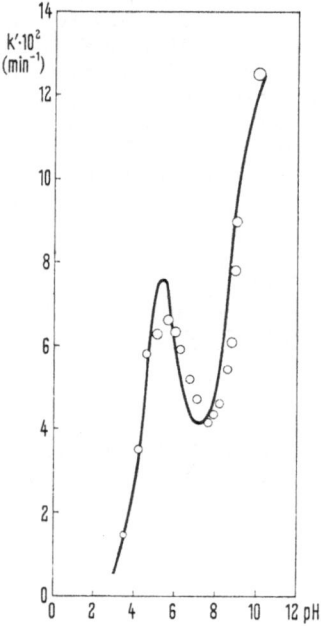

Theoretical curve of the pH-dependence of the rate constant for condensation of pyridoxal computed using equation (1). The circles denote experimental points. Buffers used: pH 3.6–5.6, 0.18 M acetate; pH 5.9–7.8, 0.18 M phosphate; pH 8.2, 0.09 M veronal; pH 8.6, 0.12 M veronal; pH 8.8–9.0, 0.18 M veronal; pH 10, 0.18 M phosphate; ionic strength $\mu = 0.5$ controlled by addition of sodium chloride. Concentration of hydroxylamine $S_N = 10^{-2} M$.

stant k_{PH^+} and at higher pH-values causes only a minor perturbation of the observed constants. Hence, e.g. at pH 5.8, a twofold increase in acetate buffer concentration causes an increase in the value of rate constant of about 15%. The major factor affecting the pH-dependence of the rate of oxime formation is thus the antecedent acid-base equilibria changing the protonation of reactants.

At pH <3 equilibrium 'between protonized form of pyridoxal (PH_2^+) and unprotonized form of hydroxylamine was studied from the dependence of limiting currents of pyridoxaloxime on hydroxylamine concentration at various pH-values. For the reaction

$$PH_2^+ + NH_2OH \rightleftharpoons \text{oxime}$$

for $pK_{NH_2OH^+} = 6.0$, $pK_{PH_2^+} = 4.23$ and for $pK_{oxime} \gg 10$ the value of $K = [\text{oxime}]/[PH_2^+][NH_2OH] = 2.10^7$ was found.

Similarly the pH-dependence of the reaction rate of pyridoxal-5-phosphate was measured. Due to the additional dissociation steps resulting from the introduction of phosphoric acid part, the interpretation of the shape of this dependence is more complicated.

This work will be described in detail in Collection of Czechoslovak Chemical Communications.

Zusammenfassung. Es wurde polarographisch die Kinetik der Reaktion des Pyridoxals mit Hydroxylamin verfolgt. Die pH-Abhängigkeit der Geschwindigkeitskonstante der Kondensierung konnte auf Grund verschiedener Reaktivität einzelner dissoziierter Formen des Pyridoxals mit nichtprotonisiertem Hydroxylamin erklärt wer:' .

P. ZUMAN and O. MANOUŠEK

Polarographic Institute, Czechoslovak Academy of Science, Prague and Central Research Institute of Food Industry, Prague (Czechoslovakia), December 30, 1963.

[1] W. P. JENCKS, J. Amer. chem. Soc. *81*, 475 (1959).
[2] E. H. CORDES and W. P. JENCKS, Biochemistry *1*, 773 (1962).

Part VIII

Polarographic Reduction of Substituted Benzonitriles

35. O. Manoušek, P. Zuman, and O. Exner: Quantitative treatment of substituent effects in polarography. VI. Polarographic reduction of benzonitriles *m*- and *p*-substituted with electronegative groups. *Collect. Czechoslov. Chem. Commun.* **33**, 3979 (1968).

36. P. Zuman, and O. Manoušek: Fission of activated carbon-nitrogen and carbon-sulphur bonds. XII. Polarographic reduction of substituted benzonitriles bearing in *para*- position a carbonyl group in acidic media. *Collect. Czechoslov. Chem. Commun.* **34**, 1580 (1969).

37. P. Čársky, and P. Zuman: The use of simple molecular orbital theory to elucidate the polarographic behaviour of some *para*- substituted benzonitriles. *Collect. Czechoslov. Chem. Commun.* **34**, 497 (1969).

In benzonitriles of the type $CN - C_6H_4 - X$, where X = COO⁻, COOR or CN, the $C - C$ bond can be reductively cleaved.[35] In the nucleophilic substitution on the benzene ring the cyanide ion is the leaving group. For X=COR in acidic media, the cyano group is reduced[36] in a four-electron process to CH_2NH_2. Quantum chemistry was used[37] to decide which of the electronegative groups in substituted benzonitriles undergoes electroreduction. The prediction was correct only in about 60% of the cases.

QUANTITATIVE TREATMENT OF SUBSTITUENT EFFECTS IN POLAROGRAPHY. VI.*

POLAROGRAPHIC REDUCTION OF BENZONITRILES *m*- AND *p*-SUBSTITUTED WITH ELECTRONEGATIVE GROUPS

O.Manoušek, P.Zuman** and O.Exner

J. Heyrovský Institute of Polarography,
Czechoslovak Academy of Sciences, Prague 1

Received March 19th, 1968

Benzonitriles bearing a COO⁻, COOR or CN group in the *meta*- or *para*-position are reduced at pH > 8 in one two-electron step in which cyanide ions are the leaving group. Effects of substituents can be expressed by Hammett equation: $\Delta E_{1/2} = \varrho_X \sigma_X^-$. The activation by electron attracting substituents is for benzonitriles less important than for aryl alkyl sulphones and benzenesulphonamides. The transition state in the benzonitrile reduction is probably more symmetrical.

The polarographic reduction wave of benzonitrile was first reported by Stackelberg and Stracke[1]. In dimethylformamide, the reduction of benzonitrile gives a one-electron wave corresponding to the formation of a radical anion, confirmed[2] by electron spin resonance. One wave was also reported[2] in this medium for 4-chloro-benzonitrile and 4-aminobenzonitrile, whereas for 4-cyanobenzoic acid, two one-electron waves were observed. They were attributed to the formation of mono and dianion radicals. For 4-amino- and 4-fluorobenzonitriles, waves corresponding to a consumption of 1·5 electrons were observed[2]. These were assigned to elimination of F⁻ or NH₂⁻, followed by dimerization, because 4,4′-cyanobiphenyl was identified as the reduction product. Similarly 1·5 electrons were reported[3] to be consumed in the reduction of aqueous and acetonitrile solutions of 1-ethyl-4-cyanopyridinium ions, with the formation of a radical of the dimer. Nevertheless in this case cyanide ions were eliminated. In 2- and 4-cyanopyridines[4-6] in aqueous media, at higher pH-values a two-electron reduction wave was observed corresponding to an elimination of cyanide ions. The reduction path was thus different from that observed in acidic solutions, where hydrogenation of the cyanide group took place giving a four-

---

*   Part V: This Journal *32*, 2066 (1967).
**   Present address: Department of Chemistry, University of Birmingham, Great Britain.

electron step. The course of the reaction in alkaline media was confirmed[7] by controlled potential electrolysis.

In the course of the study of the reaction of diphenyldiazomethane with substituted benzoic acids[8] a reduction wave was observed in the medium pH-range in buffered aqueous solutions of p-cyanobenzoic acid at about $-1\cdot8$ V (Fig. 1). As no such wave was observed for other substituted benzoic acids, it was decided to study the polarographic reduction of m- and p-cyanobenzoic acids and related derivatives $I-XIII$ (Table I) in some detail. The reduction of compounds $I-VI$ has not been reported previously, and the reducibility of dicyanobenzenes VII and VIII was mentioned only briefly earlier[6,9,10].

This paper is a part of our wider research project of the study of aromatic derivatives bearing two groups which can under certain conditions be reducible[10]. Generally, the electroreduction of both groups can take place either simultaneously (and the electrode process can be even reversible[9]) or consecutively. In the latter case it is generally not yet possible to predict which of the two groups will undergo reduction at more positive potentials. It is therefore necessary to prove in each single case which of the two groups present is reduced first. In the case of m- and p-substituted benzonitriles in this and related papers[11-13], controlled potential electrolysis and rectangular voltage polarization using a commutator method were applied for the study of the electrode process in addition to the analysis and interpretation of polarographic curves. In this paper only the polarographic behaviour of these substituted benzonitriles is discussed in which the reduction involves the cyano group. Behaviour of benzonitrile derivatives, in which the other electronegative group is reduced

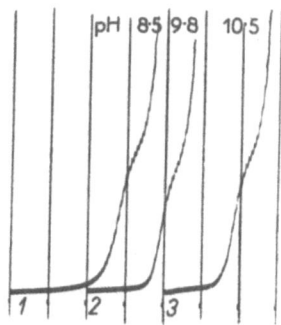

FIG. 1

Reduction of p-Cyanobenzoic Acid

$4 . 10^{-4}$M Depolarizer in Britton-Robinson buffers, pH given on the polarogram.

Curve 1 starting at $-1\cdot2$ V, 2 and 3 at $--1\cdot4$ V, 200 mV/absc., full scale sensitivity $5\cdot5$ μA.

and the cyano group has a role only of a substituent, is discussed elsewhere together with other reductions of the given electronegative group[11-13] (*e.g.* with SO_2CH_3, SO_2NH_2, or $COCH_3$).

EXPERIMENTAL

Apparatus. The polarograph, vessels, capillary, potentiostat and pH-meter were the same as in ref.[11]. Rectangular voltage polarization was followed using a Kalousek commutator[14] as constructed by Němec and Holub[15].

Substances. Substituted nitriles *I—XIII* (Table I) were prepared according to the procedure described in the literature or applying known methods so that the structure would not be in any doubt. The details of the syntheses and literature references are described elsewhere[16]. 0·01M Stock solutions of the nitriles were prepared in ethanol and were stable for at least two weeks.

TABLE I

Polarographic Reduction of Substituted Benzonitriles of the Type $X-C_6H_4-CN$ in Borax (pH 9·3)

No	X^a	M.p.,°C (corr.)	$E_{1/2}$	n^b	σ	
I	3-COO$^-$	219	$-1\cdot85^c$	—	$+0\cdot0$	
II	4-COO$^-$	218	$-1\cdot776^d$	2	$+0\cdot35$	(σ^-)
III	3-COOCH$_3$	61	$-1\cdot761^e$	2	$+0\cdot36$	
IV	4-COOCH$_3$	67	$-1\cdot527^f$	2	$+0\cdot76$	(σ^-)
V	3-COOC$_6$H$_5$	94	$-1\cdot624$	—	$\approx +0\cdot47$	
VI	4-COOC$_6$H$_5$	87	$-1\cdot416$	—	$\approx +0\cdot95$	(σ^-)
VII	3-CN	162	$-1\cdot811$	2	$+0\cdot61$	
VIII	4-CN	223	$-1\cdot612^g$	2	$0\cdot96$	(σ^-)
IX	4-CHO	102	CO^{h,13}	—	—	
X	3-COCH$_3$	99	CO$^{h\ 13}$	—	—	
XI	4-COCH$_3$	59	CO^{h,13}	—	—	
XII	3-COC$_6$H$_5$	90	CO^{h,17}	—	—	
XIII	4-COC$_6$H$_5$	112	CO^{h,17}	—	—	

^{a} In X = 3-CONH$_2$ and 4-CONH$_2$ four-electron reduction of the amidic group, in X = 3-SO$_2$CH$_3$, 4-SO$_2$CH$_3$, 3-SO$_2$NH$_2$, and 4-SO$_2$NH$_2$ reduction of the C—S bond[11,16] occurs; X = H, 3,4-(Cl)$_2$, 3,5-(Cl)$_2$, 3-CF$_3$; 4-CF$_3$, 3-N(CH$_3$)$_3^{(+)}$, and 4-N(CH$_3$)$_3^{(+)}$ do not give measurable waves. ^{b}n determined coulometrically, formation of cyanide ions proved by commutator method; ^{c}Ill-developed wave, commutator method indicates formation of cyanide ions at the potential of beginning of final current rise; $^d\alpha n = 1\cdot07$; $^e\alpha n = 1\cdot05$; $^f\alpha n = 1\cdot09$; $^g\alpha n = 1\cdot02$; ^{h}reduction of the carbonyl group.

Procedures and techniques. Polarographic *i–E* curves were recorded for $5 . 10^{-4}$M solutions containing 5% of ethanol. Half-wave potentials were measured using the three electrode system as described elsewhere[17].

The number of electrons consumed in a borate buffer pH 9·3 containing $5 . 10^{-4}$M$-1 . 10^{-3}$M substance and 5% to 10% ethanol was determined coulometrically. Measurement of the decrease in concentration of the electrolysed compound with time and integration of the current consumed in electrolysis made it possible to calculate the number of electrons consumed as described earlier[11]. The value n = 2 given in Table I is an average of several measurements. Due to the position of the waves at negative potentials fluctuations of about $\pm 10\%$ were not unexpected.

The evidence for the *formation of cyanide ions* during the electrolysis of nitriles *I— VIII* was obtained using a rectangular voltage polarization in the commutator method of Kalousek[14]. Solutions studied were 0·05M in borax (pH 9·3) and were $5 . 10^{-4}$M in substance and 5% in ethanol. The auxiliary potential at the used frequency of 6 Hz was chosen so as to correspond to the potential of the beginning of the limiting current of the studied compound *I— VIII* and the anodic wave of the cyanide ions was recorded. The curves were corrected for the blank by subtracting the current of the supporting electrolyte recorded at the same frequency. Eventually the anodic wave of the cyanide ions was recorded at various concentrations of the parent substance. To prove the identity of the anodic wave the increase of the wave recorded with the commutator was observed after addition of a 0·01M-KCN solution to make its final concentration $2 . 10^{-4}$M. As an independent proof of the cyanide formation, the method based on the formation of chlorocyan (by treatment with chloramine T) and its reaction with barbituric acid in a pyridine solution to form polymethine dyes[18] was used. The solution after electrolysis was two to five times diluted with water and neutralized by 0·1N-H_2SO_4. To the resulting solution 0·1 ml of 1% chloramine T (sodium *p*-toluenesulphochloroamide) was added and left about 1 min to react. The reagent is prepared by dissolving 3 g of barbituric acid in 15 ml of pyridine, adding 3 ml of concentrated hydrochloric acid and making up to 50 ml with water. Of this reagent solution 0·5 ml is added to the reaction mixture. A red-violet colour indicates the presence of cyanide. Because of the sensitivity allowing determination of $5 . 10^{-5}$M cyanide, it is sufficient to reduce electrolytically only about 10% of the original $5 . 10^{-4}$M benzonitrile derivative to obtain a significant proof of cyanide formation.

Determination of the *yield of cyanide ions* was carried out either using the dropping mercury or the mercury pool electrode. For the determination of this yield using a dropping electrode 0·5 to 1·0 ml of $5 . 10^{-4}$ to $1 . 10^{-3}$M solution of the nitrile *II, IV* or *VIII* was electrolysed in 0·05M borax (pH 9·3) containing 5% ethanol. The controlled potential was chosen so as to correspond to the beginning of the limiting current of the wave. Decrease of the benzonitrile reduction wave and increase in the height of the anodic wave of cyanide ions was recorded after several time intervals. Calibration of the anodic wave was carried out by comparison with the wave of a standard solution of cyanide ions of known concentration. Oxygen was removed from the solution by purging with nitrogen prior to the electrolysis. The electrolysis cell was then sealed off. During electrolysis inert gas was not introduced into or above the electrolysed solution, so as to prevent losses of cyanide ions by volatilization. Another source of errors might be a prolonged introduction of nitrogen after the addition of potassium cyanide. When all these precautions were taken, yields of 95 to 100% were found for the reduction of nitriles *II, IV* or *VIII*.

Large-scale controlled potential electrolysis was carried out in 25 ml of 0·05M borax solution $1 . 10^{-3}$M in nitrile and 10% in ethanol using a mercury pool electrode with a surface of 20 cm^2. The potential of the working cathode was controlled using a potentiostat. The potential of the pool electrode was chosen after the results of the preliminary measurements of the current–voltage curves using both a stationary mercury drop electrode and the pool electrode were obtained. For nitriles the curves obtained with both these electrodes were similar to those obtained with the dropping electrode, and differed only in the shift of the final rise of current towards more positive potentials. During electrolysis with a mercury pool electrode a stream of nitrogen was introduced above the electrolysed solution. An absorber was placed in series with the electrolytic cell. It consisted of a polarographic cell through which the escaping gas passed in a way that ensured complete absorption. The absorber contained 25 ml of 0·4M-NaOH. The electrolysis was carried out for 40 min during which time about 50% conversion was achieved. After that 1M-H_2SO_4 was added to the electrolysed solution to adjust the pH to about 1·0. The rest of the hydrogen cyanide was then transfered into the absorber by a stream of nitrogen. The absorption in 0·4M-NaOH was quantitative; no losses occurred in 0·3 to 1·0M-NaOH even after passing a stream of nitrogen for an hour. On the other hand, the losses by volatilization from borate solution at pH 9·3 are considerable, because a considerable proportion of the cyanide is present in the volatile acid form (pK_1 9·32). Yields found corresponded to about 95% of theoretical.

RESULTS AND DISCUSSION

Substances *I — VIII* in acidic solution caused only a shift of the final rise in current, but no separated wave was observed. At pH above 8 one two-electron wave was observed (Fig. 1), the number of electrons transferred being determined polarographically. Both the half-wave potential and the wave-height were pH-independent. At pH above 11 the height of the wave decreased owing to hydrolysis taking place in the bulk of the solution.

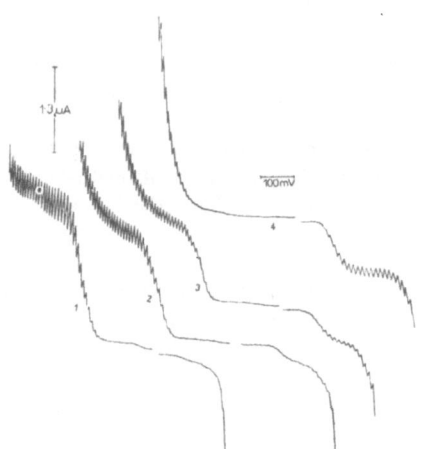

Fig. 2

Polarographic Curves of Terephthalodinitrile during Controlled Potential Electrolysis Using a Dropping Mercury Electrode

0·05M Borax pH 9·3, 2 . 10^{-4}M electrolysed substance, 1·0 ml of the solution electrolysed at the potential of the limiting current. Curves recorded after: *1* 0; *2* 30 min, *3* 90 min, *4* 12 hours. Cathodic curves starting at −1·5 V, anodic at −0·4 V. Negative potentials increase to the left.

The limiting current is governed by diffusion and the shape of the wave, as shown by the logarithmic analysis (Table I) is only slightly different from that of a reversible one-electron process. Formation of cyanide ions as product of electrolysis was confirmed by the anodic wave when the Kalousek commutator method[14] was used. Only for terephthalic acid dinitrile (*VIII*) was the formation of an unstable intermediate proved by the commutator (even when the controlled potential electrolysis yielded cyanide ions). A further proof of the formation was given by controlled potential electrolysis, carried out in the potential region in which the wave at pH 9·3 in a borax buffer was observed. The detection of cyanide ions was carried out using the reaction with barbituric acid and pyridine[17].

Determination of the yield of cyanide ions was carried out both after electrolysis with the dropping mercury electrode and a mercury pool electrode. With the dropping mercury electrode the decrease of the wave of the benzonitrile derivative and the increase in the height of the anodic cyanide wave were recorded (Fig. 2) in a solution stired by the falling off drops. Any other motions of the solution were prevented and no nitrogen was introduced above the solution to prevent removal of hydrogen cyanide. With the mercury pool electrode, however, the hydrogen cyanide formed was transferred by a stream of gas into an adjacent absorber, where cyanide was adsorbed in strongly alkaline media and determined polarographically, using its anodic wave.

Because the half-wave potentials are pH-independent, the proton transfer cannot precede the electrode process proper. From polarographic data it is, at present, impossible to distinguish whether the proton-transfer occurs in aqueous solutions prior to the second electron uptake according to the Scheme (*A*)−(*C*):

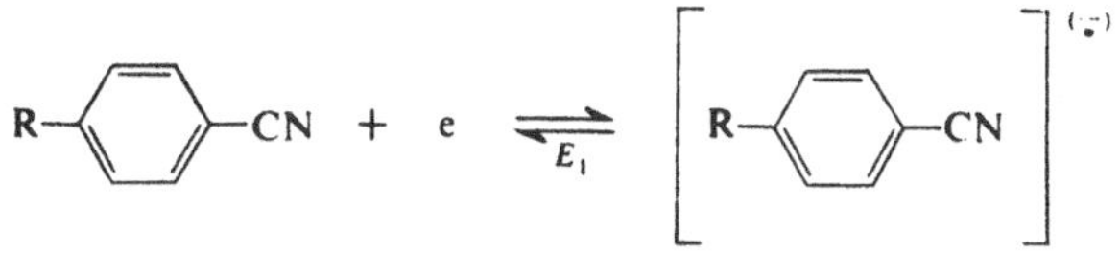

$$R-\!\!\!\left\langle\bigcirc\right\rangle\!\!\!-CN \;+\; e \;\underset{E_1}{\rightleftarrows}\; \left[R-\!\!\!\left\langle\bigcirc\right\rangle\!\!\!-CN\right]^{(\cdot-)} \qquad (A)$$

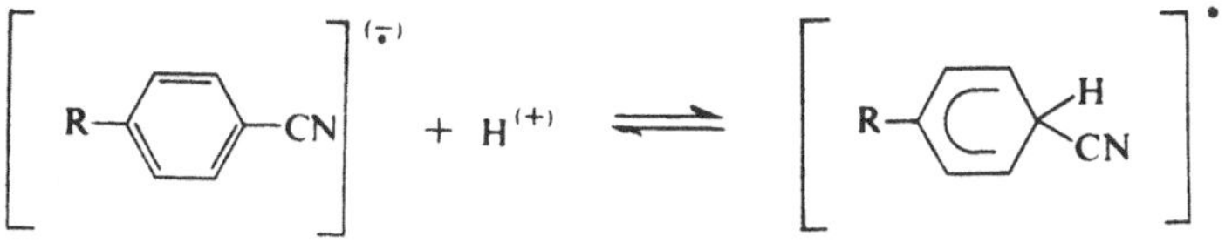

$$\left[R-\!\!\!\left\langle\bigcirc\right\rangle\!\!\!-CN\right]^{(\cdot-)} \;+\; H^{(+)} \;\rightleftarrows\; \left[R-\!\!\!\left\langle C\right\rangle\!\!\!\!\begin{smallmatrix}H\\CN\end{smallmatrix}\right]^{\cdot} \qquad (B)$$

$$\left[R-\!\!\!\left\langle C\right\rangle\!\!\!\!\begin{smallmatrix}H\\CN\end{smallmatrix}\right]^{\cdot} \;+\; e \;\xrightarrow{E_2}\; R-\!\!\!\left\langle\bigcirc\right\rangle \;+\; CN^{(-)} \qquad (C)$$

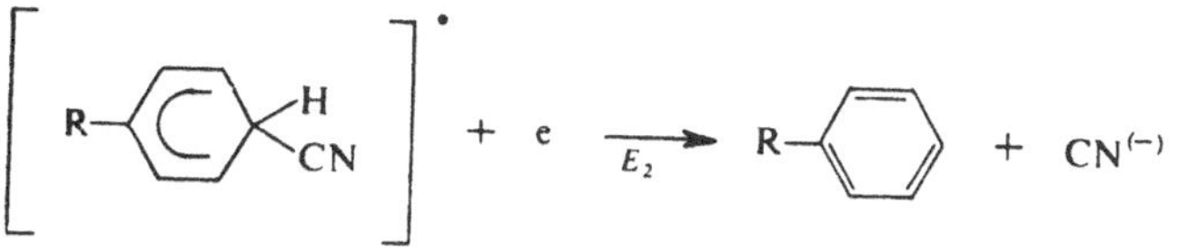

or consecutive with the second electron uptake as in (A), (D) and (E):

$$\left[R-\!\!\!\bigcirc\!\!\!-CN\right]^{(\bar{\cdot})} + \; e \; \xrightarrow[E_3]{} \; \left[R-\!\!\!\bigcirc\!\!\!-CN\right]^{(2-)} \qquad (D)$$

$$\left[R-\!\!\!\bigcirc\!\!\!-CN\right]^{(2\bar{\cdot})} + \; H^{(+)} \longrightarrow \bigcirc\!\!\!-R \; + \; CN^{(-)} \qquad (E)$$

The evidence obtained in nonaqueous solutions[2] indicated that in the absence of proton donors the uptake of the second electron takes place at more negative potentials, indicating that E_3 is more negative than E_1. Because the radical is usually reduced at more positive values than the corresponding radical anion it seems plausible that $E_2 \gtrless E_1$ which would be consistent with the finding of a single two-electron wave. Hence Scheme $(A)-(C)$ seems to be more probable.

The studied system could be considered to follow an S_N2-like mechanism only

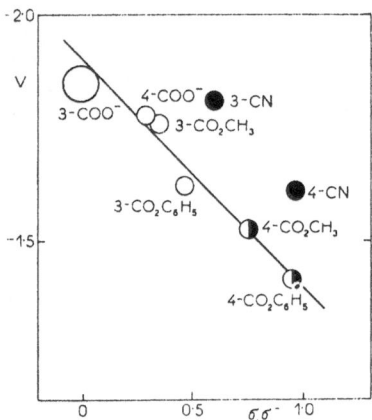

FIG. 3

Dependence of Half-Wave Potentials of Substituted Benzonitriles on Hammett Substituent Constants σ_X

Halved points: σ_{p-X}^- constants were used; full points-deviate.

if steps (B) and (C) would occur practically simultaneously, where both the bond-breaking and bond formation were the rate determining steps. A mechanism of aromatic S_N2 substitution was considered to be probably operating in the reduction of substituted phenyl methyl sulphones[11] and benzenesulphonamides[12]. These reactions have in common that the entering group is the hydrogen atom (or hydride ion); so far as we know none of these reactions has an analogy among homogeneous chemical reductions. The character of an S_N2-like substitution for phenylsulphones and benzenesulphonamides is in agreement with the extraordinary strong effect of some electronegative substituents. Both the pH-independence of the half-wave potential and the logarithmic analysis indicate that in the reduction of benzonitriles, step (A) is potential determining.

For the study of substituent effects the waves in a borax buffer pH 9·3 were measured and the half-wave potentials compared (Table I). When plotted[19] against Hammett substituent constants σ_m and σ_p^- the half-wave potentials gave a reasonable linear plot (Fig. 3). The same substituent constants were used as in our previous papers[11,12]. For substituent constants of the group $COOC_6H_5$ direct measurements are not available. Based on the approximate relation by Charton[20], it is possible to compute $\sigma_{m-COOC_6H_5} \approx 0.47$ whereas the constant $\sigma_{p-COOC_6H_5}^-$ was estimated only by an approximate guess. Based on comparison with the substituent $COOCH_3$ a value $\sigma_{p-COOC_6H_5}^- \approx 0.95$ is considered as the best guess.

The deviations for dinitriles indicated the possibility of another mechanism operating and hence these substances are being studied in more detail. The relatively large value of the reaction constant $\varrho_{\pi,CN} = 0.51$ V, indicating a high sensitivity toward substituent effect, is in agreement with the relatively negative half-wave potential of the parent compound[19] and can be considered as a further proof of the empirical equation: $\varrho_{\pi,R} = \varkappa(E_{1/2})_H + C$. The direction of the slope of the $E_{1/2}-\sigma$ plot is in agreement with a nucleophilic electrode process, probably (A). The necessity of using σ_{p-X}^- rather than σ_{p-X} constants indicates that the conjugation interaction of the substituent in the transition state is of great importance, it is in accord with the supposed mechanism of the S_N2-like type. Nevertheless the mesomeric substituent effect is not as strong as it is in the case of substituted phenyl methyl sulphones[11] and benzenesulphonamides[12]. In the latter two reaction series the values of σ_{p-X}^- constants (derived from reactions of para-substituted phenols and anilines) are not positive enough to express the operating mesomeric effect. For substituted benzonitriles, nevertheless, the use of substituent constants σ_{p-X}^- is adequate (Fig. 3). This indicates that in the case of benzonitriles the activation of electron attracting substituents is relatively less important than in the reaction series of sulphones[11] and sulphonamides[12]. The transition state in the reduction of substituted benzonitriles probably resembles more the structure of a symmetrical intermediate. Formation of radical anions of certain stability[2,21] are in agreement with this deduction.

For technical assistance we thanks Mrs H. Bilyková.

REFERENCES

1. Stackelberg M. v., Stracke W.: Z. Elektrochem. *53*, 118 (1949).
2. Rieger P. H., Bernal I., Reinmuth W. H., Fraenkel G. K.: J. Am. Chem. Soc. *85*, 683 (1963).
3. Schwarz W. H., Kosower E. M., Shain I.: J. Am. Chem. Soc. *83*, 3164 (1961).
4. Volke J., Kubíček J., Šantavý F.: This Journal *25*, 1510 (1960).
5. Volke J., Holubek J.: This Journal *28*, 1597 (1963).
6. Kardos A. M., Valenta P., Volke J.: J. Electroanal. Chem. *12*, 84 (1966).
7. Lund H.: Acta Chem. Scand. *17*, 2325 (1963).
8. Talvik A., Zuman P., Exner O.: This Journal *29*, 1266 (1964).
9. Kargin Yu., Manoušek O., Zuman P.: J. Electroanal. Chem. *12*, 443 (1966).
10. Zuman P.: Z. Anal. Chem. *224*, 374 (1967).
11. Manoušek O., Exner O., Zuman P.: This Journal *33*, 3988 (1968).
12. Manoušek O., Exner O., Zuman P.: This Journal *33*, 4000 (1968).
13. Zuman P., Manoušek O.: This Journal, in the press.
14. Kalousek M.: This Journal *13*, 105 (1948).
15. Němec L., Holub I.: Unpublished results.
16. Exner P., Boček K.: This Journal, in the press.
17. Zuman P., Exner O., Rekker R. F., Nauta W. T.: This Journal *33*, 3213 (1968).
18. Munk V., Matoušková J.: Českoslov. farm. *6*, 252 (1957).
19. Zuman P.: *Substituent Effects in Organic Polarography*. Plenum Press, New York 1967.
20. Charton M.: J. Org. Chem. *28*, 3121 (1963).
21. Zuman P., Manoušek O., Vig S. K.: J. Electroanal. Chem. *19*, 147 (1968).

Translated by the author (P. Z.).

FISSION OF ACTIVATED CARBON–NITROGEN AND CARBON–SULPHUR BONDS. XII.*

POLAROGRAPHIC REDUCTION OF SUBSTITUTED BENZONITRILES BEARING IN *para*-POSITION A CARBONYL GROUP IN ACIDIC MEDIA

P.Zuman** and O.Manoušek

*J. Heyrovský Institute of Polarography,
Czechoslovak Academy of Sciences, Prague 1*

Received March 18th, 1968

4-Cyanoacetophenone and 4-cyanobenzaldehyde are reduced in acid media in a four-electron step followed by a one-electrone reduction of the protonated carbonyl group. In the four-electron step the cyano group is reduced in a protonated species to methylenamino grouping. 4-Cyano-benzophenone is reduced in acid media in two two-electron steps. In the first step which is split into two one-electron waves, the carbonyl group is reduced. In the consecutive two-electron step 4-cyanobenzhydrol is further reduced.

In most nitriles studied so far the cyano group was attached to an aliphatic double bond[1–9] and the reduction was carried out in aqueous neutral media. When the reduction was carried out in 0·2M-HCl the substance $R^1—CH=CH—R^2$ with $R^1 = R^2 = CN$ was reported[10] (similarly as $R^1 = CN$, $R^2 = COOC_2H_5$) to give a wave twice as big as those for $R^1 = R^2 = COOH$, $COOCH_3$, or $CONH_2$. The reduction in the heterocyclic series, first reported[11] for the 2-cyano-pyridine, was further extended to 3- and 4-cyanopyridines[12–15] which were studied in consider-able detail. The authors reported[12–14] that 2- and 4-cyanopyridine in acid solution gave a four-electron wave, corresponding to formation of an aminomethyl derivative, whereas in alkaline solution a two-electron splitting off the cyanide ion was observed.

In the application of Hammett equation to the half-wave potentials of aromatic carbonyl compounds in acid media deviations were observed for *p*-cyano derivatives and a further study of these compounds was suggested[16]. On the other hand, during a study of substituted benzophenones[17] an increase of the current above the value corresponding to a two electron reduction was observed in acid media. To explain these findings and in connection with our other work on the polarographic reactivity of carbon–nitrogen bonds[18–21] a study of the course of the polarographic reduction of substituted benzonitriles bearing in *para*-position a carbonyl group, briefly report-

---

*   Part XI: This Journal *33*, 4000 (1968).
**   Present address: Department of Chemistry, University of Birmingham, Great Britain.

ed earlier[22], was carried out. Because for R = COOR, CN, CONH$_2$ etc. another course of the reduction was observed[23], the investigation dealt with in this study was restricted to 4-cyanoacetophenone (*I*), 4-cyanobenzaldehyde (*II*) and 4-cyanobenzophenone (*III*).

$$CN-\!\!\!\!\boxed{}\!\!\!\!-R$$

I, R = COCH$_3$
II, R = CHO
III, R = COC$_6$H$_5$

EXPERIMENTAL

Apparatus. Polarographic current–voltage curves were recorded with a Heyrovský polarograph (Mark VIII, 1937, produced by V. Nejedlý) and a polarograph LP 60 (Laboratorní přístroje, Prague). The electrolyzed solution was placed in a Kalousek vessel with a mercurous sulphate electrode, separated from the electrolyzed solution by liquid junction. The capillary used had a drop-time of $t_1 = 3 \cdot 2$ s and an out-flow velocity of $m = 2 \cdot 1$ mg/s.

Controlled potential electrolysis was carried out using a manual control and a system of two polarographs[24], using a magnetically stirred mercury pool electrode (surface about 13 cm^2). As the anode a graphite electrode, as reference electrode a mercurous sulphate electrode were used. Current–voltage curves were recorded using a mercury pool electrode and a potentiostat designed by Němec and Holub[25] in connection with the line recorder (0–2 mV) (Meßgeräte und Armaturwerk, Karl-Marx, Magdeburg-Buckau). The continuously increasing potential has been applied from a polarograph LP 60 (Laboratorní přístroje, Prague) with a scanning rate of 400 mV/min to the entrance of the potentiostat. On the terminal of the potentiostat the current was measured and recorded as a function of the voltage difference between the working mercury pool electrode and the auxiliary electrode.

Controlled potential electrolysis with the dropping mercury electrode was carried out in a small volume (0·5 ml) stirred by falling drops in an modified H-cell with a separated calomel electrode. A drawn-out capillary was used, with a narrow orifice and a greater out-flow velocity (of about 4–5 mg/s), but with a normal drop-time ($t_1 = 3$ s).

Values of pH were measured using a pH-meter Mark pHM 4 (Radiometer, Copenhagen) with a glass electrode type G 200 B with a reproducibility to 0·03 pH-units. Absorption spectra of the solutions were measured using a spectrophotometer SF 4 in quarz cuvettes. Infrared spectra were recorded using spectrophotometer UR 10 (Zeiss, Jena).

Substances and solutions. 4-Cyanoacetophenone(*I*) m.p. 59°C, and 4-cyanobenzaldehyde(*II*) m.p. 102°C, were commercial products. As the polarographic behaviour of 4-cyanobenzophenone (*III*) was rather anomalous, two independent ways of synthesis were used described elsewhere[17]. Stock solutions of these substances for polarography and for controlled potential electrolysis were prepared 2 . 10^{-2}M in 96% ethanol. Solutions of acids, acetate, phosphate and borate buffers and of sodium hydroxide were prepared from reagent grade chemicals.

Experimental techniques. 2 . 10^{-4}M solutions were used for polarographic analyses. Large scale controlled potential electrolysis was carried out with a mercury pool electrode using 100 ml of solution 0·2 to 1 . 10^{-3}M in the depolarizer and usually 10% ethanol. The potential of the pool electrode was controlled and kept constant up to ±10 mV. The course of electrolysis was followed by recording in regular time intervals polarographic current-voltage curves. For identification of products reduced in successive steps the electrolysis was carried out at a potential corresponding to $E_{1/4}$. Solutions after electrolysis carried out with a graphite anode (which was not separated* from the electrolyzed solution) were coloured slightly yellow. This corresponds to reactions of the graphite.

The technique used in the controlled potential electrolysis with the dropping mercury electrode and the methods used for microcoulometric determination of the number of electrons transferred were described in detail elsewhere[26]. For identification of electrolysis product the carbonyl group conjugated with a benzene ring was detected using UV spectra and polarographically. An aliquot of the electrolyzed solution was transferred into a proper solvent for spectrophotometry or into a proper supporting electrolyte for polarography. For the proof of the primary amino grouping in a test tube (after neutralization) or on the filter paper the ninhydrin reaction was carried out as follows: An aliquot of the electrolyzed solution was adjusted to pH 4—6 by addition of 1M-NaOH and 1% acetone solution of ninhydrin was added. Reaction mixture was heated on a water-bath till a blue-violet colour appears, indicating the presence of an CH_2NH_2 grouping. In a similar way a blank test was carried out with the solution prior to electrolysis. This procedure in connection with spectrophotometry was used for semiquantitative estimation of the yields of electrolysis products.

For the recording of the IR-spectra an aliquot of the electrolyzed solution was extracted three times with ether. Combined extracts were dried with calcium chloride and evaporated to dryness. The residue was dissolved in trichloromethane, purified by active charcoal, and the solvent distilled off. From the residue a 6% solution was prepared in trichloromethane which was used for spectrophotometry.

RESULTS

4-Cyanoacetophenone (I)

At pH > 4 the ketone I is reduced in a single two-electron step accompanied by an ill-developed adsorption prewave (Fig. 1**). At pH > 8 the effect of adsorption was more pronounced. Anodic wave[27] in alkaline media was much better developed for I than for unsubstituted acetophenone. At pH < 3 the height of this wave (i_1) increased to a height corresponding to a four-electron reduction (Fig. 2). This increase plotted as a function of pH has a shape of a dissociation curve with a pK' about 2. The half-wave potential at pH < 2 is shifted to more negatige values by some 53 mV/pH (Table I). Simultaneously with the increase in wave i_1, another wave (i_2) increases at more negative potentials and below pH 0 reaches a value corresponding to a one-electron transfer. The increase in current above that corresponding to a two-electron reduction is controlled by a rate of a chemical reaction. The kinetic character of this increase has been proved by the dependence on the height of the mercury column. The saddle-shaped wave in the region of the current rise (Fig. 1 , 2) indicates that the reaction takes place also at the surface of the electrode[28].

---

* It was possible to carry out the electrolysis without separation of the reference electrode, as the reduction process was irreversible.

430

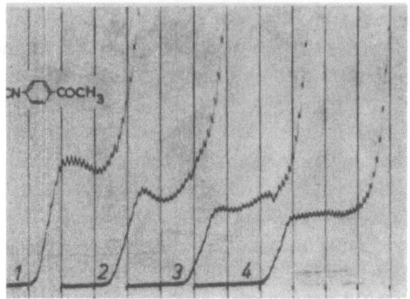

FIG. 1

Reduction of 4-Cyanoacetophenone in Buffered Solutions

$2 . 10^{-4}$M depolarizer in Britton-Robinson buffers. Curves Starting at $-0\cdot4$ V, s.c.e., 200 mV/absc., full scale sensitivity $5\cdot5$ µA. pH 1 2, 2 2·9, 3 3·55, 4 4·3.

TABLE I

Half-Wave Potentials (V *vs.* s.c.e.) of 4-Cyanoacetophenone (I) in Sulphuric Acid Solutions

$[H_2SO_4]$	3N	1N	0·3N	0·1N	0·03N	0·01N
$E_{1/2}$	$-0\cdot63_6$	$-0\cdot65_4$	$-0\cdot67_4$	$0\cdot69_8$	$0\cdot72_2$	$0\cdot74_5$

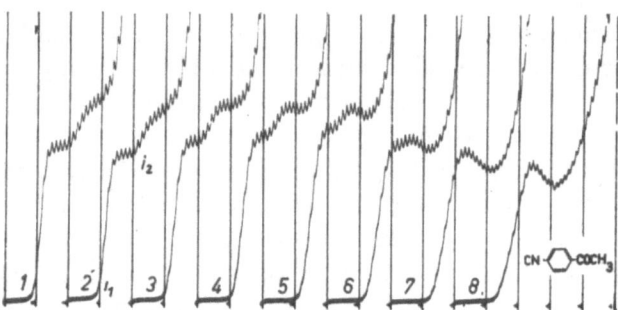

FIG. 2

Reduction of 4-Cyanoacetophenone in Sulphuric Acid

$2 . 10^{-4}$M depolarizer in sulphuric acid. Curves starting at $-0\cdot4$ V, s.c.e., 200 mV/absc., full scale sensitivity $4\cdot8$ µA Concentration of H_2SO_4: 1 5, 2 3, 3 1, 4 0·3, 5 0·1, 6 0·03, 7 0·01, 8 0·03N.

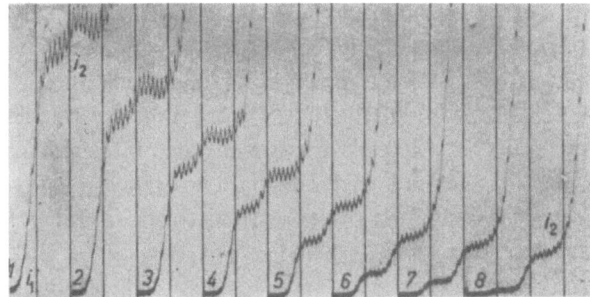

FIG. 3

Polarographic Curves of 4-Cyanoacetophenone in the Course of Controlled Potential Electrolysis Using a Mercury Pool Electrode

0·5M-H_2SO_4, 25% ethanol, 2·5 . 10^{-3}M electrolyzed substance. Reduction was carried out to 95% at the potential of the limiting current of the first wave, curves were recorded in the electrolysis cell after time intervals given on the polarogram. Curves starting at -0.6 V, 200 mV/absc., s.c.e., full scale sensitivity 17 µA.

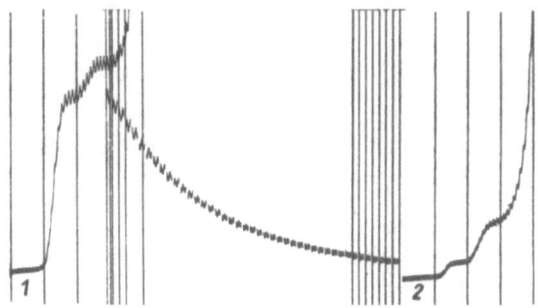

FIG. 4

Controlled Potential Electrolysis of 4-Cyanoacetophenone under Microcoulometric Conditions

0·4N-H_2SO_4, 3 . 10^{-4}M depolarizer, 1·0 ml solution. Reduction was carried out at the potential of the limiting current of the first wave; the limiting current of the first wave was recorded automatically every 1 000 s. Total time of electrolysis 44·280 s. *1* i–E curve before electrolysis; *2* after electrolysis. Curves start at -0.8 V, mercurous sulphate electrode, 200 mV/absc., full scale sensitivity 6·7 µA.

When the controlled potential electrolysis was carried out at the potential of the limiting current of the first wave (i_1) in $0.5M-H_2SO_4$ using the mercury pool electrode (Fig. 3) or in $0.2M-H_2SO_4$ with the dropping mercury electrode (Fig. 4) the height of the wave i_1 decreased during electrolysis. On the contrary, the height of the more negative one-electron wave i_2 and its half-wave potential remained practically unchanged (Figs 3 and 4). This indicates that the two waves i_1 and i_2 do not belong to either a polarographic reduction of two various forms of a depolarizer, or to two independent reduction paths, but rather to two successive electrode processes. Moreover, the product of the first reduction step does not undergo a chemical change even during the prolonged time of controlled potential electrolysis.

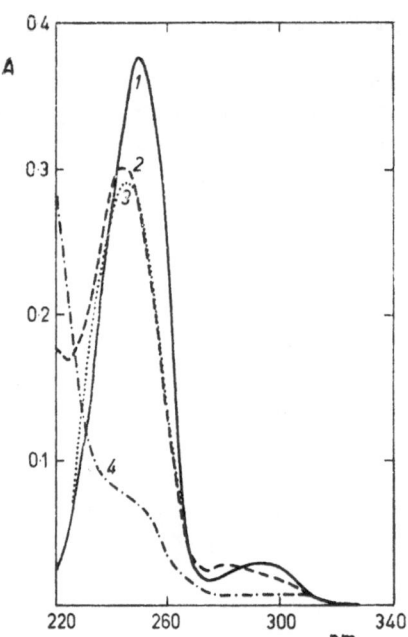

FIG. 5

UV-Absorption Spectra of 4-Cyanoacetophenone before and after Controlled Potential Electrolysis Using a Mercury Pool Electrode

$1.6 . 10^{-5}M$ studied compound in $0.012N-H_2SO_4$. *1, 2, 4* 4-cyanoacetophenone, *3* acetophenone (λ_{max} 246 nm); *1* before electrolysis (λ_{max} 250 nm); *2* product of the electrolysis in the first wave (λ_{max} 244 nm); *4* product of the electrolysis in the second wave (complete reduction).

TABLE II

Half-Wave Potentials of the Reduction Product of 4-Cyanoacetophehone (*I*)

Controlled potential electrolysis was carried out at the potential of the limiting current of the four-electron wave (i_1), solution titrated to pH 6 and diluted with buffer to $2 . 10^{-4}M$ solution; V *vs.* S.C.E.

pH	3.75	4.1	4.65	6.15	7.5	8.4	8.8	9.1	9.3
$E_{1/2}$	-1.09_4	-1.11_8	-1.147	-1.20	-1.25_4	-1.29_4	1.32_7	1.35_4	1.38_2

When comparing the absorption spectra prior to and after the reduction (Fig. 5) it can be noted that the high intensity peak at 250 nm decreased during the electrolysis only by some 20% and was shifted somewhat to shorter wave-lengths. The resulting peak at 244 nm had practically the same extinction coefficient as that of the unsubstituted acetophenone at 246 nm. This indicates that the conjugation of the carbonyl with the benzene ring was not lost during reduction.

An investigation of the polarographic behaviour of the electrolysis product has shown (Fig. 6) that its behaviour resembles that of other substituted acetophenones[29], when the pH was below 8·5. In principle, a one-electron wave of the protonated form was observed in acidic media, followed by a two-electron wave at higher pH-values. The two-electron wave corresponding to the reduction of the carbonyl group bearing no proton shows a shift of the half-wave potential (Table II), whereas for most other acetophenones bearing uncharged substituents undergoing dissociation the half-wave potential of this wave is pH-independent[29]. At higher pH-values the reduction product undergoes homogeneous chemical reactions, shown by formation of further waves (Fig. 6) and by changes in polarographic curves with time. Finally, the presence of a primary amino group in the electrolysis product was proved by ninhydrin reaction.

When the electrolysis was carried out at the potential of the limiting current of wave i_2 and spectra of the electrolysis product were recorded (Fig. 5, curve 4), the absorption band at 245 nm practically disappeared. On polarographic curves

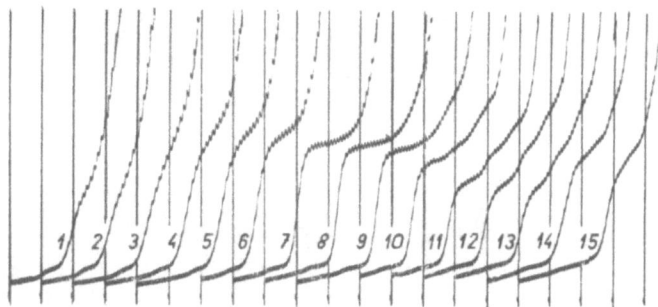

FIG. 6

Dependence of Reduction Waves of the Product of Controlled Potential Electrolysis of 4-Cyanoacetophenone on pH

Solution of 4-cyanoacetophenone in 0·4N-H$_2$SO$_4$ was electrolyzed at the potential of the limiting current of the first wave and pH adjusted by addition of 4N-NaOH to 6·0. Solution mixed with Britton Robinson buffer to give pH values given on the polarogram and final concentration of $2·5. 5 10^{-4}$M in depolarizer. Curves starting at: *1* −0·6 V; *6—9* −0·8 V; *10—15* 1·0 V; s.c.e., 200 mV/absc., full scale sensitivity 2·4 μA.

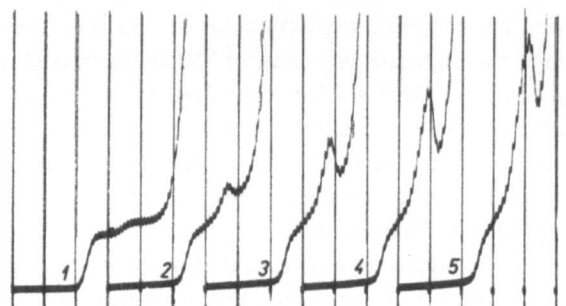

FIG. 7

Effect of Dimethylformamide on Reduction Waves of 4-Cyanoacetophenone

1M-H_2SO_4, 2 . 10^{-4}M depolarizer, concentration of dimethylformamide: *1* 0; *2* 10%; *3* 30%; *4* 50%; *5* 70%. Curves start at 0·0 V, s.c.e., 200 mV/absc., full scale sensitivity 6·9 μA.

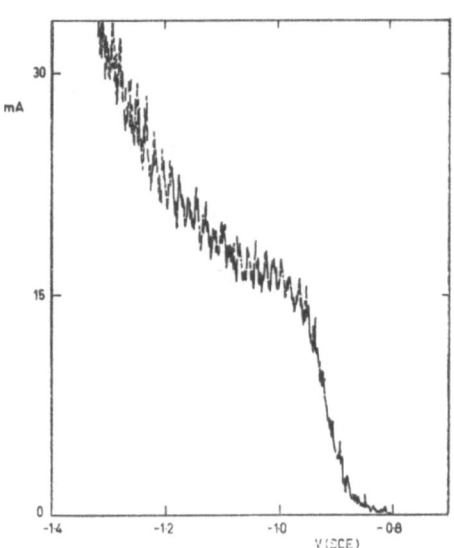

FIG. 8

Actual Record of a *i–E* Curve of 4-Cyanoacetophenone Using a Stirred Mercury Pool Electrode 0·3N-H_2SO_4, 1 . 10^{-3}M depolarizer.

no waves corresponding to a reduction of a benzoyl grouping was observed. Similarly, when the controlled potential electrolysis was carried out at the limiting current of the two-electron wave in a phosphate buffer pH 6·8 and borate buffer pH 9·3 the absorbance at 250 nm decreased in the course of the reduction. These waves thus correspond to the reduction of the carbonyl group.

When dimethylformamide was added to the solution of 4-cyanoacetophenone in 1M-H_2SO_4 an increase of the height of a catalytic wave in the potential range of the one electron wave i_2 was observed (Fig. 7). At given dimethylformamide concentration the height of this catalytic wave of hydrogen evolution increased with increasing sulphuric acid concentration.

Curves recorded with the stirred mercury pool electrode in 0·3N-H_2SO_4 (Fig. 8, 9) have shown that the four-electron reduction step can be also observed, even when at somewhat more negative potentials than with the dropping mercury electrode. The one-electron reduction step, observed with the dropping electrode was nevertheless obscured by the current oscillations (Fig. 8) and the mean current showed only a continuous increase with increasingly more negative potentials after the rise of the first wave (Fig. 9).

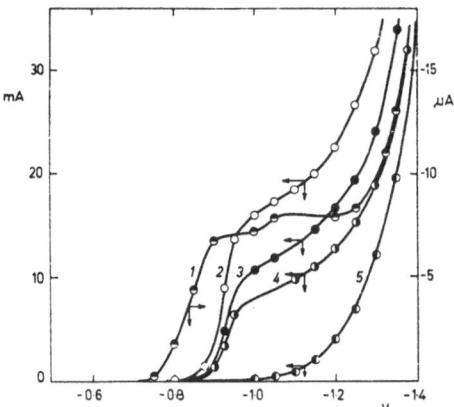

FIG. 9

Comparison of Current–Voltage Curves of 4-Cyanoacetophenone Using a Dropping Mercury and Stirred Mercury Pool Electrodes

0·3N-H_2SO_4, 1.10^{-3}M depolarizer. Curves: 1 Polarographic wave recorded with a dropping mercury electrode placed directly in the electrolyzer; 2—5 current–voltage curves with a stirred mercury pool electrode; 1 and 2 before electrolysis; 3—5 after controlled potential electrolysis at −1·05 V (s.c.e.) for: 3 5 min 4 15 min; 5 50 min. The same reference electrode (s.c.e.) was used for the dropping and pool mercury electrodes.

4-Cyanobenzaldehyde (II)

Similarly as ketone I the aldehyde II is reduced at pH > 3 in one two-electron wave, which increases at lower pH-values and reaches a height corresponding to a four-electron process in about $1N\text{-}H_2SO_4$. This wave i_1 is followed at more negative potentials by a one-electron wave i_2 which is much less well developed than for ketone I (Fig. 10). The wave i_1 is accompanied by an adsorption pre-wave. In alkaline media the separation of two one-electron waves can be observed already at pH 9, and in $1M\text{-}NaOH$ the two one-electron waves found are in ratio strictly $1:1$ and well separated. Basing on the interpretation of the reduction scheme for carbonyl com-

TABLE III

Half-Wave Potentials of 4-Cyanobenzophenone (III)
  Solutions contain 50% ethanol, V vs. s.c.e.

Solution	i_I		i_{II}
0·1N-HCl	$-0\cdot723(i_I)_1$	$-0\cdot788(i_I)_2$	$-0\cdot943$
Borax pH 9·3	$-1\cdot210$		—
0·1N-NaOH	$-1\cdot208$		—

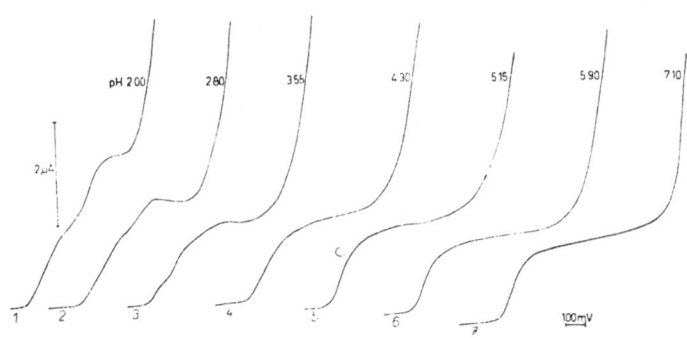

FIG. 10

Polarographic Curves of 4-Cyanobenzaldehyde in the Course of Controlled Potential Electrolysis Using a Mercury Pool Electrode

$0\cdot1M\text{-}H_2SO_4$, $1 . 10^{-2}M$ electrolyzed substance. Reduction was carried out at the potential of the peak current (1·2 V), curves were recorded after intervals of 20 minutes. Curves starting at $-0\cdot8$ V, s.c.e., full scale sensitivity $5\cdot6 . 10^{-5}A$.

pounds[30] it can be deduced that protonation of the radical anion formed in the first reduction step is slower than with most other benzaldehydes and that the radical anion $4\text{-}CNC_6H_4CHO^{(-)}$ is relatively less reactive toward interaction with alkali metal cations.

The controlled potential electrolysis at the potential of the limiting current of the four-electron wave i_1 shows a linear dependence of the logarithm of the limiting current i_1 on time[31]. This indicates that the parent compound II does not participate in any homogeneous chemical reaction consecutive to the electrode process. Nevertheless, the height of wave i_2 does not remain completely unchanged during the electrolysis. Even when the ratio $i_1 : i_2$ changes during the electrolysis in favour of i_2, the absolute height of the wave i_2 decreases (Fig. 10). The primary product of the four-electron reduction undergoes some chemical changes in the solution.

4-Cyanobenzophenone (III)

Due to the slight solubility it was necessary to study the electrochemical behaviour of ketone III in aqueous solutions containing 50% ethanol. In these solutions at pH > 5 the reduction takes place in one two-electron step the half-wave potential of which is little pH-dependent at pH < 7. A slight shift toward negative potentials was observed at higher pH-values. At pH > 9 a separation of two wave occurs, but is dependent on the kind and concentration of the cation present in the electrolyte (Table III). The reaction of the radical anion formed in the first reduction step with protons is hence not too fast, but this species is rather reactive toward cations[29,30]. At pH < 5 the first wave is split into two waves $(i_1)_1$ and $(i_1)_2$, the first of which is

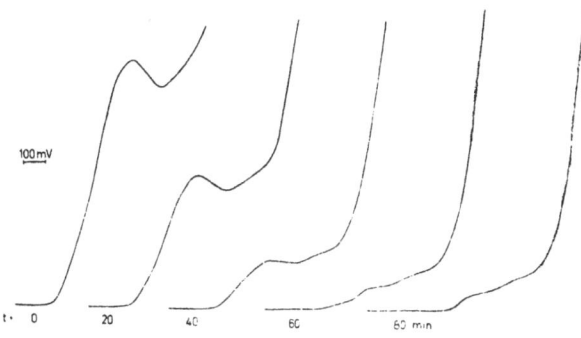

FIG. 11

Dependence of Reduction Waves of 4-Cyanobenzophenone on pH

$2 . 10^{-4}M$ depolarizer, Britton-Robinson buffers pH indicated on the polarogram. Curves 1,2 starting at -0.6 V; 3,4 at -0.7 V; 5--7 at -0.8 V, s.c.e.

affected by adsorption. Simultaneously, an increase of a wave i_{II} at more negative potentials is observed (Fig. 11). The increase in the wave-heigt i_{II} has a shape of a dissociation curve with pK' of about 2·6. At pH below 2 the more negative wave i_{II} reaches a height corresponding to a transfer of two electrons, so that the total wave-height corresponds to a four-electron process. The number of electrons transferred was verified by comparison with unsubstituted benzophenone. The total height of the wave below pH 2, which is independent of a further decrease in pH, is limited by diffusion: This was proved by linear dependence on square root of the height of mercury column and by linear dependence on concentration of substance III. The ratio of waves $(i_I)_1$ and $(i_I)_2$ is concentration independent at concentrations below $5 . 10^{-4}$M. The wave i_{II} is drawn-out and deformed, particularly at higher concentrations of the mineral acid.

Controlled potential electrolysis was carried out in 0·1N-H_2SO_4 containing 50% ethanol at -0.78 V. This potential corresponds approximately to the half-wave potential of wave $(i_I)_2$. The height of wave $(i_I)_1 + (i_I)_2$ decreased during electrolysis (Fig. 12), but the height of the wave i_{II} decreased too, until a residual wave is observed. In the UV-spectra the band at 258 nm during the electrolysis vanishes and a new band of low intensity at 245 nm is formed. The isolated product was studied in IR-spectra and has shown a decrease of the CO band at 1665 cm^{-1}, no change in CN band at 1230 cm^{-1}, increase of OH bands at 3400 cm^{-1} and 3600 cm^{-1} and of the C—O band at $1080-1100$ cm^{-1}. No bands corresponding to NH_2 or C=NH in the region between 3300 and 3500 cm^{-1} were observed.

When in the same solution the electrolysis was carried out at the potential of limiting current of i_{II}, both waves decreased with time in the same ratio until they completely disappeared. Simultaneously in the UV-spectra the band at 245 nm disappeared and two shoulders at 224 and 238 nm were observed. The CN band at 1230 cm^{-1} remained unchanged, the C—O band at $1080-1100$ cm^{-1} disappeared and the two CH_2 bands at 2850 and 2920 cm^{-1} became prominent.

DISCUSSION

Reduction processes taking place in solutions of the investigated ketones $I-III$ can be divided into two types: For 4-cyanoacetophenone (I) and 4-cyanobenzaldehyde (II) in acidic media the cyano group in the protonated form is attacked first, with a subsequent reduction of the carbonyl group, whereas for 4-cyanobenzophenone (III) the attack of the carbonyl group is followed by a consecutive reduction. Therefore these two reduction types will be discussed separately.

The change in the height of the polarographic wave i_1 with pH, appearance of the typical one-electron wave i_2 of the carbonyl reduction at more negative potentials, limited change in the UV-spectra during the controlled potential electrolysis, polarographic behaviour of the reduction product and the proof of a formation of the primary amino group in the first reduction step represent findings that are the basis of the suggested reduction scheme $(A)-(D)$ for the reduction of compounds I $(R = CH_3)$ and II $(R = H)$.

$$RCO\text{—}\langle\bigcirc\rangle\text{—}CN \underset{k_{-1}}{\overset{k_1}{\rightleftharpoons}} RCO\text{—}\langle\bigcirc\rangle\text{—}CN + H^+ \qquad (A)$$
$$\underbrace{\phantom{RCO\text{—}\langle\bigcirc\rangle\text{—}CN}}_{H^{(+)}}$$

$$RCO\text{—}\langle\bigcirc\rangle\text{—}CN + 4\,e + 4\,H^{(+)} \xrightarrow{E_1} RCO\text{—}\langle\bigcirc\rangle\text{—}CH_2NH_3^{(+)} \qquad (B)$$
$$\underbrace{\phantom{RCO\text{—}\langle\bigcirc\rangle\text{—}CN}}_{H^{(+)}}$$

It is assumed that the protonation (A) the rate of which limits the height of the wave i_1 occurs on the cyano group rather than on the carbonyl group. This conclusion can be drawn from the fact that the hydrogen ions in the reduction step (B) occur in the reduced form of the cyano group rather than in the keto group, and from the difference to other substituted benzonitriles. In these compounds either one-electron is accepted by the aromatic ring and a radical is formed[32,33], or the cyanide ion is reductively substituted[22,23]. It is assumed that in these compounds the cyano group is not protonated and hence not hydrogenated. The group RCO in *para*-

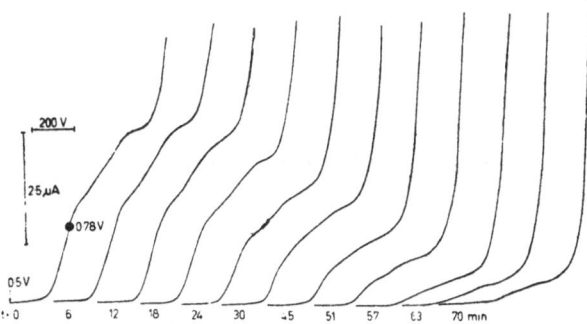

FIG. 12

Polarographic Curves of 4-Cyanobenzophenone in the Course of Controlled Potential Electrolysis Using a Stirred Mercury Pool Electrode

0.1N-H_2SO_4, 25% ethanol, $2.6 \cdot 10^{-4}$M depolarizer. Reduction was carried out at -0.78 V (S.C.E.) corresponding to the rising portion of the first wave, curves were recorded after the intervals shown in the polarogram starting at -0.5 V, S.C.E.

position is assumed to enhance the protonation. It is possible that the catalytic hydrogenation of $C\equiv N$ to CH_2NH_2 which is carried out in glacial acetic acid[34] takes place also in the protonated form of the nitrile. It seems hardly possible that the protonation at the electrode surface would affect sólely the carbonyl group, as the substituent effect of $RCOH^+$ will not be so different from that of RCO and of other electronegative substituents like RSO_2 to result in such difference in the reactivity of the cyano group. It cannot be nevertheless excluded that the second proton transfer (C) takes place prior to the electron uptake (B). If this sequence is operating it must be assumed that the form electroactive in wave i_1 is biprotonated. The rate of protonation of the carbonyl group (C) limits the height of the wave i_2. Radical formed in reaction (D) can react with mercury, with the reaction medium, undergo dimerization or be further electrolyzed at more negative potentials.

$$R{-}\underset{(+)OH}{\overset{\|}{C}}{-}\!\!\!\!\langle\,\rangle\!\!\!\!{-}CH_2NH_3{}^{(+)} \underset{k_{-1}}{\overset{k_2}{\rightleftharpoons}} RCO{-}\!\!\!\!\langle\,\rangle\!\!\!\!{-}CH_2NH_3{}^{(+)} + H^+ \quad (C)$$

$$R{-}\underset{(+)OH}{\overset{\|}{C}}{-}\!\!\!\!\langle\,\rangle\!\!\!\!{-}CH_2NH_3{}^{(+)} + e \underset{E_2}{\rightleftharpoons} R\overset{\cdot}{\underset{OH}{C}}{-}\!\!\!\!\langle\,\rangle\!\!\!\!{-}CH_2NH_3{}^{(+)} \quad\quad (D)$$

4-Cyanobenzophenone (III) is reduced in the first two-electron step on the carbonyl group. This was proved by the separation of two one-electron steps $(i_1)_1$ and $(i_1)_2$ in acidic media, which is rather characteristic for carbonyl compounds[29,30]. The half-wave potential of wave $(i_1)_1$ fits reasonably well the Hammett $E_{1/2}$–σ plot for the one-electron reduction wave of protonated benzophenones in acidic media[17]. Finally the spectroscopic evidence both in UV and in IR indicates that in the first two-electron step the carbonyl group is reduced and the cyano group remains intact. The deviations observed in the controlled potential electrolysis with a mercury pool with compound II and in particular III can be explained by reactions of some pro-products, for example radicals, at the electrode surface.

The nature of the second two-electron step i_{II} is not well understood. This reduction must be preceded by protonation either of the compound III or of the product of the first two-electron step. The site of the protonation was not distinguished. The IR spectra indicate that in the second reduction step the cyano group remains intact, but the C—OH group can be reduced to CH_2. Reductions of activated C—OH bonds are known[35], but as benzhydrol was found under conditions used to be electro-inactive, it would be necessary to consider the effect of the p-cyano group either on the reduction process or on the pre-protonation. The reduction of 4-cyanobenzophenone in acidic media hence differs both from the reduction of 4-cyanoacetophenone and from other substituted benzophenones[17,29,30].

4-Cyanocinnamic acid has been reported recently to be reduced in acid media in a four-electron wave[36], but it is not clear to which of the above two classes this reduction belongs.

The variability in the course of the reduction of benzonitrile derivatives is great, showing the above mentioned possibilities of radical formation, elimination of cyanide ions, and the reduction of the cyano group, in addition to systems in which the other electronegative group on the benzene ring is electroactive and the cyano groups acts solely as a substituent. The complete difference in the reduction course of 4-cyano-acetophenone and 4-cyanobenzophenone is nevertheless rather striking. One possible explanation is based on the consideration, that during the electrode process the phenyl rings are oriented flat at the electrode surface. For the acetophenone derivative *I* the side-chain bearing the carbonyl group can be freely rotated, the cyano group can be protonated and oriented in a position favouring its reduction. For the benzophenone derivative *III* the position of the carbonyl group toward the electrode seems to be fixed in such a way that its reduction occurs preferably to the cyano group.

This study indicates the complexity and variability of problems involved in electro-activity of aromatic compounds bearing two electronegative groups[37]

REFERENCES

1. Stackelberg M. v., Stracke W.: Z. Elektrochem. *53*, 118 (1949).
2. Rjabov A. V., Panova G. D.: Dokl. Akad. Nauk SSSR *99*, 547 (1954).
3. Bird L., Hale H.: Anal. Chem. *24*, 586 (1952).
4. Spillane L. J.: Anal. Chem. *24*, 587 (1952).
5. Matyska B., Klier K.: This Journal *21*, 1592 (1956).
6. Claver G. C., Murphy M. E.: Anal. Chem. *31*, 1682 (1959).
7. Strause S. F., Dyer E.: Anal. Chem. *27*, 1906 (1955).
8. Bargain M.: Compt. Rend. *254*, 130 (1962); *255*, 1948 (1962); *256*, 1990 (1963).
9. Marciszewski H.: Chem. Anal. (Warszawa) *5*, 119 (1960).
10. Koršunov I. A., Vodzinskij, Ju. V., Vjazanskin N. S., Kalinin A. I.: Ž. Obšč. Chim. *29*, 1364 (1959).
11. Jarvie J. M. S., Osteryoung R. A., Janz G. J.: Anal. Chem. *28*, 264 (1956).
12. Volke J., Kubíček J., Šantavý F.: This Journal *25*, 1510 (1960).
13. Volke J., Holubek J.: This Journal *28*, 1597 (1963).
14. Kardos A. M., Valenta P., Volke J.: J. Electroanal. Chem. *12*, 84 (1966).
15. Volke J., Kardos A. M.: This Journal *33*, 2560 (1968).
16. Zuman P.: This Journal *25*, 3225 (1960).
17. Zuman P., Exner O., Rekker R. F., Nauta W. T.: This Journal *33*, 3213 (1968).
18. Zuman P., Horák V.: This Journal *26*, 176 (1961).
19. Manoušek O., Zuman P.: This Journal *29*, 1432 (1964).
20. Manoušek O., Zuman P.: This Journal *29*, 1730 (1964).
21. Zuman P., Exner O.: This Journal *30*, 1832 (1965).
22. Manoušek O., Zuman P.: Chem. Commun. *1965*, 158.
23. Manoušek O., Exner O., Zuman P.: This Journal *33*, 3979 (1968).
24. Manoušek O., Zuman P.: This Journal *29*, 17 (1964).
25. Němec L., Holub I.: Unpublished results.

26. Manoušek O., Exner O., Zuman P.: This Journal *33*, 3988 (1968).
27. Heyrovský M.: This Journal *28*, 26 (1963).
28. Majranovskij S. G., Liščeta L. I.: This Journal *25*, 3025 (1960).
29. Zuman P., Barnes D., Ryvolová-Kejharová A.: Discussions Faraday Soc. *45*, 202 (1968).
30. Zuman P.: This Journal *33*, 2548 (1968).
31. Meites L.: *Proc. XXII. Conf. IUPAC,Prague 1967,* in the press.
32. Rieger P. H., Bernal I., Reinmuth W. H., Fraenkel G. K.: J. Am. Chem. Soc. *85*, 683 (1963).
33. Schwarz W. M., Kosower E. M., Shain I.: J. Am. Chem. Soc. *83*, 3164 (1961).
34. Huber W.: Z. Anal. Chem. *197*, 236 (1963).
35. Zuman P., Wawzonek S.: *Progr. Polarography*, Vol. 1, p. 303. Interscience, New York 1962.
36. Fleet B.: Private communication.
37. Zuman P.: Z. Anal. Chem. *224*, 374 (1967).

Translated by the author (P. Z.).

THE USE
OF SIMPLE MOLECULAR ORBITAL THEORY
TO ELUCIDATE THE POLAROGRAPHIC BEHAVIOUR
OF SOME *para*-SUBSTITUTED BENZONITRILES

P.ČÁRSKY and P.ZUMAN

Institute of Physical Chemistry and
J. Heyrovský Institute of Polarography,
Czechoslovak Academy of Sciences, Prague 2

Received March 18th, 1968

The nature of the polarographic reduction of the *para*-substituted benzonitriles *I—VI* depends strongly on the *para* substituent. The first reduction step could involve cleavage of the C—C bond, which leads to CN⁻ formation, or reduction of the cyano group to a —CH=NH function, or reduction of the *para*-substituted group. The experimental results obtained with compounds *I—VI* are compared with the squares of the expansion coefficients of the lowest free molecular orbital and with their bond orders before and after an uptake of two electrons.

The electrochemical properties of disubstituted benzene derivatives R^1—C_6H_4—R^2 which bear two electronegative groups R^1 and R^2 in *para* position have been recently discussed in some detail[1-3]. A mutual mesomeric interaction of R^1 and R^2 has been observed and in some cases the resulting electrochemical behaviour indicates properties which differ from the properties of both of the possible monosubstituted compounds R^1—C_6H_5 and R^2—C_6H_5. In such case of mutual interaction it is impossible to treat the electrochemical reactivity as a superimposition of the reactivities of C_6H_5—R^1 and C_6H_5—R^2. The present study tries to show the use and the limitations of the simple molecular orbital theory in the solution of these problems in the case of *para*-substituted benzonitriles[2,3].

CALCULATIONS

The standard HMO method was used[4]. Numerical calculations were carried out using the National Elliott 803b computer. The empirical parameters used, are of the form $\alpha_X = \alpha + \delta_X \beta$ and $\beta_{XY} = \varrho_{XY}\beta$ where α is the carbon Coulomb integral and β is the resonance integral of the C—C π-bond, and δ_X and ϱ_{XY} are numerical constants. The following values are used in the calculations:

Cyano group: $\qquad \alpha_N = \alpha + 0.5\beta; \ \beta_{CN} = 1.4\beta.$

Cyano group (protonated form): $\alpha_N = \alpha + 3.5\beta; \ \beta_{CN} = 1.4\beta.$

Carbonyl: $\qquad \alpha_O = \alpha + \beta; \quad \beta_{CO} = \beta.$

Carbonyl (protonated form): $\quad \alpha_O = \alpha + 4.0\beta; \ \beta_{CO} = \beta.$

Carboxyl (anion): $\qquad \alpha_O = \alpha + \beta; \quad \beta_{CO} = \beta.$

Methylsulfonyl: $\qquad \alpha_{SO_2} = \alpha + 1.9\beta; \ \beta_{SO_2CH_3} = 0.7\beta; \ \beta_{SO_2C} = 0.8\beta;$

$$\alpha_{H_3} = \alpha - 0.5\beta; \ \beta_{C \equiv H_3} = 3.0\beta.$$

(In this case the superposition of the conjugative model of methyl[5] and the model of the $-SO_2-$ group given by Gerdil and Lucken[6] was used).

Sulfamyl[6]: $\qquad \alpha_{SO_2} = \alpha + 2.0\beta; \ \beta_{SC} = 0.8\beta; \ \alpha_N = \alpha + \beta; \ \beta_{SN} = 0.5\beta$

Table I presents the expansion coefficients of the lowest free molecular orbital (LFMO) for compounds $I - VI$. For compounds III and IV the bond orders before and after an uptake of one and two electrons are given in Figs 1 and 2.

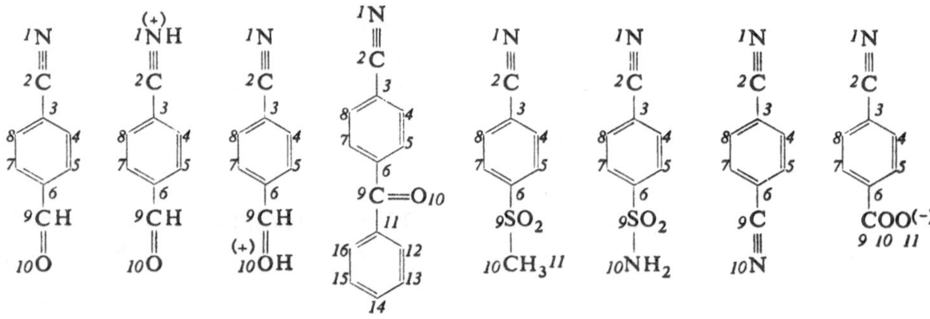

| Ia | Ib | Ic | II | III | IV | V | VI |

DISCUSSION

According to the simple molecular orbital theory the first step of the polarographic reduction can be electron transfer from the Hg-dropping electrode into the lowest free molecular orbital (LFMO) of a depolarizer. The distribution of that electron within the depolarizer's molecular structure may be regarded as given by the squares of the LFMO expansion coefficients (c_i^2). The greater the value of c_i^2 the greater the probability, that the electron remains localized at atom i. If the squares of the coefficients are remarkably large for two atoms on the same bond, this will probably indicate the reducible center of the molecule.

TABLE I

Expansion Coefficients (c_i) of the LFMO for Compounds $I-VI$

Position i	c_i	c_i^2	Position i	c_i	c_i^2
		p-Cyanobenzaldehyde (Ia)			
1	−0·2928	0·0857	6	0·2405	0·0578
2	0·1710	0·0292	7	−0·3105	0·0964
3	0·3556	0·1264	8	−0·1419	0·0201
4	−0·1419	0·0201	9	0·5447	0·2967
5	−0·3105	0·0964	10	−0·4134	0·1709
		p-Cyanobenzaldehyde (*Ib*)			
1	−0·2104	0·0443	6	0·3054	0·0933
2	0·5419	0·2936	7	−0·2073	0·0430
3	0·2373	0·0563	8	−0·2835	0·0804
4	−0·2835	0·0804	9	0·3824	0·1462
5	−0·2073	0·0430	10	−0·3458	0·1196
		p-Cyanobenzaldehyde (*Ic*)			
1	−0·2649	0·0702	6	0·1056	0·0112
2	0·1122	0·0126	7	−0·3536	0·1251
3	0·3604	0·1299	8	−0·0728	0·0053
4	−0·0728	0·0053	9	0·6975	0·4865
5	−0·3536	0·1251	10	−0·1704	0·0290
		p-Benzoylbenzonitrile (*II*)			
1	−0·2365	0·0560	9	0·5024	0·2524
2	0·1274	0·0162	10	−0·4006	0·1604
3	0·2988	0·0893	11	0·1018	0·0104
4	−0·1017	0·0103	12	−0·2641	0·0698
5	−0·2729	0·0745	13	−0·0347	0·0012
6	0·1710	0·0293	14	0·2729	0·0745
7	−0·2729	0·0745	15	−0·0347	0·0012
8	−0·1017	0·0103	16	−0·2641	0·0698
		p-Methylsulfonylbenzonitrile (*III*)			
1	−0·3798	0·1442	7	−0·3110	0·0968
2	−0·0714	0·0051	8	0·2178	0·0474
3	0·4772	0·2277	9	0·3435	0·1180
4	0·2178	0·0474	10	−0·0378	0·0014
5	−0·3110	0·0968	11	−0·0898	0·0081
6	−0·4551	0·2071			

TABLE I
(*Continued*)

Position i	c_i	c_i^2	Position i	c_i	c_i^2
		p-Sulfamylbenzonitrile (*IV*)			
1	0·2665	0·0710	6	0·3076	0·0946
2	0·0307	0·0009	7	0·2655	0·0705
3	−0·3528	0·1245	8	−0·1320	0·0174
4	−0·1320	0·0174	9	−0·4096	0·1677
5	0·2655	0·0705	10	0·6045	0·3654
		p-Dicyanobenzene (*V*)			
1	0·3850	0·1482	6	−0·3827	0·1464
2	−0·2872	0·0825	7	0·2478	0·0614
3	−0·3827	0·1464	8	0·2478	0·0614
4	0·2478	0·0614	9	−0·2872	0·0825
5	0·2478	0·0614	10	0·3850	0·1482
		p-Cyanobenzoic acid (*VI*)			
1	−0·3564	0·1270	7	−0·2793	0·0780
2	0·2481	0·0616	8	−0·2145	0·0460
3	0·3811	0·1452	9	0·3938	0·1551
4	−0·2145	0·0460	10	−0·2670	0·0713
5	−0·2793	0·0780	11	−0·2670	0·0713
6	0·3471	0·1205			

The effect of an electron uptake on bond strength can roughly be described by the change of the bond order. Of course this quantum chemical index takes into account only the π-electron part of the bond, nevertheless a large increase in its value would indicate that the bond cleavage was unlikely.

p-*Cyanobenzaldehyde* (I) *and* p-*benzoylbenzonitrile* (II). From Fig. 3 it is apparent that compounds *I* and *II* ought first to be reduced at the carbonyl group and only then at the cyano group. This statement is in agreement with the experimental data[1,2] for aldehyde *I* in neutral or alkaline media and for benzophenone *II* even in acid media[3]. In acid media the protonated form of *I* undergoes a reduction in which two of the three bonds in the cyano group are cleaved. Even though a bi-protonation of the electroactive form cannot be excluded, the presence of one proton is regarded as more probable. Nevertheless it is impossible to decide whether the proton is attached to the cyano group or to the carbonyl group of the molecule which is polarized in the field of the electrode. Of the two possible structures *Ib* and *Ic*, only the model using structure *Ib* was in agreement with the experimental data (Fig. 4).

p-Methylsulfonylbenzonitrile (III) *and p-sulfamylbenzonitrile* (IV). The LFMO expansion coefficient for the carbon in the cyano group is extremely small. So it seems probable that the reduction will occur on the sulphonyl group (Fig. 5). For *III* it was found[7] experimentally that a cleavage of the C—S bond occurs and that it leads to CH_3—SO_2^- formation. The same cleavage was also found for *IV*, but was followed by the S—N bond rupture. The remarkable decrease of the C—S and S—N bond orders (*cf.* Figs 1 and 2) is in accordance with these experimental results.

p-Dicyanobenzene (V) *and p-cyanobenzoic acid* (VI). The differences among the squares of the LFMO coefficients are not sufficiently great. Thus the LFMO cannot be treated as localized on a definite bond but as spread along the whole molecule. In this case the simple molecular orbital theory fails. Thus for compounds *V* and *VI* the experimentally found[3] cleavage of the C—C bond in alkaline solution, leading to the CN⁻ formation, cannot be explained.

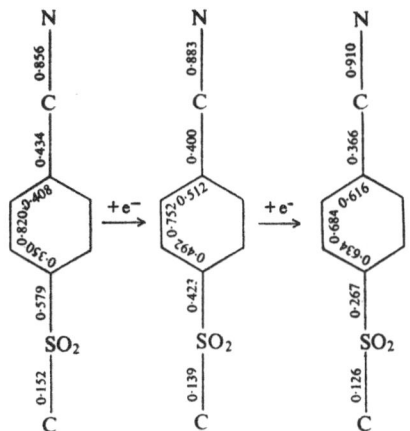

FIG. 1

Bond Orders before and after an Uptake of One and Two Electrons for *p*-Methylsulfonyl-benzonitrile (*III*)

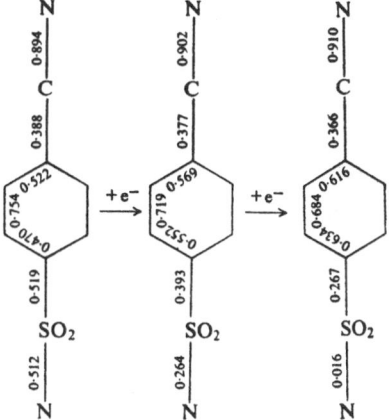

FIG. 2

Bond Orders before and after an Uptake of One and Two Electrons for *p*-Sulfamylbenzo-nitrile (*IV*)

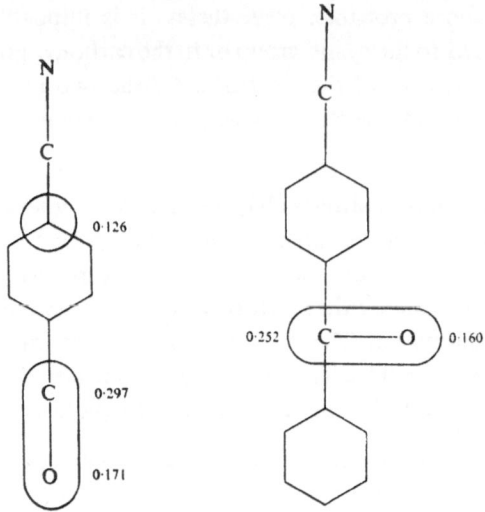

FIG. 3

The Squares of the LFMO Expansion Coefficients for Compounds *I* and *II*
(Only those coefficients are presented, for which $c_i^2 > 0.1$).

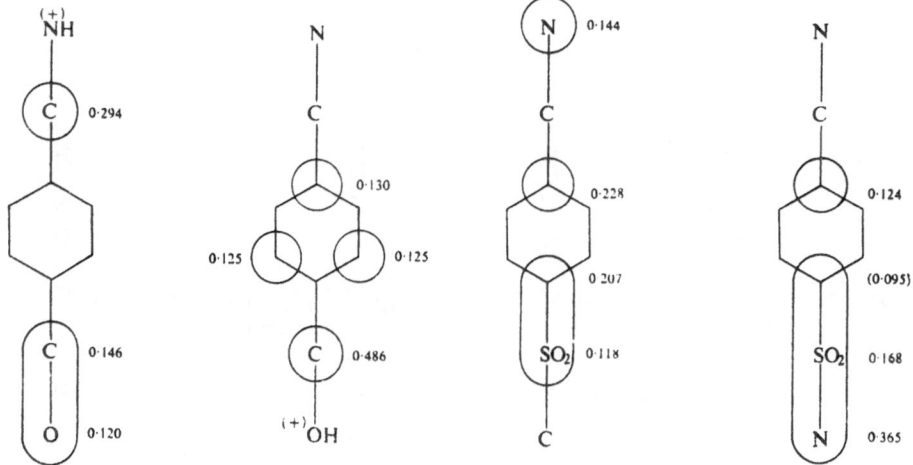

FIG. 4

The Squares of the LFMO Expansion Coefficients for Structure *Ib* ($\delta_N = 3.5$) and *Ic*
($\delta_O = 4.0$)
▸ (Only those coefficients are presented, for which $c_i^2 > 0.1$).

FIG. 5

The Squares of the LFMO Expansion Coefficients for Compounds *III* and *IV*
(Only those coefficients are presented, for which $c_i^2 > 0.1$).

The authors wish to express their thanks to Dr R. Zahradnik for his interest and for the helpfull discussions held during the course of the work.

REFERENCES

1. Zuman P.: Z. Anal. Chem. *224*, 314 (1967).
2. Manoušek O., Zuman P., Exner O.: This Journal, *33* 3979)1968).
3. Zuman P., Manoušek O.: This Journal, in the press.
4. Koutecký J., Paldus J., Zahradník R.: This Journal *25*, 617 (1960).
5. Streitwieser A., jr.: *Molecular Orbital Theory for Organic Chemists.* Wiley, New York 1961.
6. Gerdil R., Lucken E. A. C.: Mol. Phys. *9*, 529 (1965).
7. Manoušek O., Exner O., Zuman P.: This Journal, *33* 3988 (1968).

Translated by the author (P. Z.).

Part IX

Polarography of Nonbenzenoid Aromatic Substances

38. P. Zuman: Polarography of some aromatic compounds: sydnones and azulenes. (in German) *Z. Physikal Chem.* (Leipzig): Sonderheft 1958, 243.
39. P. Zuman: Polarography of nonbenzenoid aromatic and related substances. II. The course of the reduction of sydnones at the dropping mercury electrode. *Collect. Czechoslov. Chem. Commun.* **25**, 3245 (1960).
40. P. Zuman: Polarography of nonbenzenoid aromatic and related substances. III. The course of the reduction of N,N-polymethylene-bis-sydnones. *Collect. Czechoslov. Chem. Commun.* **25**, 3252 (1960).
41. P. Zuman: Polarography of nonbenzenoid aromatic and related substances. IV. Polar effects of substituents in phenylsydnones; the application of modified Hammett equation. *Collect. Czechoslov. Chem. Commun.* **25**, 3256 (1960).
42. P. Zuman, J. Chodowski, and F. Šantavý: Polarography of nonbenzenoid aromatic and related substances. VII. A polarographic study of the acid-base properties of the tropylium ion. *Collect. Czechoslov. Chem. Commun.* **26**, 380 (1961).
43. P. Zuman, and J. Chodowski: Polarography of nonbenzenoid aromatic and related substances. VIII. Adsorption processes during the electroreduction of the tropylium ion. *Collect. Czechoslov. Chem. Commun.* **27**, 759 (1962).

Sydnones are reduced[38,39] in the protonated form in a six-electron step, and in the unprotonated form in a four-electron step, to form hydrazine derivatives. Two sydnone rings separated by a methylene chain interact[40] and are reduced consecutively. Differences in substituent effects in a phenyl ring bound on 3-C or 4-N offer information about transmission of polar effects through the sydnone ring. Protonation of azulene rings can be followed.[38]

From the change in wave-height with pH,[42] when adsorption effects[43] are avoided, the rate of formation of tropyl alcohol from tropylium ion can be calculated. The rate constant ($k = 2.10^6$ l Mole^{-1} sec^{-1}) of this volume reaction is in good agreement with data obtained by relaxation techniques.

Polarographisches Institut, Tschechoslowakische Akademie der Wissenschaften, Prag

Polarographie mancher aromatischen Verbindungen; Sydnone und Azulene[1]

Von

P. Zuman[2]

Mit 20 Abbildungen und 2 Tabellen

In der organischen Chemie wurden in letzter Zeit manchen fünf- und siebengliedrigen Ringen mit aromatischen Eigenschaften Aufmerksamkeiten gewidmet [1]. Diejenigen Verbindungen, bei denen das π-Elektronensextett entweder unter Aufnahme von Elektronen von einem exocyclischen Atom oder von einem anderen Ring gebildet wird, können als pseudoaromatische bezeichnet werden.

Als Beispiele können Dimethylfulven (I), bei welchen ein Elektron von einer exocyclischen Gruppe dem Ring übergegeben wurde, weiter die Tropone (II) und Sydnone (III), bei welchen die π-Elektronensextettbildung durch die Verschiebung eines π-Elektrons des Ringes zu dem Sauerstoffatom ermöglicht ist, angegeben werden. Diejenigen von diesen Verbindungen, welche durch keine klassische Formel ausgedrückt werden können, werden auch als mesoionische Verbindungen bezeichnet [2], [3], [4], [5]. Als Beispiel der Zweiringsysteme können die Azulene angeführt werden. Hier hat das eine Elektron die Tendenz, von dem siebengliedrigen Ring in den fünfgliedrigen überzugehen. Es entsteht eine Struktur mit vollaromatischer (IV) oder polarisierter (V) Verteilung der Elektronen auf dem fünfgliedrigen Ring.

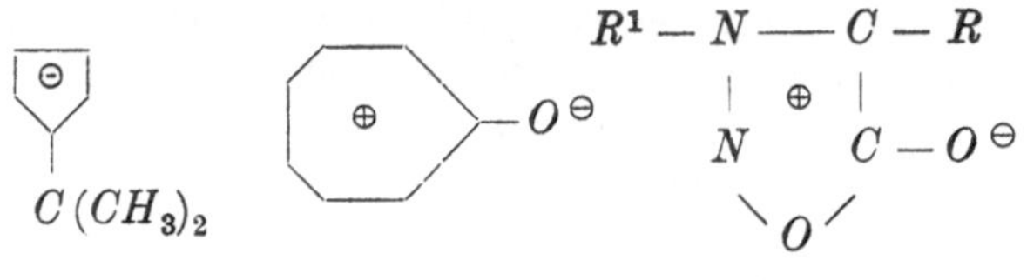

(I)   (II)   (III)

---

[1] Vortrag, gehalten anläßlich des Polarographischen Kolloquiums in Dresden vom 3. bis 7. Juni 1958.

[2] Dr. Petr Zuman, Prag XII, Blanicka 15, ČSR.

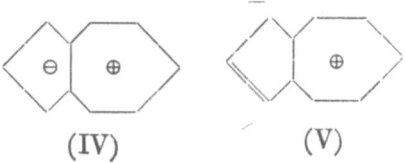

(IV)   (V)

Die Polarographie der Tropone und Tropolone wurde in dem Beitrag von ŠANTAVÝ [6] diskutiert. Hier soll zuerst das polarographische Verhalten der Sydnone besprochen werden. Die untersuchten Verbindungen wurden uns von J. C. EARL (Thurlton, England). I. M. HUNSBERGER (Fordham University, N. Y., USA.) und D. J. VOADEN (Oxford, England) in liebenswürdiger Weise zur Verfügung gestellt.

Sydnone
pH-Abhängigkeit der Sydnon-Stufen

Die Aryl-Sydnone (VI) werden in saurer Lösung in einer sechselektronigen Stufe reduziert (Abb. 1), deren Höhe mit steigendem

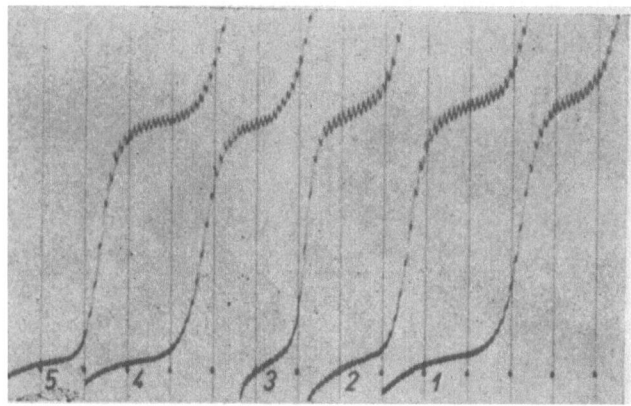

Abb. 1. Stufen der Arylsydnone in saurer Lösung.
$6,6 \cdot 10^{-}$ M 3-p-Methoxyphenylsydnon in: 1. 0,1 M-HCl; 2. 1 M-HCl; 3. 5 M-HCl; 4. 0,1 M-H_2SO_4; 5. 1 M-H_2SO_4. Von 0 V, kath. Schaltung, gegen ges. Kalomelelektrode, 200 mV/Absz., $h = 65$ cm, Empf. 1:15

pH-Wert bei pH größer als etwa 3 in der Form einer Dissoziationskurve abnimmt (Abb. 2). Bei pH größer als 5 ist auf der Kurve nur eine vierelektronige Stufe sichtbar, deren Halbstufenpotential vom

454

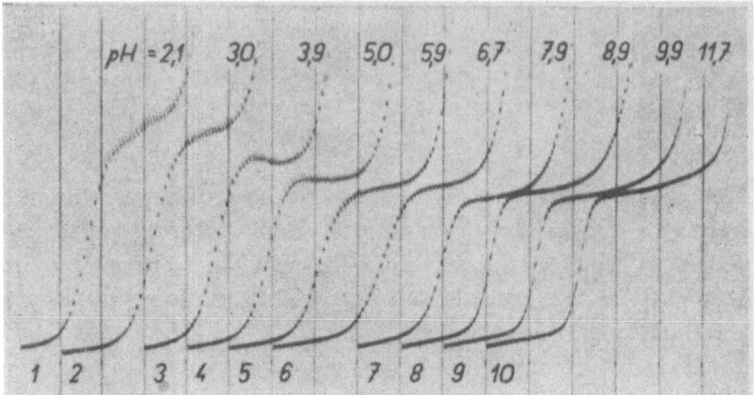

Abb. 2. pH-Abhängigkeit der Stufen des 3-Phenylsydnons in Puffergemischen nach BRITTON und ROBINSON.

5·10⁻⁵ M 3-Phenylsydnon. Die pH-Werte sind auf dem Polarogramm angegeben. Kurven 1 und 2 von 0,4 V, Kurven 3 bis 6 von 0,6 V, Kurven 7 bis 10 von 0,8 V. Kath., ges. Kalomelektrode, 200 mV/Absz., h = 57 cm, Empf. 1:7

pH-Wert der Lösung praktisch unabhängig ist. Das Halbstufenpotential der sechselektronigen Stufe verschiebt sich dagegen mit steigendem pH-Wert der Lösung zu negativeren Potentialen (Abb. 3).

$$\langle\!\rangle\!-N_3\!-\!_4C \qquad R^1\!-\!N\!-\!-\!C\!-\!R^2 \qquad \langle\!\rangle\!-N\!-\!-CH_2$$

(VI) (VII R^1 = Alkyl Benzyl) (VIII)

Bei den 3-Alkylsydnonen (VII) liegt das Halbstufenpotential bei so negativen Werten, daß die Reduktion in saurer Lösung durch die Wasserstoffabscheidung überdeckt wird (Abb. 3, 4). Auf den Kurven kann man nur die pH-unabhängige vierelektronige Stufe bei höheren pH-Werten beobachten (Abb. 4).

Als Vergleichssubstanz diente die *N*-Nitrosoanilinessigsäure (VIII). Die undissoziierte Säure sowie ihr Anion werden in einer vierelektronigen Stufe reduziert. Die Höhe dieser Stufe nimmt bei pH größer als 7,5 in der Form einer Dissoziationskurve ab (Abb. 5). Das Halbstufenpotential der zweielektronigen Stufe im alkalischen Gebiet ist vom pH unabhängig.

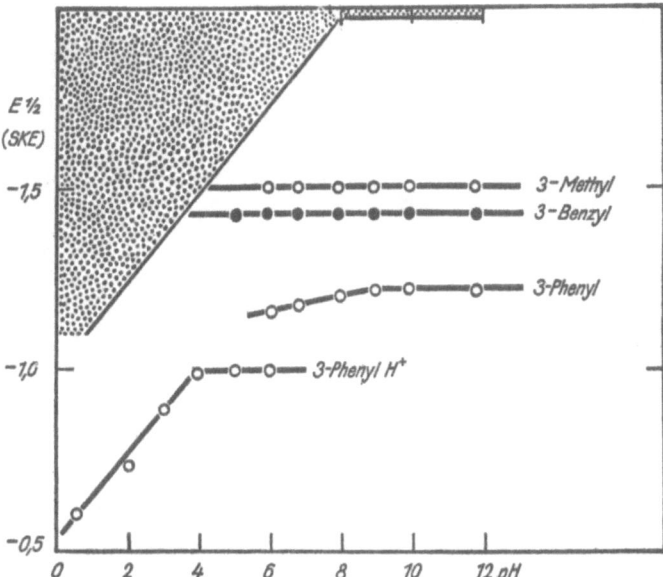

Abb. 3. pH-Abhängigkeit der Halbstufenpotentiale der in Stellung 3 substituierten Sydnone.

Die Irreversibilität aller dieser Stufen wurde aus ihrer Steilheit bestimmt, die z. B. bei den vierelektronigen Stufen der Sydnone einem Einelektronübergang entspricht. Für die Irreversibilität zeugen weiter

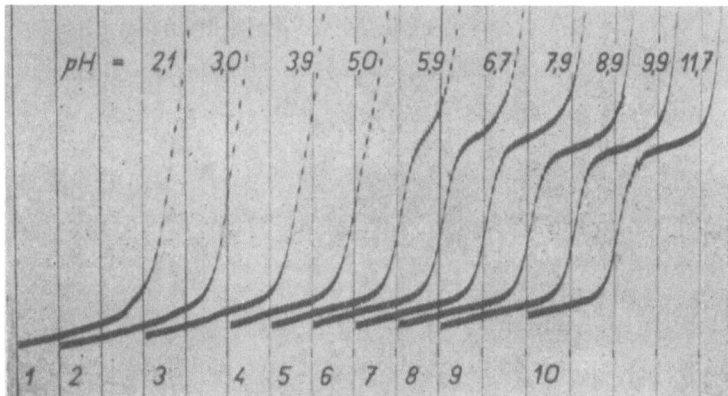

Abb. 4. pH-Abhängigkeit der Stufen des 3-Benzylsydnons in Puffergemischen nach BRITTON und ROBINSON.

$5 \cdot 10^{-5}$ M 3-Benzylsydnon. Die pH-Werte sind auf dem Polarogramm angegeben. Kurven 1 und 2 von 0,4 V, Kurve 3 von 0,6 V, Kurven 4 bis 9 von 0,8 V, Kurve 10 von 1,0 V. Kath. ges. Kalomelektrode, 200 mV/Absz., $h = 57$ cm; Empf. 1:7

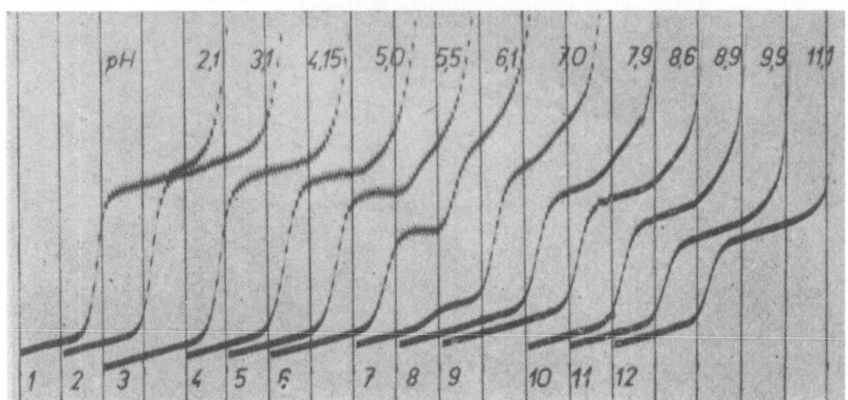

Abb. 5. pH-Abhängigkeit der Stufen der *N*-Nitrosoanilinessigsäure in Puffer-gemischen nach BRITTON und ROBINSON.

5,5 · 10⁻⁵ *M N*-Nitrosoanilinessigsäure. Die pH-Werte sind auf dem Polarogramm angegeben Kurven 1 bis 3 von 0,2 V, Kurven 4 bis 6 von 0,4 V, Kurven 7 bis 9 von 0,6 V, Kurven 10 bis 12 von 0,8 V. Kathod. Schaltung, gegen ges. Kalomelektrode, 200 mV/Absz., t_1 = 2,3 sec, h = 57 cm, Empf. 1:5

die Reduktion bei kontrolliertem Potential mit Hilfe des Potentio-staten nach PEIZKER [7], mit dem Umschalter nach KALOUSEK und RÁLEK [8] sowie die Versuche mit der Impulspolarographie mit dem Gerät von VALENTA [9] und mit dem Polaroskop nach HEYROVSKÝ [10].

Reduktionsschema und -mechanismus

Die vierelektronige Reduktion der Sydnone in alkalischer Lösung erfolgt nach dem Schema:

$$R^1 - N \overline{\quad\quad} C - R^2$$
$$\begin{array}{ccc} | & \oplus & | \\ N & & C - O^{\ominus} \\ & \searrow O \nearrow & \end{array} + 4e + 3H^+ \rightarrow \begin{array}{c} R^1 - N \overline{\quad\quad} CHR \\ | \quad\quad\quad | \\ NH_2 \quad COO^- \end{array}$$

$$(IX)$$

Die Entstehung des Hydrazins (IX) wurde mit Hilfe des Potentio-staten nach PEIZKER [7] (Abb. 6) sowie des Umschalters nach KALOU-SEK (Abb. 7) in der Schaltung nach Abb. 8, mit Hilfe der anodischen Stufe des Hydrazins (IX) nachgewiesen. Die Entstehung dieser Ver-bindung IX wurde zugleich auch mit Hilfe des anodischen Strom-

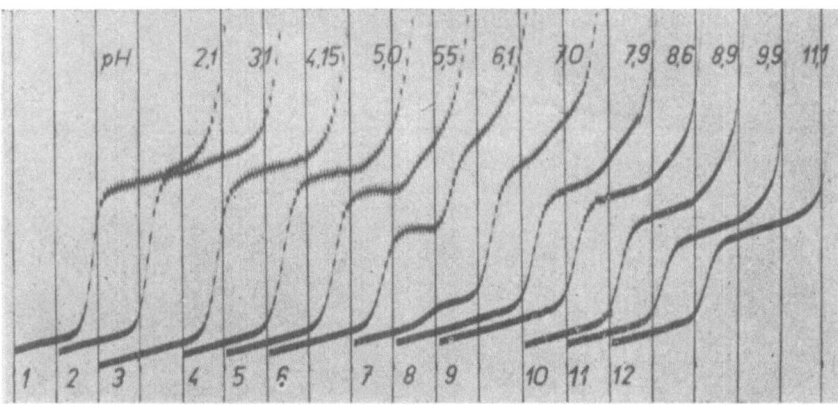

Abb. 6. Elektrolytische Sydnonreduktion.

0,005 M Benzylsydnon in Boratpuffer pH 9,3 beim Potential des Grenzstromes unter Anwendung der Quecksilbertropfelektrode mit Potentiostat nach Pelzker 30 Stunden reduziert. Zum Polarographieren wurde die Lösung 20mal mit Boratpuffer verdünnt. Kurve 1:
Vor der Elektrolyse; Kurve 2: Nach der Elektrolyse. Anodische Stufe des Hydrazins wurde
gebildet. 200 mV/Absz., Empf. 1:20

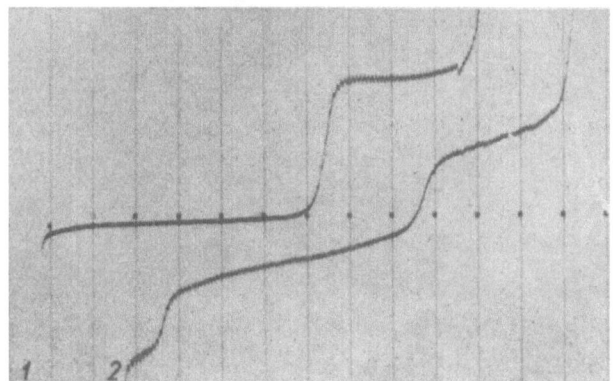

Abb. 7. Sydnonreduktion mit Umschalter nach KALOUSEK.

2,5 · $10^{-4} M$ 2-Phenylsydnon in Boratpuffer pH 9,3 mit 5% Äthanol. (1) Primitive Kurve;
(2) umgeschaltete Kurve, Hilfspotential entspricht dem Diffusionsstrom, Frequenz der
Umschaltung 3,5 Hz. Kurven von + 0,2 V, anod.-kathod. Schaltung, gegen ges. Kalomelektrode, 200 mV/Absz., h = 65 cm, Empf. 1:50

maximums ("peak") bei der Impulspolarographie mit einem einzelnen
Triangelimpuls (Abb. 9) sowie mit Hilfe der Einschnitte auf der dV/dt-
V-Kurven (Abb. 10) bestätigt.

Aus der pH-Unabhängigkeit dieser Stufe sowie aus ihrer Steilheit
folgt, daß der potentialbestimmende Schritt am wahrscheinlichsten
die reversible Aufnahme des ersten Elektrones ist. Als Folgereaktionen

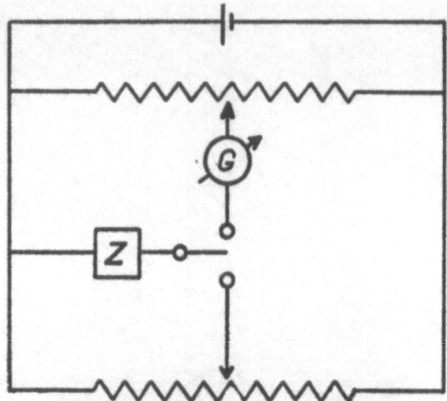

Abb. 8. Schaltschema des Umschalters nach KALOUSEK und RALEK.

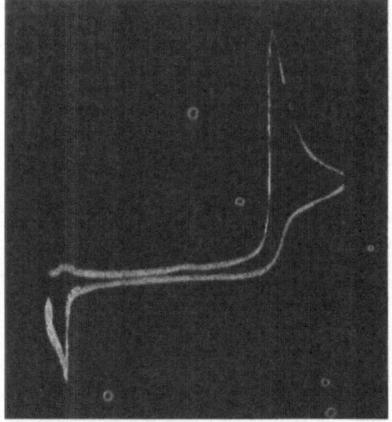

Abb. 9. i-V-Kurven des Sydnons in alkalischer Lösung mit der impulspolarographischen Einrichtung.

$1,25 \cdot 10^{-4}$ M 3-Phenylsydnon in 0,05 M Boratpuffer pH 9,3. Gerät nach Valenta,
$t_1 = 22,7$ sec, 1 Triangelimpuls, Steilheit des Impulses 0,63 V/sec, i_{max} — 1,59 A.
Anfangspannung — 0,04 V gegen Quecksilberboden, Endspannung — 1,90 V

können entweder eine langsamere Aufnahme weiterer Elektronen und
Protonen oder eine Umlagerung des entstandenen Radikals unter
weiterer Elektronen- und Protonenaufnahme stattfinden.

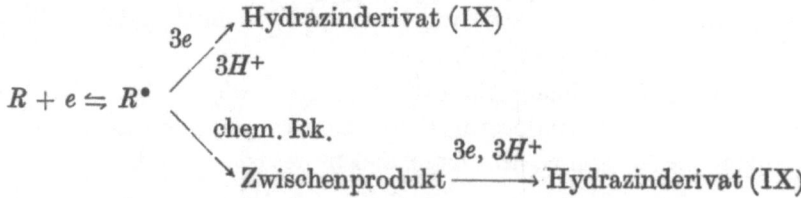

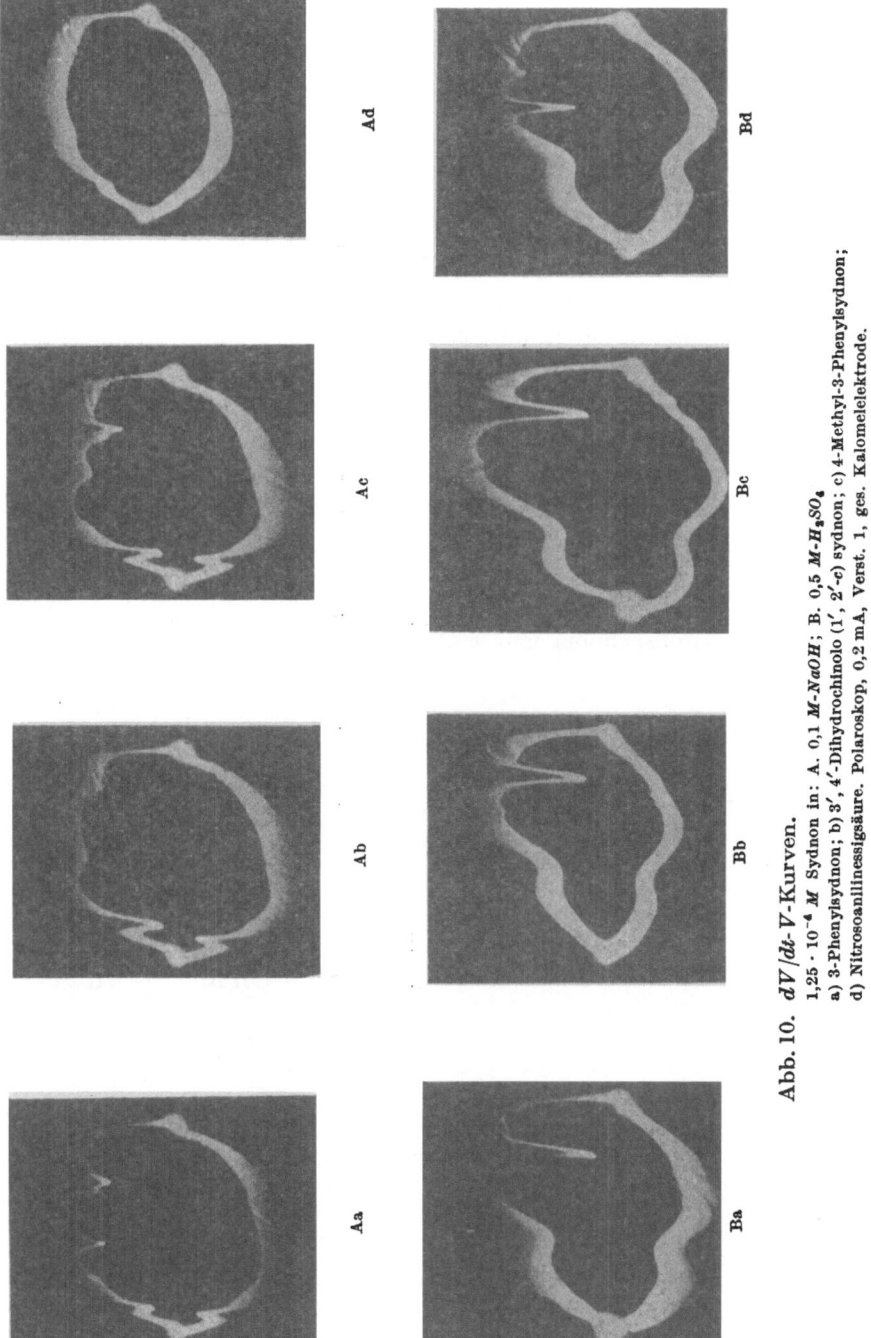

Abb. 10. dV/dt-V-Kurven.

$1{,}25 \cdot 10^{-4}$ M Sydnon in: A. 0,1 M-$NaOH$; B. 0,5 M-H_2SO_4

a) 3-Phenylsydnon; b) 3′, 4′-Dihydrochinolo (1′, 2′-e) sydnon; c) 4-Methyl-3-Phenylsydnon;
d) Nitrosoanilinessigsäure. Polaroskop, 0,2 mA, Verst. 1, ges. Kalomelelektrode.

460

Die oszillographischen Kurven verlaufen in alkalischer Lösung dermaßen kompliziert, daß sie keine weiteren Aussagen über den Mechanismus der Elektroreduktion unter polarographischen Bedingungen erlauben.

Bei sechselektroniger Reduktion in saurer Lösung konnte mit keiner der angegebenen Methoden die Entstehung des Hydrazins oder einer Verbindung, welche sich durch eine anodische Stufe äußern würde, bestätigt werden. Nach der Elektrolyse beim Potential des Diffusionsstromes des 3-Arylsydnones wurden in der Lösung Ammoniumionen nachgewiesen. Der Elektroreduktion ist eine Protonierung des Ringes vorgelagert, deren Geschwindigkeit die Stufenhöhe begrenzt. Die Stufe besitzt kinetischen Charakter. Die thermodynamischen Dissoziationskonstanten der Sydnone müssen jedoch in stark sauren Lösungen liegen, wo sie der Messung wegen eines Zerfalles der Substanz unzugänglich sind. Aus diesem Grunde konnten die Rekombinationsgeschwindigkeitskonstanten nicht berechnet werden.

Das Reduktionsschema lautet also:

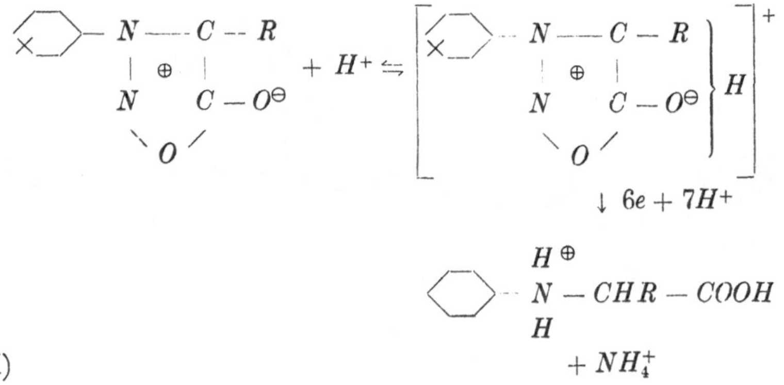

(X)

Dieselben Endprodukte wurden durch katalytische Hydrierung des 3-Phenylsydnons [2] sowie durch die Reduktion mit Zink in verdünnter Essigsäure [11] erhalten.

Über den detaillierten Reduktionsmechanismus der Elektronenübergabe können an Hand der zur Verfügung stehenden Ergebnisse keine Aussagen gemacht werden.

Bei der Reduktion des Anions der *N*-Nitrosoanilinessigsäure (VIII)
konnte in alkalischem Medium weder mit dem Umschalter, noch mit
der Elektrolyse, noch mit den oszillographischen Methoden ein anodischer Vorgang festgestellt werden. Durch den Reduktionsvorgang
in saurer Lösung entsteht die ionisierte Form des Hydrazins (XI) aus
der protonosierten Form (X). Die anodische Stufe des Hydrazins
(XI bzw. IX) kann nach der Elektrolyse in saurer Lösung und Überführung in Puffergemische von höheren pH-Werten identifiziert
werden (Abb. 11). In diesem Falle ist die Anwendung eines Potentiostaten

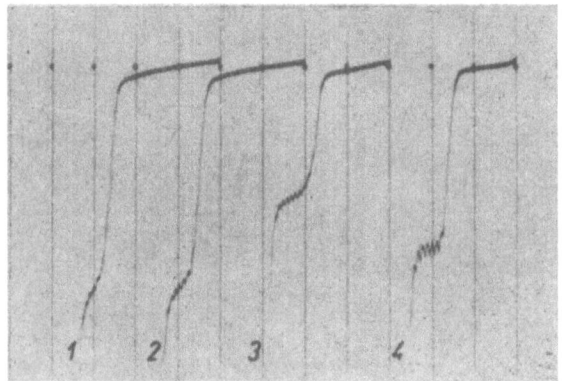

Abb. 11. Reduktionsprodukte der *N*-Nitrosoanilinessigsäure in saurem Medium.
0,01 *M N*-Nitrosoanilinessigsäure in 0,1 *M*-*H₂SO₄* bei dem Potential des Grenzstromes unter
Anwendung der Quecksilbertropfelektrode mit dem Potentiostat nach Peizker 24 Stunden
reduziert. Zum Polarographieren wurde die Lösung 20mal mit der Pufferlösung verdünnt.
Anodische Stufe des Hydrazins in 1. Acetatpuffer *pH* 4,7; 2. Phosphatpuffer 6,8; 3. Boratpuffer *pH* 9,3; 4. 0,2 *M*-*NaOH*. Kurven von — 0,4 V zurück aufgenommen, anod.-kath.
Schaltung, gegen ges. Kalomelelektrode 200 mV/Absz., *h* = 65 cm, Empf. 1:20

[7] zur Identifizierung der Reduktionsprodukte vorteilhafter als die Versuche mit der Impulspolarographie [9] und dem Umschalter [8]. Bei
diesen letztgenannten Methoden können die Produkte der Elektrolyse
nur in demselben Medium, in welchem sie hergestellt wurden, mit Hilfe
der anodischen Vorgänge studiert werden. In saurer Lösung erfolgt
die Auflösung des Quecksilbers bei negativeren Potentialen als diejenigen der anodischen Stufen des Hydrazins XI. Auf der umgeschal-
teten Kurve ist keine anodische Stufe sichtbar. Im mittleren pH-
Bereich entsteht Hydrazin bei der Umschaltung zum Grenzstrome
der freien Säure wie auch zum Grenzstrome des Anions (Abb. 12).

462

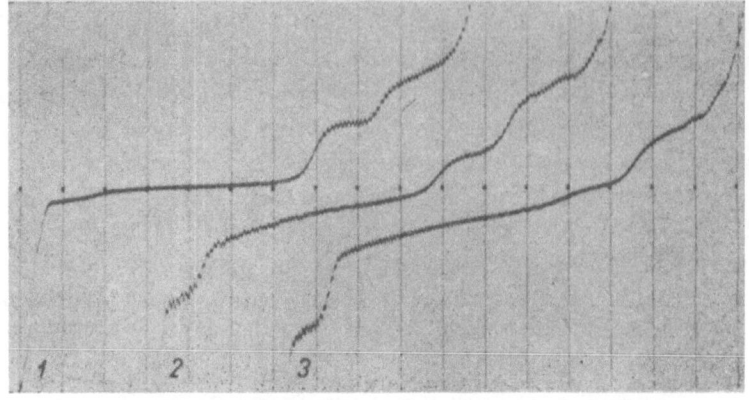

Abb. 12. Reduktion der *N*-Nitrosoanilinessigsäure mit dem Umschalter nach
KALOUSEK.

2,5 · 10⁻⁴ *M N*-Nitrosoanilinessigsäure in Britton-Robinson-Puffer *pH* 6 mit 5 % Äthanol.
1. Primitive Kurve; 2. Kurve nach der Umschaltung, Hilfspotential entspricht dem Grenzstrom der ersten Stufe; 3. Kurve nach der Umschaltung, Hilfspotential entspricht dem
Grenzstrom der zweiten Stufe, Frequenz der Umschaltung 3,5 Hz. Kurven von + 0,2 V,
anod.-kathod. Schaltung, gegen ges. Mercurosulfatelektrode, 200 mV/Absz., *h* = 65 cm,
Empf. 1 : 70

Bei den beiden Protonübergängen sind die polarographischen Grenzströme durch die Rekombinationsgeschwindigkeit begrenzt. Das
Gesamtschema lautet:

$$
\left[\begin{array}{c} CH_2COOH \\ | \\ \bigcirc-N \\ | \\ NO \end{array} \right\}H\right]^{+} \underset{\;}{\overset{-H^+}{\rightleftharpoons}} (A)
\left[\begin{array}{c} CH_2COO^- \\ | \\ \bigcirc-N \\ | \\ NO \end{array} \right\}H^+\right] \overset{-H^+}{\rightleftharpoons} (B)
\begin{array}{c} CH_2COO^- \\ | \\ \bigcirc-N \\ | \\ NO \end{array}
$$

$+4e \downarrow +4H^+ \quad (C)$ $+4e \downarrow +4H^+ \quad (D)$ $+2e \downarrow +11H^+ \ (E)$

$$
\begin{array}{c} CH_2COOH \\ | \\ \bigcirc-N \quad +H_2O \\ | \\ NH_3^+ \end{array}
\qquad
\begin{array}{c} CH_2COO^- \\ | \\ \bigcirc-N \quad +H_2O \\ | \\ NH_3^+ \end{array}
\qquad
\begin{array}{c} CH_2COO^- \\ | \\ \bigcirc-NH \\ \\ +\ \text{Produkte} \end{array}
$$

(XI)

Ein dem Gleichgewicht *B* analoger Vorgang sowie ähnliche
Reduktionsprodukte der Reaktionen *D* und *E* wurden in letzter Zeit
für die Reduktion von *N*-Nitrosoaminen vom Typus $RR'N - NO$
von ZAHRADNIK [12] vorgeschlagen.

Durch Reduktion der Sydnone in alkalischer Lösung entstehen also dieselben Verbindungen wie durch die Reduktion der entsprechenden N-Nitrosoamine in saurer Lösung. Das wurde durch den Vergleich anodischer Stufen der Hydrazinverbindungen der reduzierten Formen bewiesen.

Konstitutionseinflüsse

Wenn wir die Reduktion des 3-Phenylsydnons (XII) mit derjenigen der N-Nitrosoanilinessigsäure (VIII) vergleichen, können wir beobachten, daß die Reduktion des 3-Phenylsydnons (XII) im alkalischen Medium um etwa 80 mV positiver als die Reduktion des Anions der N-Nitrosoanilinessigsäure (XIII) verläuft.

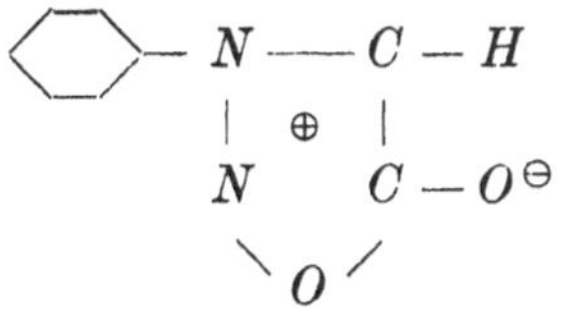

$$\langle\rangle\!-N\!-\!\!-C\!-\!H \qquad\qquad \langle\rangle\!-N\!-\!\!-CH_2$$

(XII) $E_{1/2} = -1{,}21_5\,V$ 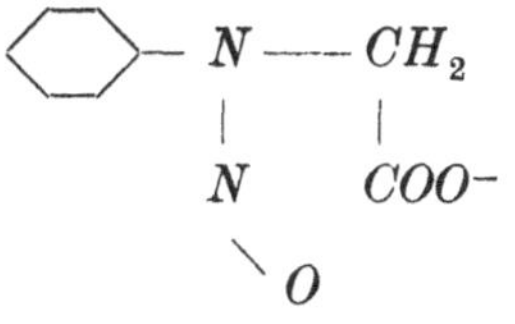   (XIII) $E_{1/2} = -1{,}29_5\,V$

Dagegen ist die ionisierte Form des 3-Phenylsydnons (XIV) um etwa 100 mV negativer reduzierbar als die freie Säure (XV).

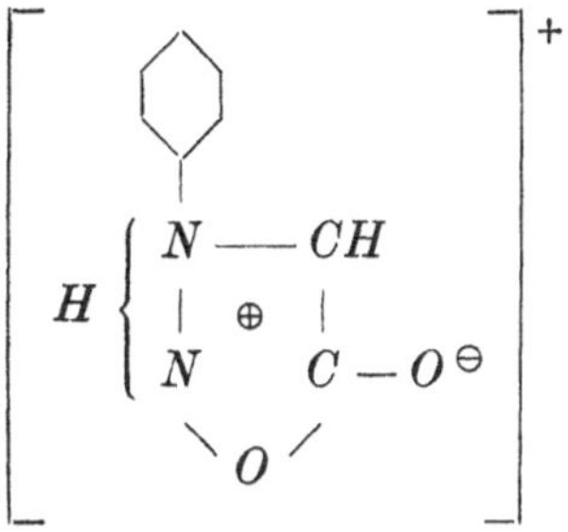

(XIV)   (XV)

$(E_{1/2} = -0{,}61\,V;\ 0{,}5\,M\ H_2SO_4)$ $(E_{1/2} = -0{,}51\,V;\ 0{,}5\,M\ H_2SO_4)$

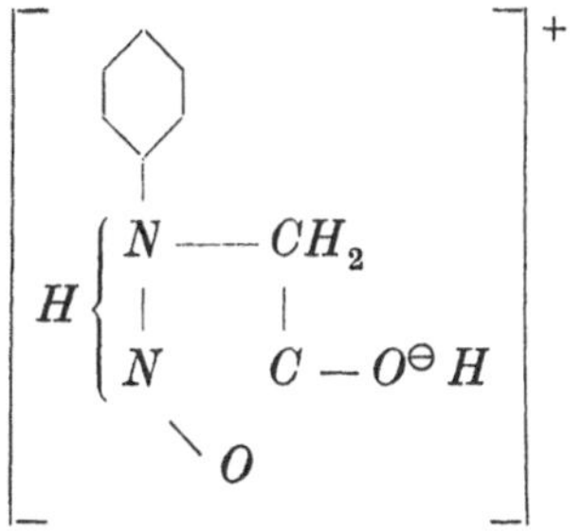

Weil die Reduktionsmechanismen in sauren wie auch in alkalischen Lösungen dieser beiden Verbindungen verschieden sind, sollen diese Werte nur einen groben Überblick über den Einfluß der Ringbildung geben.

Weil es sich in den sauren Lösungen um komplizierte Reduktionsvorgänge handelt, werden im weiteren die pH-unabhängigen Halbstufenpotentiale, die man im alkalischen Gebiete erhält, verglichen.

Zuerst wurde der Einfluß der Konjugierung bei der Substitution in Stellung 3 verfolgt (Abb. 13). Es wurden folgende Werte der Halbstufenpotentiale gefunden:

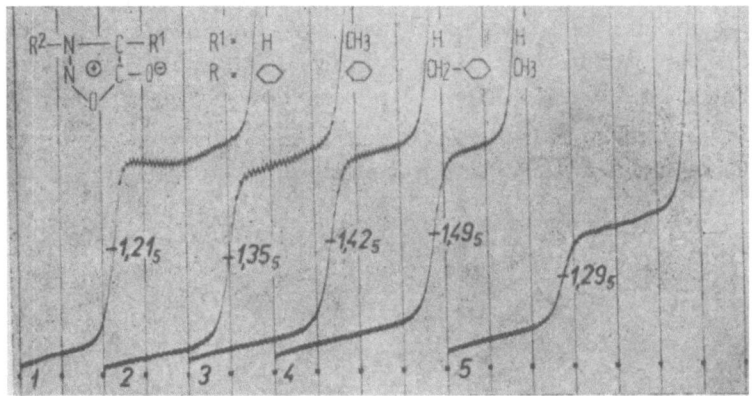

(XII)   (XVI)   (XVII)

Borax pH 9,3	$E_{1/2} = -1,23_0$ V	$E_{1/2} = -1,45_0$ V	$E_{1/2} = -1,52_0$ V
0,1 M $NaOH$	$E_{1/2} = -1,21_5$ V	$E_{1/2} = -1,42_5$ V	$E_{1/2} = -1,49_5$ V
Verschiebung im Vergleich zum Methylderivat	$+ 0,28_5$ V	$+ 0,07_0$ V	0

Abb. 13. Eintluß der Konjugierung auf die Reduzierbarkeit der Sydnone.
5 · 10⁻⁵ M Sydnonlösung in 0,1 M-$NaOH$. 1. 3-Phenylsydnon; 2. 3-Phenyl-4-Methylsydnon; 3. 3-Benzylsydnon; 4. 3-Methylsydnon; 5. N-Nitrosoanilinessigsäure. Von 0,8 V, kath., ges. Kalomelelektrode, 200 mV/Absz., $h = 65$ cm, Empf. 1:10.

Die konjugierte Phenylgruppe erleichtert wesentlich die Reduktion der Bindung $N - O$. Es handelt sich um einen günstigen Fall, in welchem — infolge des gleichen Mechanismus — der wohlbekannte

Einfluß der Phenylgruppe quantitativ ausgedrückt werden konnte. Zum Beispiel in den Paaren wie Aceton—Acetophenon, Acetaldehyd—Benzaldehyd unterscheiden sich die Reduktionsmechanismen so weit, daß ein Vergleich nicht möglich ist. Die Größe der Verschiebung, die in gewissem Maß von der Art der reduzierbaren Gruppe abhängig sein sollte, stimmt gut mit den beiden anderen Beispielen überein, die man den Tabellen [14] entnehmen konnte (Tab. 1).

Tabelle 1

Einfluß der Phenylgruppe auf die Halbstufenpotentiale

	pH 4		pH 9,2	
$CH_3COCOOH$	$-1,07$ V	$> +0,37$	$-1,55$	$> 0,26$
$C_6H_5COCOOH$	$-1,70$ V		$-1,29$	
CH_3NO_2	$-0,81$	$> +0,36$	$-0,90$	$> 0,16$
$C_6H_5NO_2$	$-0,45$		$-0,94$	

Einen ähnlichen Einfluß kann man, nach der Mitteilung von W. Rössler [15], bei der Phenylsubstitution des Tropons sowie des Azulens beobachten.

Der Einfluß der Phenylgruppe ist durch die Anwesenheit einer Methylenbrücke nicht völlig abgeschirmt. Durch die Substitution eines Wasserstoffatoms der Methylgruppe, wie es im Benzylsydnon der Fall ist, ist der $+I$-Effekt der Methylgruppe verringert.

Der Einfluß der Substitution durch eine Methylgruppe in Stellung 4 wurde an Hand zweier Paare verfolgt:

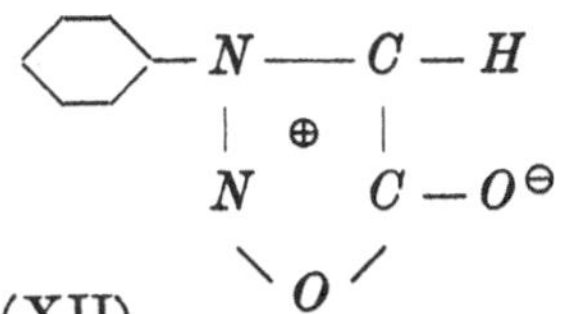

(0,1 M NaOH)
Unterschied
0,14$_0$ V
0,13 V

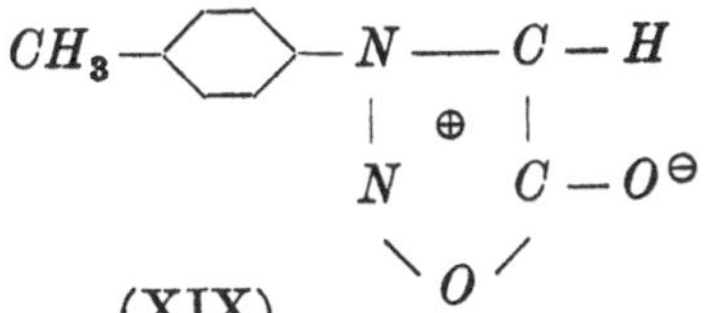

(XII) $E_{1/2} = -1,21_5$ V

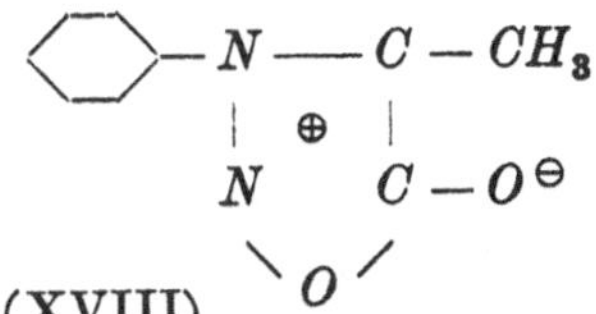

(XVIII) $E_{1/2} = -1,35_5$ V

(XIX) $E_{1/2} = -1,25$ V

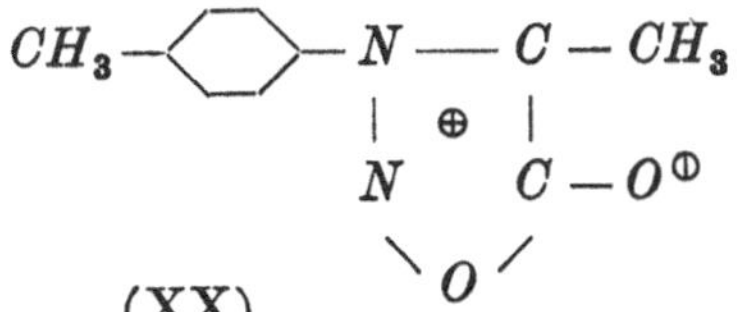

(XX) $E_{1/2} = -1,38$ V

Die Verschiebung zu negativeren Potentialen ist durch den $+I$-Effekt der Methylgruppe verursacht. Der induktive Effekt der Methylgruppe ist größer als bei den benzoiden Verbindungen, wo z. B. zwischen dem Naphthalin und seinen Methylderivaten praktisch kein Unterschied gefunden wurde [16]. Die Verschiebung ist auch größer als die größte beim Azulenen durch Methylsubstitution verursachte [17]. Eine Verschiebung von derselben Größenordnung konnte bei Methylsubstitution an der Doppelbindung in Styrol [16], Stilben [18] oder in Cumarin [19] beobachtet werden (Tab. 2).

Tabelle 2

Einfluß der Methylgruppe auf die Reduzierbarkeit

	$E_{1/2}$	Lit.		$E_{1/2}$	Lit.
Naphthalin	— 2,50	16	Styrol	— 2,39	16
α-Methylnaphthalin	— 2,50	16	β-Methylstyrol	— 2,58	16
β-Methylnaphthalin	— 2,50	16			
			Stilben	— 2,18	16
Azulen	— 1,63	17	α-Äthylstilben.....	— 2,47	18
1-Methylazulen.....	— 1,69	17			
			Cumarin	— 1,53	19
2-Methylazulen.....	— 1,75	17			
			3-Methylcumarin ..	— 1,68	19
4-Methylazulen.....	— 1,71	17			
			4-Methylcumarin ..	— 1,64	19
5-Methylazulen.....	— 1,64	17			
6-Methylazulen.....	— 1,71	17			

Weiter werden die sterischen Einflüsse sowie die Einflüsse des Ringschlusses diskutiert.

Die Substitution durch Alkylgruppen in ortho-Stellung in benzoiden Verbindungen [20], [21], [22], [23] sowie in Azulenderivaten [24], in welchen die Reduktion in der Seitenkette verläuft, verursacht eine Aufhebung der Komplanarität, was eine Verschiebung der Halbstufenpotentiale zu negativeren Werten zur Folge hat.

Eine ähnliche sterische Beeinflussung der Komplanarität wurde auch bei den Phenylsydnonen beobachtet. Der Unterschied gegenüber den obengenannten Verbindungen liegt darin, daß hier die reduzierbare Gruppe ein weiterer Ring ist (und nicht etwa eine Keto-, Aldehyd-, Azo- oder Nitrogruppe, wie es in den obenerwähnten Beispielen der Fall war).

Folgende drei Gruppen wurden verfolgt (0,1 M $NaOH$) (Abb. 14
15,16). Die erste Gruppe besteht aus XII, XXI, XXII, XXIII und
XXIV, in der zweiten befinden sich XVIII, XX und XXV, zur
dritten gehören XVII, XXVI und XXVII.

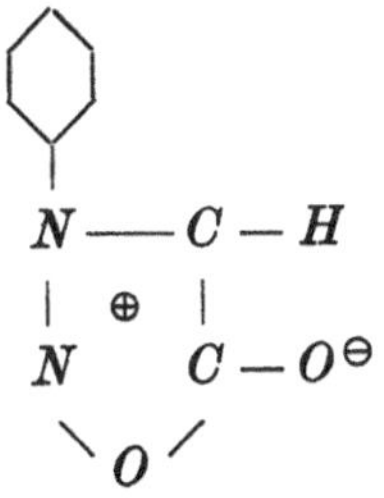

$$N\text{---}C\text{---}H \quad\quad \overset{CH_3}{N\text{---}C\text{---}H} \quad\quad \overset{CH_3CH_2}{N\text{---}C\text{---}H}$$

(XII)   (XXI)   (XXII)

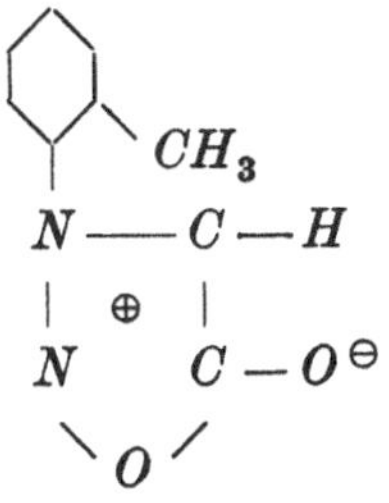

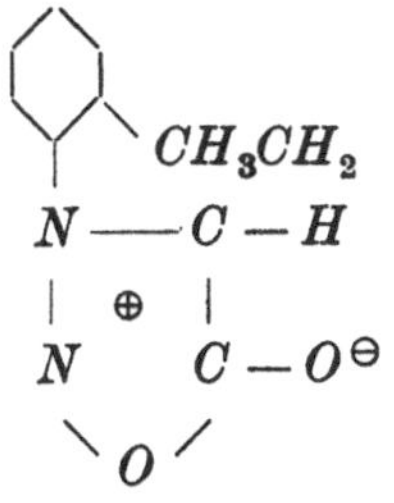

$E_{1/2} = -1{,}21_5$ V $-1{,}30$ V $E_{1/2} = -1{,}30$ V

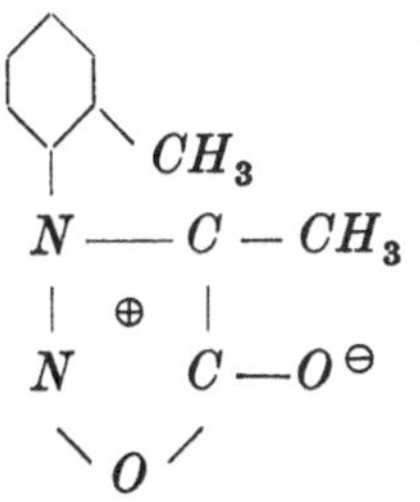

(XXIII)   (XXIV)

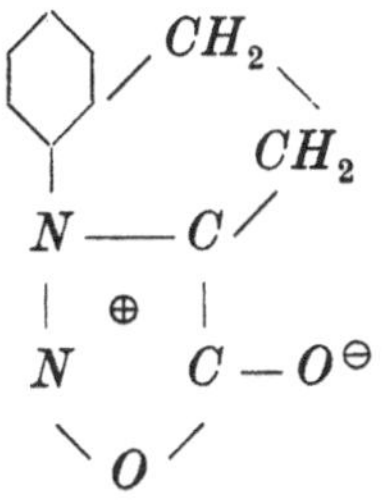

$E_{1/2} = -1{,}44$ V* $-1{,}23$ V

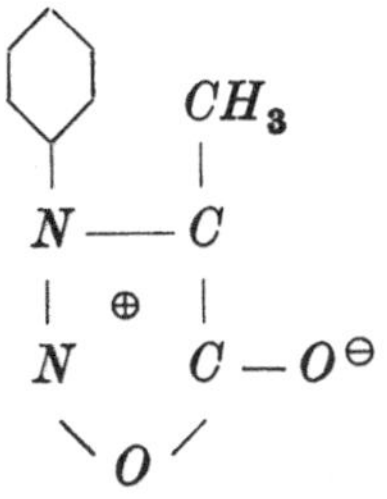

(XVIII)   (XX)

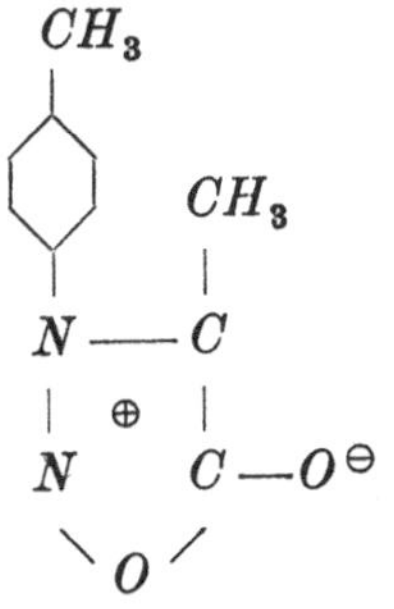

$E_{1/2} = -1{,}35_5$ V $-1{,}38$ V

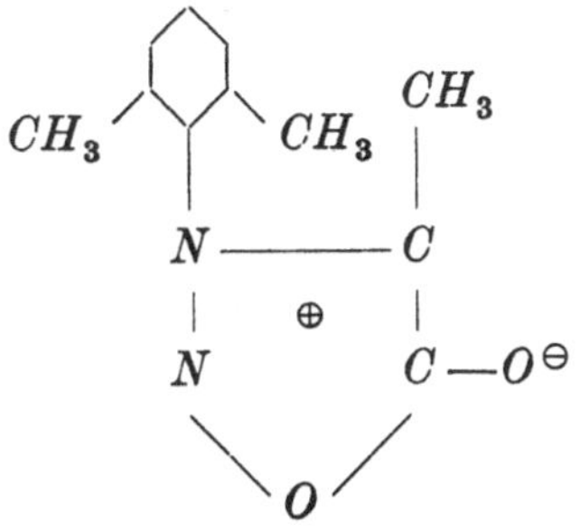

(XXV)

$-1{,}46$ V

* Dieser Wert wurde aus den Werten für 3-o-Methylphenylsydnon (XXII)
(—1,30) und aus der Differenz der Werte für 3-Phenyl-4-methylsydnon
(XVIII) (—1,35) und 3-Phenylsydnon berechnet (XII) (—1,21₅).

$$CH_3 - N - C - H$$

(structures with charges)

(XVII) — $1,49_5$ V $E_{\frac{1}{2}} =$ (XXVI) — 1,56 V   (XXVII) — 1,63** V

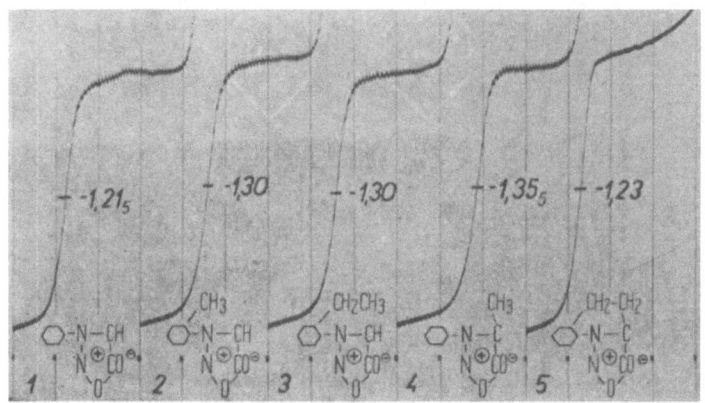

Abb. 14. Sterische Einflüsse auf die Sydnonreduktion.
$5 \cdot 10^{-5}$ M Sydnonlösung in 0,17 M-NaOH. 1. 3-Phenylsydnon; 2. 3-o-Tolylsydnon; 3. 3-o-Äthylphenylsydnon; 4. 3-Phenyl-4-methylsydnon; 5. 3', 4'-Dihydrochinolo [1', 2'-c] sydnon. Von 0,8 V, kath., ges. Kalomelelektrode, 200 mV/Absz., $h = 65$ cm, Empf. 1:10

Die Substitution durch eine Methylgruppe in para-Stellung verursacht eine Verschiebung des Halbstufenpotentials um 25 mV zu negativeren Werten. Nach den Regeln der theoretischen organischen Chemie soll die polare Wirkung einer Gruppe in ortho-Stellung der polaren Wirkung dieser Gruppe in para-Stellung ähnlich sein. An der beobachteten Verschiebung um etwa 90 mV bei der ortho-Substitution durch eine Methylgruppe in XXI muß deswegen die Aufhebung der Komplanarität beteiligt sein. Eine weitere Unterstützung dieser Annahme ist das

** Dieser Wert wurde aus den Werten für 3-Methylsydnon XVII ($-1,49_5$) und aus der Differenz der Werte für 3-Phenylsydnon XII ($1,21_5$) und 3-Phenyl-4-Methylsydnon XVIII (— 1,35) berechnet.

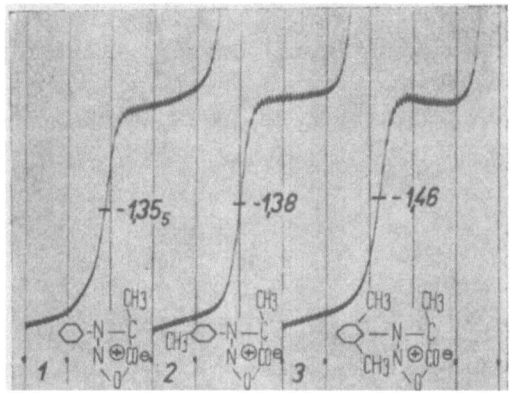

Abb. 15. Sterische Einflüsse auf die Sydnonreduktion.
5 · 10⁻⁵ M Sydnonlösung in 0,17 M-NaOH. 1. 3-Phenyl-4-methylsydnon; 2. 3-p-Tolyl-4-methylsydnon; 3. 3- (2,6-Xylyl)-4-methylsydnon. Von 1,0 V, kath., ges. Kalomelelektrode, 200 mV/Absz., h = 65 cm, Empf. 1:10

Verhalten des o-Äthylphenylsydnons XXII, bei welchem praktisch dieselbe Verschiebung wie bei dem o-Methylderivat XXI beobachtet wurde. Wäre die Potentialverschiebung in erster Reihe durch polare Effekte verursacht, so wäre eine größere Verschiebung bei dem Äthylderivat zu erwarten infolge des größeren induktiven Effektes dieser Gruppe. Dagegen kann bei der Beeinflussung der Komplanarität kein großer Unterschied zwischen der Wirkung der Methylgruppe und der Äthylgruppe erwartet werden. — Substitution durch zwei ortho-

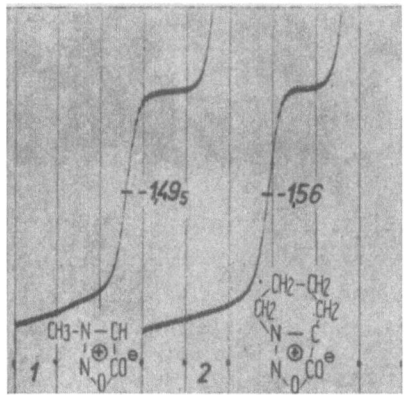

Abb. 16. Einfluß des Ringschlusses auf die Sydnonreduktion.
5 · 10⁻⁵ Sydnonlösung in 0,17 M-NaOH. 1. 3-Methylsydnon; 2. Tetrahydropyrido-[c]-sydnon. Von 1,0 V, kath., ges. Kalomelelektrode, 200 mV/Absz., h = 65 cm, Empf. 1:10

Methylgruppen verursacht eine Verschiebung von 105 mV in XXV. Es scheint, daß die zweite Gruppe nur den polaren Einfluß ausübt.

Durch den Ringschluß in XXIV wird der Phenylkern wieder näher zur Ebene des Sydnonringes herangebracht. Gleichzeitig wird auch die induktive Wirkung der Methylgruppe verringert. Diese beiden Einflüsse verursachen eine Verschiebung des Halbstufenpotentials bei der Verbindung mit anneliertem Ring (XXIV) um 210 mV gegen die positiveren Werte im Vergleich zu der Verbindung mit offener Seitenkette (XXIII).

Ähnlich verursacht der Ringschluß eine teilweise Aufhebung der induktiven Effekte der Methylgruppen im Dihydropyridosydnon XXVI.

Endlich wurden die polaren Einflüsse der Substituenten in para-Stellung am Phenylkern des 3-Phenylsydnons (XXVIII) verfolgt.

Die Halbstufenpotentiale wurden in der Weise verschoben, wie es mit Hilfe der modifizierten HAMMETTschen Gl. [13], [25]

$$(E_{\frac{1}{2}}) - (E_{\frac{1}{2}})_0 = \varrho_\pi \cdot \sigma_i$$

vorausgesagt werden konnte. Die Übereinstimmung mit der angeführten Gleichung ist auf Abbildung 17 sichtbar. Der Wert der Konstante ϱ_π, welche die Beeinflußbarkeit des reduzierbaren Skeletts XVIII durch die Substitution in para-Stellung ausdrückt, ist $\varrho_\pi =$

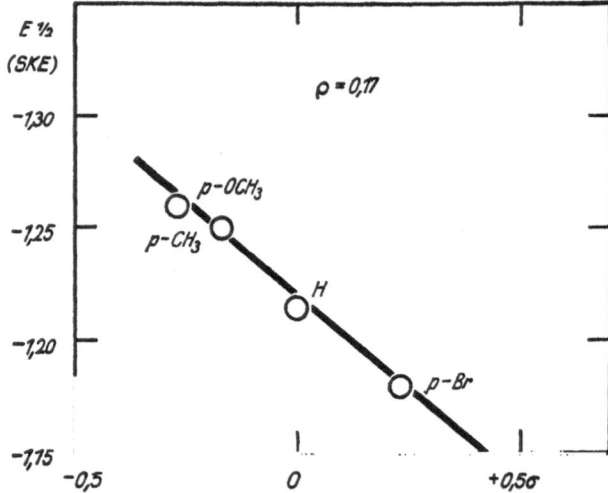

Abb. 17. Die Abhängigkeit der Halbstufenpotentiale der para-substituierten 3-Phenylsydnone in 0,1 M-$NaOH$ von der σ-Konstante nach HAMMETT

0,17. Derselbe Substituent verursacht also bei der Substitution im 3-Phenylsydnon eine größere Verschiebung des Halbstufenpotentials als z. B. bei der Substitution in Azoverbindungen, in Derivaten des Benzophenonsemicarbazons oder -oxims, jedoch eine kleinere als bei der Substitution in Derivaten des Benzophenons, Acetophenons oder Benzaldehyds.

Die Verschiebung durch eine Methylgruppe in para-Stellung ist praktisch dieselbe bei der Substitution in 3-Phenylsydnon wie in 3-Phenyl-4-Methylsydnon.

Durch Ersatz der CH-Gruppe in dem Sydnonring durch ein Stickstoffatom wurde bei der Verbindung XXIX, einem Abkömmling des Oxtriazols, eine, Verschiebung der Reduktion um etwa 0,3 V zu positiveren Potentialen beobachtet. Ein weiteres Studium der Oxatriazole ist im Gange.

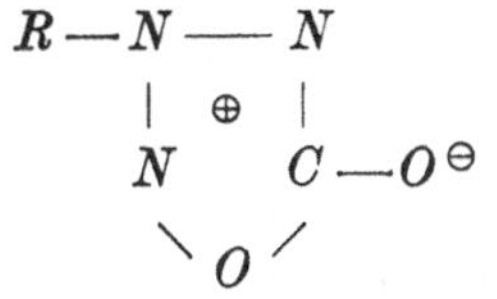

$$
\begin{array}{ll}
R-N-N & \\
\quad |\ \oplus\ | & R = \text{Isopropyl} \\
\quad N \quad C-O^{\ominus} & \quad\quad \text{Cyklohexyl} \\
\quad\ \searrow O \nearrow & \quad\quad \text{p-Bromphenyl}
\end{array}
$$

Schlußfolgerungen

Aus den angegebenen Beispielen konnten die wichtigsten Wirkungen gezeigt werden, durch welche die Konstitution das polarographische Verhalten organischer Stoffe beeinflußt. Die Reduzierbarkeit wird durch Konjugation, durch direkte induktive Effekte, durch polare Wirkungen, welche durch einen Phenylkern übertragen wurden, sowie durch sterische Hinderung der Komplanarität und durch Ringschluß beeinflußt.

Es bleibt noch übrig zu erörtern, welche Beiträge dem Endziel solcher Arbeiten — der näheren Erkenntnis des Zustandes des organischen Moleküls — von der Polarographie gebracht wurden. Bei dieser Gelegenheit ist es notwendig zu bemerken, daß die Polarographie für irreversible Systeme keine Angaben über die Polarisation im Ruhezustand sondern nur Informationen über die Polarisierbarkeit in dem Moment der Reaktion bieten kann.

Durch den Verlauf der polarographischen Reaktion wurde bestätigt, daß nicht die C-O-Bindung oder die C-N-Bindung (wie es aus manchen der benützten Strukturformeln abgeleitet wurde), sondern die N-O-Bindung am leichtesten polarisierbar ist. Der große Einfluß

der Substitution durch einen Phenylkern in Stellung 3 oder durch
eine Methylgruppe in Stellung 4 zeugt dafür, daß sich die Polarisations-
effekte über den Sydnonring leicht bis auf die reduzierte Bindung
übertragen. Ähnlich zeugt von der Beweglichkeit der π-Elektronen
die Gültigkeit der modifizierten HAMMETTschen Gleichung. Die Gültigkeit dieser Gleichung, die bisher für die Reduktion einer Bindung
in α-Stellung zum Phenylkern angewandt wurde, wurde hier für den
Einfluß der Substitution am Phenylkern auf die Reduktion der N-O-
Bindung, welche in β-Stellung zu dem Phenylkern steht, bestätigt.

Alle diese Feststellungen zeugen dafür, daß das π-Elektronen-
system im Sydnonring mobil ist, ähnlich wie bei den anderen aromatischen Verbindungen.

Azulene

Die erste Reduktionsstufe der Azulene genannten Kohlenwasserstoffe wurde von HEILBRONNER [17] in ungepufferten oder alkalischen
Lösungen der Tetraalkylammoniumsalze mit 75% Dioxan im Potentialbereich von $-1,7$ V beobachtet. Diese einelektronige Stufe ist
durch eine oder zwei weitere Stufen bei $-2,4$ V und $-2,6$ V begleitet. In gepufferten Lösungen von pH kleiner als 8 kann keine
dieser Stufen bei einem positiveren Potential als demjenigen der Wasserstoffabscheidung aus der Pufferlösung beobachtet werden.

In wäßrigen Lösungen der starken Säuren mit 10% Dioxan oder
Äthanol wurde von uns eine weitere Reduktionsstufe bei $-0,4$ V
beobachtet, begleitet von einer katalytischen Stufe bei $-1,0$ V
(Abb. 18). Die zweielektronige Stufe bei $-0,4$ V wurde der Reduktion des Azulenkations (XXX)

(XXX)

zugeschrieben. Die Höhe dieser Stufen nimmt bei sinkender Konzentration der starken Säure in der Form einer Dissoziationskurve ab.
Bei kleineren Säurekonzentrationen erfolgt eine Trübung der Lösung
durch den unlöslichen Kohlenwasserstoff.

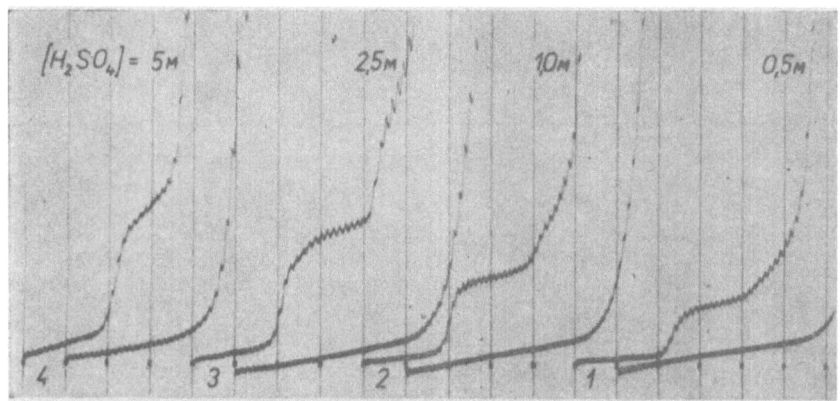

Abb. 18. Abhängigkeit der Guajazulenstufe von der Schwefelsäurekonzentration.
1 · 10⁻⁴ *M* Guajazulen in Schwefelsäurelösung folgender Konzentration: 1. 0,5 *M*; 2. 1,0 *M*;
3. 2,5 *M*; 4. 5 *M*; Von 0,4 V, *Hg*-Boden, kath., 200 mV/Absz., *h* = 65 cm, Empf. 1:15

Für die uns zugänglichen Azulene wurden für den polarographischen Dissoziationsexponenten pK' folgende annähernde Werte erhalten:

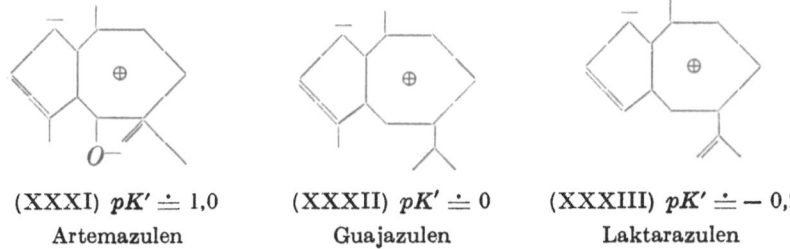

(XXXI) $pK' \doteq 1,0$ (XXXII) $pK' \doteq 0$ (XXXIII) $pK' \doteq - 0,2$
Artemazulen   Guajazulen   Laktarazulen

Die Stufen besitzen kinetischen Charakter; die Form der Dissoziationskurve entspricht auch bei dieser heterogenen Reaktion der von BRDIČKA und KOUTECKÝ [26], [27] abgeleiteten Formel, die Dissoziationskurve verschiebt sich jedoch mit steigender Azulenkonzentration zu kleineren pH- bzw. H_0-Werten.

Um den Einfluß der Inhomogenität zu vermeiden, wurde in Anwesenheit von Lösungsmitteln gearbeitet, die es ermöglichten, auch die nichtionisierten Kohlenwasserstoffe in Lösung zu halten. Zu diesen Zwecken wurden Mischungen von Schwefel- oder Salzsäure mit Eisessig oder noch besser mit wäßrigem Äthanol als geeignet gefunden. Jedoch auch in diesen Lösungen wurden Verschiebungen des pK'-Wertes der Dissoziationskurve (Abb. 19) mit der Azulenkonzentration beobachtet.

474

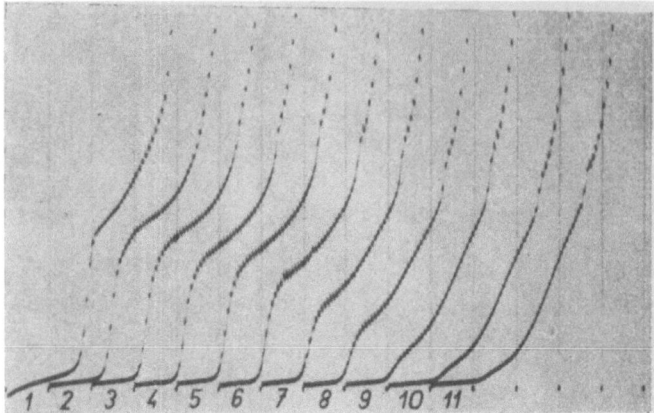

Abb. 19. Abhängigkeit der Guajazulenstufen von der Schwefelsäurekonzentration in Anwesenheit von 30 % Äthanol.

$2 \cdot 10^{-4}$ M Guajazulen in Wasser-Äthanol-Gemisch mit 30 % Äthanol mit folgenden Schwefelsäurekonzentrationen: 1. 17,8 %; 2. 8,9 %; 3. 4,45 %; 4. 3,55 %; 5. 2,67 %; 6. 1,78 %; 7. 0,89 %; 8. 0,45 %; 9. 0,18 %; 10. 0,089 %; 11. 0,044 %. Von 0,6 V, kath., Graphitelektrode, 200 mV/Absz., $h = 45$ cm, Empf. 1:5

Die Verschiebung ist am wahrscheinlichsten durch eine der Protonierung vorgelagerte Reaktion in der Lösung verursacht, die mit Hilfe der Absorptionsspektroskopie untersucht werden soll. Im Einklang mit der Verschiebung des pK'-Wertes wurde bei kleineren Säurekonzentrationen ein nichtlinearer Verlauf der Abhängigkeit der Stufenhöhe von der Azulenkonzentration beobachtet. Bei höheren Azulenkonzentrationen bildet sich eine Adsorptionsvorstufe (Abb. 20).

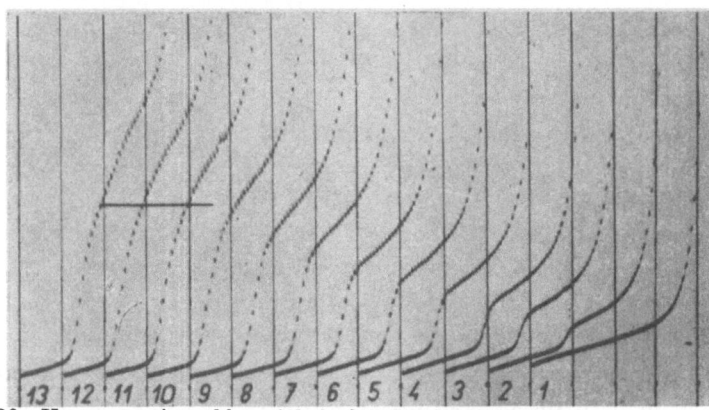

Abb. 20. Konzentrationsabhängigkeit der Guajazulenstufe.

2 % H_2SO_4, 30 % Äthanol, Konzentration des Guajazulens: 1. 0; 2. 0,12; 3. 0,24; 4. 0,36; 5. 0,476; 6. 0,59; 7. 0,70; 8. 0,91; 9. 1,11; 10. 1,30; 11. 1,49; 12. 1,67; 13. $1,84 \cdot 10^{-4}$ M. Von 0,6 V, kath., Graphitelektrode, 200 mV/Absz., $h = 45$ cm, Empf. 1:2

Die große Differenz zwischem dem Reduktionspotential des protonisierten Azulens und der freien Base, die etwa 1,3 V beträgt, ist durch den Verlust der Symmetrie des Moleküls bzw. durch die Abnahme des aromatischen Charakters verursacht. Dieser Unterschied ist viel größer als die bisher gefundenen Unterschiede zwischen dem Halbstufenpotential einer Säure und der korrespondierenden Base.

Die beobachtete Reduktionsstufe ist ein wichtiger Beweis der Existenz der protonierten Azulene, welche bisher nur spektroskopisch [28], [29], [30] nachgewiesen wurden. Es ist die positivste bisher beobachtete Reduktion eines Kohlenwasserstoffes und der erste Fall der Reduktion eines Kations eines Kohlenwasserstoffes. Es ist auch der erste unstrittige Beweis der Reduktion einer C-Säure, die bereits bei manchen Aldehyden und Ketonen, Anthocyanidinen [31], Isoflavonen [32], Δ^4-3-Ketosteroiden [33] sowie bei der Phthalsäure [34] vorausgesetzt wurde.

Durch die Bildung eines Kations kann die Erhöhung der Stufenhöhe des Laktorviolins sowie einiger Azulenketone [24] in sauren Lösungen erklärt werden. Auch der Einfluß der Jodsäure auf die Stufen mancher Kohlenwasserstoffe [35] kann durch Säurebildung verursacht sein.

Die Polarographie bietet also die Möglichkeit, die Säure-Basen-Eigenschaften auch in solchen extremen Beispielen zu verfolgen, wie es bei den Azulenen der Fall ist.

Für die Zurverfügungstellung der Präparate der Sydnone möchte ich Herrn Prof. Dr. J. C. Earl, Prof. Dr. I. M. Hunsberger und Dr. D. J. Voaden, der Oxatriazole Herrn Prof. Dr. J. H. Boyer, der Azulene Herrn Dr. V. Herout, herzlichst danken. Mein Dank gilt weiter Herrn Dr. J. Peizker für die Hilfe bei den Elektrolysen mit dem Potentiostaten sowie Herrn Dr. P. Valenta für die Ermöglichung der impulspolarographischen Aufnahmen.

Schrifttum

[1] Baker, W., The Developments of the Concept of Aromaticity in Prospectives in Organic Chemistry, herausgegeben von A. Todd. Interscience Publishers 1956, S. 28.

[2] Baker, W., W. D. Ollis und V. D. Poole, J. Chem. Soc. (1949) 307; (1950) 1542.

[3] Baker, W., und S. W. D. Ollis, Chem. Ind. (1955) 910.

[4] Bieber, T. I., Chem. Ind. (1955) 1055.

[5] Jennen, J., Chimie Industrie 67 (1952) 356.

476

[6] ŠANTAVÝ, F. Polarographisches Kolloquium 1958, 210.

[7] PEIZKER, J., Kandidatenarbeit, Prag 1957.

[8] KALOUSEK, M., und M. RALEK, Chem. listy 48 (1954) 808; Collection Czechoslov. Chem. Commun. 19 (1954) 1099.

[9] VALENTA, P., Polarographisches Kolloquium 1958, 46.

[10] HEYROVSKY, J., Anal. Chim. Acta 12 (1955) 600.

[11] EARL, J. C., Recueil Trav. Chim. Pays Bas 75 (1956) 346.

[12] ZAHRADNIK, R., Privatmitteilung.

[13] ZUMAN, P., Chem. listy 48 (1954) 94.

[14] ZUMAN, P., Collection Czechoslov. Chem. Commun. 15 (1950) 1107.

[15] RÖSSLER, W., Privatmitteilung.

[16] LAITINEN, H. A., und S. WAWZONEK, J. Am. Chem. Soc. 64 (1942) 1765, 365.

[17] CHOPARD-DIT-JEAN, L. H., und E. HEILBRONNER, Helv. Chim. Acta 36 (1953) 144.

[18] GOULDEN, F., und F. L. WARREN, Biochem. J. 42 (1948) 420.

[19] PATZAK, R., und L. NEUGEBAUER, Monatsh. Chem. 83 (1952) 776.

[20] FIELDS, M., C. VALLE und M. KANE, J. Am. Chem. Soc. 71 (1949) 421.

[21] PREVOST, CH., P. SOUCHAY und CH. MALLEN, Bull. Soc. Chim. Fr. (1953) 18.

[22] BENT, R. L., und Mitarbeiter, J. Am. Chem. Soc. 73 (1951) 3100.

[23] GOWARD, G. W., C. E. BRICKER und W. C. WILDMAN, J. Organ. Chem. 20 (1955) 378.

[24] GERDIL, R., und E. HEILBRONNER, Helv. Chim. Acta 40 (1957) 141.

[25] ZUMAN, P., Chem. listy 47 (1953) 1234; Collection Czechoslov. Chem. Commun. 19 (1954) 599; Chem. Zvesti 8 (1955) 939.

[26] KOUTECKY, J., und R. BRDIČKA, Collection Czechoslov. Chem. Commun. 12 (1947) 237.

[27] KOUTECKY, J., Chem. listy 47 (1953) 9; Collection Czechoslov. Chem. Commun. 18 (1953) 311.

[28] PLATTNER, PL. A., E. HEILBRONNER und S. WEIBER, Helv. Chim. Acta 35 (1952) 1036.

[29] HEILBRONNER, E., und M. SIMONETTA, Helv. Chim. Acta 35 (1952) 1049.

[30] CHOPARD-DIT-JEAN, L. H., und E. HEILBRONNER, Helv. Chim. Acta 35 (1952) 2170.

[31] ZUMAN, P., Chem. listy 46 (1952) 328; Collection Czechoslov. Chem. Commun. 18 (1953) 36.

[32] VOLKE, J., und V. SZABO, Chem. listy 50 (1956) 1095; Collection Czechoslov. Chem. Commun.

[33] ZUMAN, P., J. TENYGL und M. BŘEZINA, Chem. listy 47 (1953) 1152; Collection Czechoslov. Chem. Commun. 19 (1954) 46.

[34] RYVOLOVÁ, A., und V. HANUS, Chem. listy 50 (1956) 46; Collection Czechoslov. Chem. Commun.

[35] HOIJTINK, G. J., J. VAN SCHOOTEN, E. DE BOER und W. J. AALBERSBERG, Recueil Trav. Chim. Pays-Bas 73 (1954) 355.

POLAROGRAPHY OF NONBENZENOID AROMATIC AND RELATED SUBSTANCES. II.*
THE COURSE OF THE REDUCTION OF SYDNONES AT THE DROPPING MERCURY ELECTRODE

P. Zuman

Polarographic Institute, Czechoslovak Academy of Science, Prague

Received June 4th, 1960

To Academician J. Heyrovský on his 70th birthday.

Sydnones are reduced at the dropping mercury electrode at higher pH-values by a pH-independent four electron step. The formation of a hydrazine derivative in this step has been proved by a commutator method, controlled potential electrolysis, impulse polarography and oscillographic $dE/dt - E$-curves. The protonized form in acid solutions is reduced under uptake of six electrons. Ammonium ions were detected in the reaction mitutre. pH-Independent half-wave potentials obtained at pH greater than about 8 are most suitable for structural studies.

Attention recently has been paid to the chemistry of five- and seven-membered ring compounds in which the sextet of π-electrons is secured by a transfer from or to an exocyclic atom. Sydnones may be included among these compounds. These five-membered ring compounds are planar and show in some respect the behaviour of aromatic systems[2,3]. They may be classified to a group of "mesoionic compounds" that cannot be simply described by any classical covalent formula. The most common formula[2,3] (*I*) will be used in this paper.

$$R^1\!-\!N_3\!-\!_4C\!-\!R^2$$

I

In this series of papers the sydnones, their derivatives and analogues were chosen as model substances for the demonstration of structural effects on polarographic curves. Especially, the effect of substituents on half-wave potentials is discussed in detail. The reasons for the choice of sydnones as model substances can be summarized as follows: The reduction waves for sydnones are well developed and enable a precise measurement of wave-heights and of half-wave potentials. A comparison of the structural effects is further simplified by the fact that at pH values higher than about pH 7·5 the half-wave potential is independent of pH. This is preferable for the discussion of structural effects. Furthermore, the course of the reduction and the value of the transfer coefficient remained the same for all of the over thirty derivatives studied. Most of the compounds are well defined, crystalline compounds, soluble in water or in the presence of small amounts of ethanol. In spite of the apparent

---

* Part I.: Z. physik. Chem., Sonderheft *1958*, 243.

complexity, the substances are relatively simple with restricted possibilities of substitution in two positions only — one on nitrogen in position 3, one on carbon in position 4.

Having studied the course of the reduction process described in this paper, the effect of substitution with alkyl and aryl groups in positions 3 and 4 will be treated using the modified Taft equation[4]. The polar and mesomeric effects of the phenyl group were isolated. The effect of substituents on phenyl groups in positions 3 and 4, treated by the modified Hammett equation, represent another argument for the mobility of π-electrons of the sydnone ring. In a further paper the course of the reduction of N,N'-polymethylene-bis-sydnones was followed together with a study of mutual interactions of two sydnone rings, separated by a saturated aliphatic chain of different length.

Experimental

Usual polarographic circuitry was used with a Heyrovský-type polarographic apparatus with photographic registration and a Kalousek vessel with a separated calomel electrode. The galvanometer had a sensitivity of $2 \cdot 4 \cdot 10^{-9}$ A/mm. The dropping electrode used in most experiments, had the following constants: $m = 2 \cdot 3$ mg/s, $t = 3 \cdot 7$ s at $h = 63$ cm. In some experiments controlled drop time was used.

For controlled potential measurements with a dropping mercury electrode the potentiostat constructed by Peizker[5] was employed. Polarization by reactangular voltage was performed using a electronic commutator device developed by Rálek and Kalousek[6]. For measurements of polarization curves using a triangular voltage the single-sweep apparatus constructed by Valenta[7] has been applied. Measurements of the oscillographic $dE/dt-E$-curves were carried out using the apparatus Polaroskop produced by Křižík (Prague). pH-values were measured on a pH-meter Radiometer Mark 21d.

$$R^1{-}N{-}$$

II, $R^1 = CH_3$
III, $R^1 = CH_2C_6H_5$
IV, $R^1 = C_6H_5$

As representative types following substances were chosen: 3-Methyl sydnone (II), prepared by D. J. Voaden (Oxford, England), 3-benzyl sydnone (III) prepared by I. M. Hunsberger (Fordham University, New York, USA) and 3-phenyl sydnone (IV) prepared by J. C. Earl, (Haddiscoe, England), 0,01M stock solutions of these substances were prepared by dissolving the crystalline product in 50% aqueous ethanol. Buffer solutions were prepared from chemicals of reagent grade. The final concentration of ethanol in solutions polarographed was 1%.

Results

In acid solutions of 3-*aryl sydnones* a six-electron wave was observed (Fig. 1) the height of which decreased above pH about 3 in a form of the dissociation curve. At pH about 6 the height of the wave reaches the value for a four-electron process. With further increase in pH the limiting current remains unchanged. The number of electrons transferred was demonstrated by comparison with waves of p-nitrotoluene, N-nitrosodiphenylamine and 2-nitroso-1-naphthol.

The six-electron-wave in acid solution was attributed to the reduction of the protonized form, the four-electron wave was attributed to the reduction of the unprotonized form. Separation of the waves corresponding to these two processes was not observed. But the change in the slope of the curve between pH 3 and 7 shows that the current of the six-electron process decreases to zero and that the current of the four-electron process rises at the cost of the six-electron wave.

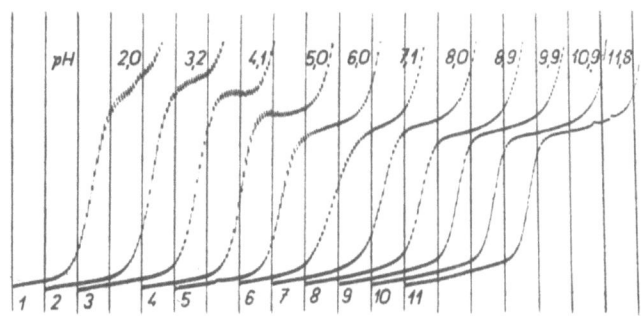

Fig. 1

pH-Dependence of Waves for 3-Phenyl Sydnone in Buffers according to Britton and Robinson. $5 . 10^{-5}$M 3-phenyl sydnone. pH-values are given on the polarogram. $1-3$ starting at -0.2 V, $4-5$ at -0.4 V, $6-11$ at -0.6 V, S.C.E., 200 mV/absc., $h = 57$ cm, sens. 1 : 7.

The six-electron-step takes place with the consumption of protons whereas in the four-electron-step no protons are transferred before the potential determining step. This was demonstrated by the shift of the half-wave potential of the six-electron wave with pH and by the independence of the half-wave potential of the four-electron-step on pH-values (Fig. 2).

The half-wave potential of the four-electron-wave is little influenced both by the cation and anion of the supporting electrolyte. The half-wave potential of this wave is practically independent of the sydnone concentration; with increasing droptime a small shift towards more positive potentials was observed.

The four-electron wave is diffusion controlled. This was proved by linear dependence of the wave height on concentration of the depolarizer, on the square root of mercury head and on one sixth power of the drop-time t_1. Similarly diffusion control was proved for the six-electron-wave in strongly acid solutions. At the intermediate pH the increase of current above the four-electron value was independent of mercury pressure and thus limited by the rate of a chemical reaction.

The slope of the log-plot of the four-electron-wave was found to be be-

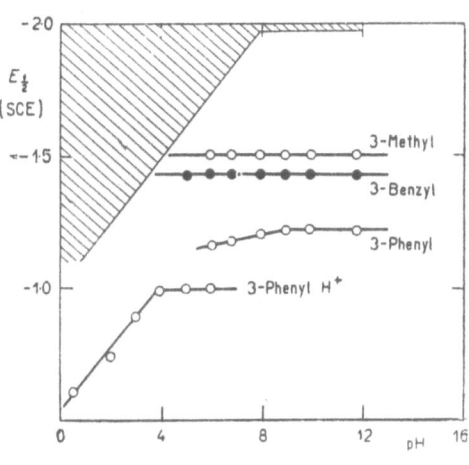

Fig. 2

pH-Dependence of the Half-Wave Potentials of Sydnones, Substituted in Position 3 For 3-phenyl derivatives potentials for protonized and unprotonized form are given.

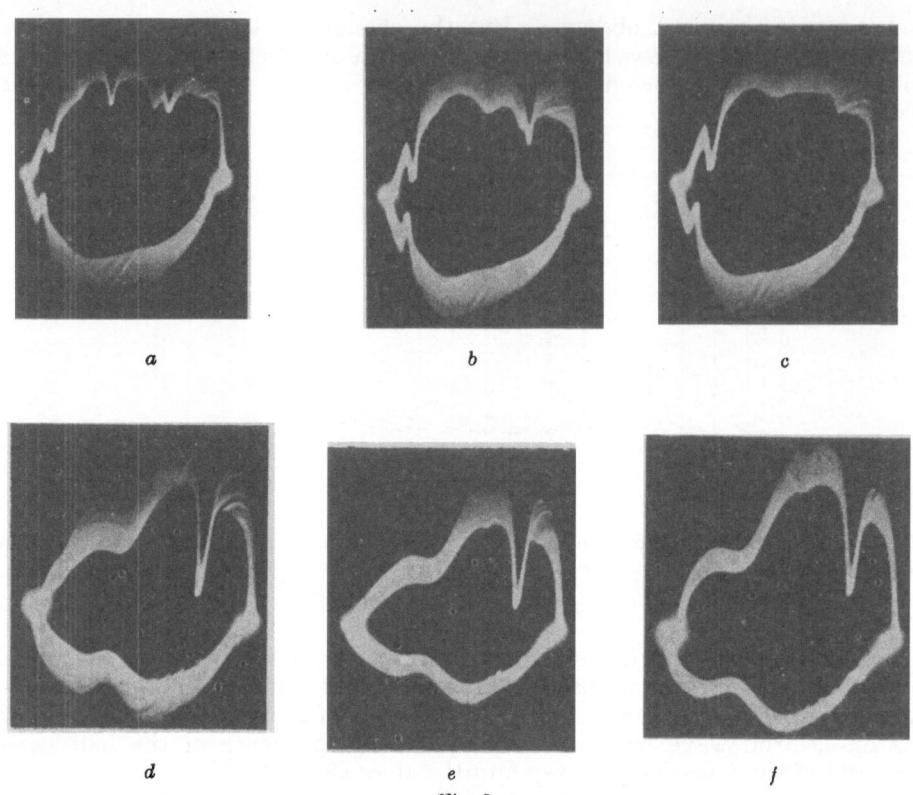

Fig. 3

Oscillographic $dE/dt - E$-Curves of Sydnone Derivatives

1. 25 . 10^{-4}M sydnone in *A* 0·1M-NaOH, *B* 0·5M-H_2SO_4. *a* phenylsydnone, *b* 3′,4′-dihydro-quinolo[1′,2′-c]sydnone, *c* 3-phenyl-4-methylsydnone. Polaroskop, 0·2 mA, amplif. 1, s.c.e.

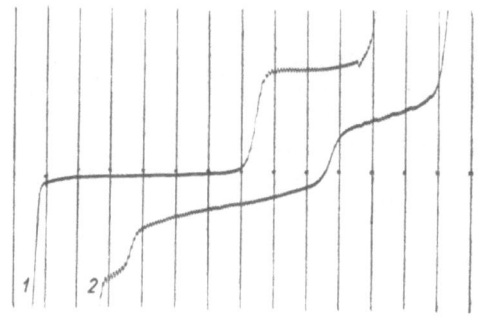

Fig. 4

Sydnone Reduction Using the Commutator Method by Kalousek

2·5 . 10^{-4}M 3-phenylsydnone, borate buffer pH 9·3 with 5% ethanol: *1* primitive curve; *2* commutated curve, auxiliary potential corresponds to the limiting current. Frequency of commutation: 3·5 c/s. Curves starting at +0·2 V, s.c.e., 200 mV/absc., $h = 65$ cm, sens. 1:50.

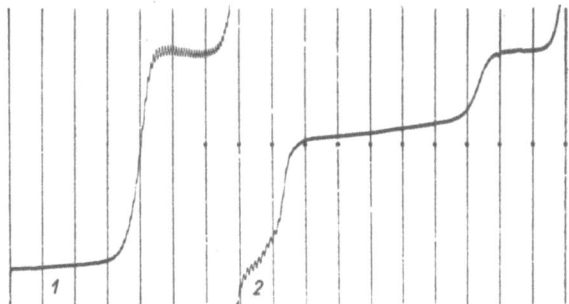

Fig. 5

Controlled Potential Electrolysis of Sydnones

0,005M benzylsydnone, in a borate buffer pH 9·3, was electrolysed using a dropping mercury electrode with potential controlled by a potentiostat by Peizker[5]. The electrolysis was performed for 30 hours. The resulting solution was diluted twentyfold using a borate buffer. 1 before electrolysis; 2 after electrolysis. 200 mV/absc., sens. 1 : 20.

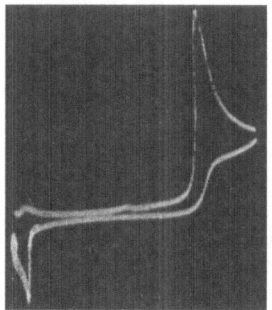

Fig. 6

Oscillographic i–E-Curves of Sydnone in Alkaline Solutions using the Technique of Impulse Polarography

1·25 . 10^{-4}M 3-phenylsydnone in 0·05M borate buffer pH 9·3. Apparatus by Valenta[7], $t_1 = 22·7$ s, one triangular impulse, the slope of the impulse 0·63 V/s, $i_{max} = 1·59$ A. Starting voltage −0·04 V (against Hg-pool), final voltage −1·90 V.

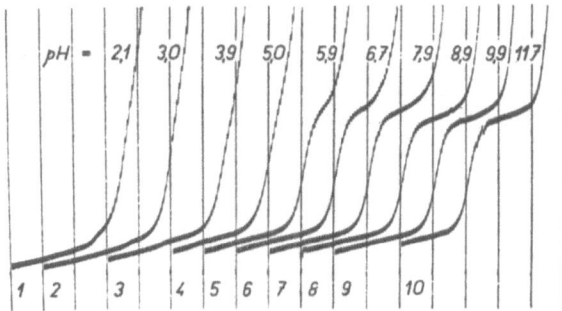

Fig. 7

pH-Dependence of Waves for 3-Benzylsydnone in Buffers According to Britton and Robinson

5 . 10^{-5}M 3-phenylsydnone. pH-values are given on the polarogram. 1 and 2 starting at −0·4 V, 3 at −0·6 V; 4—9 at −0·8 V; 10 at −1·0 V, S. C. E., 200 mV/absc., $h = 57$ cm, sens. 1 : 7.

tween 60 and 65 mV. The wave is thus considerably less steep than would correspond to a reversible four-electron process. The irreversibility of the process has been further proved by the rectangular voltage polarization and by the single sweep method, by which no anodic process at the potentials corresponding to the reduction wave could be observed. Similarly on the $dE/dt - E$ curves registered on the oscilloscope (Fig. 3) no anodic incision was observed in the potential range of the cathodic process.

Reduction products formed in the four-electron step were identified using methods mentioned in the preceding paragraph.

Thus using the commutator method of Kalousek, in which the auxiliary potential corresponded to the limiting current of the four-electron wave, an increase in an anodic wave at positive potentials was observed (Fig. 4). The wave was identified as a wave for a hydrazine derivative (corresponding to a formation of a mercury salt) by comparison with other hydrazine derivatives at several pH-values. Similar results were obtained using a controlled potential electrolysis at the potential of the limiting current of the four-electron-wave using a dropping mercury electrode. The polarographic curve of the eletrolysis product shows again the anodic wave at positive potentials (Fig. 5). The formation of a hydrazine derivative was also proved in the electrolysed solution by potentiometric acidimetric titration. In sydnone solutions no dissociation step could be found between pH 2 and 12. In the solution after electrolysis a species with the acid dissociation constant about pK 9·8 (in agreement with other hydrazines) was detected.

The formation of hydrazines was further proved by the single sweep method[7]. On the curve obtained with triangular voltage sweep, in addition to the cathodic peak (corresponding to the reduction of sydnone), there was observed an anodic peak corresponding to the formation of a mercury salt with the hydrazine derivatives formed in the reduction process at more negative potentials (Fig. 6). Finally, on the oscillographic $dE/dt - E$ curves in alkaline media (Fig. 3. $Aa-Ac$) a reversible incision at positive potentials is observed. This incision corresponds to the redox-system of mercury bound with the hydrazine derivative formed at negative potentials.

The product of the six-electron-wave in acid solutions is inactive polarographically. No anodic process could be detected either by the commutator method, or by controlled potential electrolysis or by impulse polarography, or on the $dE/dt-E$ curves (Fig. 3$Ba-Bc$). Ammonium ions could be detected in the reaction mixture after controlled potential electrolysis.

3-*Alkylsydnones* are reduced at a more negative potential. The six-electron wave is overlapped by hydrogen evolution (Fig. 2) and only the four-electron wave at higher pH-values is observed (Fig. 7). The half-wave potential of the four-electron wave is pH-independent (Fig. 2). The formation of a hydrazine derivative has been proved using all the auxiliary methods mentioned above.

Discussion

In solutions of sydnones two forms are present in acid-base equilibria. The conjugated acid form (*V*) is reduced in acid solutions. The formation of ammonium ions could be proved among the products of electrolysis. This form

(V) is in equilibrium with the unprotonized form (I), reduced by uptake of four electrons to a hydrazine derivative (VI). The presence of a hydrazine grouping in the reduction product at higher pH-values has been proved by all auxiliary methods.

Reduction products in acid solution are most likely identical with products obtained by catalytic hydrogenation[8,9] and with those isolated after reduction by zinc in a medium of diluted acetic acid[10].

For the reduction of sydnones the following scheme is proposed:

The equilibrium (A) is disturbed by the electrolytic process and restored by recombination. Due to the fact that the corresponding equilibrium constant pK_A is inaccessible potentiometrically and that in strongly acid solutions sydnones are decomposed, quantitative treatment has been abandoned. It seems to be probable that the protonized form is decomposed (as with N-nitrosoamines[11]) so that the values of dissociation constants could be determined using kinetic studies of homogeneous reactions only.

Concerning the mechanism of the electrode reaction proper in the four-electron reduction of the particle I a hypothesis can be formulated based on the slope of the wave and the independence of the half-wave potential of this wave on the pH-value. We can assume that the uptake of the first electron will be the potential determining step. This reaction is followed by the uptake of further electrons and protons in steps exerting a influence on potential. In accordance with a mechanism proceeding in several steps are the several incissions observed on the $dE/dt - E$-curves (Fig. 3).

The observed course of reduction shows that the most reactive bond for a nucleophilic attack by electrons is the N—O—bond. This is not clearly seen from the formula I. On the other hand the strongly polar character of the C—O$^{(-)}$ bond suggested by formula I could not been proved polarographically. In agreement with this deduction the IR-spectra of sydnones show a carbonyl frequency more similar to that of no lactone carbonyl than to that of a strongly polarized carbonyl group[12]. Thus the best presentation of the formula for sydnones remains, in our opinion, an open question.

Finally it should be stressed that as a result of the present investigation optimum conditions for structural comparisons were determined. Values of half-wave potentials at pH greater than about 8 that are pH-independent were selected as most suitable for these purposes.

The author expresses his thanks to Drs J. C. Earl, I. M. Hunsberger and D. J. Voaden, who kindly supported him by the gifts of samples of sydnones. The careful technical assistance of Miss N. Pejchová is acknowledged, as well as the help of Drs J. Peizker and P. Valenta in the experiments with controlled potential electrolysis and impulse polarography respectively.

References

1. Zuman P. Z.: Z. physik. Chem., Sonderheft *1958*, 243.
2. Baker W., Ollis W. D.: Quart. Revs *11*, 15 (1957), where the older literature is given.
3. Zuman P.: Chem. listy *53*, 1029 (1959).
4. Zuman P.: This Journal *25*, 3225 (1960) where further references are given.
5. Peizker J.: This Journal, in the press.
6. Kalousek M., Rálek M.: This Journal *19*, 1099 (1954); Chem. listy *48*, 808 (1954).
7. Valenta P.: Z. physik. Chem., Sonderheft *1958*, 46.
8. Baker W., Ollis W. D., Poole V. D.: J. Chem. Soc. *1949*, 307.
9. Earl J. C., Lefévre R. J. W., Wilson J. R.: J. Chem. Soc. *1949*, S 103.
10. Earl J. C.: Rec. trav. chim. *75*, 937 (1956).
11. Zahradník R.: This Journal *23*, 1529 (1958); Chem. listy *51*, 937 (1957)
12. Exner O.: Private communication.

Translated by the Author.

POLAROGRAPHY OF NONBENZENOID AROMATIC AND RELATED SUBSTANCES. III.*
THE COURSE OF THE REDUCTION OF N,N'-POLYMETHYLENE-BIS-SYDNONES

P. ZUMAN

Polarographic Institute, Czechoslovak Academy of Science, Prague

Received June 4th, 1960

To Academician J. Heyrovský on his 70th birthday.

Protonized forms of N,N'-polymethylene-bis-sydnones (V) are reduced principally in one twelve-electron wave in acid solutions. At higher pH-values four electron reduction of the two sydnone rings, in compound VII, in two successive steps were proved. Products of these reductions are hydrazine derivatives VIII and IX. The difference of the half-wave potentials of the four-electron waves is a measure of the mutual interaction of the two sydnone rings. The reduction of the first ring is shifted to a more positive potential under influence of the second ring. The half-wave potential of the reduction of the second sydnone ring is identical with that of 3-alkyl sydnones. The effect of the hydrazine part is thus screened by the polymethylene chain. The mutual interaction of the two sydnone rings was corrected for the statistical factor and explained on the basis of dynamic, polar, and orientation factors. Substitution of methyl and chloro groups on carbon in the sydnone ring does not change the reduction mechanism and cause only shifts of half-wave potentials. On the other hand in C,C'-dibromo-N,N'-polymethylene-bis-sydnones a reduction of the C—Br bond precedes the reduction of the sydnone ring.

N,N'-Polymethylene-bis-sydnones ($Ia-Ic$), prepared by Daeniker and Druey[2], which show an anticancer activity *in vitro*, give[2] UV- and IR-spectra, where little dependence on the number of methylene groupings n in I could be detected.

Ia, n = 2, R^1 = H	III, n = 2, R^1 = Cl
Ib, n = 4, R^1 = H	IVa, n = 2, R^1 = Br
Ic, n = 6, R^1 = H	IVc, n = 6, R^1 = Br
II, n = 2, R^1 = CH_3	

To show whether there is a difference in the dynamic properties of these substances their polarographic behaviour was studied. To allow discussion of the length of the polymethylene chain as well as of the influence of the

---

* Part II: This Journal *25*, 3245 (1960).

exchange of R^1 (Part V of this series), the course of the polarographic reduction of N,N'-polymethylene-bis-sydnones had to be distinguished first. The present case represents an example that has not been studied extensively in polarography, namely, the concurrence and mutual interaction of two identical groups on a saturated chain.

Experimental

The polarographic equipment, the capillary and the auxiliary apparatus were the same as in the preceding paper[1]. N,N'-Ethylene-bis-sydnone (*Ia*) (m. p. 169°, destr.), N,N'-tetramethylene-bis-sydnone (*Ib*) (m. p. 168—170°, destr.), N,N'-hexamethylene-bis-sydnone (*Ic*) (m. p. 114°, destr.), C,C'-dimethyl-N,N'-ethylene-bis-sydnone (*II*) (m. p. 182—3°, destr.), C,C'-dichloro-N,N'-ethylene-bis-sydnone (*III*) (destr.), C,C'-dibromo-N,N'-ethylene-bis-sydnone (*IVa*) (m. p. 163°, destr.) and C,C'-dibromo-N,N'-hexamethylene-bis-sydnone (*IVc*) (m. p. 120—122°, destr.) were synthetized by Daeniker and Druey[2] of CIBA (Basel, Switzerland). The 0·001 molar stock solutions of these substances were prepared by dissolving substances *I* and *II* in water, and substances *III* and *IV* in ethanol.

The Britton-Robinson, phosphate and borate buffers used were prepared from reagent grade chemicals.

Results and Discussion

The Course of the Reduction of N,N'-Ethylene-bis-sydnone (*Ia*).

The reduction of ethylene-bis-sydnone (*Ia*) proceeds in an acid solution in one wave which at sufficiently low pH-values achieves the height corresponding to a twelve-electron reduction. With increasing pH-value the height of this wave decreases in the form of a dissociation curve, until at pH about 6 it reaches the height corresponding to a transfer of eight electrons (Fig. 1). Above this pH-value the total height does not change until pH 10. At higher pH-values decomposition occurs, accompanied by changes in the polarographic curves with time. The number of electrons consumed was determined by comparison with 3-phenyl- and 3-benzylsydnone.

At pH greater than about 7 the eight-electron wave splits into two four electron waves.

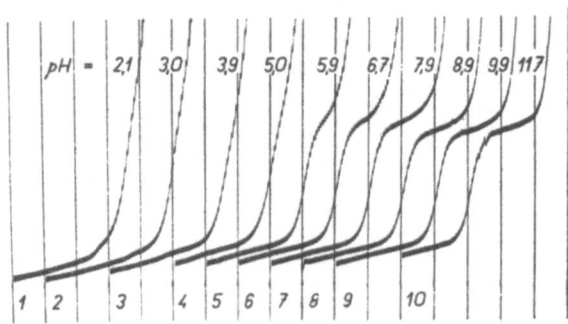

Fig. 1

Influence of pH on the Waves of N,N'-Ethylene-bis-sydnone

1 . 10⁻⁴M Depolarizer in Britton-Robinson buffers. pH-values are given on the polarogram. Curves *1—3* starting at −0·6 V, *4—11* at −0·8 V, s. c. e., 200 mV/absc., $h = 55$ cm, sens. 1 : 10.

The half-wave potential of the twelve-electron wave is shifted to more negative values with the increasing pH-value of the buffer solution (Fig. 2). The half-wave potential of the first four-electron wave at higher pH-values is pH-independent, similarly to the half-wave potentials of the reduction waves of the unprotonized form of simple 3-aryl- and 3-alkylsydnones.

For the plot of pH-dependence of the half-wave potential (for the more negative four-electron wave) two linear parts are observed (Fig. 2): At a pH lower than about 9·3 the straight part has a slope of about 60 mV/pH. At a pH higher than about 9·3 the half-wave potential was found independent of pH.

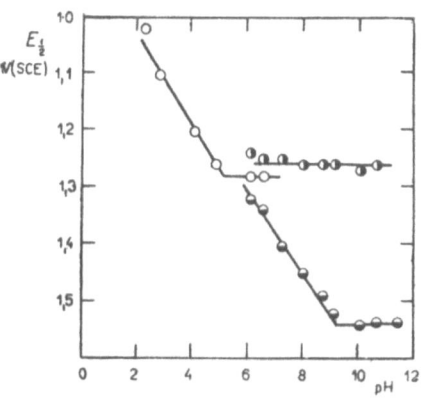

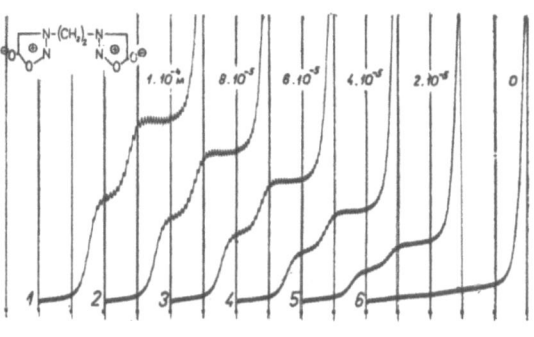

Fig. 2

Dependence of the Half-Wave Potentials of N,N′-Ethylene-bis-sydnone on pH in Britton-Robinson Buffers

1 . 10⁻⁴M depolarizer; ○ the twelve-electron wave, ◑ ◐ the two four-electron steps.

Fig. 3

Influence of Concentration on Waves of N,N′-Ethylene-bis-sydnone

Borate Buffer pH 9·3, 0·01% gelatine. Concentration of the depolarizer is given on the polarogram. Curves starting at −1·0 V, 200 mV/absc., $h = 55$ cm, sens. 1 : 20.

It was deduced from the dependence of the wave-heights on the mercury pressure that the total limiting current is governed by diffusion. Only at pH-values where the current is a little higher than the current, corresponding to the eight-electron reduction (i. e. at pH about 5·5), does the part of the current corresponding to the increase above the eight-electron process behave like a kinetic current. All the waves are a linear function of the concentration and the ratio of the two four-electron waves does not change with increasing concentration of the depolarizer (Fig. 3). As for simple sydnones[1] the irreversibility of all processes could be demonstrated. The formation of hydrazine-derivatives in both the four-electron steps at higher pH was proved using the polarization by a periodically applied rectangular voltage and from the oscillographic $dE/dt - E$ curves (cf. ref.[1]).

Based on the analogy to the behaviour of simple sydnones[1] and the results reported above the following reduction scheme is suggested (p. 3255).

The twelve-electron wave in acid solutions corresponds to the reduction of the protonized form V to the amine-derivative VII, with the formation of ammonium ions. The more positive four-electron wave in the medium pH range corresponds to the reduction of the N—O bond in the first sydnone-ring in the

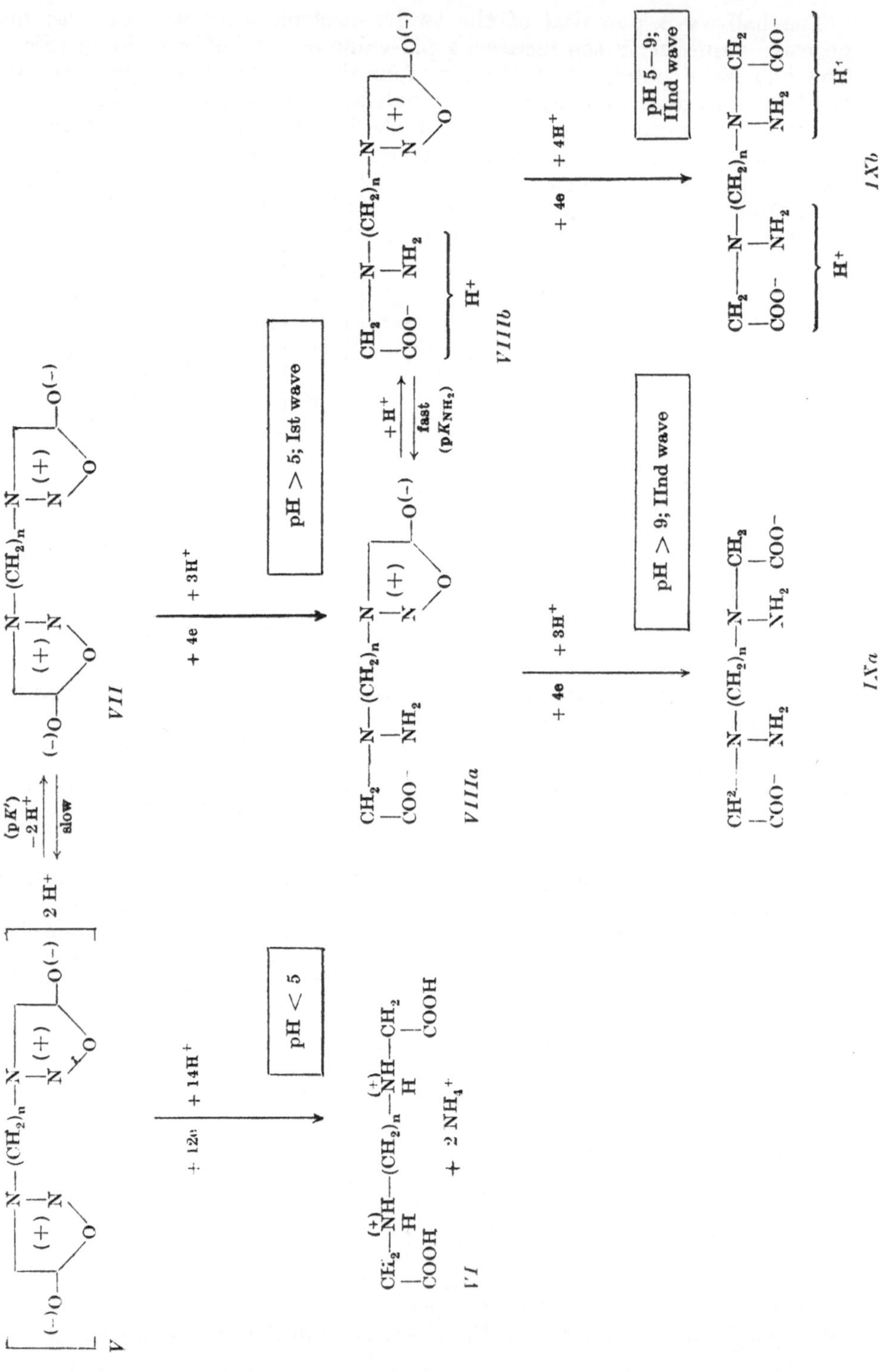

$$\text{VII}$$

$$+ 4e \quad + 3H^+$$

$$\boxed{\text{pH} > 5; \text{ Ist wave}}$$

$$\text{VIIIa}$$

$$\xrightarrow[\text{fast}]{+H^+ \atop (pK_{NH_2})}$$

$$\text{VIIIb}$$

$$+ 4e \quad + 3H^+$$

$$\boxed{\text{pH} > 9; \text{ IInd wave}}$$

$$+ 4e \quad + 4H^+$$

$$\boxed{\text{pH } 5-9; \text{ IInd wave}}$$

$$IXa$$

$$IXb$$

$$A$$

$$2H^+ \quad \xrightarrow[\text{slow}]{(pK') \atop -2H^+}$$

$$+ 12e \quad + 14H^+$$

$$\boxed{\text{pH} < 5}$$

$$\text{CH}_2\!-\!\overset{(+)}{N}H\!-\!(CH_2)_n\!-\!\overset{(+)}{N}H\!-\!CH_2$$
$$\text{COOH} \quad H \quad H \quad \text{COOH}$$
$$+ 2\,NH_4^+$$

$$VI$$

unprotonized sydnone *VII*. By this reduction a hydrazine derivative *VIII* is formed. This hydrazine is in a protolytic equilibrium (*VIIIa—VIIIb*). The more negative four-electron wave at pH between 6 and 9 corresponds to the reduction of the protonized form of hydrazine derivative *VIIIb*. The more negative four electron wave in the pH region where its half-wave potential is pH-independent (i. e. at pH-values higher than about 9) corresponds to the reduction of the N−O bond in the unprotonized hydrazine derivative *VIIIa*. The reduction product at the potentials of the more negative wave is a polymethylene-bis-hydrazine derivative (*IX*). In the middle range of pH between pH 5 and 7 the half-wave potentials of the two four-electron steps are so little different that on the curve 5 (Fig. 1) one eight-electron wave can be observed only, corresponding to the simultaneous reduction of both sydnone rings in unprotonized form *VII* directly to the compounds *IX*.

The equilibrium between the protonized (*V*) and unprotonized (*VII*) form of N,N'-ethylene-bis-sydnone (n = 2) is not established extremely fast. The height of the twelve-electron wave is limited by the rate of proton transfer. The value of the pH, where the wave-height reaches a current, corresponding to a ten-electron process, is equal to about pH 3·3. Due to the fact that the thermo-dynamic dissociation constant has a value lower than 1 and was not determined, quantitative treatment was not possible.

The equilibrium between the intermediate reduction products *VIIIa* and *VIIIb* is established immediately and does not influence the limiting currents. It causes the shift of the more negative wave with pH and enables the separation of the two reduction steps. The value pH 9·3 (pK''), corresponding to the intersection of the linear parts of the $E_{\frac{1}{2}}$-pH-plot for the second four-electron wave, most probably corresponds to the thermodynamic dissociation constant pK_a of the hydrazine group in *VIIIa*.

N,N'-Tetramethylene-bis-sydnone (*Ib*) and N,N'-Hexamethylene-bis-sydnone (*Ic*)

The polarographic curves obtained with N,N'-*tetramethylene-bis-sydnone* (Ib) as well as N,N'-*hexamethylene-bis-sydnone* (Ic) closely resemble that of N,N'-ethylene-bis-sydnone (*Ia*). They differ in that with substances *Ib* and *Ic* the wave corresponding to the reduction of the cation (*V* for n = 4 and 6) in acid solutions was not observed. This wave is shifted to more negative potentials and is obscured by the evolution of hydrogen from the buffer.

The character of the first four-electron wave (corresponding to the reduction of *VII* for n = 4 and 6) does not change with the length of the polymethyl-

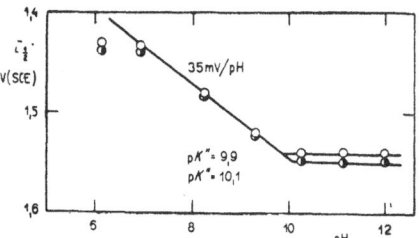

Fig. 4

Dependence of the Half-Wave Potentials of N,N'-Polymethylene-bis-sydnones on pH in Britton-Robinson Buffers

$1 \cdot 10^{-4}$M depolarizer, ○ N,N'-tetramethylene-bis-sydnone, ◑ N,N'-hexamethylene-bis-sydnone.

ene chain. The half-wave potential of this wave of the unprotonized sydnone-ring is similarly pH-independent for *Ib* and *Ic* as it is for *Ia*.

The slope of the dependence of the half-wave potential of the more negative wave on pH was found to be only 35 mV/pH for *Ib* and *Ic* (Fig. 4). The pK'' values found were 9·9 for *Ib* and 10·1 for *Ic*.

The Mutual Influence of the Sydnone Rings

For the comparison of half-wave potentials of different compounds with a view to explain the effect of constitution it is necessary that the electrode processes for the substances compared are practically the same. In addition to other criteria[3] an analogous $E_{\frac{1}{2}}$–pH-plot in all instances compared is of utmost importance. The value of the slope $dE_{\frac{1}{2}}/d\text{pH}$ must be practically the same or the half-wave potentials compared ought to be pH-independent which is even more preferable.

Because of the rather complicated behaviour of polymethylene-bis-sydnones at lower pH-values, the effects of the mutual influence of both sydnone rings are discussed first basing on the pH-independent half-wave potentials of the first four-electron wave at higher pH-values. Borate buffer of pH 9·3 was chosen for comparison, where the half-wave potential of the second wave is practically pH-independent.

The half-wave potential of the more negative wave, corresponding to a reduction of *VIIIa*, practically coincides with the half-wave potentials of 3-alkylsydnones (Table I, Fig. 5). The half-wave potential of this wave is practically independent of the length of the unprotonized side-chain. The effects of carboxy- and hydrazino groups in the side chain are completely screened by the polymethylene chain.

On the other hand, there is a distinct mutual interaction of the two unprotonized sydnone rings in *VII*. If the polymethylene chain hinders the interaction of the two rings, only one eight-electron wave, corresponding to a simultaneous reduction of both rings would be observed. Instead of such a unique wave a shift of the half wave potential of the more positive part of the wave to more positive values with the decrease in length of the polymethylene chain was observed at pH 9·3. Due to the presence of the second sydnone ring the

Table I

Comparison of Half-Wave Potentials of N,N′-Polymethylene-bis-sydnones and 3-Alkylsydnones with Dissociation Constants of Polymethylene Dicarboxylic Acids

Derivative	Sydnones				Dicarboxylic Acids		
	$(E_{\frac{1}{2}})_{\text{I}}{}^a$	$(E_{\frac{1}{2}})_{\text{II}}{}^a$	Δ^b	$\Delta_{\text{corr}}{}^c$	$pK_1{}^d$	$pK_2{}^d$	$\Delta\,pK^e$
n = 2 *Ia*	$-1\cdot28_7$	$-1\cdot52_6$	$+0\cdot23_9$	$+0\cdot17_9$	4·19	5·48	1·29
n = 4 *Ib*	$-1\cdot38_2$	$-1\cdot52_7$	$+0\cdot14_5$	$+0\cdot08_5$	4·42	5·41	0·99
n = 6 *Ic*	$-1\cdot39_4$	$-1\cdot53_0$	$+0\cdot13_6$	$+0\cdot07_6$	4·52	5·40	$0\cdot88^g$
n = ∞				$+0\cdot01_7{}^f$			0·60
3-Butyl		$-1\cdot53_0$					
3-Methyl		$-1\cdot51_5$					

^{a} V *vs.* S. C. E.; ^{b} $\Delta = (E_{\frac{1}{2}})_{\text{I}} - (E_{\frac{1}{2}})_{\text{II}}$ in volts; ^{c} $\Delta_{\text{corr}} = \Delta - 0\cdot06$ V in volts; ^{d} values from reference[6]; ^{e} $\Delta\,pK = pK_2 - pK_1$; ^{f} $(\Delta_{\text{corr}})_{n=\infty} = 2\cdot3RT/\alpha nF \log 2$; ^{g} $(\Delta\,pK)_{n=\infty} = \log 4$.

reduction of the first sydnone ring is facilitated. The effect is greater the shorter is the polymethylene chain (Table I, Fig. 5).

The difference between the half-wave potentials of the first and second wave at higher pH is a measure of how much the reduction of the first sydnone ring is influenced by the presence of the second ring. This difference is greater the shorter is the chain (Table I).

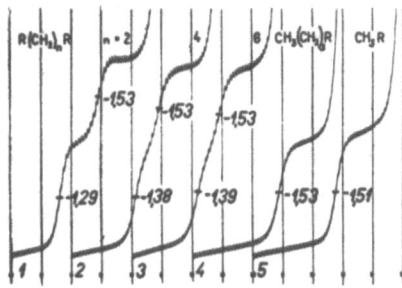

Fig. 5

The Influence of the Length of the Polymethylene Chain on the Reduction Waves of N,N'-Polymethy-lene-bis-sydnone
Borate buffer pH 9·3, 5 . 10⁻⁵M deloparizer.
1 N,N'-ethylene-bis-sydnone; 2 N,N'-tetramethyl-ene-bis-sydnone; 3 N,N'-hexamethylene-bis-sydno-ne; 4 N-n-butyl-sydnone; 5 N-methyl-sydnone. Curves starting at $-1·0$ V, s.c.e., 200 mV/absc., $h = 55$ cm, sens. 1 : 10.

For the discussion of the influence of the length of the polymethylene chain on the shift of the half-wave potentials it is necessary first to exclude the possibility of the polar effect of the chain methylene groups themselves on the reducibility of the sydnone ring. The half-wave potentials of 3-alkylsubstituted sydnones show half wave potentials limiting with the increasing length of the chain to about $-1·53$ V. This excludes the possibility that the shift of the half-wave potential of the first four-electron wave to substantially more positive values is due to the effect of methylene groups. Also consistent with this view is the practical independence of the half-wave potential of the more negative four-electron wave (similar to that of 3-alkylsydnones) on the chain-length. Thus in the more negative wave the side chain influences the half-wave potential similarly to the alkyl groups, whereas in the more positive wave the reduction potential is affected by the mutual interaction of the two sydnone rings.

To enable a quantitative correlation of half-wave potentials and chain-length it is necessary to recognize that even if both waves were to coincide, the measured half-wave-potentials $(E_{\frac{1}{2}})_I$ and $(E_{\frac{1}{2}})_{II}$ of the four-electron waves would not be identical. In the case of a one eight-electron wave (which would be found with a very long polymethylene chain, where no mutual interaction of the ω-groups would be possible), this difference would be equal to the difference of $E_{\frac{1}{4}}$ and $E_{\frac{3}{4}}$. For alkylsydnones such a difference has been found to be about 60 mV. To get a comparable value for the difference of $(E_{\frac{1}{2}})_I$ and $(E_{\frac{1}{2}})_{II}$ it is necessary to correct the measured difference $\Delta = (E_{\frac{1}{2}})_I - (E_{\frac{1}{2}})_{II}$ for this smallest difference (not due to the mutual interaction of both sydnone rings) for $n = \infty$, i.e. 60 mV and to use the corrected value $\Delta_{corr} = \Delta - 0,06$ V.

Furthermore it is necessary to consider that in a compound with two identical reducible groups the probability of the electrode reaction can be taken as twofold of that for a corresponding compound with a single electroactive group. According to the supposition, used first by Wegscheider[4] for dissociation of dibasic acids, it is possible to suppose that the rate constant of the electrode process cannot change by more than a factor of two. Then according to the equation for the half-wave potential of an irreversible process[5] a following shift in $(E_{\frac{1}{2}})_I$ compared with $(E_{\frac{1}{2}})_{II}$ (caused by a statictical factor only) could be expected:

$$(\Delta_{corr})_{n-\infty} = 2\cdot3RT/\alpha\, nF \log 2$$

In our case, where $\alpha n \doteq 1$ (estimated from the wave-shape) a shift of the first wave $(\Delta_{corr})_{n-\infty} \leqq + 0\cdot01_7$ V would be expected. The higher values of the shift Δ_{corr} observed (Table I) may be interpreted as due to the influence of other factors besides the statistical factor (if the latter has any influence at all).

An example of a system where the mutual influence of two identical groups on the polymethylene chain and similar effects besides the statistical factor, have been most thoroughly treated are the dissociation constants of dibasic acids[6,7]. It can be shown (Fig. 6) that the type of dependence of shifts of the half-wave potential of the first wave of sydnones on the number of methylene groups is similar to such a dependence observed with dissociation constants. The mechanism of the mutual interaction of two carboxylic groups on a polymethylene chain has not been unequivocally distinguished yet. Whereas Bjerrum[8] and Ingold[7,9] suppose mainly that the direct field effect be operating, the theory by Westheimer and Kirkwood[10] can be interpreted[6] as including the role of the internal inductive effect too. Thus similarly in the mutual in-

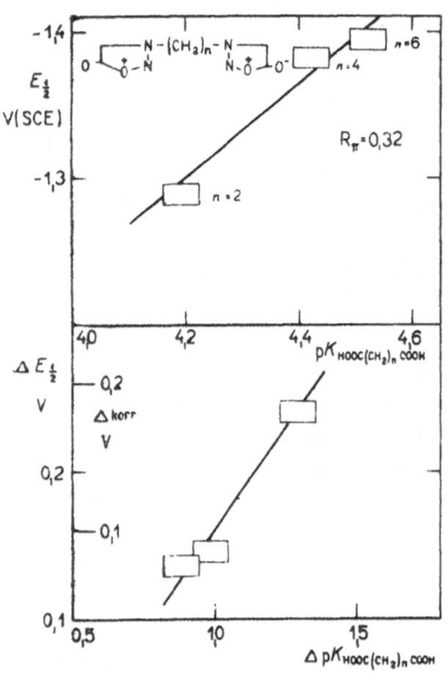

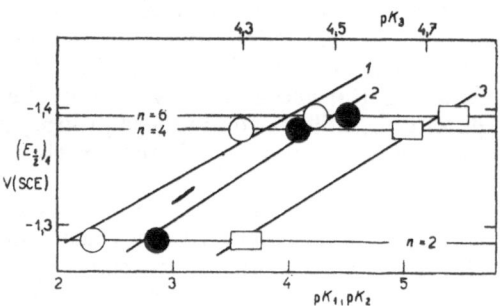

Fig. 6

Dependence of the Half-Wave Potentials of the First Four-electron Wave of N,N′-Poly-methylene-bis-sydnones at pH 9·3 on the pK_a. Values of the First Dissociation Constant of Polymethylene Dibasic Acids for Different Number of Methylene Groups (above) Dependence of the difference $\Delta E_{\frac{1}{2}}$ and corrected difference (Δ_{corr}) of the half-wave potentials of the first and second four-electron wave of polymethylene-bis-sydnones (at pH 9·3) on the difference of the first and second dissociation constants ΔpK of polymethylene dibasic acids for different number of methylene groups (below).

Fig. 7

Dependence of the Half-Wave Potentials of the First Four-electron Wave of N,N′-Poly-methylene-bis-sydnones at pH 9·3 on pK-Values of the Dissociations Constants of Acids of the Type R—$(CH_2)_{n/2}$—COOH, where R is equal to: *1* —$\overset{(+)}{N}H_3$; *2* —Cl and *3* —C_6H_5. Sydnones with n Methylene Groups were Compared with Acids with n/2 Methylene Groups. Lower Axis: pK_1, pK_2; upper axis: pK_3.

teraction of the two sydnone rings both polar effects — field and internal cannot be separated.

In Fig. 7 it is shown that the effect of the second sydnone ring on the reducibility of the first one is also comparable with the effects of some other groups along a polymethylene chain. In this figure the half-wave potentials

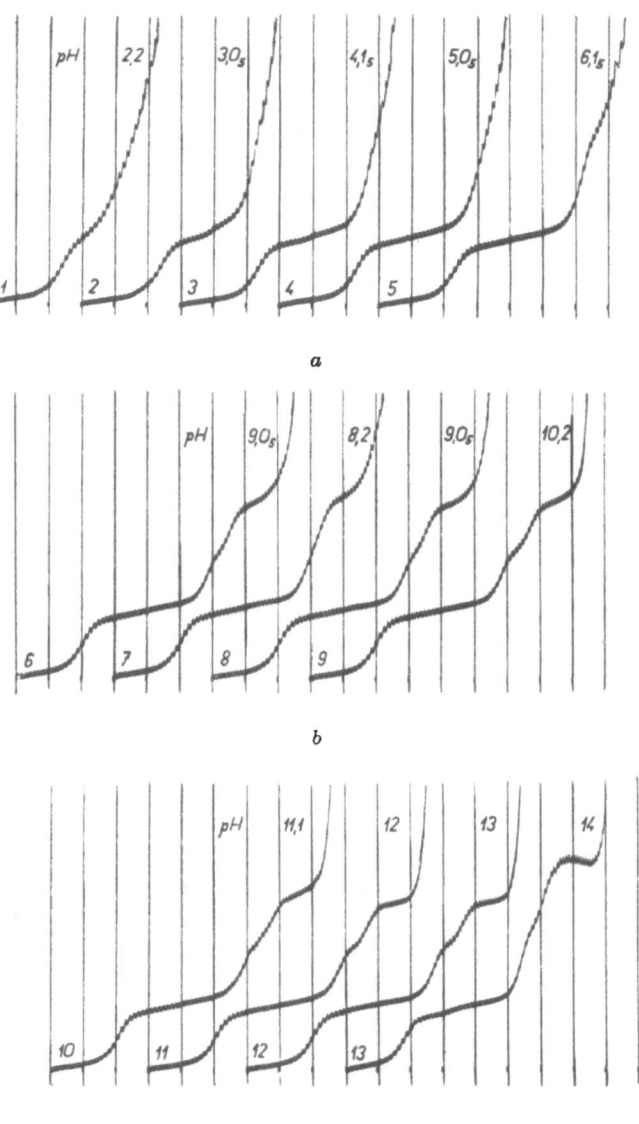

Fig. 8

The Influence of pH-Values on the Waves of C,C'-Dibromo-N,N'-hexamethylene-bis-sydnone 1 . 10⁻⁴M depolarizer, *1—11* Britton-Robinson buffers (pH-values given on the polarogram), *12* 0·1N-NaOH; *13* 1·0N-NaOH. Curves starting at −0·2 V, s. c. E., 200 mV/absc., h = 55 cm, sens. 1 : 10.

of N,N′-polymethylene-bis-sydnones were compared with pK-values of poly-methylene phenyl-, chloro- and aminoacids. To express the additivity of the effects of both sydnone rings, sydnones with n methylene groups were com-pared with acids with n/2 methylene groupings.

The correlation of the shifts of half-wave potentials with pK-values for dissociation constants of dicarboxylic acids (where the field effect is claimed to predominate) as well as of chloroacetic acids (quoted often as an example for internal polar effects) and finally of polymethylene ω-phenyl carboxylic acids (where due to the absence of dipole moment of the phenyl group no field effect can be expected) shows either that there is some inconsistency in the present interpretation of dissociation constants or may be interpreted as a proof of both field and internal inductive effects operating in polarography of polymethylene-bis-sydnones.

Two additional factors seem to be important in the step-wise reduction of the sydnone rings: the charge (and dipolar character) of the sydnone ring and the orientation at the electrode surface.

The importance of the charge is demonstrated by the already mentioned screening of the hydrazine portion of the molecule in VIII and by the fact that no separation of waves occurs with protonized forms in acid media. Even when the twelve-electron wave is shown only for Ia (and for II and III), the absence of separation of the two steps for ethylene-derivative represents a conclusive evidence that protonized forms differ substantially from unprotonized ones. This difference could be explained also by different orientation.

Another explanation would be the different conformation of the protonized and unprotonized form. To explain the actual independence of the reduction of both sydnone rings in the protonized state, a repulsion of the protonized sydnone rings could result in a streched out form. On the other hand the con-formation of the unprotonized form could allow interaction of the negative pole of one sydnone ring with the positive of the other one.

Such a conformation cannot be presumed to take place in the intermediate reduction products VIII where the side chain (containing COO⁻ and amino-group) exerts on the reduction of second sydnone ring an influence similar to that of an alkyl group.

In accordance with a theory that the difference in $(E_{\frac{1}{2}})_I$ and $(E_{\frac{1}{2}})_{II}$ is due to dynamic effects are the UV-spectra[2]. Here the position of the absorption maxima are practically identical for N,N′-polymethylene-bis-sydnones and for 3-alkylsydnones. The doubled extinction coefficient reflects the equal probability of the absorption of a light quantum by two separated chromo-phores.

Identity of the wave-length of absorption bands also enables us to elimi-nate hyperconjugation as a reason for the observed mutual influence. Effects of this kind were observed[11] in UV-spectra of compounds of the type Ar.$(CH_2)_n$ — R where n = 1 and sometimes n = 2 but never for higher values of n.

Thus dynamic polar effects — both field and inner inductive effects — and the orientation of the molecule at the electrode are supposed to cause the separat-ion of reduction waves of the two sydnone rings at higher pH-values.

The Influence of the Substituent on Carbon on the Course of the Reduction

With C,C′-*dimethyl*-N,N′-*ethylene-bis-sydnone* (II) as well as with C,C′-*dichloro*-N,N′-*ethylene-bis-sydnone* (III) the character of polarographic curves, the number of waves, their dependence on pH as well as the dependence of their half-wave potentials on pH closely resembles that found for unsubstituted N,N′-ethylene-bis-sydnone (*Ia*). Only the half-wave potentials are shifted due to the polar effects of substituents. These effects which are best compared at pH 9—10 (where both reduction processes are pH-independent) are discussed in Part V of this series[12].

Bromo-derivatives

Contrary to the findings with methyl- and chloroderivatives, the presence of bromine in the molecule of polymethylene-bis-sydnone completely changes the course of the reduction. On the polarographic curves of C,C′-dibromo-N,N′-ethylene-bis-sydnone (*IVa*) and C,C′-dibromo-N,N′-hexamethylene-bis-sydnone (*IVc*) a wave at −0,6 V is observed (Fig. 8) in addition to more negative waves. Both the height of the is more positive wave and its half-wave potential are pH-independent. The waves at more negative potentials coincide both by their shape and pH-dependence with the corresponding waves of unsubstituted N,N′-polymethylene-bis-sydnones (Fig. 9). Only at higher pH-values than about 9·5 were changes with time observed — mainly with the ethylene derivative *IVa* — which has already influenced the ratio of waves on Fig. 9.

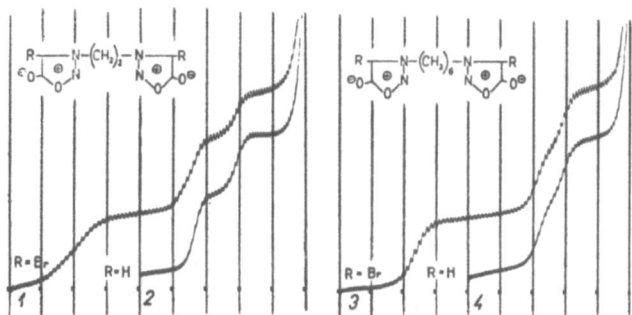

Fig. 9

Comparison of the Curves of C-Bromo-sydnones with Corresponding Curves of Unsubstituted Sydnones

Borate buffer pH 9·3, 5 . 10⁻⁵M sydnone; *1* C,C′-dibromo-N,N′-ethylene-bis-sydnone; *2* N,N′-ethylene-bis-sydnone; *3* C,C′-dibromo-N,N′-hexamethylene-bis-sydnone; *4* N,N′-hexamethylene-bis-sydnone. Curves *1* and *3* starting at −0·2 V, *2* and *4* at −1·0 V, s. c. e., 200 mV/absc., h = = 55 cm, sens. 1 : 15.

A single wave at more positive potentials was observed with *IVc*. A slight separation of two waves of equal height was found with *IVa* (Fig. 9). The total limiting current of these waves corresponds to a four-electron transfer. Bromine atoms are reduced in this wave according to the following scheme:

$$\text{Br—Sy—(CH}_2)_n\text{—Sy—Br} + 2\,e + 2\,\text{H}^+ \rightarrow \text{Sy—(CH}_2)_n\text{—Sy} + 2\,\text{Br}^-$$

(where Sy represents a sydnone ring).

With waves at more negative potentials both sydnone rings are reduced similarly to unsubstituted N,N'-polymethylene-bis-sydnones. The half-wave potentials of sydnones Ia and Ic, formed in the reduction of dibromo-derivatives IVa and IVc, are about $10-15$ mV more negative than potentials of substances Ia and Ic added to the solution. This fact can be explained[13] either by different orientation at the electrode surface or by a change accompanied with a formation of the reducible form preceding the electron transfer. The third possibility — another course of the reduction of the reactant to the same final product at potentials of the more negative wave — does not seem to apply here.

The difference in the separation of the two steps in the bromine wave in IVa and IVc shows a mutual interaction of both bromo-substituted rings of the type mentioned in the preceding paragraph. But the separation of the reductions of the two C—Br bonds is by far not so evident as for the wave of the sydnone rings. The separation of waves for two identical groups on a polymethylene chain is thus dependent also on the nature of the reducible bond. This is not in disagreement with the proposed influence of dynamic effects and orientation.

From the practical point of view it is important that polarographic measurements enable us to predict that it is possible to prepare C-chloroderivatives of hydrazine by electro-reduction of sydnones. For C-bromo sydnones the electro-reduction renders unsubstituted sydnones only, or, at sufficiently negative potentials, hydrazine derivatives of the type IX.

This example shows, how a little change in the molecule such as the substitution of chlorine for bromine can entirely change the reduction mechanism.

The author expresses his thanks to Drs H. U. Daeniker and J. Druey, who kindly supported him by the gifts of samples of sydnones, and to Dr O. Exner for the careful revision of the manuscript. The technical assistance of Miss N. Pejchová is acknowledged.

References

1. Zuman P.:.This Journal *25*, 3245 (1960).
2. Daeniker H. U., Druey J.: Helv. Chim. Acta *40*, 918 (1957).
3. Zuman P.: This Journal *25*, 3225 (1960).
4. Wegscheider R.: Monatsh. *16*, 153 (1895); *23*, 599 (1902).
5. Brdička R.: This Journal *19*, 841 (1954).
6. Hückel W.: *Theoretische Grundlagen der organischen Chemie*. Bd. 2, 7. Aufl., S. 654. Akademische Verlagesgesellschaft, Leipzig 1954.
7. Ingold C. K.: *Structure and Mechanism in Organic Chemistry*, p. 728. Cornell University Press, Ithaca, 1953.
8. Bjerrum N.: Z. physik. Chem. *106*,.219 (1923).
9. Ingold C. K.: J. Chem. Soc. *1931*, 2179.
10. Westheimer F. H., Kirkwood J. G.: J. Chem. Phys. *6*, 506, 513 (1938).
11. Gillam A. E., Stern E. S.: *An Introduction to Electronic Absorption Spectroscopy in Organic Chemistry*, 2nd Ed., p. 134. Arnold, London 1957.
12. Zuman P.: This Journal, in the press.
13. Zuman P.: *D. Sc. Thesis*. Czechoslovak Academy of Science, Prague 1959.

Translated by the Author.

POLAROGRAPHY OF NONBENZENOID AROMATIC AND RELATED SUBSTANCES. IV.*
POLAR EFFECTS OF SUBSTITUENTS IN PHENYLSYDNONES; THE APPLICATION OF MODIFIED HAMMETT EQUATION

P. ZUMAN

Polarographic Institute, Czechoslovak Academy of Science, Prague

Received June 4th, 1960

To Academician J. Heyrovský on his 70th birthday.

Half-wave reduction potentials in three reaction series: 3-phenylsydnones substituted in the 3'- and 4'-position, 3,4-diphenylsydnones substituted on the 3-phenyl group in position 4' and 3,4-diphenylsydnones substituted on the 4-phenyl group in position 4' were measured. Their values fit to the modified Hammett equation: $\Delta E_{\frac{1}{2}} = \varrho_{\pi,R}\sigma_X$. Additivity of the effects of substituents on both phenyl groups in 3,4-diphenylsydnone, on the reducibility of the N—O bond, was proved.

The half-wave potentials of organic substances can be influenced by polar, steric and meso-meric (resonance) effects of the substituents. In properly chosen groups of substances it is possible to treat these effects quantitatively[1]. The most widely used form of these relations is the modified Hammett equation (1)

$$\Delta E_{\frac{1}{2}} = \varrho_{\pi,R}\sigma_X \qquad (1)$$

where $\varrho_{\pi,R}$ is the reaction constant, characterizing the reducible group and influenced by the conditions of polarographic electrolysis, σ_X is the substituent constant, based on dissociation constants of substituted benzoic acids[2,3], characterizing the effect of the substituent. This equation applies for *meta*- and *para*-substituted benzene derivatives bearing a polarographically active group in the side chain. For reaction series, where a resonance interaction of the reducible group with the aromatic ring plays an important role, in the presence of substituents that undergo a resonance interaction with the ring, a special set of substituent constants σ_X^- is to be applied. A linear relationship between the shift of half-wave potential and substituent constant σ_X is taken as a proof[4] that polar effects of the substituent *only* have an influence on the reducibility of the polarographically active group.

It is shown in this paper that even for reaction series of *meta*- and *para*-substituted 3-phenylsydnones and 3,4-diphenylsydnones, where the whole sydnone ring represents the polarographically active group, only polar effects of the substituents have an influence on their half-wave potentials.

Experimental

Polarographic apparatus, vessels and capillaries were the same as in previous work[5].

3-Phenylsydnone, 3-(p-bromophenyl)sydnone and 3-phenyl-4-methylsydnone were kindly presented by Professor J. C. Earl (Thurlton, Haddiscoe, Norwich, England), 3-(m-methoxyphenyl)-sydnone, 3-(p-methoxyphenyl)sydnone, 3-(m-bromophenyl)sydnone, 3-(p-bromophenyl)sydnone and 3-(p-carboxyphenyl)sydnone by Dr. R. A. Eade (School of Chemistry, The University of New South Wales, Sydney), 3-(m-chlorophenyl)sydnone (m. p. 144—5°), 3-(p-chlorophenyl)-

---

* Part III: This Journal 25, 3252 (1960).

sydnone (m. p. 114–6°), 3-(*m,p*-dichlorophenyl)sydnone (m. p. 151–2°), 3-(*m*-methoxyphenyl)-sydnone (m. p. 141–3)°, 3-(*p*-methoxyphenyl)sydnone (m. p. 130–4°), 3-(*p*-carboxyphenyl)-sydnone (decomp. ca. 252°), 3-(*p*-acetoxyphenyl)sydnone (m. p. 142–3°), 3-(*p*-hydroxyphenyl)-sydnone (decomp. ca. 246°) and 3,4-diphenylsydnone (m. p. 185–8°) were kindly donated by Farbenfabrik Bayer A. G., Leverkusen, 3-(*m*-tolyl)sydnone and 3-(*p*-tolyl)-4-methylsydnone (m. p. 161°) by Dr D. J. Voaden (Chemistry Department, The University, Oxford) and 3,4-di-phenylsydnone, 3-(*p*-chlorophenyl)-4-phenylsydnone (m. p. 123–4°), 3-phenyl-4-(*p*-chlorophe-nyl)sydnone (m. p. 129°), 3,4-di(*p*-chlorophenyl)sydnone (m. p. 144°), 3-(*p*-tolyl)-4-phenylsyd-none (m. p. 151°), 3-phenyl-4-(*p*-tolyl)sydnone (m. p. 145–6°) and 3,4-di(*p*-tolyl)sydnone (m. p. 157°) were kindly donated by Professor W. Baker (Department of Organic Chemistry, The University, Bristol).

The stock solutions were usually prepared in 50% or in 96% ethanol in concentrations of $2.5 . 10^{-3}$M. The concentration of depolarizer in the polarographed solution was in most instances $5 . 10^{-5}$M. The ethanol content of the polarographed solutions was 1% or 2% in the case of 3-phe-nylsydnones and 50% in the case of 3,4-diphenylsydnones (to eliminate the adsorption pheno-mena occurring at low ethanol concentrations).

The pH-dependence of the polarographic reduction waves was followed in Britton-Robinson buffers. 0·05M borax with 2% ethanol was chosen as the most suitable medium for the comparison of substituted 3-phenylsydnones, 0·05M-NaOH with 50% ethanol for the comparison of 3,4-di-phenylsydnone and its derivatives.

The technique of measurement of half-wave potentials is demonstrated in Fig. 1. The curves were registered photographically first from positive to negative, then from negative to positive potentials (to eliminate the hysteresis of the recording system) using a small voltage span and a scanning rate of 200 mV/min. The mean value of the half-wave potentials of the two curves was de-termined and compared with the half-wave potential of thallium ions, under the same conditions, taken as an internal standard ($E_{\frac{1}{2}} = -0.445$ V, *vs.* S. C. E.).

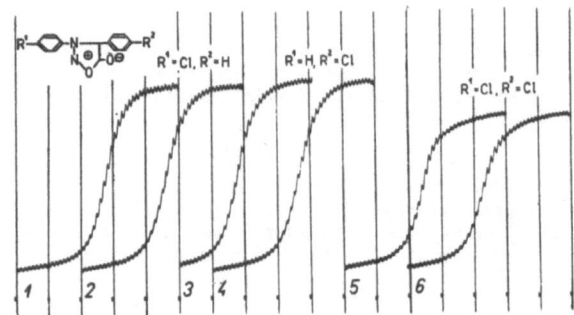

Fig. 1

Example of the Measurement of the Half-Wave Potentials of Chloroderivatives of 3,4-Diphenyl-sydnone

$5 . 10^{-5}$M depolarizer, 0·05M borax, 0·01% gelatine. Structure given on the polarogram. Pair of curves obtained by polarization from positive to negative potentials and from negative to positive potentials. Curves starting at: *1* and *5* −0·802 V; *3* −0·903 V, S. C. E., 100 mV/absc., *h* = 55 cm, sens. 1 : 4.

Results

Substituted 3-Phenylsydnones

The polarographic behaviour of 3-phenylsydnones, substituted in *para*- and *meta*-positions on the phenyl ring, is analogous to that of the unsubstituted 3-phenylsydnone[5]. Thus the half-wave potentials of these compounds were measured in a borax buffer pH 9·3, where they are pH-independent[5] (Table I). Since maxima appeared with some substances, 0·005% gelatine was added.

The waves under these conditions correspond to a four-electron reduction process with the formation of a hydrazine derivative[5].

The correlation of the half-wave potentials of substituted 3-phenylsydnones with Hammett's total polar substituent constant σ_x is given in Fig. 2. A greater deviation is observed only with the 4-hydroxy derivative for which the σ_{p-OH^-} value shows[3] a variation within a broad range.

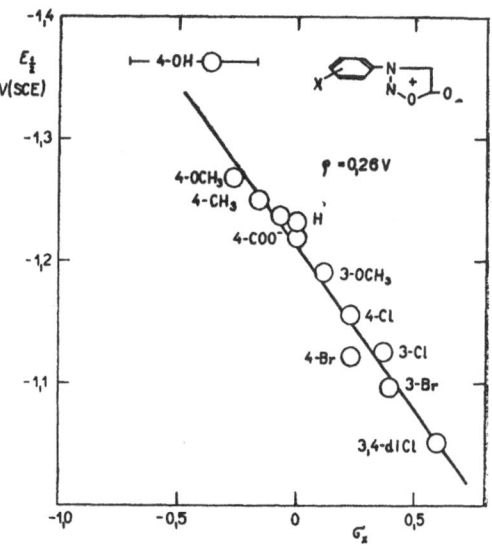

Fig. 2

Dependence of the Half-Wave Potentials of 3-Phenylsydnones, Substituted on the Phenyl Ring, on the Total Polar Substituent Constant σ_x

3,4-Diphenylsydnones

The polarographic behaviour of 3,4-diphenylsydnone is similar to that of 3-phenylsydnone in that in acid solution a six-electron wave is observed, decreasing gradually with increasing pH-value. At higher pH-values a one four-electron step was observed, corresponding to the formation of the hydrazine derivative[5]. Contrary to 3-phenylsydnones the values of the half-wave potentials for 3,4-diphenylsydnone and its derivatives have not reached at pH 9·3 the

Table I

Half-Wave Potentials of 3-Phenylsydnones Substituted in *meta*- and *para*-Positions on the Phenyl Ring

Borate buffer pH 9·3, 0,005% gelatine added, 25°C.

Substituent	$E_{\frac{1}{2}}$ V (S. C. E.)	σ_x
4′-Hydroxy	−1·36₃	−0·37[c]
4′-Methoxy	−1·26₈	−0·27
4′-Methyl	−1·26[a]	−0·17
3′-Methyl	−1·23₈	−0·07
none	−1·23₃	0·0
4-Carboxy (anion)	−1·21₉	0·0
3′-Methoxy	−1·19₁	+0·11₅
4′-Acetoxy	−1·16₂[b]	+0·30
4′-Chloro	−1·15₇	+0·23
3′-Chloro	−1·12₆	+0·37
4′-Bromo	−1·12₂	+0·23
3′-Bromo	−1·09₇	+0·39
3′,4′-Dichloro	−1·05₇	+0·60[d]

[a] Computed from the value for 3-(p-tolyl)-4-methylsydnone; [b] Hydrolyzes with formation of 3-(4′-hydroxyphenyl)sydnone; [c] A substantially uncertain value; [d] $\Sigma\sigma$;

pH-independent value yet (Fig. 3). Thus the values of comparable half-wave potentials, given in Table II, were measured in 0.05M-NaOH with $5 \cdot 10^{-5}\text{M}$ depolarizer and 50% ethanol. The half-wave potentials of the few corresponding derivatives available are correlated with the total polar substituent constant σ_x in Fig. 4.

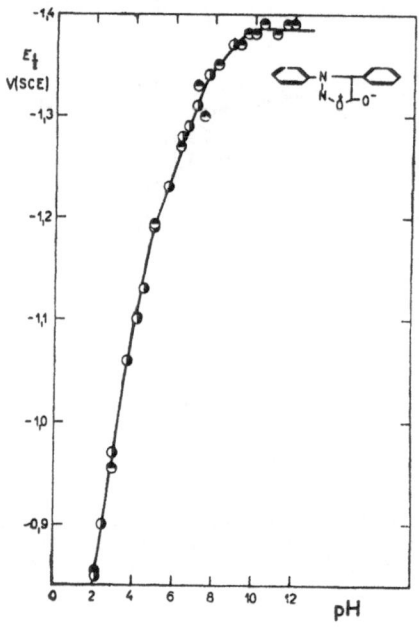

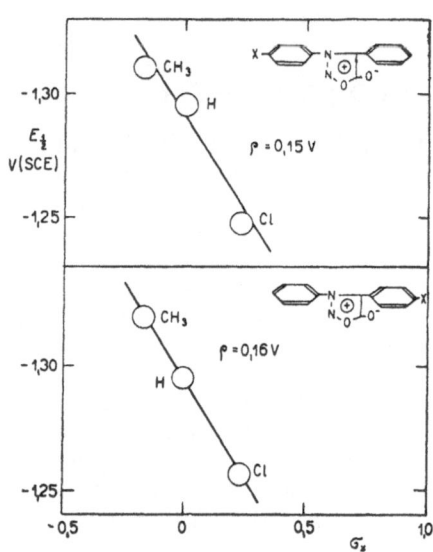

Fig. 3

The pH-Dependence of the Half-Wave Potentials of the 3,4-Diphenylsydnone in Britton-Robinson Buffers
Different sets of measurements.

Fig. 4

Dependence of the Half-wave Potentials of the 3,4-Diphenylsydnones Substituted on a 3-Phenyl Group (above) and on a 4-Phenyl Ring (below), on the Total Polar Substituent Constant σ_x

Discussion

Experimental points fit well the linear free energy relationship (1) as given in Fig. 2. This can be taken as proof[4] that substituents in m- and p-positions on the 3-phenylgroup on the sydnone ring exert principally a polar effect on the reduction of the $N-O$ bond in the ring. So far the validity of the modified Hammett equation (1) has been proved for reductions bonds of atoms bound directly on the benzene ring (as in iodobenzenes) and of bonds in α-position to the ring (e. g. in acetophenones, nitrobenzenes, stilbenes etc.).

The $N-O$ bond, reduced in sydnone rings, is in the β-position with respect to the phenyl in position 3 and in the γ-position with respect to the phenyl in position 4. The polar effects are thus easily transmitted by the sydnone ring. The easy mobility of the π- and p-electrons of the sydnone ring can be taken as further proof for the conjugated system in the sydnone ring.

Table II

Table II

Half-Wave Potentials of 3,4-Diphenylsydnones Substituted in 4'-Position on Both Rings
0·05M-NaOH, 50% ethanol, 25°C.

4'-Substituent		$E_{\frac{1}{2}}$	σ_x	ΔH^a
on 3-phenyl	on 4-phenyl	V (S. C. E.)		
H	H	$-1·29_5$	0	—
Cl	H	$-1·24_7$	$+0·23$	$0·04_8$
H	Cl	$-1·25_6$	$+0·23$	$0·03_9$
Cl	Cl	$-1·21_0$	—	$0·08_5$
CH_3	H	$-1·31_0$	$-0·17$	$-0·01_5$
H	CH_3	$-1·31_9$	$-0·17$	$-0·02_4$
CH_3	CH_3	$-1·33_7$	—	$-0·04_2$

^{a} $\Delta H = (E_{\frac{1}{2}})_H - (E_{\frac{1}{2}})_X$.

Table III

Influence of the Substitution in Position 4 on the Shift in the Half-Wave Potential, due to a Methyl
Group in the Position 4' Bound on the Phenyl Ring in Position 3 (S. C. E., in volts)

Substituent in Position 4	Substituent in Position 3		
	Phenyl	p-Tolyl	Δ^a
Methyl	$-1·36_5$	$-1·38$	$0·01_5$
Phenyl	$-1·29_5$	$-1·31_0$	$0·01_5$

^{a} Shift due to a p-methyl group, in volts.

Positive values of the reaction constants $\varrho_{\pi,R}$ were found in all three reaction series studied. This can be explained[1] as due to the nucleophilic mechanism of the potential determining step. The most probable nucleophilic reagent involved is the electron (or the electrode). The linear relationship represents further evidence that probably the same reduction mechanism is operating for all compounds studied.

Due to the small number of derivatives available the values of $\varrho_{\pi,R}$ for both reaction series given in Fig. 4 are not accurate and their difference is not significant. The comparison of the susceptibility of the 3,4-diphenylsydnones to the influence of the substituents on the 3-phenyl group with that to the influence of the substituents on the 4-phenyl group cannot thus be given. Similarly the effect of substitution of the hydrogen atom for the phenyl group in position 4, on the influence of substituents on the 3-phenyl group, even when it seems to diminish the value of $\varrho_{\pi,R}$, cannot be discussed responsibly (Figs. 2 and 4). The substitution of phenyl for methyl on the other hand seems to have a small effect (Table III), but the limitation to one substituent only (moreover having a small value of the substituent constant $\sigma_{p\text{-}CH_3} = -0·17$) invalidates once more the significance of this deduction.

An additional proof of the polar effect of substituents is given in the additivity of the effects of individual substituents. An example in the 3-phenyl series is the value of the half-wave potential of the m,p-dichloroderivative (Table I). Even more convincing are the effects of substituents in 3,4-diphenylsydnones (Table II). The sum of the shifts caused by the two chlorine atoms in the two monochloro derivatives is 0.08_7 V, in good agreement with 0.08_5 V found for the dichloro derivative. Similarly the sum of the effects of the two p-methyl groups in the two monomethyl compounds is -0.03_9 V and the shift with the dimethyl derivative is -0.04_2 V.

The author expresses his thanks to all those organic chemists, mentioned in the Experimental Part, who kindly supported him by the gifts of samples of sydnones. The careful technical assistance of Miss N. Pejchová is acknowledged.

References

1. Zuman P.: This Journal *25*, 3226 (1960), where further references are given.
2. Hammett L. P.: *Physical Organic Chemistry*, p. 184. McGraw-Hill, New York 1940.
3. Jaffé H. H.: Chem. Revs *53* 191 (1953).
4. Taft R. W. Jr.: *Steric Effects in Organic Chemistry* (Ed. M. S. Newman), p. 556. Wiley, New York 1956.
5. Zuman P.: This Journal *25*, 3245 (1960).

Translated by the Author.

POLAROGRAPHY OF NONBENZENOID AROMATIC AND RELATED SUBSTANCES. VII.*
A POLAROGRAPHIC STUDY OF THE ACID-BASE PROPERTIES OF THE TROPYLIUM ION

P. ZUMAN, J. CHODKOWSKI** and F. ŠANTAVÝ

Polarographic Institute, Czechoslovak Academy of Science, Prague and Chemical Institute, Medical Faculty, Palacký's University, Olomouc

Received July 10th, 1958***

To Academician J. Heyrovský on his 70th birthday.

At concentrations lower than about $1.5 . 10^{-4}$ M the tropylium ion gives one single polarographic reduction wave at -0.3 V, the height of which decreases with increasing pH in the form of a dissociation curve. The limiting currents are controlled by the rate of formation of tropylium ion from tropylalcohol, which is reduced at very negative potentials. Using the treatment by Koutecký the rate constant for this reaction was calculated ($k_2 = 2 . 10^6$ l mol^{-1} s^{-1}). For the rate of the formation of the base (tropylalcohol) under conditions of equilibrium the value $k_1 = 5 . 10^1$ s^{-1} was computed. Neither the form of the dissociation curve nor the pK'-values are influenced by the kind of buffer used, by the presence of pyridinium ion and by the ionic strength. The experimentally verified scheme (3) presents a new simple type of electrode mechanism.

Quantitative treatment of the polarographic study of acid-base equilibria of azulenes[1] gave rise to problems. In aqueous solutions of strong acids the cations of azulenes were soluble and gave well defined waves, but due to the low solubility of the unionised forms the equilibria measured were heterogeneous in nature. When using nonaqueous solutions and aqueous solutions of organic solvents, the quantitative expression of the acidity scale presented complications in conection with the low basicity of azulenes.

Hence we have looked for another hydrocarbon, where it would be possible to study polarographically the establishment of the homogeneous acid-base equilibria at higher pH-values. We have found that the tropylium ion (*I*), which possesses the properties of a Lewis acid, fulfils the above conditions. The study of this compound was also aimed to extend our knowledge of the chemistry of the derivatives of tropane, tropone and tropolone[2-5].

Polarographic reduction of tropylium was first described by Ždanov and coauthors[6-8]. He limited his interest to the study of unbuffered solutions only, where complicated curves were obtained due to the changes at the electrode surface during the electrode process.

An anodic process in solutions of tropylium ion has been reported recently[9].

---

* Part VI: Tetrahedron, in the press.

** Standing address: Institute of Physicochemical Analytical Methods, Polish Academy of Science, Warsaw, Poland.

*** Revised manuscript received on January 10th, 1960.

In this paper acid-base reactions, followed in a solution of low concentration of tropylium, were described. The behaviour at higher concentrations, where the influence of adsorption on polarographic curves is observed, is dealt with in a subsequent paper[10].

Experimental

Apparatus. Polarographic curves were registered on the Heyrovský type polarograph, product of Zbrojovka mark V 301 with photographic recording including galvanometer of maximum sensitivity $2·9 . 10^{-9}$ A/mm. The Kalousek electrolysis cell was used with an external saturated calomel electrode as reference electrode. The capillary used had following characteristics: $t_1 = 3·1$ s (in 1M-KCl, at 0·0 V NKE), $m = 2·78$ mg s^{-1} at $h = 60$ cm. In some instances drop-time was regulated using a device produced by Tesla-Prague.

The time-dependence of the instantaneous current ($i - t$ curves) was recorded oscillographically on the first and second drop using the apparatus designed by Němec[11]. The capillary used had $m = 1·78$ mg s^{-1} and $t_1 = 4·6$ s at $h = 55$ cm.

The impulse polarograph used was built by Valenta[12], the Polaroscope for $dE/dt - E$ curves was a product of Křižík-Prague. The commutator method according to Kalousek[13] was applied using an apparatus modified by Rálek[14].

The pH-values of buffers used were determined using a glass electrode and the Autojonometer type PMH 21d, product of Radiometer, Copenhagen.

Chemicals. Tropylium bromide was prepared by C. R. Ganellin (Queen Mary College, University of London). Materials for the preparation of buffer solutions were products of Lachema, Brno, reagent grade. Tetrabutylammonium hydroxide was prepared from tetrabutylammonium iodide, product of Fluka, Swizerland.

Solutions. In stock solutions of tropylium bromide during standing in air (and, after a longer period even in the specimen of tropylium bromide kept in a closed ampoule) a substance was formed, giving two polarographic waves at potentials more negative than that of tropylium. Moreover these waves were observed even at pH higher that about 7, where no more waves for tropylium were observed. The difference in the half-wave potentials of the two waves of equal height was about the same in the whole pH-range studied. Thus the possiblity of formation of an aromatic aldehyde was excluded. Oxidation products did not formed condensation products with ammonia[16]. This was taken as further proof that aldehydic group is not involved. The character of the waves, the pH-dependence of the wave-heights and half-wave potentials of these waves were analogous to that found with tropone and its derivatives[3]. It is therefore supposed that in the autoxidation of the tropylium ion an oxoderivative of tropane is formed by the action of air oxygen. For that reason the stock solutions were prepared always fresh.

Universal buffer according to Britton and Robinson, acetate and pyridine buffers were used. In the last two buffers the concentration of the acid component (acetic acid, pyridinium ion) was held constant and the pH changed by changing the basic component (acetate, pyridine). The ionic strength was maintained constant by addition of sodium perchlorate.

Techniques. The buffer solutions were deaerated by a stream of oxygen, stock solution of tropylium bromide was added and after a final deaeration the curve was recorded. The influence of the concentration of the buffer and depolarizer; the dependence on pH and the mercury pressure were studied in the usual way.

The reversibility of the acid-base reaction was proved in the following way: To the $2 . 10^{-4}$M solutions of tropylium bromide in buffers of pH 2·8 and 5·4 a concentrated solution of sodium hydroxide was added in order to adjust the pH-value to about 11. By addition of an equivalent amount of hydrochloric acid (with respect to the hydroxide added) the pH was readjusted to the initial value.

After adjusting the concentration of the tropylium ion on the same value by addition of water, the solutions before and after alkalifying were electrolyzed polarographically.

To determine the number of electrons transferred in the electrode process, the waves in $1\cdot5 \cdot 10^{-4}$M solutions of tropylium bromide, acetophenone (Merck) and 2-hydroxy-1,4-naphthoquinone (Light) were compared in Britton-Robinson buffer pH $4\cdot0$.

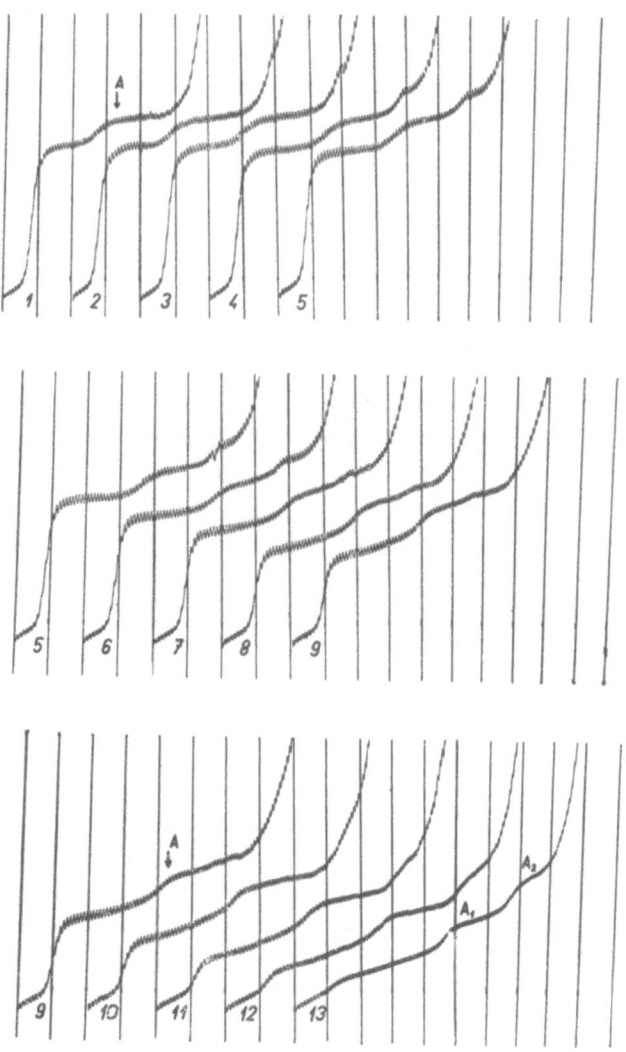

Fig. 1

Influence of pH on Reduction Waves of Tropylium Ion

$2 \cdot 10^{-4}$M tropylium bromide in buffers according to Britton and Robinson of pH: *1* 2·4; *2* 2·8; *3* 3·1; *4* 4·0; *5* 4·5; *6* 4·8; *7* 5·0; *8* 5·2; *9* 5·4; *10* 5·6; *11* 5·9; *12* 6·2; *13* 6·6₅. Curves started at 0·0 V, s. c. E, 204 mV/absc., $h = 60$ cm, sens. 1 : 5.

Little wave A and waves A_1 and A_2 in alkaline solutions respectively, correspond to the reduction of the product of autoxidation of tropylium ion.

For the study of the reversibility of the electrode process current-voltage curves using the impulse polarograph[12], the commutator method[13,14] and the $dE/dt - E$ curves as well as the $i - t$-curves[11] were registered in Britton-Robinson buffers of pH 2·4 (for 1·5, 3·0 and 12 . 10^{-4}M tropylium bromide), of pH 3·1 (for 2 . 10^{-4}M) and of pH 5·8 (for 5 . 10^{-4}M where still due to the kinetic character only one single wave, proportional to concentration of tropylium bromide, was found[10]). With the commutator method frequency 3·5 cycles/s, with the impulse polarograph a single triangular sweep with scanning rates from 0·78 V/s to 4 V/s was used.

Potentiometric titrations were performed in solutions of 0·1M and 0·05M tropylium bromide with addition of 0 to 0·5M-NaCl using a standard solution of 0·05M-NaOH.

Results

In buffered solutions containing less than 1·5 . 10^{-4}M tropylium bromide, a single one-electron wave was observed (Fig. 1). The number of electrons transferred was determined by comparison of the wave height with waves of equimolar concentrations of napthoquinone and acetophenone. The course of the electroreduction is irreversible. The slope of the wave expressed as the coefficient in logarithmic analysis is 0·040, at pH between 2·8 and 5·4. The irreversibility was further proved by the oscillographic polarography, using the curves $dE/dt - E$ (comp. Fig. 10 in the next paper in this series[10]), by impulse polarography (comp. Fig. 3 in[10]) and by the commutator method[13,14].

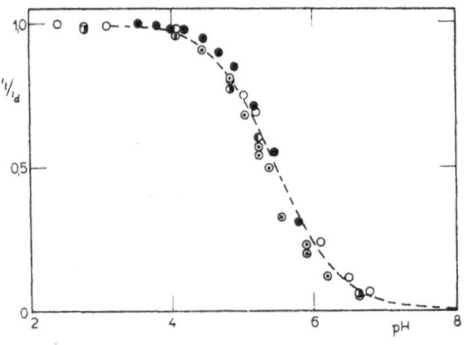

Fig. 2

pH-Dependence of the Limiting Current of Tropylium Ion

Britton-Robinson buffer, concentration of tropylium: ◑ 1 . 10^{-4}M; ⊙ 2 . 10^{-4}M; ◔ 5 . 10^{-4}M (in most instances the points coincide; ● acetate buffers, 2 . 10^{-4}M tropylium bromide; ○ pyridine buffers, 2 . 10^{-4}M tropylium bromide. Experimental points are given at $t_1 = 3·3$ s, theoretical curve for $k_2 = 1 . 10^6$ l mol^{-1} s^{-1}.

The half-wave potential is dependent on the concentration of tropylium ion. At pH 4 a shift of about 30 mV towards more positive values was observed when the concentration was changed from 2·5 . 10^{-5}M to 1·5 . 10^{-4}M tropylium.

The height of this wave decreases with increasing pH in the form of a dissociation curve (Fig. 2). From the pH-value, for which the wave-height decreased to 50% of its original height, the value $pK' = 5.64$ was determined at $t_1 = 3.3$ s. At lower pH the wave-height is proportional to a square-root of h. At higher pH-values, when the wave-height decreased to about 10% of its original height, the wave-height was independent of the mercury head.

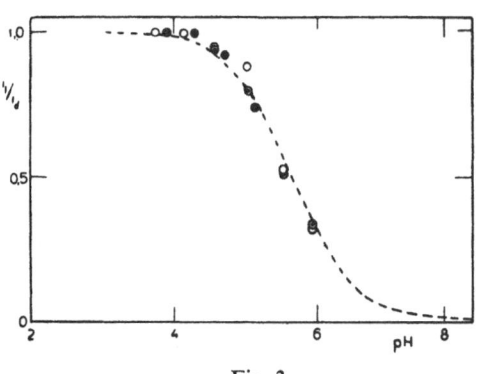

Fig. 3

Effect of Ionic Strength on Dissociation Curves

$2 . 10^{-4}$M tropylium bromide; ○ 0·1M acetic acid, concentration of sodium acetate changed from 0·016M to 1·0M, ionic strength $\mu = 1.0$; ⊙ 0·05M acetic acid, concentration of sodium acetate changed from 0·008M to 0·5M, ionic strength $\mu = 0.5$; ● 0·01M acetic acid, concentration of sodium acetate changed from 0·0016M to 0·04M, ionic strength $\mu = 0.1$. Ionic stregth regulated by addition of sodium perchlorate. For $t_1 = 3.3$ s theoretical curve is drawn for $k_2 = 2 . 10^6$ l mol^{-1} s^{-1}.

The dependence of the instantaneous current i on drop-time t_1 at potentials, where limiting current has been reached, has the form $i \sim t_1^p$. For the coefficient p following values were found: At low pH $p = 0.24$ to 0.30 (at pH 2·4), $p = 0.22$ (at pH 3·1), not markedly different from the value $p \doteq 0.19$ for a diffusion controlled current on the first drop. At higher pH, when the limiting current already decreased, a value $p = 0.60$ to 0.64 (at pH 5·8) was found in accordance with the value $(p = 0.67)$ expected for a kinetic current. The current is thus an example of the formation of an electroactive species from the electroinactive form, diffusing towards the electrode.

The changes, observed on polarographic curves with changes in pH, are reversible. This has been proved, when the solution of tropylium was alkalified in the absence of air and then reacidified. The wave obtained in this way was identical with that obtained for the original buffer. The decrease is therefore not due to a decomposition, but due to a reversible chemical reaction. This finding is in agreement with the observations made during a titration of tropylium ion by a solution of hydroxide[17].

The form of the dissociation curve corresponds closely to the theoretical treatment given by Koutecky[18] for a reversible chemical reaction of first order, preceding the electrode reaction proper. The dissociation constant used in the calculation of rate constants was determined potentiometrically and the value found $(pK = 4.6)$ was in good agreement with that given in the literature[19] $(pK = 4.75)$.

Neither the pK'-values nor the form of the dissociation curve is practically influenced by the nature of the buffer used (Fig. 2) and the ionic strength (Fig. 3). The presence of base and the conjugated acid of pyridine type has no effect either; the shape of polarographic curves is practically identical in buffers of varying composition but of the same pH-value. The half-wave potential, which is to a certain degree dependent on the concentration of tropylium (p. 383), is for pH-values lower than 4·6 independent of pH and equal to $-0·27$ V (s. c. e) (Table I). For higher pH-values the half-wave potential is shifted to more negative potentials, following the equation $E_{\frac{1}{2}} = -0·27 - 0·040$ pH (cf. Fig. 12 in[10]).

Table I

Half-Wave Potentials of $1·5 \cdot 10^{-4}$M Tropylium Bromide

Buffer	pH	$E_{\frac{1}{2}}$ (V, S. C. E.)
Britton-Robinson	2·4	—0·26
	2·8	—0·26
	3·1	—0·26$_5$
	4·0	—0·26
	4·3	—0·27
	4·5	—0·27
	4·8	—0·27
	5·2	—0·29
	5·4	—0·29
	5·8	—0·31
	6·2	—0·32
	6·6	—0·33$_5$
Acetate	3·8	—0·27$_5$
	4·1$_5$	—0·28
	4·6	—0·28$_5$
	5·0	—0·28
	5·5$_5$	—0·29
	5·9	—0·31

Discussion

Acid-base Reaction. Tropylalcohol (*II*) has not been prepared[19] on synthetic scale and the routes used[17,19] resulted in the isolation of ditropylether. The course of potentiometric[17] as well as polarographic curves, seems to enable us to conclude — in agreement with Doering[17] — that in the reaction of tropylium ion at higher pH-values (mainly at low concentrations used in electrometric methods) the tropylalcohol (*II*) is formed as primary product, being in equilibrium with tropylium ion (*I*). Tropylalcohol may further undergo a competitive reaction with ditropylether as end product[17].

For the formulation of the acid-base reaction of tropylium, two possible mechanisms can be suggested: Either a solvolysis[19] (I):

$$\text{\includegraphics{} } + 2\,H_2O \quad \underset{k_2\text{I}}{\overset{k_1\text{I}}{\rightleftarrows}} \quad \text{\includegraphics{} } + H_3O^+ \qquad (I)$$

$$\qquad\quad I \qquad\qquad\qquad\qquad\qquad II$$

or — similar to the reaction with other nucleophilic reagents, like cyanide ion[17] — a specific reaction with the hydroxyl ion[6] (2):

$$\text{\includegraphics{} } + OH^- \quad \underset{k_2\text{I}}{\overset{k_1\text{II}}{\rightleftarrows}} \quad \text{\includegraphics{} } \qquad (2)$$

$$\qquad\quad I \qquad\qquad\qquad\qquad\qquad II$$

Basing on potentiometric measurements it is impossible to decide between these two alternatives. Polarographic measurements on the other hand enable us to distinguish these two possibilities as follows:

Tropylium ion (I) is reduced at positive potentials, whereas for tropylalcohol (II) a reduction wave at $-2\cdot09$ V was claimed[8]. The limiting currents of tropylium ion are not (in the whole pH-range studied) simply proportional to the concentration of this species in the bulk of solution but rather — mainly at pH greater than about 4 — controlled by the rate of formation of this particle with rate constant k_2^I or k_2^{II}. The reaction with the constant k_2^I proceeds in buffered solutions as a reaction of the pseudo first order, as are the recombination reactions that have been studied so far[20]. The reaction with a constant k_2^{II} would be a first order reaction. The slope of the pH-dependence of the limiting current would be half that observed in the previous case. Since the experimentally found dependence the limiting current of tropylium ion on pH possess the form of a dissociation curve with a slope corresponding to a pseudo first order reaction (Figs. 2, 3), the mechanism (2) can be excluded.* Thus the view of Ždanov[6], supporti ng mechanism (2), based on qualitative observations only, could not been confirmed.

Using the treatment by Koutecký[18] and values $pK' = 5\cdot64$ at $t_1 = 3\cdot3$ s and $pK_{titr} = 4\cdot6$ the rate constant of the reaction forming the electroactive species I was computed as $k_2 = 2 \cdot 10^6\,l\,mol^{-1}\,s^{-1}$. For the rate of formation of base II the value $k_1 = 5 \cdot 10^{11}\,l\,mol^{-1}\,s^{-1}$ was found.

The conditions[21] for the application of the above treatment are fulfilled, as $k_2 t_1 > 10$ and as, in the pH-region where the kinetic current is observed, the equilibrium in the solution is shifted towards the electroinactive form II. It seems that the ten-fold excess, as encountered in the present case, is sufficient to fulfil the second condition[21].

---

* We wish to express our thanks to Dr J. Koryta for directing our attention to the possibility of deciding the mechanism based on the slope of the pH-dependence.

Since the limiting current is independent of potential in a broad potential range[21] (Fig. 1), as it is not substantially affected by the presence of small amounts of surface active substances[21] and as it is not affected by a change in the ionic strength[22], it seems that the reaction can be regarded as homogeneous. Due to the relatively low rate of the governing reaction the dimensions of the reaction layer exceed substantially that of the double layer (μ is of the order 10^2 Å). Moreover the difficulties encountered with the ordinary type of recombination acid-base reactions[21,23,24] can be at last partly due to the negative sign (or zwitterionic form) of the species reacting according to $A^- + H_3O^+ \rightleftharpoons HA + H_2O$. In the present case the species reacting with hydrogen ion bears no unit charge.

Other complicating factors, involved in the study of most of the O- and N-acids studied so far, such as the reactions with proton donors other than hydrogen ion[25-36], the influence of hydration[37,38], the rôle of enolization[38] and the effect of adsorption of proton donors[39] are in the case of tropylium ion either excluded by experimental findings (i. e. by the absence of effects of buffer nature, of buffer capacity, of the presence of pyridinium ions and of ionic strength) or by the structure of this acid-base pair.

Thus contrary to uncertainty concerning the values of the rate constants for dissociation and recombination[20-24,40] of O- and N-acids determined polarographically, the values given above can be accepted as equivalent to other rate constants of homogeneous reactions. The values for both the rate constant k_2 corresponding to a reaction of a cation with a base and k_1 referring to a unimolecular reaction of a Lewis acid, are substantially smaller than values for acid-base reactions of O- and N-acids, determined by various methods[40]. The magnitude of the measured constants is rather similar to that of rate constants found for C-acids. If the reaction course given in (1) is not substantially different from the actual mechanism of the acid-base reaction, then this similarity can be explained by comparison of bond-breaking energies of the C—H bond in C-acids and of C—OH in the reaction with constant k_2.

Electrode Process. For numerous acid-base reactions that have been studied polarographically so far a shift of the half-wave potential with pH has been observed even for pH < pK. This seems to indicate that proton transfer takes part even in the electrode process proper. Thus e. g. in the reduction of free pyruvic acid (formed by recombination from anion and hydrogen ion) further protons are consumed during the formation of lactic acid. This is shown by the value of $dE_{\frac{1}{2}}/d$ pH which corresponds to a transfer of more protons than are involved in the acid-base reaction preceding the electron transfer.

Contrary to these observations, the half-wave potential for the reduction of tropylium ion at pH smaller than 4·6 is pH independent, similarly as for other hydrocarbons[41-44]. Electrode process is thus simpler than for other reactions of the "recombination type".

On the other hand we were not able to detect for tropylium ion, whether at pH > > pK' the half-wave potential is pH-independent, as was frequently shown for other electrode processes accompanied by a proton transfer (e. g.[1,45-47]) and as required by theory[48]. At pH 5·6 and higher (even for the highest concentration where adsorption did not influence polarographic curves[10], i. e. at concentration $1·5 . 10^{-4}$M) the currents were so small that an exact measurement of half-wave potentials was impossible.

The value for the slope of the $E_{\frac{1}{2}}$−pH dependence for pH values in the interval $\overset{.}{p}K < pH < pK'$ (0·040) is identical with the value of the coefficient from the logarithmic analysis of the slope of the current-voltage curve. Such correlations were often observed[49] and an attempt was made to explain them on the basis of the theory for irreversible reduction processes[50].

Thus the dependence of both the wave-height and half-wave potentials on pH are consistent with a reduction scheme:*

$$\text{(structure)} + H_3O^+ \rightleftharpoons \text{(structure)} + 2\,H_2O \qquad (3)$$

The consumption of one electron, confirmed here by comparison with waves for other depolarizers, is in agreement with coulometric measurements by Ždanov[7,8] (even if in their first paper the authors[6] supposed a two-electron process). The reduction path *via* heptatrienyl radical under formation of ditropyl is similar to that forwarded for the reduction of tropylium ion by zinc in a solution acidified by the hydrolysis of tropylium bromide only[17].

The formation of ditropyl in the polarographic reduction of tropylium ion was suggested by Ždanov[7,8] (who neglected the rôle of the proton transfer reaction preceding the electrode process proper). He anticipated the formation of ditropyl basing on the number of electrons transferred only. The dimerization following the electrode process proper (provided that the electron transfer is not totally irreversible) could be proved by us from the shift of the half-wave potentials with increasing concentration of tropylium. The direction of this shift is in accordance with that given by the equation derived by Hanuš[51] for systems where the product of an electrode process is inactivated by dimerization.

The electrode system proper differs from most reductions of hydrocarbons in aqueous solutions[41-45] in that that in the case of tropylium no proton transfer occurs either during or subsequent to the electron uptake[52].

Polarographic reduction of acids proceeds usually at more positive potentials than the reduction of the conjugated bases. Besides other factors the fact should

---

* The symbol $\rightleftharpoons$ is used for a reversible mobile reaction, whereas $\rightleftarrows$ is used for a reversible chemical reaction, without emphasizing whether the mobility of the establishment of this equilibrium is high or not.

be borne in mind that electron is a nucleophilic reagent and that the acid is always more electrophilic than the coresponding base. For the pair tropylium ion (I) – tropylalcohol (II) the difference in half-wave potentials is over 1·8 V. Besides the positive charge of the aromatic tropylium ion which supports its electrophilic character, other factors (e. g. dimerization following the electron transfer) can play a rôle.

The last mentioned factor influences seemingly also the relatively easy reducibility of tropylium ion, manifested by its positive potential. The reduction occurs at more positive potentials than e. g. that of the azulenium cation[1]. But the comparison and discussion of this difference is excluded by the difference in the reduction mechanisms. Nevertheless it is worth mention that in the sequence: protonized benzene, azulenium ion, tropylium ion the acidity decreases and the half-wave potentials are shifted to more positive values.

Conclusions. The reaction (3) represents a *new type of chemical reaction preceding the electrode process proper.* This acid-base reaction is simpler than that observed earlier with O- and N-acids. Not only the hydrogen ions do not take part in the electrode process proper, but also the kind of the buffer used, its capacity, the presence of acids of the ammonium type and the ionic strength are without influence. This difference (besides the change in the number of charged particles taking part in the reaction that occurs with O-acids – but not with N-acids) together with the substantially slower reaction rates can be sought in the difference in mechanism. Whereas the recombination step for O- and N-acids as well as for C-acids (like azulenium ion) is normally an addition reaction, with tropylium ion an elimination occurs.

The authors express their thanks to Dr C. R. Ganellin for the specimen of tropylium bromide and to Drs V. Hanuš and J. Koryta for their critical comments of the manuscript. The careful technical assistance of Miss N. Pejchová is acknowledged.

References

1. Zuman P.: Z. physik. Chem. (Leipzig), Sonderheft *1958*, 243.
2. Šantavý F.: Chem. listy *47*, 1539 (1953); Chem. Technik *8*, 316 (1956).
3. Bartek J., Mukai T., Nozoe T., Šantavý F.: This Journal *19*, 885 (1954); Chem. listy *48*, 1123 (1954).
4. Jámbor B., Bartek J., Šantavý F.: This Journal *20*, 1244 (1955); Chem. listy *49*, 932 (1955).
5. Šantavý F., Jámbor B., Němečková A., Mollin J., Bartek J.: This Journal *22*, 1655 (1957); Chem. listy *51*, 704 (1957).
6. Volpin M. E., Ždanov S. I., Kursanov D. N.: Dokl. Akad. nauk SSSR *112*, 264 (1957).
7. Ždanov S. I.: Z. physik. Chem. (Leipzig) Sonderheft *1958*, 235.
8. Ždanov S. I., Frumkin A. N.: Dokl. Akad. nauk SSSR *122*, 412 (1958).
9. Geske D. H.: J. Am. Chem. Soc. *81*, 4145 (1959).
10. Chodkowski J., Zuman P.: This Journal, in the press.
11. Němec L.: *Thesis*. Department of Chemistry, Charles University, Prague 1958.
12. Valenta P.: Z. physik. Chem. (Leipzig), Sonderheft *1958*, 46.
13. Kalousek M.: This Journal *13*, 105 (1948); Chem. listy *40*, 149 (1946).
14. Rálek M., Novák L.: This Journal *21*, 248 (1956); Chem. listy *49*, 557 (1955).

15. Kemula W., Kublik Z.: Anal. Chim. Acta *18*, 104 (1958).
16. Zuman P.: This Journal *15*, 839 (1950); Chem. listy *46*, 599 (1952).
17. Doering W. von E., Knox L. H.: J. Am. Chem. Soc. *79*, 352 (1957).
18. Koutecký J.: This Journal *18*, 597 (1953); Chem. listy *47*, 323 (1953).
19. Doering W. von E., Knox L. H.: J. Am. Chem. Soc. *76*, 3203 (1954).
20. Brdička R.: This Journal *19*, S 41 (1954).
21. Koryta J.: Z. Elektrochem. *64*, 23 (1960).
22. Grabowski Z. R., Bartel E. T.: Roczniki chem. *34*, 611 (1960).
23. Strehlow H.: Z. Elektrochem. *64*, 45 (1960).
24. Zuman P.: Chem. listy *54*, 1244 (1960).
25. Hanuš V.: *Proc. Ist Intern. Polarograph. Congr. Prague 1951*, Part I., p. 811. Published by Přírodovědecké nakladatelství, Prague 1951.
26. Hans W., Henke K. H.: Z. Elektrochem. *57*, 595 (1953).
27. Wiesner K., Wheatley M., Los J. M.: J. Am. Chem. Soc. *76*, 4858 (1954).
28. Wheatley M. S.: Experientia *12*, 339 (1956).
29. Green J. H., Walkley A.: Australian J. Chem. *8*, 51 (1955).
30. Zahradník R., Svátek E., Chvapil M.: This Journal *24*, 347 (1959); Chem. listy *51*, 2232 (1957).
31. Bartel E. T., Grabowski Z. R., Kemula W., Turnowska-Rubaszewska W.: Roczniki Chem. *31*, 13 (1957).
32. Hrdý O.: This Journal *24*, 1180 (1959); Chem. listy *52*, 1058 (1958).
33. Zuman P.: J. Electrochem. Soc. *105*, 758 (1958).
34. Kemula W., Grabowski Z. R., Bartel E. T.: Roczniki chem. *33*, 1125 (1959).
35. Becker M., Strehlow H.: Z. Elektrochem. *64*, 42 (1960).
36. Becker M., Strehlow H.: Z. anal. Chem. *173*, 100 (1960).
37. Becker M., Strehlow H.: Z. Elektrochem. *64*, 129 (1960).
38. Ono S., Takagi M., Wasa T.: This Journal *26*, 141 (1961).
39. Volková V.: Nature *185*, 743 (1960).
40. Bell R. P.: Quart. Revs. *13*, 169 (1959).
41. Laitinen H. A., Wawzonek S.: J. Am. Chem. Soc. *64*, 1765 (1942).
42. Kolthoff I. M., Lingane J. J.: *Polarography*, II. Ed., Vol. II., p. 634. Interscience, New York 1952.
43. Hoijtink G. J., Van Schooten J.: Rec. trav. chim. *71*, 1089 (1952); *72*, 691, 903 (1953); *73*, 355 (1954).
44. Hoijtink G. J., De Boer E., Van der Meij P. H., Weijland W. P.: Rec. trav. chim. *74*, 277 (1955); *75*, 487 (1956).
45. Zuman P.: Chem. listy *48*, 94 (1954).
46. Zuman P., Tenygl J., Březina M.: This Journal *19*, 46 (1954); Chem. listy *47*, 1152 (1953).
47. Volke J., Volková V.: This Journal *20*, 1332 (1955); Chem. listy *49*, 490 (1954).
48. Tanaka N., Tamamushi R.: *Proc. Ist. Intern. Polarograph. Congr. Prague 1951*, Part I, p. 486 (532). Published by Přírodovědecké vydavatelství, Prague 1951.
49. Gardner H. J., Lyons L. E.: Revs Pure Appl. Chem. (Australia) *3*, 134 (1953).
50. Franklin J. L., Field F. H.: J. Am. Chem. Soc. *75*, 2819 (1953).
51. Hanuš V.: Chem. zvesti *8*, 702 (1954).
52. Zuman P., Chodkowski J., Potěšilová H., Šantavý F.: Nature *182*, 1535 (1958).

Translated by the Author (P. Z.).

POLAROGRAPHY OF NONBENZENOID AROMATIC AND RELATED SUBSTANCES. VIII.*

ADSORPTION PROCESSES DURING THE ELECTROREDUCTION OF THE TROPYLIUM ION

P. ZUMAN and J. CHODKOWSKI**

Polarographic Institute, Czechoslovak Academy of Science, Prague

Received July 10th, 1958***

With increasing concentrations of tropylium bromide in buffered solutions one, two and finally three types of adsorption process occur. The particular adsorption processes differ in the rate of the formation of the adsorbate and in the characteristic features of this adsorbate. This was shown by the methods of classical polarography, by measurement of the time-dependence of instantaneous and mean currents, by methods of oscillographic and impulse polarography and by the study of electrocapillary curves. For the area occupied on the surface of the electrode by one sevenmembered ring in the reduced state the value $S = 29$ Å^2 was computed. Using the covalent radii and the bond lengths the value $S = 22$ Å^2 was calculated for the area occupied by one tropylium ring.

Ždanov and his coauthors[1-3] observed effects on polarographic curves of tropylium ion which they ascribed to adsorption, which was also proved by electrocapillary curves, registered using the capillary electrometer method. They missed the first two adsorption waves and their curves have shown some irregularities. Hence, during the study of acid-base properties of tropylium[4] we paid attention to adsorption phenomena as well. The adsorption effects were studied using the methods of classical polarography, but also by measurement of the change of the mean and instantaneous current with time, by measurement of electrocapillary curves and using methods of single-sweep and oscillographic polarography as well as using the commutator method.

Experimental

Apparatus, capillary, chemicals used, solutions, and techniques were the same as in preceding part of this series[4].

The time-dependences of the instantaneous currents at a given potential (*i–t*-curves) were

---

* Part VII.: This Journal *26*, 380 (1961).

** Standing adress: Institute of Physicochemical Analytical Methods, Polish Academy of Science, Warsaw, Poland.

*** Revised manuscript received on October 10th, 1960.

recorded on the "first" and "second" drop using the apparatus designed and built by Němec[5]. The capillary used had $m = 1.78$ mg s^{-1} and $t_1 = 4.6$ s at $h = 55$ cm.

For studies of the change of the mean current with drop-time and independently, with mercury flow rate, an electronic device was used enabling the regulation of the drop-time. Measurements of the mean current as a function of the drop-time were performed in solutions of $1 . 10^{-3}$M tropylium bromide in Britton-Robinson buffer of pH 2·4 at $m = 1.38$ mg s^{-1} and $h = 30$ cm for $t_1 = 6.4$ s to 1·0 s. The influence of the rate of flow was followed in the same solution for $m = 1.38$ mg s^{-1} to 3·25 mg s^{-1} at t_1 between 1·0 s and 2·6 s.

Electrocapillary curves were measured in Britton-Robinson buffer pH 2·8, containing 0; 1; 2; 5; and $10 . 10^{-4}$M tropylium bromide. The drop-time for 20 drops was measured.

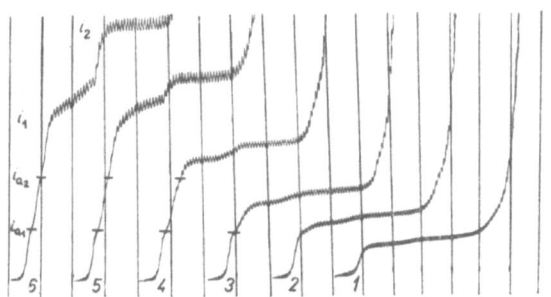

Fig. 1

The Waves of Tropylium Ion at Different Concentrations
Britton-Robinson buffer of pH 2,4. Concentration of tropylium bromide: 1 1·0; 2 2·0; 3 3·0; 4 5·0; 5 7·5; 6 $10 . 10^{-4}$M. From 0 V, s.c.e., 216 mV/absc., $h = 60$ cm, sens. 1 : 15.

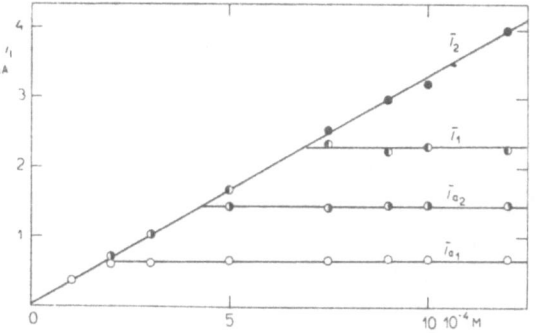

Fig. 2
Dependence of Wave-Heights on Tropylium Ion Concentration
Britton-Robinson buffer of pH 2·4; limiting current O i_{a1}; ◐ i_{a2}; ◑ i_1; ● i_2, against concentration (molar) of tropylium bromide.

Results and Discussion

Influence of Concentration and pH. When the dependence of the limiting current on the concentration of tropylium bromide is followed in buffers of pH lower than about 4, we can observe (Fig. 1) that in addition to a wave at -0.25 V (further denoted as i_{a_1}) a new wave occurs at more negative potentials, when the concentration of tropylium ion is higher than about $1.7 . 10^{-4}$M. This wave, which we shall denote as i_{a_2}, also reaches a limiting value, and does not change with further addition of tropylium at concentrations higher than about $4.3 \cdot 10^{-4}$M (Fig. 2). In solutions containing even higher concentrations of tropylium ion, another wave, denoted as i_1, occurs. In solutions of $1.2 \cdot 10^{-3}$M tropylium bromide the following half-wave potentials were measured (V, S.C.E.):

Buffer	Britton-Robinson pH 2-4	Acetate pH 4·0
$(E_{1/2})_{ia1}$	-0.16_5	-0.21
$(E_{1/2})_{ia2}$	-0.23_5	-0.25
$(E_{1/2})_{i1}$	-0.28_5	-0.31_5

Waves i_{a_2} and i_1 coincide at higher pH-values. Finally for concentrations of tropylium ion exceeding $7 \cdot 10^{-4}$M, the total height of $i_{a_1} + i_{a_2} + i_1$ ceases to increase with increasing concentration. A new wave at -0.65 V appears, the sum $(i_{a_1} + i_{a_2} + i_1 + i_2)$ being proportional to the concentration of tropylium bromide. The general picture of the dependence of the height of particular waves on concentration of the depolarizer is given in Fig. 2.

Similarly also on the $i–E$-curves obtained with the single-sweep method, recorded with the instrument designed by Valenta[6], one, two or three current peaks were observed with increasing concentration of the depolarizer (Fig. 3). Following values of peak potentials (E_n) were found in a Britton-Robinson buffer pH 2·4 containing $1.2 \cdot 10^{-3}$M tropylium bromide.

	E_n, V (S.C.E.)	$E_n - E_{1/2}$, V
E_1 corresponds to $(E_{1/2})_{ia1}$	-0.29	-0.12_5
E_2 corresponds to $(E_{1/2})_{ia2+i1}$	-0.37	-0.11
E_3 corresponds to $(E_{1/2})_{i2}$	-0.75	-0.10

Potentials, at which the current peaks arrive at their maximum height, are (in accordance with the theory) shifted towards more negative potentials in comparison with the half-wave potentials in classical polarography. The magnitude of this shift is another proof for the irreversibility of the electrode process.

With increasing pH value the total height of the tropylium wave (Fig. 4) (i. e. the sum of $i_{a_1} + i_{a_2} + i_1 + i_2$) decreases in the form of a dissociation curve, the shape of which is in a good agreement with that predicted by theory (Fig. 5) for a pseudo-first order reaction preceding the electrode process proper[7]. The value $pK' = 5.5$ does not differ from the value $pK' = 5.64$ found at lower concentrations[1].

On polarographic curves (Fig. 4) such a number of waves is observed, as for the same total current intensity at lower pH-values at a correspondingly lower concentration. E. g. in a solution of $1 \cdot 10^{-3}$M tropylium bromide at pH 5·8 (curve 8, Fig. 4), where the total current corresponds to about 50% of the original height at pH 2·8 to 4·1 (curves 1 and 2, Fig. 4), the form of the current-voltage curve, the presence of two waves $(i_{a_1}$ and $i_{a_2} + i_1)$, and their relative height are the same as in $5 \cdot 10^{-4}$M tropylium

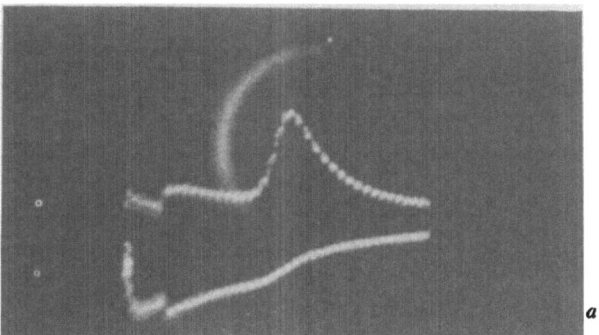

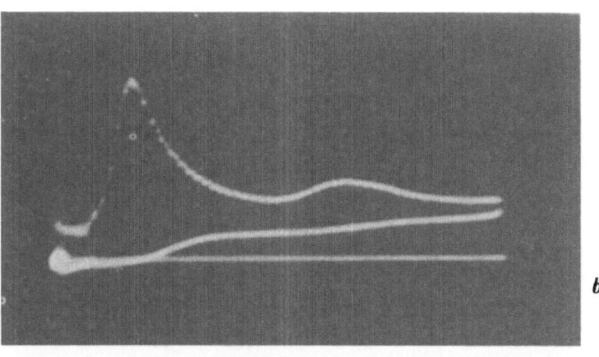

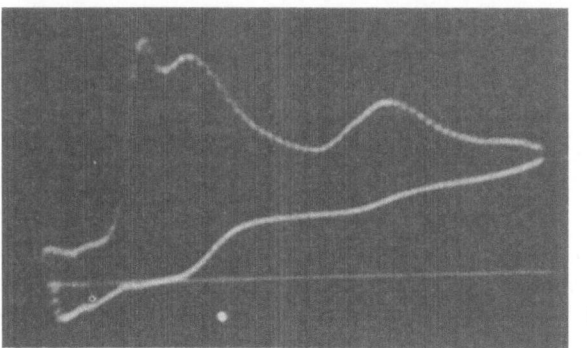

Fig. 3

Current-Voltage Curves Obtained With the Single-Sweep Method, Using Apparatus according to Valenta[5]

Britton-Robinson buffer of pH 2·4. *a*) $1·5 \cdot 10^{-4}$M; *b*) $3 \cdot 10^{-4}$M; *c*) $1·2 \cdot 10^{-3}$M tropylium bromide. Scanning rate $0·78$ V s^{-1}, triangular pulse, drop-time 21 s, pulse-time 1 s, $m = 1·62$ mg s^{-1}.

518

bromide at pH 2−4. If we thus follow the change of the curves (Fig. 4) in a $1 . 10^{-3}$M solution of tropylium bromide with pH, we observe first the decrease of the wave i_2 and at higher pH-values also of the wave $(i_{a_2} + i_1)$. At still higher pH-values only the waves i_{a_1} appear on polarographic curves.

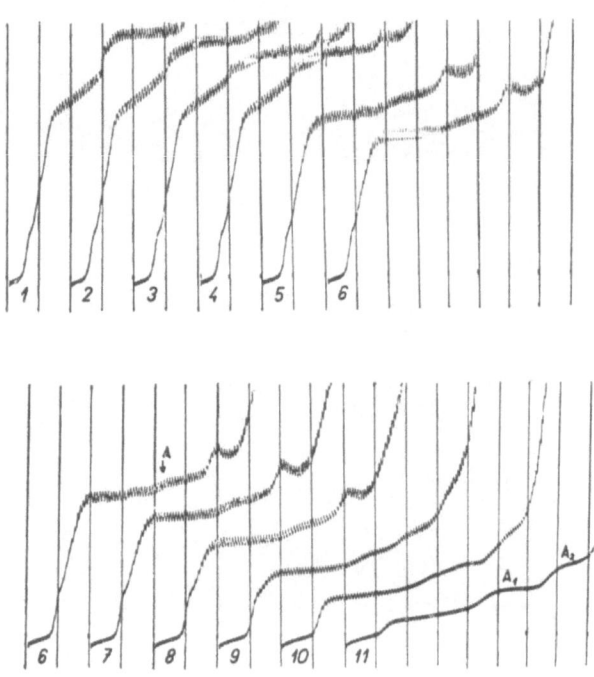

Fig. 4

Dependence of the Tropylium Waves on pH

$1 . 10^{-3}$M tropylium bromide, Britton-Robinson buffers of pH: 1 2·8; 2 4·1; 3,4 4·4₅; 5 5·1; 6 5·3; 7 5·5; 8 5·8; 9 6·0; 10 6·4; 11 6·7; Wave A respectively A_1 and A_2 at higher pH-values correspond to an oxidation product of tropylium. From 0 V, S.C.E., 208 mV/absc., $h = 60$ cm, sens. 1 : 15.

Character of the Total Limiting Current. The total height of the sum of waves is proportional to the concentration of tropylium ion in the whole concentration and pH-range studied. The total limiting current is at lower pH-values controlled solely by diffusion, at higher pH-values it is influenced also by a chemical reaction (cf.[4]). This was verified under usual polarographic experimental conditions from the dependence of the total limiting current on the mercury head h, for which at lower pH the relation $\bar{i}_{lim} \sim h^{1/2}$, and at higher pH-values than about 6 the relation $\bar{i}_{lim} \sim h^0$, were proved.

Using a capillary with a regulated drop-time in a solution of $1 . 10^{-3}$M tropylium bromide at pH 2·4 a dependence of the mean total limiting current $\left(i. \, e. \; \bar{i}_{lim} = i_{a_1} + i_{a_2} + i_1 + i_2\right)$ on the drop time t_1 possessed the form $\bar{i}_{lim} \sim t^{0.205}$ (theory $t^{0.19}$)

(at $m = 1.38$ mg s^{-1}). The dependence of this mean current on the mercury flow rate m followed the relation $\bar{i}_{lim} \sim m^{0.69}$ (theory $m^{0.67}$).

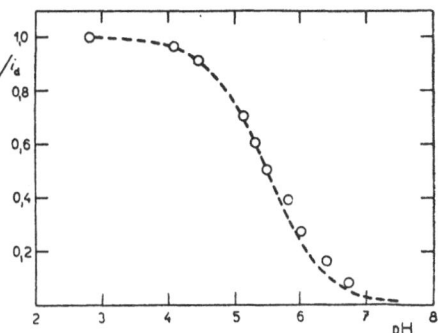

Fig. 5
Dependence of Limiting Current of Tropylium Ion on pH
Britton-Robinson buffer, $1 . 10^{-3}$M tropylium bromide. Experimental points and theoretical curve at $t_1 = 3.3$ s for $k_1 = 9.3 . 10^5$.

Table I

The Values for the Coefficient p in the Relation $i \sim t^p$ for Reduction Waves of Tropylium Ion under Different Conditions

Tropylium(M)	E^a, V	Type of Current[b]	p
		pH 2·4	
$1.5 . 10^{-4}$	0·30	$(i_{a1})_{lim}$	0·26
	0·40	$(i_{a1})_{lim}$	0·20
$1.2 . 10^{-3}$	0·70	$(i_{a1} + i_{a2} + i_1 + i_2)_{lim}$	0·21$_5$
$1.5 . 10^{-4}$	0·25	$0.75(i_{a1})_{lim}$	0·31
$3.0 . 10^{-4}$	0·20	$0.5(i_{a1} + i_{a2})_{lim}$	0·64
$1.2 . 10^{-3}$	0·15	$0.1(i_{a1} + i_{a2} + i_1 + i_2)_{lim}$	0·56
		pH 5·8	
$5.0 . 10^{-4}$	0·25	$0.7(i_{a1})_{lim}$	0·74
	0·30	$0.9(i_{a1})_{lim}$	0·62
	0·40	$(i_{a1})_{lim}$	0·53

[a] Potential (S.C.E.), at which the i–t-curves were registered [b] $(i_{a1})_{lim}$ denotes that the i–t-curve was registered at potentials corresponding to the mean limiting current of wave i_{a1}. Thus $0.5(i_{a1} + i_{a2})_{lim}$ shows that the time-change of the instantaneous current was registered at a potential where the mean current was equal to 50% of the total mean limiting current of the sum of waves i_{a1} and i_{a2}, observed at this concentration.

Further proof for the diffusion character of the total limiting current was given by measurements of the time-dependence of the instantaneous current (i–t-curves). The measurements were performed on the first drop after the appropriate polarizing

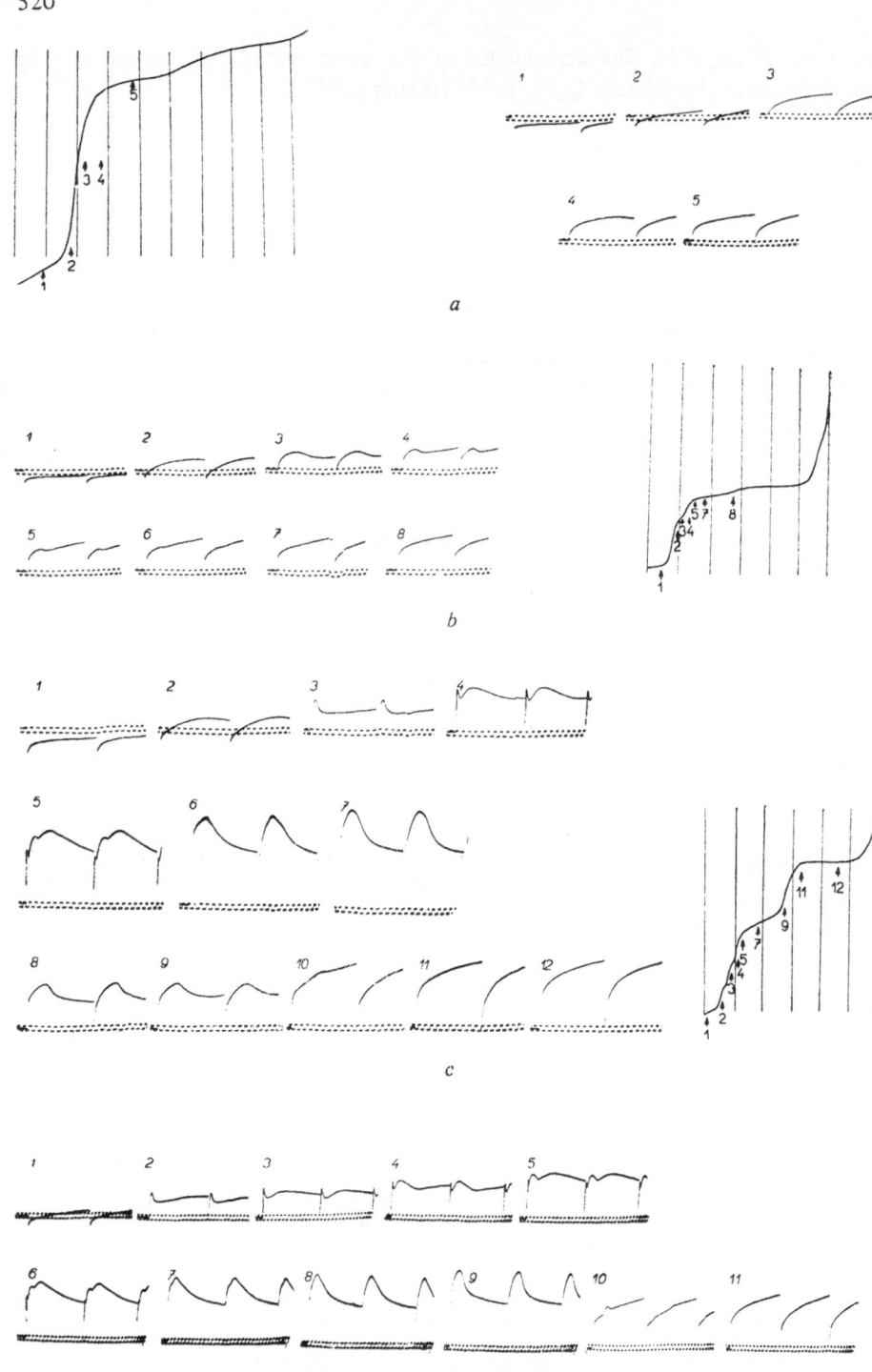

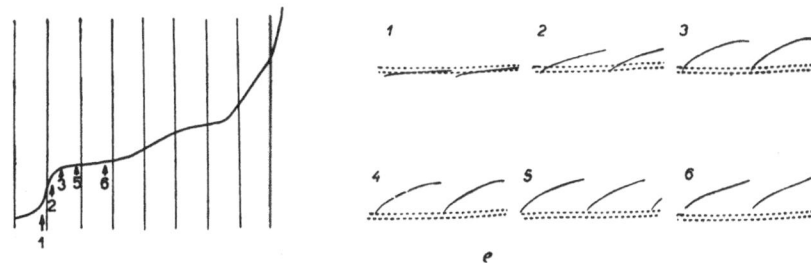

Fig. 6

Curves of the Dependence of the Instantaneous Current on Time Registered by the Apparatus According to Němec[5]

Britton-Robinson buffers. *a*) $1{\cdot}5 . 10^{-4}$M tropylium bromide, pH 2·4, $h = 55$ cm, the distance of the calibration lines 0,062 µA, time basis 5. *1* 0·1 V; *2* 0·2 V; *3* 0·25 V; *4* 0·3 V; *5* 0·4 V. *b*) 3 . $. 10^{-4}$M tropylium bromide, pH 2·4; $h = 55$ cm, the distance of the calibrations lines 0·062 µA, time basis 5. *1* 0·1 V; *2* 0·2 V; *3* 0·24 V; *4* 0·28 V; *5* 0·32 V; *6* 0·36 V; *7* 0·4 V; *8* 0·6 V. *c*) $1{\cdot}2 . 10^{-3}$M tropylium bromide, pH 2·4, $h = 55$ cm, the distance of the calibration lines 0·062 µA (curves *1—7*) and 0·16 µA (curves *8—12*) time basis 5. *1* 0·05 V; *2* 0·15 V; *3* 0·2 V; *4* 0·25 V; *5* 0·3 V; *6* 0·35 V; *7* 0·4 V; *8* 0·55; *9* 0·6 V; *10* 0·65 V; *11* 0·70 V; *12* 1·0 V. *d*) $1{\cdot}2 . 10^{-3}$M tropylium bromide, pH 2·4, $h = 35$ cm, the distance of the calibration lines 0·062 µA (curves *1—9*) and 0·16 µA (curves *10—11*) time basis. 3. *1* 0·15 V; *2* 0·2 V; *3* 0·22 V; *4* 0·25 V; *5* 0·28 V; *6* 0·30 V; *7* 0·32 V; *8* 0·35 V; *9* 0·50 V; *10* 0·65 V; *11* 1·0 V. *e*) $5 . 10^{-4}$M tropylium bromide, pH 5·8, $h = 55$ cm, the distance of calibration lines 0·062 µA, time basis 5. *1* 0·2 V; *2* 0·25 V, *3* 0·3 V; *4* 0·35 V; *5* 0·4 V; *6* 0·6 V. Arrows on diagrams show potentials at which the *i–t*-curves were recorded.

voltage has been applied. The results are given in Table I. The value of the coefficient *p* defined by the relation $i \sim t^p$ for limiting currents at lower pH-values is near to 0·2. With limiting currents at pH 5·8 (and above) and with such currents on the rising part of polarographic wave that are not yet influenced by adsorption, the value of *p* was found to be about 0·6 (Table I), typical for currents limited by the rate of chemical reaction.

For *i–t*-curves on second (and following) drops the values of the coefficient *p* were usually higher and often departures from the linear plot log *i*–log *t* were detected.

Proofs for Adsorption. Limiting currents of waves i_{a_1}, i_{a_2} and i_1 are at concentrations, at which the current, is already independent of further addition of tropylium ion controlled by the rate of growth of the mercury dropping electrode. The reduction product is adsorbed on the freshly grown surface[8]. This was demonstrated by the mentioned course of the dependence of the limiting currents on concentration of tropylium ion (Fig. 2) and by the character of the dependence of these mean limiting currents on mercury pressure. All three mean currents i_{a_1}, i_{a_2} and i_1 at these concentrations of tropylium ion are proportional to the height of mercury head *h*.

A further proof for the adsorption character of the particular currents followed from the study of dependence of instantaneous currents on time (*i–t*-curves). The course of these *i–t*-curves has in principle a shape, described for adsorption currents by Brdička[8]. In Fig. 6 *i–t*-curves are given, recorded at different concentrations of tropylium ion, at different potentials and for two different mercury pressures.

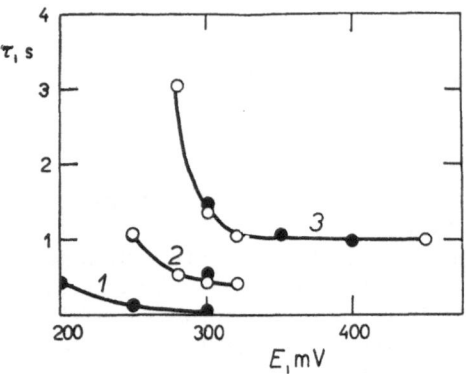

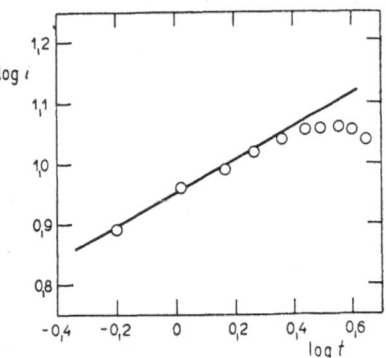

Fig. 7

Dependence of the Time (τ) in Which the *i–t*-Curve Reaches its Maximum Value on Potential and on the Height of the Mercury Head
Britton-Robinson buffer of pH 2·4, 1·2 . 10^{-3}M tropylium bromide. τ — the time interval necessary for current to reach its maximum value. $\bigcirc$ $h = $ = 35 cm, $\bullet$ $h = $ 55 cm.*1* current corresponding to wave i_{a1}, *2* to wave i_{a2}; *3* to wave i_1.

Fig. 8

Logarithmic Analysis of an *i–t*-Curve Influenced by Adsorption
Britton-Robinson buffer of pH 2·4; 1·5 . 10^{-4}M tropylium bromide, $h = 55$ cm, potential $-0·3$ V.

Three phenomena, observed on these *i–t*-curves, should be discussed here in some detail: *1*. For the adsorption current i_{a_1} the decrease of current, due to the formation of an adsorbed layer, is followed during the later phase of the drop-life by a new increase in current. This increase could be explained by the assumption that the adsorbate does not form a total hindrance of electroreduction but that in the presence of the adsorbed layer a slow electrode process can occur through the layer or in other words a hindered diffusion occurs. Such phenomena have been observed recently for metal ions in the presence of surface active substances[9,10]. *2*. The initial increase of the current (corresponding to mass transfer by diffusion or by the accompanying chemical reaction) is followed by a decrease of current due to the increased blocking of the surface by the adsorbed layer. The time-interval τ, after which the current starts to decrease, is shorter the higher is the concentration of the tropylium ion and the more negative is the potential (*i. e.* the higher is the surface concentration of the reduced form) (Fig. 7). This time-interval is assumed to correspond to a period during which the coverage of the surface of the electrode occurs. With increasing tropylium ion concentration the rate of formation of the adsorbed layer is increased and thus this time-interval shortened. Similarly also an increased rate of formation of the adsorbate can be predicted at more negative potentials (in the neighbourhood of the electrocapillary maximum). 3. For higher concentration of the depolarizer two or even three successive increases and decreases of the current with time can be observed on the i–t-curve. These particular changes correspond to the processes manifested by limiting currents i_{a_1}, i_{a_2} and i_1. The formation of different types of adsorbates is assumed. – A similar form of i–t-curve was observed for the reversible system of riboflavine[8], where it was ascribed to leucoform and to a semiquinone.

The logarithmic analysis of i–t-curves reveals the role of adsorption in the later phase of the drop-life even in cases where this effect is not visually evident from the original i–t-curves (Fig. 8).

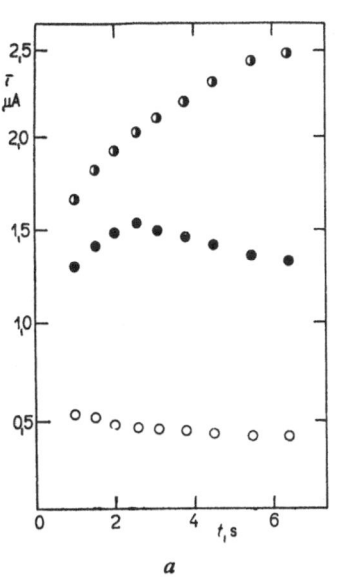

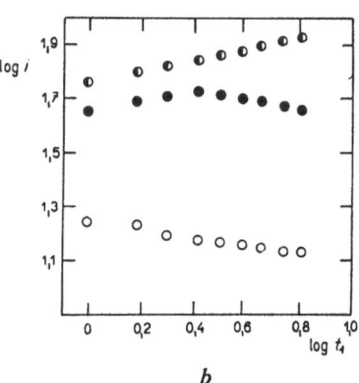

a   *b*

Fig. 9

Dependence of the Mean Current (i) on the Drop-Time

Britton-Robinson buffer of pH 2·8; $1·2 \cdot 10^{-3}$M tropylium bromide $m = 1·38$ mg/s, regulated drop-time. *a)* $i_{\lim} - t_1$; *b)* $\log i_{\lim} - \log t_1$. $\bigcirc - (i_{a_1})_{\lim}$; $\bullet - (i_a + i_{a_2} + i_1)_{\lim}$; $\mathbb{O} - $ total limiting current.

Also the dependence of the mean current values i on the regulated drop-time t_1 at given m shows (Fig. 9) a similar course to that found for the instantaneous current. The analogy is especially striking for the dependence of $\log i_{\lim}$ on $\log t_1$ for the wave ($i_a + i_1$) (compare Fig. 8 and 9b), whereas for the wave i_a the rising part with a fresh electrode surface is missing. The dependence of the mean current i on the out-flow velocity of mercury m for a constant drop-time t_1 possesses for all three kinds of limiting currents the assumed form $i \sim m^{2/3}$.

The adsorption character of waves i_{a_1}, i_{a_2} and i_1 at sufficiently high concentrations was proved also from the shape of the electrocapillary curves (Fig. 10). We suppose that at potentials more negative than the increase of current, corresponding to the wave i_1, an adsorbed insoluble film is formed. The wave i_2 is an increase in the reduction current, following the desorption of the film at -0.6 V. On the electrocapillary curve (Fig. 10) the corresponding potentials are marked by arrows. Such examples of "desorption currents" were in polarography described both in the absence[11-13] and presence[14] of depolarizers. Such a type of "desorption current" was observed in several instances for anodic waves corresponding to the formation of a slightly soluble mercury compound[15-17].

Capacity and adsorption phenomena can also be observed on the oscillographic dE/dt–E-curves (Fig. 11). The anodic in-cut corresponds to a substance formed at more positive potentials (a so called artefact).

Area Occupied by the Reduction Product. Following proofs for the adsorption characters of particular waves given above, it was possible using the equation[8] (1)

$$(i_a)_{max} = 0.85nFzm^{2/3}t_1^{-1/3} \tag{1}$$

(where z is the number of moles adsorbed on 1 cm^2, the other symbols possessing their usual meaning) to compute for $n = 1$ the area (S) covered by one sevenmembered ring of the reduced form on the surface of the dropping electrode. Equation (1), derived for reversible systems[8], is valid assuming that diffusion controls the masstransfer and the formation of a layer on the electrode surface proceeds rapidly when the concentration of the adsorbed form arrives at a given surface-concentration.

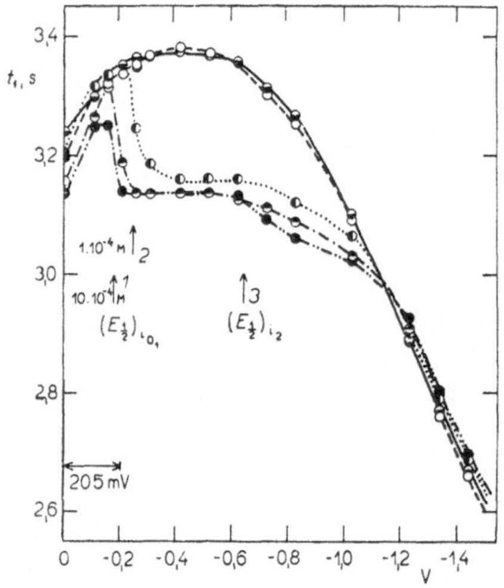

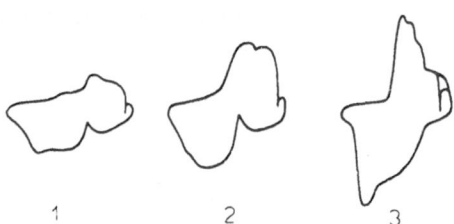

Fig. 10

Electrocapillary Curves

Britton-Robinson buffer of pH 2·8, concentration of tropylium bromide: $-$ $\bullet$ $-$ 0; $\circ$ $-\ -\ -$ 1 . 10^{-4}M; $\cdots$ $\mathbb{O}$ $\cdots$ 2 . 10^{-4}M; $\cdot - \cdot$ $\bullet$ $\cdot -$ 5 . 10^{-4}M; $-\cdot\cdot-$ $\bullet$ $-\cdot\cdot-$ 1 . 10^{-3}M. Half-wave potentials: *1* for wave i_{a1} at 1 . 10^{-3}M tropylium bromide, *2* for wave i_{a1} at 1 . 10^{-4}M tropylium bromide and *3* for wave i_2 at 1 . 10^{-3}M tropylium bromide.

Fig. 11

Oscilographic Curves dE/dt–E

Britton-Robinson buffer of pH 2·4, concentration of tropylium bromide *1* 1·5. 10^{-4}M; *2* 3 . 10^{-4}M and *3* 1·2 . 10^{-3}M. Polaroskop P 524, amplification 2, alternating component 0·2mA. The negative incision on curve *3* vanishes in the last part of drop-life. Anodic incision vanishes, if the electrode is not polarized towards more positive potentials.

With these assumptions the following values were computed for the area (S) covered by one ring in the reduction product at particular limiting currents:

Current	$S, \text{Å}^2$
i_{a1}	29
i_{a2}	13·5
i_1	8·2

These values which are substantially smaller than those found in most of the examples that have been studied so far are in good agreement with the molecular dimensions of the tropylium ion. For this symmetrical molecule it was possible using the covalent atomic radii and bond distances for C—C equal to 1·40 Å and for C—H equal to 1·1 Å to calculate the molecular radius. A value of $r = 3·5$ Å was found corresponding to an area 22 Å^2 for one sevenmembered ring.

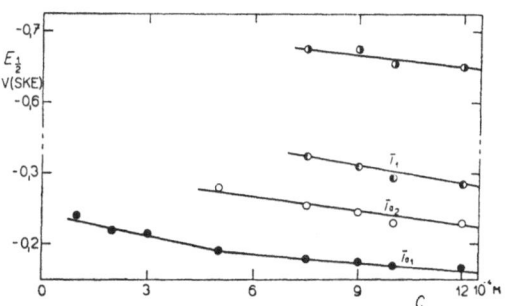

Fig. 12

Dependence of the Half-Wave Potentials on Tropylium Concentration Britton-Robinson buffer of pH 2·4. Half-wave potentials of waves: ● i_{a1}; ○ i_{a2}; ◐ i_1; ◑ i_2. c concentration of tropyliumbromide.

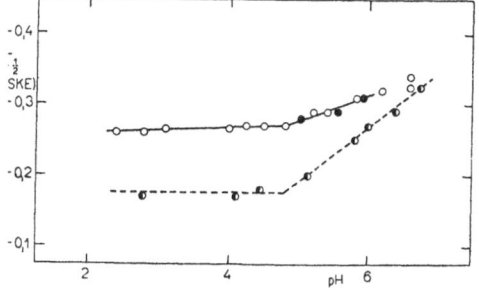

Fig. 13

Dependence of Half-Wave Potentials of Tropylium on pH Wave i_{a1} in solution of $1·5 . 10^{-4}$M tropylium bromide ○ Britton-Robinson buffers, ● acetate buffers, ◐ $1 . 10^{-3}$M tropylium bromide, Britton-Robinson buffers.

For the wave i_{a1} we can thus assume that the ring of the reduced molecule is placed parallel to the surface of the electrode; for the wave i_{a2} another orientation of the reduction produced particles in the surface of the electrode can be supposed. Finally at the potentials where the limiting current of the wave i_1 is observed, the formation of an adsorbed multilayer can be accepted as plaussible. In agreement with such an explanation is the impenetreability of this adsorbate and its stability in a relatively broad potential range.

Half-wave Potentials. The half-wave potentials of all the waves are shifted with increasing concentration of tropylium bromide towards more positive values (Fig. 12). The mercury pressure has little influence on half-wave potentials; only for wave i_2 this influence is more pronounced (the increase of the height of mercury column form $h = 30$ cm to $h = 70$ cm causes a shift of the half-wave potential of about 15 mV to more negative values).

The shift of the half-wave potentials with the change in pH is similar to that found at lower concentrations[4] (Fig. 13). In $1 . 10^{-3}$M solutions of tropylium ion the half-wave potential in the region between pH 2 and 4·6 is practically pH-independent. At pH higher than about 4·6 the half-wave potential is shifted towards more negative values with increasing pH-value, the slope of the shift $dE_{1/2}/dpH$ being equal to about 0·06 V/pH. This slope is greater than that (0·04 V/pH) observed at lower concentrations of the tropylium ion[4]. This difference is influenced by the decrease of the limiting current with increasing pH, involving also a decrease of the current at $-0\cdot2$ V. The decrease of the concentration of the electroactive form at the surface of the electrode results in a further shift of the half-wave potential to more negative potentials (cf. p. 766).

The Wave at $-1\cdot2$ V. In addition to all depolarisation effects mentioned above another steep almost discontinuous wave can be observed at $-1\cdot2$ V. This wave appears on polarographic curves at concentrations of tropylium bromide higher than about $1 . 10^{-4}$M. At concentrations above $1\cdot5 . 10^{-4}$M the limiting current does not change further with concentration of tropylium ion. The height of this wave increases somewhat with increasing pH-values. For pH above 5 due to the decrease in concentration of the tropylium ion at the surface of the electrode, the height of the wave at $-1\cdot2$ V decreases again. The half-wave potential of this wave is almost pH-independent.

The height of this wave increases steeply with increasing height of the mercury

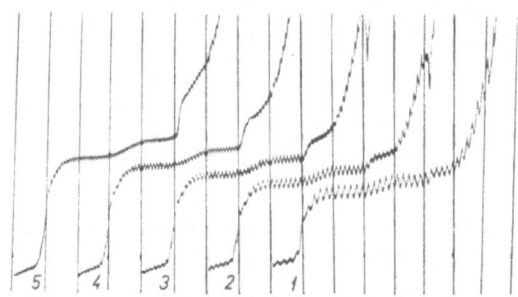

Fig. 14

Dependence of the Wave at $-1\cdot2$ V on the Height of Mercury Head

Acetate buffer of pH 3·8, $3 . 10^{-4}$M tropylium bromide, $h = 1$ 30 cm; 2 40 cm; 3 50 cm; 4 60 cm and 5 70 cm. From 0 V, S.C.E., 206 mV/absc., sens. 1 : 10.

head (h) and corresponds to a relation $i \sim h^n$, where $n > 1$ (Fig. 14). This dependence is caused by an unusual dependence on mercury outflow rate (m). This was demonstrated using a dropping mercury electrode with regulated drop-time. Whereas the dependence of the mean current $\bar{i}$ on the drop-time (t_1) possesses a form corresponding approximately to $\bar{i} \sim t_1^{0.2}$, the dependence on the mercury out-flow is nonlinear and approaches $\bar{i} \sim m^{1.0}$.

Since only at -1.2 V does the electrocapillary curve acquire the normal shape (Fig. 11), it is possible that this wave is due to a capacity phenomena, possibly combined with a maximum of the second kind. This would explain the discontinuous character of this wave and the anomalous dependence on the mercury-outflow[18]. Also the sharp in-cut at negative potentials on the dE/dt–E-curves (Fig. 11) is in accordance with a sudden change in capacity. On the other hand it cannot be ruled out that this wave represents an unusual type of catalytic hydrogen evolution, observed also for other cationic hydrocarbons, like azulenes[19]. This explanation seems to be supported by the fact that at potentials, at which this wave is observed, an increase of the current occurs in the presence of pyridine in the buffer used. The wave at -1.2 V in pyridine–pyridinium chloride buffers corresponds undoubtedly to a catalytic hydrogen evolution. Whereas the wave in the absence of pyridine forms only a small fraction of the total limiting current of tropylium ions, the wave in pyridine containing buffers is substantially higher. The height of this wave increases with increasing pyridine concentration and attains its maximum value at a pyridine concentration equimolar to that of tropylium ion. With a further increase of the pyridine buffer concentration the wave-height does not change further with increasing concentration. With a surplus of the pyridine buffer the wave at -1.2 V is a linear function of the tropylium concentration (in the studied range of concentrations) and directly proportional to the height of mercury head.

Conclusions

The theories of polarographic currents influenced by the presence of a surface active substance have been developed recently[9,10,20-24], but only limited attention has been paid to the role of the adsorption of electroactive forms.

The theory of adsorption of oxidized or reduced forms of the depolarizer was developed for reversible processes by Brdička[8]. For such mobile electrode equilibria it is possible to distinguish from the shift of the separated adsorption wave (relative to the redox potential), whether the oxidized or reduced form is adsorbed.

Recently several cases were detected (manifested by the formation of a separate wave), where adsorption phenomena accompany irreversible processes[9,14,25-27] involving no adsorptive substance other than the depolarizer.* With such irreversible

---

* These changes of i–E-curves should be distinguished from a change in the capacity current, which can be observed even for adsorption of the oxidized form e. g. with some steroids[28].

systems it is often difficult fo distinguish which of the several waves observed corresponds to a "pure" electrode process, unaffected and unhindered by adsorption.

In solutions containing the oxidized form solely, an adsorption of the *reduced* form influences the rates of electrode processes and results in the formation of a separate wave. If the retardation of the rate of the reduction process is sufficient, no more (or substantially smaller) current flows after a film of the reduced substance is formed. A decrease of the instantaneous current during a single drop to zero (or to a small value followed by a slow increase of the current) is observed on the i–t-curves. The mean current is principally governed by the rate of growth of the mercury drop. The height of the separated wave on polarograms is then proportional to $t^{-1/3}$ or h respectively.

At more negative potentials* an increase of current in the form of another wave is observed. This may be caused by two kinds of processes: *1.* At sufficiently negative potentials the rate of the electrode process becomes sufficient to cause further reduction of the oxidized form. A new wave occurs, corresponding to a reduction through a film of the adsorbed reduction product. The rate of this process differs from the rate on unoccupied surface. *2.* At sufficiently negative potentials a desorption of the film occurs, causing a rise of the current to a value, corresponding to the total amount supplied by diffusion. If the desorption would be a fast process occuring at a sharply defined potential, the rise of mean current would be discontinuous.

As the experimental material on irrevesible processes influenced by the adsorption of depolarizer available is limited, no attempts are made to use the criteria[9,10,24] enabling the distinction between the rôle of adsorption equilibria and reaction rates. It can be assumed that in cases, where separate waves are formed on polarographic curves, the equilibrium is shifted towards the adsorbed reduced form and the rate of formation of the adsorbed film is high.

The relation between the time necessary for formation of the adsorbed film[20] (ϑ) and the inverse square of concentration ($1/C^2$) was not veriefied experimentally in the present study. Several successive adsorption processes (see below) and observable rate of the slow reduction process through the adsorbed film (shown by the final increase of the i–t-curves) hindered the estimation of ϑ. As an approximate measure of this time-interval, the time τ (Fig. 7) was measured from the begining of the drop-life to the point where the instantaneous current reaches its maximum value. The time-interval τ decreases with increasing concentration of tropylium ion and with increasingly negative potentials (Fig. 6, 7). This trend is in accordance with that expected for ϑ.

The present case is unsuitable for quantitative treatment because with increasing concentrations of tropylium ion three types of adsorption phenomena occur at different potentials. (Fig. 1, 2, 6). We assume that three different types of adsorption processes take place, differing in the rate of formation of the adsorbate (possibly

---

* At concentrations higher than that sufficient to a complete surface covering.

in the equilibrium conditions) and in the properties of the film formed. The reduced form is adsorbed, either ditropyl or radical[2,3], the adsorption being either a physical or a chemisorption under formation of a compound with mercury. The available experimental material does not allow a distinction between the mechanism of the reduction of particular waves. From the comparison with electrocapillary curves we conclude that the current i_2 is most probably due to a desorption process (cf. p. 112). On the other hand we can assume from the shape of i–t-curves that in waves i_{a_2} and i_1 a slow electrode process occurs. The difference in the structure of films causing the different rates of electrode processes through films formed at limiting currents of i_{a_1} and i_{a_2} can be due to a difference in physical properties, structure or chemical composition of the two films. The impenetrability of the film, formed at limiting current i_1, can be explained by the formation of a multilayer film (cf. p. 169)

Ždanov and collaborators[1-3] recognized the adsorption character of their first wave $(i_{a_1} + i_{a_2} + i_1)$ and the rôle of the adsorption of the reduced form. They missed the waves i_{a_1} and i_{a_2}. Therefore their argument[3] that the adsorption is of another type, because the area covered by the reduced form $(5·6 \text{ Å}^2)$ calculated using equation (1) is too small, is erroneous. On the other hand our value (29 Å^2) is in good agreement with the area (22 Å^2) of an sevenmembered ring. This coincidence seems to prove that both assumptions necessary for application of equation (1) (i. e. a shift of equilibrium towards the adsorbed form and rapid formation of the adsorbed layer) are fulfilled for the reduced form of tropylium. – Further proofs or adsorption of the reduced form given by Ždanov are unconvincing, since the effect of both temperature and added surface active substance on adsorption currents is different for different depolarizers and surface active substances. The assignment[1] of the kinetic character to wave i_2 is due to an erroneous measurement of the wave height. Distortions observed at potentials more negative than $-1·0$ V are caused by unbuffered media used[1-3].

In the present case – as in numerous others[9,10] – the study of i–t-curves provides a deeper insight into adsorption processes, unveiling three successive adsorption phenomena.

The careful technical assistence of Miss N. Pejchová is acknowledged. We thank Dr C. R. Ganellin, London, for the specimen of tropylium bromide and to Drs V. Hanuš, J. Koryta, J. Kůta, S. I. Ždanov, Moscow, and Dr M. Březina for their comments on the manuscript.

References

1. Volpin M. E., Ždanov S. I., Kursanov D. N.: Dokl. Akad. nauk SSSR *112*, 264 (1957).
2. Ždanov S. I.: Z. physik. Chem. (Leipzig), Sonderheft *1958*, 235.
3. Ždanov S. I., Frumkin A. N.: Dokl. Akad. nauk SSSR *122*, 412 (1958).
4. Zuman P., Chodkowski J., Šantavý F.: This Journal *26*, 380 (1961).
5. Němec L.: *Thesis.* Charles University, Prague 1958.
6. Valenta P.: Z. physik. Chem. (Leipzig), Sonderheft *1958*, 46.
7. Koutecký J.: This Journal *18*, 597 (1953); Chem. listy *47*, 323 (1953).
8. Brdička R.: This Journal *12*, 522 (1947).

530

9. Schmid R. W., Reilley C. N.: J. Am. Chem. Soc. *80*, 2087 (1958).
10. Kůta J., Smoler I.: Z. Elektrochem. *64*, 285 (1960).
11. Heyrovský J., Šorm F., Forejt J.: This Journal *12*, 11 (1947).
12. Valenta P.: This Journal *16*, 239 (1951); Chem. listy *45*, 249 (1951).
13. Valenta P.: Chem. zvesti *8*, 767 (1954).
14. Zuman P.: Chem. zvesti *8*, 787 (1954), where previous experimental evidence and ideas are summarised.
15. Zuman P., Zumanová R., Teisinger J.: This Journal *20*, 139 (1955); Chem. listy *48*, 1499 (1954).
16. Manoušek O., Zuman P.: This Journal *20*, 1340 (1955); Chem. listy *49*, 668 (1955).
17. Fedoroňko M., Zuman P.: This Journal *21*, 678 (1956); Chem. listy *49*, 1484 (1955).
18. Dvořák J.: This Journal *19*, 39 (1954); Chem. listy *47*, 969 (1953).
19. Zuman P.: Z. physik. Chem. (Leipzig), Sonderheft *1958*, 243.
20. Koryta J.: This Journal *18*, 206 (1953).
21. Delahay P., Trachtenberg I.: J. Am. Chem. Soc. *79*, 2355 (1957).
22. Delahay P., Trachtenberg I.: J. Am. Chem. Soc. *80*, 2094 (1958).
23. Delahay P., Fike C. T.: J. Am. Chem. Soc. *80*, 2628 (1958).
24. Weber J., Koutecký J., Koryta J.: Z. Elektrochem. *63*, 583 (1959).
25. Zuman P., Tenygl J., Březina M.: This Journal *19*, 46 (1954); Chem. listy *47*, 1152 (1953).
26. Zuman P.: Kabát M.: This Journal *19*, 873 (1954); Chem. listy *48*, 368 (1954).
27. Kemula W., Cisak A.: Roczniki Chem. *31*, 337 (1957).
28. Zuman P., Černý V.: This Journal *25*, 1925 (1959); Chem. listy *52*, 1468 (1958).

Translated by the Author (P. Z.).